Patrick S.P. Wang

Pattern Recognition, Machine Intelligence and Biometrics

Patrick S.P. Wang

Pattern Recognition, Machine Intelligence and Biometrics

With 605 figures, 4 of them in color

Editor
Professor Patrick S.P. Wang
CCIS, 360 Huntington Ave
Northeastern University
Boston, MA 02110,USA
https://sites.google.com/site/mozart200
E-mails: patwang@ieee.org, pwang@acm.org

ISBN 978-7-04-033139-4
Higher Education Press, Beijing

ISBN 978-3-662-58545-0   ISBN 978-3-642-22407-2 (eBook)
DOI 10.1007/978-3-642-22407-2

Springer Heidelberg Dordrecht London New York

Patrick S.P. Wang's Brief Vitae

Patrick S.P. Wang, *Fellow of IAPR*, *ISIBM and WASE*, is a professor of Computer and Information Science at Northeastern University, USA, Shanghai East China Normal University Zi-Jiang Visiting Chair Professor, research consultant at MIT Sloan School, and adjunct faculty of computer science at Harvard University. He received PhD in C.S. from Oregon State University, M.S. in I.C.S. from Georgia Institute of Technology, M.S.E.E. from Taiwan University (Taipei) and B.S.E.E. from Chiao Tung University (Hsin-chu).

As *IEEE* and *ISIBM Distinguished Achievement Awardee*, Dr. Wang was on the faculty at University of Oregon and Boston University, and senior researcher at Southern Bell, GTE Labs and Wang Labs prior to his present position. Dr. Wang was Otto-Von-Guericke Distinguished Guest Professor of Magdeburg University, Germany, and iCORE (Informatics Circle of Research Excellence) visiting professor of University of Calgary, Canada, Honorary Advisor Professor for China's Sichuan University, Xiamen University, and Guangxi Normal University. In addition to his research experience at MIT AI Lab, Dr. Wang has been a visiting professor and invited to give lectures, do research and present papers in a number of countries and areas, from Europe, Asia and many universities and industries in the U.S.A. and Canada. Dr. Wang has published over 160 technical papers and 26 books on Pattern Recognition, A.I. Biometrics and Imaging Technologies and has 3 OCR patents by US and Europe Patent Bureaus. One of his books is so important and widely cited that the USA Department of Homeland Security(DHS) uses it as reference for Call For Proposals 2010. For details please refer to DHS website: Image Pattern Recognition—Synthesis and Analysis of Biometrics (WSP): https://www.sbir.dhs.gov/PastSolicitationDownload.asp#101005.

As IEEE senior member, he has organized numerous international conferences and workshops including *conference co-chair of the 18th IAPR ICPR* (International Conference on Pattern Recognition) in 2006, Hong Kong, China, and served as reviewer for many journals and NSF grant proposals. Professor Wang is currently *founding Editor-in-Chief of IJPRAI* (Int. J. of Pattern Recognition and A.I.), and *Machine Perception and Artificial Intelligence Book* Series by World Scientific Publishing Co. and Imperial College Press, London, UK, and elected chair of IAPR-SSPR (Int. Assoc. for P.R.). Dr. Wang has been invited to give talks in many International Conferences including AIA2007, Innsbruck, Austria; IAS2007, Manchester, UK; IEEE-SMC2007, 2009, 2010, Montreal, San Antonio, Istanbul, respectively; World-Comp2010, Las Vegas, USA; CIS2007, Harbin, China, eForensics2008, Adelaide, Australia; ISI2008, Taipei, Taiwan, China; BroadCom2008, Pretoria, South Africa; VISAPP2009, Lisboa, Portugal; UKSim2011, Cambridge, UK, and IADIS2010, 2011, Freiburg, Germany, and Roma, Italy, respectively. Dr. Wang received IEEE Distinguished Achievement Award at IEEE-BIBE2007 at Harvard Medical, for Outstanding Contributions in Bioinformatics and Bioengineering.

In addition to his technical achievements and contributions, Professor Wang has been also very active in community services, and has written several articles on Du Fu, Li Bai's poems; Verdi, Puccini, Bizet, and Wagner's operas; and Mozart, Beethoven, Schubert and Tchaikovsky's symphonies. A collection of selected proses was published in his book *Harvard Meditation Melody* by Jian-Shing Pub. Co., Taipei, which won best publication award by Taiwan, China.

Lotfi A. Zadeh (left) and Patrick S.P. Wang (right)

Dear Patrick,

Many thanks for your message and the kind words.

I appreciate very much what you wrote. As you know, I am highly impressed by your achievements. With regard to the foreword, I have a problem. After my heart attack in December 2008 my vision and my hearing have experienced a decline. Today, to read printed matter I have to use a magnifying glass. Reading messages does not present a problem but reading a book does. This is why writing a foreword — even to a book dedicated to my admired friend, K.S.Fu, would be stressful. It is a source of great regret for me not to be able to respond affirmatively to your invitation. Please keep in touch.

With my warm regards.

Sincerely,

Lotfi

Lotfi A. Zadeh
Professor in the Graduate School
Director, Berkeley Initiative in Soft Computing (BISC)

King-Sun Fu (left) and Patrick S.P. Wang (right) in 1981 Int. Conf. Advanced Automation (ICAA1981), Taipei, China

Patrick S.P. Wang (left) and King-Sun Fu (right) in 1981 Int. Conf. Advanced Automation (ICAA1981), Taipei, China

In Honor and Memory of Late Professor King-Sun Fu

The late Professor King-Sun Fu is one of the founding fathers of pattern recognition, who, with visionary insight, founded the International Association for Pattern Recognition around 1980. In the almost 30 years since then, the world has witnessed the rapid growth and development of this field.

King-Sun Fu

(10/2/1930 [Nanjing, China] – 4/29/1985 [D.C.,USA])

IEEE COMPUTER SOCIETY

IEEE TRANSACTIONS ON
PATTERN ANALYSIS AND MACHINE INTELLIGENCE

K. S. FU, Senior Editor
School of Electrical Engineering
Purdue University
West Lafayette, Indiana 47907
Telephone (317) 494-8825

November 4, 1981

Dr. Patrick S. Wang
GTE Research Laboratories
40 Sylvan Road
Waltham, MA 02154

Dear Dr. Wang:

The enclosed manuscript has been submitted for possible publication in the
IEEE Transactions on Pattern Analysis and Machine Intelligence. I should
greatly appreciate it if you or one of your associates would review the
manuscript for the Transactions. The paper is entitled: PAMI 81-8-5R,
FUZZY TREE AUTOMATA AND SYNTACTIC PATTERN RECOGNITION.

For your convenience, I enclose our standard referee form. Thank you very
much for your assistance. This paper is for your re-review.

Sincerely,

K. S. Fu
Senior Editor, IEEETPAMI

KSF:mjb

Enclosure

INSTITUTE OF ELECTRICAL AND ELECTRONICS ENGINEERS

PURDUE UNIVERSITY SCHOOL OF ELECTRICAL ENGINEERING

24 January 1980

Professor Patrick S. P. Wang
Department of Computer Science
University of Oregon
Eugene, OR 97403

Dear Patrick:

Thank you for your letter dated January 15, 1980. The following is my schedule to Eugene:

February 22, San Francisco - Eugene (UA 862)
8:25 a.m. 9:46 a.m.

February 23, Eugene - San Francisco (RW737)
1:45 p.m. 2:55 p.m.

The title of my talk is "Recent Developments in Syntactic Pattern Recognition." I will need an overhead projector and a 35mm slide projector for my talk.

Looking forward to seeing you in February.

Sincerely,

K. S. Fu
Goss Professor
of Engineering

KSF/msh

Electrical Engineering Building
West Lafayette, Indiana 47907

PURDUE UNIVERSITY
SCHOOL OF ELECTRICAL ENGINEERING

January 19, 1981

Dear Dr. Wang:

We have proposed a NSF Workshop on Structural and Syntactic Pattern Recognition, which will be held on June 22-24, 1981, in Saratoga Springs, N.Y. General Electric is also a co-sponsor of the workshop. The purpose of this workshop is to assess the progress in structural and syntactic pattern recognition in terms of its contributions and limitations to practical applications and to project the future directions of this area. Knowing your interest and past contributions in this area, we would like to invite you to participate in this workshop. Your traveling expenses will be paid by the workshop pending the final approval of our proposal by NSF.†

The major emphasis of the workshop is on group discussions rather than many formal presentations of papers. We are planning to have a series of background papers and a small number of short presentations for new research results and applications. Please do not hesitate to let me know if you would like to give a paper at the workshop.

We hope that you will accept our invitation and we are looking forward to seeing you at the workshop.

Sincerely,

K. S. Fu
Goss Professor of Engineering
on behalf of the Workshop
Organizing Committee
K.S. Fu (Co-Chairman)
T. Pavlidis (Co-Chairman)
J. L. Mundy
R. K. Aggarwal

† NSF rule may not allow us to pay participants from industry.

- -

Please return this form to K. S. Fu before February 15, 1981.

I will - ___ accept the invitation to participate in the NSF Workshop on
  will not - ___
Structural and Syntactic Pattern Recognition (June 22-24).

I would like to present a paper with title: _____

_____. Please send an abstract if possible.)

Name _____

Electrical Engineering Building
West Lafayette, Indiana 47907

GTE Laboratories Incorporated
40 Sylvan Road
Waltham MA 02154
617 890 8460

GTE Telenet Technology Center

June 12, 1981

Dr. K. S. Fu
Goss Professor of Engineering
School of Electrical Engineering
Purdue University
West Lafayette, IN 47909

Dear Dr. Fu:

Enclosed please find a copy of the GTE Seminar Announcement about your visit.

Your hotel has been reserved and prepaid at the Hotel Sonesta, 5 Cambridge Parkway, Cambridge (617-491-3600) for June 25th and 26th.

I will be waiting for you at 5:25 PM (June 25th) at Logan International Airport.

On behalf of Dr. Joel Krugler, Manager of GTE Telenet Technology Center, and GTE Laboratories, let me extend our warmest welcome to you. We believe your visit will prove to be very enjoyable and mutually beneficial.

Sincerely yours,

Dr. Patrick Wang

PW:vs
cc: Dr. J. Krugler

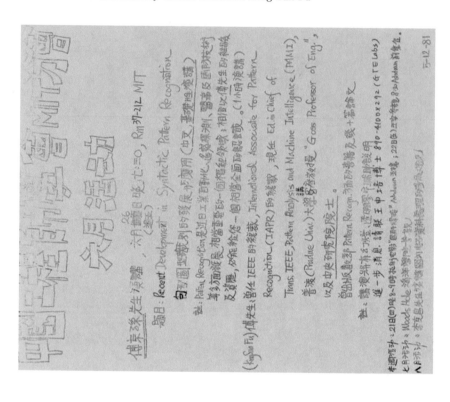

玉先生：

收到來信及電話謝：

附上照片二張，甚可用

至於寫京孫紀念文我很

感謝你們對京孫的友情

師情，我亦希望能為一文，

但我希望你能諒解我是

經過考慮，亦方知該從何說起，

所以只能作罷，請原諒。

　　　　祝　一家好

學安

傅陳綏樺敬上

一月二日

GTE

Patrick Wang

GTE Laboratories Incorporated
40 Sylvan Road
Waltham, MA 02254
617 890 4100

Date _____ 6.10

Sonesta Hotel, 在
S Cambridge Park, Cam. (Guarantee
late payment)

Foreword By J.K. Aggarwal

It is again a pleasure and an honor to introduce this collection of papers in memory of the late Professor King-Sun Fu. Professor Fu contributed significantly to the growth of pattern recognition through teaching, research, scholarship and supervision of Ph.D. students at Purdue University. It is sad that Professor Fu did not live to see the spectacular growth that we are witnessing today in the use of pattern recognition and computer vision in industry, military, and daily products. Professors Tzay Young and King-Sun Fu published a collection of papers entitled Handbook of Pattern Recognition and Image Processing. The book was published in 1986 after Professor Fu passed away in 1985.

Professor Patrick S.P. Wang has followed in the spirit of the collection published by Young and Fu.

The present book is fifth in the series published by Professor Patrick S.P. Wang and his co-editors. The earlier four collections are:
(i) Handbook of Pattern Recognition and Computer Vision
    Editors: C.H. Chen, L.F. Pau and P.S.P. Wang
    1993, World Scientific Publishing
    32 papers, 984 pages
(ii) Handbook of Pattern Recognition and Computer Vision
    Editors: C.H. Chen, L.F. Pau and P.S.P. Wang
    1999, World Scientific Publishing
    34 papers, 1019 pages
(iii) Handbook of Pattern Recognition and Computer Vision
    Editors: C.H. Chen and P.S.P. Wang
    2005, World Scientific Publishing
    33 papers, 639 pages
(iv) Pattern Recognition and Computer Vision
    Editor: P.S.P. Wang
    2010, River Publishers
    28 papers, 451 pages

The present volume again honors the contributions of Professor Fu. He was the founding editor of the IEEE Transaction on Pattern Analysis and Machine Intelligence. Today this transaction has the highest regard of researchers and developers. It is among the crown jewels of the publications of the IEEE. He also helped start the organization known as the International Association for Pattern Recognition (IAPR) in the early 1970s. Today IAPR has membership from 42 countries all over the globe. The family and friends of Professor Fu created the King-Sun Fu Prize in his memory under the auspices of the IAPR. IAPR awards this prize every two years at its International Conference on Pattern Recognition popularly known as ICPR. It is noteworthy that seven winners of the King-Sun Fu Prize (out of a total of

11 awarded so far) are among the authors in the book edited by Professors Young and Fu.

The present book consists of 30 papers clustered in five broad chapters. It consists of original contributions from a few coupled disciplines, including pattern recognition, machine intelligence, computer vision, image processing, and signal processing. Authors review existing methodologies, propose new approaches, and build systems on fairly diversified applications including face recognition, activity recognition, and biometric authentication. The book will appeal to researchers, students, and technologists.

It is a pleasure to be part of the tribute to Professor King-Sun Fu.

J.K. Aggarwal
Cullen Trust Professor
Department of Electrical and Computer Engineering
The University of Texas at Austin
May 2011

Foreword By Brian C. Lovell

This is the second in a series of books on pattern recognition to honour the memory of Professor King-Sun Fu. No one deserves more credit for the founding of the International Association for Pattern Recognition than King-Sun Fu. In 1971, King-Sun Fu invited several leading researchers to set up a committee for an international conference on pattern recognition[1]. This led directly to the First International Joint Conference on Pattern Recognition (IJCPR) which was held in Washington, D.C., from 30th October to 1st November, 1973. Discussions at the Lyngby IJCPR in 1974 recommended the formation of an entirely new international society which would be a kind of federation of national organizations in pattern recognition. The name of the organization would be "The International Association for Pattern Recognition" as suggested by Herbert Freeman.

The IAPR came into official existence in January, 1978 with King-Sun Fu as the inaugural President. Today the IAPR has 43 member organisations from all around the globe. It sponsors numerous pattern recognition conferences in many countries as well as organising the International Conference on Pattern Recognition (formerly IJCPR) conference series.

Thus it was a tremendous blow when King-Sun Fu died suddenly on 29 April 1985 aged just 55. His untimely death was a loss to the whole community and was keenly felt by his family, friends, students, and his IAPR colleagues. He, more than anyone, created the organization, serving as its inaugural president, and was a leading figure in the field of Pattern Recognition. In 1988 the IAPR awarded the first K.S. Fu Prize to commemorate his achievements. The prize was to be given no more often than biennially to a living person "in recognition of a technical contribution of far-reaching significance and impact on the field of pattern recognition or its closely allied fields made at any time in the past." The K.S. Fu prize serves as the "Nobel Prize" for pattern recognition.

It is appropriate to commemorate King-Sun Fu's achievements with this second edition of collected works representing emerging aspects of the exciting field of pattern recognition which he did so much to establish. These chapters cover diverse topics ranging from biometrics to image forensics and data mining. They give an insight into just how much the field of pattern recognition has advanced since the foundational works of King-Sun Fu in the 1970s.

Brian C. Lovell
President of the IAPR, 2008–2010
May 2011

---

1 Freeman, H., Detailed History of the IAPR, http://www.iapr.org/docs/IAPR-History. pdf, [last visited: March 2011]

Foreword By Sargur N. Srihari

My memories of King-Sun Fu date from the early 1970's when I was a graduate student beginning my studies in pattern recognition. At that time Fu's work at Purdue University was very influential for anyone interested in entering the field.

Fu was synonymous with a new area of pattern recognition called *syntactic pattern recognition*[1]. Statistical pattern recognition was already an established field with the appearance of the pre-print of Duda and Hart's textbook in 1972[2]. Syntactic pattern recognition was an alternative approach to recognizing visual patterns, such as handwritten characters, based on formal grammars. Fu was a vigorous advocate of the syntactic approach as an alternative to purely statistical models. The goal was to exploit knowledge of structure in designing pattern recognition algorithms. There existed a small body of work such as that of R. Narasimhan at the University of Illinois who proposed a full set of grammatical rules for the English alphabet[3]. Others included R. A. Kirsch of the National Bureau of Standards who proposed rules for shapes[4] and A. Joshi of the University of Pennsylvania who suggested tree grammars[5]. While Fu and Narasimhan, who both did their graduate work at the University of Illinois-Urbana/Champaign, were interested in syntactic pattern recognition , Fu took it to the next stage by proposing the use of probabilities with grammatical rules. The explanatory powers associated with such methods were attractive to the field of Artificial Intelligence which was just beginning to take off. Today pattern recognition and the associated field of machine learning have enormous applications in every field. It is quite common to see both syntactic and statistical pattern recognition being used in the same system, e.g., domain-specific rules written by linguists and maximum entropy Markov models are both used in a single commercially available information extraction system[6].

Fu also made important contributions to an area he named as *sequential pattern recognition*[7]. The goal here was to build pattern recognition algorithms which could take features as input one at a time and stop when a sufficient level of confidence was reached. He adapted Abraham Wald's se-

---

1   K.S. Fu, *Syntactic Methods in Pattern Recognition*, Academic Press, 1974.
2   R.O. Duda and P.E. Hart, *Pattern Recognition and Scene Analysis*, Wiley-Interscience, 1973.
3   R.A. Narasimhan, "Syntax-directed Interpretation of Classes of Pictures", Communications of the ACM, vol. 9, issue 6, 1966.
4   R.A. Kirsch, "Computer interpretation of English text and picture patterns", *IEEE Trans. Elect. Comp.*, EC-13 (Aug. 1964), pp. 363–376.
5   A. Joshi, S.R. Kosaraju, H. Yamada, "String Adjunct Grammars", *Proceedings Tenth Annual Symposium on Automata Theory*, Waterloo, Canada, 1969.
6   K.S. Fu, *Sequential Methods in Pattern Recognition and Machine Learning*, Academic Press, 1968.
7   R.K. Srihari, W. Li, C. Niu and T. Cornell, "InfoXtract: A Customizable Intermediate Level Information Extraction Engine", *Journal of Natural Language Engineering*, Cambridge U. Press , 14(1), 2008, pp.33–69.

quential probability ratio test, known primarily to statisticians, and brought it to the attention of the pattern recognition community. Today sequential models have developed much further and are at the heart of machine learning methods such as hidden Markov models, conditional random fields and linear dynamical systems.

Arguably, the biggest legacy of Fu was the establishment of the *IEEE Transactions on Pattern Analysis and Machine Intelligence* (IEEE-TPAMI) which began publication in 1981. It was an almost single-handed effort to build a community of researchers. Previously papers on pattern recognition would go to *IEEE Transactions on Computers* (IEEE-TC) which was dominated by digital circuits research. In fact the first I heard about IEEE-TPAMI was after I had submitted a paper to IEEE-TC, I received a call from King-Sun Fu asking whether it would be appropriate to transfer it to the newly established IEEE-TPAMI. I readily agreed and was pleased to have had a paper in the first issue of IEEE-TPAMI[8].

It was not surprising that Fu chose to use the phrase *Pattern Analysis* since that reflected both statistical methods and grammatical rules. The term *Machine Intelligence* reflected the inclusion of Artificial Intelligence (AI). At that time AI was thought of as consisting of algorithms based on logic, heuristics and expert rules. The phrase "machine intelligence" was less used but it had been the title of an influential series of volumes emerging from Edinburgh University[9]. The term "machine" avoided negative connotations of artificial intelligence such as whether natural intelligence was superior to artificial intelligence. Today IEEE-PAMI is the pre-eminent transactions of the IEEE. It has the highest impact factor of all Artificial Intelligence journals. Perhaps IEEE-PAMI itself is a tribute to King-Sun Fu's foresight.

Finally I had two other interactions with King-Sun Fu. As a graduate student I wrote to him about some fine points in the sequential models he espoused. I received a prompt response from him clarifying it and sending me a helpful pre-print. My only occasion to meet him in person was at a *Computer Vision and Pattern Recognition (CVPR)* conference in Chicago in 1979.

The present volume reflects many of the same themes and applications that were foreseen by King-Sun Fu in the early days of syntactic pattern recognition, sequential models and the pre-eminent journal that he founded.

Sargur N. Srihari
SUNY Distinguished Professor
University at Buffalo, The State University of New York
May 2011

---

8 S.N. Srihari, "Recursive Implementation of a Two-step Nonparametric Decision Rule," *IEEE Transactions on Pattern Analysis and Machine Intelligence*, IEEE-TPAMI, vol. 1, no. 1, 1981, pp. 90–94.

9 D. Michie, On Machine Intelligence, Edinburgh University Press, 1974.

Preface

In our time of rapidly changing Information Technology (IT) age, there is growing interest in Pattern Recognition and Machine Intelligence and its applications to Biometrics in academia and industries. Novel theories have been found, with new design of technology and systems, in hardware, software and mid-ware. They are extensively studied and widely used to our daily life to solve realistic problems, including science, engineering, agriculture, e-commerce, education, robotics, government, hospital, games and animation, medical imaging analysis and diagnosis, military, personal identification and verifications, and homeland security. The foundation of all these can be traced back to late Professor King-Sun Fu (10/2/1930—4/29/1985), one of the fathers of Pattern Recognition, who was visionary and founded the International Association for Pattern Recognition (IAPR) in 1978. Ever since then, after 33 years, the world has eye witnessed the rapid growth and development of this field, and most people can sense and be touched by its applications in our daily lives.

Today at the eve of his 81th birthday and 26th anniversary of unfortunate and untimely passing of Prof. Fu, we are proud to produce this volume of collected works by the world renowned professionals and experts in Pattern Recognition, Machine Intelligence, and Biometrics (PRMIB) in honor and memory of the late Professor King-Sun Fu. We hope this book will help promote further its course of not only fundamental principles, systems and technologies, but also its vast applications to help solving our daily life problems.

There have been tons of complimentary remarks and praises about Professor Fu's extraordinary achievements on PRMIB, including hundreds of paper and book publications, numerous prestigious awards, founder of IEEE-PAMI and IAPR and an NSF Research centers and so on. Professor Fu is known as one of the greatest computer scientists and engineers. But actually more than that, he is also a great artist and music lover. He believes liberal arts education, and plays volleyball well. In his Lab at Purdue University, most, if not all his team, researchers and students' popular entertainment or exercise is to play volleyball. He also enjoys reading novels and poems. I remember when he came to Boston to give a seminar, we had some discussions on his most recent work then, on "towards a unification theory of statistical and syntactic pattern recognition". He noticed on my desk that there was a paper clip of my article "On Puccini's opera 'Turandot' and Chinese folk song 'Jasmine flower' ". It immediately attracted his attention and interest to request a copy. Everywhere we met or communicated in USA, Canada, Europe or Asia, he always showed strong concerns on not only progress of pattern recognition, but also world affairs, humanity and environment. Although, regrettably, he did not finish his ambitious plan to establish a theory towards unification of statistical, syntactical and structural pattern recognition, pretty much like Einstein did not have time to finish his effort towards unification of forces, yet, Professor King-Sun Fu indeed was not only a distinguished professor, but also a great teacher from whom we do learn much. No wonder, as Professor Thomas Huang put it, in his King-Sun Fu Prize laureate speech at 2002 ICPR, Quebec City, Canada, that "There is no so-called last student of Professor Fu. We all are his students". Yes, indeed, Professor Fu, you live in our hearts forever.

This book of Pattern Recognition, Machine Intelligence and Biometrics in memory of Prof King-Sun Fu is divided into 4 parts as follows:

Part I: **Pattern Recognition and Machine Intelligence** begins with A Review of Applications of Evolutionary Algorithms in Pattern Recognition by Luis Gerardo de la Fraga and Carlos A. Coello Coello, which shows some of the most representative work regarding techniques and applications of volutionary algorithms in pattern recognition. K.C. Wong, Dennis Zhuang, Gary C.L. Li and En-Shiun Annie Lee introduce Pattern Discovery and Recognition in Sequences. In A Hybrid Method of Tone Assessment for Mandarin CALL System by Yang Qu, Xin He, Yue Lu and Patrick S.P. Wang, an approach based on forced alignment of Hidden Markov Model (HMM) is employed to train utterances for obtaining model of getting accurate syllable boundary of utterances. Zheng Liu and Wei Wu present a state-of-art review of the fusion techniques for infrared images in Fusion with Infrared Images for an Improved Performance and Perception. Janos Csirik and Horst Bunke introduce a general framework for pattern classification in wireless sensor networks that aims at increasing the lifetime of the underlying system by using a number of features as small as possible in Feature Selection and Ranking for Pattern Classification in Wireless Sensor Networks. In Principles and Applications of RIDED-2D — A Robust Edge Detection Method in Range Images, Jian Wang, et al. proposes a novel Rule-based Instantaneous Denoising and Edge Detection method (RIDED-2D) for preprocessing range images.

Part II: **Computer Vision and Image Processing** begins Lens Shading Correction for Dirt Detection by Chih-Wei Chen and Chiou-Shann Fuh, presents a novel inspection framework to detect dirt and blemish in production line of optical fabrication automatically. In Using Prototype-Based Classification for Automatic Knowledge Acquisition, Petra Perner and Anja Attig describe how prototype-based classification can be used for knowledge acquisition in image classification. Elena S'anchez-Nielsen and Mario Hern'andez-Tejera include experimental results with inside and outside video streams demonstrating the effectiveness and efficiency for real-time machine vision based tasks in unrestricted environments. In Human Extremity Detection for Action Recognition, Elden Yu and J.K. Aggarwal propose that the location of human extremities alone (including head, hands, and feet) provides an excellent approximation to body motion. Brian C. Heflin, et al. review influential works along recent topics of ensemble learning approaches devised for recognizing and tracking objects in Tracking Learning for Object Recognition and Tracking. In Depth Image Based Rendering, Michael Schmeing and Xiaoyi Jiang give an introduction to Depth Image Based Rendering and discusses some challenges including proposed solutions.

Part III: **Face Recognition and Forensics** begins with Usman Tariq, Yuxiao Hu, and Thomas S. Huang's Gender and Race Identification by Man and Machine, and details a comprehensive study on gender and race identification from different facial representations. In Common Vector Based Face Recognition, Ying Wen, Yue Lu, Pengfei Shi, and Patrick S.P. Wang study an approach for face recognition based on the difference vector plus the kernel PCA (Principal component analysis). In A Look at Eye Detection for Unconstrained Environments, Brian C. Heflin, et al. take a look at eye detection for the latter, which encompasses problems of flexible authentication, surveillance, and intelligence collection. Weishi Zheng et al. introduce recent advanced developments of using KPCA for nonlinear image preprocessing in Kernel Methods for Facial Image Preprocessing. Sangita Bharkad and Manesh Kokare give a brief survey of current fingerprint matching methods and technical achievement in this area in Fingerprint Identification—Ideas, Influences, and Trends of New Age. Hong Huang gives a Comparative Study of Face Recogition—Subspaces versus submanifolds. In Linear and Nonlinear Feature Extraction Approaches for Face Recognition, Wensheng Chen, Pong C. Yuen, Bin Fang and Patrick S.P. Wang introduce recent progress and existing challenges in the area of face recognition (FR). In Facial Occlusion Reconstruction Using Direct Combined Model, Chingting Tu and Jenn-Jier James Lien develop means to recover the occluded region(s) of the facial image such that the performance of these applications can be improved. Sargur N. Srihari and Chang Su consider generative models for forensic evidence where the goal is to describe the distributions using graphical models and to use such models to compute probabilistic metrics for measuring the degree of individuality of a forensic modality or of a piece of evidence in Generative Models and Probability Evaluation for Forensic Evidence. In Feature Mining and Pattern Recognition in Digital Image Forensics, Qingzhong Liu et al. present some recent results on detecting JPEG steganograms, doubly compressed JPEG images, and resized JPEG images based on a unified framework of feature mining and pattern recognition approaches.

Part IV: **Biometrics Authentication** begins with Biometric Authentication by Jiunn-Liang Lin, Ho-Ling Hsu, Tai-Lang Jong and Wen-Hsing Hsu, which introduces several biometric recognition technologies, including person's face, eye pupil, sound, etc., the standardization development of biometrics technology, and recent projects in all countries for various applications. In Radical-Based Hybrid Statistical-Structural Approach for Online Handwritten Chinese Character Recognition, Chenglin Liu and Longlong Ma describe a new radical-based online handwritten Chinese character recognition approach which combines the advantages of statistical methods and radical-based structural methods. In Current Trends in Multimodal Biometric System — Rank Level Fusion, Marina Gavrilova and Maruf Monwar provide an in-depth overview of traditional multimodal biometric systems and current trends in multimodal biometric fusion. In Off-line Signature Verification by Matching With a 3D Reference Knowledge Image — From Research to Actual Application, Maan Ammar introduces a method for off-line verification of signatures, which can verify signatures and detect skilled forgeries with outstanding performance. In Unified Entropy Theory on Pattern Recognition and Maximum MI Discrimination based Subspace Pattern Recognition, a unified entropy theory on Pattern Recognition is presented by Ding Xiaoqing. In Fundamentals of Biometrics — Hand Written Signature and Iris by Radhika K R and Sheela S V, they introduced a new method combining hand written signature characteristics and iris texture variations to form an occurrence vector to provide biohashing. In Recent Trends in Iris Recognition, Lenina Birgale and Manesh Kokare discusse recent trends of Iris Recognition by means of Knowledge, Possession and Reality. In Using Multisets of Features and Interactive Feature Selection to Get Best Qualitative Performance for Automatic Signature Verification, Maan Ammar reports the results obtained by using the MSF technique in case of using large number of feature sets. Last but not least, in Fourier Transform in Numeral Recognition and Signature Verification, Giovanni Dimauro presents the fundamentals of digital transforms and their use in handwriting recognition.

A short description of each chapter evidently is not adequate to present the excellent contributions by all authors. The high-quality Publication of Higher Education Press (HEP) and Springer-Verlag, however, provides a ample opportunity to let each chapter to speak for itself, and therefore readers are strongly encouraged to peruse individual chapters in detail. This monumental and milestone book is indeed very rich and stimulating of vibrating activities in theory, applications, and system technologies in Pattern Recognition, Machine Intelligence and Biometrics (PRMIB).

Working out this book cannot be done by a single individual alone. It is the result of dedicated team work. I would like to take this opportunity to show my deepest appreciations to all contributors, without whom this book can never come out. I also want to thank all foreword writers (in alphabetic order): Professors Jake Aggarwal, Brian Lovell, and Sargur Srihari, who are either current or past President, Fellows, or King-Sun Fu Prize Laureates of IAPR, the largest and most prestigious organization of its kind. I also want to show my appreciation to Mrs. Meihua Wu Fu, who kindly provides some photos of late Professor King-Sun Fu. Thanks also go to HEP and Springer, for their strong support of this book project. Jake and his student Elden

Yu's suggestion to preface is also appreciated. Finally, I like to thank my parents, sons George Da-Yuan and David Da-Wen, for their understanding, encouragement and prayer. Above all, thank God for granting me life and soul, without which I can never have today.

Patrick S.P. Wang
Northeastern University (Boston)
East China Normal University (Shanghai),
Taiwan University of Science and Technology (Taipei)
May 2011

Contents

Part I: Pattern Recognition and Machine Intelligence

Part II: Computer Vision and Image Processing

Part III: Face Recognition and Forensics

Part IV: Biometric Authentication

Contributors

J.K. Aggarwal	Department of Electrical and Computer Engineering, The University of Texas, U.S.A.
Maan Ammar	Damascus University, Syria
Quanzhang An	Changzhou EGing Photovoltaic Technology Co., Ltd., China
Anja Attig	Institute of Computer Vision and Applied Computer Sciences, IBaI, Germany
Sangita Bharkad	S.G.G.S. Institute of Engg. and Technology, India
Lenina Birgale	Shri Guru Gobind Singhji Institute of Engineering and Technology, India
Terrance E. Boult	University of Colorado at Colorado Springs, U.S.A.
Horst Bunke	University of Bern, Switzerland
Chih-Wei Chen	Taiwan University, Taipei, Taiwan, Chain
Wensheng Chen	Shenzhen University, Shenzhen, China
Zhongxue Chen	The University of Texas Health Science Center, Houston, U.S.A.
Carlos A. Coello Coello	Depto. de Computación, Instituto Politécnico Nacional, México
Janos Csirik	University of Szeged, Hungary
Luis Gerardo de la Fraga	Depto. de Computación, Instituto Politécnico Nacional, México
Giovanni Dimauro	Università degli Studi di Bari 'Aldo Moro', Dipartimento di Informatica, Bari, Italy
Xiaoqing Ding	Tsinghua University, China
Bin Fang	Chongqing University, Chongqing, China
Chiou-Shann Fuh	Taiwan University, Taipei, Taiwan, Chain
Marina L. Gavrilova	University of Calgary, Canada
Mehrtash Harandi	Australia The University of Queensland, School of ITEE, Australia
Xin He	Motorola China Research Center, China
Brian C. Heflin	University of Colorado at Colorado Springs, U.S.A.
Mario Hern'andez-Tejera	Instituto de Sistemas Inteligentes y Aplicaciones Numéricas en Ingeniería 35005 Las Palmas de G.C., Spain
Ho-Ling Hsu	STARTEK Engineering Inc., Hsinchu, China
Wen-Hsing Hsu	Tsing Hua University, Hsinchu, Taiwan, China
Yuxiao Hu	Microsoft Corporation, Redmond, U.S.A.
Thomas S. Huang	University of Illinois at Urbana-Champaign, Urbana, U.S.A.
Hong Huang	Chongqing University, China
Xiaoyi Jiang	University of Münster, Department of Mathematics and Computer Science, Germany

Tai-Lang Jong	Tsing Hua University, Hsinchu, Taiwan, China
Manesh Kokare	Shri Guru Gobind Singhji Institute of Engineering and Technology, India
Jianhuang Lai	Sun Yat-sen University, China
Gary C.L. Li	PAMI Group, Department of System Design, University of Waterloo, Canada
Enshiun Annie Lee	PAMI Group, Department of System Design, University of Waterloo, Canada
Yan Liang	Sun Yat-sen University, China
Jenn-Jier James Lien	Cheng Kung University, Tainan, Taiwan, China
Jiunnliang Lin	Tsing Hua University, Hsinchu, China
Zheng Liu	University of Ottawa, Ottawa, Canada
Qingzhong Liu	Sam Houston State University, Huntsville, U.S.A.
Chenglin Liu	Institute of Automation, Chinese Academy of Sciences, China
Brian C. Lovell	School of Information Technology and Electrical Engineering, The University of Queensland, Australia
Yue Lu	Department of Computer Science and Technology, East China Normal University, China
Fangfang Lu	Shanghai Jiao Tong University, China
Yue Lu	East China Normal University, China
Longlong Ma	Institute of Automation, Chinese Academy of Sciences, China
Lin Mei	The Third Research Institute of Ministry of Public Security, Shanghai, China
Md. Maruf Monwar	University of Calgary, Canada
Petra Perner	Institute of Computer Vision and Applied Computer Sciences, IBaI, Germany
Mengyu Qiao	New Mexico Tech, Socorro, U.S.A.
Yang Qu	Department of Computer Science and Technology, East China Normal University, China
Radhika K R	B M S College of Engineering, India
Bernardete Ribeiro	University of Coimbra, Coimbra, Portugal
Anderson Rocha	University of Campinas (Unicamp), Brazil
Elena Sánchez-Nielsen	Dpto. E.I.O. y Computación. Universidad de La Laguna 38271 La Laguna, Spain
Walter J. Scheirer	University of Colorado at Colorado Springs, U.S.A.
Patrick S.P. Wang	Northeastern University, Boston, U.S.A.
Michael Schmeing	University of Münster, Department of Mathematics and Computer Science, Germany
Sheela S V	B M S College of Engineering, India
Pengfei Shi	Shanghai Jiao Tong University, China
Sargur N. Srihari	The State University of New York at Buffalo, U.S.A.
Chang Su	The State University of New York at Buffalo, U.S.A.
Andrew H. Sung	New Mexico Tech, Socorro, U.S.A.

Javid Taheri	The University of Sydney, Australia
Usman Tariq	University of Illinois at Urbana-Champaign, Urbana, U.S.A.
Ching-Ting Tu	Cheng Kung University, Tainan, Taiwan, China
Jian Wang	The Third Research Institute of Ministry of Public Security, China
Ying Wen	Columbia University, New York, U.S.A.
Andrew K.C. Wong	PAMI Group, Department of System Design, University of Waterloo, Canada
Wei Wu	Sichuan University, Sichuan, China
Xiaohua Xie	Sun Yat-sen University, China
Zhenqiang Yao	Shanghai Jiao Tong University, Shanghai, China
Mingde Yin	Nanjing University of Aeronautics and Astronautics, Nanjing, China
Elden Yu	Department of Electrical and Computer Engineering, The University of Texas at Austin, U.S.A.
Pong C. Yuen	Hong Kong Baptist University, Hong Kong, China
Weishi Zheng	Sun Yat-sen University, Guangzhou China, and Queen Mary University of London, London UK
Yaojie Zhu	Shanghai Yanfeng Visteon Automotive Trim Systems Co., Ltd., China
Dennis Zhuang	PAMI Group, Department of System Design, University of Waterloo, Canada
Yaoxian Zou	Sun Yat-sen University, China

Part I: Pattern Recognition and Machine Intelligence

1 A Review of Applications of Evolutionary Algorithms in Pattern Recognition

Luis Gerardo de la Fraga[1] and Carlos A. Coello Coello[1,2]

Abstract This chapter presents a review of some of the most representative work regarding techniques and applications of evolutionary algorithms in pattern recognition. Evolutionary algorithms are a set of metaheuristics inspired on Darwin's "survival of the fittest" principle which are stochastic in nature. Evolutionary algorithms present several advantages over traditional search and classification techniques, since they require less domain-specific information, are easy to use and operate on a set of solutions (the so-called population). Such advantages have made them very popular within pattern recognition (as well as in other domains) as will be seen in the review of applications presented in this chapter.

1.1 Introduction

We will start by providing a very general description of a pattern recognition process [1, 2] in order to make this chapter self-contained.

The pattern recognition process consists of the three stages shown in Fig. 1.1: (1) segmentation, (2) feature selection and (3) classification. In Fig. 1.1, the input is a set of pixels (i.e., an image), but other types of input data are also possible (e.g., three dimensional scanned points of a human face).

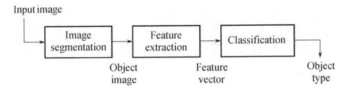

Fig. 1.1 The three phases of a pattern recognition system.

Segmentation is the partition of an input image, or data, into its constituent parts or objects. It is known that, in general, automatic segmentation is a very difficult task. The output of the segmentation stage is usually

1 CINVESTAV-IPN (Evolutionary Computation Group) Departamento de Computación, Av. IPN No. 2508, Col. San Pedro Zacatenco, Mexico, D.F. 07360, MEXICO. E-mails: {fraga, ccoello}@cs.cinvestav.mx.

2 The second author is also with the UMI LAFMIA 3175 CNRS at CINVESTAV-IPN.

another image with raw pixel data, constituting either the boundary of a region or all the points in the region itself.

Feature selection deals with extracting features for differentiating one class of objects from another. The output of this stage is a vector of values of the measured features.

The last stage is called classification, or recognition. This is the process that assigns a label to each object based on the information provided by their descriptors.

As will be seen in this chapter, Evolutionary Algorithms (EAs) have been applied to all three stages of the pattern recognition process [3]. In fact, some authors such as Rizki et al. [4] have applied EAs to each stage of a pattern recognition system, with minimum human intervention, with the aim of obtaining the best possible (overall) pattern recognition system.

From the several metaheuristics that currently exist (see for example [5, 6]) for solving optimization and classification problems, EAs, which emulate the natural selection mechanism, have become one of the most popular, mainly because of their ease of implementation, their flexibility and their high effectivity in a wide variety of tasks [7 – 10]. Thus, this chapter aims to provide a general (although not comprehensive due to obvious space limitations) overview on the use of EAs for solving pattern recognition tasks.

The remainder of this chapter is organized as follows. Section 1.2 presents a short introduction to evolutionary algorithms, including two examples that illustrate their use. Within these two examples, two specific evolutionary algorithms (namely, genetic algorithms and differential evolution) are discussed in more detail. Then, in Section 1.3 we provide a brief review of several applications representative of the work done regarding the use of EAs in any of the three previously indicated stages of a pattern recognition task. Some possible paths for future research regarding the use of EAs in pattern recognition are briefly discussed in Section 1.4. Finally, the main conclusions of this chapter are drawn in Section 1.5.

1.2 Basic Notions of Evolutionary Algorithms

Evolutionary algorithms are bio-inspired metaheuristics that attempt to emulate Charles Darwin's natural selection mechanism (i.e., the "survival of the fittest" principle) with the purpose of solving (mainly optimization) problems [8]. Metaheuristics are high-level frameworks that combine basic (low-level) heuristic methods to explore the search space of a problem in a more efficient and effective way [11]. A *heuristic* is a technique that searches for good solutions at a reasonable computational cost, but without guaranteeing optimality. In fact, in some cases, heuristics cannot even determine how far is a certain solution from the optimum [12].

The idea of getting inspiration from the natural selection mechanism to solve problems is not new, since it can be traced back to the 1930s [13]. Nevertheless, it was until the 1960s that these early ideas were actually implemented. Three are the main paradigms considered within EAs: genetic algorithms [14, 15, 7], evolution strategies [16–18] and evolutionary programming [19, 20]. Each of them was developed independent from the others, and with different motivations (e.g., genetic algorithms were originally developed to solve machine learning problems, whereas evolution strategies were originally developed to solve optimization problems). There have been also variations of some of these approaches. The most remarkable is genetic programming [21–24], which is a variation of the genetic algorithm that uses tree-based encoding, instead of the original fixed-length binary strings adopted in genetic algorithms.

EAs, in their different versions and variations have been found to be very effective for solving a wide variety of optimization and classification problems [7, 18, 20]. The main reasons for their popularity are their generality (they require little domain-specific information), ease of implementation and use, combined with their high effectiveness in solving highly complex problems [7].

The basic operation of an EA can be summarized as follows. First, they generate a set of possible solutions (called a "population") to the problem that is being solved. Such a population is normally generated in a random manner. Each solution in the population (called an "individual") encodes all the decision variables of the problem. In order to assess the suitability of such individuals, a fitness function must be defined. Such a fitness function is a variation of the objective function of the problem that we wish to solve and is used as a relative measure of performance among individuals (i.e., solutions that represent better objective function values are the fittest). Then, a selection mechanism must be applied in order to decide which individuals will "mate". This selection process is normally based on the fitness contribution of each individual (i.e., the fittest individuals have a higher probability of being selected). Upon mating, a set of children or "offspring" are generated. Such offspring are "mutated" (this operator produces a small random change, with a low probability, on the contents of an individual), and become the population to be evaluated at the following iteration (called a "generation"). This process is repeated until reaching a stopping condition (normally, a maximum number of generations).

The previous description corresponds to a general EA, but each specific EA adopts variations of this procedure. For example, a genetic algorithm will perform crossover (several possible crossover schemes exist), whereas evolutionary programming uses only mutation. It is also worth indicating that there are a few recent metaheuristics that are also normally considered as EAs. From them, the two most popular are differential evolution (DE) [25] and particle swarm optimization (PSO) [26].

Next, we will illustrate two application examples of the use of EAs in pattern recognition. In these examples, two specific types of EAs will be briefly described (namely, genetic algorithms and differential evolution).

1.2.1 Example 1: Clustering

Clustering can be seen as a previous stage to classification, in which groups of objects are labeled with an integer value and assigned into subsets (the so-called *clusters*), so that objects in the same cluster are similar according to some measure. Clustering is considered a method of unsupervised learning, and is commonly used for statistical data analysis in many disciplines, including data mining, pattern recognition, image analysis, machine learning and bioinformatics [27, 28].

Perhaps the most popular clustering algorithm in current use is k-means [29]. The k-means algorithm produces a partition of n observations $(x_1, x_2, \cdots, x_n)$, where x_i is a vector of features of dimension d, into k $(k < n)$ sets $S = \{S_1, S_2, \cdots, S_k\}$ such that the distances from each observation to the nearest partition are minimized:

$$\min_S \sum_{i=1}^{k} \sum_{x_j \in S_i} \|x_i - \mu_i\|^2, \tag{1.1}$$

where μ_i is the mean of the observations in the set i.

Very recently, it was proved that the complexity of the problem tackled by the k-means algorithm is NP-hard [30] in a general Euclidean space of dimension d, even for 2 clusters. It is also NP-hard even in 2 dimensions [31] for a general number of clusters. NP-hard complexity of a problem implies that, at least up to now, there is no procedure that can solve it in a polynomial time. In other words, this is really a very difficult task, which therefore justifies the use of a metaheuristic to solve it. Here, we will use a genetic algorithm to solve this problem.

A Simple Genetic Algorithm

Genetic algorithms (GAs) emphasize the importance of sexual recombination (which is the main operator) over the mutation operator (which is used as a secondary operator). They also use probabilistic selection (like evolutionary programming and unlike evolution strategies). The basic operation of a GA is illustrated in Fig. 1.2.

First, an initial population is randomly generated, as indicated in the description of a general EA. It is worth indicating, however, that deterministic or semi-deterministic procedures can also be used for generating the initial population (using, for example, a greedy procedure).

The individuals of the population of a GA will be a set of strings of characters (letters and/or numbers) called *chromosomes* that represent all the possible solutions to the problem. Chromosomes are made of *genes* that correspond to each of the decision variables of the problem. Finally, genes are

Generate initial population of size M in a random manner;
Repeat
    Evaluate fitness of each individual within M;
    Select (normally in a probabilistic way) individuals
        (called *parents*) based on fitness;
    With probability P_c, apply the crossover operator
        between pairs of parents.
    Apply the mutation operator to all the children
        generated from the previous step with a probability P_m.
    Apply elitism (i.e., the best individual in the population
        remains intact for the following generation).
Until the stop condition is reached

Fig. 1.2 Pseudocode of a simple genetic algorithm.

made of *alleles* which correspond to characters or symbols in the alphabet used as a basis for the encoding. This notation is graphically depicted in Fig. 1.3.

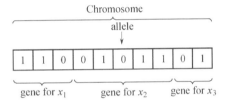

Fig. 1.3 Example of the binary encoding traditionally adopted with the genetic algorithm.

One aspect that has great importance in the case of the genetic algorithm is the encoding of solutions, since GAs normally use an indirect representation of the solutions of the problem (unlike evolution strategies that typically adopt a direct representation of solutions, by operating over vectors of real numbers). Traditionally, GAs adopt a binary encoding regardless of the type of decision variables of the problem to be solved, mainly because of the fact that this sort of encoding can be considered as universal [7], but biological arguments have also been provided to favor this sort of encoding [32]. Nevertheless, other types of encodings are also possible and have been used with genetic algorithms (see for example [33–35]).

After defining the encoding to be adopted by the GA, a *fitness function* value must be computed for each of the chromosomes in the population. These fitness values measure the quality of the solution encoded by the chromosome and represent a relative performance measure that will allow us to know which solutions are preferable over others. Knowing each chromosome's fitness, a *selection* process takes place to choose the individuals (presumably, the fittest) that will become the parents of the following generation. A variety of selection schemes exist [36], including *roulette wheel* selection [37], *stochastic remainder* selection [38, 39], *stochastic universal* selection [40, 41], *ranking* selection [42] and *tournament* selection.

After being selected, *crossover* takes place. During this stage, the genetic material of a pair of individuals is exchanged in order to create the population of the next generation. This operator is applied with a certain probability P_c

to pairs of individuals selected to be parents (P_c is normally set between 60% and 100%). When using binary encoding, there are three main ways of performing crossover:

1) *Single-point crossover*   A position of the chromosome is randomly selected as the crossover point as indicated in Fig. 1.4.

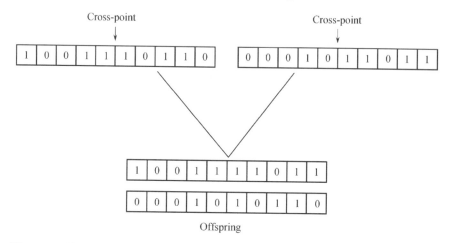

Fig. 1.4 Use of a single-point crossover between two chromosomes. Notice that each pair of chromosomes produces two descendants for the next generation.

2) *Two-point crossover*   Two positions of the chromosome are randomly selected for exchanging chromosomic material, as indicated in Fig. 1.5.

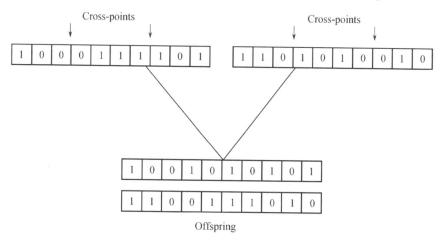

Fig. 1.5 Use of a two-point crossover between two chromosomes. In this case the genes at the extremes are kept, and those in the middle part are exchanged.

3) *Uniform crossover*   This operator was proposed by Syswerda [43] and can be seen as a generalization of the two previous crossover techniques. In

this case, for each bit in the first offspring it decides (with some probability P_c) which parent will contribute its value in that position, as indicated in Fig. 1.6. The second offspring will receive the bit from the other parent. Although for some problems uniform crossover presents several advantages over other crossover techniques [43], in general, one-point crossover seems to be a bad choice, but there is no clear winner between two-point and uniform crossover [44, 34].

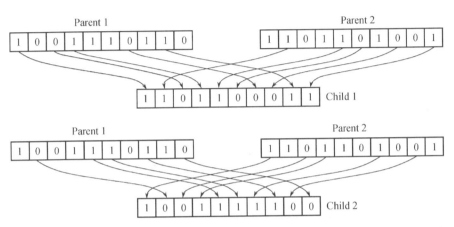

Fig. 1.6 Use of 0.5-uniform crossover (i.e., adopting a 50% probability of crossover) between two chromosomes. Notice how half of the genes of each parent go to each of the two children. First, the bits to be copied from each parent are selected randomly using the probability desired, and after the first child is generated, the same values are used to generate the second child, but inverting the source of procedence of the genes.

Other crossover operators are, of course, possible, mainly when adopting other encodings (see for example [45]).

The offspring generated by the crossover operator are subject to *mutation*, which is a genetic operator that randomly changes a gene of a chromosome. If we use a binary representation, a mutation changes a 0 to 1 and viceversa. This operator is applied with a probability P_m to each allele in a chromosome (P_m normally adopts a low probability that goes from 1% up to 10% as a maximum). The use of this operator allows the introduction of new chromosomic material to the population and, from a theoretical perspective, it assures that—given any population—the entire search space is connected [46].

Finally, the individual with the highest fitness in the population is retained, and it passes intact to the following generation (i.e., it is not subject to either crossover or mutation). This operator is called *elitism* and its use is required to guarantee convergence of a simple GA, under certain assumptions (see [47] for details).

The previous procedure must be repeated a certain number of times, until reaching a stopping criterion. The most commonly adopted stopping criterion is to use a (pre-defined) maximum number of iterations (or *generations*), but other criteria are also possible [7].

Although GAs can benefit from self-adaptation mechanisms that allow their parameters to be defined in an automated way [48], their users normally fine-tune their parameters (population size, crossover and mutation rates, etc.) by hand, using a trial-and-error process.

The Example

To show how this simple GA works, we will use the k-means problem. Let's assume that we have five observations and three clusters. Thus, an individual can be represented as $(1, 2, 1, 3, 2)$ meaning that the first and third observations form the first cluster. Analogously, the second and fifth observations form the second cluster, and the third cluster is formed with only the fourth observation.

The binary encoding corresponding to the above individual would be the following: $([0, 1], [1, 0], [0, 1], [1, 1], [1, 0])$. Any of the previously discussed crossover operators can be used in this case. Mutation is also applied as indicated before (using a NOT operator). The fitness value of each individual is obtained using Eq. (1.1). In this case, and considering the encoding adopted by the GA, the means must be computed from the code sets, each time we need to obtain the fitness value of an individual.

The use of a population, combined with the application of mutation, reduces the probability that the GA gets trapped in local minima. When using a GA, the use of a randomly generated population (using a uniform distribution) to start the search, solves the initialization of the k-means algorithm, which is known to have a significant impact on its performance [49].

It is worth indicating that GA-based clustering algorithms [50, 51] have shown superior performance than traditional clustering algorithms [49].

1.2.2 Example 2: Robust Ellipse Fitting

A silhouette of an ellipse, or a data set with elliptical form, can be detected by fitting. Given a function D that calculates the distance from an ellipse $\boldsymbol{x}$ to a point $\boldsymbol{p}$, the best ellipse is that which minimizes the function:

$$g : \mathbb{R}^5 \to \mathbb{R},$$

$$g = \sum_{i=1}^{n} D^2(\boldsymbol{x}, \boldsymbol{p}_i), \tag{1.2}$$

where $\boldsymbol{p}_i = [x_i, y_i]^{\mathrm{T}}$ is the vector representing the coordinates of each point $i, i = 1, 2, \cdots, n$, and the ellipse is represented by a vector of five parameters $\boldsymbol{x} = [a, b, x_c, y_c, \alpha]^{\mathrm{T}}$, where a and b are the semimajor and semiminor axes of the ellipse in a canonical form (centered at the origin of the coordinate system), (x_c, y_c) are the coordinates of the new origin for the translated ellipse, and α is the rotation angle between the semimajor axis and the original x axis.

To solve the problem stated in Eq. (1.2) is relatively simple by using the least squares method. The least squares procedure is based on the observation that the minimum of Eq. (1.2) is located at the place where its derivative (or its gradient) is equal to zero. Using the algebraic distance, the problem is linear and generates the most efficient algorithm to solve it [52]. Considering the real orthogonal distance between each point and the ellipse, this becomes a nonlinear problem which can be solved using the Gauss-Newton method [53].

The problem here is that using the algebraic distance generates distorted ellipses if points are not provided with sufficient accuracy. Also, the least squares method is not resistant to the presence of outliers in the data: points far away from the ellipse will be considered using the square of their distances. Furthermore, in the case of the nonlinear problem, an initial point, located very near to the actual optimum is required in order to solve it using mathematical programming techniques.

In order to obtain better results, we will use the linear solution to the problem as a starting point for the nonlinear algorithm. It is worth noting, however, that by doing this, the outliers affect even more to the linear solution.

Taking the sum of the distances, instead of their squared values, generates a new problem that will not present the same difficulties as the least squares procedures previously described. In this case, the problem consists of minimizing the following function:

$$g_1 : \mathbb{R}^5 \to \mathbb{R},$$

$$g_1 = \sum_{i=1}^{n} D(\boldsymbol{x}, \boldsymbol{p}_i), \tag{1.3}$$

The minimization of Eq. (1.3) is robust to the presence of outliers because every distance will be taken as it is, rather than adopting its squared value. The problem now is that Eq. (1.3) cannot be solved using any conventional least squares approach. Thus, the use of a metaheuristic is now appropriate. Something interesting here is that using an EA to solve this problem is much simpler [54, 55] than adopting a nonlinear least squares procedure, because we only need a procedure that computes the value of the function g_1 (we do not need to compute gradients or to perform any matrix inversion).

In this case, it is not possible to use the simple GA described in Section 1.2.1, because the most suitable representation for the solutions of this problem are vectors of real numbers. Although it is possible to use GAs with real-numbers encoding (see for example [56, 57]), we will adopt in this case an EA that has been found to be very powerful when dealing with this sort of problems (i.e., those in which the decision variables are real numbers): differential evolution [58, 25].

Differential Evolution

Differential Evolution was proposed by Kenneth Price and Rainer Storn in the mid 1990s [59, 60, 25]. DE is an evolutionary (direct-search) algorithm which has been mainly used to solve continuous optimization problems. DE shares similarities with traditional EAs. Unlike simple GAs [7], DE does not adopt binary encoding. Also, it does not use a probability density function to self-adapt its parameters as done with Evolution Strategies [61]. Instead, DE performs mutation based on the distribution of the solutions in the current population. Thus, the search directions and any possible step sizes depend on the location of the individuals that were selected to calculate the mutation values.

The most popular nomenclature adopted to refer to the different DE variants is called "DE/rand/1/bin", where "DE" means Differential Evolution, the word "rand" indicates that individuals selected to compute the mutation values are chosen at random, "1" is the number of pairs of solutions chosen and finally "bin" means that a binomial recombination is used. The corresponding algorithm of this variant ("DE/rand/1/bin") is shown in Fig. 1.7 and is the most popular in the specialized literature.

The "CR" parameter controls the influence of the parent in the generation of the offspring. Higher values mean less influence of the parent. The "F" parameter scales the influence of the pairs of solutions that are selected to obtain the mutation value (only one pair in the case of the algorithm in Fig. 1.7). The stopping criterion of DE is normally a maximum number of iterations (as in the case of the simple GA), but other, more elaborate criteria are also possible (see for example [62]).

Several DE variants are possible. To exemplify this point, we took from the paper by Mezura et al. [63] the eight DE variants adopted, each of which will be briefly described next. The modifications from variant to variant are in the recombination operator used (Steps 9 to 15 in Fig. 1.7) and also in the way individuals are selected to calculate the mutation vector (Step 7 in Fig. 1.7). The variants adopted by Mezura et al. [63] are the following:

1) Four variants whose recombination operator is discrete, always using two individuals: the original parent and the DE mutation vector (step 11 in Fig. 1.7). Two discrete recombination operators: binomial and exponential. The main difference between them is that for binomial recombination, each variable value of the offspring is taken at each time from one of the two parents, based on the value of "CR". On the other hand, in the exponential recombination, the value of each variable that forms the offspring is taken from the first parent until a random number surpasses the "CR" value. From this point, all the remaining offspring variable values will be taken from the second parent. These variants are called: "DE/rand/1/bin", "DE/rand/1/exp", "DE/best/1/bin" and "DE/best/1/exp" [64]. The "rand" variants select at

```
1   Begin
2       G=0
3       Create a random initial population x_{i,G} ∀i, i = 1, ⋯ , NP
4       Evaluate f(x_{i,G}) ∀i, i = 1, ⋯ , NP
5       For G=1 to MAX_GEN Do
6           For i=1 to NP Do
7  ⇒            Select randomly r_1 ≠ r_2 ≠ r_3 :
8  ⇒            j_{rand} = randint(1, D)
9  ⇒            For j=1 to D Do
10 ⇒                If (rand_j[0, 1) < CR or j = j_{rand}) Then
11 ⇒                    u_{i,j,G+1} = x_{r_3,j,G} + F(x_{r_1,j,G} − x_{r_2,j,G})
12 ⇒                Else
13 ⇒                    u_{i,j,G+1} = x_{i,j,G}
14 ⇒                End If
15 ⇒            End For
16             If (f(u_{i,G+1}) ⩽ f(x_{i,G})) Then
17                 x_{i,G+1} = u_{i,G+1}
18             Else
19                 x_{i,G+1} = x_{i,G}
20             End If
21         End For
22         G = G + 1
23     End For
24  End
```

Fig. 1.7 "DE/rand/1/bin" algorithm. randint(min,max) is a function that returns an integer between min and max. rand[0, 1) is a function that returns a real number between 0 and 1. Both are based on a uniform probability distribution. "NP", "MAX_GEN", "CR" and "F" are user-defined parameters. "D" is the dimensionality of the problem. Steps pointed with arrows change depending on the DE version adopted.

random to all the individuals to compute mutation and the "best" variants use the best solution in the population besides the random ones.

2) Two variants with arithmetic recombination, which, unlike discrete recombination, are rotation invariant. These are "DE/current-to-rand/1" and "DE/current-to-best/1" [64]. The only difference between them is that the first selects the individuals for mutation at random and the second one uses the best solution in the population besides random solutions.

3) "DE/rand/2/dir" [65], which incorporates objective function information to the mutation and recombination operators. The aim of this approach is to guide the search to promising areas faster than traditional DE. Their authors argue that the best results are obtained when the number of pairs of solutions is two [65].

4) Finally, a variant with a combined discrete-arithmetic recombination, the "DE/current-to-rand/1/bin" [64].

Each variant's implementation details are summarized in Table 1.1.

The Example

In our example, the vectors of real numbers for each individual will be of size 5, since we have five decision variables (a, b, x_c, y_c, α). The population is randomly initialized within the allowable range of the decision variables. In

Table 1.1 DE variants adopted by Mezura et al. [63]. j_r is a random integer number generated within the range $[0, n]$, where n is the number of decision variables of the problem. $U_j(0, 1)$ is a real number generated at random between 0 and 1. Both numbers are generated using a uniform distribution. In their experiments, Mezura et al. [63] used $p = 1$

Nomenclature	Variant
rand/p/bin	$u_{i,j} = \begin{cases} x_{r_3,j} + F \cdot \sum_{k=1}^{p}(x_{r_1,j}^p - x_{r_2,j}^p) & \text{if } U_j(0,1) < CR \text{ or } j = j_r \\ x_{i,j} & \text{otherwise} \end{cases}$
rand/p/exp	$u_{i,j} = \begin{cases} x_{r_3,j} + F \cdot \sum_{k=1}^{p}(x_{r_1,j}^p - x_{r_2,j}^p) & \text{from } U_j(0,1) < CR \text{ or } j = j_r \\ x_{i,j} & \text{otherwise} \end{cases}$
best/p/bin	$u_{i,j} = \begin{cases} x_{best,j} + F \cdot \sum_{k=1}^{p}(x_{r_1,j}^p - x_{r_2,j}^p) & \text{if } U_j(0,1) < CR \text{ or } j = j_r \\ x_{i,j} & \text{otherwise} \end{cases}$
best/p/exp	$u_{i,j} = \begin{cases} x_{best,j} + F \cdot \sum_{k=1}^{p}(x_{r_1,j}^p - x_{r_2,j}^p) & \text{from } U_j(0,1) < CR \text{ or } j = j_r \\ x_{i,j} & \text{otherwise} \end{cases}$
current-to-rand/p	$u_i = x_i + K \cdot (x_{r_3} - x_i) + F \cdot \sum_{k=1}^{p}(x_{r_1}^p - x_{r_2}^p)$
current-to-best/p	$u_i = x_i + K \cdot (x_{best} - x_i) + F \cdot \sum_{k=1}^{p}(x_{r_1}^p - x_{r_2}^p)$
current-to-rand/p/bin	$u_{i,j} = \begin{cases} x_{i,j} + K \cdot (x_{r_3,j} - x_{i,j}) + F \cdot \sum_{k=1}^{p}(x_{r_1,j}^p - x_{r_2,j}^p) & \text{if } U_j(0,1) < CR \text{ or } j = j_r \\ x_{i,j} & \text{otherwise} \end{cases}$
rand/2/dir	$v_i = v_1 + \dfrac{F}{2}(v_1 - v_2 + v_3 - v_4)$ where $f(v_1) < f(v_2)$ and $f(v_3) < f(v_4)$

our example, a and b can vary between 1 and one half of the size of the input image (if we consider that every point also represents a pixel position in an image), (x_c, y_c) can be inside the image, and $\alpha \in [0 : \pi]$.

For the fitness value, we use in this case the value of the function g_1 indicated in Eq. (1.3). The results obtained by DE are shown in Fig. 1.8, in which they are compared with respect to those obtained with the linear algorithm. It can be clearly seen that the ellipse fitted by DE is much better than the other one.

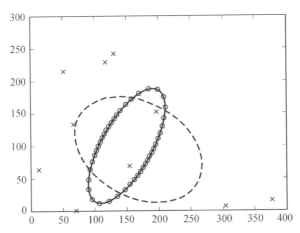

Fig. 1.8 45 ellipse points and 10 outliers. We show with a solid line to the ellipse fitted by differential evolution, using the sum of distances. With a dashed line we show the ellipse fitted with the linear algorithm, according to the sum of the squared distances. Clearly the last one is very distortioned and is not resistant to the presence of outliers.

1.3 A Review of EAs in Pattern Recognition

In this section, we will briefly review some representative work on the use of EAs for pattern recognition tasks. Our review will provide examples of all three stages of the pattern recognition process: (1) segmentation, (2) feature selection and (3) classification. Each of these groups of applications are discussed in the following subsections.

1.3.1 Segmentation

Image segmentation denotes a process by which a raw input image is partitioned into non-overlapping regions such that each region is homogeneous and connected. A segmented image is often considered connected if there exists a connected path between any two pixels within the region. A region is considered homogeneous if all of its pixels satisfy a homogeneity criterion defined over one or more pixel attributes such as intensity, texture, color, range, etc. Computing such an image partition is a problem of very high combinatorial complexity. Given the astronomical size of the search space, an exhaustive or near-exhaustive enumeration of all possible image partitions to arrive at a segmented image is usually infeasible. This motivates the use of EAs in this problem.

In [66] Bhandarkar and Zhang use three hybrid EAs for the segmentation of gray level images: (1) a GA hybridized with simulated annealing (SA) [67] called SA-GA, (2) a GA hybridized with microcanonical annealing (MCA) [68] called MCA-GA, and (3) a GA hybridized with a random cost algorithm (RCA) [69] called RCA-GA. It should be noted that SA, MCA and RCA are all stochastic hill-climbing search techniques (i.e., they are local search engines). The problem was defined as one of minimization, in which the fitness function incorporated both region contour (or edge information) and region gray-scale uniformity (such uniformity was quantified using gray-level variance values). Clearly, the desired segmentation corresponded to the global minimum of the cost function proposed by the authors. The authors noted, however, that their fitness function presented many local minima. The results of the three hybrid schemes were compared with respect to those of a simple GA using several images, both with and without (Gaussian) noise. RCA-GA was the fastest approach, closely followed by MCA-GA. However, it was SA-GA (the slowest among the hybrid approaches) the one that obtained the best overall results in terms of both visual quality and cost value of the final segmentation. All the hybrid approaches had a better performance than the simple GA, which clearly showed the usefulness of local search in this case.[3]

Jiang and Yang proposed in [71] a hybrid approach that combines features of genetic algorithms [7] and tabu search [72]. The proposed approach, which was called "evolutionary tabu search" (ETS) was used for the segmentation of cell images. Knowing that most cells in the human body have ellipse-like boundaries, the authors used an ellipse equation to describe the boundary of a cell. Since the definition of an ellipse requires five parameters (two to determine its location and three more to determine its size and orientation), the authors used these five values as the decision variables to be obtained. The goal was, evidently, to fit the cell boundary as well as possible, and, therefore, the fitness function was defined in such a way that it counted the number of points that were within a certain (fixed) distance from the ellipse. In order to find the edge points, the authors adopted the Canny operator [73]. The proposed hybrid was compared with respect to its two components considered separately (i.e., the genetic algorithm and tabu search), showing to be superior to both of them. The authors reported that their pro-

---

3 Population-based approaches (e.g., EAs) hybridized with a local search procedure are also known as memetic algorithms [70].

posed approach was able to find consistently (i.e., there was little variation of results over several independent runs) near-optimal results for several images with red blood cells.

Bocchi et al. presented in [74] an EA for image segmentation. The main idea of this approach is to perform a colonization of a bidimensional world (i.e., the image) by a certain number of populations, each of which represents a different region of the image. The individuals of each population, competed in order to occupy all the available space and to adapt to the specific local environmental characteristics of the world. This approach was inspired on the famous game of "Life" [75]. The authors validated their proposed approach using several synthetic images. Their results produced significant improvements with respect to those found by the well-known fuzzy c-means clustering algorithm [76]. The authors indicated that their approach can be used for the segmentation of gray-scale, color and textural images. Indeed, they indicated that their approach can be extended to any vector-valued parametric images, regardless of their number of components.

Krawiec et al. provided in [77] a review of work on the use of genetic programming (GP) [22] for object detection and image analysis. As indicated before, GP refers to a variation of the genetic algorithm in which a tree-encoding is adopted. This special kind of encoding evidently requires of different alphabets and specialized operators, since we are, in fact, evolving programs rather than simple vectors of decision variables. The trees used in GP consist of both functions and terminals. The most commonly adopted functions are the following [22]:

```
Arithmetic operations (e.g., +, -, ×, ÷ )
Mathematical functions (e.g., sine, cosine, logarithms, etc.)
Boolean Operations (e.g., AND, OR, NOT)
Conditionals (IF-THEN-ELSE)
Loops (DO-UNTIL)
Recursive Functions
Any other domain-specific function
```

Terminals are typically variables or constants, and can be seen as functions that take no arguments. GP incorporates fitness-based selection, crossover and mutation (obviously, all of them are modified so that they can properly deal with trees), but also adds special procedures for generating the initial population (e.g., a maximum tree height is normally enforced to avoid an excessive memory use), as well as other operators for a variety of tasks (e.g., to destroy a certain percentage of the population in order to improve diversity, or to "encapsulate" or protect certain subtrees that we want to keep). Because of its nature, GP provides a powerful tool for building scalable and adaptive image analysis systems, which use raw image data as input and produce complete recognition systems as their output. Krawiec et al. [77] point out, however, that the main drawback of GP is that (as any other EA) it normally requires a considerable number of fitness function evaluations, and each of these evaluations are quite expensive (computationally speaking) in this case. They also indicate that most of the applications that they reviewed are focused on feature-based recognition, in which the evolved system is able to discriminate between positive and negative examples based on certain features of the image. However, they found very few applications focused on model-based recognition in which the underlying assumption is that a database of models of recognized objects is available and comparisons are then done between the input image and the object models. This sort of approach allows for the recognition of more complex (even compound) objects, but has been rarely adopted with GP.

1.3.2 Feature Selection

Feature selection refers to the selection of the most relevant features (e.g., of a class of objects) so that we can build robust learning models (e.g., for classifying classes of objects). By robustness we refer to finding the minimum subset of features that are useful to keep in order to obtain an efficient and improved solution to the problem of our interest (e.g., to classify a set of objects). This task is important, since normally not all the available features are useful and keeping any unnecessary features increases the computational cost required to solve the problem. Performing an optimal feature selection requires an exhaustive search of all possible subsets of features. This makes this problem suitable for using EAs, because of the very large search space normally involved.

In [78], Muni et al. presented an online feature selection algorithm based on GP. In fact, the use of the tree-based encoding of GP also allows for the design of the classifiers, which is something they do for a multicategory classification problem. The authors generated the initial population in such a way that the initialization process itself, generated classifiers that had a high probability of using smaller feature subsets. The fitness function adopted by the authors assigns a higher fitness value to any classifier that is able to classify more samples using less features. Indeed, this can be seen as a multiobjective fitness function [79], since it performs both feature selection and the design of classifiers at the same time. The authors also proposed two crossover operators specially designed to perform feature selection. The authors compared results with respect to those reported in the specialized literature for several data sets that go from low (four) to high (over 7000) dimensionality. Their proposed approach provided better performance for both two-class and multiclass problems, even in problems having redundant or bad features (which were artificially added). The main limitation of this approach is only its applicability, which is constrained to numerical attributes only, because their GP implementation adopted arithmetic functions to design the classifiers.

Watchareeruetai et al. [80] adopted a variation of Linear Genetic Programming (LGP)[4] [81] for extracting features from images. The authors indicated that the main source of problems when using LGP for image feature extraction is the excessive redundancy associated with the encoding adopted by this technique. Such redundancy can substantially increase the computational cost of LGP, since the same costly solutions are being evaluated more than once. In order to deal with this problem, the authors proposed a transformation of the LGP representation into a canonical form that has no redundancies. The experiments conducted by the authors indicated reductions in computational time that go from 7% up to 62% with respect to the use of the original LGP encoding.

Kowaliw et al. [82] adopted cartesian GP[5] [84] to define a set of transforms on the space of grayscale images which were meant to facilitate a further classification process. The idea is that these transforms (which are really programs) emphasize distinguishing characteristics. The authors applied their proposed approach for detecting muscular dystrophy-indicating inclusions in cell images. In their experiments, the authors were able to discover a set of features that could achieve 91% recognition of healthy cells, and an 88% recognition of sick cells. These accuracies provided a 38% improvement over predefined features alone. However, the authors considered as more important the fact that this approach may constitute an important step towards having a GP-based recognition system that automatically adapts to a given database without any human intervention.

Guo et al. [85] proposed an automatic image pattern recognition system which was used for classification of medical images. This system adopts (apparently, for the first time, as claimed by the authors) a generalized primitive texture feature extraction technique based on a histogram region of interest by thresholds (HROIT), which is used to characterize the Oculopharyngeal Muscular Dystrophy (OPMD) disease. They also proposed a new technique, based on the integration of genetic programming (GP) and the expectation maximization (EM) algorithm [86] (called GP-EM), for generating feature functions automatically, based on the primitive features obtained from HROIT. The GP-EM system makes simpler the learning task by generating a single feature (which is actually a program generated by GP). The authors showed that this approach led to higher classification accuracies (90%, on average, in diagnosing the OPMD disease) and lower standard deviations, being able to outperform another classification system based on a Support Vector Machine (SVM) [87].

Raymer et al. [88] proposed an approach in which a GA was used to perform, simultaneously, feature selection, feature extraction and classifier training. The GA was used to reduce the dimensionality of the feature set.

---

4 LGP is a GP [22] variant that evolves sequences of instructions from an imperative programming language. The term *linear* is used in this case to denote the structure of the imperative program representation [81].

5 Cartesian Genetic Programming was proposed in [83] with the purpose of evolving digital circuits and represents programs as directed graphs.

Basically, the GA transformed a set of patterns into a lower dimensionality (a set of weight vectors that scale the individual features of the original pattern vectors was used for this sake). The aim was to reduce the dimensionality as much as possible, while maximizing the classification accuracy. The authors also adopted a binary masking vector to perform a selection of a subset of the features under consideration. This was done while performing the dimensionality reduction, since the GA used both vectors (the weight vectors and the binary masking vector) in its chromosomes. Each of the resulting subsets of features were evaluated in terms of their classification accuracy on some test data using a nearest neighbor classifier. They also used this approach in combination with the k nearest neighbor classification rule to allow linear feature extraction (instead of binary vectors, real numbers were used in this case). The proposed approach was tested on medical and on biochemical data, producing very competitive results in all cases, while using less features than the other classifiers with respect to which it was compared.

1.3.3 Classification and Clustering

As indicated before, clustering can be seen as a previous stage to classification, and even as a simpler form of classification, which makes it very difficult to distinguish one task from the other. Because of that, both of them are considered in this section.

Iglesia et al. [89, 90] used a multi-objective GA called *Nondominated Sorting Genetic Algorithm-II* (NSGA-II) [91] for doing partial classification (this is called the *nugget discovery task*). The aim was to maximize both confidence and coverage of the rules. The initial population of the GA was not randomly generated. Instead, the authors used a special procedure that looked at the data to ensure that no rules with zero coverage were produced (as happens when a randomly generated population is adopted). In order to achieve this, a default rule was adopted (in this default rule, all limits were maximally spaced and all labels were included), and the remainder of the population were just mutations of the default rule. The authors compared results with respect to another approach which was developed by some of the same authors. For their validation, the authors used data sets taken from the UCI repository [92]. The proposed approach was able to produce sets of rules of similar quality as the other approach, and was even able to outperform it in some cases.

EAs have been widely used for clustering, as made evident in the survey on GA-based clustering written by Sheikh et al. [93]. In this review, the authors indicated that one of the main advantages of EAs, when used for clustering, is that, unlike classical clustering techniques (e.g., k-means [29], fuzzy c-means [76], etc.), they do not require the number of clusters as an input parameter, since this value can be produced during the search. This review also showed a rich variety of applications, including image compression, microarray data analysis, document clustering and text clustering, among others. Finally, the authors indicated that in their review they found that GAs had only being applied to distance-based clustering algorithms (e.g., k-means) but not with other types of clustering algorithms.

Bandyopadhyay and Maulik presented in [94] a GA for the automated clustering of data sets. The proposed approach can evolve the number of clusters while performing the clustering of the data. The authors adopted a special encoding that contains both real numbers (which encode the centers of the clusters) and "don't" care symbols (which are used to encode a variable number of clusters). The fitness of individuals was computed using the Davies-Bouldin index [95]. The proposed approach was validated using four artificial and two real-world data sets in which the number of clusters goes from two to six, and the number of dimensions goes from two to nine. In their experiments, the authors considered both overlapping and non-overlapping data sets. In all cases, the results were found to be quite competitive.

In a further paper, Bandyopadhyay et al. [96] proposed the constrained elitist multiobjective genetic algorithm based classifier (CEMOGA-Classifier) for developing nonparametric classifiers. The proposed approach can overcome problems common to traditional (single-objective) classifiers, such as overfitting/overlearning and ignoring smaller classes. In the proposed classifier, the authors considered three objectives: minimize (1) the number of misclassified training points and (2) the number of hyperplanes, and maximize (3) the product classwise correct recognition rates. Since CEMOGA-Classifier generates several solutions (called *nondominated* [79]), and it is desirable to have only one (which corresponds to the desired classifier), the authors adopted a validity function to select only one solution from the non-dominated set. This validity function is really an aggregating function that combines the values achieved by the classifier in each of the three objectives considered during the optimization process. This validity function allows the user to weight the importance that he/she wishes to give to each of these objectives. Results were compared with respect to those produced by two well-established multi-objective evolutionary algorithms: NSGA-II [91] and the Pareto Archived Evolution Strategy (PAES) [97]. In general, CEMOGA-Classifier was able to approximate the class boundaries of the data sets adopted using a smaller number of hyperplanes, which indicates a superior generalization capability.

1.4 Future Research Directions

As has been shown in the previous sections, the use of EAs in pattern recognition is a very active research area. There are, however, several other topics within this area that still represent interesting research topics. The following are a few examples:

Integration

One of the long-term goals of using EAs in pattern recognition must be the development of fully automated systems that can be applied to different databases with minimum (or no) human intervention. This may require to combine EAs with other approaches (e.g., fuzzy logic and/or machine learning techniques) as well as the design of new architectures that allow an efficient and effective integration of different types of approaches throughout the different stages involved in a pattern recognition process. The use of multiobjective optimization techniques (which are designed to solve problems in which we aim to optimize two or more (normally conflicting) objectives) may be useful for this task [79] and have, indeed, been used (and have been raising increasing interest) in a variety of pattern recognition tasks (see for example [98 – 100]).

Efficiency

As indicated before, a critical aspect that could certainly limit the applicability of EAs in pattern recognition is their computational cost. Although EAs by themselves are computationally inexpensive, they normally require to evaluate the fitness of several hundreds or thousands of solutions in order to obtain the necessary information to guide the search. In many pattern recognition tasks (e.g., in segmentation) these fitness evaluations are computationally expensive. There has been little work until now regarding the incorporation of fitness approximation techniques [101] to EAs adopted for pattern recognition tasks. Such techniques could provide an interesting alternative to improve the efficiency of an EA, at the expense of sacrificing some accuracy, if this is affordable at the application at hand.

Use of other Metaheuristics

In recent years, other bio-inspired metaheuristics have become increasingly popular in a wide variety of applications [102]. These metaheuristics also have a lot of potential in pattern recognition tasks, but their use in this domain still remains relatively scarce. Representative examples of these new metaheuristics are the following:

– Particle Swarm Optimization: This is actually another type of EA which was originally proposed by James Kennedy and Russell C. Eberhart in the mid-1990s [103, 26]. This metaheuristic simulates the movements of a flock of birds which aim to find food. In this approach, the behavior of each individual (called *particle*) is affected by either the best local (i.e., within a certain neighborhood) or the best global individual. As other EAs, it also uses a population as well as a fitness measure. However, because of its design, this metaheuristic allows individuals to benefit from their past experiences, which is a feature that is normally inexistent in traditional EAs. This technique has been widely applied in a variety of problems [26].

-   Artificial Immune Systems: Our immune system can be seen as a highly parallel intelligent system that is able to learn and retrieve previous knowledge (in other words, it has "memory"), while solving complex recognition and classification tasks (namely, detecting antigens that invade our body). These features make immune systems quite attractive from a computational point of view, and has motivated a lot of research on the development of mathematical and computational models that emulate its operation. Artificial immune systems were introduced in the mid-1990s and since then, have attracted increasing interest from researchers who have used it for a variety of tasks, including a few classification and pattern recognition problems [104–106].
-   Ant Colony Optimization: This is a metaheuristic inspired by the behavior shown by colonies of real ants which deposit a chemical substance on the ground called *pheromone* [107, 108]. The pheromone influences the behavior of the ants: they tend to take those paths in which there is a larger amount of pheromone. Pheromone trails can thus be seen as an indirect communication mechanism used by the ants. This system also presents several interesting features from a computational perspective, and has triggered a significant amount of research. The first of these ant-based systems, called *ant system* was originally proposed for the traveling salesman problem. However, over the years, this approach (and its several variations, which are now collectively denominated *ant colony optimization* algorithms) has been applied to a wide variety of combinatorial optimization problems [108].

1.5 Conclusions

In this chapter, we have provided a short introduction to EAs and some of their applications within pattern recognition. This introduction included a summarized description of a few EAs (genetic algorithms, genetic programming and differential evolution). Then, several applications of them in segmentation, feature selection and classification (including clustering) were reviewed. This review indicated a great interest in using EAs for pattern recognition tasks, and also pointed out to the possible use of EAs combined with other approaches for the development of fully automated pattern recognition systems.

In the last part of this chapter, some further ideas to extend the research in this area were provided, including the use of other bio-inspired metaheuristics and the incorporation of fitness approximation techniques to reduce the (normally high) computational cost associated to the use of EAs.

We really hope that the information provided in this chapter increases the interest from researchers working in pattern recognition to use EAs, for that has been its main goal.

Acknowledgements

The second author acknowledges support from CONACyT project no. 103570.

References

[1] Gonzalez R C, Woods R E (1992) Digital Image Processing. Addison-Wesley, New York

[2] Castleman K R (1996) Digital Image Processing. Prentice Hall, New Jersey

[3] Pal S K, Wang P P (eds) (1996) Genetic Algorithms for Pattern Recognition. CRC Press, Boca Raton

[4] Rizki M M, Zmuda M A, Tamburino L A (2002) Envolving pattern recognition systems. IEEE Transactions on Evolutionary Computation, 6(6): 594–609

[5] Glover F, Kochenberger G A (eds) (2003) Handbook of Metaheuristics. Kluwer Academic Publishers, Norwell

[6] Ibaraki T, Nonobe K, Yagiura M (eds) (2005) Metaheuristics. Progress as Real Problem Solvers. Springer, New York

[7] Goldberg D E (1989) Genetic Algorithms in Search, Optimization and Machine Learning. Addison-Wesley, New York

[8] Fogel D B (1995) Evolutionary Computation. Toward a New Philosophy of Machine Intelligence. The Institute of Electrical and Electronic Engineers, New York

[9] Eiben A E, Smith J E (2003) Introduction to Evolutionary Computing. Springer, Berlin

[10] Sivanandam S N, Deepa S N (2008) Introduction to Genetic Algorithms. Springer, Berlin

[11] Blum C, Roli A (2003) Metaheuristics in Combinatorial Optimization: Overview and Conceptual Comparison. ACM Computing Surveys, 35(3): 268–308

[12] Reeves C B (ed) (1993) Modern Heuristic Techniques for Combinatorial Problems. Wiley, Chichester

[13] Fogel D B (ed) (1998) Evolutionary Computation. The Fossil Record. Selected Readings on the History of Evolutionary Algorithms. The Institute of Electrical and Electronic Engineers, New York

[14] Holland J H (1962) Concerning Efficient Adaptive Systems. In: Yovits M C, Jacobi G T, Goldstein G D (eds) (1962) Self-Organizing Systems, pp 215–230. Spartan Books, Washington D C

[15] Holland J H (1962) Outline for a Logical Theory of Adaptive Systems. Journal of the Association for Computing Machinery, 9: 297–314

[16] Schwefel H P (1965) Kybernetische Evolution als Strategie Der Experi-Mentellen Forschung in Der Strömungstechnik. Dipl-Ing Thesis

[17] Schwefel H P (1977) Numerische Optimierung von Computer-Modellen mittels der Evolutionsstrategie. Birkhäuser, Basel, Alemania

[18] Schwefel H P (1981) Numerical Optimization of Computer Models. Wiley, Chichester

[19] Fogel L J (1966) Artificial Intelligence Through Simulated Evolution. Wiley, New York

[20] Fogel L J (1999) Artificial Intelligence Through Simulated Evolution. Forty Years of Evolutionary Programming. Wiley, New York

[21] Koza J R (1989) Hierarchical genetic algorithms operating on populations of computer programs. In: Sridharan N S (ed) Proceedings of the 11th International Joint Conference on Artificial Intelligence, pp 768–774. Morgan Kaufmann, San Mateo

[22] Koza J R (1992) Genetic Programming. On the Programming of Computers by Means of Natural Selection. MIT Press, Cambridge

[23] Koza J R (1994) Genetic Programming II: Automatic Discovery of Reusable Programs. MIT Press, Cambridge

[24] Koza J R, Bennet F H, III, Andre D et al (1999) Genetic Programming III: Darwinian Invention and Problem Solving. Morgan Kaufmann, Sna Mateo

[25] Price K V, Storn R M, Lampinen J A (2005) Differential Evolution. A Practical Approach to Global Optimization. Springer, Berlin

[26] Kennedy J, Eberhart R C (2001) Swarm Intelligence. Morgan Kaufmann, San Francisco

[27] Xu R, Wunsch D (2009) Clustering. IEEE Press and Wiley, Hoboken

[28] Gan G, Ma C, Wu J (2007) Data Clustering: Theory, Algorithms, and Applications. Society for Industrial and Applied Mathematics. Philadelphia, Pennsylvania

[29] MacQueen J B (1967) Some Methods for Classification and Analysis of Multivariate Observations. In: Proceedings of the 5th Berkeley Symposium on Mathematical Statistics and Probability, 2: 281–297. University of California Press, Berkeley

[30] Aloise D, Deshpande A, Hansen P et al (2009) NP-hardness of Euclidean Sum-of-squares Clustering. Machine Learning, 75(2): 245–249

[31] Mahajan M, Nimbhorkar P, Varadarajan K (2009) The Planar k-means Problem is NP-hard. Lecture Notes in Computer Science, 5431: 274–285

[32] Holland J H (1975) Adaptation in Natural and Artificial Systems. University of Michigan Press, Ann Arbor

[33] Ronald S (1995) Genetic Algorithms and Permutation-encoded Problems: Diversity Preservation and a Study of Multimodality. PhD Thesis, The University of South Australia

[34] Michalewicz Z (1996) Genetic Algorithms + Data Structures = Evolution Programs, 3rd Edn. Springer, New York

[35] Rothlauf F (2002) Representations for Genetic and Evolutionary Algorithms. Physica-Verlag, New York

[36] Goldberg D E, Deb K (1991) A Comparison of Selection Schemes used in Genetic Algorithms. In: Gregory J E Rawlins (ed) Foundations of Genetic Algorithms, pp 69–93. Morgan Kaufmann, San Mateo

[37] De Jong K A (1975) An Analysis of the Behavior of a Class of Genetic Adaptive Systems. PhD Thesis, University of Michigan, Ann Arbor, Michigan, USA

[38] Booker L B (1982) Intelligent Behavior as an Adaptation to the Task Environment. PhD Thesis, Logic of Computers Group, University of Michigan, Ann Arbor, Michigan, USA

[39] Brindle A (1981) Genetic Algorithms for Function Optimization. PhD Thesis, Department of Computer Science, University of Alberta, Alberta, Canada

[40] Baker J E (1987) Reducing Bias and Inefficiency in the Selection Algorithm. In: John J Grefenstette (ed) Genetic Algorithms and Their Applications: Proceedings of the Second International Conference on Genetic Algorithms, pp 14–22. Lawrence Erlbaum Associates, Hillsdale

[41] Grefenstette J J, Baker J E (1989) How Genetic Algorithms work: A critical look at implicit parallelism. In: David Schaffer J (ed) (1989) Proceedings of the Third International Conference on Genetic Algorithms, pp 20–27. Morgan Kaufmann Publishers, San Mateo

[42] Baker J E (1985) Adaptive Selection Methods for Genetic Algorithms. In: John J Grefenstette (ed) Proceedings of the First International Conference on Genetic Algorithms, pp 101–111. Lawrence Erlbaum Associates, Hillsdale

[43] Syswerda G. Uniform Crossover in Genetic Algorithms. In: Schaffer J D (ed) (1989) Proceedings of the Third International Conference on Genetic Algorithms, pp 2–9. Morgan Kaufmann, San Mateo

[44] Mitchell M (1996) An Introduction to Genetic Algorithms. MIT Press, Cambridge

[45] Dumitrescu D, Lazzerini B, Jain L C et al (2000) Evolutionary Computation. CRC Press, Boca Raton

[46] Buckles B P, Petry F E (eds) (1992) Genetic Algorithms. Technology Series. IEEE Computer Society Press, New York

[47] Rudolph G (1994) Convergence Analysis of Canonical Genetic Algorithms. IEEE Transactions on Neural Networks, 5(1): 96–101

[48] Eiben A E, Hinterding R, Michalewicz Z (1999) Parameter Control in Evolutionary Algorithms. IEEE Transactions on Evolutionary Computation, 3(2): 124–141

[49] Peña J M, Lozano J A, Larrañaga P (1999) An Empirical Comparison of Four Initialization Methods for the k-means Algorithm. Pattern Recognition Letters, 20: 1027–1040

[50] Maulik U, Bandyopadhyay S (2000) Genetic Algorithm-based Clustering Technique. Pattern Recognition, 33: 1455–1465

[51] Krishna K, Narasimha Murty M (1999) Genetic k-means Algorithm. IEEE Trans on Systems, Man and Cybernetics Part B, 29(3): 433–439

[52] Fitzgibbon A, Pilu M, Fisher R B (1999) Direct Least Square Fitting of Ellipses. IEEE Pattern Analysis and Machine Intelligence, 21(5): 476–480

[53] Ahn S J, Rauth W, H-J Warnecke (2001) Least-squares Orthogonal Distances Fitting of Circle, Sphere, Ellipse, Hyperbola, and Parabola. Pattern Recognition, 34(12): 2283–2303

[54] de la Fraga L G, Vite Silva I, Cruz-Cortes N (2009) Euclidean Distance fit of Conics Using Differential Evolution, pp 171–184. Springer, Heidelberg

[55] de la Fraga L G, Lopez G M Dominguez (2010) Robust Fitting of Ellipses with Heuristics. 2010 IEEE Congress on Evolutionary Computation, CEC 2010, (AC-CEPTED)

[56] Herrera F, Lozano M, Verdegay J L (1998) Tackling Real-Coded Genetic Algorithms: Operators and Tools for Behavioural Analysis. Artificial Intelligence Review, 12(4): 265–319

[57] García-Martínez C, Lozano M, Herrera F et al (2008) Global and Local Real-Coded Genetic Algorithms Based on Parent-Centric Crossover Operators. European Journal of Operational Research, 185(3): 1088–1113

[58] Chakraborty U K (2008) Advances in Differential Evolution. Studies in Computational Intelligence. Springer, Heidelberg

[59] Storn R, Price K (1995) Differential Evolution: A Simple and Efficient Adaptive Scheme for Global Optimization over Continuous Spaces. Technical Report TR-95-012. International Computer Science Institute, Berkeley

[60] Storn R, Price K (1997) Differential Evolution: A Fast and Efficient Heuristic for Global Optimization over Continuous Spaces. Journal of Global Optimization, 11(4): 341–359

[61] Schwefel H P (1995) Evolution and Optimum Seeking. Wiley, New York

[62] Zielinski K, Laur R (2008) Stopping Criteria for Differential Evolution in Constrained Single-objective Optimization. In: Chakraborty U K (ed) Advances in Differential Evolution. Studies in Computational Intelligence. Springer, Heidelberg

[63] Efrén Mezura-Montes, Jesús Velázquez-Reyes, Carlos A Coello Coello (2006) Comparing Differential Evolution Models for Global Optimization. In: Maarten Keijzer et al (ed) (2006) 2006 Genetic and Evolutionary Computation Conference (GECCO'2006), 1: 485–492, Seattle, Washington, USA, July 2006. ACM Press, New York

[64] Price K V (1999) An Introduction to Differential Evolution. In: David Corne, Marco Dorigo, Fred Glover (eds) New Ideas in Optimization, pp 79–108. McGraw-Hill, London

[65] Feoktistov V, Janaqi S (2004) Generalization of the Strategies in Differential Evolution. In: Proceedings of the 18th International Parallel and Distributed Processing Symposium (IPDPS 2004), 2004, Santa Fe, New Mexico, USA, p 165a, New Mexico, USA, April 2004. IEEE Computer Society.

[66] Bhandarkar S M, Zhang H (1999) Image segmentation using evolutionary computation. IEEE Transactions on Evolutionary Computation, 3(1): 1–21

[67] Kirkpatrick S, Gellatt C D, Vecchi M P (1983) Optimization by Simulated Annealing. Science, 220(4598): 671–680

[68] Creutz M (1983) Microcanonical monte-carlo simulation. Physical Review Letters, 50(19): 1411–1414

[69] Wang Y H, Prade R A, Griffith J et al (1994) A Fast Random Cost Algorithm for Physical Mapping. Proceedings of the National Academy of Sciences of the United States of America, 91(23): 11094–11098

[70] Moscato P (1999) Memetic Algorithms: A Short Introduction. In: David Corne, Fred Glover, Marco Dorigo (eds) New Ideas in Optimization, pp 219–234. McGraw-Hill, New York

[71] Tianzi Jiang, Faguo Yang (2002) An Evolutionary Tabu Search for Cell Image Segmentation. IEEE Transactions on Systems, Man and Cybernetics Part B – Cybernetics, 32(5): 675 – 678

[72] Glover F, Laguna M (1997) Tabu Search. Kluwer Academic Publishers, Boston

[73] Canny J (1986) A Computational Approach to Edge-Detection. IEEE Transactions on Pattern Analysis and Machine Intelligence, 8(6): 679 – 698

[74] Bocchi L, Ballerini L, Hässler S (2005) A New Evolutionary Algorithm for Image Segmentation. In: Franz Rothlauf et al (ed) Applications of Evolutionary Computing. Evoworkshops 2005: EvoBIO, EvoCOMNET, EvoHOT, EvoIASP, EvoMUSART, and EvoSTOC, pp 264 – 273. Springer. Lecture Notes in Computer Science, Vol 3449. Lausanne, Switzerland, March/April 2005

[75] Gardner M (1970) The fantastic combinations of John Conways new solitaire game "life". Scientific American, 223: 120 – 123

[76] Bezdek J C (1981) Pattern Recognition with Fuzzy Objective Function Algorithms. Kluwer Academic Publishers, Norwell

[77] Krawiec K, Howard D, Zhang M (2007) Overview of Object Detection and Image Analysis by Means of Genetic Programming Techniques. In Proceedings of the 2007 Frontiers in the Convergence of Bioscience and Information Technologies, pp 779 – 784. IEEE Computer Society Press

[78] Muni D P, Pal N R, Das J (2006) Genetic Programming for Simultaneous Feature Selection and Classifier Design. IEEE Transactions on Systems, Man and Cybernetics Part B – Cybernetics, 36(1): 106 – 117

[79] Coello Coello C A, Lamont G B, Van Veldhuizen D A (2007) Evolutionary Algorithms for Solving Multi-Objective Problems, 2nd edn. Springer, New York

[80] Watchareeruetai U, Takeuchi Y, Matsumoto T et al (2008) Transformation of Redundant Representations of Linear Genetic Programming into Canonical Forms for Efficient Extraction of Image Features. In: 2008 IEEE Congress on Evolutionary Computation (CEC 2008), pp 1996 – 2003, Hong Kong, June 2008. IEEE Service Center

[81] Brameier M F, Banzhaf W (2007) Linear Genetic Programming. Springer, New York

[82] Kowaliw T, Banzhaf W, Kharma N et al (2009) Evolving Novel Image Features Using Genetic Programming-based Image Transforms. In 2009 IEEE Congress on Evolutionary Computation (CEC'2009), pp 2502 – 2507. IEEE Press, Trondheim

[83] Miller J F, Thomson P, Fogarty T (1998) Designing Electronic Circuits Using Evolutionary Algorithms. Arithmetic Circuits: A Case Study. In: Quagliarella D, Périaux J, Poloni C et al (eds) Genetic Algorithms and Evolution Strategy in Engineering and Computer Science, pp 105 – 131. Morgan Kaufmann, Chichester

[84] Julian F Miller, Peter Thomson (2000) Cartesian Genetic Programming. In: Riccardo Poli, Wolfgang Banzhaf, William B Langdon, Julian Miller, Peter Nordin, Terence C Fogarty (eds) Genetic Programming, European Conference, EuroGP 2000, pp 121 – 132, Edinburgh, Scotland, UK, April 2000. Springer. Lecture Notes in Computer Science, vol 1802

[85] Guo P F, Bhattacharya P, Kharma N (2009) An Efficient Image Pattern Recognition System Using an Evolutionary Search Strategy. In Proceedings of the 2009 IEEE International Conference on Systems, Man and Cybernetics. IEEE Press, San Antonio

[86] Mitchell T M (1997) Machine Learning. McGraw-Hill, London

[87] Vapnik V N (1999) The Nature of Statistical Learning Theory, 2nd edn. Springer, New York

[88] Raymer M L, Punch W F, Goodman E D et al (2000) Dimensionality Reduction Using Genetic Algorithms. IEEE Transactions on Evolutionary Computation, 4(2): 164 – 171

[89] de la Iglesia B, Reynolds A, Rayward-Smith V J (2005) Developments on a Multi-objective Metaheuristic (MOMH) Algorithm for Finding Interesting Sets of Classification Rules. In: Carlos A Coello Coello, Arturo Hernández Aguirre, Eckart Zitzler (eds) Evolutionary Multi-Criterion Optimization. Third International Conference, EMO 2005, pp 826 – 840, Guanajuato, México, March 2005. Springer. Lecture Notes in Computer Science, vol 3410

[90]   de la Iglesia B, Richards G, Philpott M S et al (2006) The Application and Ef-
        fectiveness of a Multi-objective Metaheuristic Algorithm for Partial Classification.
        European Journal of Operational Research, 169: 898 – 917

[91]   Deb K, Pratap A, Agarwal S et al (2002) A Fast and Elitist Multiobjective Genetic
        Algorithm: NSGA – II. IEEE Transactions on Evolutionary Computation, 6(2): 182 –
        197

[92]   Newman D J, Hettich S, Blake C L et al (1998) UCI Repository of machine learning
        databases http://www.ics.uci.edu/~mlearn/MLRepository.html. Accessed 12 Octo-
        ber 2010

[93]   Rahila H, Sheikh M M, Raghuwanshi et al (2008) Genetic Algorithm Based Cluster-
        ing: A Survey. In First International Conference on Emerging Trends in Engineering
        and Technology, pp 314 – 319. IEEE Press, Nagpur

[94]   Bandyopadhyay S, Maulik U (2002) Genetic clustering for automatic evolution of
        clusters and application to image classification. Pattern Recognition, 35(6): 1197 –
        1208

[95]   Davies D L, Bouldin D W (1979) Cluster separation measure. IEEE Transactions
        on Pattern Analysis and Machine Intelligence, 1(2): 224 – 227

[96]   Bandyopadhyay S, Pal S K, Aruna B (2004) Multiobjective GAs, Quantitative
        Indices, and Pattern Classification. IEEE Transactions on Systems, Man and Cy-
        bernetics – Part B: Cybernetics, 34(5)

[97]   Knowles J D, Corne D W (2000) Approximating the Nondominated Front Using
        the Pareto Archived Evolution Strategy. Evolutionary Computation, 8(2): 149 – 172

[98]   Das R, Mitra S, Banka H, Mukhopadhyay S (2007) Evolutionary Biclustering with
        Correlation for Gene Interaction Networks. In: Ashish Ghosh, Rajat K De, Sankar
        K Pal (eds) Pattern Recognition and Machine Intelligence. Second International
        Conference (PReMI' 2007), pp 416 – 424. Springer, Lecture Notes in Computer Sci-
        ence, Vol 4815, Kolkata, India, December 18 – 22 2007

[99]   Radtke P V W, Wong T, Sabourin R (2009) Solution Over-Fit Control in Evolution-
        ary Multiobjective Optimization of Pattern Classification Systems. International
        Journal of Pattern Recognition and Artificial Intelligence, 23(6): 1107 – 1127

[100]   Chatelain C, Adam S, Lecourtier Y et al (2010) A Multi-model Selection Framework
        for Unknown and/or Evolutive Misclassification Cost Problems. Pattern Recogni-
        tion, 43(3): 815 – 823

[101]   Jin Y (2005) A Comprehensive Survey of Fitness Approximation in Evolutionary
        Computation. Soft Computing, 9(1): 3 – 12

[102]   Corne D, Dorigo M, Glover F (eds) (1999) New Ideas in Optimization. McGraw-Hill,
        London

[103]   Kennedy J, Eberhart R C (1995) Particle Swarm Optimization. In Proceedings of
        the 1995 IEEE International Conference on Neural Networks, pp 1942 – 1948. IEEE
        Service Center, Piscataway

[104]   Dasgupta D (eds) (1999) Artificial Immune Systems and Their Applications.
        Springer, Berlin

[105]   de Castro L N, Timmis J (2002) Artificial Immnue System: A New Computational
        Intelligence Approach. Springer, London

[106]   Wang W, Gao S, Tang Z (2009) Improved pattern recognition with complex artificial
        immune system. Soft Computing, 13(12): 1209 – 1217

[107]   Dorigo M, Di Caro G (1999) The Ant Colony Optimization Meta-Heuristic. In:
        David Corne, Marco Dorigo, Fred Glover (eds) New Ideas in Optimization. McGraw-
        Hill, London

[108]   Dorigo M, Stützle T (2004) Ant Colony Optimization. MIT Press, Cambridge

2 Pattern Discovery and Recognition in Sequences

Andrew K.C. Wong, Dennis Zhuang, Gary C.L. Li and En-Shiun Annie Lee[1]

Abstract   Today, a huge amount of DNA and protein sequences are available, but the growth of biological knowledge has not kept pace with the increasing data. Hence much more effective computational methods are required to reveal the inherent functional units in these sequences in the form of patterns that could be related back to the biological world and life science applications. Pattern discovery provides such a computational tool. This chapter provides a brief review of known pattern discovery techniques for sequence data and later presents a new pattern discovery framework capable of discovering statistically significant sequence patterns effectively without relying on any prior domain knowledge. In response to the "too many patterns" problem, our algorithm is able to remove redundant patterns including those attributed by their strong statistical significant sub-patterns. It hence renders a compact set of quality patterns making interpretation and further model development much easier. When applying to transcription factor binding site data, it obtains a relatively small set of patterns — 14 out of 18 consensus binding sites are associated with our top ranking patterns.

2.1 Introduction

This chapter is to provide a brief review of known pattern discovery techniques for sequence data and to present a new pattern discovery framework developed by our research team.

Sequence data is an important type of data acquired in different forms for various applications. From commercial initiative such as supermarket transaction sequences to scientific research such as DNA and protein sequences, enormous amount of data are gathered. Today, most pattern discovery methods are developed by two research communities: bioinformatics and data mining. Biological sequences are the main targets of the former and transaction sequences are the hot subjects of the latter [1]. Compared to transaction sequences, biological sequences typically have a smaller number of sequences and alphabet size but each of them is much lengthier. Furthermore, their element contains a single item. Elements of transaction sequences on the other hand are usually itemsets which could contain multiple items. This chapter

---

1   PAMI Group, Department of System Design, University of Waterloo.

will focus on biological sequences.

With the advent of high-throughput sequencing techniques, a huge number of DNA and protein sequences is available nowadays, but the increase of biological knowledge has not kept pace with its growth. This enormous amount of data makes manual analysis impossible and demand much more effective computational methods to reveal the functional meaning inherent in these sequences. Pattern discovery provides such a computational tool. The underlying patterns of residue associations are believed to be related the to functional units that share or conserve for structural and functional requirement of the biological molecules. It might provide multiple backups to ensure that one corruption would not result in a lethal consequence. Recent findings suggest that even once considered as repeated junks may also have a role to play for the organism. Hence discovering substring patterns that occur more frequently than expected is the underlying assumption for pattern discovery. However, most of the current pattern discovery methods produce a huge set of patterns among which many are only resulted from coincidences. To reduce these redundant and false positive patterns, statistical assessments are often introduced to evaluate the quality of the output patterns so as to provide a basis for removing them.

Section 2.2 presents a brief literature review of pattern discovery in sequences. Note that pattern and motif are used interchangeably. Section 2.2.1 defines the input sequences and gives some preliminary definitions. Section 2.2.2 summarizes various types of pattern definitions. Section 2.2.3 shows some current state of art techniques in discovering patterns. We will here concentrate on the problem of discovering previously unknown patterns. This is in response to a recent concern arisen due to the complexity, availability and the difficulty of getting the required domain knowledge. This also provides a base for the classical pattern matching or pattern recognition approach where known patterns have to be first identified. Pattern discovery attempts to discover patterns which can be related back to knowledge inherent in the data. Once patterns are discovered, biologists could devise experiment to explore the reasons of their occurrences. It is a pathway to move data a step closer to models and knowledge. This is the spirit of scientific research. Furthermore, the discovered patterns can be used as functional features to characterize and further classify the input sequences. Thus, one of the important applications of pattern discovery is sequence classification.We do not intend to compare the performance of individual methods in finding relevant patterns since different methods hold quite different assumptions in reporting patterns. As pointed out in [2], despite much effort to date, detection of biological motifs remains a great challenge for computational biologists. Performance comparison among pattern discovery tools has been proven to be a very difficult task because we lack an absolute standard to evaluate the correctness of the tools. There are many pattern discovery tools in the literature and it is impossible to report all of them exhaustively. Nevertheless, a simple yet detailed comparison within certain framework could be found in [3]. Here, we try to present some representative methods. Moreover, since pattern discovery tools would usually produce a large set of patterns, to address such problem is more imminent. We will review and discuss some of the commonly used measures (such as statistical significance) to assess the quality of patterns and pattern pruning methods in Section 2.2.4.

Section 2.3 presents our pattern discovery framework. Our method first discovers statistical patterns in terms of character strings whose character association deviates significantly from the default random model such as character independence. Because of their statistical significance, their occurrences imply functional necessity or selection. In other words, their non-random existence suggests that their occurrence has a role or multiple roles to play — from enormous examples found in natural languages to biomolecular sequences. Recognizing that some of seemingly statistically significant patterns may be attributed by their embedded strong statistical significant subpatterns, we use the concept of statistically induced characteristics to capture these fake patterns. Hence, by extracting or retaining non-induced patterns (patterns not induced by their strong significant subpatterns), a more compact set of output patterns could be obtained. If interpretation and further analysis is the goal of pattern discovery, such reduction is very significant. Furthermore, an efficient and exact algorithm instead of heuristic ones is developed to discover and retain such patterns. Our algorithm is based on the data structure known as generalized suffix tree. Non-induced patterns are directly identified in the suffix tree without generating frequent or significant patterns and applying post processing afterward. In other word, our algorithm is able to quickly discover patterns or motifs related to functional structures without relying on any prior domain knowledge. It renders a very generic and care-free analytical platform for pattern discovery. This pattern discovery framework has been tested by long sequence of natural language where all the punctuations and spaces are removed [4]. Over 18 000 functional units such as words and short phrases are discovered as patterns from the entire book of "Pride and Prejudice" in less than 3 seconds. It is then applied to the transcription factor binding site data [4] to demonstrate its effectiveness in identifying biologically meaningful patterns without any prior knowledge.

Section 2.4 closes our discussion with concluding remarks.

2.2 Sequence Patterns and Pattern Discovery — A Brief Review

Pattern discovery in biological sequences is one of the most challenging problems in molecular biology and computer science. The problem can be formulated as follows: given a set of input sequences, find unknown patterns that occur frequently or unexpectedly. It is hoped that these patterns may reveal the regularities (functional units) hidden in the input data.

First we need to define pattern. After pattern is defined, further constraints (i.e., number of occurrences) can be applied to render desirable patterns by various pattern discovery tools. Scoring functions that assess the quality or interestingness of the patterns can be built into the pattern discovery process or used in the post-processing to a subset of high score discovered patterns. The scoring function can be based on pattern properties (i.e., pattern length, support, number of occurrences, degree of ambiguity) or statistical significance. Section 2.2.4 discusses more about scoring functions based on statistical significance.

2.2.1 Introduction to Input Sequences and Simple Counting Stati-stics

Let Σ be a set of distinct elements $\{e_1, e_2, \cdots, e_{|\Sigma|}\}$, called the alphabet, and $|\Sigma|$ be its size. A sequence S over Σ is an ordered list of elements $s_1 s_2 \cdots s_n$. In general, the input data might come as multiple sequences $S_1, S_2, \cdots, S_m$ with lengths $L_1, L_2, \cdots, L_m$, respectively. Let L be their overall length. For DNA sequences, the alphabet $\Sigma = \{A, C, G, T\}$ and for protein sequences the alphabet contains 20 amino acids. Most pattern discovery methods can in principle be adapted to any finite alphabet. However, the implementation might not be easy and the computation time might greatly increase for large alphabet size.

Here we introduce the definition of number of occurrences and support of a pattern for better illustrating pattern discovery methods.

The **number of occurrences** of a pattern P in a set of multiple sequences is denoted by k_P. The *occurrence list* is $L_P = \{\cdots, (i, j), \cdots\}$ where the ordered pair (i, j) is a position denoting that P occurs at position j in sequence i. The *support* of P denoted by q_P in the multiple sequences is the number of sequences in which P occurs at least once. A pattern is said to be frequent if its number of occurrences is not less than a specified minimum requirement min_{occ}, i.e., $k_p \geqslant min_{occ}$.

Example 1

Let the input sequences be $S_1 = \mathrm{A\,T\,C\,G\,A\,T}$ and $S_2 = \mathrm{T\,C\,G\,A\,T\,C}$. The number of occurrences of pattern A T and G A T C is 3 and 1 with their occurrence list $f(1,1); (1,5); (2,4)g$ and $f(2,3)g$ and their supports 2 and 1, respectively.

Support can be an important property for a pattern, especially when input sequences are thought to share conserved regions with structural or functional importance. However, it is only meaningful if the input contains multiple sequences.

The number of occurrences of a pattern sometimes is confused with the support in literature especially when each input sequence is short which makes it unlikely to contain two or more pattern occurrences. In the practice, each sequence could be long enough to contain multiple occurrences of a pattern.

2.2.2 Two Types of Sequence Pattern

Definitions of sequence pattern vary in literatures. They can be chiefly summarized into two categories: deterministic pattern and probabilistic pattern. Deterministic pattern either occurs in (or matches) the input sequences or not, that is, we can determine whether a deterministic pattern occurs in a given position in the input sequences. For probabilistic pattern, we can compute the probability that it occurs in a given position of the input sequences. The higher the probability, more probably it occurs in that position.

Deterministic Pattern

The simplest deterministic pattern is just a short sequence of characters from alphabet. It is called solid pattern. TACGTA and TATATA are examples of solid patterns. More complex deterministic patterns can be formed by incorporating into solid patterns certain features as described below: ambiguous character, wild card, flexible gap.

An ambiguous character represents (and thus can match) any of the characters from a subset of the original alphabet. It is usually denoted by a list of distinct characters enclosed in square brackets or a code outside of the original alphabet. For example, [A G], [C T], [A C T] are ambiguous character and can be represented by R, Y, H, respectively in IUPAC Nucleotide Code. TA[C T]GTA or equivalently TAYGTA is a pattern containing ambiguous character indicating that either C or T can be in the 3rd position when matching this pattern in the input sequences.

Wild card (or also known as do not care character) is a special kind of ambiguous character that matches any character from the alphabet. Wild card is denoted by the letter N for DNA sequences, X for protein sequences and most often it is denoted by a dot ".". Pattern containing one or more wild cards are called gapped pattern. A fixed length gap denoted by $x(i)$ contains i consecutive wild cards. For example, TA. CG or TANNCG or TA-x(2)-CG is a gapped pattern. When we match this pattern in the input sequences, we can skip matching for the 3rd and 4th position, i.e., any sequence with 6 characters beginning with TA and ending with CG can be its occurrence.

A flexible gap is a gap with variable length and is usually denoted by $x(i, j)$ where i and j are the lower and the upper bound length of the gap, respectively. For example, $x(1, 3)$ represents any fixed length gap from $\{x(1),$ $x(2), x(3)\}$. Gap with any length including 0 is denoted by character*. TA-x(1,3)-CG is a pattern with flexible gap matching sequence beginning with TA and then CG starting in the 4th, 5th or 6th position.

Complex deterministic pattern can contain all of the above features. The pattern TA-x(2)-[C T]G-x(3, 5)-CA is an example. Although complex pattern can be composed of all of the above mentioned features, many pattern discovery methods are restricted to some of these features. For example, some methods may allow only ambiguous character and some only fixed length gap.

Pattern with mismatches can further extend the expressive power of deterministic pattern by allowing certain mismatches. It is composed of two parts: 1) the pattern description often called the consensus pattern which is a deterministic pattern (usually a solid pattern) and 2) the part that states the degree of tolerance or the number of mismatches in matching the consensus pattern in input sequences. Given a subsequence from the input, if three edit operations (substitution, insertion and deletion) are allowed, the minimum number of edit operations (edit distance) needed to transform the subsequence into the consensus is called the number of mismatches. In the common case where the consensus is a solid pattern and only substitution is allowed, the number of mismatches is the hamming distance between the subsequence and the consensus pattern.

For example, given a consensus pattern TACGTA and the allowed number of mismatches is 2. The subsequence GACGTA has 1 edit distance from consensus and is a match for the consensus. The sequence GACGTC is 2 edit distance from consensus and is still considered as an occurrence for the consensus. GATGTC is 3 edit distance from consensus and hence does not considered as a match with the consensus.

Mismatch is different from ambiguous character. Mismatch can happen in any position of a pattern whereas ambiguous character allows certain characters in a fixed position.

Probabilistic Pattern

One drawback of deterministic pattern is that the pattern description does not show compositions of characters in each position nor use this information in the pattern discovery process. For example, from the deterministic pattern TA[C T] GT, we only know that C or T is allowed in the 3rd position, but in the actual case, their frequency of occurrences might be different, say only 30% and 70%, respectively. Probabilistic pattern utilizes this piece of information which might be important in distinguishing weak pattern from random noise.

The simplest probabilistic pattern is position weight matrix (PWM), sometimes called position specific score matrix (PSSM) or a profile (although profile is often used for more complicate probabilistic pattern, i.e., pattern with gaps). PWM is a table representing the pattern with l positions. The column describes the position and the row describes each character from the alphabet. Each entry of the matrix is the probability of observing the given character (relative frequency) at given position. Figure 2.1 shows the example PWM with 6 positions and its sequence logo for visualization.

Given the PWM and the background distribution of characters from al-

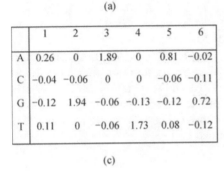

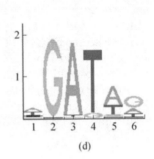

	1	2	3	4	5	6
A	0.40	0	0.98	0	0.62	0.24
C	0.22	0.02	0	0	0.02	0.05
G	0.06	0.98	0.01	0.08	0.06	0.59
T	0.32	0	0.01	0.92	0.30	0.12

(a)

	1	2	3	4	5	6
A	0.67	$-\infty$	1.95	$-\infty$	1.31	-0.07
C	-0.17	-3.98	$-\infty$	$-\infty$	-3.98	-2.39
G	-1.98	1.97	-3.98	-1.65	-1.98	1.23
T	0.34	$-\infty$	-3.98	1.88	0.27	-0.98

(b)

	1	2	3	4	5	6
A	0.26	0	1.89	0	0.81	-0.02
C	-0.04	-0.06	0	0	-0.06	-0.11
G	-0.12	1.94	-0.06	-0.13	-0.12	0.72
T	0.11	0	-0.06	1.73	0.08	-0.12

(c)

(d)

Fig. 2.1 Probabilistic pattern for transcription factor GATA3 binding sites from JASPAR database [5]. (a) First, define the Position Weight Matrix A, $A(c, j)$ as the relative frequency for character c at position j; (b) Next, define the Log-odd score matrix A' for PWM as, $A'(c, j) = \log_2 \dfrac{A(c, j)}{f(c)}$, here we assume $f(c) = \dfrac{1}{4}$ for $c \in \Sigma_{DNA}$; (c) Relative entropy matrix R for PWM, $R(c, j) = A(c, j) \cdot A'(c, j)$ and (d) The corresponding sequence logo of PWM for visualization. The y coordinate for sequence logo is bits.

phabet, we can calculate a score which indicates how well a sequence segment $s_1 s_2 \cdots s_l$ matches the PWM. The score is given by the following formula.

$$\prod_{j=0}^{l} \frac{A(s_j, j)}{f(s_j)}, s_j \in \Sigma,$$

where $A(s_j, j)$ is an entry of PWM and $f(c)$ is the background frequency of character c. The product is the odd score that the sequence segment matches the probability distribution represented in PWM. The higher the score, the better is the match between the sequence segment and the probabilistic pattern.

We can replace the product in the formula with summation through logarithm transform. With the log-odd value stored in matrix A', the computation of matching score can be simplified as shown in the formula below.

$$\sum_{j=0}^{l} \log_2 \frac{A(s_j, j)}{f(s_j)} = \sum_{j=0}^{k} A'(s_j, j), s_j \in \Sigma.$$

Example 2

Suppose that the sequence segment is TGATTA and the background frequency is 0.25 for all characters, using the matrix $\boldsymbol{A}$ and $\boldsymbol{A}'$ in Fig. 2.1, the matching odd score is

$$\frac{0.32}{0.25} \times \frac{0.98}{0.25} \times \frac{0.98}{0.25} \times \frac{0.92}{0.25} \times \frac{0.30}{0.25} \times \frac{0.24}{0.25} = 83.38.$$

The log-odd score is $0.34 + 1.97 + 1.95 + 1.88 + 0.27 - 0.07 = 6.34$. Note that if the background frequency is 0.25 for all characters, then the denominator will be the same and can be neglected.

The above score corresponds to the degree of matching between a given sequence segment and PWM. It is not the scoring function that assesses the quality of the probabilistic pattern itself. As the scoring function is essential in the probabilistic pattern discovery process, we describe one of the common measures, relative entropy or sometimes called information content. The relative entropy of a given PWM is defined to be

$$\sum_{j=0}^{l} \sum_{c \in \Sigma} \boldsymbol{A}(c,j) \log_2 \frac{\boldsymbol{A}(c,j)}{f(c)} = \sum_{j=0}^{l} \sum_{c \in \Sigma} \boldsymbol{R}(c,j) = \sum_{j=0}^{l} R_j,$$

where R_j is the relative entropy for position j. The relative entropy can be calculated by summing all entries in the matrix $\boldsymbol{R}$.

The relative entropy provide a measure that indicates how well conserved a pattern is with respective to background distribution. In a particular position, the more the distribution of the random variable differs from the background distribution, the higher is the relative entropy for that position. It is used as the optimization objective (search guide) in many probabilistic pattern discovery methods.

The relative entropy is a good measure for two probabilistic patterns that have close number of occurrences (in many cases, close support) as this measure does not take into account the number of occurrences but instead depends on the relative frequency in PWM. In fact, many methods that use relative entropy depend on the assumption that the probabilistic pattern occurs exactly once in all or nearly all of the input sequences (have the same or close support).

The sequence logo in Fig. 2.1 is a visualization of the probabilistic pattern (PWM). Each column of the sequence logo corresponds to one position of the probabilistic pattern. The height of each column is the relative entropy R_j. The characters in a position are displayed sorted according to their relative frequency with the most frequent character on the top. The height of each character at a position is proportional to its relative frequency. A quick look at the sequence logo reveals the most conserved (consensus) character for each position.

2.2.3 Sequence Pattern Discovery Techniques

There are many approaches for discovering deterministic patterns and probabilistic patterns. Here we just select a few representative approaches for each
type of pattern and discuss their specific problems and insights. We do not
attempt to present a comprehensive survey of all sequence pattern discovery
methods.

Deterministic pattern discovery methods would often produce a set of
patterns that meet some predefined level of support, number of occurrences
and score. Scoring function can be built into the pattern discovery process
directly or used in the post-processing stage. The output set generally contains a large number of patterns. Although this may be a drawback, pattern
discovery usually conducts exhaustive search ensuring all patterns that meet
the minimum requirement are reported. Further structures could be extracted
from this set of discovered patterns by removing redundant patterns, extracting motif base and pattern clustering.

Probabilistic pattern discovery methods would usually find one or several
best scoring patterns. Scoring function is essential for probabilistic pattern
discovery as it would serve as the optimization objective in the pattern discovery process. These methods would take the risk of missing other patterns
that are statistically as meaningful as those reported probabilistic patterns.

Direct comparison between deterministic and probabilistic pattern discovery methods is uncommon because most deterministic patterns contain gaps
or even flexible gaps while ungapped PWM is often adopted as probabilistic
pattern.

Exhaustive Enumeration

The simplest approach is to enumerate all possible patterns that satisfying
user defined constraints such as pattern length, number of occurrences, support, and number of mismatches. After enumeration, score can be computed
for each pattern and patterns with highest scores or scores above certain
threshold can be outputted. Pruning the exhaustive search could add some
improvements over the running time. For example, the pattern that does not
meet the support requirement can be pruned without further expansion such
as proposed by Pratt [6].

The advantage of such approach is that we are guaranteed to find the
best scoring patterns and we may output an arbitrary number of high scoring
patterns. However, this method is only good for short and simple pattern, as
the running time is exponential to the pattern length (the number of possible
candidates is even much larger if mismatches, gaps and flexible gaps are
allowed) for deterministic pattern and could be $O(L^L)$ (the number of PWMs
computed from all possible occurrence sets) for probabilistic pattern beyond
feasible computational resource. With such high computational complexity,
the probabilistic pattern discovery methods always output locally optimal
patterns instead global ones. In comparison, deterministic pattern discovery

methods could manage to produce a set of patterns satisfying constraints defined by user. The YMF [7] uses such approach to detect transcription factor binding sites in the promoter regions of Yeast.

Discovery of Deterministic Patterns

Most deterministic pattern discovery methods output patterns with number of occurrences not below min_{occ}. min_{occ} should be at least 2 to make the pattern nontrivial. This output set often contains many redundancies. Some elegant methods reduce redundancies in their output, which not only improves the output quality but also improve the running time. Some would also use suffix tree or suffix array to achieve better efficiency.

1) *Maximal pattern* Here we examine one type of redundancies for gapped pattern (solid pattern with do not care character). Given the input sequences, the pattern P is said to be maximal if it cannot be made more specific by replacing any do not care character with other characters or appending an arbitrary character to its left/right without decreasing the number of occurrences.

For example, given the input sequence $s = CACACATACATAC$, the pattern $P_1 = ACA$. A having 2 occurrences is not maximal as $P_2 = ACATA$ has the same number of occurrences. Pattern $CA. ACATAC$ having 2 occurrences is the maximal pattern as it cannot be made more specific while keeping the same number of occurrences.

Therefore those patterns that can be made more specific are redundant with respective to some maximal patterns for they have the same set of occurrences. It would be desirable to avoid generating these redundant patterns in the pattern discovery process. TEIRESIAS [8] algorithm is the representative method in discovering maximal patterns.

2) *Irredundant patterns*   Although the definition of maximal pattern produces a less redundant output, the set of maximal patterns could contain other redundancy defined in [9]. A pattern is called irredundant if it is not covered by other maximal patterns, or more formally, the occurrence list of the pattern could not be obtained as the union of the occurrence lists of other maximal patterns. Given min_{occ}, all irredundant patterns could be extracted to form a pattern basis from which all other maximal patterns could be generated. There are also other similar irredundant definitions [10, 11], resulting in different pattern bases.

For example, consider again the input sequence $s = CACACATACATAC$, the pattern $P_1 = CACA$. $A.A.A.A$ with occurrence list $L_{P_1} = \{0, 2\}$ is irredundant for it is a maximal pattern and it is not covered by other maximal patterns. $P_2 = CA. AC$ with $L_{P_2} = \{0, 4, 8\}$ is not covered by other maximal patterns and hence is an irredundant pattern. For pattern $P_3 = CA. A$ with $L_{P_3} = \{0, 2, 4, 8\}$ is maximal but is not irredundant for it is covered by patterns P_1 and P_2 (i.e., $L_{P_3} = L_{P_1} \bigcup L_{P_2}$).

Discovery of Probabilistic Patterns

It is hard to obtain optimal probabilistic patterns. Most of the algorithms for probabilistic pattern discovery employ Expectation Maximization (EM) or Gibbs Sampling techniques, which can converge to a local maximum, rather than to the global one. Gibbs sampling can be viewed as a stochastic implementation of Expectation Maximization. Here we will describe these two methods in a very basic form. They assume that the probabilistic pattern occurs exactly once in each input sequence.

Expectation Maximization   Probabilistic pattern discovery using Expectation Maximization is first proposed by [12]. This method gives directly the probabilistic pattern (PWM) without explicitly obtaining the set of its occurrences. It consists of two steps, E step and M step in the iteration. EM algorithm iterates between calculating the probability for each l-mer (subsequence of length l) based on the current PWM and computing a new PWM based on the probabilities. It can be shown that this procedure converges to a local maximum of the log likelihood of the resulted PWM.

Initialization step: Set the initial PWM randomly

Iteration step

E step: For each l-mer in the input sequences, the probability that it is generated by the current PWM, rather than by the background distribution, is computed. The sum of the probabilities of all l-mers in each sequence is normalized to one.

M step: The expected count of an observed character in a certain position of PWM is calculated as the sum of the probabilities of each l-mer whose corresponding position contains this character. The probability of observing a certain character in a given position (an entry) of PWM is then updated by normalizing the expected count of that character with respective to the total expected count (the sum of the probabilities of all l-mers).

Repeat iteration step, until the stop condition is met. (i.e., no further improvement)

Gibbs Sampling   Unlike EM algorithm, Gibbs Sampling [13] explicitly maintain a set of l-mers, which represents the occurrences of the probabilistic pattern. PWM is directly computed from this set of occurrences. In the iteration of Gibbs Sampling, the set of occurrences is updated and its corresponding PWM is computed.

Initialization step:
The initial set of occurrences is formed by randomly select one l-mer from each sequence. Denote the occurrence in sequence i by o_i.

Iteration step:
(1) Pick randomly one sequence i;
(2) Compute PWM based on the set of occurrences except o_i;
(3) For each l-mer in sequence i, compute the probability that it is generated by the PWM, rather than by the background distribution;
(4) Choose randomly the new occurrence o_i' among all l-mers of sequence i according to their corresponding scores (it is more probable to be chosen with higher score);
(5) Replace o_i with o_i' in the set of occurrences.

Repeat iteration step, until the stop condition is met.

EM algorithm takes the weighted average across all l-mers whereas Gibbs Sampling takes a weighted sample from all l-mers. Given sufficient iterations, the algorithm will efficiently sample the joint probability distribution of the likelihood of probabilistic pattern PWM. It will converge to a local maximum as EM.

MEME [14] is an improved version of EM algorithm: (1) it removes the assumption that the probabilistic pattern occurs exactly once in each input sequence; (2) it can be forced to report several patterns instead of only one best pattern and (3) it increases the chance to find the globally optimal pattern.

2.2.4 Assessment of Pattern Quality

Assessment of pattern quality is important in ranking the discovered patterns for users where most deterministic pattern discovery tools may produce a large set of patterns and even essential as a guide (optimization objective) for probabilistic pattern discovery. For deterministic pattern, the commonly used scoring function to evaluate pattern quality is z-score. For probabilistic pattern, one of the scoring functions, relative entropy has been introduced in Section 2.2.2, so here we will show other scoring methods that allows for more sophisticated background models and addresses some of its issues.

z-score

z-score evaluates the statistical significance of the pattern. It measures how unlikely that the same number of occurrences of the pattern is observed in the input sequences that are generated by the random background model.

For example, suppose we have 100 sequences, each of length L_i and a solid pattern l of length l with the number of occurrences k_P. The number of possible occurrences of P in each sequence is $L_i - l + 1$, so the total number of potential occurrences is $\sum_{i=0}^{100} (L_i - l + 1)$. We want to be able to decide whether the observed number of occurrences k_P is what we might expect if the input sequences are generated by random background model.

There are many background models for generating random sequences. The simplest one is to generate each character in the sequences independently with given probabilities of individual characters. This is called Bernoulli scheme. Another commonly used model is to generate the sequences using markov chain of order r. Transition probabilities of the markov chain are estimated by the $(r + 1)$-mer frequencies from the input sequences or some larger background sequences (i.e., all promoter regions of a species).

With the background model, we can compute the probability $P_{r_{ij}}$ of a pattern P occurring in j position of i sequence. Let X_{ij} be the Bernoulli variable indicating whether P occurs in j position of i sequence. So the ex-

pected number of occurrences of P in the random sequences is $E(X_P) =$
$$E\left(\sum_i \sum_j X_{ij}\right).$$

Now we have the expected number of occurrences, we can ask how significant the observed number of occurrences k_P is different from $E(X_P)$. z-score
is given by the following formula to measure the significance.

$$z_P = \frac{k_P - E(X_P)}{\sigma(X_P)},$$

where $\sigma(X_P) = \sqrt{Var(X_P)}$ is the standard deviation. It is the number of
standard deviations by which the observed value k_P differs from the expected
value. This normalized score allows for comparing different patterns. Usually
we require $z_P \geqslant t$ for a pattern to be statistically significant, t is the minimum
threshold.

The problem of measuring the pattern significance is related to the hypothesis testing in statistics. The significance value helps us to decide whether
or not to reject the null hypothesis (the observed number of occurrences is
by chance) under the random assumption. The standard criterion for reject-
ing the null hypothesis is when we obtain a p-value $\geqslant 0.05$ (probability that
the pattern is generated by random model). This value suggests that if we
do 100 experiments, in each generating 100 random sequences, we would
expect to see that there are 5 experiments in which P occurs k_P times. It
can also interpret as the probability of making the mistake to reject the null
hypothesis.

Log likelihood and MAP Score

In Section 2.2.2, one of the scoring functions for probabilistic pattern
relative entropy was introduced. The disadvantage of relative entropy is that
only the simplest background model is allowed (the Bernoulli scheme) and it
does not account for the number of occurrences of the pattern.

Log likelihood [15] addresses the problem of not accounting for the number
of occurrences. Log likelihood is roughly proportional to the relative entropy
of the pattern times to the number of occurrences. Likelihood of a model
is the probability that the observed data could be generated by the model.
Typically the logarithm of this probability is used instead for computation
convenience. It is given by

$$\log Pr(data|model) = \sum_i \log P_r(data_i|model).$$

Here for example, the model could be PWM. We can calculate the probability
that all observed l-mers are generated by the PWM as follows.

$$\log Pr(data|PWM) = \sum_i \log Pr(l - mer_i|PWM)$$

However, one might not want to know only the probability that the l-mer is generated by the PWM but also the probability that it is generated by the PWM rather than the background random model. Therefore, we want to incorporate the background model for scoring the probabilistic patterns.

This is achieved by MAP (maximum a priori probability) score. It allows for more complicated background models (i.e., markov chain of order 3 that makes it easier to rule out the low complexity repeats such as AAAAA, TTTT). MAP is the probability of the model given the data and is given below.

$$\log Pr(model|data) = \log \frac{Pr(data|model) \cdot Pr(model)}{Pr(data)}$$
$$= \log Pr(data|model) + \log Pr(model) - \log Pr(data).$$

If all models are a priori equally likely and we do not incorporate the background model, the MAP score is the same with the Log likelihood.

In Section 2.2.2, we also gave a score for measuring how well a sequence segment matches PWM. It is the special case of the MAP where we assume all models are a priori equally likely and the background model is Bernoulli scheme. It is given by the following formula.

$$\log Pr(PWM|\text{sequence segment}) = \log \frac{Pr(\text{sequence segment}|PWM)}{Pr(\text{sequence segment})}$$
$$= \log \frac{\prod_{i=0}^{k} \boldsymbol{A}(s_i, i)}{\prod_{i=0}^{k} f(s_i)}.$$

2.3 Our Pattern Discovery Framework

2.3.1 Our Traditional Random Sequence Synthesis and Recognition

We would like to bring back an early work on pattern discovery of biomolecule sequences using a probabilistic technique called random sequence [16]. We presented a methodology which combines a probabilistic definition, the synthesis of random sequences, as well as representation and recognition aspects. This technique is able to synthesize a class of short biomolecular sequences into a probabilistic pattern known as a random sequence and then use this random sequence to recognize subsequences belonging to that class from a much longer sequence. The final detection is achieved through an optimal matching of the random sequence against the subsequences of the target sequence. Figure 2.2 gives a schematic of the framework.

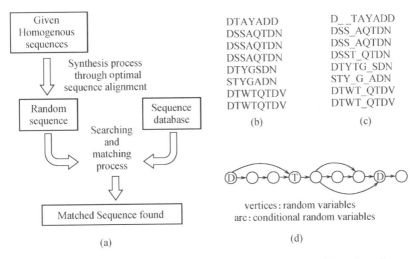

DTAYADD   D _ _TAYADD
DSSAQTDN   DSS_AQTDN
DSSAQTDN   DSS_AQTDN
DSSAQTDN   DSST_QTDN
DTYGSDN   DTYTG_SDN
STYGADN   STY_G_ADN
DTWTQTDV   DTWT_QTDV
DTWTQTDV   DTWT_QTDV

(b)   (c)

vertices : random variables
arc : conditional random variables

(d)

(a)

Fig. 2.2 Random Sequence: (a) The complete framework of Random Sequence Synthesis and Recognition Schema. Given a set of homologous biological sequences, a random sequence for patterns is generated from an iterative algorithm which uses an optimal sequence alignment. Once the random sequence is created, it is used to scan along a longer sequence database to match for a distance-based entropy score. The final result presents significance subsequences along the segment with minimal entropy. (b) A given collection of raw unsynthesized homogeneous sequences as initial input; (c) Their Optimal Alignment ("_" denotes a "null character") and (d) The Synthesized Random Sequence (the blank nodes represent a random vertex with probabilistic distribution of its composed characters).

Thus a random sequence for the pattern S can be defined by the following probabilities below. The columns represent each element in the sequence and the rows represent each alphabet element. The cell represents the probability of an alphabet element occurring as that position in the pattern sequence.

$$r_1^P = \begin{bmatrix} p_{01} & p_{02} \cdots & p_{0|s|} \\ \vdots & \ddots & \vdots \\ p_{|\Sigma|1} & p_{|\Sigma|2} \cdots & p_{|\Sigma||s|} \end{bmatrix}.$$

Alignment of Two Random Sequences

Given sequence s, let's say we have two random sequences, $r_1^s r_2^s$. Perform alignment on these two sequences locally using the classical algorithm, dynamic programming. For the purposes of a simple demonstration, use the score where matches are scored as $+1$, mismatches are scored as 0, and gaps are scored as -1.

$$D(i,j) = min \begin{cases} 0 \\ D(i-1,j-1) + matched \ or - mismatch \\ D(i,j-1) + gapped \ at \ j \\ D(i-1,j) + gapped \ at \ i \end{cases}.$$

For more complex scoring systems, probability is calculated. Let's assume that the characters in the sequence are independent and identically distributed, in this case the characters in the sequence can be considered as independent events and thus by the multiplication rule for probabilities their probability multiplied. In order to convert the probability to entropy, manipulation of the probability using log is used.

$$Pr(S) = Pr(s_1)Pr(s_2)\cdots Pr(s_n) = \prod_i Pr(s_i)$$

$$\log Pr(S) = \log \prod_i Pr(s_i) = \sum_i \log Pr(s_i)$$

$$H(S) = -\sum_i P(s_i)\log P(s_i).$$

In the application of random sequence, a complex Entropy distance score is used to measure the similarity between two sequences. In addition, the known base distribution of the existing data is incorporated as PAM and BLOSSUM matrices.

$$d(s_i,s_j) = \sum_{u=0}^{n}\sum_{t=0}^{n} p_{ui}p_{tj}BLOSSUM(e_u,e_t)[H(s_k) - q_1 H(s_i) - q_2 H(s_j)].$$

Iterative Algorithm

The algorithm of creating a finalized random sequence involves iteratively combining existing random sequences until a finalized random sequence is found.

Initialization:
Let $R = \{r_1, r_2, \cdots, r_N\}$ and let $D = \{\ \}$ be the list of distances that is initially empty
Iteration step:
For $i = 1 \ldots N - 1$
    For $j = i + 1$ to N
        Using Dynamic Programming technique, find the distance between d_{ij}
        between r_i and r_j Store d_{ij} in D
Combination Step:
For $i = 1 \ldots N - 1$
    Find minimum d_{ij}
    Synthesize r_i and r_j into new random sequence r'_{ij}

Continued

Label or Re-label the new r'_{ij}
Insert new r'_{ij} into R and remove r_i and r_j from R and d_{ij} from D
Update D with r'_{ij}
Repeat
Exit Condition:
Output final r'_{ij}

Results of Random Sequences — Detecting Subsequence Patterns in a Long Sequence

The result of the final algorithm is run on 19prokaryotic 5SRNA sequences and the same pattern can be found in 30 out of 34 sequences. The part "cgaac" pattern found is reportedly interacts with the tRNA common arm sequence.

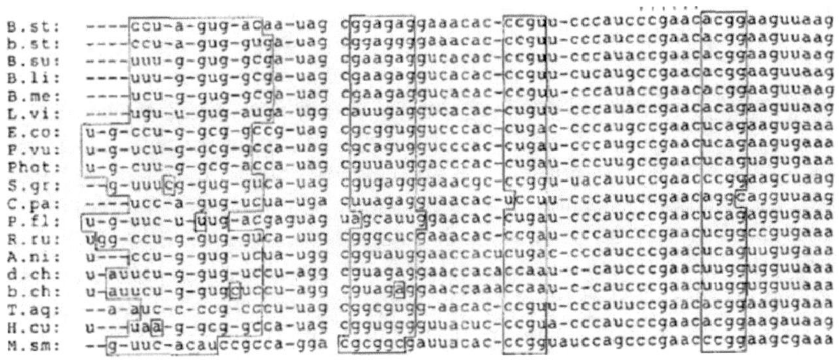

```
B.st:   ----ccu-a-gug-acaa-uag cggagaggaaacac-ccguu-cccaucccgaacacggaaguuaag
b.st:   ----ccu-a-gug-guga-uag cggaggggaaacac-ccguu-cccaucccgaacacggaaguuaag
B.su:   ----uuu-g-gug-gcga-uag cgaagaggucacac-ccguu-cccauaccgaacacggaaguuaag
B.li:   ----uuu-g-gug-gcga-uag cgaagaggucacac-ccguu-cucaugccgaacacggaaguuaag
B.me:   ----ucu-g-gug-gcga-uag cgaagaggucacac-ccguu-cccauaccgaacacagaaguuaag
L.vi:   ----ugu-u-gug-auga-ugg cauugaggucacac-cuguu-cccauaccgaacacagaaguuaag
E.co:   u-g-ccu-g-gcg-gccg-uag cgcgguggucccac-cugac-cccaugccgaacucagaagugaaa
P.vu:   u-g-ucu-g-gcg-gcca-uag cgcaguggucccac-cugau-cccaugccgaacucagaagugaaa
Phot:   u-g-cuu-g-gcg-acca-uag cguuuauggacccac-cugau-cccuugccgaacucaguagugaaa
S.gr:   --g-uuucg-gug-guca-uag cgugagggaaacgc-ccggu-uacauuccgaacccggaagcuaag
C.pa:   ---ucc-a-gug-ucua-uga cuuagagguaacac-uccuu-cccauuccgaacaggcagguuaag
P.fl:   u-g-uuc-u-uug-acgaguag uagcauuggaacac-cugau-cccaucccgaacucagaggugaaa
R.ru:   ugg-ccu-g-gug-guca-uug cgggcucgaaacac-ccgau-cccaucccgaacucggccggugaaa
A.ni:   u---ccu-g-gug-ucua-ugg cgguauggaaccacucgugac-cccauccgaacucaguugugaaa
d.ch:   u-auucu-g-gug-uccu-agg cguagaggaaccacaccaau-c-caucccgaacuuggugguuaaa
b.ch:   u-auucu-g-gugcucu-agg cguagaggaaccaaaccaau-c-caucccgaacuugguguuaaa
T.aq:   --a-auc-c-ccg-cgcu-uag cggcgugg-aacac-ccguu-cccauuccgaacacggaagugaaa
H.cu:   u---uaa-g-gcg-gcga-uag cggugggguuacuc-ccgua-cccauccugaacacggaagauaag
M.sm:   --g-uuc-acauccgcca-gga cgcggcgauuacac-ccgguauccagcccgaacccggaagcgaaa
```

Fig. 2.3 A comparison between 2 alignments of 19 5RNAs.

The chapter presents both the basic notion and an algorithm of the synthesis process. Since the random sequence contains probabilistic characteristics of many sequences in the class, detection of signal in the target sequence is much more reliable. The random sequence synthesized from the above algorithm is used as a template to determine probabilistic variations on the DNA sequence. As the random sequence is shifted along the DNA sequence, its distance value within that frame of segment in the string is recorded. The successful detection is based on the optimal matching of the DNA sequence segments with the random sequence. Figure 2.4 shows the synthesis of the 12 transfer RNA sequences of 70 bases and the recognition of these signals from a long DNA sequence obtained from bovine mitochondrial genome. The experiment successfully shows the minimum distance value, occurring at base position 364, is the beginning of the tRNA. In conclusion, the distance obtained by comparing synthesized random sequence and subsequence segments in the search genome can be used to detect hidden patterns similar to the sequences of a class.

As a specific objective of this chapter, we introduce a new pattern dis-

DNA Strand

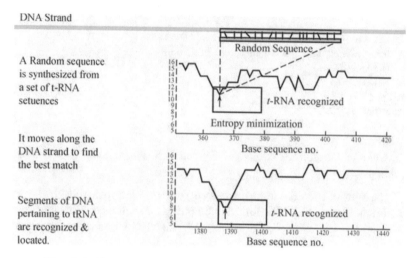

Fig. 2.4 Recognition of t-RNA along a long DNA sequence.

covery method recently developed by our research team. First, we introduce
the generalized suffix tree as the data structure for representing strings. It
can be constructed in linear time and space. The details of it and its linear time and space construction algorithms can be found in Gusfield [17].
These implementation details are left to the readers, however it should be
noted that the concept of a suffix link from the linear time construction algorithm is incorporated in our framework. We next establish the connection
between consecutive patterns and path labels in the suffix tree. We then link
significant representative patterns with the nodes in the suffix tree [4].

2.3.2 Our New Pattern Discovery Framework

Formal definition of Generalized Suffix Tree

Given a collection of strings $S_1, S_2, \ldots, S_m$ over Σ, the generalized suffix
tree T for these multiple strings is a rooted directed tree with the following
properties:

1) Each leaf node is labeled by a position (i, j) or a set of positions
$\{\ldots, (i, j), \ldots\}$ where (i, j) indicates a suffix of string S_i starting at the position j.

2) Each internal node has at least two outgoing edges each of which is
labeled with a non-empty substring of one of the input strings. No two edges
out of a node can have the edge-label starting with the same character.

Most often, a termination character $\$ \in \Sigma$ is appended to each string
to ensure that T exists for this set of multiple strings. Figure 2.5 gives an
example for two input strings.

We will use an example that is revisited throughout the examples in this

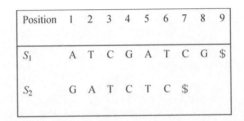

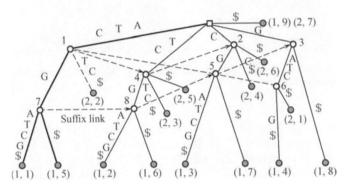

Fig. 2.5 Generalized suffix tree representing S_1 = ATCGATCG\$ and S_2 = GATCTC\$.

section. The generalized suffix tree T is constructed for the multiple strings

$$S_1 = \text{ATCGATCG\$ and } S_2 = \text{GATCTC\$}.$$

Three types of nodes are represented in the suffix tree diagram: The square node is the root, the empty hallowed circles are the internal nodes, and the solid green circles denote the leaf nodes. The pair (i, j) at the end of a leaf node indicates that a suffix of a string starting at position j of the sequence i is represented by the path label of this leaf node. For example, $(1, 1)$ gives the suffix ATCGATCG\$, which is found in sequence 1 starting at position 1. The edges are labeled with substrings.

Looking for Patterns and Consecutive Patterns in Suffix Tree

In Fig. 2.6, from the root to node 1 the edge is labeled "ATC". Then from node 1 to node 7 the edge is labeled "G". Finally at node 7 there are two edges leading to leaf nodes: ATCG\$ and \$. Concatenating the path from the root to node 1 to node 7 and down to $(1, 1)$ gives the final pattern ATCGATCG\$, which is found in sequence 1 starting at position 1. Similarly, Concatenating the path from the root to node 1 to node 7 and down to $(1, 5)$ gives the final pattern ATCG\$, which is found in sequence 1 starting at position 5 as ATCG\$.

To define formally, the label of the path from the root ending at node v is the string resulted from concatenation of the substrings that label the edges along that path. The label of a path from root ending inside an edge (v, w),

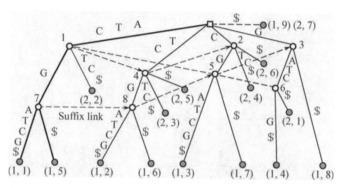

Fig. 2.6 A path concatenated from root node to leaf node in the generalized suffix tree represents a pattern.

is the label of the path from the root ending at node v, concatenating with the remaining characters of the label of the edge (v, w) down to the end of the path. For convenience, the label of a path ending at node v is represented by $pl(v)$, the path label of v. Accordingly, a consecutive pattern that occurs at least once in the input strings has its unique path in suffix tree and is represented by the path label.

Counting the Number Occurrences of a Pattern in Suffix Tree

Let's consider the pattern **ATC**, which ends at node 1. There are three occurrences of this pattern in the two sequences: the first one and second one is displayed respectively in the first sequence $S_1 = ATCGATCG\$$, and the third one is displayed in the second string $S_2 = GATCTC\$$. These pattern occurrences correspond to the three leaf nodes that are found in the subtree of *node 1*: the leaf nodes (1, 1), (1, 5), (2, 2), as shown in Fig. 2.7. These index pairs (1, 1), (1, 5), and (2, 2) indicate the positions of occurrences of pattern **ATC**. Thus **ATC** has 3 occurrences.

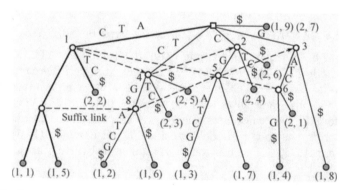

Fig. 2.7 Occurrences of a pattern "ATC" is found in the subtree rooted at the pattern's nod.

Frequent Patterns in Suffix Tree

The number of occurrences of a consecutive pattern is the number of positions found in the subtree under its path. By storing into each node x the number of positions $k(x)$ in the sub tree rooted by it, the number of occurrences of a consecutive pattern can be easily obtained by finding the node which is at or below the end of the pattern path. For example, the number of occurrences of pattern **ATC** whose path ends at node 1 is given by $k(node\ 1) = 3$. Hence frequent patterns are represented by labels of paths that end at or above a node x where $k(x) \geqslant min_{occ}$.

Pattern **ATC** occurs at positions $(1, 1)$, $(1, 5)$, and $(2, 2)$ (see Fig. 2.8) thus having $k_P = 3$ and pattern **TC** occurs at positions $(1, 2)$, $(1, 6)$, $(2, 3)$, and $(2, 5)$ (see Fig. 2.9) thus having $k_P = 4$. Next, these occurrence counts will be used to determine if the pattern is statistically significant.

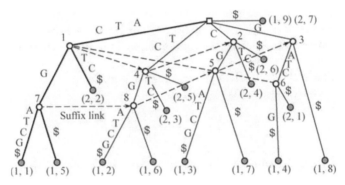

Fig. 2.8 Occurrences of the pattern "ATC" and the number of occurrence is counted as the number of leaves in the subtree rooted under the pattern's node, which is 3.

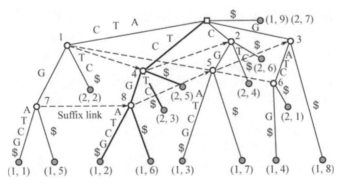

Fig. 2.9 Occurrences of the pattern "TC" and the number of occurrence is counted as the number of leaves in the subtree rooted under the pattern's node, which is 4.

Statistically Significant Patterns

To define statistically significant patterns, a background random model for determining the expected frequency of P is needed. Without being given specific domain knowledge for the background, we adopt the simplest model: the Bernoulli scheme. With this scheme, the probability of P occurring in a position of a random sequence is $pr(P) = \prod_{i=1}^{m} pr(p_i)$, where $p_i \in \Sigma$. Let X_i be a Bernoulli variable that indicates whether P occurs in position i of a random sequence. The total number of possible positions is $L_P = \sum_{i=1}^{n}(l_i - m + 1)$, so the number of occurrences of P is a random variable $X_P = \sum_i X_i$ which follows a binominal distribution. Its expected number of occurrences is $E(X_P) = pr(P) \cdot L_P$.

To measure how k_P of P deviates from its expected one if the given sequences are generated from the random model, we use the standard residual [18].

Definition 2.1   Statistically significant pattern.
The standard residual is defined as

$$z_P = \frac{k_P - E(X_P)}{\sqrt{E(X_P)}}$$

A pattern is statistically significant or over-represented [11] if $z_P \geq t$ where t is the predefined minimum threshold.

Representative Pattern in a Suffix Tree

Patterns having the same list of occurrence positions can be clustered into an equivalent group. Representative pattern in the group is the pattern having the highest statistical significance or order, i.e., it cannot be extended either to the left or to the right without decreasing its frequency. Representative patterns can be identified from the GST fast and effectively using suffix links.

A Generalized Suffix Tree can be constructed in linear-time by utilizing suffix links. Suffix links are used to extend the suffix and prevent the algorithm from restarting at the root to traverse down the entire tree. In Fig. 2.10, the suffix link from node 7 to node 8 indicates that the path label at node 8 (TCG) is the second suffix of the path label (ATCG) at node 7. It is used to locate representative patterns. Formally a suffix link of v points to u if $pl(v)$ is an one character left extension of $pl(u)$. The node u is called the suffix node of v because $pl(u)$ is the suffix of $pl(v)$.

Next, we need to identify nodes corresponding to representative patterns in the tree. This can be efficiently achieved by utilizing the suffix links. Note that the paths of representative patterns end at nodes instead of within edges. A pattern with path ending within an edge can be further extended by at

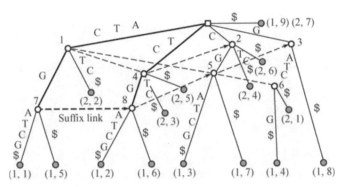

Fig. 2.10 Suffix link indicates that the path label at a second node is the second suffix of the path label at the first node.

least one character to the right without decreasing the number of occurrences and thus by definition cannot be a representative pattern. For example, the smaller subpattern has a superpattern with the same number of occurrences. However, it is not a one-to-one mapping; not all nodes correspond to representative patterns. As we might notice that the pattern associated with the path ending at the node is not a representative pattern because it has a superpattern with the same number of occurrences which has higher statistical significance. We demonstrate this in a series of diagrams.

In Figs. 2.11 and 2.12, a representative pattern (say ATCG at 7) is the path label of an internal node (7) such that: 1) No suffix link pointing to it. (i.e, no extension to the left; extension to the right could only have lower frequency) Note that node 8 (TCG) with a suffix link pointing to it has an extension A to the left while the frequency remains the same.

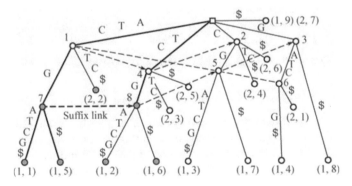

Fig. 2.11 Requirement one of representative pattern is that no suffix link is pointing to the node.

2) Though with suffix link pointing to it, say (TC) at node 4, yet whose position frequency (i.e., 4) is greater than that (i.e., 3) of its left hand

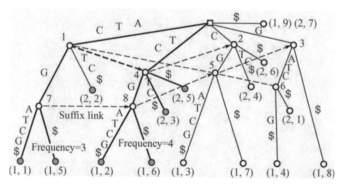

Fig. 2.12 Requirement two of representative pattern is that the suffix link pointing to it is greater than that of its left hand extension.

extension (path label ATC at node 1).

The representative patterns $pl(x)$ (A T C G, T C, ...) (7 and 4) is statistically significant if $z_{pl(x)} \geq t$.

Definition 2.2   Significant representative pattern.

Significant representative pattern is both statistically significant and representative (see Fig. 2.13).

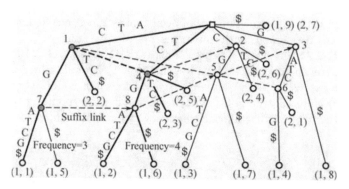

Fig. 2.13 Final representative pattern satisfies its two requirements.

Removal of Statistical Fake Patterns Induced by Their Sub-patterns

Although significant representative patterns contain fewer redundancies, there could still be too many of them for human experts to interpret. Some significant representative patterns can be statistically induced by other significant representative patterns and are better to be removed.

Definition 2.3   Statistically induced pattern.

Let P' be a subpattern of P. The conditional statistical significance of P given P' is defined as

$$z_{P|P'} = \frac{k_P - E(X_P|P')}{\sqrt{E(X_P|P')}}$$

where $E(X_P|P') = pr(P|P') \cdot k'_P = \dfrac{pr(P)}{pr(P')} \cdot k_{P'}.$

Given a set of significant representative patterns, a pattern P in it is said to be statistically induced if there exists a proper subpattern P' of P such that $z_{P|P'} < t$. A proper subpattern P' is not statistically induced.

This conditional statistical significance is used to evaluate how strongly the statistical significance of a pattern is actually attributed by the occurrences of one of its proper subpatterns. Patterns whose significances are due to their proper subpatterns by mere chance are fake patterns. Hence removing them would render a more succinct set of patterns.

Here we describe an efficient algorithm to find non-induced patterns. If a pattern P is not induced by its proper subpattern P' that has the smallest conditional statistical significance $z_{P|P'}$ among all its proper subpatterns known as valid patterns for P, then P is not statistically induced. In other word, if P is non-induced, then $z_{P|P'} \geqslant t$ for any valid pattern of it, including the one with smallest $z_{P|P'}$. Thus, we develop Algorithm 1 to efficiently discover non-induced patterns by associating each significant representative pattern with its valid pattern. More specifically, we identify valid nodes representing the valid patterns for each node corresponding to significant representative pattern from the lowest order to highest order.

Algorithm 1   — Discovery of non-induced patterns
1) Construct a generalized suffix tree T for the input sequences.
2) Annotate $k(v)$ the number of positions under each node v of T.
3) Extract a set of nodes whose $k(v) \geqslant min_{occ}$.
4) Sort the above nodes in ascending order according to order of $pl(v)$ using counting sort.
5) For each node v

> Find the valid node w for v using Procedure 1
> If v is not a suffix node and $z_{pl(v)} \geqslant t$
> and $pl(v)$ is not induced by $pl(w)$

      Output $pl(v)$
    End if
End for

Procedure 1 (for Algorithm 1)   — Find valid node
1) Let v_S and v_P be the suffix node and parent node of v respectively.
2) If $pl(v_S)$ is non-induced
      Let w_1 be v_S
    Else
      Let w_1 be the valid node of v_S
    End if
3) If $pl(v_P)$ is non-induced
      Let w_2 be v_P
    Else
      Let w_2 be the valid node of v_P
    End if

4) Pick one node with the smallest conditional statistical significance out of w_1 and w_2 to be the proper node of v.

Running Time Analysis

Steps $1-3$ can be done in linear time. Step 4 uses counting sort to sort the nodes according to the path length and can thus be done also in linear time as the maximal path length is at most the length of the longest input string. Steps 5a and 5b take constant time. Step 5 can be done in $O(L)$ time. Therefore, non-induced patterns can be found in linear time.

Application of Pattern Discovery to Transcription Factor Binding Site Data

We apply our method to identify transcription factor (TF) binding sites on Yeast using the widely studied SCPD database [19] with many of its TFs known along with their regulated genes. They are taken from the upstream (promoter) regions of genes regulated by one or more TFs. Each set of genes is called regulon and is associated with one or more TFs. The genes are believed to be co-regulated by specific TFs and the binding sites for them are experimentally determined in the database. Three conditions are imposed when choosing the regulon: (1) the number of genes in it should be at least three, (2) the consensus binding sites are available, and (3) the number of gaps or "do not care" characters in the consensus should be at most two. The condition (3) is imposed since we discover only consecutive patterns in the current stage. There are totally 18 such regulons. For each regulon of the TF(s), the upstream sequences of genes are extracted from position $+1$ to -800 relative to the ORF (translation start site).

We design a score combining the statistical significance and support to rank the discovered patterns since the former is based only on its number of occurrences and no information of its support is used. However, to find transcription binding sites amongst multiple sequences, the number of supports is important. These genes are regulated by one or more TFs, and thus we expect that each upstream region of the input gene sequence contains one or more binding sites. Hence, patterns with higher support should be considered more important than those with less support. For example, if we have discovered two patterns TTTAAA and CTTCCT with close statistical residual but different support, say 2 and 7, then the latter will be more important and more likely to correspond to binding sites. Hence, a combined score is defined as

$$score = \frac{\text{Support}}{\text{No. of genes}} \cdot \text{Standard residual}.$$

In DNA sequences tandem, repeats are common. For example, in a sequence like AAAAAATTTTTT, the pattern AAAA occurs at positions 1, 2 and 3 which overlap multiple times. Hence, a post-processing step is applied to further remove patterns whose occurrences overlap in the original sequences.

We discovered the non-induced patterns for each dataset and compared the result with YMF [7] and Weeder [20], respectively (see Figs. 2.14, 2.15 and Table 2.1). We ranked the discovered patterns according to the combined score and chose the top 15 ones for comparison. For YMF, we used its webserver and obtained 5 best motifs through FindExplanators [21] for motif length from 6 to 8 (all available parameters), resulting a total number of 15 patterns. 0 spacers and 2 degenerate symbols are allowed in the motif definition. For Weeder, we downloaded the standalone platform and used the medium mode. All motifs recommend in the final output are used for comparison. For each motif reported by Weeder, we use only the best occurrences with the percentage threshold greater than 90 as binding site predictions. We use the measures nSn (sensitivity), nPPV (positive predictive value), nPC (performance coefficient) and nCC (correlation coefficient) defined in [3] in comparison.

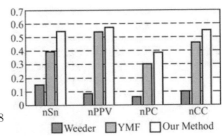

Fig. 2.14 Combined measures over 18 datasets.

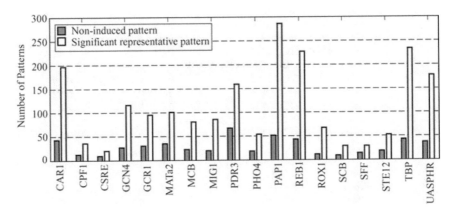

Fig. 2.15 Number of reported patterns after removing induced patterns among significant representative patterns.

Among the 18 datasets, our discovered patterns within rank 13 match the consensus binding sites in 14 datasets and 4 of them are ranked top. The patterns in bold do not match the binding sites in the remaining 4 datasets CPF1, CSRE, MATalpha2 and SFF. The reason why our discovered patterns

Table 2.1 Comparison of our method, YMF and Weeder SPCD datasets (the patterns among top 15 that achieve the best nSn is used for comparison). IUPAC Nucleotide Code is used

TF		Motif/Pattern	nSn	nPPV	nPC	nCC	Rank
CAR1	Consensus	AGCCGCCR					
	Weeder	CCTAGCCG	0.23	0.09	0.07	0.14	
	YMF	GCCGCCG	0.7	1	0.7	0.84	
	Our Method	AGCCGCC	0.88	1	0.88	0.94	6
CPF1	Consensus	TCACGTG					
	Weeder	CACGTGGC	0	0	0	−0.01	
	YMF	YCACGWG	1	0.5	0.5	0.71	
	Our Method	TTC	0.29	0.01	0.01	0.04	10
CSRE	Consensus	YCGGAYRRAWGG					
	Weeder	GCGGTCGG	0	0	0	−0.01	
	YMF	CGGATRRA	0.58	0.22	0.19	0.35	
	Our Method	CCGG	0.33	0.08	0.07	0.15	1
GCN4	Consensus	TGANT					
	Weeder	TGACTC	0.07	0.13	0.05	0.08	
	YMF	TGWCTR	0.18	0.51	0.15	0.29	
	Our Method	TGACT	0.34	1	0.34	0.57	13
GCR1	Consensus	CWTCC					
	Weeder	TCTGGCATCC	0.1	0.2	0.07	0.13	
	YMF	TCTYCCY	0.3	0.48	0.23	0.37	
	Our Method	TTCC	0.68	0.39	0.33	0.5	9
MATalpha2	Consensus	CRTGTWWWW					
	Weeder	GGAAATTTAC	0.13	0.14	0.07	0.13	
	YMF	ACGCGT	0	0	0	0	
	Our Method	GAAAAAAG	0	0	0	−0.01	1
MCB	Consensus	WCGCGW					
	Weeder	AGACGCGT	0.19	0.08	0.06	0.1	
	YMF	ACGCGT	0.68	1	0.68	0.82	
	Our Method	ACGCGT	0.68	1	0.68	0.82	1
MIG1	Consensus	CCCCRNNWWWWW					
	Weeder	CCCCAG	0.39	0.1	0.09	0.19	
	YMF	CCCCRS	0.5	0.21	0.18	0.32	
	Our Method	CCCCAG	0.33	0.29	0.18	0.3	2
PDR3	Consensus	TCCGYGGA					
	Weeder	GTCTCCGCGG	0.32	0.14	0.11	0.19	
	YMF	TCCGYGGA	1	1	1	1	
	Our Method	TCCGCGGA	0.64	1	0.64	0.8	1
PHO4	Consensus	CACGTK					
	Weeder	GAAACGTG	0.07	0.02	0.02	0.02	
	YMF	CACGTGSR	0.71	0.75	0.58	0.73	
	Our Method	CACGTG	0.71	1	0.71	0.84	1
RAP1	Consensus	RMACCCA					
	Weeder	AGCACCCA	0.13	0.23	0.09	0.17	
	YMF	CACCCA	0.64	0.86	0.58	0.74	
	Our Method	CACCCA	0.64	0.86	0.58	0.74	8
REB1	Consensus	YYACCCG					
	Weeder	ACCCGC	0.14	0.05	0.04	0.08	
	YMF	TTACCCG	0.7	1	0.7	0.84	
	Our Method	TTACCCG	0.7	1	0.7	0.84	7

TF		Motif/Pattern	nSn	nPPV	nPC	nCC	Continued Rank
ROX1	Consensus	YYNATTGTTY					
	Weeder	CCTATTGT	0.28	0.05	0.04	0.07	
	YMF	TTGTTS	0.48	0.29	0.22	0.35	
	Our Method	ATTGTT	0.6	0.63	0.44	0.6	6
SCB	Consensus	CNCGAAA					
	Weeder	AGTCACGAAA	0.47	0.26	0.2	0.31	
	YMF	CACGAA	0.61	1	0.61	0.78	
	Our Method	CACGAAA	0.71	1	0.71	0.84	1
SFF	Consensus	GTMAACAA					
	Weeder	CTGTTTAG	0.13	0.02	0.02	0.04	
	YMF	TAAWYA	0.38	0.08	0.07	0.17	
	Our Method	AAAGG	0.13	0.04	0.03	0.06	2
STE12	Consensus	TGAAACA					
	Weeder	ATGAAACA	0.2	0.05	0.04	0.07	
	YMF	ACATGS	0.06	0.1	0.04	0.07	
	Our Method	TGAAAC	0.86	0.7	0.63	0.77	3
TBP	Consensus	TATAWAW					
	Weeder	CCGCTG	0	0	0	−0.02	
	YMF	CRCATR	0.01	0.02	0.01	0	
	Our Method	ATATAAA	0.43	0.89	0.41	0.62	13
UASPHR	Consensus	CTTCCT					
	Weeder	TGTCAGCG	0	0	0	−0.01	
	YMF	CCTCGTT	0.14	0.21	0.09	0.17	
	Our Method	CTTCCTC	0.71	0.86	0.64	0.78	9
Average	Weeder		0.16	0.09	0.05	0.09	
	YMF		0.48	0.51	0.37	0.47	
	Our Method		0.54	0.65	0.44	0.56	

have no match in CPF1, CSRE and SFF is that their consensus binding sites have fewer than 2 occurrences. As for MATalpha2, the consensus has 6 occurrences, but it has many substitutions. Because our program runs with $min_{occ} = 5$ and is confined to discover consecutive patterns, these consensuses are not discovered.

The overall performances of Weeder, YMF and our method across 18 datasets are evaluated by the combined measures in [3] and shown in Fig. 2.14. It indicates that the overall performance of our method is better than YMF. That Weeder does not perform well comparatively might be that the percentage threshold 90 is too strict. However, Weeder does not provide a good strategy in choosing this parameter.

Figure 2.15 shows the number of reported patterns in terms of non-induced pattern and significant representative patterns. After removing induced patterns among significant representative patterns, our method produces a relative small set of patterns of which the number of reported patterns ranges from 8 to 67. The result shows that our method is able to retain those patterns associated with conserved functional units in the promoter regions while reducing the number of patterns.

2.4 Conclusion

It has been a long journey since Professor K.S. Fu pioneered the new field of stochastic syntactic pattern recognition. It was through the developing of distance measures between biomolecular sequences by optimal alignment [22] that Professor Fu and the first author (A. Wong) of this chapter came to know each other when bioinformatics was in its infancy. Through much discussions and encouragement, A.Wong has launched into the new field of bioinformatics, pattern recognition on discrete and mixed-mode valued data as well as structural data. This has led to Wong's later development.

In this chapter, some of Wong's early works on discrete-valued data and random graphs [23, 24] in IEEE Transacations on PAMI are cited. That later led to random sequences synthesis and analysis [25, 26] and other works [18, 27, 28]. This chapter also provides a brief review of known pattern discovery techniques for sequence data and presents a new sequence pattern discovery framework developed by our research team that yields very important results. It has solved a very difficult problem of discovering statistical significant patterns in multiple sequences without relying on prior domain knowledge and reducing redundant patterns whose statistical significance is induced by their strong subpatterns to render a more compact pattern set.

The results from the transcription factor binding sites experiment confirm the algorithm's ability to acquire a relatively small set of patterns that reveal interesting, unknown information inherent in the sequences. While the algorithm drastically reduces the number of patterns, it is still able to retain patterns associated with conserved functional units in the promoter regions without relying on prior knowledge.

Our future work will advance in the following directions: (1) Extending our method to discover patterns with gaps; (2) Discovering distance patterns in protein sequences and relating the discovered patterns to three-dimensional conformation and low sequence similarity.

References

[1]   Zhao Q, Bhowmick S S (2003) Sequential Pattern Mining: A Survey, Technical Report, CAIS. Nanyang Technological University, Singapore, No 2003118

[2]   Das M K, Dai H K (2007) A Survey of DNA Motif Finding Algorithms. BMC Bioinformatics, 8 (Suppl 7): S21

[3]   Tompa M, Li N, Bailey T L et al (2005) Assessing Computational Tools for the Discovery of Transcription Factor Binding Sites. Nature Biotechnology, 23(1): 137 – 144

[4]   Wong A K C, Zhuang D, Gary C L Li et al (2010) Discovery of Non-induced Patterns from Sequences. In: Pattern Recognition in Bioinformatics, pp 149 – 160

[5]   http://jaspar.genereg.net/. Accessed 3 October 2010

[6]   Jonassen I (1996) Efficient Discovery of Conserved Patterns Using a Pattern Graph. Technical Report 118, Department of Informatics, University of Bergen, Norway

[7] Sinha S, Tompa M (2002) Discovery of Novel Transcription Factor Binding Sites by Statistical Overrepresentation. Nucleic Acids Research, 30(24): 5549–5560

[8] Rigoutsos I, A Floratos (1998) Combinatorial Pattern Discovery in Biological Sequences: The TEIRESIAS Algorithm. Bioinformatics, 14(1): 55–67

[9] Parida L, Rigoutsos I, Floratos A et al (2000) Pattern Discovery on Character Sets and Real-valued Data: Linear Bound on Irredundant Motifs and an Efficient Polynomial Time Algorithm. In: Proceedings of the eleventh ACM-SIAM Symposium on Discrete Algorithms, SODA 2000, pp 297–308

[10] Pisanti N, Crochemore M, Grossi R et al (2005) Bases of Motifs for Generating Repeated Patterns with Wild Cards. IEEE/ACM Trans on Computational Biology and Bioinformatics, 2(1): 40–49

[11] Pisanti N, Crochemore M, Grossi R et al (2004) A Comparative Study of Bases for Motif Inference. In String Algorithmics. KCL Publications, London

[12] Lawrence C E, Reilly A A (1990) An Expectation Maximization (EM) Algorithm for the Identification and Characterization of Common Sites in Unaligned Biopolymer Sequences. PROTEINS: Structure, Function, and Genetics, 7, 41–51

[13] Lawrence C E, Altschul S F, Boguski M S et al (1993) Detecting Subtle Sequence Signals: A Gibbs Sampling Strategy for Multiple Alignment. Science, 262(5131): 208–214

[14] Bailey. T L, Elkan C (1995) Unsupervised Learning of Multiple Motifs in Biopolymers Using Expectation Maximization. Machine Learning, 21(1/2): 51–80

[15] D'haeseleer M (2006) How does DNA Sequence Motif Discovery Work? Nature Biotechnology, 24, 959–961

[16] Wong A K C, Reichert T A, Aygun B (1974) A Generalized Method for Matching Informational Macromolecular Cod Sequences. Journal of Computers in Biology and Medicine, 4, 43–57

[17] Dan Gusfield (1997) Algorithms on Strings, Trees, and Sequences. Computer Science and Computational Biology

[18] Wong A K C, Wang Y (1997) High-Order Pattern Discovery from Discrete-Valued Data. IEEE Trans On Knowledge Systems, 9(6): 877–893

[19] SCPD Database, http://rulai.cshl.edu/SCPD/. Accessed 1 October 2010

[20] Eskin E, Pevzner P (2002) Finding Composite Regulatory Patterns in DNA Sequences. Bioinformatics, 18(1): S354–S363

[21] Blanchette M, Sinha S (2001) Separating Real Motifs from Their Artifacts. Bioinformatics, 17(1): S30–S38

[22] Wong A K C, Reichert T A, Aygun B (1974) A Generalized Method for Matching Informational Macromolecular Cod Sequences. Journal of Computers in Biology and Medicine, 4: 43–57

[23] Wong A K C, Wang C C (1979) DECA-A Discrete-Valued Ensemble Clustering Algorithm. IEEE Trans on Pattern Analysis and Machine Intelligence, PAMI-1(4): 342–349

[24] Wong A K C, You M L (1985) Entropy and Distance of Random Graphs with Application to Structural Pattern Recognition. IEEE Trans on Pattern Analysis and Machine Intelligence, PAMI-7(5): 599–609

[25] Chan S C, Wong A K C (1991) Synthesis and Recognition of Sequences. IEEE Trans on PAMI-13(12): 1245–1255

[26] Wong A K C, Chiu D K Y, Chan S C (1995) Pattern Detection in Biomolecules Using Synthesis Random Sequence. Journal of Pattern Recognition, 29(9): 1581–1586

[27] Zhang, Wong A K C (1997) Towards Efficient Multiple Molecular Sequence Alignment. IEEE Trans on SMC, pp 918–932

[28] Wong A K C, G Li (2008) Simultaneous Pattern Clustering and Data Grouping. IEEE Trans Knowl Data Eng, 20(7): 911–923

3 A Hybrid Method of Tone Assessment for Mandarin CALL System

Yang Qu[1], Xin He[2], Yue Lu[1,*], Patrick S P Wang[1,3]

Abstract Mandarin is known as a tonal language, so the tone is a distinctive discriminative feature in Mandarin. However, accurate segmentation of utterances has a great effect on tone recognition. In this chapter, an approach based on forced alignment of HMM (Hidden Markov Model) is employed to train utterances for obtaining model of getting accurate syllable boundary of utterances. Moreover, for the purpose of getting more objective tone evaluation, a competing models based measure is used to get tonal syllable and tone assessments. We combine two scoring functions to acquire the overall tone scoring results. The experimental results demonstrate that this proposed hybrid method, using forced alignment based tone model and competing models based tone assessment, gives a relative accurate tone assessment.

3.1 Introduction

Overview of CALL

Language learning makes people easier to communicate with each other, so it becomes a hot field in education. The traditional face-to-face teaching, which is based on interpersonal communication between the teacher and student, is crucial in oral language learning. This teaching method is highly effective, but it is not enough. Compared with the face-to-face teaching method, the self-teaching method is obviously a useful supplement. It is not subject to the restrictions of time and place, varies with each individual, and thereby we can develop different learning strategies. However, because of the limitations of the level of self-learners, they are hard to find errors and correct the incorrect pronunciations on their own.

Along with the popularization of computers and networks, computers have become so widespread in schools and homes and their uses have expanded so dramatically that the majority of language teachers now think

1 Department of Computer Science and Technology, East China Normal University, Shanghai 200241, China.

2 Motorola China Research Center, Shanghai, China.

3 College of Computer and Information Science, Northeastern University, Boston, MA 02115, USA.

* Corresponding author. Department of Computer Science and Technology, East China. Normal University, Shanghai 200241, China. E-mail: ylu@cs.ecnu.edu.cn (Y. Lu).

about the implications. Technology can bring about changes in the teaching methodologies of foreign language beyond simply automating fill-in-the-gap exercises [1].

Computer Assisted Language Learning (CALL) is a form of computer-based accelerated learning which can offer many potential benefits for both learners and teachers, because they offer learners the chance to practice extra learning material at their convenient time in a stress-free environment [2, 3]. It can be used to reinforce what has been learned in the classrooms. Simultaneously, it can also be used as remedial to help learners with limited language proficiency.

The origin and development of CALL can be trace back to the 1960s. Early CALL systems focus mainly on training reading, writing and listening skills, but rarely on oral language skill. Since the early 1970s, a number of different approaches to automatic speech recognition (ASR) have been proposed and implemented by researchers, including dynamic time warping (DTW), template matching, knowledge-based expert systems, neural nets, and Hidden Markov Modeling (HMM) [4, 5]. HMM-based modeling applies sophisticated statistical and probabilistic computations to the problem of pattern matching at the sub-word level. The generalized HMM-based approach has been proved that it is an effective approach utilized in speech recognition. Even if it is not the most effective one, method for creating high-performance speaker-independent recognition engines that can cope with large vocabularies. The vast majority of today's commercial systems deploy this HMM-based technique [6]. At the moment, using current CALL technology, even with its current limitations, for the development of speaking abilities has gained much attention. In real life, some success on the application of CALL has been made. Particularly, the computer-mediated communication, which can help speaking skills closely linked to "communicative competence" (ability to engage in meaningful conversation in the target language) and provide controlled interactive speaking practice outside the classroom, is a successful application by using current CALL technology [6].

Speech recognition plays a key role in the pronunciation learning. With the rapid development of speech processing and recognition technologies, more and more researchers focus on the progress in the area of CALL systems to assist pronunciation learning [7 – 9]. Many research groups have already undertaken such studies, and even developed the more mature applications, such as the University of Cambridge, the FLUENCY of the Carnegie Mellon University Language Technology Research Institute [10], the EduSpeak of the SRI International research institute [7, 11], the Academic Center for Computing and Multimedia Studies of Kyoto University [12 – 14], the Tsinghua University, the ThinkIT Laboratory of the Institute of Acoustics, Chinese Academy of Sciences [15], the Hong Kong Polytechnic University and the PLASER of the Hong Kong University of Science and Technology [16].

Chapelle [17] suggests that some design features and evaluation criteria for multimedia CALL might be developed on the basis of hypotheses about ideal conditions for second language acquisition (SLA). In his paper, he also outlines a relevant theory of SLA and enumerates seven hypotheses it implies for ideal conditions such as input saliency, opportunities for interaction, and learner focus on communication. These evaluation criteria relate to many areas of the speech signal processing and speech recognition, such as computer-aided pronunciation quality assessment, natural language processing and so on.

Specificity of Mandarin CALL System

Mandarin is a monosyllabic structure language. Each character is pronounced as a monosyllable, and has its own meaning. In our daily conversation, consonant and vowel are generally used to distinguish different characters and words. Moreover, tone is also integrant. Although the syllable may consist of the same consonant and vowel, the meaning of the syllable will not the same when the tone is different. Therefore, the meaning discriminating of the tone makes it equally important to consonant and vowel. Tone, as a basilic aspect of distinguishing features in Mandarin speech, makes Mandarin obviously different from the Indo-European languages. In fact, considering Mandarin itself, the tone is also a crucial benchmark to tell apart different dialects. In Mandarin, consonant, vowel and tone are the three essential elements of the Mandarin speech.

The main speech technologies used in the CALL system include high-quality speech compression, pronunciation assessment, speech synthesis, etc. The pronunciation assessment within CALL plays a vital role. Tone is usually described by pitch, and it represents the pitch changes in words. Since Mandarin is a tonal language, the accurate pronunciation is relevant not only to phoneme, but also to the tone of each word. Especially, the tone recognition of the isolated word often can get good recognition results. But in continuous speech, two neighboring tones will interact with each other commonly, so that the tone recognition rate can not remain at a high level, and the final pronunciation assessment result will be influenced. Meanwhile, various environments, diverse dialects of Mandarin, and the speaker's gender, age, emotion, attitude can also cause a lot of difficulties in recognition. Take different people as an example, the pitch of a woman is usually higher than that of a man, young people mostly have higher pitch than old people. However, even if the same person, he would also have a dissimilar pitch in various circumstances. As a rule, when a person is excited, his pitch will rise higher, but if a person is down in the mouth, his pitch will descend slightly lower.

It is very important to pronounce tone precisely in live communication, and therefore, evaluating tone pronunciations is crucial for a Mandarin CALL system [18]. Yet, there are only a few works have been done in tone assessment of Mandarin CALL system [19 – 22]. For English, tone is only relevant to prosody, and it is unrelated to the word itself. Emotions and attitudes are reflected by the intonation that people use when they speak. Depending on the intonation, the same sentence may show different attitudes. In [23], an automatic intonation assessment system for second language is proposed based on a top-down scheme. But in a tonal language, pitch of voice is used to distinguish one word from another [24]. In Mandarin, each character is associated with one out of five possible tones. The tone of a given character is also context-dependent according to Tone Sandi [25]. To be effective, a Mandarin CALL system requires the ability to accurately measure tone quality.

Our work

The aim of our work described in this chapter is to study a method for automatic tone assessment in a Mandarin CALL system. Figure 3.1 shows the block diagram of our tone evaluation system. Initially, a speech voice input into the system. Then the front-end feature extraction converts the input speech waveform to a sequence of mel-frequency cepstral coefficients (MFCC) parameters. Subsequently, Viterbi decoding of HMM is applied to do forced alignment for obtaining syllables and performing the automatic segmentation to acquire the position of each syllable. For getting a relatively objective tone evaluation results, the competing model based tone assessment is used to each syllable in our system. Here we adopt two models to receive two scores respectively, a 39-dimension tonal model for tonal syllable score and a tone model we proposed for tone score. Finally, considered the importance of tone in Mandarin, the overall tone score is obtained by combining tonal syllable score and tone score. The system described here is focused on measuring the tone pronunciation quality of speech in Mandarin. In our system, the tone assessment can be mainly divided into two parts: one is tone model based on forced alignment of HMM; the other one is tone assessment based on competing models.

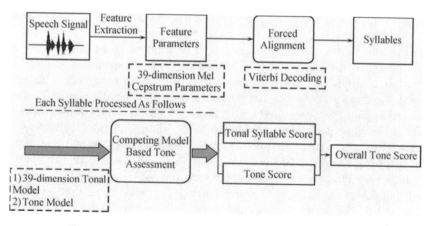

Fig. 3.1 Block-diagram of our tone evaluation system.

Accurate segmentation of syllables has great impact on the accuracy of tone recognition. Therefore, a training measure based on HMM forced alignment is used to improve the accuracy of segmentation in Mandarin. Moreover, for getting an objective tone evaluation, the competing models based approach was introduced in this chapter. Especially, the tonal syllable score is obtained by using tonal syllable competing models, and the tone score is computed by using tone competing models.

The outline of the chapter is as follows: Section 3.2 gives related works in phonetic segmentation, automatic pronunciation assessment and tone assessment. In Section 3.3, a new method of tone assessment using the forced alignment based tone model and combining the competing models is proposed. Section 3.4 gives a detailed description of the experimental procedure and analysis for the proposed approach. Conclusions are given in Section 3.5.

3.2 Related Work

For the purpose of synthesis and recognition, speech often needs to be segmented into phonetic units firstly. Research about automatic phonetic labeling has been investigated extensively in TTS (text-to-speech) [26–30]. At the same time, automatic segmentation for natural speech is also meaningful for automatic speech recognition. Automatic segmentation techniques described in the literature can roughly be classified into two major categories, implicit segmentation and explicit segmentation [31]. In the explicit case, the segmentation algorithm split up the incoming utterance into segments as already known, so it is linguistically constrained to an a priori known phonetic sequence, while in the implicit case there is no prior knowledge of the corresponding phonetic sequence. At present, HMMs, which are able to take the statistical variability of speech into account, are widely used in speech recognition. Simultaneously, the HMM-based segmentation approaches are also frequently used in speech synthesis [32–35]. In [32], the paper described an automatic procedure for the segmentation of speech, and an acoustic phonetic unit HMM recognizer is used in their system. Makashay et al. compare the impact of several phonetic segmentation techniques on TTS synthesis, and the CWT (Speaker Dependent Cross Word Triphone HMMs) method provides the best segmentation in their experiment [33]. Adell et al. give a comparative study of automatic phone segmentation methods for TTS [34]. Mporas et al. proposed a hybrid HMM-based architecture for speech segmentation [35]. Most of the researchers do the phonetic segmentation by Viterbi forced alignment using HMM and analyze the phonetic boundaries by boundary checking rules. However, the forced alignment based tone model is little reported in tone assessment.

In previous work, many researchers have investigated on automatic pronunciation quality assessment by speech recognition techniques [12, 36–42]. Franco et al. introduced HMM-based scores by using average phone segment posterior probabilities to automatic assess pronunciation quality in a computer-aided language instruction system [7, 36, 37]. A method of automatic pronunciation scoring, which utilizes a likelihood-based Goodness of Pronunciation (GOP) measure, is investigated for use in CALL systems [39]. An English pronunciation learning system for Japanese students is introduced in [12]. Molina et al. proposed an ASR based pronunciation evaluation method to automatically generate the competitive lexicon [40]. Cincarek et al. describe an approach for automatic pronunciation scoring of words and sentences for non-native speech [41]. De Wet et al. described an attempt to automate the large-scale assessment of oral language proficiency and listening comprehension for fairly advanced students of English as a second language [42]. Most of them focus on the English speech. In Mandarin, some improvements also have been developed in pronunciation evaluation [8, 15, 43, 44]. Among them, an approach based on the ranking among 411 syllables in Mandarin was presented to measure confidence [43]. Furthermore, a relative measure based on the ranking among all competing biphone models was also used [8].

Tone is one of the important components in Mandarin. Particularly in the assessment of Mandarin speech, tone plays a vital role. Although considerable research effort has been invested in the development of CALL systems, little attention has been paid to tone assessment of Mandarin CALL system. In [45], Y Cao et al. took a decision tree based approach to obtain the quantitative result of tone variation patterns in continuous Chinese speech. They used the Mandarin Database collected under 863 Project and their recognition accuracy is 69.1%. In [46], the real-context model was proposed as a new concept to be used in the tone unit selection. And a method based on hierarchical clustering is performed to generate a more refined tone model. It is still a challenging problem for Mandarin to evaluate the tone of characters in a sentence correctly. In Surendran's paper, he first proposed an information-theoretic measure to analyse the importance of tones in recognition of Mandarin Chinese, and then investigates the automatic recognition of tones in Mandarin speech [47]. Jiang et al. mainly study an effective CALL system for strongly accented Mandarin speech, and three technologies, speaker CMN (speaker-dependent Cepstrum Mean Normalization), HLDA (Heteroscedastic Linear Discriminate Analysis) and MAP (Maximum a Posteriori), are used to resolve three outstanding problems as follows: the channel distortion in application environment, low discrimination ability of the acoustic model and the mismatch between the speaker-independent acoustic model and the assessed speech, accordingly the accuracy of the phonetic pronunciation quality evaluation is improved [48].

3.3 Proposed Approach

In this section, we propose an approach, which uses the forced alignment based tone model and combines the competing models based tone assessment to improve the tone recognition accuracy in Mandarin. A detailed description of the approach will be described in this section.

3.3.1 Preprocessing of Segmentation Using HMM-Based Forced Alignment

The higher segmentation accuracy implies that the better assessment can be achieved. Accordingly, speech signals should be segmented first. As showed in the upper half of Fig. 3.1, before the tone assessment, we preprocess the input speech voice for initial segmentation by using HMM-based forced alignment. Definitely, during the forced alignment, the acoustic model plays a vital role. The acoustic model for the HMM forced alignment is constructed by using the 863 Mandarin corpus [49]. The model is based on context-dependent triphone modeling, and the acoustic feature is 39 dimensional Mel-frequency cepstral coefficients (MFCCs, include 13 static, 13 delta, and 13 delta delta) tonal model. The model is constructed by using HTK tools [50]. We use phones as basic units. A system that uses phones as basic units should consider the context dependency in modeling those phones [51]. In our system, 90 phones are adopted to construct 1505 triphones HMM model. Then we create the tied model and train it using the same training data. In our HMMs, each triphone is modeled in 5-state topology, and each state has 8 Gaussians. In our experiment, the input speech waveform is first extracted into 39-dimension MFCC parameters. Then Viterbi algorithm is used to do forced alignment. As a result, each input speech is segmented into several tonal syllables.

Take the Chinese sentence "偶然的机会 (Ouran de Jihui)" as an example, two different segmentation results are given in Fig. 3.2 and Fig. 3.3, respectively.

Figure 3.2 shows the initial segmentation results of the sentence "偶然的 机会 (Ouran de Jihui)" by using HMM-based forced alignment. The pronunciation of the sentence in Mandarin can be expressed as "ou3 ran2 de0 ji1 hui4". In the vertical direction, lines in the waveform plot indicate the boundaries of tonal syllables. The pronunciation of each syllable is labeled under the waveform. Among the labeled area, "sil" represents the silence in the head and tail parts of this sentence. The bottom panel of Fig. 3.2 shows the spectrogram of the sentence. Moreover, on the spectrogram, it also plots the corresponding pitch vector of this utterance by using the blue irregular line.

For Comparison, Fig. 3.3 is the manually segmented results of the same sentence of Fig. 3.2. In the pronunciation labeled area, "pau" is labeled between syllable "de0" and "ji1". Here "pau" indicates the pause between two syllables. Compared with Fig. 3.2, the location of syllable boundaries of Fig. 3.2 are generally before the boundaries of Fig. 3.3.

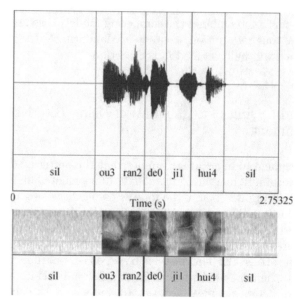

Fig. 3.2 An example of initial segmentation results for the test utterance "偶然的机会 (Ouran de Jihui)" segmented by HMM-based forced alignment.

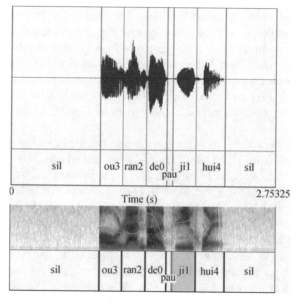

Fig. 3.3 An example of segmentation results for the test utterance "偶然的机会 (Ouran de Jihui)" segmented by manual.

3.3.2 Forced Alignment Based Tone Model

In this section, we partitioned training samples into five tonal classes, the high flat (tone 1), the low rising (tone 2), the high low rising (tone 3), the high falling (tone 4), and the neutral tone (tone 5).

For getting accurate tone assessment, we first use forced alignment of HMM to identify each syllable. Accordingly, accurate segmentation is particularly important. In order to segment accurately, forced alignment based tone model was proposed in our system.

According to the forced alignment results, each utterance was segmented into several words. In real world, speech is time-varying and non-stationary, so that shot-term techniques are always used to process speech signals. Pitch feature is often used to represent the vibration rate of audio signals, which can be represented by the fundamental frequency, or equivalently, the reciprocal of the fundamental period of voiced audio signals.

There are many different functions in time-domain short-term based pitch tracking algorithms [52 – 57]. Among them, the autocorrelation function (ACF) and the average magnitude difference function (AMDF) are more typical functions. ACF is relatively noise immune, and it can work well in many contexts. But, in order to cover the F0 ranges encountered in speech, the window over ACF is relatively large. So the amount of heavy calculation impacts its application of real-time processing. Compared with the autocorrelation method, AMDF has the advantage of low computation and high precision, and is employed by the U.S. government in the speech coding standard LPC-10 [36]. The defect of the AMDF is that its peak values decrease with time lag.

In this section, for all the segmented utterances, we employed the normalized sum magnitude difference square function (SMDSF) and Viterbi post-processing algorithm [58] to extract pitch features, which are 4-dimension, five tone triphone features. The SMDSF combines the ACF and the ADMF effectively, so it can be used as a standard measure of non-periodic level of signals, and it can accurately estimate speech pitch at any sample rate in real time. Details of the pitch tracking method can be found in [59].

Figure 3.4 shows the pitch tracking results of one frame speech signal by using different pitch tracking algorithms. As shown by the graph, the top panel is the original speech signal of one frame, other three panels plot the difference among AMDF, SMDSF and normalized SMDSF functions.

The normalized SMDSF-based pitch tracking combined with the Viterbi post-processing algorithm can get less pitch estimation errors. Take the utterance "偶然的机会 (Ouran de Jihui)" for example, Fig. 3.5 illustrates the difference among three functions, the top one uses the normalized SMDSF pitch tracking algorithm, the second one uses the normalized SMDSF-based pitch tracking algorithm combined with the medium smoothing post-processing algorithm, and the third one uses the normalized SMDSF-based pitch tracking algorithm combined with the Viterbi post-processing algorithm.

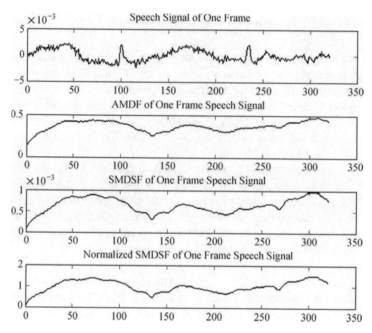

Fig. 3.4 Pitch tracking results of one frame by using different pitch tracking algorithms.

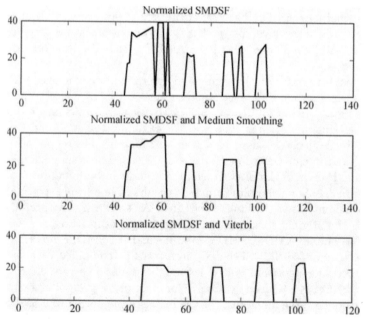

Fig. 3.5 Pitch tracking results of the utterance "偶然的机会 (Ouran de Jihui)" by using different pitch tracking algorithms.

Before training the model, each training utterance needs to be segmented into several syllables by using HMM forced alignment. The acoustic model for the HMM forced alignment is based on context-dependent triphone modeling, and the model is a 39-dimension Mel-frequency cepstral coefficients (MFCC) tonal model. After forced alignment, the segmental results are used as parameters in training our tone model.

3.3.3 Competing model based tone assessment

With the purpose of getting objective tone evaluation, we used a relative measure based on competing models. Basically, the whole hybrid evaluation procedure contains two parts: tonal syllable evaluation and tone evaluation.

In tonal syllable evaluation, a simplified INITIAL/FINAL net based on the linguistic knowledge of Mandarin pronunciation is proposed to form legal competing models.

Syllable is the smallest unit of the Mandarin pronunciation. It is the basic unit of speech structure. A Mandarin syllable has either a structure form of CV (Consonant-Vowel or INITIAL-FINAL [25, 60]) or a single structure form of V (Vowel). The beginning of a syllable is called as consonant, and the INITIAL is the initial consonant of a syllable. Here FINAL is the vowel. Table 3.1 (a) and (b) show that the total number of INITIALs and FINALs are, respectively, 21 and 37.

Table 3.1 21 INITIALs of Mandarin syllables and 37 FINALs of Mandarin syllables

(a) 21 INITIALs of Mandarin syllables
b, p, m, f, d, t, n, l, g, k, h, j, q, x, z, c, s, zh, ch, sh, r
(b) 37 FINALs of Mandarin syllables
a, ai, an, ang, ao, e, ei, en, eng, er, i, ia, ian, iang, iao, ie, ii, iii, in, ing, iong, iou, o, ong, ou, u, ua, uai, uan, uang, uei, uen, uo, v, van, ve, vn

In our system, the INITIAL/FINAL format based tonal syllable assessment proceeds as follows:

1) For a syllable of "(INITIAL) + FINAL" format, we define the set of competing models as "(INITIAL)*+FINAL*+TONE*" where the part between brackets is optional and the * is a wildcard representing all the possible INITIALs/FINALs/TONEs. For example, the Chinese character "偶" is pronounced "ou3", it is composed of FINAL "ou" and TONE "3", and it has no INITIAL. Thus its competing models can be expressed by "FINAL* + TONE" format. If the full FINAL net listed in Table 3.1 (b) is used, we will have 37×5 competing models at most, such as "a0".

2) We then send the syllables to the competing models for a log probability evaluation and find the rank (zero-based) of syllable in the competing models.

3) Each syllable has a different set of competing models. We therefore divide the rank of syllable by the size of its competing models to obtain a

rank ratio between 0 and 1. Then the syllable score of the ith syllable in an utterance can be expressed by

$$Score_i = \frac{100}{1 + \left|\dfrac{r_i}{a}\right|^b} \tag{3.1}$$

where r_i is the rank ratio of the ith syllable, and a and b are the tunable parameters of this scoring function.

Since the total number of INITIALs, FINALs and TONEs are, respectively, 21, 37 and 5, so each syllable has $21 \times 37 \times 5$ competing models at most. It is too large. Accordingly, the simplified net is introduced to form legal competing models. Table 3.2 lists the simplified net, which includes 2 INITIAL nets and 3 FINAL nets.

Table 3.2 Simplified INITIAL/FINAL net based on the linguistic knowledge

(a) 2 INITIAL nets

INITIAL net 1 c, ch, s, sh, z, zh, j, q, x
INITIAL net 2 d, t, g, h, k

(b) 3 FINAL nets

FINAL net 1   en, eng, iong , ong, ou, u, uen
FINAL net 2   ao, e, o, ou, u , uo
FINAL net 3   in, ing, iong

Take the phrase "斗争 (Douzheng)" (it means fight) for example, it can be pronounced as "dou4 zheng1". For the first word "dou4", INITIAL "d" is in the second INITIAL net "d, t, g, h, k", while FINAL "ou" is in the first FINAL net "en, eng, iong , ong, ou, u, uen" and the second FINAL net "ao, e, o, ou, u , uo". Therefore, the competing models of "dou4" can be represented in the form as "INITIAL from INITIAL net 1 + FINAL from FINAL net 1 and FINAL net 2 + TONE".

In tone evaluation, we employed the forced alignment based tone model to extract pitch features of a given utterance, and send each syllable into five tone competing models for a log probability evaluation and find the tone rank (zero-based) of syllable in the competing models. In like manner, the tone competing models is defined as "INITIAL+FINAL+TONE*" where the INITIAL/FINAL is constant and the * is a wildcard representing all the possible TONEs. Similarly, Eq. (3.1) was brought to compute the ith correct tone score. After previous two evaluations, we have obtained two scores. The overall tone scoring function is designed as a weighted average of two scores:

$$Score_{overall} = w_s \cdot Score_{tonalsyllable} + w_t \cdot Score_{tone}, \tag{3.2}$$

where w_s and w_t are the tunable parameters, and $w_s + w_t = 1$.

3.4 Experimental Procedure and Analysis

Speech Database

The speech corpus used in the experiments is the Mandarin Database collected under 863 Project (863 Database), which is a speaker-independent, large vocabulary, continuously read Mandarin Chinese speech database [49]. The text material was selected from the Chinese newspaper *People's Daily*, telescripts and dictionaries. The corpus covers 2 185 sentences and 388 phrases. The database is recorded by 80 speakers, which include 40 males and 40 females, and each speaker read 520 utterances. 60 speakers' (30 males and 30 females) records are used as training set and the rest are used as testing set. All the Mandarin speeches of the 863 Database are in the format of 8 000 Hz sampling rate, 16 bits bit rate, mono channel and PCM audio format.

Mandarin Speech Tone Recognition with Forced Alignment Based Tone Model

To evaluate the impact of tone recognition of different segmentation results, we used three models for comparison.

The configurations of models for experiments are listed as follows:

1) ToneBaseline: baseline, 4-dimension, five tone triphone models;
2) ToneAligned: trained after forced alignment;
3) ToneCheckedmanually: trained after segmentation checked manually.

All of the three are 4-dimension, five tone triphone models. ToneBaseline is considered as the baseline. ToneAligned and ToneCheckedmanually are the enhanced version of ToneBaseline. ToneAligned is the forced alignment based tone model which is introduced in previous section. ToneCheckedmanually was generated after segmentation checked manually. For ToneAligned and ToneCheckedmanually, the detailed introductions are carried out in the following.

In the process of generating ToneAligned model, forced alignment is performed to align all the training utterances into tonal syllables. The 39 dimension tonal acoustic model for the HMM forced alignment is trained by HTK tools [50]. And the model was trained by 42 693 utterances in 863 Database. The segmental results of forced alignment are referred in training process.

As for the ToneCheckedmanually model, a set of manually adjusted labels generated by a single human labeler were used as references for utterances segmentation. Then, the results of segmentation were used in training.

In our experiment, TEST SET 1 and TEST SET 2 were used in testing. Each TEST SET has 100 sentences. The results of our experiments are presented in Tables 3.3 and 3.4, and the corresponding line charts are in Figs. 3.6 and 3.7. In the two line charts, Models of ToneBaseline, ToneAligned, and ToneCheckedmanually are represented by TB, TA, and TC respectively. The vertical direction represents the tone recognition rate.

Table 3.3   Tone recognition accuracy of three configurations for TEST SET 1

	Segmentation	
	HTK	Manually
ToneBaseline (TB)	27.33	27.33
ToneAligned (TA)	59.49	59.82
ToneCheckedmanually (TC)	58.18	59.57

Table 3.4   Tone recognition accuracy of three configurations for TEST SET 2

	Segmentation	
	HTK	Manually
ToneBaseline (TB)	71.50	69.94
ToneAligned (TA)	73.33	71.36
ToneCheckedmanually (TC)	73.25	72.77

Tables 3.3 and 3.4 list the tone recognition accuracy attained by using three different configurations of models. In TEST SET 1, all of the sentences were read by one male reader, while TEST SET 2 was recorded by one female tester. After segmentation, 1 222 and 1 200 tonal syllables were respectively obtained in TEST SET 1 and TEST SET 2. As expected, the tone recognition accuracy with ToneAligned model and ToneCheckedmanually model is higher than the accuracy with ToneBaseline model separately. As shown in Tables 3.3 and 3.4, the segmentations based on HTK and Manually give the similar accuracy of tone recognition with the same test databases. The results, showed in Figs. 3.6 and 3.7, are particularly distinct. The tone recognition accuracy of female reader was significantly higher than that of male reader. After analysis, we found that it is because utterances read by the female reader are clearer than utterances read by the male reader. The tone recognition accuracy of ToneAligned is close to the accuracy of ToneCheckedmanually. From the experiment, we found that the performance of using ToneAligned is better than ToneBaseline.

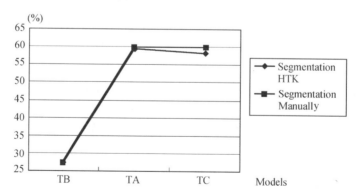

Fig. 3.6 Tone recognition accuracy of three different models for TEST SET 1.

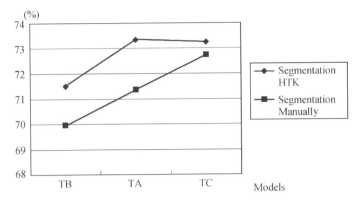

Fig. 3.7 Tone recognition accuracy of three different models for TEST SET 2.

3.4.1 Competing model based tone assessment

After getting the forced alignment based tone model, we shows the competing model based tone evaluation experiment in this part.

In tonal syllable evaluation, three methods were adopted for tone evaluation in comparing.

1) METHOD 1: use syllable-based competing models; baseline without tone.

2) METHOD 2: use syllable-based competing models; with tone.

3) METHOD 3: use INITIAL/FINAL-based competing models; with tone.

The simplified net was used in all of the three methods. METHOD 1 is considered as the baseline without tone. The syllable-based competing models of METHOD 1 are in form "(INITIAL) + FINAL". For METHOD 2 and METHOD 3, tone is considered in their competing models. METHOD 2 is the method we introduced. Consequently, INITIAL/FINAL-based competing models were applied in METHOD 3. In this method, INITIAL score and FINAL score were obtained separately, after that the two scores were combined to construct a syllable score.

For instance, the word "斗 (Dou)" (means fight) can be pronounced as "dou4". In our experiment, we compared the three methods, and in each method we read "dou4" as "dou4", "dou1", "dou3", "tou4", "yi1", respectively.

The simplified INITIAL/FINAL net is used in our experiment. Table 3.5 shows the tonal syllable scores attained by using three methods. All the test utterances were read by one tester. By using METHOD 1, we obtained 29 syllable-based competing models. 102 syllable-based competing models which got by METHOD 2 were with different tones. In METHOD 1 and METHOD 2, Eq. (3.1) was used to calculate the tonal syllable score. In METHOD 3,

4 INITIAL competing models and 21 FINAL competing models were generated from this approach. Similarly, Eq. (3.1) was applied to get the ith correct INITIAL score and the ith correct FINAL score. The syllable scores were computed by INITIAL and FINAL scores. Since tone was considered in METHOD 2 and the number of competing models got from METHOD 2 was larger than that from METHOD 1 and METHOD 3, tonal syllable scores acquired from METHOD 2 were more objective.

Table 3.5   Tonal syllable score of "dou4" by three methods

Pronunciation	METHOD 1	METHOD 2	METHOD 3
dou4	100	100	89.55
dou1	98.13	96.98	87.69
dou3	59.35	55.14	58
tou4	28.44	31.62	33.38
yi1	32.28	29.42	55.05

In tone evaluation part, we employ a hybrid method, in which we use the ToneAligned model to extract pitch features, and send the syllables to the five tone competing models for a log probability evaluation , so that we can find the tone rank (zero-based) of syllable in the competing models. Table 3.6 shows the tone ranks and corresponding tone scores of "dou4", and combined the preceding syllable scores from METHOD 2 to obtain the overall score.

Compared with Table 3.5 and Table 3.6, the results of our hybrid method are better than that of using only competing models method. For the wrong pronunciation "dou1", "dou3", "tou4" and "yi1", our hybrid method gives a lower score than only competing models method (METHOD 2 in Table 3.5). The hybrid method can show the difference between correct pronunciations and wrong pronunciations obviously.

Table 3.6   Tone evaluation results of "dou4" by hybrid method

Pronunciation	Rank	Tone Score	Syllable Score	Overall Score
dou4	0	100	100	100
dou1	3	40.98	96.98	68.98
dou3	3	40.98	55.14	48.06
tou4	1	86.21	31.62	58.92
yi1	4	28.09	29.42	28.76

Moreover, in order to get more intuitive results, we use the 863 Database to test in our compared experiment. The test set includes 4 male utterances data and 4 female utterances data. There are 800 sentences, 9 692 syllables in all. Table 3.7 gives the tone recognition results of syllables. By using our hybrid method, the tone recognition accuracy of syllables reaches 83.92%. The tone recognition rate of only using tone models is 66.41% and the results of only using competing models is 83.89%. In a related work, [45] reported a recognition rate of 66.3% for using HMM tone models with a decision tree based approach, and it also reported an accuracy of 69.1% for using stochastic polynomial tone models with a decision tree based approach. Compare with [45], the tone recognition accuracy of syllables by using our hybrid method improves about $15\% - 17\%$.

Table 3.7 Tone recognition accuracy (%) of syllables

	Syllables
ToneAligned Model	66.41
Competing Model	83.89
Hybrid Method (Tone Aligned Model + Competing Model)	83.92

We average the tone score of syllables in one sentence as an average result of a sentence. Table 3.8 shows the average results, it is obvious that our hybrid method is superior to the only competing model method, and its tone accuracy reaches 82%.

Table 3.8 Average tone score (%)

Competing Model	80.75
Hybrid Method (Tone Aligned Model + Competing Model)	82

Table 3.9 tabulates the specific tone recognition results of five tones. The first line lists the results of the decision tree based approach reported in [45]. The next two lines show the results of the only competing model method and the hybrid method. Compared with the first line, the only competing model method and the hybrid method have been improved significantly. By using our hybrid method, the tone accuracy of five tones increases about 20% in average. Particularly, the accuracy of Tone 5 was improved by more than 40%, and the accuracy of Tone 3 was improved about 25%. For Tone 3 and Tone 5, the accuracy of the hybrid method is also better than the accuracy of only competing model method. On average, the result of the hybrid method is slightly better than that of the only competing model method.

Table 3.9 Tone recognition accuracy (%) of five tones

	Tone 1	Tone 2	Tone 3	Tone 4	Tone 5	Average
Approach in [45]	73.4	64.2	59.1	68.8	54.1	66.3
Competing Model	84.42	82.21	78.21	86.4	97.59	85.77
Hybrid Method	82.22	83.94	84.84	84.37	98.08	86.69

3.5 Conclusions

In this chapter, we aims to get improved tone modeling, in the form of better treatment for syllable boundary segmentation, and also to attain objective pronunciation evaluation. Experiments show that the forced alignment of HMM based training technique is a feasible measure to solve the problem of getting better segmentation. It is further shown that the hybrid approach of using forced alignment based tone model with competing models based tone assessment is indeed able to give better performance in tone evaluation. The tone recognition accuracy of syllables in the hybrid approach is 83.92%, and the average tone accuracy reaches 82%. By using this hybrid measure, the tone assessment result is more objective than the result of only using competing models. In particular, the tone recognition accuracy of Tone 3 and Tone 5 is increased significantly.

In our work, although the tone accuracy has been ameliorated generally, but for Tone 1 and Tone 4, the results of our hybrid method are slightly worse than the only competition model method. So in the future, we would like to find a suitable method to further improve the tone accuracy.

References

[1]   Noemi D (2007) Computer-assisted Language Learning: Increase of Freedom of Submission to Machines? http://www.terra.es/personal/nostat. Accessed 10 December 2010

[2]   Witt S M, Young S J (1998) Performance Measures for Phone-level Pronunciation Teaching in Call. Paper Presented at the Speech Technology in Language Learning, Marholmen, Sweden, 25 – 27 May 1998

[3]   Zinovjeva N (2005) Use of Speech Technology in Learning to Speak a Foreign Language. Speech Technology

[4]   Levinson S, Liberman M (1981) Speech Recognition by Computer. Scientific American, pp 64 – 76

[5]   Weinstein C, McCandless S, Mondshein L et al (1975) A System for Acoustic-phonetic Analysis of Continuous Speech. IEEE Transactions on Acoustics, Speech and Signal Processing, 23(1): 54 – 67

[6]   Ehsani F, Knodt E (1998) Speech Technology in Computer-aided Language Learning: Strengths and Limitations of a new Call Paradigm. Language Learning and Technology, 2(1): 54 – 73

[7]   Franco H, Abrash V, Precoda K et al (2000) The Sri Eduspeak (tm) System: Recognition and Pronunciation Scoring for Language Learning. In: Integrating Speech Technology in Language Learning, Dundee, Scotland, 2000, pp 123 – 128

[8]   Chen J C, Jang J S R, Tsai T L (2007) Automatic Pronunciation Assessment for Mandarin Chinese: Approaches and System Overview. Computational Linguistics and Chinese Language Processing, 12(4): 443 – 458

[9]   Wang H C, Waple C J, Kawahara T (2009) Computer Assisted Language Learning System Based on Dynamic Question Generation and Error Prediction for Automatic Speech Recognition. Speech Communication, 51(10): 995 – 1005

[10]   Eskenazi M, Hansma S (1998) The Fluency Pronunciation Trainer. In: Proceedings of Speech Technology in Language Learning

[11]   Teixeira C, Franco H, Shriberg E et al (2000) Prosodic features for Automatic Text-independent Evaluation of Nativeness for Language Learners. In: Proc ICSLP

[12]   Tsubota Y, Kawahara T, Dantsuji M (2004) Practical use of English Pronunciation System for Japanese Students in the call Classroom. Paper Presented at the Proc. INTERSPEECH 2004, Jeju Island, Korea, 4 – 8 October 2004

[13]   Imoto K, Tsubota Y, Raux A et al (2002) Modeling and Automatic Detection of English Sentence Stress for Computer-assisted English Prosody Learning System. In: ICSLP-2002, 2002, pp 749 – 752

[14]   Wang H, Kawahara T (2008) Effective Error Prediction Using Decision Tree for asr Grammar Network in call System. In: IEEE International Conference on Acoustics, Speech and Signal Processing 2008, pp 5069 – 5072

[15]   Pan F P, Zhao Q W, Yan Y H (2008) Mandarin vowel Pronunciation Quality Evaluation by a Novel Formant Classification Method and its Combination with Traditional Algorithms. Paper Presented at the Proc. ICASSP 2008, Las Vegas, Nevada, USA 30 March – 4 April 2008

[16] Mak B, Siu M, Ng M et al (2003) Plaser: Pronunciation Learning via Automatic Speech Recognition. In: Proceedings of the HLT-NAACL 03 workshop on Building Educational Applications Using Natural Language Processing, 2003. Association for Computational Linguistics, pp 23 – 29

[17] Chapelle C (1998) Multimedia call: Lessons to be Learned from Research on Instructed sla. Language Learning & Technology, 2(1): 21 – 39

[18] Zhang Y B, Chu M, Huang C et al (2008) Detecting tone Errors in Continuous Mandarin Speech. In: Proceedings of IEEE International Conference on Acoutics, Speech and Signal Processing, Las Vegas, NV, USA, January 2008, pp 5065 – 5068

[19] Yao Qian F K S, Tan Lee (2008) Tone-enhanced Generalized Character Posterior Probability (gcpp) for Cantonese lvcsr. Computer Speech & Language, 22(4): 360 – 373

[20] Qian Y, Soong F K (2009) A Multi-space Distribution (msd) and Two-stream tone Modeling Approach to Mandarin Speech Recognition. Speech Communication, 51(12): 1169 – 1179

[21] Wei S, Wang H, Liu Q et al (2007) Cdf-matching for Automatic Tone Error Detection in Mandarin Call System. In: IEEE International Conference on Acoustics, Speech and Signal Processing

[22] Hussein H, Wei S, Mixdorff H et al (2010) Development of a Computer-aided Language Learning System for Mandarin: Tone Recognition and Pronunciation Error Detection. In: Proceedings of the Speech Prosody

[23] Arias J, Yoma N, Vivanco H (2010) Automatic Intonation Assessment for Computer Aided Language Learning. Speech Communication, 52(3): 254 – 267

[24] Pike K L (1948) Tone Languages. University of Michigan Press, Ann Arbor

[25] Lee L S (1997) Voice Dictation of Mandarin Chinese. IEEE Signal Processing Magazine, 14(4): 63 – 101

[26] Ljolje A, Hirschberg J, Santen JPHv (1996) Automatic Speech Segmentation for Concatenative Inventory Selection. In: Santen JPHv, Sproat RW, Olive JP, Hirschberg J (eds) Progress in Speech Synthesis. Springer, New York, pp 305 – 312

[27] Demuynck K, Laureys T (2002) A Comparison of Different Approaches to Automatic Speech Segmentation. In: Proceedings of the 5th International Conference on Text, Speech and Dialogue, 2002. Springer, pp 277 – 284

[28] Kawai H, Toda T (2004) An Evaluation of Automatic Phone Segmentation for Concatenative Speech Synthesis. In: Proceedings of IEEE International Conference on Acoustics, Speech, and Signal Processing

[29] Malfrere F, Deroo O, Dutoit T et al (2003) Phonetic Alignment: Speech Synthesis-based vs. Viterbi-based. Speech Communication, 40(4): 503 – 515

[30] Mporas I, Lazaridis A, Ganchev T et al (2009) Using Hybrid Hmm-based Speech Segmentation to Improve Synthetic Speech Quality. In: 13th Panhellenic Conference on Informatics, 2009, pp 118 – 122

[31] Van Hemert J (1991) Automatic Segmentation of Speech. IEEE Transactions on Signal Processing, 39(4): 1008 – 1012

[32] Brugnara F, Falavigna D, Omologo M (1993) Automatic Segmentation and Labeling of Speech Based on Hidden Markov Models. Speech Communication, 12(4): 357 – 370

[33] Makashay M, Wightman C, Syrdal A et al (2000) Perceptual Evaluation of Automatic Segmentation in Text-to-speech Synthesis. In: Proc ICSLP, 2000, pp 431 – 434

[34] Adell J, Bonafonte A, Gómez J et al (2005) Comparative Study of Automatic Phone Segmentation Methods for Tts. In: Proceedings of ICASSP

[35] Mporas I, Ganchev T, Fakotakis N (2008) A Hybrid Architecture for Automatic Segmentation of Speech Waveforms. In: IEEE International Conference on Acoustics, Speech and Signal Processing, 2008, pp 4457 – 4460

[36] Franco H, Neumeyer L, Kim Y et al (1997) Automatic Pronunciation Scoring for Language Instruction. Paper Presented at the IEEE International Conference on Acoustics, Speech, and Signal Processing, Munich, Germany, April 21 – 24

[37] Franco H, Neumeyer L, Digalakis V et al (2000) Combination of Machine Scores for Automatic Grading of Pronunciation Quality. Speech Communication, 30(2 – 3): 121 – 130

[38] Neumeyer L, Franco H, Digalakis V et al (2000) Automatic Scoring of Pronunciation Quality. Speech Communication, 30(2 – 3): 83 – 93

[39] Witt S M, Young S J (2000) Phone-level Pronunciation Scoring and Assessment for Interactive Language Learning. Speech Communication, 30(2−3): 95−108

[40] Molina C, Yoma N B, Wuth J et al (2009) Asr Based Pronunciation Evaluation With Automatically Generated Competing Vocabulary and Classifier Fusion. Speech Communication, 51(6): 485−498

[41] Cincarek T, Gruhn R, Hacker C et al (2009) Automatic Pronunciation Scoring of Words and Sentences Independent from the Non-native's First Language. Computer Speech & Language, 23(1): 65−88

[42] De Wet F, Van der Walt C, Niesler T (2009) Automatic Assessment of Oral Language Proficiency and Listening Comprehension. Speech Communication, 51(10): 864−874

[43] Chen J C, Jang J S R, Li J Y et al (2004) Automatic Pronunciation Assessment for Mandarin Chinese. Paper Presented at the IEEE International Conference on Multimedia and Expo, Taipei, Taiwan, China, June 2004

[44] Ge F, Pan F, Liu C et al (2009) An Svm-based Mandarin Pronunciation Quality Assessment System. In: The Sixth International Symposium on Neural Networks (ISNN 2009), 2009. Springer, pp 255−265

[45] Cao Y, Huang T Y, Xu B (2004) A Stochastically-based Study on Chinese Tone Patterns in Continuous Speech (in Chinese). Acta Automatic Sinica, 30(2): 191−198

[46] Liu Z J, Shao J, Zhang P Y et al (2007) Research on Tone Recognition in Chinese Spontaneous Speech (in Chinese). Acta Physica Sinica, 56(12): 7064−7069

[47] Surendran D R (2007) Analysis and Automatic Recognition of Tones in Mandarin Chinese. The University of Chicago, Chicago

[48] Jiang T, Tang M, Ge F et al (2009) An Effective Call System for Strongly Accented Mandarin Speech. In: International Conference on Research Challenges in Computer Science, 2009, pp 92−95

[49] Zu YQ (1999) The Text Design for Continuous Speech Database of Standard Chinese (in Chinese). Acta Acustica, 24(3): 236−247

[50] Young S, Evermann G, Gales M et al (2006) The htk book (for htk version 3.4). Cambridge University, London

[51] Lin C Y, Wang H C (2010) Speech Recognition Based on Attribute Detection and Conditional Random Fields. In: Wang PSP (ed) Pattern Recognition and Machine Vision: In Honor and Memory of Late Prof. King-Sun Fu. River Publishers, Denmark, pp 405−415

[52] Ross M, Shaffer H, Cohen A et al (1974) Average Magnitude Difference Function Pitch Extractor. IEEE Transactions on Acoustics, Speech and Signal Processing, 22(5): 353−362

[53] Mei X, Pan J, Sun S (2001) Efficient Algorithms for Speech Pitch Estimation. In: Proceedings of 2001 International Symposium on Intelligent Multimedia, Video and Speech Processing, pp 421−424

[54] Shimamura T, Kobayashi H (2001) Weighted Autocorrelation for Pitch Extraction of Noisy Speech. IEEE Transactions on Speech and Audio Processing, 9(7): 727−730

[55] Boril H, Pollak P (2004) Direct Time Domain Fundamental Frequency Estimation of Speech in Noisy Conditions. In: Proceedings of European Signal Processing Conference, 2004, pp 1003−1006

[56] Amado R (2008) Pitch Detection Algorithms Based on Zero-cross Rate and Autocorrelation Function for Musical Notes. In: International Conference on Audio, Language and Image Processing, 2008, pp 449−454

[57] K. Abdullah-Al-Mamun F S, G. Muhammad (2009) A High Resolution Pitch Detection Algorithm Based on Amdf and acf. Journal of Scientific Research, 3(1): 508−515

[58] Liu J, Zheng T F, Deng J et al (2005) Real-time Pitch Tracking Based on Combined Smdsf. Paper Presented at the Proc. INTERSPEECH 2005, Lisbon, Portugal, 4−8 September 2005

[59] Liu J, Zheng F, Wu W H (2006) Real-time Pitch Tracking Based on Sum of Magnitude Difference Square Function. Journal of Tsinghua University (Science & Technology), 46(1): 74−77

[60] Chou F C, Tseng C Y, Lee L S (2002) A Set of Corpus-based Text-to-speech Synthesis Technologies for Mandarin Chinese. IEEE Transactions on Speech and Audio Processing, 10(7): 481−494

4 Fusion with Infrared Images for an Improved Performance and Perception

Zheng Liu[1] and Wei Wu[2]

Abstract   Infrared sensors are being widely used to detect the infrared radiation beyond the human visible spectrum from varied objects in different applications. As the infrared sensor is not subject to poor illuminations, complementary information to visual perception is available from the acquired infrared images. Thus, the applications may benefit from fusing infrared images with results of other sensing modalities. This chapter presents a state-of-art review of the fusion techniques for infrared images. Three fusion levels, i.e., pixel level, feature level, and decision level, are considered. For each category, representative fusion algorithms are described and discussed. Applications, which take advantage of these fusion techniques, are also briefly summarized in this chapter.

4.1 Introduction

With the development of sensor technologies, it is possible for heterogeneous sensor modalities to perform across different wavebands of the electromagnetic spectrum [1, 2]. The information acquired from these wavebands can be combined with a so-called sensor/data fusion technique, in which an enhanced single view of a scene with extended information content is achieved as the final result. The application of sensor/data fusion techniques can be found in a wide range of applications including multi-focus imagery, concealed weapon detection (CWD), intelligent robots, surveillance systems, medical diagnosis, remote sensing, non-destructive testing (NDT), etc. [3 – 14]. When the data is in the form of two-dimensional image, the fusion is also known as image fusion.

Infrared or thermal imaging cameras can detect radiation in the infrared range $0.9 - 14$ μm. As all objects near room temperature emit infrared radiation, it is possible to see the environment with or without visible illumination, especially for warn objects like humans and warm-blooded animals. Thus, infrared imaging is being extensively used for military and civilian applications. For a complexed environment, fusion with infrared images may lead to a better analyzing performance. This chapter is devoted to the fusion with infrared images, but the fusion methods can be applied to other sensor modalities as

---

1   University of Ottawa, Ottawa, Canada. E-mail: zheng.liu@ieee.org.
2   Sichuan University, Sichuan, China. E-mail: wuwei@scu.edu.cn.

well [15, 16].

This chapter is organized as follows. Section 4.2 describes the principle of infrared imaging. Detailed review of fusion algorithms with infrared images is presented in Section 4.3. Section 4.4 introduces various applications. Summary of this chapter can be found in Section 4.5.

4.2 The Principle of Infrared Imaging

All the possible electromagnetic radiation consists of the electromagnetic spectrum as shown in Fig. 4.1 and corresponding wavelengths are listed in Table 4.1. The wavelength of the visible light ranges approximately from 390 nm to 770 nm. After the visible light comes the infrared (IR) , which ranges from 770 nm to 1 mm and is further divided into five parts, e.g., near IR, short IR, mid-wave IR, long-wave IR, and far IR.

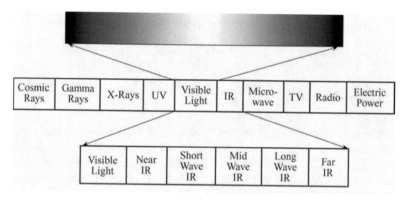

Fig. 4.1 The whole electromagnetic spectrum [1].

Table 4.1 The electromagnetic wavelength table [17]

Electromagnetic Wave	Wavelength $\lambda(\mu m)$
Cosmic Rays	$\lambda < 10^{-7}$
Gamma Rays	$10^{-4} > \lambda > 10^{-8}$
X-Rays	$0.1 > \lambda > 10^{-7}$
UV	$0.39 > \lambda > 0.01$
Visible Light	$0.77 > \lambda > 0.39$
IR	$10^3 > \lambda > 0.77$
Microwave	$10^6 > \lambda > 10^3$
TV and Radio Wave	$10^{11} > \lambda > 10^6$
Electric Power	$\lambda > 10^{10}$

Objects having temperature over 0 K $(-273.15°)$ can generally emit infrared radiation across a spectrum of wavelengths. The intensity of an object's emitted IR energy is in proportion to its temperature. The emitted energy

measured as the target's emissivity, which is the ratio between the emitted
energy and the incident energy, indicates an object's temperature. At any
given temperature and wavelength, there is a maximum amount of radiation
that any surface can emit. If a surface emits this maximum amount of radiation, it is known as a blackbody. Planck's law for blackbody defines the
radiation as [18]

$$I_{\lambda,b}\left(\lambda, T\right) = \frac{2hc^2}{\lambda^5} \frac{1}{e^{\frac{hc}{\lambda k T}} - 1}, \tag{4.1}$$

where $I(\lambda, T)$ is the spectral radiance or energy per unit time, surface area,
solid angle, and wavelength (Unit: W m^2 μm^{-1} sr^{-1}). The meaning of each
symbol in above equation is listed below [19]:

  λ: *wavelength (meter)*;
  T: *Temperature (kelvin)*;
  h: *Planck's constant (joule/hertz)*;
  c: *speed of light (meter/second)*;
  k: *Boltzmann's constant (joule/kelvin)*.

Usually, objects are not blackbodies. According to Kirchhoff's law, there is
$R + \varepsilon = 1$, where ϵ is the emissivity and R is the reflectivity. Emissivity
is used to quantify the energy-emitting characteristics of different materials
and surfaces. The emitted energy of an object reaches the IR sensor and is
converted into an electrical signal. This signal can be further converted into a
temperature value based on the sensor's calibration equation and the object's
emissivity. The signal can be displayed and presented to the end users. Thus,
thermography can "see in the night" without an infrared illumination. The
amount of radiation increases with temperature; therefore, the variations in
temperature can be identified by thermal imaging. The IR cameras can generally be categorized into two types: cooled infrared detectors and uncooled
infrared detectors. They can detect the difference in infrared radiation with
insufficient illumination or even in total darkness. The use of thermal vision techniques can be found in numerous applications such as military, law
enforcement, surveillance, navigation, security, and wildlife observation [15].
The IR image can provide an enhanced spectral range that is imperceptible to
human beings and contribute to the contrast between objects of high temperature variance and the environment. Compared with a visual image, the IR
image is represented with a different intensity map. The same scene exhibits
different features existing in different electromagnetic spectrum bands.

4.3 Fusion with Infrared Images

This section discusses a general implementation or procedure to fuse with
infrared images. Varied fusion algorithms at different levels are described in
detail.

4.3.1 Implementation

The fusion with infrared images can be carried out at three levels: pixel-, feature-, and decision-level. The procedures for the three types of image fusion are illustrated in Fig. 4.2 [2], which assumes a visible image is fused with an infrared image. First, a pre-processing step is adopted to align/register the input images and remove the noises so that pixels from one image modality are associated with corresponding pixels from the other. Feature-level fusion needs to extract features of interest from the inputs and fuse those features together. The extracted features are further identified or classified. For decision-level fusion, the identification and classification results are combined. A higher level conclusion is drawn from the fused result.

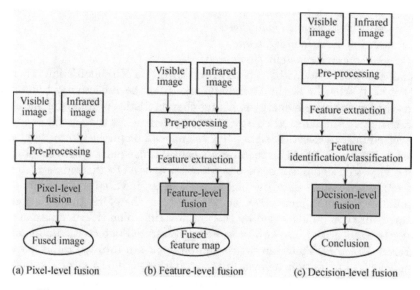

Fig. 4.2 Fusion of visible and infrared images at different levels.

Usually, the pixel-level fusion generates a fused image, which is more appropriate for human perception or further automated analysis. The result of the feature-level fusion could be a feature map, which characterizes the objects of interest in detail or highlights the features of the objects. A conclusion is derived from the decision-level fusion through the combination of classification results.

4.3.2 Algorithms

Pixel-Level Fusion

1) Multiresolution analysis based fusion

The basic idea of multiresolution analysis (MRA) based fusion is to represent an image with some basis functions, which make image features like edges and boundaries easily accessed and manipulated. Since the fused image is a weighted sum of basis functions, choosing a basis function similar to the original signal will facilitate the analysis.

A flowchart of the fusion process is given in Fig. 4.3. Input image $A(x, y)$ and $B(x, y)$ are first represented by a MRA algorithm as $\sum_i y_i^A g_i(x, y)$ and $\sum_i y_i^B g_i(x, y)$. The basis function is $g_i(x, y)$. y_i^A and y_i^B are the transform coefficients for image A and B respectively and combined by a fusion rule. The new coefficient $y_i^C = \mathrm{fusion}(y_i^A, y_i^B)$ is used to reconstruct the fused image $C(x, y)$ through an inverse transforming.

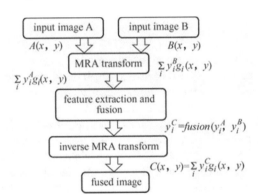

Fig. 4.3 The multiresolution analysis based pixel-level image fusion.

The research on MRA-based image fusion is focused on two aspects. One is the MRA algorithm and the other is the fusion rule. Numerous solutions have been proposed so far. The MRA algorithm and fusion rule need to be tailored for the requirements of a specific application. No one-size-fits-all solution is available. Table 4.2 summaries the MRA algorithms and fusion rules applied to pixel-level image fusion. Corresponding references can be found in the second and forth column of the table.

An example of pixel-level fusion of visible and infrared images is given in Fig. 4.4. In the visible image, the background is relatively clear while the human object is dim. On the contrary, the contrast between the human object and background is sharp in the IR image. Therefore, the fusion of these two images will enhance both the human object and background at the same time. The fused results of six MRA algorithms are shown. The discussion of the performance of these algorithms is beyond the scope of this chapter and can be found in other literatures [32−34].

Table 4.2 Multiresolution based pixel-level image fusion.

Multiresolution algorithm	References	Fusion rule	References
Laplacian pyramid	[3, 20]	Genetic algorithm	[29]
Ratio of low-pass pyramid	[21]	Expectation maximization	[30]
Steerable pyramid	[22]	Match and salience measures	[3]
Wavelet	[23]	Maximum selection and consistency verification	[23]
Multi-wavelet	[24]	Region activity measurement	[27]
Contourlet	[25]	Hidden Markow quadtree	[31]
Multi-contourlet	[26]	Model-based weighted average	[32]
Discrete wavelet frame	[27]		
Complex wavelets	[28]		

2) Empirical Mode Decomposition Based Fusion

The empirical mode decomposition (EMD) was proposed by Huang et al. to analyze non-linear and non-stationary signals [35]. It decomposes the signal into intrinsic mode functions (IMFs). In contrast with image pyramid and wavelets, EMD uses basis functions derived from the data and this is why EMD is known as a fully data-driven method. The algorithm has been extended to two-dimensional image data analysis. The input visible and infrared images are vectorized into lexicographical order and EMD is applied to the vectors and converted to the reconstructed matrices [37]. At this pre-reconstruction stage, the IMFs are multiplied by a set of weights to decrease the mutual information between the inputs as illustrated in Fig. 4.5. With the decrease in mutual information, the features from the input images are emphasized and fused together. Different weighting schemes have been applied based on the nature of the IMFs. An automated process to minimize the mutual information is expected from the future study of this fusion method. This fusion approach has been applied to the application of face recognition [38].

Further developments include using rotation-invariant EMD, which operates fully in the complex domain [39]. The weighting coefficients for the fusion are determined by the relative values of the local variance at each scale. Pyramidal empirical mode decomposition (PEMD) was introduced in [40]. As illustrated in Fig. 4.6, the EMD decomposes an image into an IMF image and a residue image. The residual image is further processed with a "RE-DUCE" operation, which consists of Gaussian filtering and down-sampling to get a approximate image. A difference of Gaussian (DOG) image is derived from the subtraction of residual image and expanded approximate image. The "EXPAND" is an inverse operation of "REDUCE". The two operations are exactly the same as those in Laplacian pyramid [20]. The approximate image can be further decomposed with the repeated operations. The fusion rules

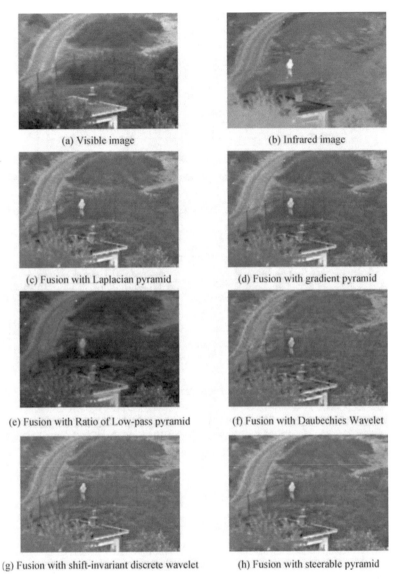

(a) Visible image

(b) Infrared image

(c) Fusion with Laplacian pyramid

(d) Fusion with gradient pyramid

(e) Fusion with Ratio of Low-pass pyramid

(f) Fusion with Daubechies Wavelet

(g) Fusion with shift-invariant discrete wavelet

(h) Fusion with steerable pyramid

Fig. 4.4 An example of multiresolution visible and infrared image fusion.

from the MRA-based fusion can be applied directly.

3) Principal Component Analysis Based Fusion

Principal component analysis (PCA) is also known as Karhumen-Loève transform, which converts a number of possibly correlated variables into a small number of uncorrelated variables, i.e., principal components. Thus, PCA-based fusion integrates the complementary information of the largest variation from the inputs [41].

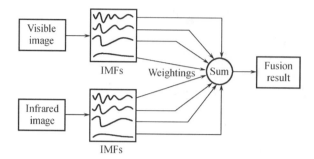

Fig. 4.5 The empirical decomposition model based image fusion [37].

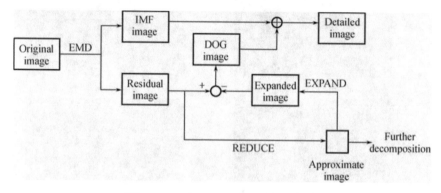

Fig. 4.6 The pyramidal EMD [40].

The stacked n input images (image size: $M \times N$) are reorganized into an array $\mathbf{X}$ consisting of $M \times N$ n-dimension vectors ($\boldsymbol{x}$) as illustrated in Fig. 4.7. For each vector $\boldsymbol{x}$, expected values $E\{\boldsymbol{x}\}$ are calculated by

$$m_x = E\{\boldsymbol{x}\} = \frac{1}{MN} \sum_{k=1}^{MN} \boldsymbol{x}_k, \qquad (4.2)$$

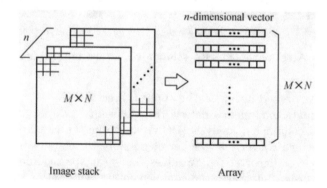

Fig. 4.7 The re-arrangement of images in PCA.

The covariance matrix C_x is defined as

$$C_x = \mathrm{E}\{(\boldsymbol{x} - m_x)(\boldsymbol{x} - m_x)^{\mathrm{T}}\}. \tag{4.3}$$

As vector $\boldsymbol{x}$ is of n dimension, matrix C_x is real and symmetric with order $n \times n$.

Let $\boldsymbol{e}_i$ and λ_i $(i = 1, 2, \ldots, n)$ denote the eigenvector and eigenvalue of matrix C respectively and the eigenvalue λ_i is organized in descending order, i.e., $\lambda_{i-1} \geqslant \lambda_i$. A matrix A can be constructed with the eigenvector $\boldsymbol{e}_i$ as

$$A = \begin{pmatrix} \boldsymbol{e}_1 \\ \boldsymbol{e}_2 \\ \vdots \\ \boldsymbol{e}_n \end{pmatrix}.$$

The vector $\boldsymbol{x}$ is mapped to a vector $\boldsymbol{y}$, which constitutes a matrix Y as follows:

$$y = A(\boldsymbol{x} - m_x). \tag{4.4}$$

The inverse operation is

$$\boldsymbol{x} = A^{\mathrm{T}}\boldsymbol{y} + m_x. \tag{4.5}$$

If only q eigenvectors are used, $\boldsymbol{x}$ can be approximated by the reconstruction of $\widehat{\boldsymbol{x}}$:

$$\widehat{\boldsymbol{x}} = Aq^{\mathrm{T}}\boldsymbol{y} + m_x. \tag{4.6}$$

where A_q, containing first q eigenvectors, denotes part of A.

The visible (I_{vi}) and infrared (I_{ir}) images are first filtered by a Gaussian filter. There is

$$S_{ir} = G(I_{ir}) = I_{ir} \otimes \exp\left(-\frac{x^2 + y^2}{\sigma^2}\right), \tag{4.7}$$

$$S_{vi} = G(I_{vi}) = I_{vi} \otimes \exp\left(-\frac{x^2 + y^2}{\sigma^2}\right), \tag{4.8}$$

where $\otimes$ refers to the convolution operation. And the deviations of the filtered images are

$$D_{vi} = I_{vi} - S_{vi}, \tag{4.9}$$

$$D_{ir} = I_{ir} - S_{ir}. \tag{4.10}$$

Then, D_{vi} and D_{ir} are used to compute the principal components. The first two principal components P_{vi} and P_{ir} are filtered with a Gaussian function to get the weighting coefficients.

$$w_{ir} = G(P_{ir}), \tag{4.11}$$

$$w_{vi} = G(P_{vi}). \tag{4.12}$$

The fused image F is obtained by the sum of two items. The first item is the average of the smoothed visible and infrared images. The second item is the weighted and normalized sum of the deviations.

$$F = \frac{S_{ir} + S_{vi}}{2} + \frac{(D_{ir} \cdot w_{ir}) + (D_{vi} \cdot w_{vi})}{w_{ir} + w_{vi}}. \tag{4.13}$$

4) Fusion with polyharmonic local sine transform

There are many other transforms for image fusion. The polyharmonic local sine transform is first proposed as a model for image compression [42]. The transform first decomposes an image into a polyharmonic component (p) and the residual (r) [43]. Then, the contrast C is defined with p and r: $C(x, y) = r(x, y)/\bar{p}$. $\bar{p}$ is the mean of p. The fusion is implemented as

$$\begin{cases} p_f = p_{vi} w_{vi} + p_{ir} w_{ir}, \\ r_f(x, y) = C_f(x, y) p_f, \end{cases} \tag{4.14}$$

where w_{vi} and w_{ir} are two non-negative weighting coefficients derived by

$$\begin{cases} w_{vi} = \dfrac{E_{vi}}{E_{vi} + E_{ir}}, \\ w_{ir} = 1 - w_{vi}. \end{cases} \tag{4.15}$$

E_{vi} and E_{ir} are the entropy of visible and infrared images respectively. The contrast of the fused image is obtained by

$$C_f(x, y) = \begin{cases} C_{vi}(x, y), & \text{if } |C_{vi}(x, y)| \geqslant |C_{ir}(x, y)|, \\ C_{ir}(x, y), & \text{otherwise}. \end{cases} \tag{4.16}$$

and $r_f(x, y) = C_f(x, y) p_f$. With the derived p_f and $r_f(x, y)$, the fused image can be obtained.

5) A total variation based fusion

A total variation (TV) approach is implemented to fuse visible and infrared images at pixel level in Ref. [44]. The relation between the true image $f_0(x, y)$ and multiple sensors' observation $f_i(x, y)$ $(i = 1, 2, \cdots, n)$ is modeled as

$$f_i(x, y) = \beta_i(x, y) f_0(x, y) + \eta_i(x, y), \ 1 \leqslant i \leqslant n, \tag{4.17}$$

where $\beta_i(x, y)$ and $\eta_i(x, y)$ are the gain and noise respectively. The purpose of fusion is to estimate $f_0(x, y)$ from $f_i(x, y)$. Eq. 4.17 can be written in another form as

$$\begin{pmatrix} f_1 \\ \vdots \\ f_n \end{pmatrix} = \begin{pmatrix} \beta_1 \\ \vdots \\ \beta_n \end{pmatrix} f_0 + \begin{pmatrix} \eta_1 \\ \vdots \\ \eta_n \end{pmatrix}, \tag{4.18}$$

$$\Rightarrow \boldsymbol{f} = \boldsymbol{\beta} f_0 + \boldsymbol{\eta}, \tag{4.19}$$

And Eq. (4.19) can be further re-written as

$$\beta^N \boldsymbol{f} = \boldsymbol{f}_0 + \beta^N \eta, \tag{4.20}$$

where there is $\beta^N = \left(\beta^T \beta\right)^{-1} \beta^T$. To obtain f_0, we need to estimate β^N so that the cost function $\|\beta^N \boldsymbol{f} - \boldsymbol{f}_0\|$ is minimized. A principal eigenvector based approach is employed to estimate the sensor gains $\beta = [\beta_1, \cdots, \beta_n]$ [44]. As the spatial variation of the sensor gain is small and β_i $(1 \leqslant i \leqslant n)$ is constant over a small region of the image. Therefore, the input images are split into small regions and sensor gains are computed for each region. The $f_0(x, y)$ is estimated through the minimization of its TV seminorm under suitable constraints [44].

6) Fusion with Non-negative Matrix Factorization

The non-negative matrix factorization (NMF) is an algorithm, which can find two non-negative factor matrices W and H for a given non-negative matrix V such that $V_{n \times m} \approx W_{n \times r} H_{r \times m}$. And r is a number smaller than n and m. If v and h are the columns of V and H, the factorization can be rewritten as $v \approx Wh$. Matrix V can be considered as the input multi-sensor images, which are represented by a linear combination of r basis vectors/images from matrix W. When the number r is set to one, the input images are linear combinations of the W, which is the fused image with the global features of the input images [45, 46].

Usually, NMF is realized by iterative updates on W and H. A cost function is defined as [47]

$$F = \|V - WH\|^2 = \sum_{i,j}^{n,m} (V_{ij} - (WH)_{ij})^2 \tag{4.21}$$

or

$$D(V \| (WH)) = \sum_{ij} \left(V_{ij} \log \frac{V_{ij}}{(WH)_{ij}} - V_{ij} + (WH)_{ij} \right). \tag{4.22}$$

A local maximum of this objective function F can be achieved by an iterative updating scheme:

$$H_{i+1} \leftarrow H_i \frac{(W^T V)_i}{(W^T WH)_i}, \quad W_{i+1} \leftarrow W_i \frac{(VH^T)_i}{(WHH^T)_i} \tag{4.23}$$

or

$$H_{i+1} \leftarrow H_i \frac{\sum\limits_i W_i V_i / (WH)_i}{\sum\limits_k W_{ki}}, \quad W_{i+1} \leftarrow W_i \frac{\sum\limits_\mu H_i V_i / (WH)_i}{\sum\limits_\nu H_{i\nu}}. \tag{4.24}$$

Matrices W and H are randomly initialized under the non-negativity constraints.

7) Statistical signal processing approach based fusion

The fusion of visible and infrared images can be formulated as a statistical estimation problem, in which the visible and infrared images are considered as an observation of the true scene. The mean square error between the fused image and the true scene is minimized by incorporating covariance intersection into the expectation maximization process [48].

In Ref. [49], a fusion scheme based on statistical image formation model is described. An iterative bootstrapping non-parametric expectation-maximization (BNEM) algorithm is proposed. The whole procedure is depicted in Fig. 4.8.

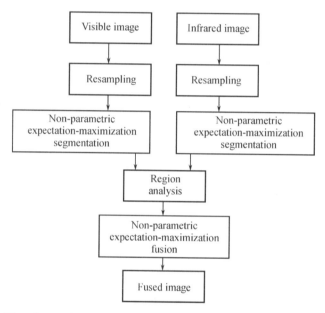

Fig. 4.8 Flowchart of bootstrapping non-parametric expectation-maximization fusion.

The image re-sampling is implemented with a random sampling model. Then, an unsupervised Bayesian segmentation based on bootstrapping NEM algorithm is applied to generate a region map for each image. As input images may contain different objects appearing with different shapes, a joint region map is determined from the region maps of input images [50]. Similarly, the image formation model can be written as

$$z_i(j) = \beta_i(j)s(j) + \varepsilon_i(j), \ i = 1, 2, \cdots, q, \tag{4.25}$$

where i and j are the sensor and image pixel index respectively and there are:

$z_i(j)$: the observed sensor image;
$s(j)$: the true scene;

$\beta_i(j)$: sensor selectivity factor;

$\varepsilon_i(j)$: random distortion.

The BNEM is then used to estimate the model parameters and produce the fused image. A set of iterative equations that produce approximate maximum-likelihood estimates are developed with the BNEM algorithm in Ref. [49]. Computational efficiency and better image quality are pursued for future work on BNEM-based image fusion.

8) Image fusion in color space

Another way to fuse visible and infrared images is in the color space as human visual system is sensitive to colors. A new technique to fuse a color visual image with an infrared image for concealed weapon detection was proposed in Ref. [9]. The fusion algorithm is illustrated in Fig. 4.9. Input images are manipulated in three color spaces, i.e., RGB (R:red, G:green, B:blue), HSV (H:hue, S:saturation, V:brightness value) and LAB (L:brightness, A:red-green chrominance, B:yellow-blue chrominance). The original infrared image and its reverse polarity are fused with the V component from the HSV space respectively. The discrete wavelet frame (DWF) is used, which is a shift invariant transform. Following the flowchart in Fig. 4.9, seven fused images, $\{F_{1\text{RGB}}, F_{1\text{LAB}}, F_{2\text{LAB}}, F_{2\text{RGB}}, F_{2\text{HSV}}, F_{3\text{HSV}}, F_{3\text{RGB}}\}$, can be obtained. Among the fused results, $F_{3\text{RGB}}$ is most similar to the original color image.

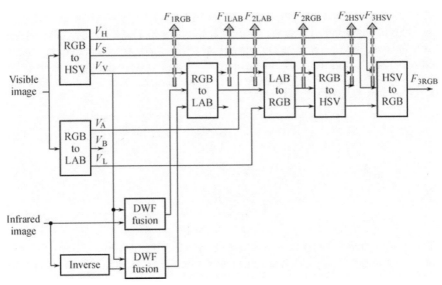

Fig. 4.9 Fusion of a color visual image and an infrared image [9].

Another color fusion scheme is shown in Fig. 4.10 [51]. The reverse polarity of the infrared (IIR) image is also used in this method. A curvelet transform is employed to fuse the visible and infrared image as well as the IIR image. The fused results (F_1 and F_3) are fed to the R (red) and B (blue) channel respectively. The G (green) channel is the linear combination result (F_2) of visible $(I_{vi}(i,j))$ and infrared $(I_{ir}(i,j))$ image:

$$F_2(i,j) = \alpha(i,j) \cdot I_{vi}(i,j) + (1 - \alpha(i,j)) \cdot I_{ir}(i,j),$$
$$\alpha(i,j) = \frac{I_{ir}(i,j) - \min(I_{ir}(i,j))}{\max(I_{ir}(i,j)) - \min(I_{ir}(i,j))}. \tag{4.26}$$

The RGB channels compose the fused result.

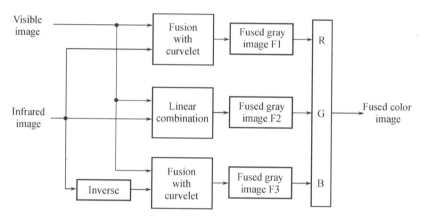

Fig. 4.10 Fusion in color space.

The flowchart of linear image fusion is given in Fig. 4.11 [52]. The visible and infrared images are firstly fused with false color fusion scheme. Then, a color transfer operation is performed to obtain the final fusion result. The fusion is implemented in the YUV (luminance/chrominance) color space. The relation between RGB and YUV space is

$$\begin{bmatrix} Y \\ U \\ V \end{bmatrix} = \begin{bmatrix} 0.299 & 0.587 & 0.114 \\ -0.1471 & -0.2888 & 0.4359 \\ 0.6148 & -0.5148 & -0.1000 \end{bmatrix} \begin{bmatrix} R \\ G \\ B \end{bmatrix}. \tag{4.27}$$

The false color fusion in YUV space is as

$$\begin{cases} Y_C = I_{vi}, \\ U_C = I_{vi} - I_{ir}, \\ V_C = I_{ir} - I_{vi}. \end{cases} \tag{4.28}$$

The visible image (I_{vi}) is mapped to the achromatic channel and therefore high-frequency information of the visible image is kept in the fused result. The difference of visible and infrared image is mapped to U and V channel respectively. Color transfer in YUV color space is as

$$P_F = \frac{\sigma_R^P}{\sigma_C^P}\left(P_C - \mu_C^P\right) + \mu_R^P, \ (P = Y, U, V), \tag{4.29}$$

where C, R, and F refer to the false color fused image, the reference image, and the final fused image respectively. P denotes the three color channels in YUV space. μ and σ refer to the mean and standard deviation of channel P.

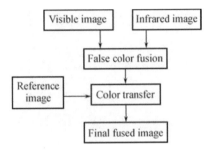

Fig. 4.11 Linear image fusion in color space.

In order to maintain or improve the contrast of the fused image, a color contrast enhancement was further implemented by enhancing the V channel with a ratio η. Thus, Eq. (4.29) changes to

$$
\begin{cases}
P_F = \dfrac{\sigma_R^P}{\sigma_C^P}\left(P_C - \mu_C^P\right) + \mu_R^P, \ (P = Y, U), \\[2mm]
V_F = \eta \dfrac{\sigma_R^V}{\sigma_C^V}\left(V_C - \mu_C^V\right) + \mu_R^V, \\[2mm]
\eta = D/\mu_D, \\[2mm]
D = |I_{ir} - \mu_{ir}|.
\end{cases} \tag{4.30}
$$

Here, D is the divergence of the intensity of each pixel from the mean intensity of the IR image and η is the ratio of local to global divergence of the IR image. A 3×3 window is used to get the local mean divergence to eliminate noise in the IR image.

Feature-Level Fusion

1) Union Operation on Image Contours and Silhouettes

Different from the pixel-level operation, feature-level fusion extracts and fuses high-level features from the input images for a higher level understanding of the captured information, which helps avoid the limitations of low-level fusion. As the low-level fusion is not guided by the desired characteristics for the output, the fused image is subject to the loss of information from the inputs. In Ref. [53], contours derived from visible and infrared images are fused for robust object detection. The process follows the flowchart in Fig. 4.12.

The first step is the background subtraction to identify the initial regions-of-interest (ROIs). A statistical background model using signal Gaussian function is applied to each pixel in the infrared image. Color and intensity information is used within these areas to obtain the corresponding ROI in the visible image [53]. Then, a contour saliency map is derived from each ROI based on the input and background gradient information. After thinning and thresholding operation, the most salient contour segments for the same image region are obtained for visible and infrared images respectively. A simple union of these segments fuses these contours. An A^* path-constrained search along watershed boundaries of the ROI is used to complete and close

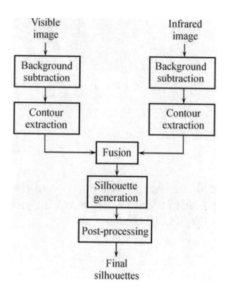

Fig. 4.12 Contour-based fusion for background subtraction.

broken segments [54]. The contour image is flood-filled to produce the final silhouette.

Similarly, human body is detected by fusing the silhouettes derived from color and infrared images in Ref. [55]. Let $c(X, Y)$ and $t(X, Y)$ represent the color and infrared silhouette vectors respectively. The corresponding mean background values are μ_c and μ_t. The estimate of probability is

$$\begin{cases} P\left(S|c(X,Y)\right) = 1 - \mathrm{e}^{-\|c(X,Y)-\mu_c(X,Y)\|^2}, \\ P\left(S|t(X,Y)\right) = 1 - \mathrm{e}^{-\|t(X,Y)-\mu_t(X,Y)\|^2}, \end{cases} \tag{4.31}$$

where S refers to the human silhouette and (X, Y) is the coordinate. The threshold values in Eq. (4.31) are chosen as the smallest values so that shadows in both color and infrared images are eliminated.

$$\begin{cases} \text{Product rule:} & P\left(S|c\left(X,Y\right)\right) P\left(S|t\left(X,Y\right)\right) > \tau_{product}, \\ \text{Sum rule:} & P\left(S|c\left(X,Y\right)\right) + P\left(S|t\left(X,Y\right)\right) > \tau_{sum}, \\ \text{Max rule:} & \max\{P\left(S|c\left(X,Y\right)\right), P\left(S|t\left(X,Y\right)\right)\} > \tau_{max}, \\ \text{Min rule:} & \min\{P\left(S|c\left(X,Y\right)\right), P\left(S|t\left(X,Y\right)\right)\} > \tau_{min}. \end{cases} \tag{4.32}$$

2) Fusion with logic operations

Feature-level fusion always employs a feature extraction process to derive useful information from input images. In Ref. [56], a seed detection algorithm was used to extract the targeted features, such as object's centroid, size, aspect ratio and angular direction, from both visible and infrared images. With this information, a rectangle can be drawn to show the detected objects for each input. The detected objects may have three different relationships as illustrated in Fig. 4.13. In intersect condition, both tracking information contains information of the same object; thus, an OR operation is applied to obtain the union of infrared tracking I and visual tracking V. In disjoint condition, each detected object contains information for different targets, so the same OR operation is applied. For the include condition, an AND operation is suggested to obtain the intersection of I and V.

$$\begin{cases} \text{OR operation}: & F = I \cup V, \\ \text{AND operation}: & F = I \cap V. \end{cases} \tag{4.33}$$

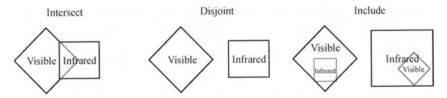

Fig. 4.13 Different relationships between detected objects.

3) Fusion with dual-ν support vector machine

The amplitude and phase features of a facial image are independent of changes of illumination and contrast [57]. Thus, these features are employed to improve the performance of face recognition. The extracted amplitude features F_A and phase features F_P are divided into 3×3 regions to calculate the weighted average region based activity. The resulted n activity levels are used to train a dual ν-support vector machine (2ν-SVM) to identify if the F_A feature is acceptable or not [57, 58]. If the SVM classification margin of F_A is greater than that of F_P, the output $O(x,y)$ of the SVM is 1; otherwise it is -1. The fused feature vector $FF(x,y)$ is obtained by

$$FF(x,y) = \begin{cases} F_A(x,y) \text{ if } O(x,y) > 0, \\ F_P(x,y) \text{ if } O(x,y) < 0. \end{cases} \tag{4.34}$$

Then, a match score is obtained by matching the fused feature vectors.

Actually, a two-level hierarchical fusion scheme was proposed for multi-spectral face images in Ref. [57]. At the first level, short and long wave infrared images are fused with DWT at pixel level. The fused image is then used as the input in Fig. 4.14. The fused result at the feature level will be use for face recognition through a matching process.

4) Genetic algorithm to fuse kernel-based features

Genetic algorithm (GA) was employed to combine eigenfeatures from visible and infrared images [29]. Suppose there are R face images in the training set and each face image is represented as a vector Γ_i (of size $N \times N$). The average face Ψ and difference Φ_i are computed as

$$\Psi = \frac{1}{R} \sum_{i=1}^{R} \Gamma_i, \tag{4.35}$$

$$\Phi_i = \Gamma_i - \Psi. \tag{4.36}$$

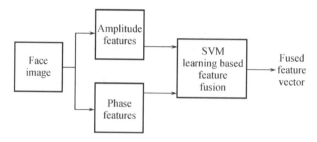

Fig. 4.14 Support vector machine for feature-level fusion.

Then, the covariance matrix is

$$C = \frac{1}{R} \sum_{i=1}^{R} \Phi_i \Phi_i^{\mathrm{T}} = A A^{\mathrm{T}}, \tag{4.37}$$

where $A = [\Phi_1 \Phi_2 \cdots \Phi_R]$. The eigenspace is defined by computing the eigenvector μ_i of C. Actually, the eigenvector ν_i of $A^{\mathrm{T}} A$ ($R \times R$ matrix) is computed first and μ_i can be computed from ν_i as

$$\mu_i = \sum_{j=1}^{R} \nu_{ij} \Phi_j. \tag{4.38}$$

Only a small number (R_k) of eigenvectors corresponding to the largest eigenvalues are kept. For a new input, Γ, the project is computed as

$$\widetilde{\Phi} = \sum_{i=1}^{R_k} w_i \mu_i \tag{4.39}$$

$$w_i = \mu_i^{\mathrm{T}} \Gamma \tag{4.40}$$

Herein, $\{w_i\}$ are the eigenfeatures. As illustrated in Fig. 4.15, the eigenspaces for visible and infrared face images are firstly built. The new inputs are projected to the two eigenspaces respectively. GAs are applied to decide eigenfeature selection from each eigenspace. Recognition is performed in the fused eigenspace. Another approach uses kernel face subspaces to implement feature-level fusion of visible and infrared images [58].

Decision-Level Fusion

1) Averaging

In Ref. [60], fusion of visible (RGB) and thermal videos is implemented to track pedestrian in surveillance area. Each pixel in the image is modeled as two dynamically growing vectors of codewords. For the RGB input, a codeword is represented by the average RGB value and the luminance range $[I_{low}, I_{hi}]$. If an incoming pixel p is with in $[I_{low}, I_{hi}]$ and the dot product of

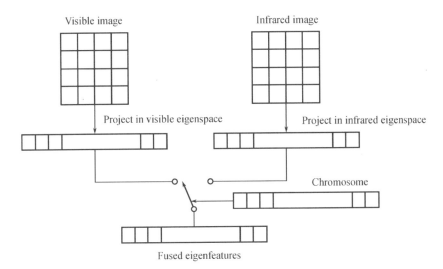

Fig. 4.15 Fusion with GA in eigenspace domain.

p_{RGB} with the codeword RGB is less than a predefined threshold, the pixel p will be classified as background. For the thermal monochromatic input, a codeword is represented by an intensity range $[T_{low}, T_{hi}]$. The incoming pixel p_T is classified by comparing the ratios of p_T/T_{low} and p_T/T_{hi}. The back group maps are computed for color and thermal images respectively:

$$p_C = \frac{\langle x_C, v_C \rangle}{|x_C||v_C|}, \tag{4.41}$$

$$p_T = \min\left(1, \frac{|x_T - v_T|}{\sigma_T}\right), \tag{4.42}$$

where v_C and v_T are the color and thermal codeword respectively and σ_T is the normalizing constant for thermal mode. The fusion is implemented by

$$p_{BG} = \frac{p_C + p_T}{2} > \text{Threshold}. \tag{4.43}$$

The fusion operation is robust when one modality is performing poorly.

2) Match Score Fusion

A match score fusion of visible and infrared face images for face recognition is proposed in Ref. [60]. First, two match scores are computed by matching the global features and local features extracted from probe and gallery face images. The global facial features are extracted with the 2D log polar Gabor transform while the local facial features are extracted with local binary pattern [61, 62]. The match scores are obtained by matching the corresponding extracted features. The two match scores for global and local features are fused with Dezert Smarandache (DSm) theory (Fig. 4.16) [64]. The fused match scores for visible and infrared face images are combined again to get a composite match score for the multi-modal input images. A final decision, accept or reject, is made with this composite score.

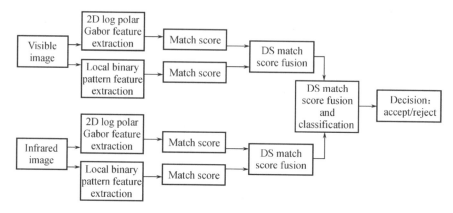

Fig. 4.16 Dezert Smarandache theory based match score fusion.

A hybrid multi-level fusion shown in Fig. 4.17 is also suggested in Ref. [61]. At the first level, visible and infrared face images are fused with a $2\nu - GSVM$ (2ν-granular support vector machine) algorithm. Then, the 2D log polar Gabor transform and local binary pattern feature extraction are applied to the fused image at the second level. The DSm is used to fuse the match score for the extracted features. Experimental results demonstrated that the integrated fusion scheme achieved the best results [61].

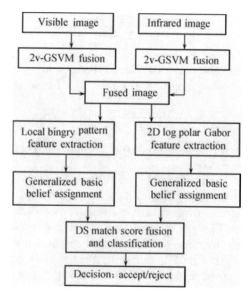

Fig. 4.17 Fusion at the pixel and decision level for face recognition.

3) Adaptive Neuro-fuzzy Inference Based Fusion

The adaptive neuro-fuzzy inference system (ANFIS) is an integration of neural network architectures and fuzzy inference system, which consists of a fuzzy layer, product layer, normalized layer, defuzzy layer, and summation layer. The ANFIS has a great potential in modeling nonlinear functions. An object tracking strategy is illustrated with the flowchart in Fig. 4.18. The targets are tracked by both the radar and infrared sensor. Then, the sensor precision values, states, and local estimation errors are fed into the ANFIS sensor confidence estimators (ASCE) to get the relevant sensor confidence degree. The ASCE takes two inputs, i.e., the contextual information (CI) and the normalized variables of the measurement square error (NVMSE). The CI describes the usability of sensor data while the NVMSE determines the relative sensor measurement precision. The output of ASCE is the sensor confidence degree.

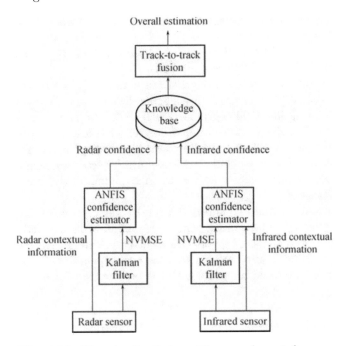

Fig. 4.18 The adaptive fusion with neuro-fuzzy inference.

The knowledge base chooses a suitable track-to-track fusion algorithm according to some reasoning rules about the sensor confidence degrees [65]. The overall estimation is finally obtained from the selected track-to-track fusion process.

4.4 Applications

Varied applications have benefited from the fusion with infrared images. Some representative applications are briefly described and discussed in this section, but this is not a complete list of the applications.

4.4.1 Surveillance Application

An important observation of pixel-level fusion for object tracking is described in Ref. [66]. For tracking purpose, IR mode is most useful while pixel-level fusion does not guarantee a better performance. However, when multiple objects are most separable from the background in both imaging modalities, the inclusion of complementary and contextual information from the inputs through video fusion makes the fused results more suitable for a human observer and computer-based analysis [66]. In other words, pixel-level fusion is not always beneficial to applications like object tracking. In this case, feature- or decision-level fusion is more appropriate.

The fusion of multiple spatiogram trackers for object tracking was described in Ref. [67]. Spatiograms are a generalization of histogram containing the same information as histograms as well as additional spatial information for each bin [67]. Pixel-based features for target tracking were split into N spatiogram model trackers. All the trackers evaluated a series of potential object position hypotheses and return a similarity score for each tracker. The score for each hypothesis is obtained by multiplying all the similarity scores from each tracker.

In Ref. [68], visible and infrared video sequences were fused for target detection purpose. The source sequences were first segmented into target and background regions. Different fusion rules were applied to the target and background regions respectively. The fused coefficients were finally combined to reconstruct the fused sequence with the inverse transform.

4.4.2 Nondestructive Inspection and Evaluation

In Ref. [69], color visual image and infrared image were used for the diagnosis of facade isolation. Information of spatial relationships from visual image, information of temperature distributions from infrared image, and information of the component's structure were used as inputs to a knowledge-based inference engine as illustrated in Fig. 4.19.

A Gaussian pyramid is first built for the high resolution visual image. From the pyramid, a visual image of the same resolution as the IR image is selected for further analysis with IR image. Features are extracted from the preprocessed visual and IR images. Then, the two images are registered and

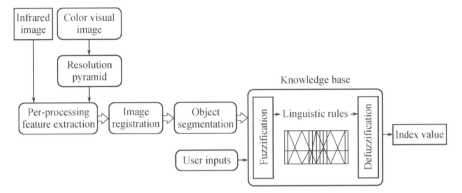

Fig. 4.19 Diagnosis of facade isolation from visual and IR images.

the correspondence between pixels is established. Objects of interest, such as windows, doors, and ventilation openings, are segmented manually. Three types of information derived from the visual and IR images are fed into the knowledge based inference engine together with the user inputs. An index value ranging from 10 to 1 represents the degradation of thermal properties [69].

Another use of pixel-level fusion can be found in the application of pressure vessel inspection [70]. In the fused image, the location of defect is highlighted, which will reduce the ambiguity and workload for an inspector.

The detection of fruit also employed image fusion technique, i.e., Laplacian pyramid and fuzzy logic, to improve the overall detection performance, especially when the fruits in the visible image are over-exposed and the fruits in the thermal image were warmer than the canopy [71].

4.4.3 Biometric Application

Due to its insensitivity to illumination changes, infrared imagery is also used for face recognition to improve the performance under varying lighting conditions. In Ref. [31], the empirical mode decomposition was applied to implement pixel-level fusion of visible and infrared face images. The cumulative match characteristics between galleries and probes were employed to measure the performance of face recognition. With the EMD fusion, the face recognition performance under poor illumination and different expressions was improved.

A weighted power series model for face verification score fusion was proposed in Ref. [71]. A linear parametric power series mode was adopted to minimize an approximated total error rate to fuse multi-modal face verification scores [71]. A useful information gain can be achieved from the fusion of visible and infrared face images.

In Ref. [72], fusion was implemented at pixel level and decision level. Pixel-level fusion generated illumination-invariant face image while decision-level fusion refined the classification results based on the average matching scores or the highest matching score from individual face recognition modules. The application framework is illustrated in Fig. 4.20. Experiments showed that fusion-based methods achieved reliable recognition results.

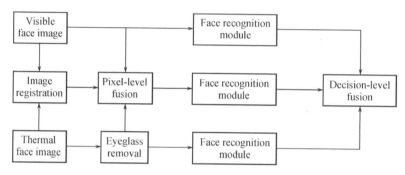

Fig. 4.20 Face recognition with pixel- and decision-level fusion.

As the multi-modal image fusion contributes to the face recognition, the design of a camera, which can simultaneously acquire a near-infrared image and visible image, was described in Ref. [73].

4.4.4 Remote Sensing

The remote sensing has found its applications in resource investigation, environmental monitoring, disaster prevention, etc. [27, 74]. Remote sensing images, such as optical, thermal, multispectral, and hyperspectral, reflect different characteristics of the same object. Fusion of multi-modal images will facilitate the further analysis of the remote sensing images. A multicontourlet transform was proposed to fuse visible and infrared images [27]. The synthetic aperture radar (SAR) and infrared images were fused to reduce noises and achieve a better visual quality as reported in Refs. [75, 76].

4.5 Summary

This chapter presents a review of fusion with infrared images for an improved performance and perception. Generally, fusion is implemented at three different levels, i.e., pixel level, feature level, and decision level. Varied algorithms are described for each fusion methodology. And the fusion techniques have benefited a diver range of applications, like surveillance, industrial inspection, biometric application, remote sensing, etc. The requirements of a specific application determine how to process and fuse the acquire data and images. Meanwhile, the assessment of the fusion of visible and infrared image raises another challenge for research. Again the evaluation of fusion performance depends on the specific application. There is no uniform solution for this.

Acknowledgements

The visible and infrared images in Fig. 4.4 are provided by Dr. Alexander Toet and available from http://www.imagefusion.org.

References

[1]   http://www.physics.gatech.edu/academics/tutorial/phys2122/Chapter.htm   (2007). Accessed 12 October 2010

[2]   Moria I, Heather J P (2005) Review of Image Fusion Technology in 2005. In: Peacock GR et al (eds) Proceedings of SPIE, vol 5782: 29–45. Bellingham, WA

[3]   Burt P J, Kolczynski R J (1993) Enhanced Image Capture Through Fusion. In: Proceedings of International Conference on Image Processing, pp 248–251

[4]   Slamani M A, Ramac L, Uner M et al (1997) Enhancement and Fusion of Data for Concealed Weapons Detection. In: Proceedings of SPIE, vol 3068: 20–25

[5]   Uner M K, Ramac L C, Varshney P K et al (1996) Concealed Weapon Detection: An Image Fusion Approach. In: Proceedings of SPIE, vol 2942: 123–132

[6]   Varshney P K, Ramac L, Slamani M A et al (1998) Fusion and Partitioning of Data for the Detection of Concealed Weapons. In: Proceedings of the International Conference on Multisource-Multisensor InformationFusion

[7]   Varshney P K, Chen H, Uner M (1999) Registration and Fusion of Infrared and Millimetre Wave Images for Concealedweapon Detection. In: Proceedings of International Conference on Image Processing, vol 13: 532–536

[8]   Aggarwal J K (1993) Multisensor Fusion for Computer Vision, NATO ASI Series F: Computer and Systems Science, vol 99

[9]   Xue Z, Blum R, Li Y (2002) Fusion of Visual and IR Images for Concealed Weapon Detection. In: Proceedings of ISIF 2002: 1198–1205

[10] Foresti G L, Snidaro L (2002) A Distributed Sensor Network for Video Surveillance of Outdoors. In: Foresti G L, Regazzoni C S, Varshney P K (eds) Multisensor Surveillance Systems. Kluwer Academic Publishers

[11] Matsopoulos G K, Marshall S, Brunt J N H (1994) Multiresolution Morphological Fusion of MR and CT Images of the Humanbrain. IEE Proceedings on Vision, Image and Signal Processing, 141(3): 137–142

[12] Koren I, Laine A, Taylor F (1998) Enhancement via Fusion of Mammographic Features. In: Proceedings of International Conference on Image Processing, pp 722–726

[13] Pohl C, Genderen J L V (1998) Multi-sensor Image Fusion in Remote Sensing: Concepts, Methods and Applications. International Journal of Remote Sensing, 19(5): 823–854

[14] Gros X E, Liu Z, Tsukada K et al (2000) Experimenting with Pixel-level NDT Data Fusion Techniques. IEEE Transaction on Instrumentation and Measurement, 49(5): 1083–1090

[15] Blum R S, Liu Z (eds) (2005) Multi-Sensor Image Fusion and Its Applications. Signal Processing and Communications. Taylor

[16] Stathaki T (ed) (2008) Image Fusion: Algorithms and Applications. Elsevier, Amsterdam

[17] http://www.siliconfareast.com/emspectrum.htm. Access 12 November 2010

[18] Maldague X P V (2001) Theory and Practice of Infrared Technology for Nondestructive Testing. Wiley Series in Microwave and Optical Engineering. Wiley, New York

[19] Rybicki G B, Lightman A P (1979) Radiative Processes in Astrophysics. Wiley, New York

[20] Adelson E H, Anderson C H, Bergen J R et al (1984) Pyramid Methods in Image Processing. RCA Engineer, 29(6): 33 – 41

[21] Teot A (1989) Image Fusion by a Ratio of Low-pass Pyramid. Pattern Recognition Letters, 9, 245 – 253

[22] Liu Z, Tsukada K, Hanasaki K et al (2001) Image Fusion by Using Steerable Pyramid. Pattern Recognition Letters, 22, 929 – 939

[23] Li H, Manjunath B S, Mitra S K (1995) Multisensor Image Fusion Using the Wavelet Transform. Graphical Models and Image Processing, 57(3): 235 – 245

[24] Wang H H (2004) A New Multiwavelet-based Approach to Image Fusion. Journal of Mathematical Imaging and Vision, 21, 177 – 192

[25] Liu K, Guo L, Li H et al (2009) Fusion of Infrared and Visible Light Images Based on Region Segmentation. Chinese Journal of Aeronautics, 22(1): 75 – 80

[26] Zhang Z, Blum R S (1999) A Categorization of Multiscale Decomposition Based Image Fusion Schemes with a Performance Study for a Digital Camera Application. In: Proceedings of IEEE, 87(8): 1315 – 1326

[27] Chang X, Jiao L, Liu F et al (2010) Multicontourlet-based Adaptive Fusion of Infrared and Visible Remote Sensing Images. IEEE Geoscience and Remote Sensing Letters, pp 549 – 553

[28] Lewis J J, O'Callaghan R J, Nikolov S G et al (2007) Pixel- and Region-Based Image Fusion with Complex Wavelets. Information Fusion, 8(2): 119 – 130

[29] Bebis G, Gyaourova A, Singh S et al (2006) Face Recognition by Fusing Thermal Infrared and Visible Imagery. Image and Vision Computing, 24(7): 727 – 742

[30] Liu G, Jing Z, Sun S (2005) Image Fusion Based on an Expectation Maximization Algorithm. Optical Engineering, 44(7): 077, 001

[31] Flitti F, Collet C, Slezak E (2009) Image Fusion Based on Pyramidal Multiband Multiresolution Markovian Analysis. Signal, Image and Video Processing, 3(3): 275 – 289

[32] Loza A, Bull D, Canagarajah N et al (2010) Non-gaussian Model-based Fusion of Noisy Images in the Wavelet Domain. Computer Vision and Image Understanding, 114(1): 54 – 65

[33] Toet A, Franken E M (2003) Perceptual Evaluation of Different Image Fusion Schemes. Displays, 24(1): 25 – 37

[34] Li S, Yang B, Hu J (2010) Performance Comparison of Different Multi-resolution Transforms for Image Fusion. Information Fusion. DOI 10.1016/j.inffus.2010.03.002. Article in Press

[35] Liu Z, Forsyth D S, Laganière R (2008) A Feature-based Metric for the Quantitative Evaluation of Image Fusion. Computer Vision and Image Understanding, 109(1): 56 – 68

[36] Huang N E, Shen Z, Long S R et al (1998) The Empirical Mode Decomposition and the Hilbert Spectrum for Nonlinear and Non-stationary Time Series Analysis. In: Proceedings of the Royal Society of London. Series A: Mathematical, Physical and Engineering Sciences, 454(1971): 903 – 995. DOI 10.1098/rspa.1998.0193. URL http://dx.doi.org/10.1098/rspa.1998.0193

[37] Hariharan H, Gribok A, Abidi M A et al (2006) Image Fusion and Enhancement via Empirical Mode Decomposition. Journal of Pattern Recognition Research pp 16 – 31

[38] Hariharan H, Koschan A, Abidi B et al (2006) Fusion of Visible and Infrared Images Using Empirical Mode Decomposition to Improve Face Recognition. In: Proceedings of International Conference on Image Processing, pp 2049 – 2052

[39] Rehman N, Looney D, Rutkowski T et al (2009) Bivariate EMD-based Image Fusion. In: IEEE 15th Workshop on Statistical Signal Processing, Cardiff, UK pp 57 – 60

[40] Li H, Zheng Y (2009) Image Fusion Algorithm Using Pyramidal Empirical Mode Decomposition. In: Proceedings of the 9th International Conference on Hybrid Intelligent Systems, pp 152 – 157. Washington, DC, USA. DOI http: //dx.doi.org/10.1109/HIS.2009.38. Accessed 3 November 2010

[41] Kumar S S, Muttan S (2006) PCA-based Image Fusion. SPIE. p. 62331T. DOI 10.1117/12.662373. URL http://link.aip.org/link/?PSI/6233/62331T/1. Accessed 3 November 2010

[42] Saito N, Remy J F (2006) The Polyharmonic Local sine Transform: A New Tool for Local Image Analysis and Synthesis Without Edge Effect. Applied and Computational Harmonic Analysis, 20(1): 41 – 73

[43] Liu S, Han J, Liu B et al (2009) Novel Image Fusion Algorithm Based on Human Perception. Optical Engineering, 48(4): 047002-047002-6

[44] Kumar M, Dass S (2009) A Total Variation-based Algorithm for Pixel-level Image Fusion. IEEE Transactions on Image Processing, 18: 2137 – 2143

[45] Zhang J, Wei L, Miao Q et al (2004) Image Fusion Based on Non-negative Matrix Factorization. In: Proceedings of International Conference on Image Processing. Singapore

[46] Tsagaris V, Anastassopoulos V (2005) Fusion of Visible and Infrared Imagery for Night Color Vision. Displays, 26(4-5): 191 – 196

[47] Lee D D, Seung H S (2001) Algorithms for Non-negative Matrix Factorization. In: Leen T K, Dietterich T G, Tresp V, (eds.) Advances in Neural Information Processing, vol 13. MIT Press, Cambridge, MA, USA

[48] Chen S, Leung H (2009) An EM-CI Based Approach to Fusion of IR and Visual Images. In: 12th International Conference on Information Fusion, pp 1325 – 1330. Seattle, WA, USA, 2009

[49] Zribi M (2010) Non-parametric and Region-based Image Fusion with Bootstrap Sampling. Information Fusion, 11(2): 85 – 94

[50] Yang J, Blum R S (2006) A Region-based Image Fusion Method Using the Expectation-Maximization Algorithm. In: Proceedings of Conference on Information Science and Systems

[51] Sun F, Li S, Yang B (2007) A New Color Image Fusion Method for Visible and Infrared Images. In: IEEE International Conference on Robotics and Biomimetics(ROBIO)

[52] Yin S, Cao L, Ling Y et al (2010) One Color Contrast Enhanced Infrared and Visible Image Fusion Method. Infrared Physics & Technology, 53(2): 146 – 150

[53] Davis J W, Sharma V (2007) Background-subtraction Using Contour-based Fusion of Thermal and Visible Imagery. Computer Vision and Image Understanding, 106, 162 – 182

[54] Russell S, Norvig P (eds) (2003) Artificial Intelligence: A Modern Approach. Prentice Hall

[55] Han J, Bhanu B (2007) Fusion of Color and Infrared Video for Moving Human Detection. Pattern Recognition, 40(6): 1771 – 1784

[56] Zhou Y, Mayyas A, Qattawi et al (2010) Feature-level and Pixel-level Fusion Routines when Coupled to Infrared Night-vision Tracking Scheme. Infrared Physics & Technology, 53(1): 43 – 49

[57] Singh R, Vatsa M, Noore A (2008) Hierarchical Fusion of Multi-spectral Face Images for Improved Recognition Performance. Information Fusion, 9(2): 200 – 210

[58] Chew H G, Lim C C, Bogner R E (2004) Optimization and Control with Applications, Chap. An Implementation of Training dual-nu Support Vector Machine. Kluwer, Dordrecht, 2004

[59] Desa S M, Hati S (2008) IR and Visible Face Recognition Using Fusion of Kernel Based Features. In: 19th International Conference on Pattern Recognition, Tampa, FL, USA pp 1 – 4

[60] Leykin A, Hammoud R (2010) Pedestrian Tracking by Fusion of Thermal-visible Surveillance Videos. Machine Vision and Applications, 21(4): 587 – 595

[61] Singh R, Vatsa M, Noore A (2008) Integrated Multilevel Image Fusion and Match Score Fusion of Visible and Infrared face Images for Robust Face Recognition. Pattern Recognition, 41(3): 880 – 893

[62] Singh R, Vatsa M, Noore A (2009) Face Recognition with Disguise and Single Gallery Images. Image and Vision Computing, 27(3): 245 – 257

[63] Ahonen T, Hadid A, Pietikainen M (2006) Face Decription with Local Binary Patterns: Application to Face Recognition. IEEE Transactions on Pattern Analysis and Machine Intelligence, 28(12): 2037 – 2041

[64] Smarandache F, Dezert J (2004) Advances and Applications of DSmT for Information
      Fusion. American Research Press, Champaign
[65] Yuan Q, Dong C, Wang Q (2009) An Adaptive Fusion Algorithm Based on ANFIS
      for Radar/Infrared System. Expert Systems with Applications, 36(1): 111 – 120
[66] Cvejic N, Nikolov S G, Knowles H D et al (2007) The Effect of Pixel-level Fusion
      on Object Tracking in Multi-sensor Surveillance Video. In: IEEE Conference on
      Computer Vision and Pattern Recognition, Minneapolis, MN pp 1 – 7
[67] Conaire C O, O'Connor N E, Smeaton A (2008) Thermal-visual Feature Fusion for
      Object Tracking Using Multiple Spatiogram Trackers. Machine Vision and Applica-
      tions, 19: 483 – 494
[68] Xiao G, Wei K, Jing Z (2008) Improved Dynamic Image Fusion scheme for Infrared
      and Visible Sequence Based on Image Fusion System. In: 11th International Confer-
      ence on Information Fusion, Cologne, pp 1 – 6
[69] Ribari S, MarEti D, Vedrina D S (2009) A Knowledge-based System for the Non-
      destructive Diagnostics of Faade Isolation Using the Information Fusion of Visual
      and ir Images. Expert Systems with Applications, 36(2): 3812 – 3823
[70] Tian Y P, Zhou K Y, Feng X et al (2009) Image Fusion for Infrared Thermography
      and Inspection of Pressure Vessel. Journal of Pressure Vessel Technology, 131L 021,
      502
[71] Bulanon D, Burks T, Alchanatis V (2009) Image Fusion of Visible and Thermal
      Images for Fruit Detection. Biosystems Engineering, 103(1): 12 – 22
[72] Toh K A, Kim Y, Lee S et al (2008) Fusion of Visual and Infra-red Face Scores by
      Weighted Power Series. Pattern Recognition Letters, 29(2): 603 – 615
[73] Heo J, Kong S, Adidi B et al (2004) Fusion of Visual and Thermal Signatures with
      Eyeglass Removal for Robust Face Recognition. In: Proceedings of the IEEE Work-
      shop on Object Tracking and Classification beyond the Visible Spectrum in Conjunc-
      tion with CVPR, pp 94 – 99
[74] Hizem W, Allano L, Mellakh A et al (2009) Face Recognition from Synchronised
      Visible and Near-infrared Images. IET Signal Processing, 3(4): 282 – 288
[75] Simone G, Farina A, Morabito F C et al (2002) Image Fusion Techniques for Remote
      Sensing Applications. Information Fusion, 3(1): 3 – 15
[76] Ye Y, Zhao B, Tang L (2009) SAR and Visible Image Fusion Based on Local Non-
      negative Matrix Factorization. In: 9th International Conference on Electronic Mea-
      surement & Instruments, Beijing, China, pp 4 – 266
[77] Zhang Y, Li Y, Zhang K et al (2009) SAR and Infrared Image Fusion Using Non-
      subsampled Contourlet Transform. In: International Joint Conference on Artificial
      Intelligence, Hainan Island, China, pp 398 – 401

5 Feature Selection and Ranking for Pattern Classification in Wireless Sensor Networks

Janos Csirik[1] and Horst Bunke[2]

Abstract Feature selection is a classical problem in the discipline of pattern recognition, for which many solutions have been proposed in the literature. In the current paper we consider feature selection in the context of pattern classification in wireless sensor networks. One of the main objectives in the design of wireless sensor networks is to keep the energy consumption of sensors low. This is due to the restricted battery capacity of today's sensors. Assuming that the features of a pattern recognition systems are acquired by the network's sensors, the objective of keeping the energy consumption of the sensors low becomes equivalent to minimizing the number of features employed in object classification. In fact, this objective is related with, but not identical to, classical feature selection, where one wants to optimize the recognition performance of a system by detecting and eliminating noisy, redundant, and irrelevant features. This paper introduces a general framework for pattern classification in wireless sensor networks that aims at increasing the lifetime of the underlying system by using a number of features as small as possible in order to reach a certain recognition performance. In experiments with data from the UCI repository, we demonstrate the feasibility of this approach. We also compare a number of classical procedures for feature subset selection in the context of pattern classification in wireless sensor networks.

5.1 Introduction

The problems of feature subset selection and feature ranking have been of paramount importance in the discipline of pattern recognition [1]. It is well known that given features are often noisy, irrelevant, or redundant. Hence, eliminating such features may lead to a better performing pattern recognition system. Also, with less features the risk of overfitting is reduced and one gets the benefits of faster execution time and less memory consumption. Principal component analysis [2] and its newer version, kernel principal component analysis [3], are two well-known methods to reduce the number of features. They compute a linear transformation from the original pattern space into a new space of lower dimensionality. Another well-known family of algorithms is based on search. Particular instances of these algorithms include forward,

---

1 Institute of Informatics, University of Szeged, Hungary.
2 Institute of Informatics and Applied Mathematics, University of Bern, Switzerland.

backward, and floating search [4, 5]. These methods iteratively add new elements to, or remove them from, a given set of features with the aim of locally optimizing the performance of the underlying system. Another classical algorithm for feature subset selection is Relief which provides a ranking of the features with respect to their ability to discriminate between different classes of patterns [6]. Other approaches to the feature subset selection problem are based on support vector machine [7], genetic algorithms [8, 9], and information gain [10]. For a more detailed discussion of feature selection we refer to the surveys provided in [1, 11 – 15].

In most of the previous papers on feature subset selection, the objective was to find a subset of features that lead to a high performance of the resulting classifier. In the current paper we address the problem of feature selection from a different perspective, namely in the context of wireless sensor networks [16]. The field of wireless sensor networks has become a focus of intensive research in recent years and various theoretical and practical questions have been addressed. One of the most critical issues faced in this domain is the restricted lifetime of the individual sensors, caused by limited battery capacity. Thus keeping the energy consumption of the individual sensors low is a key issue in wireless sensor networks. Assuming that the individual features of a pattern recognition problem to be solved in a wireless sensor network context are provided by the network's sensors, minimizing the energy consumption becomes equivalent to minimizing the number of features to be used. Clearly this problem is related to, but not identical with, previous work on feature selection because in wireless sensor networks we aim at keeping the number of features low while maintaining a certain level of classification accuracy, rather than optimizing the classification accuracy through the elimination of features.

Various approaches to minimizing energy consumption and maximizing the lifetime of sensors have been proposed. Berman et al. [17] have investigated the efficient energy management in a theoretical model. They assume that each sensor has its own monitored region and that the number of sensors largely exceeds the number of sensors necessary to monitor the required full region R. A set of sensors C covering R will be called sensor cover. Given an energy supply b_i for each sensor the task is to find a schedule of sensor covers with the maximum length such that for any sensor the total active time does not exceed b_i. The authors handle the case where the coverage area is less than 100%.

Krause et al. [18] consider the problem of monitoring spatial phenomena, such as road speed on a highway, using wireless sensors with limited battery life. The main question is to decide where to locate these sensors to best predict the phenomenon at the unsensed locations. However, given the power constraints, one also needs to determine when to selectively activate these sensors in order to maximize the performance while satisfying lifetime requirements.

Wang and Xiao [19] provide a survey on energy-efficient scheduling mechanisms in sensor networks that have different design requirements than those in traditional wireless networks. They classify these mechanisms based on their design assumptions and design objectives. Different mechanisms may make different assumptions about their sensors including detection model, sensing area, transmission range, failure model, time synchronization, the ability to obtain location and distance location, network structure and sensor deployment strategy. While all the mechanisms have a common design objective to maximize network lifetime, they may also have different objectives determined by their target applications.

Sun and Qi [20] present a concept of dynamic target classification in wireless sensor networks. The main idea is to dynamically select the optimal combination of features and classifiers based on the probability that the target to be classified might belong to a certain category. The authors also use two datasets to validate the hypothesis and derive the optimal combination of sets by minimizing a cost function. Moreover, they show that this approach can significantly reduce the computational time while at the same time achieve better classification accuracy, compared with traditional classification.

Chatterjea et al. [21] observed that in some applications very large amounts of raw data need to be transported through the wireless sensor network. This leads to high levels of energy consumption and thus diminished network lifetime. The authors allow certain nodes in the network to aggregate data by taking advantage of spatial and temporal correlations of various physical parameters and thus eliminating the transmission of redundant data. They present a distributed scheduling algorithm that decides when a particular node should perform this novel type of aggregation.

Tan and Georganas [22] propose a node-scheduling scheme, which can reduce the overall energy consumption of the underlying system, and therefore increase system lifetime, by turning off some redundant nodes. Their coverage-based off-duty eligibility rule and backoff based node-scheduling scheme guarantee that the original sensing coverage is maintained after turning off redundant nodes.

Duarte and Hu [23] classify the type of moving vehicles in a distributed, wireless sensor network. Specifically, based on an extensive real word experiment, they have compiled a data set which consists of 820 MB raw time series data and 70 MB of preprocessed, extracted spectral feature vectors, together with baseline classification results obtained by the maximum likelihood classifier.

The current paper is based on the assumption that sensors in a wireless network are only activated upon request from the base station of the underlying system. Each sensor measures one particular feature from the environment and returns its value to the base station, where the classification algorithm is executed.[3] Thus, in order to minimize energy consumption, we aim at minimizing the number of features used, or needed, for classification.

---

3   We suppose that classifier training and validation are executed on the base station as well.

In pattern recognition, it is well known that the reduction of the number of features used by a classifier does not necessarily lead to a deterioration of the recognition rate, and the literature is quite rich in methods for feature selection and feature set reduction [1, 11–13]. However, the main objective of these approaches is to increase the recognition performance of a system, through the elimination of redundant, noisy, or irrelevant features. By contrast, in this paper we reduce the number of features for the purpose of minimizing energy consumption of the sensors, and thus want to increase the lifetime of the system, while maintaining a certain level of classification accuracy. So we do not necessarily expect that we get a higher recognition performance by reducing the number of features, but aim at extending the lifetime of the classifier.

A sequential classifier combination approach to minimizing the overall classification cost has been proposed in [24]. The procedures for features set reduction applied in the current paper are loosely related to this approach. However, while the overall objective of [24] is to minimize the number of computational operations to be executed by a classifier, we aim at using the minimum number of features, i.e., sensor measurements, while maintaining a certain level of classification accuracy.

The rest of this paper is organized as follows. In Section 5.2, our general approach to feature selection for pattern classification in wireless sensor networks is outlined. Then, in Section 5.3, a number of sensor ranking strategies are introduced. Next, a series of experiments are described in Section 5.4. Finally, in Section 5.5, we present a summary and discussion, and draw conclusions from this work.

5.2 General Approach

We follow the traditional notation of statistical pattern recognition [25] and assume that a pattern x is represented by an N-dimensional feature vector, i.e., $x = (x_1, \dots, x_N)$, where x_i is the value of the i-th feature; $i = 1, \dots, N$. Let $S = \{s_1, \dots, s_N\}$ be the set of available sensors, where each sensor s_i measures exactly one particular feature $f(s_i) = x_i$ to be used by the classifier. Hence, the maximal set of features possibly available to the classifier is $\{x_1, \dots, x_N\}$. Furthermore, let $\varphi : S \to \mathbb{R}$ be a function that assigns a utility value $\varphi(x_i)$ to each feature $f(s_i) = x_i$. Concrete examples of utility functions will be discussed in Section 5.3. For the moment, let us assume that the utility of a feature x_i is proportional to its ability to discriminate between the different classes an unknown object may belong to. If feature x_i can better discriminate between the classes under consideration than feature x_j, we have $\varphi(x_i) > \varphi(x_j)$. If the two features have the same discriminating power, then $\varphi(x_i) = \varphi(x_j)$.

The basic structure of the algorithm for object classification proposed in this paper is given in Fig. 5.1. The system uses a base classifier. This base classifier can be a classifier of any type, in principle. For the purpose of simplicity, however, we assume in this paper that the base classifier is a k-nearest neighbor (k-NN) classifier [25]. Nearest neighbor classifiers are popular in pattern recognition and machine learning for at least two reasons. First, their asymptotic error has been shown to be bounded by twice the error of the optimal classifier [25]. Second, nearest neighbor classifiers do not need any training.

```
 1: begin
 2:   rank sensors s_1, ..., s_N according to the utility of the their features such that
      φ(x_1) ⩾ φ(x_2) ⩾ ... ⩾ φ(x_N)
 3:   F = ∅
 4:   for i = 1 to N do
 5:       if sensor s_i is available then
 6:           read feature f(s_i) = x_i
 7:           F = F ∪ {x_i}
 8:           classify(F)
 9:           if confidence(classify(F)) ⩾ θ then
10:               output result of classify(F) and terminate
11:           end if
12:       end if
13:   end for
14:   output result of classify(F)
15: end
```

Fig. 5.1 Algorithm for object classification with limited number of sensor measurements. (This algorithm is to be executed on the network's base station.)

Having a base classifier at its disposition, the algorithm starts with ranking the sensors in line 2. After this step, the sensors $s_1, \ldots, s_N$ are ordered according to the utility of their features $x_1, \ldots, x_N$, such that $\varphi(x_1) \geqslant \varphi(x_2) \geqslant \ldots \geqslant \varphi(x_N)$. That is, the first sensor yields the most discriminating feature, the second sensor the second most, and so on. Then the algorithm initializes the set F of features to be used by the classifier to the empty set (line 3). Next it iteratively activates one sensor after the other, reads in each sensor's measurement, and adds it to feature set F (lines 4 to 7). Once a new feature has been obtained, statement $classify(F)$ is executed, which means that the base classifier is applied, using feature set F (line 8). Note that a k-NN classifier is particularly suitable for such an incremental mode of operation where new features are iteratively added, because it needs only to compute distances of the unknown object to the training instances, and the distance computations can be performed in an incremental fashion, processing one feature after the other and accumulating the individual features' distances. In line 9, it is checked whether the confidence of the classification result is equal to or larger than a threshold θ. If this is the case the classification result is considered final. It is output and the algorithm terminates (line 10). Otherwise, if the confidence is below the given threshold θ, the next sensor is activated.

Obviously, in order to classify an unknown object, the base classifier uses nested subsets of features $\{x_1\}$, $\{x_1, x_2\}, \ldots, \{x_1, x_2, \ldots, x_i\}$ until its confidence in a decision becomes equal to or larger than threshold θ. While running through the for-loop from lines 4 to 13, it may happen that a sensor s_i becomes unavailable due to battery exhaustion or some other cause. In this case, sensor s_i will be simply skipped and the algorithm continues with sensor s_{i+1}. In case none of the considered feature subsets leads to a classification result with enough confidence, the classifier outputs, in line 14, the result obtained with the set F of features considered in the last iteration through the for-loop, i.e., for $i = N$.

An important issue in the algorithm of Fig. 5.1 is how one determines the confidence of the classifier. Many solutions to this problem can be found in the literature [26 – 28]. In the current paper, our base classifier is of the k-NN type. This means that it determines, for an unknown object represented by a subset of features $\{x_1, \ldots, x_i\}$, the k nearest neighbors in the training set, i.e., those k instances in the training set that have the smallest distance from the unknown object. Then it assigns the unknown object to that class that is represented most often among the k nearest neighbors. In case of a tie, a random decision is made. Let $k' \leqslant k$ be the number of training samples that are among the k nearest neighbors and belong to the majority class. Then we can say that the larger k', the more confident is the classifier in its decision. Consequently, we can define $confidence(classify(F)) = k'$. That is, if there are $k' > \theta$ nearest neighbors from the majority class, then the classification result is output and the system terminates. Otherwise, if the number of nearest neighbors belonging to the majority class is less than or equal to θ, the next sensor is activated. If M denotes the number of classes to be distinguished, then the range of feasible thresholds is the set of integers from the interval $[\lceil k/M \rceil, \ldots, k - 1]$.[4]

The algorithm of Fig. 5.1 is based on the assumption that many objects can be correctly classified without having access to the full set of features. An important consideration in the current paper is that the features are ordered according to their utility, such that the most salient features will be used first. Clearly, if a result with high confidence can be obtained using only feature subset $\{x_1, \ldots, x_i\}$, it is not necessary to activate any of the following sensors, $s_{i+1}, \ldots, s_N$ in order to obtain features $\{x_{i+1}, \ldots, x_N\}$. Taking such an early decision will increase the overall lifetime of the system. Obviously, there is a trade-off between recognition performance and system lifetime. On the average, one would expect that with an increasing value of θ more sensors will be activated and a higher recognition accuracy can be obtained at the expense of a shorter lifetime.

The algorithm of Fig. 5.1 takes into account that one or several features may not be available. There are several possible causes for such a case, for example, that a sensor has exhausted its battery or has become faulty for some reason. In this case, the corresponding feature x_i is skipped, and the

---

4   $\lceil X \rceil$ denotes the smallest integer equal to or larger than X.

system continues with sensor s_{i+1}.

5.3 Sensor Ranking

This section describes four possible methods for ranking the features acquired by the sensors before they are used by the algorithm of Fig. 5.1. That is, we order the features $x_1, ..., x_N$ and their corresponding sensors $s_1, ..., s_N$, such that $\varphi(x_1) \geqslant \varphi(x_2) \geqslant ... \geqslant \varphi(x_N)$. As a matter of fact, there exist many procedures in the literature that can be used for this purpose. Rather than providing an exhaustive overview here we refer to the surveys [1, 11–13] on feature selection and ranking.

5.3.1 Sensor Ranking by Relief

The first method for feature ranking considered in this paper is Relief [6]. It is based on the idea that, for feature x_i being relevant or having a high utility, elements in the training set from the same class should be closer than elements from a different class. Consequently, the utility $\varphi(x_i)$ of feature x_i is defined as

$$\varphi(x_i) = (1/M) \sum_{j=1}^{M} \left(-diff(x_{ij}, near\_hit_{ij}) + diff(x_{ij}, near\_miss_{ij})\right).$$

(5.1)

In this equation, M denotes the number of training samples and x_{ij} is the value of the i-th feature of sample x_j from the training set. Function $diff(x_{ij}, near\_hit_{ij})$ measures the distance between the value of feature i of sample x_j and the value of the same feature of the sample that is closest to x_j and from the same class. Analogously, $diff(x_{ij}, near\_miss_{ij})$ measures the distance between the value of feature i of sample x_j and the value of the same feature of the sample in the training set that is closest to x_j and from a different class. If a feature is useful, it should be able to help in discriminating different classes from each other. Therefore, this feature should have large values of $diff(x_{ij}, near\_miss_{ij})$ and small values of $diff(x_{ij}, near\_hit_{ij})$. In other words, it should give a large value of $\varphi(x_i)$.

Clearly, for a given set of labeled training instances, it is straightforward to compute the utility $\varphi(x_i)$ of each individual feature in $O(M^2)$ time. Then we reorder the features $x_1, ..., x_N$ and their corresponding sensors $s_1, ..., s_N$ such that $\varphi(x_1) \geqslant \varphi(x_2) \geqslant ... \geqslant \varphi(x_N)$.

5.3.2 Sensor Ranking by a Wrapper Approach

In the fields of Pattern Recognition and Machine Learning, feature ranking
and feature selection algorithms are traditionally subdivided into filter-based
and wrapper-based approaches [12]. A filter-based approach is characterized
by applying some formal criterion in the features space to evaluate how well
a feature can discriminate between different classes. The Relief algorithm
introduced in Section 5.3.1 may be regarded as an instance of a filter-based
algorithm. Wrapper algorithms (WAs), by contrast, use the error rate of a
full-fledged classifier to determine the quality of the individual features. As
the final goal of feature ranking or reduction is often classification, wrapper-
based approaches may be considered more useful. However, as they need a
completely developed and trained classifier, they are usually computationally
more demanding than filters.

Under a wrapper-based approach, the higher the classification accuracy
of a single feature is, the higher is its utility. Based on this observation, we
propose to divide the training set S into two disjoint subsets, S_1 and S_2, such
that $S = S_1 \cup S_2$. Subset S_1 will be used as the training, and subset S_2 as the
validation set to compute the features' utility. To determine $\varphi(x_i)$, we use
only feature x_i to represent all instances from S. In order to avoid classifier
training, it is proposed to apply a k-nearest neighbor classifier (with some
suitably chosen value of k), using S_1 as the training and S_2 as the test set.
Let κ be the number of correctly classified samples from S_2, using S_1 as the
training set. Then the utility (i.e., correct recognition rate) of feature x_i is
given by

$$\varphi(x_i) = \kappa/|S_2|. \qquad (5.2)$$

Once we have computed $\varphi(x_i)$ for each feature x_i, we rank the features
$x_1, ..., x_N$ and the corresponding sensors $s_1, ..., s_N$ such that $\varphi(x_1) \geqslant \varphi(x_2) \geqslant$
$... \geqslant \varphi(x_N)$.

Note that there is a close relation between the method of Section 5.3.1
and that of Section 5.3.2. Whenever we get a positive contribution to the
sum in Eq. 5.1 from element x_j, element x_j will also contribute to increasing
the value of the expression in Eq. 5.2 and vice versa. Therefore, Eq. 5.2 may
be interpreted as a binary version of Eq. 5.1.

5.3.3 Sensor Ranking by a Wrapper Approach Using Sequential
          Forward Search

A shortcoming of the two feature ranking methods considered until now is
that they are not taking into account any possible interdependencies between
pairs of features. For example, the fact that one feature is just a multiple or
an identical copy of another one would remain undetected by either method.
A possible way to overcome this problem is sequential forward search (SFS),

where subsets of features rather than individual features are evaluated [4]. For a set consisting of M features, there exist 2^M different subsets. Therefore, exhaustive investigation of all these subsets is usually impossible due to computational complexity problems. What SFS attempts is to focus on a small fraction of all possible subsets by starting with feature sets of size one and iteratively enlarging the best subset found until a certain moment. SFS is just a simple representative of a larger family of related algorithms. For additional algorithms belonging to this family we refer to [4, 5].

Similarly to the approach described in Section 5.3.2, SFS makes use of a division of the training set into disjoint subsets S_1 and S_2. Furthermore, it is based on a k-NN classifier that uses S_1 as the training and S_2 as the validation set, based on different feature subsets. A pseudo-code description of the algorithm is shown in Fig. 5.2. The algorithm uses two sets, F and F', of features. Initially set F includes all features (line 2) and set F' is empty (line 3). Note that set F is unordered, i.e., the x_i's may be in any random order that does not reflect their utility. In each run through the for-loop from line 4 to 8, the feature x_j is selected from F that increases the utility most when added to the features selected before (line 5). That is, x_j is the feature out of those not yet used that leads to the highest recognition accuracy when added to F'. After having identified feature x_i, it is removed from F (line 6) and added to F' (line 7). Upon termination, set F is empty and set $F' = \{x_{\pi(1)}, ..., x_{\pi(N)}\}$ is output, where the order $\pi(1), ..., \pi(N)$ of the features reflects their utility.

```
1: begin
2:   F = {x_1, ..., x_N}
3:   F' = ∅
4:   for i = 1 to N do
5:       x_π(i) = arg max_{x_j ∈ F} accuracy(F' ∪ {x_j})
6:       F = F − {x_π(i)}
7:       F' = F' ∪ {x_π(i)}
8:   end for
9:   output F' = {x_π(1), ..., x_π(N)}
10: end
```

Fig. 5.2 Feature ranking by wrapper approach using sequential forward search.

In contrast with the procedures of Sections 5.3.1 and 5.3.2, the utility of a feature under this forward search strategy is not measured in isolation, but in conjunction with all features selected before. Although this procedure is computationally more demanding than the procedures described before (it is of order $O(N^2)$ in terms of the number of features, while the previous methods are of order $O(N)$), it is motivated by the observation that the selection process of features resembles the iterative extension of the set of features in the algorithm of Fig. 5.1. Therefore, one may expect this ranking strategy particularly suitable for being used in conjunction with the algorithm of Fig. 5.1.

5.3.4 Sensor Ranking Including the Lifetime of Sensors

The aim of the sensor ranking schemes introduced in Sections 5.3.1 to 5.3.3 is to measure the discrimination power of features, such that a relatively small subset of the best ranked features is able to classify an unknown object with a sufficiently high confidence. Clearly, this will increase the lifetime of the sensors that are not used often. A possible shortcoming is, however, that the best ranked features are likely to be used frequently and thus run out of battery power rather quickly. Then the system necessarily has to resort to features with a lower utility. As these features have less discriminatory power, it may be expected that the performance of the system significantly deteriorates once the highly ranked sensors are no longer available. In order to avoid such deterioration, we propose a more general sensor ranking scheme, which takes the expected remaining lifetime of each sensor into consideration. The basic idea is to assign a weight to each sensor that is proportional to its expected remaining lifetime. This weight is then combined with any of the utility measures $\varphi(x_i)$. As a matter of fact, any of the sensor ranking measures discussed before can be used in conjunction with the method introduced in the current section.

More formally, let $T(s_i) \geqslant 0$ be an integer number denoting the expected number of measurements sensor s_i can deliver before its battery is exhausted. Given this number for each sensor $s_1, ..., s_N$, we propose using the modified utility function

$$\psi(s_i) = T(s_i) \cdot \varphi(s_i), \tag{5.3}$$

where $\varphi(s_i)$ is any of the utility functions introduced in Sections 5.3.1 to 5.3.3. Together with the modified sensor ranking strategy, we need to slightly modify the algorithm in Fig. 5.1, such that sensor ranking in line 2 is based on $\psi(s_i)$ rather than $\varphi(s_i)$. That is, sensor s_1 is the one that ranks highest under ranking function ψ, sensor s_2 is the one that ranks second highest, a.s.o. Furthermore, after the feature value delivered by sensor s_i has been read in line 6, the lifetime $T(s_i)$ of sensor s_i has to be decreased. This can be accomplished by adding, between lines 6 and 7, an additional statement

$$\text{6a:} \quad T(s_i) = T(s_i) - 1.$$

Here we assume that all $T(s_i)$'s have been properly initialized before the algorithm starts and sensor s_i becomes unavailable in line 5 if $T(s_i) = 0$.

5.4 Experiments

All algorithms described in Sections 5.2 and 5.3 were implemented and experimentally evaluated. In the field of wireless sensor networks, there are not many data sets publicly available, especially not for pattern recognition problems. In [29], a dataset for activity recognition of humans equipped with body worn sensors as well as environmental sensors is described, and in [30] the PlaceLab datasets are introduced, which were acquired in the context of research on ubiquitous computing in a home setting. However, the authors of these papers do not mention any use of the data sets for pattern classification problems. Moreover, no classification benchmarks have been defined for any of these data sets. For this reason, the authors of the current paper decided to use datasets from the UCI Machine Learning Repository [31]. The sensors were simulated by assuming that each feature in any of these datasets is delivered by a sensor. The selection of the datasets used in the experiments was guided by several considerations. First, the datasets should pose classification problems with only numerical and no missing feature values. These conditions are necessary for the k-NN classifier being applicable in a straightforward way. Moreover, in order to allow for large scale experiments, we selected only datasets including a rather large number of instances and features. An overview of the datasets used in the experiments together with their most important characteristics appears in Table 5.1. We observe that the number of instances ranges from 1593 up to 7797 and the number of features from 256 up to 5000. More details about these datasets can be found in [31].

Table 5.1 Datasets and some of their characteristic properties

Data Sets	# Instances	# Features	# Classes	Training Set	Test Set
Isolet	7 797	617	26	6 237	1 560
Gisette	7 000	5 000	2	6 000	1 000
Semeion	1 593	256	10	1 193	400
Multiple Features	2 000	649	10	1 500	500

5.4.1 First Experiment

The purpose of the first experiment was to find out how the value of threshold θ (see line 9 of the algorithm in Fig. 5.1) influences the classification *accuracy* and the total number of sensor measurements (i.e., features) needed to classify the objects. The classification *accuracy* is measured in the standard way according to

$$accuracy = correct/total$$

where *correct* and *total* denote the number of correctly classified instances and the total number of instances in the test set, respectively. Rather than reporting the total number of sensor measurements directly, we give the *efficiency*, which is defined as

$$efficiency = 1 - (used\_features/total\_features)$$

with *used_features* and *total_features* denoting the number of features actually used and the total number of features, respectively. Clearly, the effi-

ciency is bounded from below by 0 and from above by 1, corresponding to
the case that all and no features are used, respectively.

Figures 5.3–5.6 show the accuracy and the efficiency of the algorithm
for the four datasets. In these experiments, a k-NN classifier with $k = 10$
was used. Whenever a tie between different classes occurred, it was randomly
resolved.

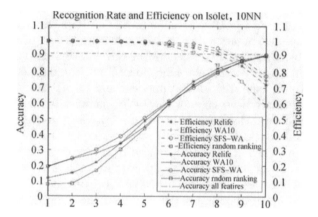

Fig. 5.3 Recognition rate and efficiency on Isolet.

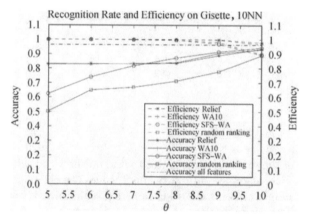

Fig. 5.4 Recognition rate and efficiency on Gisette.

In the first experiment, only the sensor ranking strategies of Sections
5.3.1 to 5.3.3 were investigated. In order to determine the feature ranking in
line 2 of the algorithm of Fig. 5.1 under the Wrapper approaches described
in Sections 5.3.2 and 5.3.3, the training set was randomly divided into two
disjoint sets TR_1 and TR_2 at a proportion of 75% and 25%, respectively.
The elements of TR_1 were used as the prototypes of a 10-NN classifier, while
TR_2 served as the test set. The ranking was determined according to the
performance of the 10-NN classifier on TR_2. To determine the feature ranking

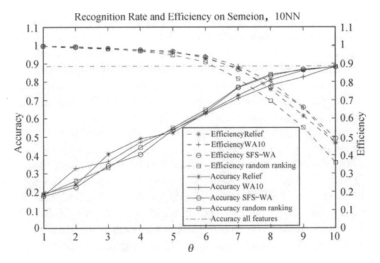

Fig. 5.5 Recognition rate and efficiency on Semeion.

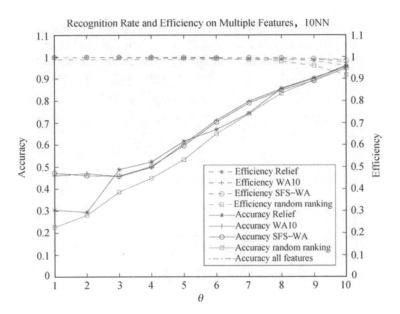

Fig. 5.6 Recognition rate and efficiency on Multiple Features.

for the method described in Section 5.3.1 (Relief), the whole training set, i.e., $TR_1 \cup TR_2$, was used to compute the ranking. In order to objectively assess the benefit that is obtained from the sensor ranking strategies, we also run the experiment under a random ranking procedure. Under this procedure, the features in line 2 of the algorithm of Fig. 5.1 were randomly ranked. As a result we obtained, for each data set, four curves that show the efficiency of the proposed approach as a function of threshold θ, and another four curves

representing the accuracy.[5] For each value of θ and each of the eight curves,
the complete test set was classified once.

In Figs. 5.3–5.6 we observe that, as one can expect, the accuracy increases
with an increasing value of threshold θ, while the efficiency decreases. As a
reference value, the accuracy obtained with the full set of features without
a reject option is also shown (accuracy all features). This accuracy value is
independent of the threshold θ. Note that it corresponds to an efficiency value
equal to zero. Obviously, for all four data sets, the accuracy of the proposed
system does not reach the accuracy obtained when all features are used and
no reject option exists, but for larger values of θ it gets quite close.

In Figs. 5.7–5.10, we represent the results of Figs. 5.3–5.6 by means of
ROC curves. In these ROC curves the *efficiency* is plotted as a function of
the *accuracy* for all considered sensor ranking strategies, Relief, WA, WA-
SFS (Wpproach Using Sequential Forward Search) and Random. One can
clearly observe the trade-off between *efficiency* and *accuracy* as with an
increasing recognition rate the efficiency decreases. It becomes also evident
that a quick deterioration of the efficiency happens only for higher recognition rates. In these curves it is particularly easy to compare the different
feature selection strategies to each other. Obviously, on data set Isolet, WA-
SFS has the highest and random ranking the lowest performance. The same
observation holds true for data set Multiple Features. On data set Gisette,
random ranking is inferior to all other strategies and Relief performs better

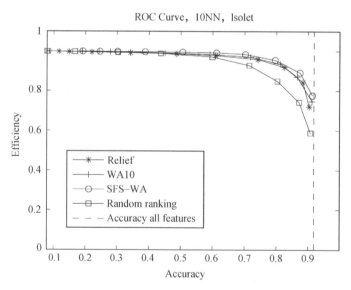

Fig. 5.7 ROC curve on Isolet.

---

5   Each of the four curves corresponds to one of the sensor ranking strategies Relief
    (Section 5.3.1), Wrapper approach (Section 5.3.2), Wrapper approach with sequential
    forward search (Section 5.3.3), and random ranking, respectively.

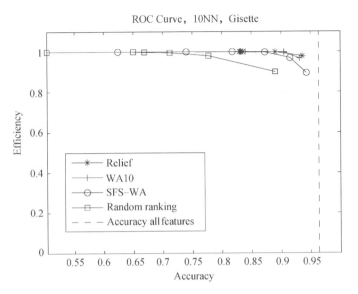

Fig. 5.8 ROC curve on Gisette.

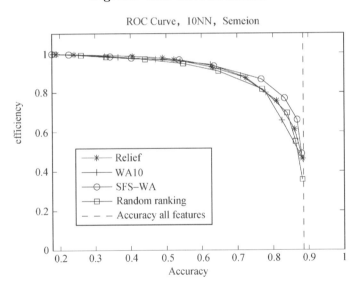

Fig. 5.9 ROC curve on Semeion.

than WA-SFS. Finally, on Semeion, WA-SFS performs best, but there is no clear looser. Depending on the value of threshold θ, either random ranking or WA has a lower performance.

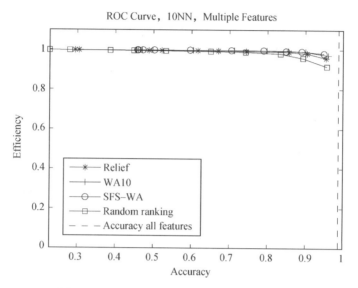

Fig. 5.10 ROC curve on Multiple Features.

5.4.2 Second Experiment

The purpose of the first experiment was to analyze the trade-off between *efficiency* and *accuracy*, depending on the value of threshold θ. However, in this experiment the life-time of the considered system, i.e., the number of classifications a system is able to perform before the sensors are no longer available because of battery exhaustion, was not directly taken into account. In the second experiment we address this issue. The aim of the experiment is to explicitly consider the lifetime of the system and analyze the trade-off between lifetime and accuracy.

We assume that the test set consists of M patterns and each feature x_i can be used exactly M times before the battery of its sensor is exhausted. This means that with a conventional pattern recognition system, which uses the full set of features for each pattern to be classified, the test set can be classified exactly once before all sensors become unavailable.[6] By contrast, with the system proposed in this paper, not all features will be used in each classification step, which allows one to classify the test set multiple times.

In the second experiment, we classify the test set multiple times until all sensors become unavailable. Let $M' \geqslant M$ be the number of pattern instances actually classified, where we count an element of the test set as often as it has

---

6 In a real world scenario, one would expect that the sensors last much longer. However, such a scenario can be easily accounted for by just multiplying all related quantities by some constant factor.

been classified. Now we define the *lifetime_ extension_ factor* as follows:

$$lifetime\_ extension\_ factor = M'/M.$$

Clearly, the *lifetime_ extension_ factor* is bounded by 1 from below. According to our assumption that each feature can be used exactly M times before it becomes unavailable, the case *lifetime_ extension_ factor* = 1 occurs if the underlying system always uses all features in each classification step. However, if less than N features are used, the value of the *lifetime_ extension_ factor* will be greater than 1.

The setup of Experiment 2 is identical to Experiment 1, but this time we measure the *accuracy* and the *lifetime_ extension_ factor* both as a function of threshold θ. A representation of the results in terms of ROC curves appears in Figs. 5.11 – 5.14. Similarly to Figs. 5.7 – 5.10 where a tradeoff between *efficiency* and *accuracy* is reported, we observe a trade-off between *accuracy* and *lifetime_ extension_ factor*. The representation in Figs. 5.11 – 5.14 may be more useful than the one in Figs. 5.7 – 5.10 from the practical point of view because it directly relates the *accuracy* to the actual lifetime of the underlying system. From such a plot a system developer can immediately see the lifetime extension gained by running the classifier at a somewhat lower accuracy. Comparing Figs. 5.7 – 5.10 with Figs. 5.11 – 5.14 with respect to the performance of the feature ranking strategies, we notice some differences. In particular, random ranking performs better in Figs. 5.11 – 5.14 than in Figs. 5.7 – 5.10. For example, in Fig. 5.11, it is the best out of all four feature ranking strategies.

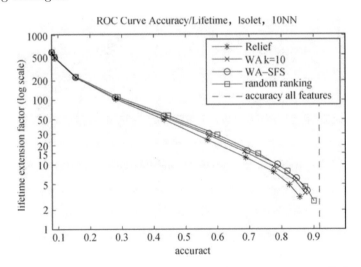

Fig. 5.11 ROC curve of accuracy and lifetime extension on Isolet.

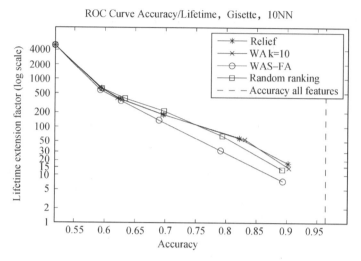

Fig. 5.12 ROC curve of accuracy and lifetime extension on Gisette.

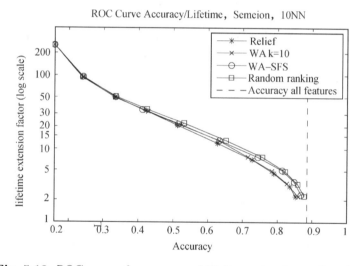

Fig. 5.13 ROC curve of accuracy and lifetime extension on Semeion.

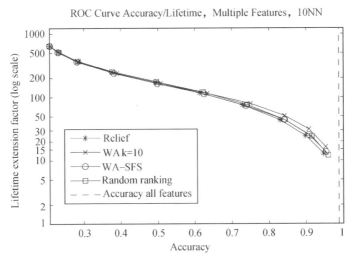

Fig. 5.14 ROC curve of accuracy and lifetime extension on Multiple Features.

5.4.3 Third Experiment

In this experiment we use the sensor ranking strategy described in Section 5.3.4, where the lifetime of the sensors is taken into account. Otherwise the experiment is identical to the second experiment. The results are presented in terms of ROC curves in Figs. 5.15 – 5.18. It is interesting to compare Figs. 5.15 – 5.18 to Figs. 5.11 – 5.14 in order to find out whether in-

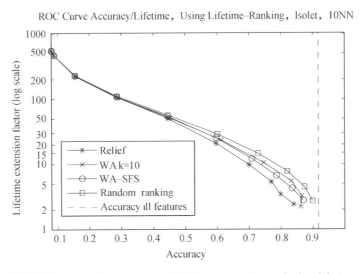

Fig. 5.15 ROC curve of accuracy and lifetime extension on Isolet, lifetime-ranked.

cluding the lifetime of the sensors has any impact on the *accuracy* or the
lifetime_ extension_ factor. On datasets Isolet, Multiple Features, and
Gisette only minor differences can be observed. On Semeion, the accuracies
for small values of the *lifetime_ extension_ factor*) slightly drop. From the
results one can conclude that including the remaining lifetime in the sensor
ranking strategy does not lead to an improvement in performance.

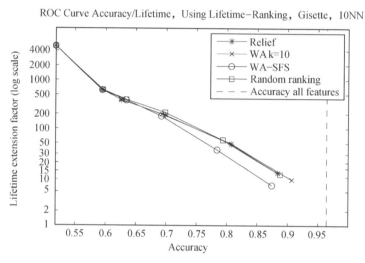

Fig. 5.16   ROC curve of accuracy and lifetime extension on Gisette, lifetime-
ranked.

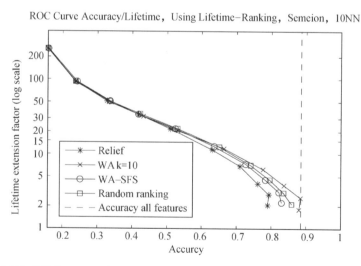

Fig. 5.17   ROC curve of accuracy and lifetime extension on Semeion, lifetime-
ranked.

ROC Curve Accuracy/Lifetime, Using Lifetime–Ranking, Multiple Features, 10NN

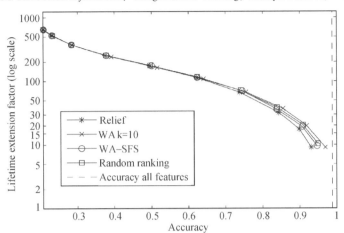

Fig. 5.18 ROC curve of accuracy and lifetime extension on Multiple Features, lifetime-ranked.

5.4.4 Fourth Experiment

From the first three experiments one can conclude that the lifetime of a system can be increased at the cost of decreased *accuracy*. However, we did not quantitatively analyze how the decrease in *accuracy* takes place over time. In this experiment we proceed similarly to Experiments 2 and 3, i.e., we classify the test set several times. But we do not report the accuracy in the global sense, i.e., in one number for all runs together, but want to see how it changes as the systems evolves over time and more sensors become unavailable.

In this experiment the ranking strategies of Sections 5.3.1 – 5.3.3 were considered. The test set was divided into smaller portions of size one tenth of the original test set size. Then the algorithm of Fig. 5.1 was applied until all sensors became unavailable. For each portion of the test data the recognition rate and the number of sensors used were recorded separately. For the sake of brevity, we show only results for the threshold $\theta = 10$ and the feature ranking strategy WA-SFS.

In Figs. 5.19 – 5.22, the results of the fourth experiment are shown. The x-axis depicts the number of rounds through the partitions of the test set, while on left and right y-axis the *accuracy* and the number of sensors actually used is given, respectively. For all four datasets we observe a similar behavior. Over many rounds the *accuracy* does not remarkably deteriorate. Only towards the right end of the plot, when more and more sensors are no longer available, it drops significantly. On the other hand, the number of sensors used for the classification fluctuates over time, depending on the underlying data set.

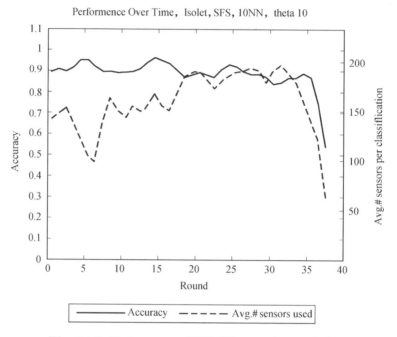

Fig. 5.19 Performance of SFS-WA over time on Isolet.

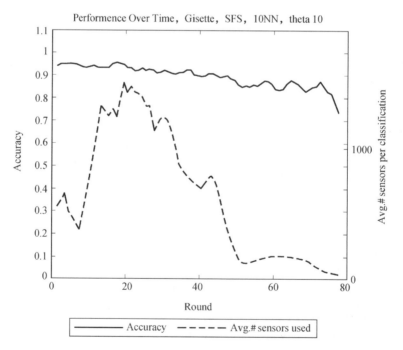

Fig. 5.20 Performance of SFS-WA over time on Gisette.

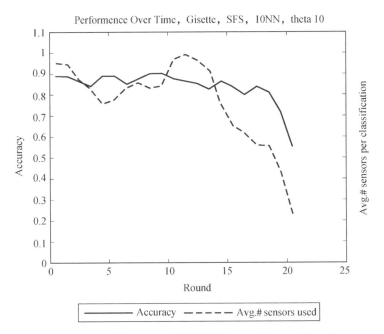

Fig. 5.21 Performance of SFS-WA over time on Semeion.

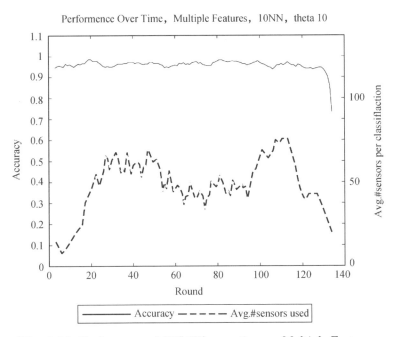

Fig. 5.22 Performance of SFS-WA over time on Multiple Features.

Towards the end, we note a remarkable drop.

There are a few more interesting phenomena to observe. On data set Isolet, the accuracy does not decrease much until about round 35. Afterwards it decays very rapidly. The number of features fluctuates remarkably, but shows an upward trend until round 32. Afterwards it quickly declines. The second phase of this decline, when only very few sensors are left, is paralleled by a steep decline of the *accuracy*. From the qualitative point of view, a similar behavior can be observed on data sets Multiple Features and Semeion. On Gisette, we note that after round 50 only few features remain to be used. Surprisingly, the recognition rate nevertheless remains quite stable for some time. It is only after round 73, when it drops more rapidly.

5.4.5 Fifth Experiment

The sensor ranking strategy described in Section 5.3.4 was motivated by preventing a steep decline of the system towards the end of the life time of the sensors. In order to measure the effect of including the remaining life time of the sensors in the ranking procedure, the fourth experiment was repeated, but this time using the ranking strategy of Section 5.3.4. The results are shown in Figs. 5.23–5.26. In these and the following figures, a sliding window of width $\lceil (\#Rounds)/20 \rceil$ was used to make the curves smoother. Apparently, there

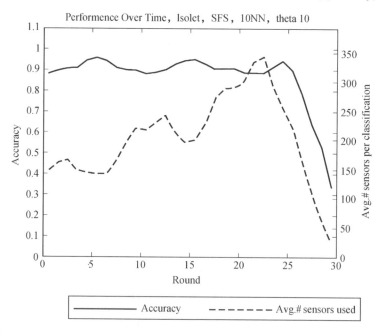

Fig. 5.23 Performance over time of lifetime-ranking and SFS-WA on Isolet.

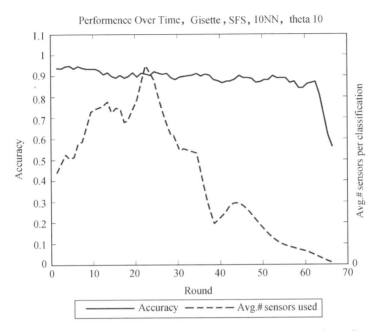

Fig. 5.24 Performance over time of lifetime-ranking and SFS-WA on Gisette.

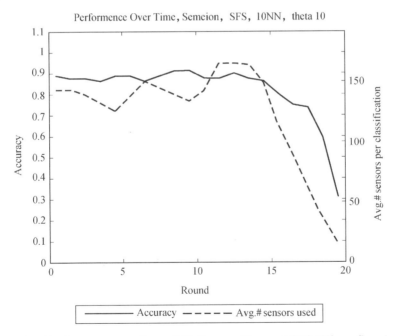

Fig. 5.25 Performance over time of lifetime-ranking and SFS-WA on Semeion.

are no significant differences to Figs. 5.19 – 5.22. Therefore we can conclude
that taking the remaining lifetime of the sensors into account does not lead
to a noticeable improvement of the system's stability towards the end of the
lifetime of its sensors.

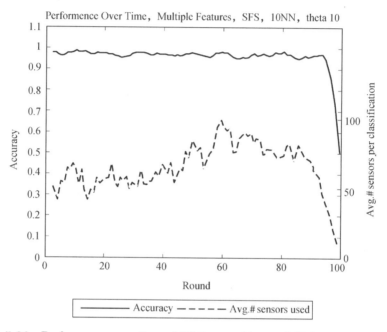

Fig. 5.26 Performance over time of lifetime-ranking and SFS-WA on Multiple
Features.

5.5 Summary, Discussion and Conclusions

In this paper we have addressed the feature subset selection problem from
the perspective of wireless sensor networks. Traditionally, methods for feature selection in pattern recognition aim at increasing the recognition rate
of a classifier by elimination of noisy, redundant, or irrelevant features. By
contrast, in this paper we propose a procedure that reduces the number of
features in order to increase the lifetime of the individual sensor, and thus
the lifetime of the overall system. This is motivated by the fact that the
most critical issue in wireless sensor networks is the limited lifetime of the
individual sensors, caused by restricted battery capacity.

In this paper, a general procedure is proposed for reducing the number
of features used by a classifier. This procedure can be applied in conjunction with any known method for feature selection or ranking. In the current
paper three well known methods, viz. Relief, a wrapper approach based on

evaluating each feature individually with a k-nearest neighbor classifier, and a wrapper approach in conjunction with sequential forward search are applied in addition to random feature ranking. The underlying base classifier is a k-nearest neighbor classifier. The basic idea of the proposed procedure is to rank all available features first and then use nested subsets of the features, starting with the best ranked individual feature until a certain confidence level is reached in classification. The confidence level of our k-nearest neighbor classifier is defined as the number of nearest prototype patterns from the majority class.

The proposed procedure was implemented and experimentally tested. As test data, four datasets from the UCI Machine Learning repository were used. A wireless sensor network scenario was simulated by assuming that the individual features are delivered by independent sensors. The results of the experiments revealed that the system behaves very well. The recognition rate monotonically increases with an increasing number of features used by the system. On the other hand, decreasing the number of features leads to an extension of the lifetime of the system. Comparing the four feature ranking strategies with each other we note that there is no clear winner. Which of the strategies works better depends on the underlying data set. An important observation made from the experiments is that during most of its lifetime, the system behaves quite stable. That is, the recognition rate slightly decreases over the system's lifetime, but a drastic drop typically happens only towards the very end when all but a few sensors ran out of power, and particularly the high ranked sensors are no longer available.

The proposed system can be applied for pattern classification in real wireless sensor networks provided that the objects or events to be classified behave in some stationary way. Because features are acquired in a sequential fashion and the decision of the classifier about the class label of an unknown object or event is only available after the first i features ($1 \leqslant i \leqslant N$) have been processed, it is required that the object or event to be recognized does not change until sensor s_i has delivered feature $f(s_i) = x_i$. This may be a problem when quickly moving objects or rapidly changing events are to be classified. However, there are many potential applications of wireless sensor networks, where this stationary assumption is typically satisfied. Examples include environment monitoring and surveillance.

There are many ways in which the work described in this paper can be extended. First of all one can think of investigating classifiers other than the k-nearest neighbor classifier. As one particular example, the naive Bayes classifier [32] seems to be a promising candidate as it processes features in a sequential way and handles nested subsets of features in an efficient way. Similarly, in addition to the feature selection and ranking strategies considered in this paper, there are many alternative methods known from the literature [1, 11–13]. It would be certainly worthwhile to extend the experiments to these methods and compare them to the ones applied in this paper. Moreover, an extension of the experiments to more datasets would be desirable, in particular datasets obtained from real wireless sensor networks. A different approach could consist in designing a multiple classifier system where the classifiers are sequentially activated and each classifier uses only a single feature [33]. From the combination of the first i individual classifiers, corresponding to features $x_1, ..., x_i$, one could derive a confidence value that in turn is used to decide whether the next classifier, which processes features x_{i+1}, should be activated.

Acknowledgements

This research has been supported by the TÁMOP-4.2.2/08/1/2008-0008 program of the Hungarian National Development Agency.

Furthermore, the authors want to thank Peter Bertholet for implementing the algorithms proposed in this paper, conducting the experiments described in Section 5.4, and preparing this manuscript.

References

[1]   Gyon I, Gunn S, Nikravesh M et al (eds) (2006) Feature Extraction. Foundations and Applications. Springer, Heidelberg

[2]   Jolliffe I (1986) Principal Component Analysis. Springer, Heidelberg

[3]   Schoelkopf B, Smola A, Mueller K R (1998) Nonlinear Component Analysis as a Kernel Eigenvalue Problem. Neural Computation, 10: 1299 – 1319

[4]   Pudil P, Novovicova J, Kittler J (1994) Floating Search Methods in Feature Selection. Pattern Recognition Letters, 15(11): 1119 – 1125

[5]   Somol P, Novovicova J, Grim J et al (2008) Dynamic Oscillating Search Algorithm for Feature Selection. Proc 19th Int Conf on Pattern Recognition, pp 1 – 4

[6]   Kira K, Rendell L A (1992) The Feature Selection Problem Traditional Methods and a New Algorithm. In: Proceedings of the 9th National Conference on Artificial Intelligence, pp 129 – 134

[7]   Guyon I, Weston J, Barnhill S et al (2002) Gene Selection for Cancer Classification Using Support Vector Machines. Machine Learning, 46(1 – 3): 389 – 422

[8]   Ferri F, Kadirkamanathan V, Kittler J (1993) Feature Subset Search Using Genetic Algorithms. In Proceedings of the IEE/IEEE Workshop on Natural Algorithms in Signal Processing, vol 740

[9]   Oh I S, Lee J S, Moon B R (2004) Hybrid Genetic Algorithms for Feature Selection. In: IEEE Transactions on Pattern Analysis and Machine Intelligence, pp 1424 – 1437

[10]   Cover T M, Thomas J A (1991) Elements of Information Theory. Wiley, New York

[11]   Siedlecki W, Sklansky J (1988) On Automatic Feature Selection. Int Journal of Pattern Recognition and Art Intelligence, 2: 197 – 220

[12]   Kohavi R, John G (1997) Wrappers for Feature Subset Selection. Artificial Intelligence, 97(1 – 2): 273 – 324

[13]   Guyon I, Elisseeff A (2003) An Introduction to Variable and Feature Selection. Journal of Machine Learning Research, 3: 1157 – 1182

[14]   Liu H, Yu L (2005) Towards Integrating Feature Selection Algorithms for Classification and Clustering. IEEE Trans KDE, 17: 491 – 502

[15]   Zhao Z, Morstatter F, Sharma S et al (2007) Advancing Feature Selection Research: ASU Feature Selection Repository. TR-10-007, School of Computing, Informatics, and Decision Systems Engineering, Arizona State University, Tempe

[16] Xiao Y, Chen H, Li F H (2010) Handbook on Sensor Networks, World Scientific

[17] Berman P, Calinescu G, Shah C, Zelinovsky A (2006) Efficient Energy Management in Sensor Networks. In: Pan Y, Xiao Y (eds) Ad hoc Sensor Networks. Nova Science Publisher

[18] Krause A, Rajagopal R, Gupta A et al (2009) Simultaneous Placement and Scheduling of Sensors. IPSN'09, ACM Press, San Francisco, pp 181 – 192

[19] Wang L, Xiao Y (2006) A Survey of Energy-efficient Scheduling Mechanisms in Sensor Networks. Mobile Networks and Applications, 11: 723 – 740

[20] Sun Y, Qi H (2008) Dynamic Target Classification in Wireless Sensor Networks. Proceedings 19th Int Conf on Pattern Recognition

[21] Chatterjea S, Nieberg T, Meratnia N et al (2008) A Distributed and Self-organizing Scheduling Algorithm for Energy-efficient Data Aggregation in Wireless Sensor Networks. ACM Trans on Sensor Networks, 4(4), Article 20

[22] Tian D, Georganas N D (2002) A Coverage-preserving Node Scheduling Scheme for Large Wireless Sensor Networks. ACM Press WSNA, 32 – 41

[23] Duarte M F, Hu Y H (2004) Vehicle Classification in Distributed Sensor Networks. Parallel Distr. Comput, 64: 826 – 838

[24] Last M, Bunke H, Kandel A (2002) A feature-based serial approach to classifier combination. Pattern Analysis and Applications, 5: 385 – 398

[25] Duda R O, Hart P E, Stork D G (2001) Pattern Classification, 2nd edn. Wiley-Interscience

[26] Chow CK (1970) On optimum error and reject trade-off. IEEE Trans on Information Theory, IT-16(1): 41 – 46

[27] Fumera G, Roli F, Giacinto G (2000) Reject Option with Multiple Thresholds. Pattern Recognition, 33 (12): 2099 – 2101

[28] Hanczar B, Dougherty E R (2008) Classification with Reject Option in Gene Expression Data. Bioinformatics, 24(17): 1889 – 1895

[29] Zappi P, Lombriser C, Farelle E et al (2009) Experiences with Experiments in Ambient Intelligence Environments. IADIS Int Conference Wireless Applications and Computing

[30] Intille S, Larson K, Tapia E M et al (2006) Using a Live-in Laboratory for Ubiquitous Computing Research, Proceedings PERVASIVE 2006, Springer LNCS 3968: 349 – 365

[31] Frank A, Asuncion A (2010) UCI Machine Learning Repository [http://archive.ics.uci.edu/ml]. Irvine, CA: University of California, School of Information and Computer Science

[32] Mitchell T M (1997) Machine Learning. Mc Graw-Hill, New York

[33] Kuncheva L I (2004) Combining Pattern Classifiers: Methods and Algorithms. Wiley, New York

6 Principles and Applications of RIDED-2D — A Robust Edge Detection Method in Range Images

Jian Wang[1,2], Zhenqiang Yao[3], Mingde Yin[4], Lin Mei[5], Yaojie Zhu[6], Quanzhang An[7,8], Fangfang Lu[9]

Abstract In computer vision field, edge detection is often regarded as a basic step in range image processing by virtue of its crucial effect. Due to huge computational costs, majorities of existing edge detection methods cannot satisfy the requirement of efficiency in several industrial or biometric applications. This Chapter proposes a novel Rule-based Instantaneous Denoising and Edge Detection method (RIDED-2D) for preprocessing range images. First of all, a new classification is proposed to categorize silhouettes of 2D scan line into eight types by defining a few new coefficients. Subsequently, several discriminant criteria on large noise denoising and edge detection are stipulated based on qualitative feature analysis on each type. By selecting some scan points as feature point candidates, a practical parameter learning method is provided to train and refine the threshold set. RIDED-2D is implemented with three mode algorithms, fastest of which is an integrated algorithm by merging calculation steps to the most extent. Since all the coefficients are established based on distances among the points or their ratio, RIDED-2D is inherently invariant to translation and rotation transformations. For refining the edge lines, a forbidden region approach is proposed to eliminate interference of the mixed pixels. Furthermore, key performances of RIDED-2D are evaluated detailedly, including computational complexity,

1 State Key Laboratory of Mechanical System and Vibration, Shanghai Jiao Tong University, Shanghai 200240, China.

2 Cyber Physical System R&D Center, The Third Research Institute of Ministry of Public Security, Shanghai 201204, China. E-mail: wjconan@gmail.com.

3 State Key Laboratory of Mechanical System and Vibration, Shanghai Jiao Tong University, Shanghai 200240, China. E-mail: zqyao@sjtu.edu.cn.

4 College of Mechanical and Electrical Engineering, Nanjing University of Aeronautics and Astronautics, Nanjing, Jiangsu Province 210016, China. E-mail: ymd501@sina.com

5 Cyber Physical System R&D Center, The Third Research Institute of Ministry of Public Security, Shanghai 201204, China. E-mail: meilin@mail.trimps.ac.cn.

6 Shanghai Yanfeng Visteon Automotive Trim Systems Co., Ltd. Shanghai 200238, China. E-mail: xiaozhuwzx@126.com.

7 Changzhou EGing Photovoltaic Technology Co., Ltd. Jintan, Jiangsu province 213213, China.

8 Shanghai Key Laboratory of Materials Laser Processing and Modification, Shanghai Jiao Tong University, Shanghai 200240, China. E-mail: aqz771612@yahoo.com.cn.

9 Institute of Image Processing and Pattern Recognition, Shanghai Jiao Tong University, Shanghai 200240, China. E-mail: free-enid@163.com.

time expenditure, accuracy and stability. The results indicate that RIDED-2D can detect edge points accurately from several real range images, in which large noises and systematic noises are involved, and the total processing time is less than 0.1 millisecond on an ordinary PC platform using the integrated algorithm. Comparing with other state-of-the-art edge detection methods qualitatively, RIDED-2D exhibits a prominent advantage on computational efficiency. Thus, the proposed method is qualified for real-time processing in stringent applications. Besides, another contribution of this chapter is introducing CPU clock counting technique to evaluate the performance of the proposed algorithm, consequently, the technique suggests a convenient and objective way to estimate algorithm's time expenditure in other platforms.

6.1 Introduction

Over the years, range images are widely used to obtain useful descriptions of 3D scenes by many computer vision techniques [1, 2]. Containing distance measurements from a selected reference point or plane to surface points of objects within a scene, range images provide precise 3D representations about the scenes and it allows more information to be recovered [3]. In a profile image, edges, which appear discontinuously in depth values, surface normal, or curvatures etc., convey much important information about an object's structure. Therefore, edge detection is often regarded as a basic step in image processing due to its crucial effect on image segmentation [4].

6.1.1 Traditional Template-Based Edge Detection on 2D Intensity Images

Earlier and recent developmentson edge detection methods have been reviewed by Kunt [5] and Basu [6], respectively. Traditional edge detection approaches such as Roberts, Prewitt, Sobel, Marr and Canny operators commonly employ specific templates to extract edges followed by combining with smoothing functions. Limited by orientation and scope, however, traditional template-based edge detection methods mostly do not work when rotation or scaling transformation occurs, and perform poorly when the edges are blurred and noisy. Since then, two or more directional operators are combined to approximate the performance of a rotationally invariant operator.

Recently, more sophisticated operators have been developed to provide a certain degree of immunity to noise, invariant to limited rotation or scaling transformation, and to detect a more accurate location of edges [6−8]. Based on the flexibility of the finite element framework, Coleman et al. [9] formulated scalable derivative edge detectors for 2D intensity images, and then presented an adaptive first-order gradient operator that can automatically modulate shape to accommodate irregular data distribution of range image data without re-sampling[10].

6.1.2 Extension from 2D to 3D Range Image Processing

Progress has been achieved in the extensions of the existing 2D edge detectors into 3D space during recent 20 years. Krishnapuram and Gupta [11] introduced morphological operations to the range image processing, and developed two-edge detection and surface structure classification schemes, i.e., edge operators using opening and closing residues, and derivative operators using erosion and dilation residues. Jiang and Bunke [12] proposed an edge detection algorithm for range images based on a scan line approximation technique. Inspired by Jiang's algorithm, a simple merging step was introduced to overcome the scan line over-segmentation problem, and then a procedure for fine localization of edge points was proposed, enhancing the robustness and accuracy [13]. Bhattacharya et al. [13] developed an approach to detect families of parallel lines in 3D space at a moderate computational cost by using a (2+2)–D Hough space. They first found peaks in 2D slope parameter space as well as the peaks in the intercept parameter space for each of these peaks. For processing anisotropic image data, Brejl and Sonka [7] reported a directional 3D edge detector which is based on interpolating the image intensity function in a small neighborhood of every voxel by a tri-cubic polynomial. Based on a finite element approach, Suganthan et al. [15] developed a shape adaptive 3×3 gradient operator for feature extraction on range image data directly without the need of image pre-processing. Liu et al. [16] proposed a new spatial clustering algorithm TRICLUST based on Delaunay triangulation. This new boundary detection function was valuable for many real world applications such as geo-spatial data processing, point-based computer graphics, etc. Sharifi et al. [17] introduced a new classification of most important and commonly used edge detection algorithms, and discussed the advantages and disadvantages of some available algorithms.

6.1.3 Modern Edge Detection on 3D Range Images

In the past thirty years, range image segmentation techniques have risen to some indirect ways of edge detection. Varieties of clustering and region growing algorithms have been developed to segment the surface to many patches, boundaries of which were classified into different edge types [18]. Hoffman and Jian [19] segmented the range image into "surface patches", which were categorized into planar, convex, or concave based on some planarity tests. Boundaries between adjacent surface patches were classified as crease or non-crease edges by a square error criterion clustering algorithm. By replacing intensity values with depth information, Bellon et al. [20] proposed a clustering algorithm to compute the normal coefficients of each image pixel for edge detection. Proposing edge strength measures that had a straightforward geometric interpretation and supported a new classification of edge points into several subtypes, Jiang and Bunke [12] provided a definition of optimal edge detectors, and presented a robust method for edge detection based on a scan line approximation technique. Subsequently, based on region growing method, they presented a novel algorithm for fast segmentation of range images into both planar and curved surface patches using high-level features (curve segments) as segmentation primitives instead of individual pixels [21]. By using a multi-resolution surface model, Maeda et al. [22] proposed a method for detecting feature points that are the local maxima of Gaussian curvatures on a 3D object. Their method could determine a desirable level by using an absolute resolution to perform the efficient matching in the geometric hashing method. Demarsina et al. [23] applied a modified region growing method with normal estimation to cluster the points in a range image, built and manipulated a graph of these clusters, and then presented an algorithm to extract closed sharp edges from the point cloud without estimating the curvature and without triangulating the point cloud. Contemporaneously, comparison and evaluation of range image segmentation algorithms have been carried out by a large number of experiments [24 – 26]. Furthermore, some methods that are popularly used in other fields have been introduced or adapted to edge detection or image segmentation, e.g., robust trimmed methods [27], Bayesian inference [28], fuzzy logic [29], Boolean derivatives [30], and case-based reasoning [31] etc.

6.1.4 Intention and Outline of the Chapter

Although traditional approaches are improved by a variety of novel ideas, the practical detection methods with low computational cost and high adaptability are still being pursued. In many industrial or biometric occasions, vision systems have to process vast range data in limited time. Usually, a few accurate edge points can bring a satisfactory answer to a vision task, but an algorithm has to output the results in a moment. Efficiency has become into one of vital factors to determine the practicality of an approach. For acquisition of higher efficiency, stability and scalability, our studies focus on denoising and edge detection for 2D scan lines in range images. However, none of the template type is employed due to their weak adaptability to rotation transformation and large noise.

This chapter is organized as follows. Section 6.1 surveys traditional template-based 2D detection operators, traces their extension in 3D space, and reviews range image segmentation techniques as well as some state-of-the-art methods on edge detection in range images. In Section 6.2, a new classification of silhouettes in scan lines is proposed after introducing several novel coefficient definitions, in order to provide mathematical description to geometrical distribution of points in a scan line. On this basis, these typical silhouettes are analyzed qualitatively using the proposed coefficients. In Section 6.3, several key ideas behind the algorithm philosophy are described on denoising and edge detection, and a highly integrated algorithm is discussed in detail. The parameters in the algorithm are determined by a new learning approach using edge point candidates. The troublesome mixed pixel problem is handled by proposing a forbidden region method to quarantine these pixels during edge detection. Section 6.4 presents a set of experiments on five real range images, and evaluates the method deliberately in terms of computational complexity, time expenditure, accuracy and stability. Section 6.5 discusses some related issues including parameters determination, preprocessing and postprocessing. Finally, Section 6.6 draws a few conclusions and suggests potential research in the future.

6.2 Definitions and Analysis

6.2.1 Coefficient Definitions

Some new coefficients are defined as Fig. 6.1. θ_i denotes the angle between +X axis and line $\overline{P_i P_{i+1}}$. Euclidean distance between P_i and P_{i-1} is named as $d_{i,i-1}$. Besides, several derived coefficients give mathematical descriptions on distribution and local features of points in a scan line.

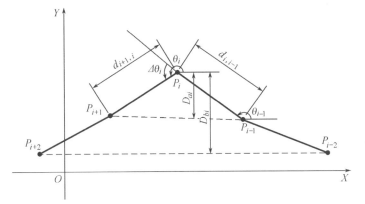

Fig. 6.1 Coefficient definitions.

Swerve Angle

Swerve angle $\Delta\theta_i$ represents the extent that scan line swerves at the point

P_i. P_i will be a candidate for a large noise or an edge point when $\Delta\theta_i$ is large enough.

$$\Delta\theta_i = \theta_i - \theta_{i-1}. \tag{6.1}$$

Point-Point Distance family

At a sharp edge, two scan points appears as an acute step and the distances between contiguous points vary quickly. Such a salient feature can be characterized by combination two of the following coefficients.

1) Unsigned difference of point-point distance (ΔDPP)

ΔDPP indicates absolute extent of distance variation among three contiguous points.

$$\Delta DPP_i = |d_{i+1,i} - d_{i,i-1}|, \tag{6.2}$$

where $d_{i,i+1}$ and $d_{i-1,i}$ are unsigned distances among P_{i-1}, P_i, and P_{i+1} (see Fig. 6.1).

2) Unsigned ratio of point-point distance (RPP)

The relative extent of distance variation among three contiguous points can be represented by an unsigned ratio of point-point distance. Note that RPP is always positive and larger than 1.0.

$$RPP_i = \max\left\{ \left|\frac{d_{i,i+1}}{d_{i-1,i}}\right|, \left|\frac{d_{i-1,i}}{d_{i,i+1}}\right| \right\}. \tag{6.3}$$

Point-chord Distance Family

At a crease edge, a scan point appears as a local peak or valley and the neighboring points depart from it at identical directions without wave or retracement. This feature can be illustrated using position relationship among them. Two coefficients are defined to give a mathematical representation:

1) Signed point-chord distances

As Fig. 6.1 shown, signed D_{ai} is the distance between P_i and $\overline{P_{i-1}P_{i+1}}$, and signed D_{bi} refers to distance between P_i and $\overline{P_{i-2}P_{i+2}}$. Along the line $P_{i-1} \to P_i \to P_{i+1}$, D_{ai} is positive if chord $\overline{P_{i-1}P_{i+1}}$ lies in the left, otherwise negative. The same rule of signs is applicable to D_{bi}. These two coefficients provide good evidence to analyze the local marching direction of points in a scan line.

2) Unsigned ratio of point-chord distance (RPC)

Ratio of point-chord distance (RPC) indicates relative extent that P_i apart from its neighbors.

$$RPC_i = \left|\frac{D_{bi}}{D_{ai}}\right|. \tag{6.4}$$

6.2.2 Distribution Analysis on Typical Silhouettes

Based on the signs of the mean and Gaussian curvature at each point, Besl and Jian classified the surface patches surrounding any point on a smooth surface into eight types: peak, pit, ridge, valley, saddle ridge, saddle valley, flat (planar), and minimal [18, 32]. During scanning a target, a 2D range finder can acquire a transverse silhouette (the so-called scan line) in a scanning period. Inspired by Besl and Jian's method, we classify local silhouettes in a scan line into eight types: smooth line, pulse, peak, pit, ridge, valley, step and jump. These typical silhouettes are illustrated in Fig. 6.2, in which some interesting points (edges, extreme points etc.) are marked with circles. If a qualified discriminant algorithm has enough capability to deal with any rotation and translation transformations, orientation of silhouettes cannot be taken into consideration. In this situation, peak and pit can be merged into one class, same to ridge and valley. Smooth line refers to a straight line or a low-curvature curve without large noise points.

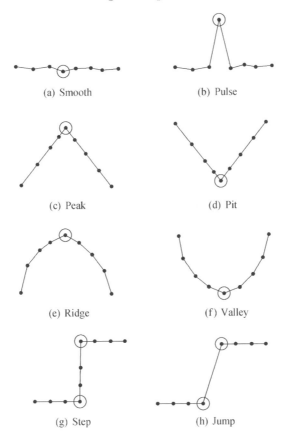

(a) Smooth   (b) Pulse

(c) Peak   (d) Pit

(e) Ridge   (f) Valley

(g) Step   (h) Jump

Fig. 6.2 Basic types of silhouettes in a scan line.

A pulse, which may occurs randomly at any position of the scan line, often is caused by a large noise in the scan line. According to common classification of edges in range images [12, 33], peak, pit, ridge and valley can be regarded as crease edges. Table 6.1 lists the qualitative feature analysis on interesting points in each type of silhouettes using aforementioned coefficients. Detailed discriminant criteria based on this table is involved in each criterion in Section 6.3.

Due to a fixed viewpoint of range finder, these interesting points cannot accurately overlap physical feature points. Nevertheless, a high-resolution range finder is often employed to scan a large-scale target in many industrial applications. These approximate feature points can satisfy the requirement in low accuracy occasions, and it is feasible to refine positions based on these points to obtain a higher accuracy.

6.3 Principles of Instantaneous Denoising and Edge Detection

Based on coefficients definitions and analysis in Section 6.2, a few determinant criteria can be stipulated for a given task. An edge recognition task usually consists of two principal steps: denoising and edge detection.

6.3.1 Large Noise Denoising

As for a scan line acquired by a laser range finder, noises can be classified to stochastic large noises (LNs) and system noises. In Table 6.1, all coefficients of a smooth line are comparatively small in that there are not distinct variances in direction or distance. Low-magnitude system noises only bring tiny waves to the scan line, and result in uncertainty of sign of $D_{ai}D_{bi}$. As long as moderate thresholds are selected, the proposed method is qualified to discriminate features from a scan line slightly polluted by system noise. Here, we only discuss the approach to eliminate the large noise.

Table 6.1 Qualitative feature analysis on each type of silhouettes

Items	Smooth	Pulse	Peak/Pit	Ridge/Valley	Step	Jump
$\Delta\theta_i$	S/M	VL	L	L	L	L
ΔDPP_i	S	S/M	S/M	S/M	L	VL
RPP_i	S	S/M	S/M	S/M	L/VL	VL
D_{ai}	S	VL	M/L	M/L	L	L
D_{bi}	S	VL	L	L	L/VL	L/VL
$\text{sgn}(D_{ai}D_{bi})$	+/-	+	+	+	+	+
RPC_i	S	S/M	L/VL	L	M/L	M/L

Notes:
a) S – Small; M – Moderate; L – Large; VL – Very large.
b) $\text{sgn}(D_{ai}D_{bi})$: Positive if D_{ai} has a same sign with D_{bi}, negative otherwise.

In the case of indoor applications, LNs are often caused by dark spots or specular points on the surface. Snowflakes, raindrops, dust or spots on the window of the scanner can bring large noise outdoors. A LN gives rise to an obvious pulse in a smooth line, as shown in Fig. 6.2 (b). Referring to Table 6.1, a pulse can be distinguished by very large $\Delta\theta_i$ and D_{ai}. In order to eliminate pulses at fastest speed and obtain the least infection to neighbors, a median filter is preferred. Denoising procedure is summarized in a flowchart as Fig. 6.3 shown, where Φ_{noise} is a swerve angle threshold for large noise, and DA_{noise} is a point-chord distance threshold. Computation of $\Delta\theta_i$ and D_{ai} involves two close neighbors only, so the traversal scope is defined as $i = 2, \cdots, n-1$ and the computing window covers 3 contiguous points.

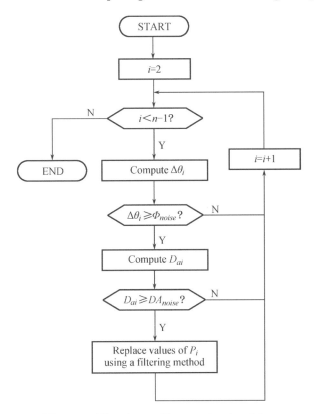

Fig. 6.3 Flowchart of large noise denoising.

6.3.2 Edge Point Detection

Generally, edges in range images are classified into three basic types: jump, crease and smooth. Smooth edges detection has been widely ignored, because they relatively seldom occur in range images [12]. According to feature descriptions in Table 6.1, peak, pit, ridge and valley have similar parameters, while the parameters of step and jump also have a high similarity. Consequently, we can classify these silhouettes into two basic edge subtypes: step edge and crease edge. In fact, the majority of jump edges often degenerate to step edges in range image due to the mixed-pixel effect.

Each edge point (EP) has a big swerve angle but less than that of LN. Therefore, swerve angle $\Delta\theta_i$ is regarded as a common discriminant coefficient.

1) Step edge points (SEPs), e.g., Fig. 6.2 (g) and (h), have large ΔDPP_i and RPP_i.

2) Crease edge points (CEPs), e.g., Fig. 6.2 (c) and (d), have large D_{ai} and RPC_i.

Figure 6.4 illustrates a procedure for detecting these two types of edges. Where Φ_{edge}, ΔDPP_{edge}, RPP_{edge}, DA_{edge} and RPC_{edge} are discriminant thresholds, noting that $\Phi_{edge} < \Phi_{noise}$, and $DA_{edge} < DA_{noise}$. Traversal scope is defined as $i = 3, \ldots, n-2$, because computation of RPC_i involves 4 close neighbors. Therefore, the size of the computing window is expanded to five contiguous points.

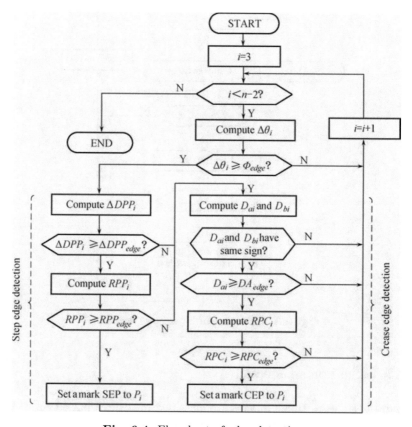

Fig. 6.4 Flowchart of edge detection.

6.3.3 Integrated Algorithm

For obtaining highest efficiency, denoising and edge detection phases are merged into one algorithm. As mentioned above, these two steps utilize computing windows with different sizes, 3 points in denoising and 5 points in edge detection. When merging, the first and the last 2 points in scan line should be excluded from main loop. As compensation, P_2 and P_{n-1} are denoised in advance. P_1 and P_n are never regarded as target points in two phases.

A detailed flowchart illustrates the integrated algorithm as Fig. 6.5. Denoising and edge detection, which are enclosed with dashed panes, are sequential within one loop.

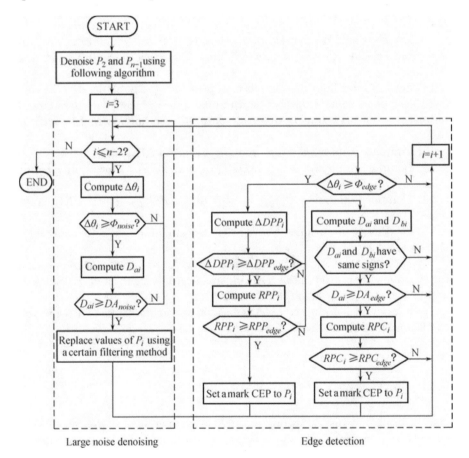

Fig. 6.5 Flowchart of integrated algorithm.

6.3.4   Parameter Determination

Seven discriminant coefficients are used in the proposed method: Φ_{noise}, DA_{noise}, Φ_{edge}, ΔDPP_{edge}, RPP_{edge}, DA_{edge} and RPC_{edge}. In a given task, they are determined by several factors, such as resolution and accuracy of range finder, task requirement, shape and orientation of objects. On the other hand, robustness, accuracy and efficiency of the method are subject to these coefficients. Besl and Jian [32] presented a noise estimation method to set some thresholds dependent on the image noise variance, so that the segmentation algorithm can adapt automatically to the noise conditions of different range images [34]. We do not adopt this method because it will increase computational cost and perform poorly in some complex detection tasks.

A new threshold determination method is proposed which learns parameter from selected target point candidates. The basic ideas are briefed as follows:

1) Several scan lines are chosen to represent typical silhouettes. In each fixed scan line, some target point candidates are picked out manually. A unique attribute is given to each candidate for discrimination: LN, CEP or SEP.

2) Compute minimum Φ_{noise} and D_{noise} for all of LN candidates.

3) Compute minimum Φ_{edge}, ΔDPP_{edge} and RPP_{edge} for all SEP candidates.

4) Compute minimum Φ_{edge}, DA_{edge} and RPC_{edge} for all CEP candidates.

5) Synthesize all results above with certain allowances to generate a threshold set.

6) Perform denoising and edge detection with the parameter set. If some target points are missed in the resulting image, select them as new candidates and return step b) to refine the threshold set.

7) Parameter learning completes as soon as current threshold set is able to discriminate all the feature points.

6.3.5   Forbidden Region of Edge Detection

Due to imperfect distinguishing capability of range finder, the mixed pixel effect, often occurs at the sharp edges, complicates the edge detection and localization in range images [35, 36]. After preliminary denoising, these mixed pixels usually appear smooth in the scanning plane but still sharp in the orthogonal plane. As a result, these pixels often locate on the steep slopes and lead to pseudo edge points (PEPs) during edge detection, see Fig. 6.6 (a). Because the PEPs are mixed with other genuine EPs randomly, it is very difficult to distinguish and remove them from edge map by conventional methods, e.g., morphological operators, and traditional filters. In order to handle the problem of mixed pixels, a forbidden region (FR) approach is proposed.

The surface of an object is surrounded by many EPs, which can be classified into row EPs and column EPs, according to two orthogonal directions. By observing a large number of real-range images, two characters on EP distribution are concluded. First character is that an EP usually has one unique identity, row EP or column EP, except for few multiple intersection points of several lines. The second important character is that a column EP seldom occurs between two neighboring row EPs in a row line, and vice versa. Based on these characters, a column FR can be generated to cover several neighboring points of a row EP. During edge detection in column lines, all points in the column FR bypass the detection procedure. Same rule is used to set up a row FR surrounding a column EP. Figure 6.6(b) illustrated an edge map with row FRs and column FRs. Edge points of row lines and column lines are marked with hollow cube and solid torus, respectively. Small cones denote FRs in columns and rows. It is obviously observed that the FRs with suitable size cover a majority of troublesome mixed pixels and quarantine their disruptive effect. Significant effects on pseudo edge points removing can be found by comparing subgragh (b) with (c) in Figs. 6.8 – 6.12. If a sharp step, on which a few mixed pixels locate, comprises k points including two endpoints, the minimum unilateral width of FR is $(k - 2)/2$ for covering all the mixed pixels. Empirically, the unilateral width of FR is 2 or 3 in our experiments.

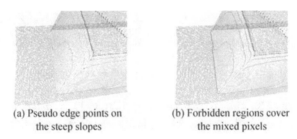

(a) Pseudo edge points on the steep slopes

(b) Forbidden regions cover the mixed pixels

Fig. 6.6 Forbidden regions.

6.4 Experiments and Evaluations

To provide comprehensive verification of applied effects, key performances of RIDED-2D are evaluated detailedly. For the sake of reconstruction actual application environment, a large number of real range images have been acquired through scanning a variety of objects with a TOF range finder. All the aforementioned algorithms are implemented in our self-developed software with embedded testing modules.

6.4.1 Experiments

We set up a vision system and develop the processing software named LMRS —
RB[10] (Lidar 3D Measuring and Recognizing System — Rules Brain) and
LMAS[11] (Lidar Measuring and Analyzing System). A SICK LMS-291 laser
range finder is hung on a spreading girder, facing to the ground (see Fig. 6.7).
Its technical parameters in the experiments are provided:

1) Measurement range: 8 m.
2) Measurement resolution: 10 mm.
3) System error: typ. ± 35 mm.
4) Statistical error standard deviation(1 sigma): typ. 10 mm (mm-mode).
5) Angular range: 100°.
6) Angular resolution: 0.25°.

Fig. 6.7 Range finder platform.

For the sake of noise reduction, each pixel in our experimental images is an
average of five scans of the range finder. By this approach, the system noises is
decreased from ±35 mm to ±20 mm approximately. In order to produce large
noise points, two small mirrorses are placed on the objects, because the range
finder cannot receive the reflective pulses from the mirrors. Using the trained
threshold sets listed in Table 6.2, five of over 50 real range images are tested
as the following set of images (Figs. 6.8 – 6.12) shown. For the convenience of
illustration and evaluation, each image has been clipped to a uniform size of
200 x 100 pixels.

---

10   LMRS v3.02 was registered software copyright of China with No. 2010SR002362 on
January 14, 2010.
11   LMAS v2.09 was registered software copyright of China with No. 2009SR039116 on
September 14, 2009.

Table 6.2 Threshold sets in experiments

Item	Unit	Example 1	Example 2	Example 3	Example 4	Example 5
Φ_{noise}	deg	120.0000	120.0000	120.0000	120.0000	120.0000
D_{noise}	mm	200.0000	200.0000	200.0000	200.0000	200.0000
Φ_{edge}	deg	16.0483	24.9562	23.9862	16.4279	27.9167
ΔDPP_{edge}	mm	10.6582	9.8641	10.0045	10.4951	27.8153
RPP_{edge}		1.5325	1.5024	1.4272	1.4752	3.3489
DA_{edge}	mm	8.8524	8.0248	5.9557	12.1254	5.5234
RPC_{edge}		1.2185	1.1962	1.4875	1.5008	1.1018

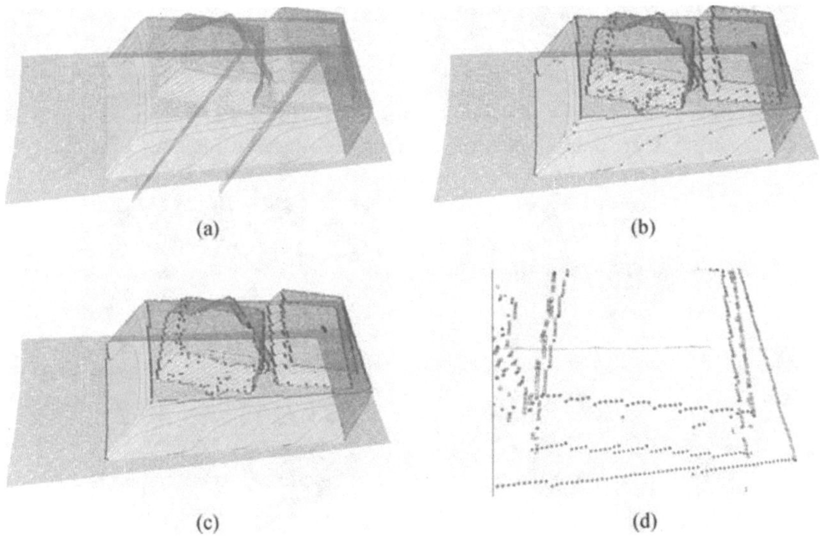

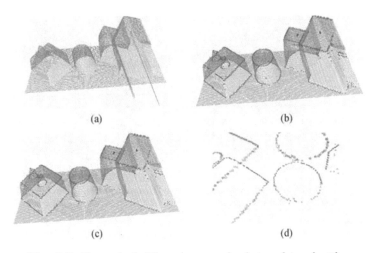

(a)

(b)

(c)

(d)

Fig. 6.8 Example 1: Three flat boxes on a desk.

(a)

(b)

(c)

(d)

Fig. 6.9 Example 2: Three boxes, a bucket and two bottles.

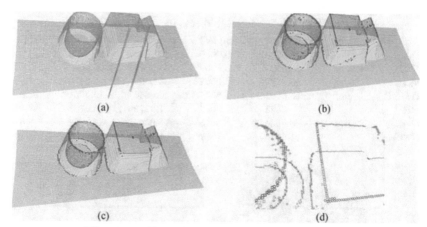

Fig. 6.10 Example 3: Two boxes and a bucket.

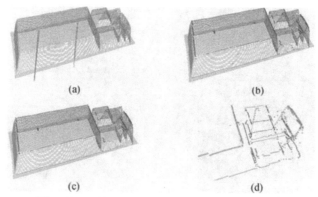

Fig. 6.11 Example 4: Two chairs and a desk.

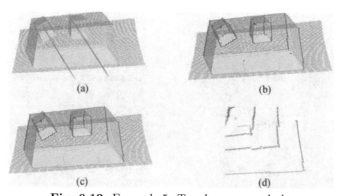

Fig. 6.12 Example 5: Two boxes on a desk.
In legends of Figs: 6.8 – 6.12: (a) Original range image (upper left). (b) Processed image when FR is disabled (upper right). (c) Processed image when FR is enabled (lower left). (d) Refined edge map (lower right).

6.4.2 Evaluations

Since accuracy, efficiency and stability are of importance in applications, and all the key performances of RIDED-2D are evaluated detailedly, including computational complexity, time expenditure, accuracy and stability. In addition, we also introduce CPU clock counting technique, which suggests a convenient and objective way to estimate algorithm's time expenditure in other platforms.

Computational Complexity

There are three main phases in the method: attributes calculation, denoising and edge detection. Generating an independent attribute array, which is used to store point attributes, has computational complexity of $3O(n)$. It comprises following sequential tiny steps, three of which, a) – c), are indispensable loops:

1) Compute $d_{i-1,i}$.
2) Compute $\Delta\theta_i$ and D_{ai}.
3) Compute D_{bi}.
4) Other parameters are computed dynamically in edge detection phase if needed.

An independent denoising phase only compares $\Delta\theta_i$ and D_{ai} with given thresholds, while without computing any parameters. If a point is distinguished as a large noise, the median filtering process will recompute attributes of the point and its four close neighbors. Since the number of large noise is often comparatively rare, computational complexity (CC) of denoising phase can be regarded as $O(n)$.

Comparing $\Delta\theta_i$ with Φ_{edge} is the first step for an independent edge detection phase, and computing ΔDPP_{edge}, RPP_{edge}, DA_{edge} and RPC_{edge}.is an optional successive step. Therefore, its CC is $O(n)$ at best and $O(3n)$ at worst.

A normal mode algorithm can be obtained by concatenating aforementioned three independent phases without any integration. Further, a fast mode algorithm is achieved when merging the attribute array calculation into other phases and maintaining independence of denoising and edge detection. With best performance, an integrated mode algorithm, which has least redundant calculation, comes into being as attribute computation and denoising which are completely merged into edge detection.

Experimentally, our algorithm involves two optional phases, which are excluded from computational complexity calculation due to their unnecessariness for RIDED-2D method. A smoothing procedure follows three main phases using median filter that introduces an additional CC of $O(n)$. Commonly, the CC of FR generation is regarded as $O(n)$, but it can be omitted because FRs prevent a large number of points in them from high-intensity computation of edge detection on a certain direction. Actually, testing data indicate that FRs can promote computational speed of the whole algorithm more than 10%.

Time Expenditure

Our testing platform has Pentium IV 2.66 GHz CPU and 1 GB DDRII 667 MHz memory. Firstly, a single scan line is tested and the results are presented in Fig. 6.13 (a). The scan line contains 401 points, including 5 step edge points, 2 large noise points that are marked with ellipses in Fig. 6.13 and 10 crease edge points. It also can be known from Fig. 6.13, that region A and C are intersected corners between ground and walls, meanwhile region B contains main targets — two boxes and a desk. As for region B, the original, edge detection, and edge detection with smoothing are illustrated in (b), (c) and (d) of Fig. 6.13 respectively. Rectangles and triangles are employed to severally mark step edge points and the large noise points (detected or smoothed) in Fig. 6.13(c) and (d). Table 6.3 lists performances of the proposed algorithm in three modes by 20 independent measures. It is worthy of noting that the time of points attribute calculation has been merged into other phases, and FR generation is included in computational time statistics. It is evidently observed that, compared with normal mode and fast mode, the integrated mode has a tremendous promotion ratio of executing speed

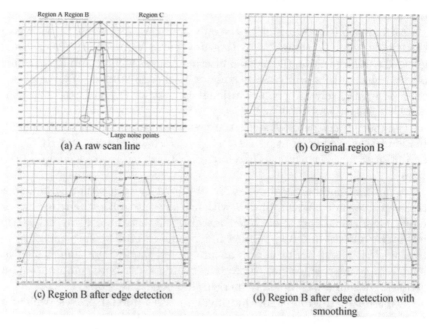

(a) A raw scan line

(b) Original region B

(c) Region B after edge detection

(d) Region B after edge detection with smoothing

Fig. 6.13 Algorithm test on a 2D scan line.

at 18.30 and 4.46, respectively. In this work, clock periods of CPU (*CP*) are computed using "Online Assembly" technique and CPU occupancy rate (*COR*) is estimated according to the monitoring graph of "Process Explorer". Based on net clock periods (CP_n) and *COR*, it is convenient and objective to estimate algorithm's time expenditure in other platforms. Table 6.4 tabulates results of performance evaluation on these images. Occasionally, a point has a "double identity" when it is classified as an edge point both in row and column scan line.

Table 6.3   Computational performance of three modes on a single scan line

Item	Unit	Normal mode	Fast mode	Integrated mode
CC_{best}		5O(n)	O(2n)+2O(n)	O(2n)+O(n)
CC_{worst}		4O(n) +O(3n)	O(2n)+O(n)+O(3n)	O(2n)+O(3n)
$T_{denoise}$	ms	4.684869	0.147672	–
$T_{detection}$	ms	6.8734014	2.738030	0.590401
T_{sl}	ms	11.558271	2.885702	0.590401
CP		31327845	7936406	1712203
COR		14.93%	14.93%	13.43%
T_{sl_net}	ms	1.725650	0.430835	0.079291
CP_n		4677247	1184905	229949

Notes:
  1) CC_{best} — computational complexity at best; CC_{worst} — computational complexity at worst; $T_{denoise}$ — time of denoising; $T_{detection}$ — time of edge detection; T_{sl} — time of total computation on a single scan line; CP — clock periods; COR — CPU occupancy rate; T_{sl_net} — net time of total computation on a single scan line; CP_n — net clock periods.
  2) $T_{denoise}$, $T_{detection}$, T_{sl}, and CP are arithmetic averages of 20 measures.
  3) Net clock periods are calculated by $CP_n = CP \times COR$, net time of total computation by $T_{sl\_net} = T_{sl} \times COR$.

Table 6.4   Computational performance of test images

Item	Unit	Example 1	Example 2	Example 3	Example 4	Example 5
N_{se}		1597	922	825	1642	1204
N_{ce}		11	60	132	56	70
N_{ln}		9	10	12	9	8
N_{me}		22	18	19	12	15
N_{ml}		0	0	0	0	0
A_{ed}	%	98.64%	98.09%	97.75%	99.27%	98.77%
A_{ln}	%	100.00%	100.00%	100.00%	100.00%	100.00%
T_{row}	ms	165.820577	68.823962	53.456499	78.812578	87.857111
T_{col}	ms	172.902259	62.194946	62.940142	84.649897	99.498611
T_{total}	ms	338.722836	131.018908	116.396640	163.462475	187.355723
CP_{total}		985501445	381894417	37837537	471540032	542845798
COR_{avg}	%	16.92%	22.49%	21.56%	21.47%	24.29%
T_{total_net}	ms	57.311904	29.466152	25.095116	35.095393	45.508705
CP_{n_total}		166746845	85888054	72837773	101239645	131857244

Notes:
  1) N_{se} — amount of step edge points; N_{ce} — amount of crease edge points; N_{ln} — amount of large noise points; N_{me} — amount of missed edge points; N_{ml} — amount of missed large noise points; A_{ed} — accuracy ratio of edge detection; A_{ln} — accuracy ratio of large noise detection; T_{row} — time of row scan line processing; T_{col} — time of column scan line processing; T_{total} — time sum of T_{row} and T_{col}; CP_{total} — total CPU periods, i.e., sum of CP_{row} and CP_{col}; COR_{avg} — COR average during row and column processing; CP_{n_total} — total net clock periods.
  2) T_{row}, T_{col}, T_{total} and CP_{total} are arithmetic averages of 20 measures. All the operations irrelevant to computation, e.g., update of view and message procedure, have been excluded.
  3) Computation is delimited in a clip region involves 200×100 pixels in each image.

Accuracy

By training and testing for several times, a good result can be usually obtained. Occasionally, there are some ambiguous points that are difficult to be distinguished, e.g., two adjacent points locating at a step edge or a corner edge with smoothing transition. As an example, an enlargement of region C in Fig. 6.12 (a) is plotted in Fig. 6.14, which illustrates a smooth corner. Detection results often include non-point or more than two points, if inexact parameters are employed. Actually, only one physical edge point exists at the corner. This ambiguous problem can be solved by selecting a best edge point as a crease edge point candidate accompanying with the parameters refinery step. In this work, more than five row scan lines are chosen to train parameters in each range image, and then a same procedure is carried out with several column scan lines. In order to obtain a high accuracy, moderately tight parameters are selected to perform denoising and edge detection. It is evidently observed from Table 6.4 that the mean recognition accuracy is higher than 99% in our experiments.

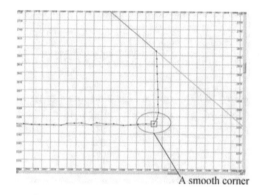

A smooth corner

Fig. 6.14 A smooth corner edge in region C of Fig. 6.13 (a).

Stability

According to analysis in Section 6.3.1, a system noise point probably introduces a large swerve angle $(\Delta\theta_i)$, but only appears tiny waves with small magnitude relatively. After selecting a suitable parameter set, it is easy to distinguish these system noises from large noise and edge points. Figure 6.13 (d) displays the same portion as Fig. 6.13 (c), but its range data represents one scan of range finder without average filtering preprocessing. Observing subgraph (c) and (d) of Fig. 6.13, it is apparent that slight stochastic system noises have no evident influence on the proposed method.

All the coefficients in the proposed method are established based on the distances between the points or their ratio rather than absolute coordinates of points. Therefore, the method is inherently invariant to translation and rotation transformation. Moreover, it is practical to design self-adapting algorithms invariant to scaling transformation by adjusting parameters according to dimensional range of scene and objects.

6.4.3 Comparison with other approaches

Since most of details of other approaches are unavailable, the comparison is limited to the data acquired from the literature. Furthermore, the majority of papers did not report the detailed evaluation data on amounts of points or edges, computing time, etc. The results are tabulated in Table 6.5 to conduct qualitative comparison with few other methods in terms of principles, main steps, major operations, as well as computational complexity estimation. It can be evidently observed that, the proposed method has a prominent advantage in computational efficiency, because it consists of a little primary arithmetic on discriminant criteria with a very low computational complexity.

6.5 Discussions and Applications

For simplicity, Section 6.3 only demonstrates the basic principles and primary algorithm of RIDED-2D neglecting some practical issues. In this section, we will discuss a few important improvements, such as operation window expansion, parameters determination, preprocessing and postprocessing as well. According to the mechanism of RIDED-2D, all the coefficients are established based on distances among the points or their ratio. Consequently, RIDED-2D is inherently invariant to translation and rotation transformations, and is easy to be expanded to process a variety of range images and intensity images after intensity-to-depth mapping. Some tips on biometrics application are briefly exemplified in Section 6.5.4.

6.5.1 Operation Window Expansion

In aforementioned sections, operation windows are preset with their minimal size in order to simplify the basic principles of our method: 3 points for large noise denoising and 5 points for edge detection. Sometimes, the size of windows has to be expanded to meet a given task, because a constant size cannot provide enough robustness and adaptability in special industrial applications.

Occasionally, in-real range images, a few LNs maybe conglomerate when big blind spots are scanned. Figure 6.15 exemplifies a range image in which each blind spot consists of more than 3 large noises, which are highlighted with red rectangles. A 7-point window is employed to detect these LNs successfully with a natural extension from the rule mentioned in Section 6.3.1. Specifically, if $\Delta\theta_i$ and D_{ai} cannot meet the LN denoising rule in the first round, P_{i-2}, P_i and P_{i+2} are concerned as a new 3-tuple then starts the second round. Once the new $\Delta\theta_{i,b}$ and D_{bi} still cannot dissatisfy the rule,

Table 6.5 Qualitative comparison on different edge detection algorithms

Item	Proposed (Integrated)	Ref. [21]	Ref. [11] (Edge operators)	Ref. [13]
Principle	Discriminant criteria	Scan line grouping; Region growing; Quadrics approximation.	Morphological residue analysis	Scan line approximation
Main steps	a) Large noises filtering; b) Edge points detection	a) Extraction of curve segments; b) Neighboring relationships determination; c) Region growing.	a) Positive roof edges detection; b) Negative roof edges detection; c) Crease edges detection; d) Jump edges detection.	a) Quadratic polynomials approximation; b) Scan lines splitting; c) Segments merging; d) Noisy segments discarding; e) Edge points recovering; f) Edge strength calculation.
Major operations	a) Basic arithmetic b) Square root	a) Integral; b) Inverse trigonometric function; c) Quadrics approximation.	a) Opening operators; b) Closing operators; c) Blur minimum operator.	a) Quadrics approximation; b) Inverse trigonometric function; c) Square root
Control variables	7 thresholds	8 thresholds	10 or more operators	6 thresholds
Robustness	Very good	Very good	Moderate	Moderate
CC_{avg}	$2O(2n)$	$3O(n) + O(2n)$	$12O(3n)$	$6O(n)$

Notes:
1) All the steps unrelated to edge detection in each method has been excluded, e.g., quadric surface approximation.
2) CC_{avg} – average computational complexity;
3) CC_{avg} estimation of ref. [21]: $O(n)$ in extraction of curve segments, $O(n)$ in neighboring relationships determination between the curve segments, and $O(n) + O(2n)$ in region growing.
4) CC_{avg} estimation of ref. [11]: $2O(3n)$ in positive roof edges detection, $2O(3n)$ in negative roof edges detection, $4O(3n)$ in crease edges detection, and $4O(3n)$ in jump edges detection.
5) CC_{avg} estimation of ref. [13]: the average computational complexity in each one of six steps is $O(n)$.

another 3-tuple consists of P_{i-3}, P_i and P_{i+3} will be selected in the third round. A 7-point window covers 7 consecutive points, so three rounds will be executed at most. It is worthy of noting that the upper limit of the operation window size has a strong constrain, because the LN denoising rule within a large LN operation window will lead to fatal conflicts with the EP detection rule or miss some small surface features, e.g., an erect rod with a small top will miss being detected.

Similarly, an EP detection window can be expanded in order to enhance the reliability of EPs or improve the detection performance on special local features. However, size and discriminant criteria of a window expansion are largely dependent on some practical conditions in a given task.

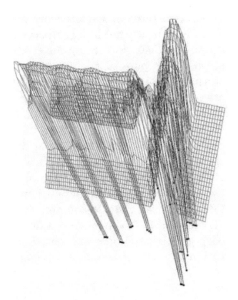

Fig. 6.15 Large noise detection with a 7-point window.

6.5.2 Parameters Determination

Seven coefficients are employed to characterize the features of large noise and edge points in this work. Investigating from the procedure of our algorithm, it is apparent that algorithm's efficiency is influenced most significantly by swerve angle thresholds (Φ_{noise} and Φ_{edge}), which look like two gates of the internal computation in the flowchart. A multitude of feature points will be missed detection when these thresholds are too high. On the contrary, a large number of points will have chances to enter internal computation resulting in huge computational costs if the thresholds are very low. Threshold selection is always a dilemma on account of the complexity and uncertain factors of a vision task. In order to achieve equilibrium between efficiency and accuracy, a compromise approach is to select several scan lines to perform parameter learning; while these scan lines include a few salient features. In consideration of adaptability, each trained parameter should be multiplied by a margin factor whose empirical value is assumed 0.90 in our experiments.

6.5.3 Preprocessing and Postprocessing

Prior to detection, many researchers often execute a filtering process, e.g., median filter and Gaussian filter on raw data to reduce stochastic noise. It is known that these classic filters are very difficult to detect subtle features from smooth surfaces, and hardly reserve local details of objects, e.g., sharp edges or crease edges. Regarding an average of two or more scan lines acquired at the same pose of the range finder as the raw datum, the influence of stochastic noises can be reduced apparently; meanwhile, the local features are maintained well.

Observation on Figs. 6.8–6.12 indicates the edge point maps are capable of illustrating each visible edge of objects. Subsequently, some refinement procedures have to be conducted to filter out thin and clear edges. For instance, a few isolated edge points are removed and the final edges are refined to fit the physical edges of the objects by the morphological methods, e.g., combination of dilation and erosion operators. However, as the main intention of this work focuses on detecting edges and denoising large noises from 2D scan line in 3D space with a high performance, refinement and linkage on edge points based on graph-based approach [4] are intended to be discussed within following research of this work.

6.5.4 Potential Applications

Recent years, synthetical pattern recognition on human physiological or behavioral characteristics has become one of most attractive research issues in biometric field, owing to its wide range of applications in commerce and law enforcement [37]. In research on facial expression analysis and face synthesis, one of recent trends is capturing subtle details such as wrinkles, creases and minor imperfections [38]. High-precision recognition and large-volume analysis have given rise to exigent requirement of high-efficiency processing method. In past decade, some modern approaches have been developed to solve this problem, e.g., non-tensor product wavelets [39, 40], geometry-based methods [38], photometric-stereo-based method [41], etc. On the other hand, for avoidance of huge computational costs, several dimensionality reduction techniques, e.g., PCA [42], LDA [14], QFDA [22] and VDE [37], have been popularly used by transforming the high dimensional data into meaningful low dimensional representation. However, it is still difficult to convert task-oriented or object-related prior-knowledge to definitive criteria with comprehensible forms.

Although the range finder in our experiments is a typical industrial sensor, the proposed method in this paper can be adopted in a variety of applications, such as geodesy, photogrammetry, biometrics, astronautics etc. Due to its robustness to noise and invariance to translation and rotation transformations, RIDED provides a novel approach to process the original range images and intensity images (they can be convert to depth images by intensity-to-depth mapping) and then output edge maps instantaneously. For instance, local features in facial range images are still satisfied with traditional classification [18], though their high-level features have apparent differences from those of industrial range images. Therefore, RIDED can be applied to edge detection and noise removal on facial images with minor changes. Having obvious low brightness contrast to surrounding skin, the wrinkles can be treated as silhouette features by brightness-to-depth mapping. The thin or wide wrinkles correspond to large noises or step edges, respectively. After the similar mapping procedure, the small imperfections can be detected and smoothed as large noises in range images. Naturally, facial creases would be recognized by the crease edges detection method given as Section 6.3.2. Of course, analysis on subtle details is strongly dependent on precision of the original images, so that the facial range images must be acquired by one or more high-resolution scanners or their combination.

6.6 Conclusions and Prospects

In this chapter, a novel rule-based method named RIDED-2D, in which none of templates is required unlike most previous methods, is proposed for denoising large noise and edge detection for 2D scan line in range image. In our method, silhouettes of 2D scan line are classified into eight types according to definitions of a few new coefficients. Sequentially, several discriminant criteria on large noise and edge detection are stipulated based on qualitative feature analysis on each type. Selecting some feature point candidates to train the threshold set, a practical parameters determination approach is provided. Then, an integrated algorithm, in which calculation steps are merged to compress computational complexity to the utmost extent, is implemented for the RIDED-2D. The proposed method is inherently invariant to translation or rotation transformation, since all the coefficients are established on the basis of distances between the points or their ratio. Furthermore, a forbidden region approach is proposed to quarantine the mixed pixels in edge detection procedure to obtain a near-perfect edge map. Detailed evaluations on computational complexity, time expenditure, detection accuracy and running stability are performed in our experiments on real range images. The results indicate that the proposed integrated algorithm can process a 2D scan line within 0.1 milliseconds on an ordinary PC. Therefore, it is qualified to real-time processing in stringent industrial occasion.

Although many problems are well-solved, some open questions can be addressed in the future:

1) How to promote accuracy of the proposed method? Generally, the edge points detected by RIDED method are not the objects' physical edge points. The virtual edge points, intersection points of two neighboring fitting curve, will be introduced to approximate actual edge points. Moreover, in order to obtain thin and clear edges without pseudo edge points in the processed edge maps, a hierarchical threshold framework can be adopted in different stages, e.g., edge detection, edge linkage, as well as edge refinement. The recognition accuracy of the multiscale discriminant criteria should be investigated based on combined classifier evaluation method [43].

2) The RIDED method is only applied to 2D scan line processing in this chapter. However, application of the relationship among scan lines to process more features in 3D space with high efficiency needs to be explored. It is hypothesized to be a potential solution that multidimensional coefficients and criteria are employed in the procedure.

3) The proposed algorithm can process a scan line within 0.1 millisecond. However, it is doubtable to process the hundreds of scan lines simultaneously within a few seconds to satisfy the stringent requirements in industrial applications. It seems to be a promising solution to overcome the problem by introducing parallel computing method to edge detection. For example, one process for a scan line might be divided into many tiny parallel threads; meanwhile, a large amount of scan lines can be processed simultaneously in a parallel algorithm.

Acknowledgements

Our research was supported by following projects:
- Major program of Science and Technology Commission of Shanghai Municipality (No. 09DZ1122202);
- National Natural Science Foundation projects (No. 51075271, No. 50821003, No. 50805094);
- National Basic Research Program of China (No. 2006B705400);
- National Science and Technology Support Projects of China (No. 2006BAH02A17);
- Science and Technology Commission of Shanghai Municipality (No. 06DZ11310).

The authors would like to thank Mr. Zhengsong Xu for helping us to design and install the range finder platform, and are grateful to some colleagues at our laboratory, Dr. Yun Xie and Dr. Kangmei Li, for providing a lot of advice.

References

[1]   Marshall D, Lukas G, Martin R (2001) Robust Segmentation of Primitives from Range Data in the Presence of Geometric Degeneracy. IEEE Transaction on Pattern Analysis and Machine Intelligence, 23(2): 304 – 314

[2]   Chen C H, Wang P S P (2005) Handbook of Pattern Recognition and Computer Vision. Third Edition EDN. World Scientific, Singapore

[3]   Bellon O R P, Silva L (2002) New Improvements to Range Image Segmentation by Edge Detection. IEEE Signal Processing Letter, 9(2): 43 – 45

[4]   Sappa A D (2006) Unsupervised Contour Closure Algorithm for Range Image Edge-Based Segmentation. IEEE Transaction on Image Processing, 15(2): 377 – 384

[5]   Kunt M (1982) Edge Detection: A Tuttorial Review. In: the IEEE International Conference on Acoustics, Speech, and Signal Processing, Palais des Comgrès, Paris, France, IEEE, pp 1172 – 1175

[6]   Basu M (2002) Gaussian-Based Edge-Detection Methods — A Survey. IEEE Transactions on Systems, Man, and Cybernetics, Part C: Applications and Reviews, 32(3): 252 – 260

[7]   Brejl M, Sonka M (2000) Directional 3D Edge Detection in Anisotropic Data: Detector Design and Performance Assessment. Computer Vision and Image Understanding, 77(2): 84 – 110

[8]   He Z, You X, Tang Y Y et al (2006) Texture Image Retrieval Using Novel Non-Separable Filter Banks Based on Centrally Symmetric Matrices. In: the 18th International Conference on Pattern Recognition, Hong Kong, China, pp 161 – 164

[9]   Coleman S A, Scotney B W, Herron M G (2005) Content-Adaptive Feature Extraction Using Image Variance. Pattern Recognition, 38(12): 2426 – 2436

[10] Coleman S A, Suganthan S, Scotney B W (2010) Gradient Operators for Feature Extraction and Characterisation in Range Images. Pattern Recognition Letters, 31(9): 1028 – 1040

[11] Krishnapuram R, Gupta S (1992) Edge Detection in Range Images Through Morphological Residue Analysis. In: the 1992 IEEE Computer Society Conference on Computer Vision and Pattern Recognition, Champaign, IL, USA, IEEE Computer Society, pp 630 – 632

[12] Jiang X, Bunke H (1999) Edge Detection in Range Images Based on Scan Line Approximation. Computer Vision and Image Understanding, 73(2): 183 – 199

[13] Katsoulas D, Werber A (2004) Edge Detection in Range Images of Piled Box-Like Objects. In: the 17th International Conference on Pattern Recognition, Cambridge, UK, IEEE Computer Society, pp 80 – 84

[14] Bhattacharya P, Liu H, Rosenfeld A et al (2000) Hough-Transform Detection of Lines in 3D Space. Pattern Recognition Letters, 21(9): 843 – 849

[15] Suganthan S, Coleman S, Scotney B (2007) Range Image Feature Extraction with Varying Degrees of Data Irregularity. In: the 2007 International Machine Vision and Image Processing Conference, Kildare, Ireland, pp 33 – 40

[16] Liu D, Nosovskiy G V, Sourina O (2008) Effective Clustering and Boundary Detection Algorithm Based on Delaunay triangulation. Pattem Recognition Letters, 29(9): 1261 – 1273

[17] Sharifi M, Fathy M, Mahmoudi M T (2002) A Classified and Comparative Study of Edge Aetection Algorithms. In: the International Conference on Information Technology: Coding and Computing, Las Vegas, Nevada, USA, IEEE Computer Society, pp 117 – 120

[18] Besl P J, Jain R C (1986) Invariant Surface Characteristics for 3D Object Recognition in Range Images. Computer Vision, Graphics, and Image Processing, 33(1): 33 – 80

[19] Hoffman R, Jian A K (1987) Segmentation and Classification of Range Images. IEEE Transaction on Pattern Analysis and Machine Intelligence, PAMI – 9(5): 608 – 620

[20] Bellon ORP, Direne AI, Silva L (1999) Edge Detection to Guide Range Image Segmentation by Clustering Techniques. In: the 1999 International Conference on Image Processing, Kobe, Japan, IEEE, pp 725 – 729

[21] Jiang X, Bunke H, Meier U (2000) High-Level Feature-Based Range Image Segmentation. Image and Vision Computing, 18(10): 817 – 822

[22] Maeda M, Kumamaru K, Inoue K (2006) Detection of 3D Feature Points by Using Multiresolution Surface model. In: The 2006 SICE-ICASE International Joint Conference, Busan, South Korean, pp 2564 – 2569

[23] Demarsin K, Vanderstraeten D, Volodine T et al (2007) Detection of Closed Sharp Edges in Point Clouds Using Normal Estimation and Graph Theory. Computer-Aided Design, 39(4): 276 – 283

[24] Hoover A, Jean-Baptiste G, Jiang X et al (1996) An Experimental Comparison of Range Image Segmentation Algorithms. IEEE Transaction on Pattern Analysis and Machine Intelligence, 18(7): 673 – 688

[25] Jiang X, Bowyer K W, Morioka Y et al (2000) Some Further Results of Experimental Comparison of Range Image Segmentation Algorithms. In: the 15th International Conference on Pattern Recognition, Barcelona, Spain, pp 877 – 881

[26] Min J, Powell M, Bowyer K W (2004) Automated Performance Evaluation of Range Image Segmentation Algorithms. IEEE Transaction on System, Man, and Cybernetics — Part B: Cybernetics, 34(1): 263 – 271

[27] Chang I S, Park R-H (2001) Segmentation Based on Fusion of Range and Intensity Images Using Robust Trimmed Methods. Pattern Recognition, 34(10): 1951 – 1962

[28] Boccignone G, Napoletano P, Ferraro M (2008) Embedding Diffusion in Variational Bayes: A Technique for Segmenting Images. International Journal of Pattern Recognition and Artificial Intelligence, 22(5): 811 – 827

[29] Melin P, Mendoza O, Castillo O (2010) An Improved Method for Edge Detection Based on Interval Type-2 Fuzzy Logic. Expert Systems with Applications, 37(12): 8527 – 8535

[30] Agaian S S, Panetta K A, Nercessian S C et al (2010) Boolean Derivatives With Application to Edge Detection for Imaging Systems. IEEE Transactions on Systems, Man, and Cybernetics, Part B: Cybernetics, 40(2): 371 – 382

[31] Frucci M, Perner P, Baja G S D (2008) Case-Based-Reasoning for Image Segmentation. International Journal of Pattern Recognition and Artificial Intelligence, 22(5): 829 – 842

[32] Besl P J, Jian R C (1998) Segmentation Through Variable-order Surface Fitting. IEEE Transaction on Pattern Analysis and Machine Intelligence, 10(2): 167 – 192

[33] Sappa A D, Devy M (2001) Fast Range Image Segmentation by an Edge Detection Strategy. In: the Third International Conference on 3D Digital Imaging and Modeling, Quebec City, Que, Canada, pp 292 – 299

[34] Jiang X, Bunke H (1994) Fast Segmentation of Range Images into Planar Regions by Scan Line Grouping. Machine Vision and Applications, 7(2): 115 – 122

[35] Tang P, Huber D, Akinci B (2007) A Comparative Analysis of Depth-Discontinuity and Mixed-Pixel Detection Algorithms. In: the Sixth International Conference on 3D Digital Imaging and Modeling, Montreal, Quebec, Canada, IEEE Computer Society, pp 29 – 38

[36] Godbaz J P, Cree M J, Dorrington A A (2008) Mixed Pixel Return Separation for a Full-Field Ranger. In: the 23rd International Conference on Image and Vision Computing, Christchurch, New Zealand, IEEE, pp 1 – 6

[37] Chen X, Zhang J (2010) Maximum Variance Difference-Based Embedding Approach for Facial Feature Extraction. International Journal of Pattern Recognition and Artificial Intelligence, 24(7): 1047 – 1060

[38] Luo Y, Gavrilova M L, Wang P S P (2008) Facial Metamorphosis Using Geometrical Methods for Biometric Applications. International Journal of Pattern Recognition and Artificial Intelligence, 22(3): 555 – 584

[39] You X, Zhang D, Chen Q et al (2006) Face Representation by Using Non-Tensor Product Wavelets. In: the 18th International Conference on Pattern Recognition, Hong Kong, China, pp 503 – 506

[40] You X, Chen Q, Wang P et al (2007) Nontensor-Product-Wavelet-Based Facial Feature Representation. In: Yanushkevich SN, Wang PSP, Gavrilova ML, Sargur N S (eds) Image Pattern Recognition: Synthesis and Analysis in Biometrics. Machine Perception and Artificial Intelligence, vol 67. World Scientific Publishing, Singapore, pp 207 – 224

[41] Lee S W, Wang P S P, Yanushkevich S N et al (2008) Noniterative 3D Face Reconstruction Based on Photometric Stereo. International Journal of Pattern Recognition and Artificial Intelligence, 22(3): 389 – 410

[42] Shih F Y, Cheng S, Chuang C F et al (2008) Extracting Faces and Facial Features from Color Images. International Journal of Pattern Recognition and Artificial Intelligence, 22(3): 515 – 534

[43] Feng X, Ding X, Wu Y et al (2008) Classifier Combination and Its Application in Iris Recognition. International Journal of Pattern Recognition and Artificial Intelligence, 22(3): 617 – 638

Part II: Computer Vision and Image Processing

7 Lens Shading Correction for Dirt Detection

Chih-Wei Chen[1] and Chiou-Shann Fuh[1]

Abstract   We present a novel inspection framework to detect dirt and blemish in production line of optical fabrication automatically. To detect dirt defect from vignette surface, we use B-Spline to generate an ideal vignette surface and subtract the contaminated surface from the generated vignette surface to locate the defect. Besides, we apply some image processing methods to optimize our inspection framework. Experimental results show the effectiveness and robustness of the proposed approach regarding dirt detection in automatic optical inspection.

7.1 Introduction

There are recently been massive growth in the application of camera module. The following Sections 7.1.1 and 7.1.2 describe the concept of image-based biometrics and illustrate some practical issues of camera module respectively.

7.1.1 Biometrics

In recent years there has been renewal of interest in the application of biometrics. Although there are some privacy issues, the potential application of biometrics are still immense. In biometrics, the physiological biometric characteristic includes iris recognition [1–4], fingerprint [5–7], facial recognition [8–10], retina [11, 12] and so on.

The biometrics features can be extracted in various ways. Some of the biometrics data are stored as image format for the purpose of post processing. The iris image [4] shows in Fig. 7.1 (a) is shoot in visible wavelength environment. Figure 7.1 (b) to (d) are fingerprint image [13], diagram of facial recognition [9], and retinal image [14] respectively. We may need to use image sensor to obtain the desired images. Therefore, the quality of image directly affects to the accuracy of the post processing of biometrics. Figure 7.2 shows the comparison between favorable image and blemished image.

Generally, we cannot just apply image de-noise approach to reduce the effect of small blemish on input image because the area of blemish is unpredictable. In this paper, we introduce a framework to detect the defect

---

1   Department of Computer Science and Information Engineering, Taiwan University, Taipei 106, Taiwan, China. E-mails: {d95013, fuh}@csie.ntu.edu.tw.

from input image. While the defect been detected, manufacturer can apply the proposed method on production line to decrease the influence caused by blemish.

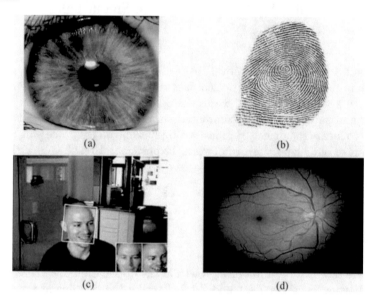

(a)   (b)

(c)   (d)

Fig. 7.1 (a) The captured iris image. (b) Fingerprint image. (c) Facial recognition in biometrics. (d) Retinal image.

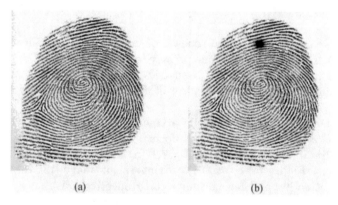

(a)   (b)

Fig. 7.2 (a) Favorable fingerprint image. (b) Blemished fingerprint image.

7.1.2 Issues of Camera Module

Camera module can be applied to acquire the required biometrics image because the related devices are obtainable and inexpensive. There are some practical issues should be dealt with before applying camera module over biometrics. This section will briefly introduce camera module and illustrate the issues caused by blemish.

Recently, the application of camera module becomes more widespread. To provide more portability and feasibility of application, the developing trend of camera module is manufactured toward small dimension and low cost.

The conventional approaches to achieve the above mentioned object is to use fewer optical lenses to lower camera module dimension and substitute plastic material for glass material to decrease cost. Unfortunately, when we use fewer optical lenses, some aberration effects become conspicuous. One of the effects shown in Fig. 7.3 is called lens shading. Lens shading phenomenon is introduced in Section 7.2.1. Simultaneously, dirt and blemish detection is needed in production line during camera module fabrication. The red circle in Fig. 7.4 shows contaminated region.

Fig. 7.3 The vignette effect shows up while shooting the Kodak Gray Card.

Occasionally, the difference between defect and vignette of lens shading is not conspicuous. It is difficult to perceive even through manual verification. Figure 7.5 gives an example of inconspicuous defect; the dirt region in red circle is similar to the background.

In this article, we propose a systematic framework to achieve the objective of automatic optical dirt detection. The remainder of this paper is organized as follows. In Section 7.2, we introduce some background knowledge of our proposed framework. Section 7.3 presents our system framework and describes the framework step-by-step. In Section 7.4, the realistic defect image and experimental results are presented. Finally, the conclusion is drawn in Section 7.5.

Fig. 7.4 Lens shading image with contaminated region.

Fig. 7.5 Lens shading image with inconspicuous contaminated region.

7.2 Background

Background knowledge include lens shading phenomenon and color filter array is introduced in Section 7.2.1 and Section 7.2.2 respectively. By integrating a series of techniques into our proposed framework, the following Sections 7.2.3 to 7.2.6 show how it is possible to overcome the limitations of each individual processing step.

7.2.1 Lens Shading Phenomenon

The phenomenon of lens shading is the reduction in light falling on the image sensor away from the center of the optical axis caused by physical obstructions. Figure 7.3 shows a phenomenon of lens shading while we shoot the Kodak Gray Card. In general, the vignette effect of lens shading may be asymmetric.

The effect of relative illumination can be calibrated by well optical design. But it becomes a challenge to calibrate vignette effect without increasing the dimension of optical design. There are many patents proposed to correct the shading effect by image post-processing. X. Y. Wang et al. proposed a lens shading correction approach by using B-spline curve fitting [15]. In their patent, the inventors reveal that the feasibility of conventional symmetric method for lens shading correction was limited by optical alignment and radial symmetry. Furthermore, the inventors also indicate that the two-direction multiplication method often inadequately corrects the shading effect. In this research, we apply B-spline curve to simulate the shading surface. The operating process is presented in Section 7.3.

7.2.2 Color Filter Array

In photography, since the typical image sensors cannot distinguish specific wavelength, color filter arrays are needed [16]. Commonly, color filter arrays have been placed above the image sensor. Figure 7.6 shows the diagram of the assembling organization about microlens, color filter array, and image sensor.

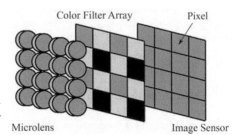

Fig. 7.6 This schematic diagram illustrates the relation between color filter array and image sensor.

There are many different types of color filter arrays proposed. Figure 7.7 shows some types of color filter arrays.

In this research, we use conventional RGB (Red, Green, Blue) images as input data. The conventional RGB image is demosaiced from raw image by using the color interpolation algorithm [17]. The color filter array of the input images in this article is Bayer filter. By observation, the green channel has occupied fifty percent of the whole image; the red channel and blue channel have occupied twenty five percent respectively. In other words, the green channel has higher chance to collect the integral information from real scene. Therefore, we only use the green channel in each input image to save computation cost.

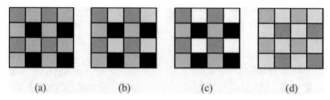

(a)   (b)   (c)   (d)

Fig. 7.7 Various types of color filter arrays. (a) Bayer filter. (b) RGBE (Red, Green, Blue, Emerald) filter. (c) RGBW (Red, Green, Blue, White) filter. (d) CYYM (Cyan, Yellow, Yellow, Magenta) filter.

7.2.3 Image De-Noise

After taking green channel of input image, the next step is to lower this image's noise. In this research, the objective of noise reduction is to make the sampling point (control point) in B-Spline more generalized.

Many of noise reduction approaches have been proposed to decrease the defect (or distortion data) from given image without affecting the image detail. For instance, A. Buades et al. proposed non-local means algorithm [18]; C. Tomasi et al. proposed bilateral filter approach [19]; and so on.

By observation, the noise patterns shown in images can be effectively reduced by using conventional 9×9 pixels median filter. The following figures show the result of noise reduction after applying the 9×9 pixels median filter.

Although the visual difference is slight, the process of image de-noise is indispensable for the following processes. Figure 7.9 shows the result of Sobel edge detection in Fig. 7.8. After image de-noising, the intensity variation in Fig. 7.9 (b) becomes stable and more suitable for the sampling control point of the B-Spline algorithm.

(a)   (b)

Fig. 7.8 The visual difference between input image and de-noised image. (a) Input image. (b) Result of 9×9 pixels median filter in (a).

(a)   (b)

Fig. 7.9 (a) The result of Sobel filtering the image in Fig. 7.8 (a). (b) Result of Sobel filtering the image in Fig. 7.8 (b).

7.2.4 Histogram Equalization

To enhance the contrast of whole image, we apply histogram equalization to make the slight variance of given image prominent. For discrete application, the probability of occurrence of intensity r_k in an image can be denoted by [20]

$$p_r(r_k) = \frac{n_k}{n}, \quad k = 0, 1, 2, \ldots, L - 1,$$

where n_k means the total pixels of specific intensity r_k; n means total pixels of whole image; and L means the number of intensity gray levels in the image ($L = 256$ for 8 bit/pixel image). Let r means the normalized gray level (from [0, 255] to [0, 1]), Fig. 7.10 shows the concept of histogram equalization.

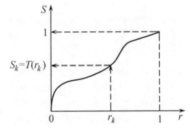

Fig. 7.10 Concept of transformation from gray level r to level s.

The transformation function of histogram equalization can be defined by

$$s_k = T(r_k) = \sum_{j=0}^{k} p_r(r_j) = \sum_{j=0}^{k} \frac{n_j}{n}, \quad k = 0, 1, 2, \ldots, L - 1.$$

And, the inverse transformation from s back to r is defined by:

$$r_k = T^{-1}(s_k), \quad k = 0, 1, 2, \ldots, L - 1.$$

Fig. 7.11 shows the effect of the image enhancement by histogram equalization.

Fig. 7.11 (a) An image with slight variance of intensity. (b) Result of histogram equalization in (a). Note: The bright spot at the lower right corner is emphasized.

By observing Fig. 7.11 (b), a bright spot appears at the lower right corner of the image. Besides, there are many sparse spots spread in the image. To increase the accuracy of our blemish detection, we apply morphological image processing to decrease the effect of those sparse spots.

7.2.5 Morphological Operation

The theory of morphology is based on set theory [20], and can be applied to achieve many image processing techniques. For instance, image component extracting, image thinning, image pruning, and so on. In this research, we use mathematical morphological opening to lower the sparse spots mentioned in Fig. 7.11 (b). Opening can be denoted by

$$A \circ B = (A \ominus B) \oplus B,$$

where $A \circ B$ means the opening operation of set A by B; $\ominus$ means the erosion operation; and $\oplus$ means the dilation operation. We use the following rule to implement erosion during realistic implementation:

$$I(x, y) = \arg \min I(x + x', y + y'),$$

where $I(x, y)$ means the intensity after erosion at position (x, y). By using 3×3 pixels square structuring element, we select the minimum intensity from current position (x, y) and its eight neighboring pixels, denoted by (x', y'), as current intensity. Similarly, we use the following rule to implement dilation during implementation:

$$I(x, y) = \arg \max I(x + x', y + y').$$

We select the maximum intensity from position (x, y) and its eight neighboring pixels as current intensity. The structuring element is fixed in this article; we apply 3 times of erosion before 3 times of dilation to achieve preferable result. Figure 7.13 (b) shows the result of opening operation in the Fig. 7.12 (a).

(a)   (b)

Fig. 7.12   (a) An image with many sparse spots. (b) Result of morphological operation based on 3 times of erosion operation and 3 times of dilation operation.

The iteration of erosion and dilation can be modified to suit the requirement of specific inspection.

7.2.6 B-Spline

The B-Spline curve is proposed by Isaac Jacob Schoenberg in 1946 [21 – 24]. B-Spline curve is a generalization of the Bézier curve, and can be applied to many fields, such as, computer-aid design, computer graphics, engineering mechanics, and so on [25, 21, 23].

We fit the lens shading image by using the B-Spline curve, because the B-Spline curve can avoid the Runge phenomenon. In this article, we implement the uniform quadratic B-Spline to generate the ideal vignette surface. The uniform quadratic B-Spline is denoted by [24]

$$S_i(t) = \begin{bmatrix} t^2 & t & 1 \end{bmatrix} \begin{bmatrix} 0.5 & -1 & 0.5 \\ -1 & 1 & 0 \\ 0.5 & 0.5 & 0 \end{bmatrix} \begin{bmatrix} p_{i-1} \\ p_i \\ p_{i+1} \end{bmatrix},$$

for $t \in [0, 1], i = 1, 2, \ldots, m - 2$, where p_{i-1}, p_i, and p_{i+1} mean the set of control points; S_i means the ith B-Spline segment. The detail of the generated surface depends on the number of sampling control points. The more control points we sample, the more exquisite surface we get. We use vertical control points to generate vertical B-Spline. Then, we apply the generated vertical B-Spline to generate whole vignette surface. The diagram of vertical B-Spline generation is shown in Fig. 7.13.

In this article, we extract 31 control points in each vertical sample, and 31 control points in each horizontal sample. Thus, the total number of control

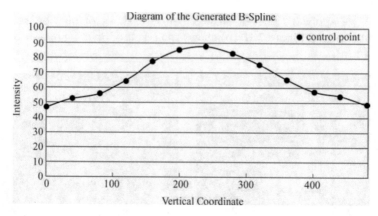

Fig. 7.13 The diagram is a concept of vertical B-Spline generation using 13 control points. After vertical B-Spline generation, we can use them to generate horizontal B-Spline.

points is 961 (=31×31). Figure 7.14 (b) shows the result of B-Spline generated surface in Fig. 7.14 (a).

Fig. 7.14 (a) The given image. (b) Result of generated B-Spline surface in (a).

7.3 Our Proposed Method

To detect dirt from vignette image, we use the backgrounds mentioned in Section 7.2. Our proposed framework is shown in Fig. 7.15.

We use the green channel of the input image to lower the issues caused by image demosaicing process. To generate an appropriate B-Spline surface, we have to moderate the intensity variation in the extracted green channel image. Thus, we smooth the green channel image by using 9×9 pixels median filter. The original image and smoothed result are shown in Figs. 7.16 and 7.17.

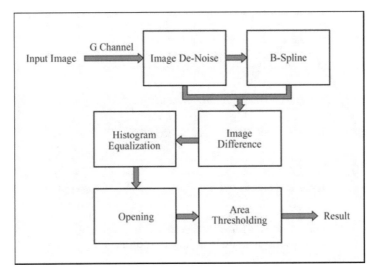

Fig. 7.15 The proposed system flowchart.

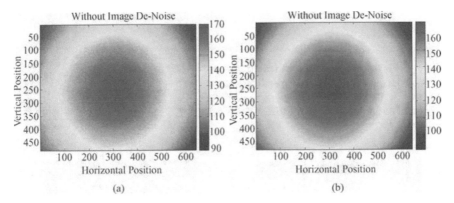

Fig. 7.16 (a) Scalar image of original image. (b) Scalar image of de-noised image.

After image de-noise, we extract M pixels' intensities as control points from N vertical sampling. Figure 7.18 shows the concept of the above mentioned sampling process; red points in Fig. 7.18 are sampled control points.

Then, we use the $M \times N$ control points to first generate vertical direction B-Spline curves. The hollow blue rectangles shown in Fig. 7.19 are the required B-Spline curves, and the solid rectangles shown in Fig. 7.20 are the generated B-Spline curves. Finally, we generate horizontal B-Spline curves to complete whole B-Spline surface by using existing vertical B-Spline information. The hollow green rectangle shown in Fig. 7.20 is one of the required horizontal B-Spline curves, and the blue parts inside green rectangle are control points. The above mentioned B-Spline curve generation can be accelerated by using parallel computation.

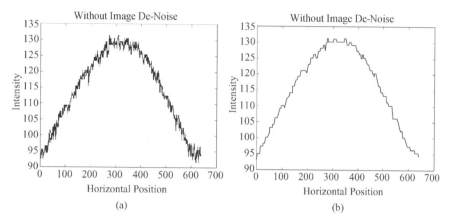

Fig. 7.17 Compare the same image line between the given image and the smoothed image. (a) Image without de-noised process. (b) Result of 9×9 pixels median filter process in (a).

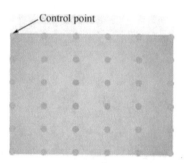

Fig. 7.18 The red points in this image are sampled control points. In this diagram, we extract 6 pixels ($M{=}6$) from each sampled column, and 6 columns ($N{=}6$) from the whole image.

Fig. 7.19 The hollow blue rectangles are the required B-Spline curves.

Fig. 7.20 The solid blue rectangles are the generated B-Spline curves from Fig. 7.19, and the hollow green rectangle is one of the required B-Spline curves. We can use blue parts inside the green rectangle as control points to compute its B-Spline curve.

Fig. 7.21 (b) shows the performance of the generated B-Spline curve, and Fig. 7.22 (b) shows the performance of the generated B-Spline surface.

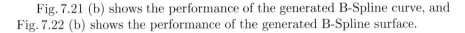

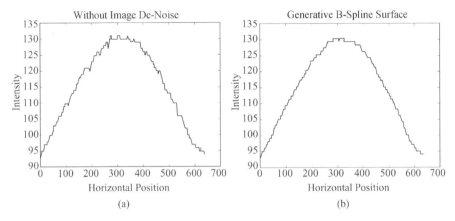

Fig. 7.21 Compare the same image line between the de-noising image and the generated B-Spline surface. (a) This curve shows the intensity variation of the de-noising image. (b) This curve means the generated B-Spline image.

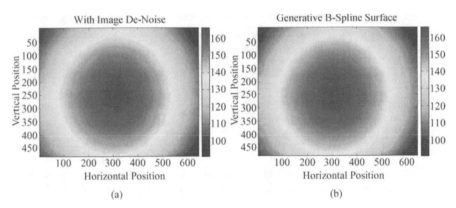

Fig. 7.22 (a) Scalar image of de-noised image. (b) Scalar image of generated B-Spline surface.

Next, we subtract the generated B-Spline surface from de-noised image to compute image difference. Figure 7.23 shows the result of image difference. To prevent the intensity variation of image difference from over-smoothing, we apply histogram equalization to enhance the result of image difference.

Unfortunately, some conspicuous perturbations in input image may be enhanced simultaneously. To lower the effect of the perturbations of input image, we use morphological operation to remove small isolated areas. Here, we use 3×3 pixels structuring element mentioned in Section 7.2.5 to process 3 times of erosion before 3 times of dilation to achieve preferable result.

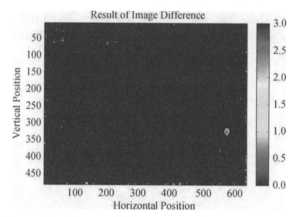

Fig. 7.23 The result of image difference. We subtract the generated B-Spline surface from de-noised image to obtain this result.

After the morphological opening, most of small isolated areas caused by conspicuous perturbations have been removed. To preserve the flexibility of application, we add an area threshold function as the final step. The objective of area threshold function is to fill the area whose pixel number is larger than or equal to threshold as a specific color, and to fill the area whose pixel number is smaller than threshold as another color. Thus, the dirt detection result can be easily rechecked by user if necessary.

7.4 Experimental Results

In this article, we use real 640×480 pixels inspection image as input image. The median filter kernel is set as 9×9 pixels. We use $961 (= 31 \times 31)$ control points to generate ideal B-Spline surface. Moreover, we apply 3 times of erosion before 3 times of dilation with 3×3 pixels structuring element mentioned in Section 7.2.5 to achieve preferable result. Besides, we set the area threshold as 85 pixels. Here, if the area of dirt particle is larger than or equal to 85 pixels, then the dirt particle will be drawn with green color, otherwise, the dirt particle will be drawn with red color.

The experimental environment is on Intel Core 2 CPU Q6600 2.4 GHz (our program only uses single core), and 2 GB RAM. The average processing time is about 2.8 seconds without using any code optimization or hardware acceleration.

The respective results of our proposed framework are shown in Figs. 7.24 – 7.32. In our experiment of 13 images, we achieve 0 misdetections and 0 false alarms.

A special case is shown in Fig. 7.28 where many noises distribute over input image. The sparse spots appear in enhanced result in Fig. 7.28 (f)

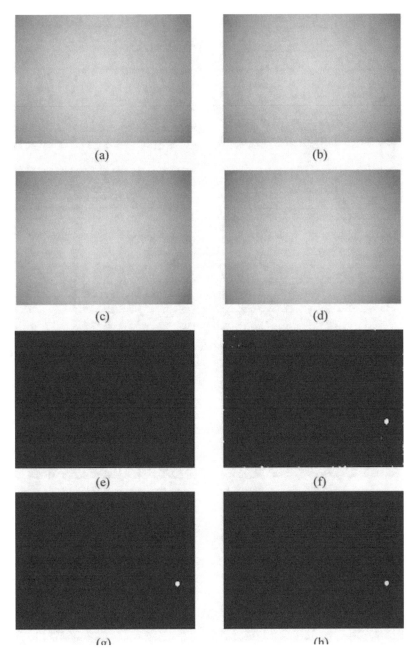

Fig. 7.24 Test Image 1. (a) Input image. (b) The green channel of input image. (c) De-noised result in (b). (d) The generated B-Spline surface. (e) Result of image difference. (f) Result of histogram equalization in (e). (g) Result of Opening in (f). (h) Result of area threshold in (g).

Fig. 7.25 Test Image 2. (a) Input image. (b) The green channel of input image. (c) De-noised result in (b). (d) The generated B-Spline surface. (e) Result of image difference. (f) Result of histogram equalization in (e). (g) Result of Opening in (f). (h) Result of area threshold in (g).

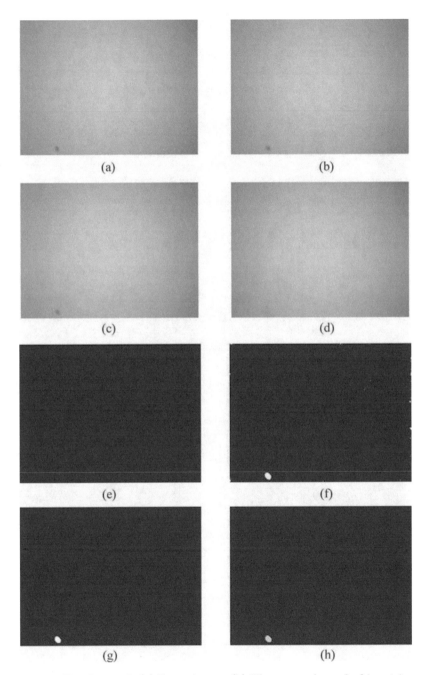

Fig. 7.26 Test Image 3. (a) Input image. (b) The green channel of input image. (c) De-noised result in (b). (d) The generated B-Spline surface. (e) Result of image difference. (f) Result of histogram equalization in (e). (g) Result of Opening in (f). (h) Result of area threshold in (g).

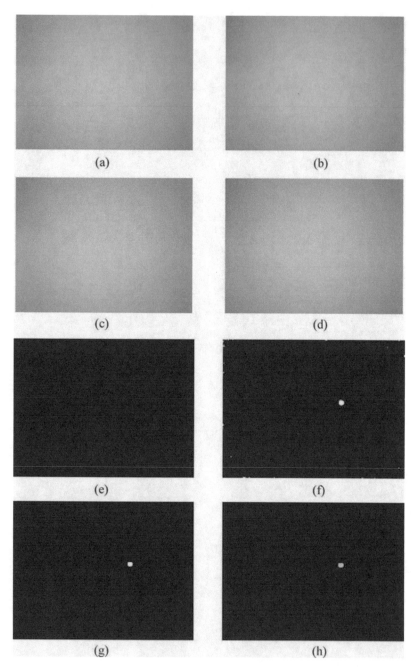

Fig. 7.27 Test Image 4. (a) Input image. (b) The green channel of input image. (c) De-noised result in (b). (d) The generated B-Spline surface. (e) Result of image difference. (f) Result of histogram equalization in (e). (g) Result of Opening in (f). (h) Result of area threshold in (g).

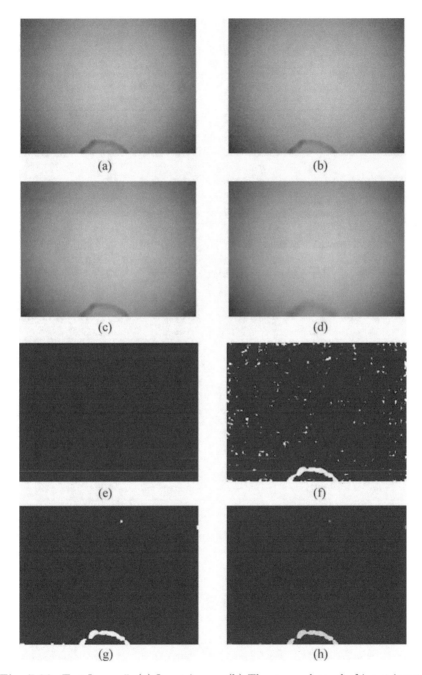

Fig. 7.28 Test Image 5. (a) Input image. (b) The green channel of input image. (c) De-noised result in (b). (d) The generated B-Spline surface. (e) Result of image difference. (f) Result of histogram equalization in (e). (g) Result of Opening in (f). (h) Result of area threshold in (g).

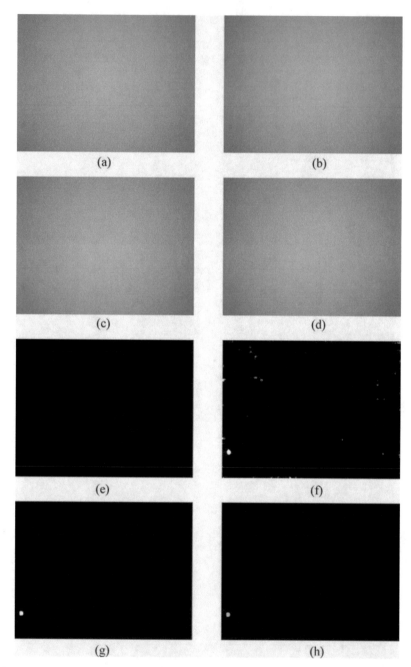

Fig. 7.29 Test Image 6. (a) Input image. (b) The green channel of input image. (c) De-noised result in (b). (d) The generated B-Spline surface. (e) Result of image difference. (f) Result of histogram equalization in (e). (g) Result of Opening in (f). (h) Result of area threshold in (g).

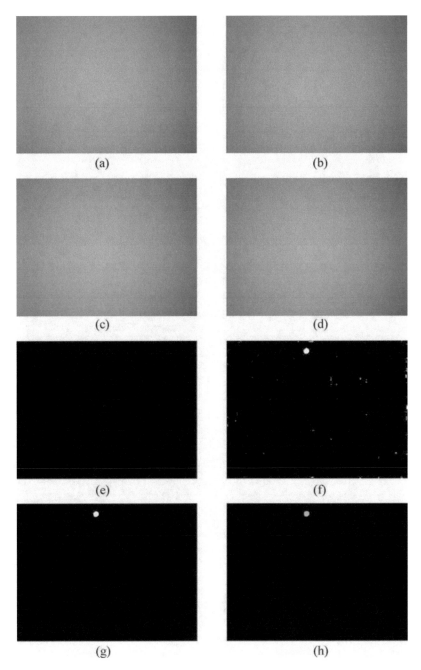

Fig. 7.30 Test Image 7. (a) Input image. (b) The green channel of input image. (c) De-noised result in (b). (d) The generated B-Spline surface. (e) Result of image difference. (f) Result of histogram equalization in (e). (g) Result of Opening in (f). (h) Result of area threshold in (g).

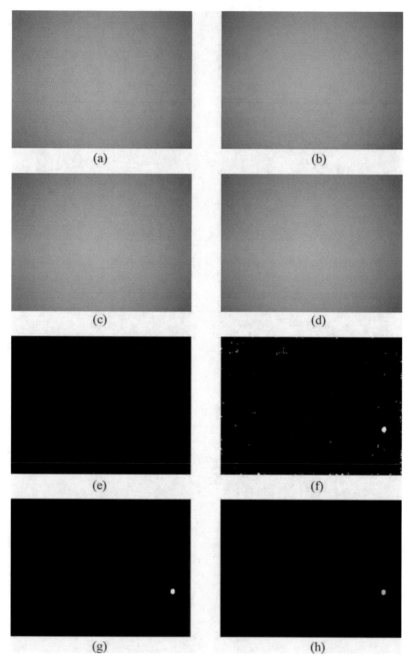

Fig. 7.31 Test Image 8. (a) Input image. (b) The green channel of input image. (c) De-noised result in (b). (d) The generated B-Spline surface. (e) Result of image difference. (f) Result of histogram equalization in (e). (g) Result of Opening in (f). (h) Result of area threshold in (g).

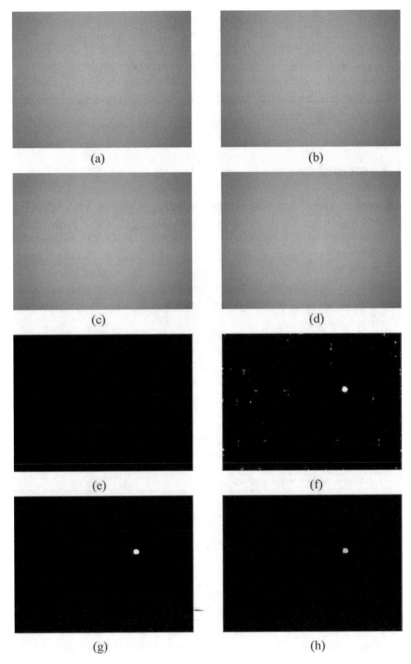

Fig. 7.32 Test Image 9. (a) Input image. (b) The green channel of input image. (c) De-noised result in (b). (d) The generated B-Spline surface. (e) Result of image difference. (f) Result of histogram equalization in (e). (g) Result of Opening in (f). (h) Result of area threshold in (g).

seriously. By using mathematic morphological opening, most of the sparse spots are removed. But, there still are some noises after opening. Therefore we apply area threshold function to emphasize the region with larger area. Users can easily distinguish the obvious or the slight defect region.

Besides, when the size of input image is large, user can apply down-sampling to lower processing cost. Of course, the parameters can be modified to fit different requirements.

7.5 Conclusions

In automatic optical inspection, the technology to detect the defect parts (or interesting parts) in given image automatically is needed. In this article, we propose a novel approach to extract defect regions from lens shading image. By applying the proposed method to inspect the optical component, we can make the optical device more reliable and more accurate for its future application. Although, the proposed method in this article tends to deal with specific problem, the concept of the approach can be extended to other inspection.

There still are many topics, including automatic defect detection for LCD (Liquid Crystal Display) panel surface, and so on.

Acknowledgements

This research was supported by the Science Council of Taiwan, China, under Grants NSC 98-2221-E-002 -150 -MY3 and NSC 95-2221-E-002-276-MY3, by EeRise Corporation, Jeilin Technology, Alpha Imaging, Winstar Technology, Test Research, Faraday Technology, Vivotek, Lite-on, and Syntek Semiconductor.

References

[1] Daugman J (2002) How Iris Recognition Works. In: Proceedings of the 2002 International Conference on Image Processing, vol 31: pp I-33 – I-36

[2] Daugman J (2003) The Importance of Being Random: Statistical Principles of Iris Recognition. Transactions on Pattern Recognition, 36(2): 279 – 291

[3] Hosseini M S, Araabi B N, Soltanian-Zadeh H (2010) Pigment Melanin: Pattern for Iris Recognition. IEEE Transactions on Instrumentation and Measurement, 59(4): 792 – 804

[4] Wikipedia (2010) Iris Recognition. http://en.wikipedia.org/wiki/Iris_recognition. Accessed 6 November 2010

[5] Maltoni D, Maio D, Jain A K et al (2009) Handbook of Fingerprint Recognition. Springer, New York

[6] Ratha N K, Bolle R (2003) Automatic Fingerprint Recognition Systems. Springer, Heidelberg

[7]   Wikipedia (2010) Fingerprint. http://en.wikipedia.org/wiki/Fingerprint. Accessed 2 November 2010

[8]   Brunelli R, Poggio T (1993) Face Recognition: Features Versus Templates. IEEE Transactions on Pattern Analysis and Machine Intelligence, 15(10): 1042 – 1052

[9]   Jain A K, Li S Z (2005) Handbook of Face Recognition. Springer, New York

[10] Wright J, Yang A Y, Ganesh A et al (2009) Robust Face Recognition via Sparse Representation. Pattern IEEE Transactions on Analysis and Machine Intelligence, 31(2): 210 – 227

[11] Jain A, Bolle R, Pankanti S (2002) Introduction to Biometrics. In: Jain A, Bolle R, Pankanti S (eds) Biometrics, pp 1 – 41. Springer, New York

[12] Wikipedia (2010) Retinal Scan. http://en.wikipedia.org/wiki/Retinal_scan. Accessed 5 November 2010

[13] Department Y P (2010) Fingerprinting. http://www.yakimapolice.org/police/pages/fingerprinting.htm. Accessed 4 April 2010

[14] Jamblichus (2010) A Man's Right Eye as Shown in a Retinal Scan. http://jamblichus.wordpress.com/tag/michel-foucault. Accessed 2 October 2009

[15] Wang X Y, Shan J, Wu D (2008) System and Method for Lens Shading Correction of an Image Sensor Using Splines. US Patent US20090268053

[16] Nakamura J (2005) Image Sensors and Signal Processing for Digital Still Cameras. Optical Science and Engineering Series. Taylor & Francis

[17] Davies A, Fennessy P (2001) Digital Imaging for Photographers, 4th edn. Focal Press, Boston

[18] Buades A, Coll B, Morel J M (2005) A Non-local Algorithm for Image Denoising. In: Computer Vision and Pattern Recognition, 2005. CVPR 2005. IEEE Computer Society Conference on 20 – 25 June 2005, vol 62: 60 – 65

[19] Tomasi C, Manduchi R (1998) Bilateral Filtering for Gray and Color Images. In: Computer Vision, 1998. Sixth International Conference on 4 – 7 January 1998: 839 – 846

[20] Gonzalez R C, Woods R E (2002) Digital Image Processing, 2nd edn. Prentice Hall, New York

[21] Weisstein E W (2010) B-spline. http://mathworld.wolfram.com/B-Spline.html. Accessed 1 January 2010

[22] Wikipedia (2010) Isaac Jacob Schoenberg. http://en.wikipedia.org/wiki/Isaac_Jacob_Schoenberg. Accessed 1 October 2010

[23] Wikipedia (2010) Spline (Mathematics). http://en.wikipedia.org/wiki/Spline_(mathematics). Accessed 1 November 2010

[24] Wikipedia (2010) B-spline. http://en.wikipedia.org/wiki/B-spline. Accessed 1 November 2010

[25] Shene C K (2010) Introduction to Computing with Geometry Course Notes. http://www.cs.mtu.edu/~shene/COURSES/cs3621/NOTES/notes.html. Accessed 1 July 2010

8 Using Prototype-Based Classification for Automatic Knowledge Acquisition

Petra Perner[1] and Anja Attig[1]

Abstract In this paper we describe how prototype-based classification can be used for knowledge acquisition in image classification. We describe the necessary functions a prototype-based classifier should have. These functions need to be developed for a low number of samples and for a large number of samples; they are feature subset selection, similarity learning and prototype selection. The classifier is applied to the internal mitochondrial movement of cells as an example. The aim was to discover the different dynamic signatures of mitochondrial movement. Our results and an outlook on future work are presented here.

8.1 Introduction

Cellular assays are highly recommended scientific tools in modern drug research and examination of pathological processes at the cellular level. Based on a cellular assay, the internal mitochondrial movement of cells is studied, for example, for two reasons: Firstly, it serves as a reference for normal organelle (endosome/phagosome) movement to allow comparison with organelle movement in the virally or bacterially infected disease states. Second, mitochondrial movements and distribution of mitochondria are sensitive markers for cell health and subtle differences in movement and distribution may serve as surrogate markers for a number of disease states. For example, mitochondria change their characteristic distribution and morphology during apoptosis, Huntington's disease, during neuronal toxicity and Alzheimer's disease. In the light of these observations, a goal of the current application is to identify "dynamic signatures" involving key parameters of mitochondrial movement and distribution as a rapid screening means for neuronal degeneration or neurodegenerative diseases.

We have solved the knowledge acquisition problem as a classification problem based on a prototype-based classifier and image analysis. Such a method needs to take into account a lot of different cases that can occur when samples are collected on-line. Negative samples can heavily outnumber positive examples and vis versa. The data set for the different classes might be heavily imbalanced. In a first consideration we have addressed the problem of having

---

1 Institute of Computer Vision and Applied Computer Sciences, IBaI. PSF 301114, 04251 Leipzig, Germany. E-mails: pperner@ibai-institut.de, www.ibai-institut.de.

only a few prototypes and later on we will have more samples. We describe in this paper what kind of methods we have developed for this type of problems.

The image analysis is based on the case-based reasoning method for image segmentation and on random sets for feature description developed by Perner [1]. Other methods for image segmentation [2, 3] and feature extraction [3, 4] are also applicable for the proposed method.

The proposed prototype-based classification method is not only essential for the biological application we describe here. It is necessary methods for all application where pattern similarity is the key concept in order to acquire the concept description from unknown data. Biometrical and Security systems where we usually have to deal with a lot of ambiguity and not formerly known concepts might also profit from these kind of method [10].

The classifier is set up based on prototypical cell appearances in the image such as for e.g., "healthy cell", "cell dead", and "cell in transition stage". For these prototypes image features are calculated based on random set theory that describes the texture of the cells. The prototype is represented then by the attribute-value pair and the class label.

These settings are taken as initial classifier settings in order to acquire the knowledge about the dynamic signatures.

The importance of the features and the feature weights are learned by the prototype-based classifier based on selected prototypes. Once the classifier is set up, each new cell is then compared with the prototype-based classifier and the similarity with the prototypes is calculated. If the similarity is high, the label of the prototype is assigned to the new cell. If the similarity with the prototypes is too low then there is evidence that the cell is in a transition stage and that a new prototype has been found. With this procedure we can learn the dynamic signature of the mitochondrial movement.

In Section 8.2 we present the methods for our prototype-based classifier. The material is described in Section 8.3 for the internal mitochondrial movement of cells.

In Section 8.4 the methodology for prototype-based classification is presented. Results are given in Section 8.5 and, finally, in Section 8.6 conclusions are presented.

8.2 Prototype-Based Classification

A prototype-based classifier classifies a new sample according to the prototypes in a database and selects the most similar prototype as output of the classifier. A proper similarity measure is necessary to perform this task but in most applications there is no a-priori knowledge available that suggests the right similarity measure. The method of choice to select the proper similarity measure is therefore to apply to the problem a subset of the numerous similarity measures known from statistics and to select the one that performs best according to a quality measure such as, for example, the classification accuracy. The other choice is to automatically build the similarity metric by learning the right attributes and attribute weights. The latter one we have chosen as one option to improve the performance of our classifier.

When people collect prototypes to construct a dataset for a prototype-based classifier it is useful to check if these prototypes are good prototypes. Therefore, a function is needed to perform prototype selection and to reduce the number of prototypes used for classification. This results in better generalization and a more noise tolerant classifier. If an expert selects the prototypes, this can result in bias and possible duplicates of prototypes causing inefficiencies. Therefore a function to assess a collection of prototypes and identify redundancy is useful.

Finally, an important variable in a prototype-based classifier is the value used to determine the number of closest cases and the final class label.

Consequently, the design-options for improving the performance of the classifier are prototype selection, feature-subset selection, feature weight learning and the 'k' value of the closest cases (see Fig. 8.1). The developed classifier is named ProtoClass.

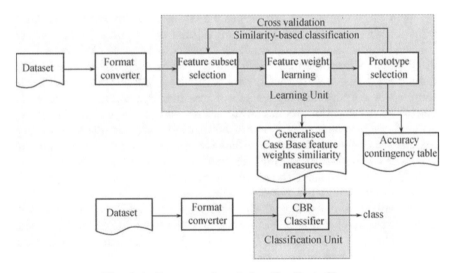

Fig. 8.1 Prototype-based classifier ProtoClass.

We assume that the classifier can start in the worst case with only one prototype per class. By applying the classifier to new samples the system collects new prototypes. During its lifetime, the system will change its performance from an oracle-based classifier that classifies the samples roughly into the expected classes, to a system with high performance in terms of accuracy.

In order to achieve this goal we need methods that can work on a small number of prototypes and on a large number of prototypes (see Fig. 8.2). As long as we have only a few numbers of prototypes, feature subset selection and learning the similarity might be the important features the system needs. If we have more prototypes we also need prototype selection.

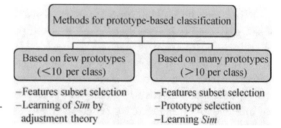

Fig. 8.2 Methods for prototype-based classification.

For the case with a small number of prototypes we chose methods for feature subset selection based on the discrimination power of attributes. For learning the similarity we use the feature-based calculated similarity and the pairwise similarity rating of the expert and apply the adjustment theory to fit the curve more to the expert's rating by adjusting the weights of the attributes.

For a large number of prototypes we chose a decremental redundancy-reduction algorithm proposed by Chang [5] that deletes prototypes as long as the classification accuracy does not decrease. The feature-subset selection is based on the wrapper approach [6] and an empirical feature-weight learning method [7] is used. Cross validation is used to estimate the classification accuracy. A detailed description of our classifier ProtoClass is given in [1]. The prototype selection, the feature selection, and the feature weighting steps are performed independently or in combination with each other in order to assess the influence these functions have on the performance of the classifier. The steps are performed during each run of the cross-validation process.

The classifier schema shown in Fig. 8.1 is divided in the design phase (Learning Unit) and the normal classification phase (Classification Unit). The classification phase starts after we have evaluated the classifier and determined the right features, feature weights, the value for 'k' and the cases.

Our classifier has a flat data base instead of a hierarchical one and this makes it easier to conduct the evaluations.

8.2.1 Classification Rule

Assume we have n prototypes that represent m classes of the application. Then, each new sample is classified based on its closeness to the n prototypes. The new sample is associated with the class label of the prototype that is closest to the sample.

More precisely, we call $x'_n \in \{x_1, x_2, \ldots, x_i, \ldots, x_n\}$ a closest case to x if $\min d\,(x_i, x) = d\,(x'_n, x)$, where $i = 1, 2, \ldots, n$.

The rule chooses to classify x into category C_l, where x'_n is the closest case to x and x'_n belongs to class C_l with $l \in \{1, \ldots, m\}$.

In the case of the k-closest cases we require k-samples of the same class to fulfill the decision rule. As a distance measure we can use any distance metric. In this work we used the city-block metric.

The pair-wise similarity measure among our prototypes demonstrates the discrimination power of the chosen prototypes based on the features (see Table 8.1).

Table 8.1 Pairwise Similarity as Indicator for Discrimination Power

Prototype	P_1	P_2	$\ldots$	P_j	$\ldots$	P_n
P_1	0	$Sim_{12}(SimE_{12})$	$\ldots$	$Sim_{1j}(SimE_{1j})$	$\ldots$	$Sim_{1n}(SimE_{1n})$
P_2	$Sim_{21}(SimE_{i1})$	0		$Sim_{2j}(SimE_{2j})$	$\ldots$	$Sim_{2n}(SimE_{2n})$
$\vdots$	$\vdots$	$\vdots$		$\vdots$		$\vdots$
P_i	$Sim_{i1}(SimE_{i1})$	$Sim_{i2}(SimE_{i2})$	$\ldots$	$Sim_{ij}(SimE_{ij})$	$\ldots$	$Sim_{in}(SimE_{in})$
$\vdots$	$\vdots$	$\vdots$		$\vdots$		$\vdots$
P_n	$Sim_{n1}(SimE_{n1})$	$Sim_{n2}(SimE_{n2})$	$\ldots$	$Sim_{nj}(SimE_{nj})$	$\ldots$	0

The calculated feature set must not be the optimal feature subset. The discrimination power of the features must be checked later. For a low number of prototypes we can let the expert judge the similarity $Sim\ E_{ij}$ with $i, j \in \{1, \ldots, n\}$ between the prototypes. This gives us further information about the problem which can be used to tune the designed classifier.

8.2.2 Feature Selection for Low Number of Prototypes

For only one prototype in each class c the discrimination power of each feature f_i with $i \in \{1, \ldots, N\}$ is checked based on the intra and inter class variance:

$$T_{f_i} = \sum_{c=1}^{m} (f_{ic} - \overline{f_i})^2, \tag{8.1}$$

where $\overline{f_i}$ is the mean value of attribute i.

We rank the features according to the value of T_{f_i} and the select the t features with the highest value. The choice of the value t has been verified by the system developer. If we have more than one prototype per class the discrimination power of a feature is calculated according to:

$$T_{f_i} = \sum_{c=1}^{m} \left(\frac{1}{n_c} \sum_{l=1}^{n_c} (f_{il} - \overline{f_{ic}})^2 + (\overline{f_{ic}} - \overline{f_i})^2 \right), \tag{8.2}$$

where n_c is the number of prototypes in class c and f_{ic} is the mean value of attribute i in class c.

In the same manner as before we rank the features according to the value of T_{f_i} and the select the t features with the highest value.

8.2.3 Using Expert's Judgment on Similarity and the Calculated Similarity to Adjust the System

Humans can judge the similarity $SimE_{ij}$ among objects on a rate between 0 (identical) and 1 (dissimilar). We can use this information to adjust the system to the true system parameters [8].

Using the City-Block Distance as distance measure, we get the following linear system of equations:

$$SimE_{ij} = \frac{1}{N} \sum_{l=1}^{N} a_l |f_{il} - f_{jl}|, \tag{8.3}$$

with $i, j \in \{1, \ldots, n\}$, f_{il} the feature l of the ith prototype and N the number of attributes. The attribute a_l is the normalization of the feature to the range $\{0, 1\}$ with

$$a_l = \frac{1}{|f_{\max,l} - f_{\min,l}|} \tag{8.4}$$

that is calculated from the prototypes. That the kind of normalization has an important impact on the results is clear [11] but since we do not have many samples we choose this simple normalization as a first choice. That the chosen values also not represent the true range of the feature value is also clear since we have too small a number of samples. The factor a_l is adjusted closer to the true value by the least square method using expert's $SimE_{ij}$

$$\sum_{i=1}^{n-1} \sum_{j=i+1}^{n} \left(SimE_{ij} - \frac{1}{N} \sum_{l=1}^{N} a_l |f_{il} - f_{jl}| \right)^2 \Rightarrow Min, \tag{8.5}$$

with the restriction $0 \leqslant a_l \leqslant \frac{1}{|f_{\max,l} - f_{\min l}|}$.

8.2.4 Feature-Subset Selection and Feature Weighting for a Large Number of Samples

The wrapper approach [6] is used for selecting a feature subset from the whole set of features and for feature weighting. This approach conducts a search for a good feature subset by using the k-NN classifier itself as an evaluation function. By doing so the specific behavior of the classification methods is taken into account. The leave-one-out cross-validation method is used for estimating the classification accuracy. Cross-validation is especially suitable for a small data set. The best-first search strategy is used for the search over the state space of possible feature combination. The algorithm terminates if we have not found an improved accuracy over the last k search states.

The feature combination that gave the best classification accuracy is the remaining feature subset. We then try to further improve our classifier by applying a feature-weighting tuning technique in order to get real weights for the binary weights.

The weights of each feature w_i are changed by a constant value $\delta : w_i :=$ $w_i \pm \delta$. If the new weight causes an improvement of the classification accuracy, then the weight will be updated accordingly; otherwise, the weight will remain as is. After the last weight has been tested, the constant δ will be halved and the procedure repeats. The process terminates if the difference between the classification accuracy of two interactions is less than a predefined threshold T.

8.2.5 Prototype Selection by Chang's Algorithm

For the selection of the right number of prototypes we used Chang's algorithm [5]. The outline of the algorithm can be described as follows: Suppose the set T is given as $T = \{t_1, \ldots, t_i, \ldots, t_n\}$ with t_i as the ith initial prototype. The idea of the algorithm is as follows: We start with every point in T as a prototype. We then successively merge any two closest prototypes t_1 and t_2 of the same class by a new prototype t, if the merging will not downgrade the classification of the patterns in T. The new prototype t may simply be the average vector of t_1 and t_2. We continue the merging process until the number of incorrect classifications of the pattern in T starts to increase.

Roughly, the algorithm can be stated as follows: Given a training set T, the initial prototypes are just the points of T. At any stage the prototypes belong to one of two sets – set A or set B. Initially, A is empty and B is equal to T. We start with an arbitrary point in B and initially assign it to A. A point p in A and a point q in B are selected such that the distance between p and q is the shortest among all distances between points of A and B. Then merging of p and q is attempted. That is, if p and q are of the same class, a vector $p*$ is computed in terms of p and q. If replacing p and q by $p*$ does not decrease the recognition rate for T, merging is successful. In this case, p and q are deleted from A and B, respectively, and $p*$ is put into A, and the procedure is repeated once again. In the case that p and q cannot be merged, i.e., if either p or q are not of the same class or merging is unsuccessful, q is moved from B to A, and the procedure is repeated. When B is empty, the whole procedure is recycled by letting B be the final A obtained from the previous cycle, and by resetting A to be the empty set. This process stops when no new merged prototypes are obtained. The final prototypes in A are then used in the classifier.

8.3 Methodology

How to use prototype-based classification for concept learning based on a cell-image classification tasks is described in this Section.

8.3.1 General Overview

The general methodology is represented in Fig. 8.3. Based on image analysis and feature description we can take some measurements from the image. With the help of automatic image analysis we find the cell nucleous and its cytoplasma in the image. The software automatically takes some measurements from this area of the cell in regard to the area, the shape, grey level, texture, movement of objects or displacement of objects.

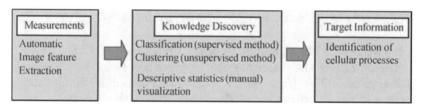

Fig. 8.3 Problem description.

This information must be summarized into more meaningful information about the cellular processes. It is preferable to come up with a single number or a few numbers or a category instead of having a bunch of numerical values.

This knowledge acquisition process can be done in different ways. Most common is the usage of descriptive statistics in order to summarize the data. In the simplest case mean, standard deviation or the distribution of a single feature or n-dimensional features are calculated and the results are visualized for the biologist in different ways for further evaluation. Correlations between features are other meaningful information that allows the biologist to manually study the data in different ways and to derive some information from the observations.

The summarization into categories is the highest level of summarization and can lead directly to the final decision the biologist has to make. It requires knowledge of what categories are underlying the data. If the categories are known a-priori this information can be taken as a starting point for classification. If the categories are not known a-priori clustering can be used to discover the categories.

In this paper, we provide a methodology how prototype-based classification can be used to summarize the data into meaningful categories that give the biologist the targeted information.

8.3.2 Prototype-Based Classification as Knowledge Acquisition Method

Figure 8.4 summarizes the knowledge acquisition process based on prototype-based classification.

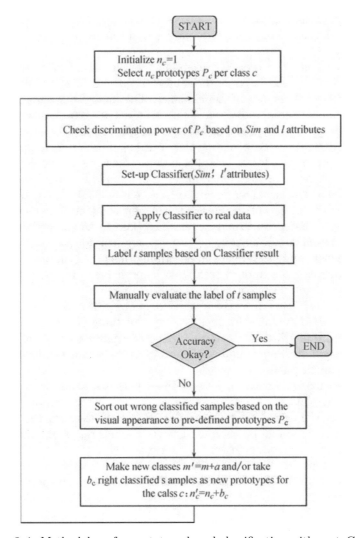

Fig. 8.4 Methodology for prototype-based classification with protoClass.

We start with one prototype for each class. This prototype is chosen by the biologist based on the appearance of the cells. It requires that the biologist has enough knowledge of the processes going on in cell-based assays and is able to decide what kind of reaction the cell is showing.

The discrimination power of the prototypes is checked first based on the attribute values measured from the cells and the chosen similarity measure. Note that we calculated a large number of attributes for each cell. However the use of many attributes does not mean that we will achieve a good discrimination power between the classes. This triggers the curse of dimensionality and it is better to come up with one or two attributes for small sample sizes in order to ensure a good performance of the classifier. The result of this process is the selection of the right similarity measure and the right number of attributes.

With this information a first classifier is set up and applied to real data. Each new data is associated with the label of the classification. We manually evaluate the performance of the classifier. The biologist assigns the true or gold label for the sample seen so far. This is stored in a data base and serves as gold standard for further evaluation. During this process the expert will sort out wrongly classified data. Wrong classification of data may happen because of too few prototypes for one class or because the samples should be divided into more classes. The decision what kind of technique should be applied is made based on the visual appearance of the cells. Therefore, it is necessary to display the prototypes of the class and the new samples. The biologist sorts these samples based on visual appearance (known as categorizing or card sorting in psychology). That this is no an easy task for a human is evident; this task needs some experience in describing image information [9].

However, it is a standard technique in psychology in particular gestalt psychology known as categorizing or card sorting.

As a result of this process we arrive at more prototypes for one class or with new classes and at least one prototype for these new classes.

The discrimination power must be checked again based on this new data set. New attributes, new number of prototypes, or a new similarity measure might be the output.

The process is repeated as long as the expert is satisfied with the result.

As result of the whole process we get a data set of samples with true class labels, the settings for the prototype-based classifier, the important attributes, and the real prototypes. The class labels represent the categories of the cellular processes going on in the experiment.

As result of our knowledge acquisition process we obtain the relevant categories for the application and a classifier that now can be used for routine work on-line.

8.4 Application

After the assay has been set up it is not quite clear what the appearances of the different phases of a cell are. This has to be learnt during use of the system.

Based on their knowledge, the biologists set up several descriptions for the classification of the mitochondria. They grouped these classes as follows: tubular cells, round cells and dead cells. For the appearance of these classes see images in Fig. 8.5.

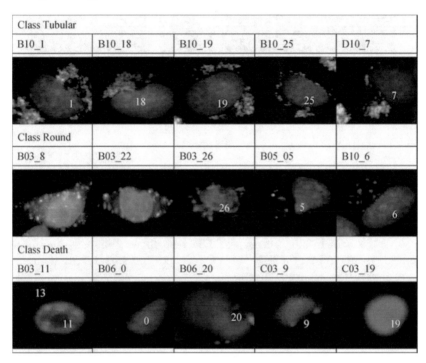

Fig. 8.5 Sample images for three classes.

Then prototypical cells were selected and the features were calculated with the software tool *CellInterpret* [1].

The expert rated the similarity between these prototypical images.

Our data set consist of 223 instances with the following class partition:

– 36 instances of class Death;
– 120 instances of class Round;
– 47 instances of class Tubular, and 114 features for each instance.

The expert chose for each class a prototype shown in Fig. 8.6 The test data set for classification has then 220 instances. For our experiments we also selected 5 prototypes pro class, respectively, 20 prototypes pro class. The associate test data sets do not contain the prototypes.

Prototype Death (B6_23)	Prototype Round (B3_22)	Prototype Tubular (F10_2)

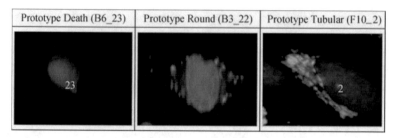

Fig. 8.6 The prototypes for the class Death, Round and Tubular.

8.5 Results

Figure 8.7 (a) shows the accuracy for classification based on different numbers of prototypes for all attributes and Fig. 8.7 (b) shows the accuracy for a test set based on only the three most discriminating attributes. To run the classifier with $k = 3$ for 3 prototypes is mean less. Therefore, the curve for $k = 3$ starts by 15 prototypes. The test shows that the classification accuracy is not so bad for only three prototypes but with increasing number of prototypes the accuracy increases.

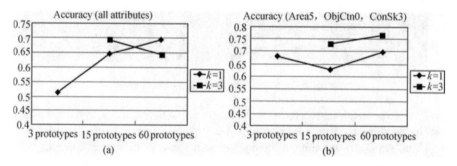

Fig. 8.7 Accuracy versus prototypes for two different feature subsets. (a) Accuracy for different numbers of prototypes using all attributes. (b) Accuracy for different numbers of prototypes using 3 attributes.

The selection of the right subset of features can also improve the accuracy and can be done based on the method presented in Section 8.2 for a low number of samples. The right chosen number of closest cases k can also help to improve accuracy but cannot be applied if we only have three prototypes or less prototypes in the data base.

Figure 8.8 shows the accuracy according on the number of attributes. In case of three prototypes the accuracy decreases as more attributes are taken for classification while in case of more prototypes the accuracy increases if more attributes are taken for classification. However, for the number of six attributes and larger the accuracy stay much stable. This observation nicely demonstrates the effect of the curse of dimensionality.

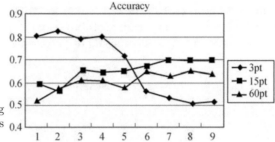

Fig. 8.8 Accuracy depending on the number of attributes $(k = 1)$.

Table 8.2 shows the discrimination power between three prototypes using the three attributes (ObjCnt0, ArSig0, ObjCnt1). It shows the best discrimination power between "death" and the two others "round" and "tubular" while there is more less discrimination power between "round" and "tubular". This observation is not surprising since it is also visually more harder to distinct "round" from "tubular" and vise-versa.

Table 8.2   Difference between 3 prototypes using the 3 attributes (ObjCnt0, ArSig0, ObjCnt1)

	B6_23	B03_22	F10_2
B6_23	0	0,669503257 (0,8)	0,989071038 (0,6)
B03_22	0,669503257 (0,8)	0	0,341425705 (0,9)
F10_2	0,989071038 (0,6)	0,341425705 (0,9)	0

Figure 8.9 shows the classification results for the 220 instances without adjustment (labelled as FB), meaning the weights a_l are equal to one $(1, 1, 1)$, and with adjustment (labeled as EB) based on expert's rating where the weights are $(0.005\,464\,48,\ 0.005\,025\,79,\ 0.002\,026\,21)$ as an outcome of the minimization problem. The result shows that accuracy can be improved by applying the adjustment theory; especially the class specific quality is improved by applying the adjustment theory.

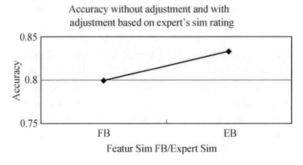

Fig. 8.9 Accuracy with and without adjustment theory.

The application of the methods for a larger sample set did not bring any significant reduction in the number of prototypes (see Fig. 8.10) or in the feature subset (see Fig. 8.11). The prototype selection method reduced the number of prototypes only by three prototypes. We take this as an indication that we have not yet enough prototypes and that the accuracy of the classifier can be improved by collecting more prototypes.

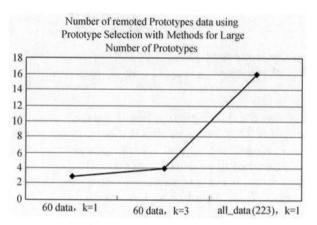

Fig. 8.10 Number of removed Prototypes.

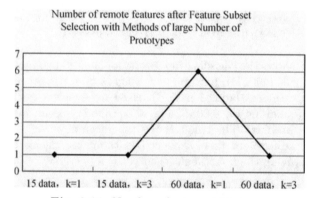

Fig. 8.11 Number of removed Features.

In Summary, we have shown that the chosen methods are valuable methods for a prototype-based classifier and can improve the classifier performance. In future work we will investigate the adjustment theory as a method to learn the importance of features based on a smaller number of features and a feature subset selection for a smaller number of samples.

8.6 Conclusion

We have presented our results on a prototype-based classification. Such a method can be used for incremental knowledge acquisition and classification. Therefore the classifier requires methods that can work on smaller numbers of prototypes and on a large number of prototypes. Our first result shows that feature subset selection based on the discrimination power of a feature is a good method for smaller numbers of prototypes. The adjustment theory in combination with an expert similarity judgment can be used to learn the true feature range in case of fewer prototypes. If we have a large number of prototypes an option for prototype selection that can check for redundant prototypes is necessary.

The system can start to work on a low number of prototypes and can instantly collect samples during the use of the system. These samples get the label of the closest case. The system performance improves with more prototypes being provided in the data base of the system. This means that an iterative process of labeled sample collection based on prototype-based classification is necessary that is followed by a revision of these samples after some time in order to sort out wrongly classified samples until the system performance has been stabilized.

References

[1] Perner P (2008) Novel Computerized Methods in System Biology-Flexible High-Content Image Analysis and Interpretation System for Cell Images. In: Perner P, Salvetti O (eds) Advances in Mass Data Analysis of Images and Signals in Medicine, Biotechnology, Chemistry and Food Industry, MDA 2008, LNAI, vol 5108: 139 – 157, Springer, Heidelberg

[2] Pham T (2007) Geo-Thresholding for Segmentation of Fluorescent Microscopic Cell Images. In: Perner P, Salvetti O (eds) Advances in Mass Data Analysis of Signals and Images in Medicine Biotechnology and Chemistry Lnai, vol 4826: 15 – 26, Springer, Heidelberg

[3] Colantonio S, Perner P, Salvetti O (2009) Diagnosis of Lymphatic Tumors by Case-Based Reasoning on Microscopic Images. Trans on Case-Based Reasoning, 2(1): 29 – 40

[4] Pham T (2007) A Novel Image Feature for Nuclear-Phase Classification in High Content Screening. In: Perner P, Salvetti O (eds) Advances in Mass Data Analysis of Signals and Images in Medicine Biotechnology and Chemistry Lnai, vol 4826: 84 – 93, Springer, Heidelberg

[5] Chang C L (1974) Finding Prototypes for Nearest Neighbor Classifiers. IEEE Trans on Computers, C-23 (11)

[6] Perner P (2002) Data Mining on Multimedia Data. LNCS, vol 2558, Springer, Heidelberg

[7] Perner P (2008) Prototype-Based Classification. Applied Intelligence, 28: 238 – 246

[8] Niemeier W (2008) Ausgleichsrechnung. de Gruyter, Berlin, New York

[9] Sachs-Hombach Kl (2002) Bildbegriff und Bildwissenschaft. In: Gerhardus D, Rompza S (eds) Kunst – Gestaltung – Design, vol 8: 1 – 38, Verlag St Johann, Saarbrücken

[10] Wang P (2010) Concept of Ambiguity and Applications to Security and Transportation Safety. In: Proceedings IEEE International Conference on System Science and Engineering 2010, pp 179 – 183

[11]   Perner J (2010) Characterizing Cell Types through Differentially Expressed Gene Clusters Using a Model-based Approach. ibai-Publishing, Leipzig

9 Tracking Deformable Objects with Evolving Templates for Real-Time Machine Vision

Elena Sánchez-Nielsen[1] and Mario Hernández-Tejera[2]

Abstract A tracking approach not only needs to matching target objects in dynamic scenes, it also needs to update templates when it is required, computing occlusions, and processing multiple-objects with real-time performance. In this Chapter, we present an heuristic search algorithm with target dynamics to match target objects with real-time performance on general purpose hardware. The results of this heuristic search are combined with the more common views of target objects, and intensity information in order to update the templates. As a result, the updating process will be computed only when the target object has evolved to a transformed shape dissimilar with respect to the current shape, providing robust tracking, and multi-object tracking because accurate template updating is performed. The paper includes experimental results with inside and outside video streams demonstrating the effectiveness and efficiency for real-time machine vision based tasks in unrestricted environments.

9.1 Introduction

One of the basic tasks in image processing and machine vision is the visual tracking of deformable objects with a wide range of applications such as surveillance, activity analysis, classification and recognition from motion, and human-computer interfaces. The tracking approaches can be classified into three main categories: contour-based tracking, feature-based tracking, and region-based tracking. The central challenge of all these approaches is to determinate the image position of a target and/or multiple targets in a dynamic vision context given one or several templates that represent the targets. According to the approach used templates can be represented by a representation of the bounding contour of the object such as active contour models [1], sub-features such as distinguishable points or lines on the target [2], and blobs associated with a connected region in the image such as mean-shift and integral histogram approaches [3, 4]. If an active vision approach is considered, it is also desirable that the tracking approach keeps the target centered in the image, moving the sensor adequately [5, 6].

---

1 Dpto. E.I.O. y Computación. Universidad de La Laguna, 38271 La Laguna, Spain. E-mail: enielsen@ull.es.
2 Instituto de Sistemas Inteligentes y Aplicaciones Numéricas en Ingeniería, 35005 Las Palmas de G.C., Spain. E-mail: mhernandez@iusiani.ulpgc.es.

At present, there are still obstacles in achieving all-purpose and robust
tracker approaches. Four main issues should be addressed in order to carry
out an effective template tracking approach:

1) *Real-time performance.* Real-time template tracking is a critical task
in many computer vision applications such as visual surveillance [7], traffic
control [8],vision based interface tasks [9], navigation tasks for autonomous
robots [10], gesture based human-computer-interaction [11], perceptual intelligence applications [12], virtual and augmented reality systems [13], or applications from the "looking and people" domain [14]. Moreover, in real-time
applications not all system resources can be allocated for tracking processes
because other high-level tasks such as trajectory interpretation and reasoning
can be demanded. Therefore, it is desirable to adjust the requirements of the
computational cost of a tracker approach to be as low as possible to make
feasible real-time performance over general purpose hardware.

2) *Matching.* Template matching is the process in which a reference template $T(k)$ is searched for in an input image $I(k)$ to determine its location
and occurrence. Over the last decade, different approaches based on searching the space of transformations using a measurement similarity have been
proposed for template based matching. Some of them explicitly establish
point correspondences between two shapes and subsequently find a transformation that aligns these shapes [15, 16]. The iteration of these two steps
involves the use of algorithms such as iterated closest points (ICP) [17, 18] or
shape context matching [16]. However, these methods require a good initial
alignment in order to converge, particularly whether the image contains a
cluttered background. Other approaches are based on searching the space of
transformations using Hausdorff matching [19], which are based on an exhaustive search that works by subdividing the space of transformations in
order to find the transformation that matches the template position into the
current image. Also, similar techniques have been used for tracking selected
targets in natural scenes [10] and for people tracking using an autonomous
robot [20]. However, no heuristic functions and no target dynamics have been
combined in these search processes. This situation leads to an increase of the
computational costs in the tracking process.

3) *Updating.* The underlying assumption behind several template tracking
approaches is that the appearance of the object remains the same through
the entire video [21 – 23, 3]. This assumption is generally reasonable for a
certain period of time and a naive solution to this problem is updating the
template every frame [10, 20, 24] or every n frames [25] with a new template
extracted from the current image. However, small errors can be introduced
in the location of the template each time the template is updated and this
situation establishes that the template gradually *drifts* away from the object
[26]. Matthews, Ishikawa and Baker in [26] propose a solution to this problem. However, their solution approach only addresses the issue related to
objects whose visibility does not change while they are being tracked. Other
approaches, fix the template at the first frame and it is not updated during

the sequence, limiting the method to work with occluding objects [21].

4) *Occlusion and multiple object tracking.* Tracking multiple objects in different scenarios involves considerable difficulty because of occlusions, working with different objects which deform differently, and/or move with different speeds. At the same time, real-time performance is reduced with the increase of targets to be tracked.

In this paper, we address these four issues (real-time performance, matching, updating, occlusions and multiple object tracking) by using a contour-based tracking approach, where the template tracking problem of objects in 3D space from 2D images is formulated in terms of decomposing the transformations induced by moving objects between frames into two parts: 1) a 2D motion, corresponding to the change of the target position in the image space, which is referred to as the template position matching problem and 2) a 2D shape, corresponding to a different aspect of the object becoming visible or an actual change of shape in the object, which is referred to as the template updating problem.

The main contributions are: (1) an A* search algorithm that uses the Kullback-Leibler measurement as heuristic to guide the search process for efficient matching of the target position, (2) dynamic update of the space of transformations using the target dynamics, dramatically reducing the number of possible alternatives, (3) updating templates only when the target object has evolved to a new shape change significantly dissimilar with respect to the current template in order to solve the drift problem, (4) representation of the most common views of the target shape evolution through a short-time visual memory, and (5) the use of intensity information to confirm the best matching position.

As a result, the first two contributions provide a fast algorithm to apply over a space of transformations for computing target 2D motion, and the other three contributions provide robust tracking because accurate template updating can be performed. In addition to these contributions, the paper also contains a number of experimental evaluations and comparisons:

- A direct comparison of the performance of conventional search approaches and the proposed A* search approach, demonstrating that the proposed approach is faster.
- An empirical comparison of updating templates using a continuous updating approach and the updating template approach that is proposed in this paper, demonstrating that no updating templates in every frame and using a dynamic short-time visual memory, lead to a more robust tracking approach.
- Results of tracking and multi-object tracking with deformable objects.
- Temporal cost for the proposed approach, illustrating that the time to track targets in video streams is lower than real-time requirements.

The structure of this paper is as follows: the problem formulation is illustrated in Section 9.2. In Section 9.3, the A* search algorithm for computing target position is described. The updating reference template problem is detailed in Section 9.4. Multi-object tracking is described in Section 9.5. Experimental results are provided in Section 9.6 and Section 9.7 concludes the chapter.

9.2 Problem Formulation

For the sake of subsequent problem formulation, some definitions are introduced first:

Definition 9.1 (Template)   Let $T(k) = \{t_1, \cdots, t_r\} \subseteq \mathbb{R}^2$ be a set of points that represent a template in step time k.

Definition 9.2 (Image)   Let $I(k) = \{i_1, \cdots, i_s\} \subseteq \mathbb{R}^2$ be another set of points that denote an input image in step time k. It is assumend that each new step time k corresponds to a new frame k of the video stream.

Definition 9.3 (Set of transformations)   Let a translational transformation g be parameterized by the x displacement g_x and the y displacement g_y. That is, $g = (g_x, g_y)$.

Let a bounded set of translational transformations be a set of transformations $\mathbb{G} = \big[g_{xmin}, g_{xmax}\big] \times \big[g_{ymin}, g_{ymax}\big] \subseteq \mathbb{R}^2$ and let $g^c = (g_x^c, g_y^c)$ denote the transformation that corresponds to the center of $\mathbb{G}$. It is defined as

$$g^c = \left(\left(\frac{1}{2}\big(g_{xmin} + g_{xmax}\big)\right), \left(\frac{1}{2}\big(g_{ymin} + g_{ymax}\big)\right)\right). \qquad (9.1)$$

Where the lower and upper bounds of $\mathbb{G}$ in x and y dimension are represented by $(xmin, xmax)$ and $(ymin, ymax)$.

9.2.1 Template Position Matching Problem

Given a step time k, a template $T(k)$, an input image $I(k)$ and an error bound ε, the template position matching problem can be viewed as the search process in the space of transformations $\mathbb{G}$ in order to find the transformation g_{opt} that satisfies the quality of match $Q(g)$.

The quality of match $Q(g)$ assigned to a transformation g is defined by the allowed error ε when template points are brought to image points using the transformation g. It is expressed as

$$Q(g) = h_l\left(g\big(T(k)\big), I(k)\right) < \varepsilon, \qquad (9.2)$$

where $h_l\big(g\big(T(k)\big), I(k)\big)$ is the partial directed Hausdorff distance [27] between the translated points set of the template $T(k)$ by the translation g (it is denoted as $g(T(k))$), and the input image $I(k)$. It is defined as

$$h_l\left(g\big(T(k)\big), I(k)\right) = \underset{t \in T}{L^{th}} \min_{i \in I} \|g\big(t(k)\big) - i(k)\|, \qquad (9.3)$$

where $\| \cdot \|$ denotes the Euclidean distance. The partial directed Hausdorff distance ranks each point of the translated template T based on its distance to the nearest point in I and uses the lth quartile ranked point as the measure of the distance.

9.2.2 Template Updating Problem

Once the new position of the target has been computed, the template is updated to reflect the change in its shape. Since the template matching position problem determines translational motion between consecutive frames, all the non-translational transformations of the image motion and three-dimensional shape change are considered to be a change in the 2D shape of the object.

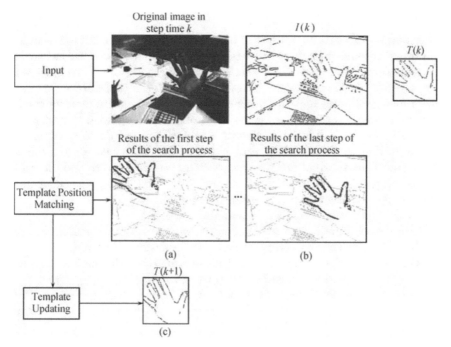

Fig. 9.1 A schematic overview of the template tracking problem. The template position matching consists of the search process in the space of transformations $\mathbb{G}$ in order to find the transformation g_{opt} that satisfies the quality of match $Q(g)$. Figure 9.1 (a) illustrates the result of applying the initial transformation $g = (g_x, g_y)$ of the search process to every point in template $T(k)$, where $Q(g)$ is not satisfied. Figure 9.1 (b) shows the result of applying the transformation that best matches the current template points $T(k)$ in the image $I(k)$, after the search process has been computed and where $Q(g)$ is satisfied. Once the template matching problem has been computed, the template is updated in order to represent the new view of the target such is illustrated in Fig. 9.1 (c).

Given $g_{opt}(T(k))$ and $I(k)$, new 2D shape changes between successive
images are computed as the measure of the discrepancy between $g_{opt}(T(k))$
and $I(k)$ under a certain error bound δ. That is, the new template $T(k+1)$ is
represented by all those points of input image $I(k)$ that are within distance
δ of some point of $g_{opt}(T(k))$ according to the following expression:

$$T(k+1) = \sum_{i \in I} \left\| g_{opt}(T(k)) - i \right\| < \delta. \tag{9.4}$$

Figure 9.1 illustrates a graphical summary of the problem formulation.
The set of points that represent $I(k)$ and $T(k)$ are edge-based features extracted from real images using a Canny edge detector [28].

9.3 Search Framework for Computing Template Position

Formulation of problem solving under the framework of heuristic search is
expressed through a *state space-based representation approach* [29], where the
possible problem situations are considered as a *set of states*. The *start state*
corresponds to the initial situation of the problem, the *goal state* corresponds
to problem solution, and the transformation between states can be carried
out by means of *operators*. As a result, problem solving is addressed as the
sequence of operators which transform the start state to the goal state.

According to the heuristic search framework described above, the A*
search problem is formulated as: the search process oriented to find the transformation g_{opt} in $\mathbb{G}$ that satisfies the quality of match $Q(g)$. The elements of
the problem solving according to this formulation are:
- Initial state: corresponds to a bounded set of $\mathbb{G}$.
- State. each search state n is associated with a subset of transformations
  $G_n \subseteq \mathbb{G}$. Each state is represented by the transformation g^c.
- Goal state is the transformation that best matches the current template
  $T(k)$ in the current image $I(k)$, according to $Q(g)$.
- Operators are the functional elements that lead the transformation of one
  state to another. For each current state n, the operators A and B are
  computed:

1) *Operator A*. Each set of transformations from the current state is partitioned into four new states by vertical and horizontal bisections. Each one
of these new states is represented by the transformation g^c.

2) *Operator B*. This operator computes the quality of match (Expression
9.3) for each one of the four new states generated by *operator A*. It is denoted
as $h_l\big(g^c(T(k)), I(k)\big)$.

Splitting each current state into four new states leads to the representation
of the search tree to be a quaternary structure, where each one of the four
states computed are referred to as *NW, NE, SE,* and *SW* cells. The heuristic
search is initiated by the association of the set of tranformation $\mathbb{G}$ with

the root of the search, and subsequently the best state at each tree-level l is expanded into four new distinct states, which are non-overlapping and mutually exclusive. The splitting operation is finished when the quadrisection process computes a translational motion according to the quality of match $Q(g)$ or all the different states have been partitioned in cells of unit size.

The best state to expand from NW, NE, SW, and SE cells at each tree-level l is computed by the evaluation function of the A* search algorithm [16], which combines features of the uniform-cost search and pure heuristic search. The corresponding cost value assigned to each state n by the evaluation function $f(n)$ is given as

$$f(n) = c(n) + h^*(n), \tag{9.5}$$

where $c(n)$ is the estimated cost of the path from the initial state s_0 to the current state n, and $h^*(n)$ is the heuristic estimate of the cost of a path from state n to the goal state.

9.3.1 Heuristic Evaluation Function $h^*(n)$

The heuristic value of the cost of a path from state n to the goal is estimated evaluating the quality of the best solution reachable from the current state n. Desirability of the best state is computed measuring the similarity among the distribution functions that characterize the current state and the goal state.

Let P denote the distribution function that characterizes the current state n and let Q denote the corresponding distribution function that characterizes the goal state. Since the quality of match $Q(g)$ is denoted by the partial directed Hausdorff distance, the distribution function P can be approximated by a histogram of distances $\{H_{g^c}\}_{i=1\cdots r}$, which contains the number of template points $T(k)$ at distance d_j with respect to the points of the input image $I(k)$, when the transformation g^c of the current state n is applied on the current template $T(k)$.

Likewise, since the quality of match $Q(g)$ is denoted by the partial directed Hausdorff distance, the distribution function Q that characterizes the goal state corresponds to the largest number of ranked template points $T(k)$ at distance zero, and closest to zero with respect to the points of the input image $I(k)$, when the transformation g^c of the current state n is applied on the current template $T(k)$. This distribution function Q can be modelled by an exponential distribution function $f(n) = ke^{-an}$, where parameter a checks the falling of the exponential function. Therefore, if the quality of match $Q(g^c)$ assigned to the transformation g^c is satisfied, the distributions P and Q, respectively, will show an appearance similar to the illustration of Fig. 9.2 (a) and (b).

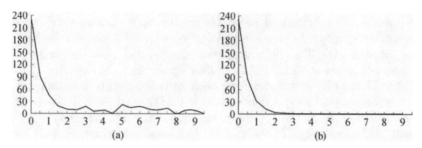

Fig. 9.2 *Distribution functions*: (a) Histogram of distances $\{H_{g^c}\}_{i=1\cdots r}$ associated
with a transformation g^c that verifies the quality of match $Q(g^c)$. (b) Distribution
function $f(n) = ke^{-an}$ with parameter $a = 1$ that characterizes the final state. The
horizontal axis represents distance values and the vertical axis denotes the number
of transformed template points with g^c at distance d_j with respect to the current
points scene $I(k)$.

The similarity between distributions P and Q is measured using the
Kullback-Leibler distance (KLD) [30, 31]:

$$D(P\|Q) = \sum_{i=1}^{R} p_i \log \frac{p_i}{q_i}, \tag{9.6}$$

where R is the number of template points. According to [30], $D(P\|Q)$ has
two important properties:
1) $D(P\|Q) \geqslant 0$;
2) $D(P\|Q) = 0$ iff $P = Q$.

The values of KLD will be non-zero and positive when the distributions
P and Q are not similar because the template points do not match the input
image points. Otherwise, the value of KLD is equal or close to zero when the
template points match the input image points.

9.3.2 Estimated Cost Function $c(n)$

An estimated cost function $c(n)$ is added to the evaluation function $f(n)$ in
order to generate a backtracking when the heuristic function leads the search
process towards no promising solutions. This depth term is based on the
number of operators of type A applied from the initial search state to the
current state n.

9.3.3 Initial State Computation

The dimension $M \times N$ of the set of the transformation $\mathbb{G}$ is dynamically
computed as an adjustable size-based set of transformations in every frame k

by means of incorporating an alpha-beta predictive filtering [32] into the A*
search algorithm. It is focused on the assumption that there is a relationship
between the size of the set of transformations and the resulting uncertainty
of the alpha-beta predictive filtering. The basic hypothesis underlying this
assumption is that the regularity of the movement of mobile objects is related
to small uncertainties in the prediction of future target locations, which leads
to a reduced set of transformations. Otherwise, the set $\mathbb{G}$ is increased when
the motion trajectory exhibits any deviation from the assumed temporal
motion model.

The dimension of $\mathbb{G}$ is computed from the parameters to be estimated
by the filtering approach. The location vector and the corresponding velocity
vector of the filtering approach [32] are jointly expressed as a state vector
$x = [\theta^{\mathrm{T}}, \dot{\theta}^{\mathrm{T}}]^{\mathrm{T}}$. The dynamic state equation assuming a constant velocity
model is formulated as

$$x(k+1) = \Phi x(k) + v(k). \tag{9.7}$$

The state vector estimation is computed as

$$\hat{x}(k+1 \mid k+1) = \hat{x}(k+1 \mid k) + v(k+1) \left[\alpha \frac{\beta}{\Delta T} \right]^{\mathrm{T}}, \tag{9.8}$$

where $\Phi = \begin{bmatrix} 1 & \Delta t \\ 0 & 1 \end{bmatrix}$, Δt is the sampling interval, $v(k)$ denotes the process
noise vector, $\hat{x}(k+1 \mid k+1)$ represents the updated estimate of x, $\hat{x}(k+1 \mid k)$
corresponds to the predicted value at time step $k+1$, and $v(k+1)$ denotes the
residual error, called innovation factor. The parameters α and β respectively
are used for weighting the target position and velocity in the update state
computation.

Since the innovation factor $v(k)$ represents a measure of the error of $\hat{z}(k+1)$,
a decision rule focused on the uncertainty measurement can be computed
in order to determine the dimension of $\mathbb{G}$. Two criteria are considered in
the decision rule design. The first one is that small values of the innovation
factor indicate low uncertainty about its estimate and therefore, a reduced
size of $\mathbb{G}$. The second criterion is that the dimension of $\mathbb{G}$ must be a $2^p \times 2^q$
value in order to assure that each terminal cell of $\mathbb{G}$ will contain only a
single transformation after the last quadrisection operation of the A* search
algorithm had been applied. Assuming these criteria, $M \times N$ is bounded by
$2^{min} \leqslant v(k) \leqslant 2^{max}$. From this, the dimension $M \times N$ of $\mathbb{G}$ can be computed
according to the following decision rule:

$$M, N = \begin{cases} 2^{min}, & \text{if } w + 2^{min} \leqslant v(k), \\ 2^{max}, & \text{if } w + 2^{min} > v(k), \end{cases} \tag{9.9}$$

where w is computed according to the following expression:

$$w = \phi(2^{max} - 2^{min}) \tag{9.10}$$

and ϕ is a parameter that weights the influence of the difference between both
bounds in the selection of the appropriate value. Figure 9.3 illustrates the
integration of the alpha-beta filtering with the computation of the adjustable
set of transformations.

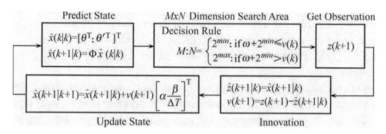

Fig. 9.3 *Adjustable size-based set of transformations:* Computation of $M \times N$
dimension of $\mathbb{G}$.

Once M and N have been computed, the low and upper bounds $[g_{xmin},$
$g_{xmax}] \times [g_{ymin}, g_{ymax}]$ of $\mathbb{G}$ in each step time k are calculated as follows:

$$g_{xmin}(k) = g_x(k-1) - \frac{M}{2}, \tag{9.11}$$

$$g_{xmax}(k) = g_x(k-1) + \left(\frac{M}{2}\right) - 1, \tag{9.12}$$

$$g_{ymin}(k) = g_y(k-1) - \frac{N}{2}, \tag{9.13}$$

$$g_{ymax}(k) = g_y(k-1) + \left(\frac{N}{2}\right) - 1, \tag{9.14}$$

where $\big(g_x(k-1), g_y(k-1)\big)$ represents the components of the solution transformation g computed in previous step time $k-1$.

9.3.4 A* Search Algorithm

In this section, we describe the steps of the A* search algorithm for the
template matching problem.

Input

$\mathbb{G}$: initial set of transformations.
ε: distance error bound allowed.
$v(k)$: uncertainty measurement computed by the alpha-beta filtering
approach.
ω : estimate which determines from what value the 2^{max} or 2^{min} bound
value should be selected for computing $M \times N$ dimension of $\mathbb{G}$.
2^{max} and 2^{min}: upper and lower values for the innovation factor $v(k)$.
$D(P\|Q)$: value of Kullback-Leibler distance between the distribution
functions that characterize the search state that is being evaluated, P,

and the goal state, Q.

η : number of operators of type A applied from the initial search state to the current state n.

Output

$M \times N$ dimension of $\mathbb{G}$
$g_{opt} = [g_{xopt}, g_{yopt}]$

Algorithm

Step 1: Compute $M \times N$ dimension of $\mathbb{G}$:

$$M, N = \begin{cases} 2^{min}, & \text{if } w + 2^{min} \leqslant v(k), \\ 2^{max}, & \text{if } w + 2^{min} > v(k). \end{cases}$$

Step 2: Find g_{opt} such that $Q(g) = h_l\big(g(T(k)), I(k)\big) < \varepsilon$ is satisfied:

While $Q(g) > \varepsilon$ **or** (all states are not made up by a single translation)
Do
    1) Split current state n into four new states $\{n\}_{i=1\cdots4}$
    2) Compute $Q(g_c) \leftarrow h_l\big(g_c(T(k)), I(k)\big)$ for each new state n_i
    3) Expand the best state n_i according to the evaluation function $f(n) = c(n) + h^*(n)$:
        (1) $c(n) \leftarrow c(n-1) + \eta$
        (2) $h^*(n) \leftarrow D(P\|Q)$
End While
    Step 3: Output $g_{opt} = [g_{xopt}, g_{yopt}]$ or no solution if all states conform a single translation and $Q(g^c) = h_l\big(g^c(T(k)), I(k)\big) > \varepsilon$ for each g^c.

9.3.5 Recovering Exceptions

Four different situations are presented in shape tracking when an alpha-beta predictive filtering is integrated with an A* heuristic search:

- The new measurement computed is non-zero and the alpha-beta prediction is zero: this situation corresponds to the start of the movement of target shape and therefore, an initialization process of the alpha-beta filtering must be computed.
- The new measurement is zero and similar to the non-zero prediction: this situation is related to an accurate estimation of the alpha-beta filtering and the A* search algorithm.
- The new measurement is non-zero and noticeably different from the non-zero prediction: this situation denotes the presence of a motion discrepancy due to erratic direction changes or extreme deformations introduced by the target shape.
- The new measurement is zero and the prediction is non-zero: stopping conditions are introduced by the target shape in its trajectory motion.

Since there is no reason to assume that the motion after a discontinuity can be accurately predicted from the motion history of the tracked object, a new initialisation of the alpha-beta filtering is computed for the last two situations.

Introduction of additional mechanisms are required in the third situation when the heuristic algorithm cannot provide a solution g_{opt}, such that the condition $Q(g) = h_q\big(g(T(k)), I(k)\big) < \varepsilon$ is verified, because erratic direction changes of the object or extreme deformations by the target shape have been introduced. With the purpose of computing target position in this situation, the search algorithm is carried out according to the next steps:

- Search Area Increase. A new dimension of $\mathbb{G}$, $M \times N$, is proportionally computed for the initial search area established to match the target position.
- Use of an Iconic Memory Stack including different significant views of the object. With the aim to guarantee the best matching of the target object, the heuristic search is computed with all the different views of the target that are stored in the short-term memory of the tracking system.
- Compute weak solutions. A weak solution is a solution where the error bound ε of the quality of function match is increased in one unit until a maximum value of 10. New A* searches using the current template and the different views that represent the target object are computed for each new value of ε until a solution is achieved.
- Alert State Establishment. The loss of the target is established when the object cannot be matched with the previous actions. These actions are achieved in the following images until new regularities are detected. A disappearance of the object from the scene represents the main reason of the alert situation.

9.4 Updating Framework for Computing Template Changes

The new template is only updated when the target object has evolved significantly to a new shape. Since the A* search algorithm defines the 2D motion component, a solution g that satisfies $Q(g)$ with a error bound ε denotes that target object is moving according to a translational motion and no different aspect of the target is becoming visible. On the other hand, new 2D shape changes take place if the A* search algorithm computes a weak solution, that is a solution g that satisfies $Q(g)$ with a maximum error bound ε_{max}.

9.4.1 Short-Time Visual Memory

The different templates that compose Short-Time Visual Memory $STVM$

represent the more common views of the target object. With the purpose of minimizing redundancies, the problem about what *representative* templates should be stored is addressed as a dynamic approach that removes the less distinctive templates in order to leave space to the new ones. In order to determine the weight of every template, we introduce a *relevance index*, which is defined according to

$$R(k, i) = \frac{T_p(i)}{1 + k - T_s(i)}, \tag{9.15}$$

where k represents the time step, i corresponds to the identification symbol template, $T_p(i)$ represents the frequency of use as template, and $T_s(i)$ corresponds to the time from the last instant it was used as the current template. A new template is inserted into $STVM$, when the quality of match $Q(g)$ ranges from ε to ε_{max}. The template with the lower index of relevance is removed from $STVM$ when the stack of templates is full and the new template is included into $STVM$.

9.4.2 Template Updating Algorithm

According to the value of $Q(g_{opt})$ computed by the A* search algorithm, each current template $T(k)$ is updated as $T(k + 1)$ based on one of the steps of the following algorithm:

Step 1 If $Q(g_{opt}) \leqslant \varepsilon$, the new template in time step $k + 1$, $T(k+1)$, is equivalent to the best matching of $T(k)$ in $I(k)$. That is the edge points of $I(k)$ that are directly overlapping on some edge point of the best matching, $g_{opt}(T(k))$, represent the new template $T(k + 1)$:

$$T(k + 1) \leftarrow \{i \in I(k) \mid \min_{t \in T(k)} \|g_{opt}(t) - i\| = 0\}.$$

Step 2 If some template of $STVM$ computes the best matching when the current template $T(k)$ cannot match the target object with an inferior or equivalent distance value ε, $Q(g_{opt}) > \varepsilon$, this template of $STVM$ is selected as the new template $T(k+1)$. Otherwise, the current template, $T(k)$ is updated by incorporating the shape change by means of the partial directed Hausdorff distance measure according to Expression 9.4. In this context, we denote $STVM = \{T(STVM)_i\}_{i=1}^{N}$ as the different templates that integrate $STVM$, $Q(g; T(STVM)_i, I(k), \varepsilon)$ as the best error bound distance ε computed for the ith template of $STVM$. The updating process is expressed as

$$T(k + 1) \leftarrow \left\{ \begin{array}{l} \{i \in I(k) \mid \min_{t \in T(k)} \|g_{opt}(t) - i\| \leqslant \delta\}, \text{if} \\ Q(g_{opt}; T(k), I(k), \varepsilon) \leqslant Q(g_{opt}; T(STVM)_i, I(k), \varepsilon) \\ T(STVM)_i, \text{if } Q(g_{opt}; T(STVM)_i, I(k), \varepsilon) \\ \qquad < Q(g_{opt}; T(k), I(k), \varepsilon) \end{array} \right\}$$

Step 3 If the best matching computed using the current template $T(k)$
and all templates of $STMS$ is superior to the error bound distance ε_{max}, no
template is updated:

$$T(k+1) \leftarrow \{\phi \ \text{if} \ Q(g_{opt}; T(k), I(k), \varepsilon) \geqslant \varepsilon_{max} \ \text{and}$$
$$Q(g_{opt}; T(STVM)_i, I(k), \varepsilon) \geqslant \varepsilon_{max}\}.$$

Figures 9.4, 9.5, 9.6, and 9.7 illustrate the results obtained with the updating template approach using rigid and non-rigid targets. Each figure shows:
1) the current edge template updated from frame k, $T(k)$, and used for matching the target object in frame $k + 1$; 2) the new template $T(k + 1)$, which is
updated from the best matching results between $T(k)$ and frame $k + 1$ and
the use of STMS when it is required. Edges were extracted using a Canny
edge detector [26].

Fig. 9.4 Template updating: The target motorcycle is moving according to
a translational motion trajectory. Therefore, $Q(g_{opt}) \leqslant \varepsilon_{min}$ and the new template $T(k + 1)$ is updated from the edge contours of frame $k + 1$ that are directly overlapping on some edge contour of the best matching $g_{opt}(T(k))$. That is
$T_{k+1} \leftarrow \{i \in I_k \mid \min\limits_{t \in T_k} \|G_{max}(t) - i\| = 0\}.$

Fig. 9.5 Template updating: A shape change of the hand target has taken place
between frame k and frame $k+1$ due to a fast motion. Therefore, $\varepsilon_{min} \leqslant Q(g_{opt}) \leqslant$
ε_{max} and the new template $T(k + 1)$ is updated from the edge contours of frame
$k+1$ that are at δ distance from some edge contour of the best matching $g_{opt}(T(k))$.
That is $T(k + 1) \leftarrow \{i \in I(k) \mid \min\limits_{t \in T(k)} \|g_{opt}(t) - i\| \leqslant \delta\}.$

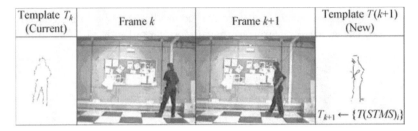

Fig. 9.6 Template updating: Since the target person has rotated around himself according to a non-translational motion from frame k to frame $k + 1$, the best matching computed is obtained by means of a template of $STMS$. Therefore, this template $T(STMS)_i$ is selected as the new template $T(k + 1)$.

Fig. 9.7 Template updating: No template is updated in step time $k + 1$, since the target car has been disappeared from the scene. All the templates of $STMS$ are used in the following images until new regularities are detected.

9.5 Multiple Object Tracking and Intensity Information

The tracking approach described may work for sequences containing multiple moving objects. In such cases each target to be tracked is modeled as a visual tracking process. Using tracking processes, each target in image sequences such as the *People* sequence shown in Fig. 9.8 is associated with an individual process. In this way, we will have as many processes as targets to be tracked. The steps for each tracking process are as follows: (1) computing template position using the search framework; (2) computing updates oftemplates, and short-time visual memory.

The tracking approach described thus far does not work well when the template being tracked becomes a template with a reduced set of edge points. For instance, frames in which much of the object is dissapearing from view such as the *Car* sequence shown in Fig. 9.9. One means of dealing with this problem is to combine our approach with more perceptual pathways such as the use of intensity information. In such cases targets are located by sliding intensity templates through the search intensity window when the bounding box area which includes the template is below a margin error. Similarity between overlapped images is measured using a sum of squared differences

(SSD), although any other measure can be used. After the intensity template is located, the heuristic search framework is computed in the best matching

Fig. 9.8 Indoor sequences: *People* Sequence, *Hand* Sequence (frames 150 and 250 are shown).

Fig. 9.9 Outdoor sequences: Car Sequence (frames 150 and 250 are shown) and Motorcycle Sequence (frames 10 and 50 are shown).

position. If the results of both approaches confirm a matching, a new template is updated according to the approach described in Section 9.4. Otherwise, the tracking approach enters in a "suspicious mode" in which additional templates from the short-time visual memory are used in the best matching position computed by the SSD approach. Once the tracking process goes into suspicious mode, it stays that way for several frames, waiting for the re-appearing of the target.

9.6 Experiments and Results

This section illustrates the performance and average runtime of the tracking approach under a variety of circumstances, noting particulary the effects of the use of arbitrary shapes and the use of mobile and fixed acquisition cameras. All experiments were performed by an Intel Core2 Duo 3.0 GHz.

In order to compare the approach proposed with previous works, we use the same values of parameters that have previously been used. In the experiments performed, the goal state of the A* search algorithm is defined as the translation g that verifies that 80% (parameter $l = 0.8$) of template edge points are at maximum 2 pixels distance (error bound distance $\varepsilon = 2.0$) from the current explored image $I(k)$. The dimension of STVM is settled up to 8. That is the limit of the different reference templates that constitute STVM is eight. This dimension can be adjusted dinamically according to tracking conditions.

9.6.1 Experimental Environment

An experimental evaluation of the proposed framework is included to show its use in practical applications for vision based interface tasks such as gesture based human-computer-interaction, navigation tasks for autonomous robots, and visual surveillance.

Thirty different sequences, which contain 750 frames as average rate, have been used for experimental evaluations, achieving the same behavior for all of them. Particularly, indoor and outdoor video streams which are labeled "People", "Hand", "Cars" and "Motorcycle" are illustrated. Each one of these sequences contains frames of 280×200 pixels that were acquired at 25 fps. Edges were extracted from frames using a Canny edge detector [28].

Different features are taken into account in the experiments: 1) the use of arbitrary shapes, such as deformable targets and rigid objects, 2) mobile and fixed acquisition cameras, 3) presence of similar objects related to the target shape, and 4) multi-object tracking. The main aspects of each sequence used are:

− *People sequence*: contains 745 frames (that is 30 seconds) that are characterized by three people of 70×148 average size that are moving through an indoor corridor surrounded by several static and moving objects near the targets shape. The people movement is based on speeding up, slowing down and erratic motions. Also, diverse turns, stretching and deformations are introduced by the target object.

– *Hand sequence*: consists of 512 frames (20 seconds) that are characterized by a flexible hand of 108×116 average size that twists, bends and rotates on a desk with different objects around it. The tracker was tested for dramatic changes in appearance (frame 200) and rotations (frame 250).

– *Car sequence*: comprised of 414 frames (17 seconds) which were acquired through a visual surveillance system. Different parked cars and a mobile car are present. The average size of the template is 60×40 pixels. The tracker was tested for scale changes (frame 100, 250).

– *Motorcycle sequence*: includes 70 frames (3 seconds) of a motorcycle of 184×149 average size that is moving fast in a traffic environment. The video stream was acquired through a mobile camera with zoom features.

9.6.2 Tracking Sequence Results

Figure 9.8 illustrates two sample frames of the indoor sequences (People sequence and Hand sequence). People sequence is from the CAVIAR database [33]. This sequence shows a clip with multiple people in an inside corridor. The processing results of this sequence in the following sections include the data of the tracking of three people through the sequence.

Figure 9.9 shows other two sample frames of the outdoor sequences (Car sequence and Motorcycle sequence). In both figures the first column shows original frames; the second column illustrates the edge-based image using Canny detector, and the third column shows edge located template.

9.6.3 Comparative Analysis Between Search Strategies

Table 9.1 shows the performance measured in seconds between the proposed A* search algorithm with the conventional blind search strategy [19] that works by subdividing the transformations space and no heuristic function is used to guide the search process. The result shows that the A* heuristic search algorithm is faster than the blind search strategy in an average rate of four times better.

Table 9.1 Comparative runtime between search strategies

Sequence	People	Hand	Car	Motorcycle
N° of frames	745	512	414	70
Seconds Sequence	30	20	17	3
Time required to process A* Search Strategy	16.2	5.48	3.96	0.6

Continued

Sequence	People	Hand	Car	Motorcycle
Time required to process Blind Search Strategy	64	21.98	15.8	2.6
Average rate between search strategies	4	4	4	4.3

9.6.4 Comparative Analysis Between Template Updating Approaches

Figure 9.10 illustrates a qualitative comparison of template updating approaches
with the *Hand* sequence. This figure shows for each approach the original
frame, the appearance of the template to be matched at $I(k)$, the result of
the matching at $I(k)$, and the step time k from which was computed the
template. For example, $T(74)$ denotes that the template was computed from
step time $k = 74$.

In order to test the robustness of the template updating algorithm, we
evaluate the performance when a short-time memory is incorporated and we
evaluate the number of template updates for each sequence, testing: (1) our
approach based on updating templates only when the target has evolved to a
new shape and (2) the template updating approach used in [10, 19, 24] based
on continuous updating at every frame using the Hausdorff measurement.

Table 9.2 summarizes the results showing that the number of required updates is minimized in relation to other updating approaches based on Hausdorff measurement [10, 19, 24]. No target object was drifted using the proposed approach; however, templates were drifted in sequences *People*, *Hand*,
and *Motorcycle* using continuous updating approach at every frame. Moreover, the use of a short-term memory avoids the loss of the target in certain
situations such as illustrated in Table 9.2. These templates from STVM are
used when: (1) an imprecision error of the edge detection process takes place
(*People* and *Hand* sequence), (2) disappearance and reappearance conditions
of the target object along the video stream (*Car* and *People* sequence), and
(3) occlusion conditions of the target (*People*, *Hand*, *Car*, and *Motorcycle*
sequence).

Table 9.2 Template updating: Number of template updates, total number of
different templates stored in STVM during the tracking process for each sequence,
total number of templates used from STVM during the tracking process and number
of frames where the target was retrieved using STVM.

Sequence	People	Hand	Car	Motorcycle
N° of frames	745	512	414	70
Number of updates	380	300	200	60
Number of different templates stored in STVM	124	19	6	6
Number of templates used from STVM	36	12	6	4
Number of frames where the target was retrieved using STVM	18	10	8	4

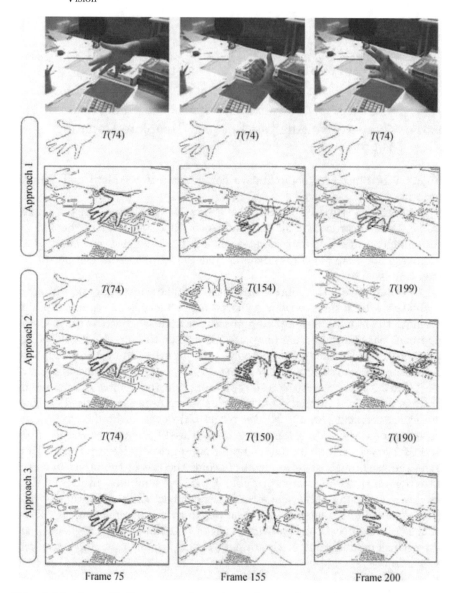

Fig. 9.10 A qualitative comparison of template updating Approaches 1, 2, and 3. With Approach 1, the template is not updated and the tracking process fails in frames 155 and 200, because the template does not reflect the change of the hand shape. With Approach 2, the template is updated every frame. In this situation, the tracking process fails because the template has also been updated in those situations where the target hand has not been evolved to a new shape and therefore, the template was constructed from background edges such as $T(154)$ and $T(199)$. With Approach 3, the template is only updated when the target hand has evolved to a new shape. With this approach, the target hand is successfully followed and the template is appropriately updated across the sequence.

9.6.5   Average Runtime Behaviour

Visual systems with real-time restrictions require the processing of each frame in a maximum time of 40 milliseconds. Table 9.3 shows the runtime in seconds for computing the A* search strategy and the template updating process for each video stream.

The time for processing each frame of sequences *People, Hand, Car, and Motorcycle*, is respectively: 36 ms, 17 ms, 11 ms, and 14 ms. The results of the procesing time for *People* sequence corresponds to the time for tracking three people through an inside corridor. The experimental results reported confirm the adaptation of the tracker proposed to real-time restrictions. Moreover, the computational cost to carry out the visual tracking is lower than the processing latency (40 milliseconds). This feature allows using the remaining time for other processes that integrate the visual system.

Table 9.3   *Runtime* for processing each process of each sequence evaluated.

Sequence	People	Hand	Car	Motorcycle
N° of frames	745	512	414	70
Seconds	30	20	17	3
Time required to process A* Search Strategy	16.2	5.48	3.96	0.6
Time required to process Template Updating	10.6	3.29	0.7	0.4
Total (seconds) to process the sequence	26.8	8.78	4.65	1.0

9.7   Conclusions and Future Work

This Chapter is concerned with robust tracking and multiple-object tracking with deformable shapes in video streams without any assumption on the speed and trajectory of the objects. An A* search framework is computed in the space of transformations to provide efficient target motion, that uses the Kullback-Leibler measurement as heuristic to guide the search process. The most promising initial search alternatives are computed through the incorporation of target dynamics. Combining heuristic information and target dynamics in the search process provides faster computations of target motion solutions than previous search algorithms with similar features.

2D shape change is captured with 2D templates that evolve with time. These templates are only updated when the target object has evolved to a new shape change. The representative temporal variations of the target shape are enclosed in a short-time visual memory. Intensity information is used to confirm the best matching position when the templates are a reduced set of edge points.

The proposed template based tracking system has been tested and has been empirically proved that: (1) the A* search proposed is faster than previous search strategies in an average rate of four times better, allowing real-time performance on general purpose hardware; (2) although abrupt motions cannot be predicted by an alpha-beta filtering approach, the tracker performance was well adapted to the non-stationary character of the person's movement which alternates abruptly between slow and fast motion such as the *People* sequence; (3) updating templates using combined results focused on the value of the quality of match, the use of a short-term memory and information intensity lead to accurate template based tracking.

This work leaves a number of open possibilities that may be worth further research, among others it may be interesting to consider: further processing on input images in order to reduce the illumination sensitivity by means of the use of an anisotropic diffusion filter instead of the Gaussian filter, that is used by the classical Canny detector in order to obtain better edge detection, and the proposal of a dynamic reformulation of the Canny edge detector by means of replacing the hysteresis step by a dynamic threshold process in order to reduce blinking effect of edges during successive frames, and as consequence generate more stable edge sequences such is proposed in Ref. [34]. The study of known models for varying illuminations such as [22] should be also interesting in order to supplement the previous approach.

Acknowledgements

This work was supported by the Spanish Government through the Ministry of Science and Innovation under the project TIN2008-06570-C04-03.

References

[1]   Isard M, Blake A (1998) Condensation-conditional Density Propagation for Visual Tracking. International Journal of Computer Vision, 29(1): 5–28
[2]   Beymer D, McLauchlan P, Coifman B et al (1997) A Real-time Computer Vision System for Measuring Traffic Parameters. In: Proc IEEE Conf on Computer Vision and Pattern Recognition (CVPR)
[3]   Adam A, Rivlin E, Shimshoni I (2006) Robust Fragments-based Tracking Using the Integral Histogram. In: IEEE Conf Computer Vison and Pattern Recognition, CVPR
[4]   Adam A, Rivlin E, Shimshoni et al (2006) Robust Fragments-based Tracking Using the Integral Histogram. Proceedings of IEEE Conference On Computer Vision and Pattern Recognition (CVPR), 2006
[5]   Aloimonos Y (ed) (1993) Active Perception. Lawrence Erlbaum Assoc, Pub, N J
[6]   Bajcsy R (1988) Active Perception. In Proceedings of IEEE Workshop on Computer Vision, 76(8): 996–1005
[7]   Collins R, Lipton A, Fujiyoshi H et al (2001) Algorithms for Cooperative Multisensor Surveillance. Proc IEEE, 89(10): 1456–1477
[8]   Koller D, Daniilidis H, Nagel H (1993) Model-based Object Tracking in Monocular Sequences of Road Traffic Scenes. In: International Journal of Computer Vision, 10: 257–281
[9]   Turk, Mathew (2004) Computer Vision in the Interface. Communications of the ACM, 47(1): 61–67
[10]   Parra R, Devy M, Briot M (1999) 3D Modelling and Robot Localization from Visual and Range Data in Natural Scenes. LNCS, 1542: 450–468, Springer
[11]   Lange C, Hernmann T, Ritter H (2003) Holistic Body Tracking for Gestural Interfaces. In Lectures Note in Computer Science, 2915: 132–13. Springer, Heidelberg

[12] Pentland A P (2000) Perceptual Intelligence. Communications of the ACM, 43(3): 35 – 44

[13] Cham T, Rehg J (1999) A Multiple Hypothesis Approach to Figure Tracking. IEEE Computer Vision and Pattern Recognition, 2: 219 – 239

[14] Gavrila D M (1999) The Visual Analysis of Human Movement: A Survey. Computer Vision and Image Understanding, 73: 82 – 89

[15] Baker S, Matthews I (2004) 'Lucas Kanade 20 Years On: A Unifying Framework'. International Journal of Computer Vision, 56(3): 221 – 255

[16] Belongie S, Malik J, Puzicha J (2002) Shape Matching and Object Recognition using shape context. IEEE Transactions on Pattern Analysis and Machine Intelligence, 24(4): 509 – 522

[17] Besl P J, McKay N (1992) A Method for Registration of 3D Shapes. IEEE Transactions on Pattern Analysis and Machine Intelligence, 14(2): 239 – 256

[18] Chen Y, Medioni G (1992) Object Modelling by Registration of Multiple Range Images. Image and Vision Computing, 10(3): 145 – 155

[19] Rucklidge W J (1996) Efficient computation of the minimum Hausdorff distance for visual recognition. LNCS 1173, Springer

[20] Schlegel C, Illmann J, Jaberg H et al (1999) Integrating Vision based behaviours with an autonomous robot. LNCS 1542: 1 – 20, Springer

[21] Comaniciu D, Ramesh V, Meer P (2000) Real-time Tracking of Non-rigid Objects Using Mean Shift. Proceedings of IEEE Conference on Computer Vision and Pattern Recognition, vol 2: 142 – 149

[22] Hager G D, Belhumeur P N (1998) Efficient Region Tracking with Parametric Models of Geometry and Illumination. IEEE Transactions on Pattern Analysis and Machine Intelligence, 20(10): 1025 – 1039

[23] Liu T L, Chen H T (2004) Real-Time Tracking Using Trust-Region Methods. IEEE Transactions on Pattern Analysis and Machine Intelligence, 26(3): 397 – 401

[24] Sánchez Nielsen E, Hernández Tejera M (2001) Tracking Moving Objects Using the Hausdorff Distance. A Method and Experiments. Frontiers in Artificial Intelligence and Applications: Pattern Recognition and Applications, IOSPRess, pp 164 – 172

[25] Reynolds J (1998) Autonomous Underwater Vehicle: Vision System. PhD thesis, Robotic Systems Lab. Department of Engineering, Australian National University

[26] Matthews I, Ishikawa T, Baker S (2004) The Template Update Problem. IEEE Transactions on Pattern Analysis and Machine Intelligence, 26(6): 810 – 815

[27] Huttenlocher D P, Klanderman G A, Rucklidge W J (1993) Comparing Images Using the Hausdorff Distance. IEEE Transactions on Pattern Analysis and Machine Intelligence, 15(9): 850 – 863

[28] Canny J (1986) A Computational Approach to Edge Detection. IEEE Transactions on Pattern Analysis and Machine Intelligence, 8(6): 679 – 698

[29] Pearl J (1984) Heuristics. Intelligent Search Strategies for Computer Problem Solving. Addison-Wesley, New York

[30] Kullback S (1959) Information Theory and Statistics. Wiley, New York

[31] Cover Thomas M, Thomas Joy A (1991) Elements of Information Theory, Wiley, New York

[32] Bar-Shalom Y, Xiao-Rong Li (1993) Estimation and Tracking: Principles. Techniques, and Software. Artech House Press, Boston

[33] Robert Fischer (2007) Caviar Tests Case Scenarios HTTP. http://groups.inf. ed.ac.uk/vision/CAVIAR/CAVIARDATA1/. Accessed 30 September 2010

[34] Antón-Canalís L, Hernández-Tejera M, Sánchez-Nielsen E (2006) AddCanny: Edge Detector for Video Processing. In: Blanc-Talon J, Philips W, Popescu D et al (eds) ACIVS 2006: LNCS, vol 4179, Springer

10 Human Extremity Detection for Action Recognition*

Elden Yu[1] and J.K. Aggarwal[2]

Abstract Johansson showed that locations of internal body joints are sufficient visual cues to characterize human motion and activities. The position of such body joints is difficult to estimate and track automatically due to clothing and self-occlusion. In this paper, we propose that the location of human extremities alone (including head, hands and feet) provides an excellent approximation to body motion. Human extremities are modeled and detected via two different methodologies. When human contours are available, extremities are modeled as points on the contour and detected using a variable-star-skeleton representation. An extremely compact posture descriptor is further defined on these precise extremities. When it is difficult to obtain contours, extremities are modeled as image patches and detected by searching the image directly. Each pixel is associated with a patch containing the pixel and a probability of the patch forming a specific class of extremities. The histogram of probable extremities is defined on such a probable extremity map as another posture descriptor. Both posture descriptors are tested on two public data sets for action recognition. The results validated the effectiveness of extremities to estimate human body motion and recognize actions.

10.1 Introduction

A critical step in human action recognition is to effectively and efficiently represent human body posture. Johansson [1] demonstrated that the location of human body joints is an effective visual cue for human recognition of actions. He attached lights on human body joints and took videos of human actions in the dark. As shown in Fig. 10.1, the set of points in an image does not really follow any Gestalt principle, which is often used in psychology to group scattered cues. But when viewing the points in an image sequence, observers can quickly find a vivid human figure in action. In essence, the human visual system can recover object information from very sparse inputs such as a set of points in motion. This phenomenon is known as biological

* Portions reprinted, with permission, from [39] © 2009 IEEE.
1 Computer and Vision Research Center, Department of Electrical and Computer Engineering, The University of Texas at Austin. E-mail: elden.yu@gmail.com.
2 Computer and Vision Research Center, Department of Electrical and Computer Engineering, The University of Texas at Austin. E-mail: aggarwaljk@mail.utexas.edu.

motion in biological vision literature [2, 3, 4].

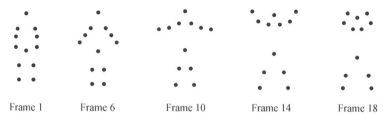

Frame 1   Frame 6   Frame 10   Frame 14   Frame 18

Fig. 10.1 Biological motion. Human visual systems can recognize actions from inputs as sparse as a set of body joints. The sequence is synthesized from motion capture.

Stimulated by Johansson's experiments, Webb and Aggarwal [5] proposed to estimate the structure of jointed objects from motion, where jointed objects have two visible points on each rigid part. Consistent with Johansson's method, modern motion capture systems generally have performers wear suits with distinct markers to identify such body joints. In the past, there have been various studies on Johansson's moving light displays (MLD), as reviewed by Cedras and Shah [6].

In general, body joints are not easily available from videos or images directly, due to self-occlusion and clothing. Can one replace body joints with other points to represent a human body? In Fig. 10.2, we display a few images of another set of points, instead of body joints. For human observers, it is as easy to identify the action as the "jumping jack" as shown in Fig. 10.1 with body joints. This new set of points including head, hands and feet is called human extremities.

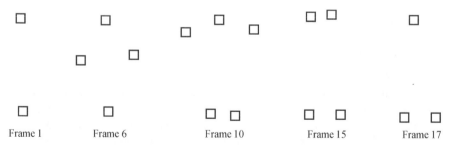

Frame 1   Frame 6   Frame 10   Frame 15   Frame 17

Fig. 10.2 Extremities for action recognition. Extremities only, including head, hands and feet, are sufficient to recognize jumping jacks. Here extremities are detected as points along the contour, and drawn as squares for display. The sequence is from the Weizmann data set [7].

In this chapter, we propose that human extremities are sufficient for approximating human motion. We validate our proposal by detecting extremities, building a corresponding extremity based human posture representation, and utilizing such a posture representation for action recognition. There are two different methods to model the extremities. When a human contour is available, we model the extremities as points on the contour. When the contour is not available, we model the extremities as image patches. For each extremity model, we have an approach detecting extremities, building posture descriptors, and recognizing actions.

The rest of this chapter is organized as follows. We review relevant works in Section 10.2. We present the detection and usage of extremities as points on a contour in Section 10.3. Then we present the detection and usage of probable extremities in Section 10.4. We report, compare, and discuss the corresponding experimental results in Section 10.5. Finally the conclusion is given in Section 10.6.

10.2 Relevant Works

Works in human action recognition are too many to review in this paper. Here we focus on those that directly inspired our own approach.

10.2.1 Star-skeleton

Fujiyoshi and Lipton [8] proposed a star-skeleton model (SS) to analyze human motion. The center of mass of a human silhouette is extracted as the star. Distances from contour points to the star are computed as a function of indices of clockwise sorted contour points. Their initial goal is to use such a representation for feature extraction to recognize cyclic human actions such as walking and running. Their features include the angle between the left leg and the vertical axis passing through the human blob centroid and the angle between the line from the head to the star and the vertical axis, as shown in Fig. 10.3.

Fig. 10.3 Star-skeleton model demonstration. Displayed from left to right are the source image, the segmented contour with the star and skeletons, and the features.

Petkovic et al. [9] used the star-skeleton to find out parts of the human body that stick out; however, they only considered those parts that fall within a pre-defined portion centered around the body, to emphasize the hand movements.

Peursum et al. [10] used a modified star-skeleton in each of the multiple views of an action and fused the two-dimension star-skeletons into a three-dimension one. They modified the star so that it is no longer the blob centroid but the "shoulder"' point of the body, which is defined as the point at one third from the head to the centroid. This creates the problem of finding the head, which is solved by designating the highest extreme point from the star-skeleton as the head; however, the head is not always the highest point.

To utilize structure information available in the star-skeleton, Chen et al. [11] defined a distance function between two star-skeletons. Each star-skeleton is converted into a vector of five extremities. If less than five are detected, they fill the rest with zero. If more than five are detected, they increase the noise smoothing level to remove extra extremities. The distance function is defined as the sum of Euclidean distances between five matched pairs of skeletons. Such a distance function is used in their HMM-based action recognition system.

In order to accurately find extremities for detecting a fence climbing action, Yu and Aggarwal [12, 13] proposed a two-star-skeleton representation (2SS) by adding the highest contour point as the second star. Two sets of local peaks are estimated, and their indices are simply paired up and averaged to find more precise extreme points; however, it is not explained in [12, 13] why two stars should perform better than a single star.

10.2.2 Body Parts

Fischler and Elschlager [14] proposed the pictorial structure model. The basic idea is to model an object through a collection of parts arranged in a deformable configuration. The appearance of each part is modeled separately, and the deformable configuration is represented by spring-like connections between pairs of parts. These models allow for qualitative descriptions of visual appearance, and are suitable for generic recognition problems.

Felzenszwalb and Huttenlocher [15] presented a statistical framework for modeling the appearance of objects with the pictorial structure models. Their contribution is to present efficient algorithms in both finding instances of an object in an image and training tree structured object models from training images. Later, Felzenszwalb et al. [16] considered detecting and localizing objects of a generic category in static images. Their models cover both entire object with a coarse global template and object parts with higher resolution part templates.

To combine the flexibility of non-parametric methods such as kernel density estimation and the tractability of a pictorial structure model, Sapp et al. [17] proposed a simple semi-parametric approach by expressing a subset of model parameters as kernel regression estimates from a learned sparse set of exemplars.

Bourdev and Malik [18] annotated human joints, extracted 3D pose, keypoint visibility and region labels over a database of 1 000 persons. They then applied such labels to develop a novel concept called poselets for part based detection and segmentation.

Ferrari et al. [19] worked on uncontrolled videos to estimate a human pose, by finding configurations of Body Parts. They use a generic weak model of a pose to cut most search space off, and rely on the detected regions as initialization to further refine search space for the parts.

Ju et al. [20] proposed the "cardboard people" model, where the limbs of a person are represented by a set of connected planar patches. The motion of the limb is estimated from optical flow fields by treating the limb as a chain structure of rigid objects. Their experiments are conducted on "walking" with only two legs visible.

Haritaoglu et al. [21] developed a real time system to estimate a human body pose and detect Body Parts from silhouettes. The system uses a silhouette-based body model which consists of six primary Body Parts (head, hands, feet, and torso) and ten secondary parts(elbows, knees, shoulders, armpits, hip, and upper back). It first compares the human body contour with predefined templates to estimate body posture. Then the head position is detected and other Body Parts are estimated with the topology of the estimated body posture. Their work was later included in the W^4 system [22].

Park and Aggarwal [23] used a hierarchical human body model, where a body is divided into the head, the upper body and the lower body. Furthermore, the head has hair and a face, the upper body has hands and torso, and the lower body has legs and feet. A maximum a posterior (MAP) classifier is employed to assign each blob into a body part. Ryoo and Aggarwal [24] followed a similar human body part model while emphasizing the high level modeling of continued and recursive activities.

10.3 Extremities as Points on a Contour

Human extremities refer to human heads, hands and feet. They provide useful information about human movement. In the simplest case, we model extremities as points along the body contour. For image sequences taken by a stationary camera, the common approach to obtain contours is to do background subtraction followed by thresholding, binary morphological operations, connected component analysis, and then a border following algorithm.

10.3.1 On Number and Positions of Stars

If there is only one star in a star-skeleton representation, the position of the star is very important. It greatly effects whether a contour point could be a local peak in the distance function from the star to the contour points, hence be a possible human extremity.

An example is given in Fig. 10.4 showing a person climbing. We focus
on detecting the left hand. The part of the contour around the left hand is
highlighted with a green solid line, while the other parts are shown with a
black dash line. For illustration purposes, four stars are chosen as shown with
blue asterisks and numbered. The detected hand from each star is shown with
a solid red square and numbered accordingly in Fig. 10.4. From this example,
it is obvious that the fourth star provided the best approximation, the second
star made a close one, and the other two produced incorrect extremities.

Fig. 10.4 Inspiring observation. The four stars
are shown with blue asterisks and their respective detections of left hand shown with solid red
squares with corresponding numbers.

Before we proceed, we briefly introduce some concepts from the computer
graphics community to make the paper self-contained.

Given a human contour represented by a set of clockwise sorted contour
points, we treat it as a simple polygon P. In geometry [25], a simple polygon
is a polygon whose sides do not intersect unless they share a vertex. A point
in the polygon (including interior and boundary) is visible with respect to
another point in the polygon if their line segment falls completely within the
polygon. For example, in Fig. 10.5, point D is visible to point E,F,G while
not visible to H.

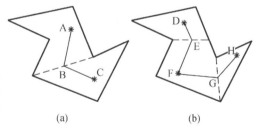

Fig. 10.5 Non-unique decomposition. On the same simple polygon, there may exist a lot of decompositions.

(a)   (b)

Shapira and Rappoport [25] defined the star polygon as follows. If inside
polygon P there exists a point v visible from every point on the contour, then
P is called star polygon and v is called star point. They further proposed that
a simple polygon can always be decomposed into a set of star polygons. From
the definition, a convex polygon is naturally a star polygon.

Shown in Fig. 10.5 is the same simple polygon decomposed into two star polygons in (a) and into three in (b). In Fig. 10.5 (a), points A, C are star points. In Fig. 10.5 (b), points D, F, H are star points. Obviously the star-polygon decomposition is not unique.

Why do different stars produce different approximations of the left hand in Fig. 10.4? There are many possible explanations such as distance, scale, and visibility. Among all the factors, we regard visibility as the most important one. The reason that the second and fourth star perform better is because the hand is visible to them, while the hand is not visible to the other two stars.

Fujiyoshi and Lipton [8] considered the human centroid as a single star. As human contours are usually not star polygons, a single star cannot be visible to all contour points. Hence their method will easily miss true human limbs or produce false alarms. In extreme conditions, the centroid may not even be inside the human silhouette.

Yu and Aggarwal [13] added the highest contour point as the second star. It can be interpreted as an intention to make all those points not visible to the center of mass visible to the second star. In this way it is hoped that most contour points will be visible to at least one of the two stars. This strategy is intuitive and reasonable; however, its practical effect is weakened in two aspects. First, it is a problem whether or not to treat the highest contour point as an extremity. In most human postures, the highest contour point is the head, hence it is desirable to include the second star as one of the detected extremities. When the assumption is violated, the inclusion might produce false alarms. Second, two detected limbs (each from a different star) are paired up and averaged, which means a good detected extremity is compromised by a bad one. We would rather have the algorithm to select the good ones and discard the bad ones.

Using a frame of a person climbing a fence, we show in Fig. 10.6 the visibility of each contour point with respect to the center of the mass and the highest contour point. Details of computing such a visibility are described later. It is obvious that with only the center of mass as the single star, a

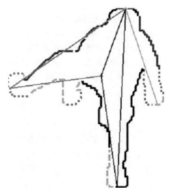

Fig. 10.6 Visibility illustration. The two stars are in blue and green respectively. Contour points visible only to the center of mass are shown with a blue solid line, visible only to the highest contour point shown with a green dash line, visible to both shown with a black solid line, and visible to neither shown with a red dotted line. Best viewed in color.

considerable portion of the contour is not visible. With the addition of the second star, more contour pieces are covered, while there is still a significant portion not visible.

10.3.2 Variable-star-skeleton Representation

Now, we develop a variable-star-skeleton (VSS) representation, motivated by observing that more and well positioned stars make contour points more visible. Although built upon previous works [8, 13], our new representation is different in two aspects, including finding stars and producing extremities out of multiple sets of candidates. We take as stars, junction points on the medial axis of the human silhouette, which may be regarded as a rough approximation of human body joints. Each star will produce a set of extreme points, as previously done in SS and 2SS. As a candidate, each extreme point will be processed according to its robustness to noise smoothing, visibility to the generating star, and proximity to its neighbors.

Detecting Junctions on a Medial Axis

For contours, a medial axis is the union of all centers of inset circles that are tangent to at least two contour points. In order to compute the medial axis, we choose the augmented Fast Marching Method by Telea and Wijk [26] among many existing algorithms such as [27, 28]. There is a threshold t controlling how short each branch of the medial axis may be. In Fig. 10.7 the computed medial axis is shown as a magenta dotted line with t=10 and t=30 respectively.

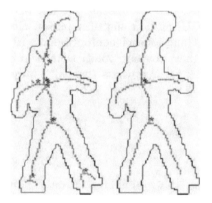

Fig. 10.7 Finding stars. The left image shows a magenta line indicating the medial axis obtained with t=10, while the right one with t=30. Each detected junction point is annotated with a black asterisk.

In order to find the junction points, we employ a lookup table (LUT) in the 3 × 3 neighborhood of every pixel on the medial axis. As each cell in the neighborhood take binary values, we have 256 total possible combinations of the 8 connected neighbors. For each combination, we determine if the center pixel is a junction point, as denoted by a black asterisk in Fig. 10.7. One may notice in the figure that sometimes two junctions are too close together; in such cases, we merge those junctions that are closer than a threshold (w) and use their mean as the estimated junction. In rare cases, the parameter t is too strict to produce any junction from the medial axis; we opt to use the center of the mass as the single star, although we can also choose to reduce t until there is at least one junction point.

Generating candidate extreme points

Suppose there are N stars denoted as $star_j (j = 1, 2, ..., N)$. Starting with the highest contour point, each point in the contour of length NC is sorted clockwise, and denoted as $P_i (i = 1, 2, ..., NC)$. As in previous works, we compute the Euclidean distance from $star_j$ to P_i as a function $dist_j(i)$. The function is then smoothed by a one-dimension Gaussian kernel with standard deviation delta. Contour points with a local peak are chosen as candidate extreme points.

In order to find the local peaks from the smoothed distance function, we proceed as follows.

1) Modify the computed distance $dist_j(i)$ to a refined version denoted as $D_j(k)$, by removing repeating values so that there are no identical values adjacent to each other in $D_j(k)$. Now the length of $D_j(k)$ should be reduced from NC to another number denoted as NK. Keep the indices $Ind_k (k = 1, 2, ..., NK)$ updated, so that for each chunk of identical distance values, their common index is the middle of the interval. The main purpose of this step is to accommodate those contour pieces where every point has the same distance to the star.

2) For each k, check if $D_j(Ind_k) > D_j(Ind_{k-1})$ and $D_j(Ind_k) > D_j (Ind_{k+1})$. If both are satisfied, it is output as a candidate extremity. Note here $k - 1$ and $k + 1$ are both modulo NK arithmetic.

Using the contour and junctions from Fig. 10.8, Fig. 10.9 shows the plots of a distance function and its smoothed version with respect to the top red star. Those with respect to the bottom green star are similar. The detected candidates are drawn as red or green crosses in Fig. 10.8 accordingly.

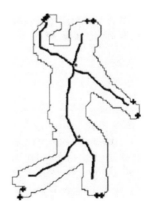

Fig. 10.8 Locating extremity candidates. The medial axis is shown with a magenta line (enhanced for display purpose), junctions as asterisks, and candidate extremities as crosses in the same color as the corresponding star.

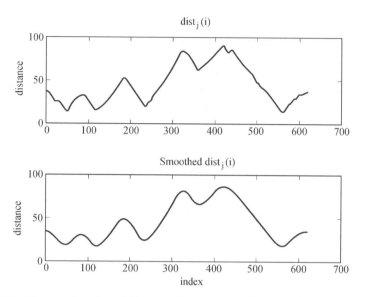

Fig. 10.9 Distance functions. The top plot shows the distance from sorted contour points to the red star, while the bottom plot shows its smoothed version.

Filtering

In this section, we determine if a candidate extreme point is kept, discarded or merged with a nearby candidate. We first associate each candidate with two properties, including robustness to the smoothing parameter and visibility to the generating star.

The robustness R may be viewed as a measurement of how much a possible human limb protrudes out of the torso. As we have located all the local peaks from the distance function D_j (ignoring j in this subsection for simplicity) as described above, we can easily modify it to locate all the local valleys as well. Given a local peak with value $D(x)$ at index x, it must have two adjacent valleys. One is on the left and the other is on the right. Suppose the higher adjacent valley has value $D(y)$ at index y, we define robustness R associated with the candidate extreme point P_x in the following equation, also illustrated in Fig. 10.10

$$R = \frac{D(x) - D(y)}{|x - y|}.$$

We connect from the candidate to the star generating it, to form a line segment. The visibility V is computed as a proportion of the line segment that lies inside a silhouette. Given two points, we use the basic raster algorithm [29] on line drawing to produce the set of points between them. Then the intersection of the set with the binary human silhouette produces line points inside the silhouette.

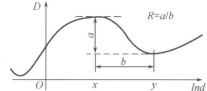

Fig. 10.10 Robustness definition. It is essentially determined by the ratio between the relative peak and its width.

With these properties, we proceed with the following procedure where the input is all those candidates as generated in Sect.10.3.2.2 and the output is the detected extremities.

1) Select candidates chosen by more than one star. Group all those candidates by hierarchical agglomerative clustering with single linkage, so that any two candidates whose indices are closer than w are put into one group. The means of all those clusters with more than one member form set $\boldsymbol{A}$, and all the single member clusters form set $\boldsymbol{B}$.

2) Select candidates with better visibility and robustness. Select from $\boldsymbol{B}$ all those candidates with R bigger than threshold $MaxR$ and V bigger than $MaxV$ into set $\boldsymbol{A}$.

3) Discard bad candidates from $\boldsymbol{B}$ with R smaller than threshold $MinR$ or V smaller than $MinV$.

4) Make at most 5 extremities. Denote the number of elements of $\boldsymbol{A}$ as $|\boldsymbol{A}|$. If $|\boldsymbol{A}| > 5$, sort $\boldsymbol{A}$ by product of R and V, stop and output the top 5 only. If $|\boldsymbol{A}| \leqslant 5$, sort $\boldsymbol{B}$ by product of R and V. Select the top $\min\{|\boldsymbol{B}|, 5 - |\boldsymbol{A}|\}$ candidates from $\boldsymbol{B}$ into set $\boldsymbol{A}$, stop and output $\boldsymbol{A}$.

10.3.3 Performance Comparison

In order to compare the performance of detecting extremities from contours with the three kinds of star-skeleton representations, we built a data set from 50 sequences of persons climbing fences. Shown in Fig. 10.11 are sample frames of a sequence. We collect 20 frames evenly distributed from each sequence to form a data set of 1 000 frames. It is checked manually to test if the proposed VSS performs better than previous methods including SS and 2SS.

Fig. 10.11 Sample frames of a fence climbing sequence.

For each frame in the data set, we have all three star skeleton representations performing detection of extremities as an approximation of head and human limbs. We manually check the results and determine the number of ground truth extreme points, true positives and false alarms. To empirically validate the relative importance of visibility and robustness criteria for human extremities, we also did experiments on the data set without the visibility or robustness criteria. Comparisons are shown in Table 10.1.

Table 10.1   Results from the three representations on the data set with 3 691 extremities

	True positive	False alarm
SS	3107/84.2	779/21.1
2SS	3381/91.6	146/4.0
VSS w/o robustness	3617/98.0	705/19.1
VSS w/o visibility	3580/97.0	384/10.4
VSS	3440/93.2	98/2.7

There are several parameters involved in all three star skeleton representations. The common parameter among the three is the Gaussian smoothing factor *delta*. There is a tradeoff between detecting more global or more local extreme points when selecting different scales of smoothing parameters. We used *delta* = 10. The t threshold is set as 30, which yields a reasonable medial axis for most binary blobs. We usually get one, two or three junctions from a medial axis. We set $w = 10$ for both merging junctions and clustering candidates. The two thresholds for R in the filtering process are set as 0.6, 0.1, and the two thresholds for V are set as 0.9, 0.5. All the parameter values are chosen empirically and used throughout the experiments.

From Table 10.1, we conclude that the two-star-skeleton (2SS) considerably improve detection accuracy from the single star skeleton. The variable star skeleton (VSS) performs best. From detection results over the 1 000 frames, we have the following observations.

When there is no junction point detected, the VSS is reduced to the single-star-skeleton, except that there is the filtering process. When there is only one junction point, the VSS is different from the single SS. An example is shown in Fig. 10.12 (a), where the VSS can successfully detect the two hands while both SS and 2SS fail. The difference lies in that the single star is usually closer to human body joints instead of being the center of mass. Hence it has better visibility to the ground truth extreme points including

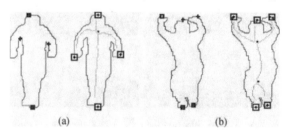

(a)   (b)

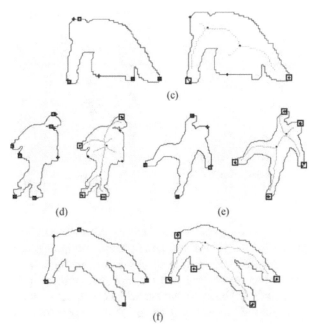

Fig. 10.12 Comparison of SS, 2SS and VSS. For each pair of images, the image on the left shows the result of SS in red crosses, the result of 2SS in blue squares; the image on the right shows the result of VSS in blue squares. In the images on the right, stars are shown in colors and their associated extremity candidates are shown in the same color crosses.

the head and limbs.

The 2SS implicitly assumes the highest contour point being the head. When the assumption does not hold, the VSS easily wins over the 2SS. Figure 10.12 (b, e) shows the hand is higher than the head, and Fig. 10.12 (f) shows the back is higher than the head.

10.3.4 Indicator of Extremities

After extracting precise extremities from contours, we develop a human posture descriptor for action recognition purposes. Motivated by the shape context descriptor proposed by Belongie et al. [30], we use a simple planar indicator to build a feature vector for each frame. As shown in Fig. 10.13, we find the relative coordinates of each extremity with respect to the center of mass of the human silhouette. The entire plane is evenly divided into $N(= 12)$ sectors, and the indicator is a N-element vector with each element indicating if there is an extremity in the sector. This is an extremely compact posture representation, since each vector occupies only 12 bits.

With such a posture representation, we recognize a variety of common

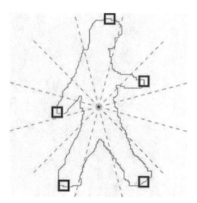

Fig. 10.13 Indicator of extremities. The 12-bit posture representation is compact yet effective.

human actions by using the discrete Hidden Markov Model (HMM) technique [31]. In particular, we use the HMM Matlab implementation by Kevin Murphy [32]. The toolbox is simple to use, however, one has to take care of the special case when an observation has not ever shown up in the training set. If left unattended, the observation will corrupt the entire sequence to produce zero likelihood[3]. A small trick to solve the problem is to always add an arbitrarily small constant to the trained HMM models. As each action is represented with an image sequence or video, the key procedure is to convert each frame to an observation symbol so that each action may be represented by an observation sequence. In order to reduce the number of observation symbols, vector quantization is commonly employed to cluster the feature vectors. The cluster label of each feature vector acts as the observation symbol for HMM usage; however, it is not always necessary if there are a limited number of unique features, as in our case.

10.4 Extremities as Image Patches

We model human extremities as points along a contour in Sect 10.3; however, an extremity is not really a single point in practice. Instead, an extremity often covers a region and is more appropriately modeled as an image patch.

10.4.1 Representing Patches

Given an image patch, there are various ways to represent it as a feature vector. The simplest way is to concatenate all the pixel intensities. For example, Turk and Pentland [33] proposed the Eigenfaces approach for face

---

3 In naïve Bayes classifier for text classification, the same phenomenon occurs and the jargon is "smoothing".

recognition, with the face images represented by concatenations of all pixels in the images. What exact representation one should use for patches depends on the specific types of patches.

Among the three types of extremities, heads are in general upright, feet are mostly upright, while hands may have various orientations. The histogram of oriented gradients (HOG) descriptor [34] can capture edge orientations at different spatial local neighborhoods. As shown in Fig. 10.14, given an extremity patch, its gradients are computed first. With a set of $m \times n$ spatial cells imposed on the gradient image, the occurrences of each of the $nBin$ edge orientations ($nBin = 9$) in the local cells are counted. All the local histograms are concatenated and normalized to form the final representation as a feature vector of length $m \times n \times nBin$.

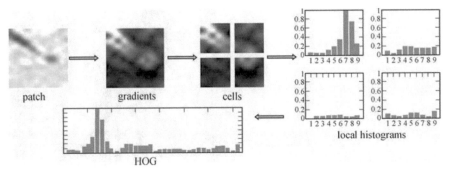

Fig. 10.14 The HOG representation of an extremity patch.

10.4.2 Predicting a Patch as an Extremity Class

Now the task is to train a classifier so that one can predict how likely a testing patch is from one of the extremity classes. There are mainly two steps, including collecting the training set of extremities and training a classifier to predict a test patch.

In order to identify a patch with a classifier, one needs both positive and negative extremity patches. We designate a set of images as the training set, and extract positive and negative patches from all images in the set. For positive extremity patches, we manually collect example patches at a fixed size and label them as heads, hands or feet. For negative examples, we have written a program to automatically collect two sets of patches according to the locations of positive patches from the images.

As illustrated in Fig. 10.15, given an image in (a), we manually collect four patches in (b), including the head, two hands, and the feet. The corresponding masks are displayed in (c) as white squares. The next points are randomly selected out of those contours of white squares as patch centers. The resulting patches may cover both a part of the true extremity and the background, as shown in (d). Then points are randomly selected out of the black regions so that the resulting patches entirely cover the background, as shown in (e). Both sets of patches in (d) and (e) are treated as negative examples, for better identification of extremities from backgrounds. In Fig. 10.16 some samples of the three extremities and negative classes are shown.

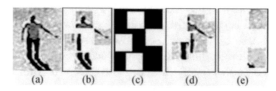

(a)   (b)   (c)   (d)   (e)

Fig. 10.15 To collect extremity patches from frames for training.

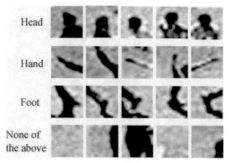

Head

Hand

Foot

None of
the above

Fig. 10.16 Samples of the collected patches for training.

Given the training set of image patches represented by feature vectors, we train and validate with a Support Vector Machine (SVM) classifier to see if a patch can be reliably classified as having positive or negative extremities.

It is a multi-class classification problem. In our work, there are three extremity classes plus the negative (not an extremity) class. The name "probable extremities" comes from the fact that the classifier produces a probability over each extremity patch. To predict the probability of an image patch as one of the extremity classes or negative class, we utilize the algorithm proposed by Wu et al.[35]. The basic flow is shown in Fig. 10.17.

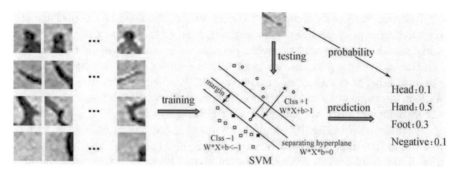

Fig. 10.17 Probability estimate of an image patch as extremities or negative.

10.4.3 Detection of Probable Extremities

We built a probable extremity map by searching over possible locations, as shown by the flow chart in Fig. 10.18. Instead of cropping a rectangle region and computing the histogram repeatedly, there is a faster method to compute histograms for all possible locations over an image. The method is called integral histogram, as first proposed by Porikli [36] and Viola and Jones [37].

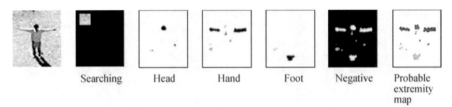

Searching   Head   Hand   Foot   Negative   Probable extremity map

Fig. 10.18 Building the probable extremity map, which includes three channels for heads, hands, and feet.

10.4.4 Histogram of Probable Extremities

With the probable extremity map built, we propose to build a histogram out of it hence the name "Histogram of Probable Extremities" and its abbreviation HOPE. No blob centroid is available here and the histogram is built in the style of the HOG descriptor.

Briefly speaking, a grid of M by N cells are imposed on the probable extremity map. In each cell, the average of probabilities across the cell is computed for each of the three extremity classes. Hence each histogram in the cell is a feature vector of length 3 (note the probability of the negative class is discarded). All the histograms from cells are concatenated and normalized to form a feature vector of length $M \times N \times 3$.

With each image represented as a HOPE descriptor of length $M \times N \times 3$, we can employ any classifier for action recognition. Each block of T consecutive frames are treated as the most basic action unit, and their HOPE descriptors are further combined to form a feature of length $M \times N \times 3 \times T$. Adjacent blocks may have O frames overlapping. Considering that the block descriptor may be very long, we use Principle Component Analysis for dimension reduction when necessary. Beyond the SVM for estimating extremity probabilities in Sect 10.4.2, we train another multi-class SVM to classify whether a block of consecutive frames belongs to one of the predefined action classes. The sequence label is assigned to the class that gets the most votes from the block based classification.

10.5 Experimental Results

In this section, we apply the two kinds of extremity models on two public data sets, including the Weizmann data [7] and the Tower data [38]. In all experiments, leave-one-out cross-validation is utilized to produce the overall classification accuracy. As a comparison, we report the performance of both extremity models in Table 10.2.

Table 10.2 Comparison of the two extremity models on two public data sets

Classification accuracy	Weizmann	Tower
Based on precise extremities	93.6	86.7
Based on probable extremities	95.7	98.3

10.5.1 Precise Extremities on Weizmann Data

In the Weizmann data, there are 93 sequences of 9 persons performing 10 different actions, as shown in Fig. 10.19. Using the provided human silhouette, we extracted the human extremities for all 5,687 frames. To get a flavor of the accuracy of the proposed VSS on this particular data set, we manually checked all 701 frames from 10 sequences performed by one person (Daria). The VSS detected 1,889 (96.1%) out of 1,966 ground truth extremities, while making only 18 false alarms.

Fig. 10.19 Weizmann data. Sample images are shown for the 10 actions, including bend, jumping jack, jump, pjump, run, side, skip, walk, wave1 and wave2.

There are only 179 unique feature vectors. After 93 iterations, the confusion matrix is produced as in Fig. 10.20. There are 2 misclassifications between jump and pjump, as they are essentially the same action taken in different views. The overall accuracy is 93.6%.

	bend	jack	jump	pjump	run	side	skip	walk	wave 1	wave 2
bend	9									
jack		9								
jump			7	1			1			
pjump			1	8						
run					10					
side						9				
skip					1		9			
walk						1		9		
wave 1									8	1
wave 2										9

Fig. 10.20 The confusion matrix of action recognition on the Weizmann data set.

10.5.2 Probable Extremities on Weizmann Data

For those images in Fig. 10.19, the corresponding vector images of probable extremities are shown in Fig. 10.21. Our best result is achieved with $M = 8, N = 6, T = 15, O = 5$.

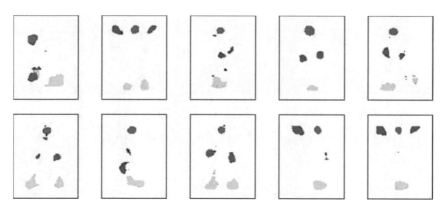

Fig. 10.21 Probable extremity map. Corresponding vector images of the probable extremities in the Weizmann data with head in red, feet in green and hands in blue. Best viewed in color.

The overall accuracy is improved to 95.7%. This improvement is due to the more complete and powerful representation of a human body posture with the HOPE descriptor. As seen in Fig. 10.21, the human bodies are visualized very clear by those extremities and there is no false alarm at all.

10.5.3 Precise Extremities on Tower Data

In comparison with the Weizmann data set, the Tower data are in lower resolution. For this data set, the camera is mounted on a tower around 70 meters tall and actors perform in a garden under the tower. Human figures in the frames are only around 40 pixels tall. There are 6 actors each performing 5 actions twice, including carrying, running, jumping, waving one hand, and waving both hands. There are 60 sequences and 2,406 frames in total. We show five sample frames of each action in Fig. 10.22.

Fig. 10.22 Tower data. From top to bottom, the actions are carrying, jumping, running, waving one hand and waving two hands.

As the resolution is low, the segmentation is not as good as in the Weizmann data set. The overall classification accuracy is 86.7%.

10.5.4 Probable Extremities on Tower Data

The five probable extremity maps corresponding to those frames in Fig. 10.22 are shown in Fig. 10.23. Since the human figures are much smaller than in the Weizmann data, the extremity detection is not as perfect. For example, the right foot in the third running frame is wrongly detected as hands; however,

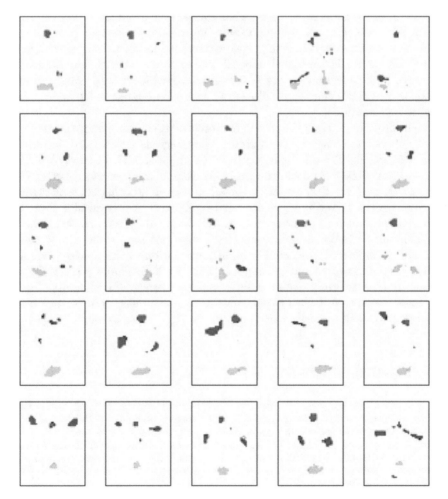

Fig. 10.23 Probable extremity map. Five sample probable extremity maps of each action in the Tower data are shown. From top to bottom, the actions are carrying, jumping, running, waving one hand and waving two hands. Best viewed in color.

even with such inevitable errors due to low resolution, we still achieve significant improvement over the approach based on the precise extremity model. The overall accuracy is 98.3%, just one error out of 60 sequences.

10.6 Conclusion

In this chapter, we presented our unique perspective on classifying human actions with extremities only.

For precise extremities, the detection improved from the simple star-skeleton to the most advanced variable-star-skeleton, with more and more comprehensive understanding on what can be seen as an extremity on a contour. After detection, the human body information is essentially reduced to only a few points. It is amazing that with so little information, we can do so much. In particular, we used only 12 bits to represent each frame and still achieve 93.6% accuracy on the Weizmann data, which is surprisingly good to support our claim that extremities are powerful enough cues for approximating human motions.

For probable extremities, the detection is more or less relaxed to detect the extremities and its nearby Body Parts. These regions are the most descriptive regions for identifying human posture, in comparison with torsos.

On the Weizmann data, the probable extremities performed better than the precise extremities, also with the advantage of avoiding contour segmentation. On the Tower data, we obtained significantly better results. From our observation, when the extremities are well detected by the precise extremity model, the probable extremity approach does not significantly improve the classification accuracy, since it is an approximation to the precise extremity after all. This is exactly what happens with the Weizmann data set. For the Tower data, since there is quite a bit of shadow associated with human figures in lower resolution, contour segmentation is not an easy job. In such cases, the probable extremities clearly outperform the precise extremities.

Acknowledgements

It is pleasure to acknowledge the support of Chia-Chih Chen for providing useful discussions and suggestions during development of techniques described in this paper and of Selina Keilani for editorial help by proofreading various versions of the manuscript. This work was supported by the Defense Advanced Research Projects Agency (DARPA) under Contract No. HR0011-08-C-0135. Any opinions, findings and conclusions or recommendations expressed in this material are those of the author(s) and do not necessarily reflect the views of DARPA.

References

[1] Johansson G (1973) Visual Perception of Biological Motion and a Model for its Analysis. Perception and Psychophysics, 14(2): 201–211

[2] Dittrich W (1993) Action Categories and the Perception of Biological Motion. Perception, 22(1): 15–22

[3] Shipley T (2003) The Effect of Object and Event Orientation on Perception of Biological Motion. Psychol Sci, 14(4): 377–380

[4] Troje N F, Westhoff C (2006) Inversion Effect in Biological Motion Perception: Evidence for a "life detector". Current Biology, 16: 821–824

[5] Webb J A, Aggarwal J K (1982) Structure From Motion of Rigid and Jointed Objects. Artificial Intelligence, 19: 107–130

[6] Cedras C, Shah M (1994) A Survey of Motion Analysis From Moving Light Displays. In: IEEE Conference on Computer Vision and Pattern Recognition, pp 214–221, 21–23 June 1994, Seattle, USA

[7] Blank M, Gorelick L, Shechtman E et al (2005) Actions as Space-time Shapes. In: IEEE International Conference on Computer Vision, pp 1395–1402. Beijing

[8] Fujiyoshi H, Lipton A (1998) Real-time Human Motion Analysis by Image Skeletonization. In: IEEE Workshop on Applications of Computer Vision, pp 15–21, 19–21 October 1998, Princeton, USA

[9] Petkovic M, Jonker W, Zivkovic Z (2001) Recognizing Strokes in Tennis Videos Using Hidden Markov Models. In: International Conference on Visualization, Imaging and Image Processing, Marbella, Spain

[10] Peursum P, Bui H, Venkatesh S et al (2004) Human Action Recognition With an Incomplete Real-time Pose Skeleton. Technical Report, January 2004

[11] Chen H S, Chen H T, Chen Y W et al (2006) Human Action Recognition Using Star Skeleton. In: International Workshop on Visual Surveillance & Sensor Networks, pp 171–178, Santa Barbara

[12] Yu E, Aggarwal J K (2006) Detection of Fence Climbing from Monocular Video. In: International Conference on Pattern Recognition, pp 375–378, 20–24 August 2006, Hong Kong, China

[13] Yu E, Aggarwal J K (2009) Detecting Persons Climbing Fences. International Journal of Pattern Recognition and Artificial Intelligence, 23(7): 1309–1332

[14] Fischler M, Elschlager R (1973) The Representation and Matching of Pictorial Structures. IEEE Transactions on Computer, 22(1): 67–92

[15] Felzenszwalb P F, Huttenlocher D P (2003) Pictorial Structures for Object Recognition. International Journal of Computer Vision, 61: 2005

[16] Felzenszwalb P F, McAllester D, Ramanan D (2008) A Discriminatively Train-ed, Multiscale, Deformable Part Model. In: IEEE Conference on Computer Vision and Pattern Recognition, pp 24–26 June 2008, Anchorage, Alaska, USA

[17] Sapp B, Jordan C, Taskar B (2010) Adaptive Pose Priors for Pictorial Structures. In: IEEE Conference on Computer Vision and Pattern Recognition

[18] Bourdev L, Malik J (2009) Poselets: Body part Detectors Trained Using 3D Human Pose Annotations. In: International Conference on Computer Vision

[19] Ferrari V, Marin-Jiminez M, Zisserman A (2008) Progressive Search Space Reduction for Human Pose Estimation. In: IEEE Conference on Computer Vision and Pattern Recognition

[20] Ju S X, Black M J, Yacoob Y (1996) Cardboard People: A Parameterized Model of Articulated Image Motion. In: International Conference on Automatic Face and Gesture Recognition

[21] Haritaoglu I, Harwood D, Davis LS (1998) Ghost: A Human Body Part Labeling System Using Silhouettes. In: International Conference on Pattern Recognition, pp 77–82, August 17–20 1998, Brisbane, Australia

[22] Haritaoglu I, Harwood D, Davis LS (2000) W4: Real-time Surveillance of People and Their Activities. IEEE Trans on Pattern Analysis and Machine Intelligence, 22(8): 809–830

[23] Park S, Aggarwal J K (2006) Simultaneous Tracking of Multiple Body Parts of Interacting Persons. Computer Vision and Image Understanding, 102(1): 1–21

[24] Ryoo M S, Aggarwal J K (2009) Semantic Representation and Recognition of Continued and Recursive Human Activities. International Journal of Computer Vision, 82(1): 1–24

[25] Shapira M, Rappoport A (1995) Shape Blending Using the Starskeleton Representa-

tion. IEEE Computer Graphics and Applications, 15(2): 44–50

[26] Telea A, van Wijk J J (2002) An augmented Fast Marching Method for Computing Skeletons and Centerlines. In: ACM Symposium on Data Visualization, pp 251–260, Aire-la-Ville, Switzerland

[27] Blum H, Nagel R N (1978) Shape Description Using Weighted Symmetric Axis Features. Pattern Recognition, 10(3): 167–180

[28] Ogniewicz R L, Kubler O (1995) Hierarchic Voronoi Skeletons. Pattern Recognition, 28(3): 343–359

[29] O'Rourke J (1994) Computational Geometry in C. Cambridge University Press, London

[30] Belongie S, Malik J, Puzicha J (2002) Shape Matching and Object Recognition Using Shape Contexts. IEEE Transacations on Pattern Analysis and Machine Intelligence, 24(4): 509–522

[31] Rabiner L R (1989) A Tutorial on Hidden Markov Models and Selected Applications in Speech Recognition. Proceedings of the IEEE, 77(2): 257–286

[32] Murphy K (1998) Hidden Markov Model (HMM) Toolbox for Matlab. http://www.cs.ubc.ca/~murphyk/Software/HMM/hmm.html. Accessed 1 Janaury 1998

[33] Turk M, Pentland A (1991) Face Recognition Using Eigenfaces. In: Proceedings of the IEEE Conference on Computer Vision and Pattern Recognition, pp 586–591, 3–6 June 1991, Hawaii, USA

[34] Dalal N, Triggs B (2005) Histograms of Oriented Gradients for Human Detection. In: IEEE Conference on Computer Vision and Pattern Recognition, vol 2 pp 886–893

[35] Wu T F, Lin C J, Weng R C (2004) Probability Estimates for Multi-class Classification by Pairwise Coupling. Journal of Machine Learning Research, 5: 975–1005

[36] Porikli F (2005) Integral Histogram: A Fast way to Extract Histograms in Cartesian Spaces. In: IEEE Conference on Computer Vision and Pattern Recognition, 1: 829–836

[37] Viola P, Jones M (2002) Robust Real-time Object Detection. International Journal of Computer Vision, 57(2): 137–154

[38] Chen C C, Aggarwal J K (2009) Recognizing Human Action From a far Field of View. In: IEEE Workshop on Motion and Video Computing 9 December 2009, Utah, USA

[39] Yu E, Aggarwal J K (2009) Human Action Recognition with Extremities as Semantic Posture Representation. In: IEEE CVPR Workshop on Semantic Learning and Applications in Multimedia, 21 June 2009, Miami Beach, USA

11 Ensemble Learning for Object Recognition and Tracking

Mehrtash Harandi[1], Javid Taheri[2] and Brian C. Lovell[3]

Abstract   Nowadays, a simple query in Google image search will return millions of results. Apparently this is not because existing vision technologies for object recognition are performing at such a high-level[4]; on the contrary, while human beings can recognize objects with little difficulty, artificial vision systems are far from matching the accuracy, speed and generality of human vision [1]. Potential applications enabled by solving the object recognition problem are enormous, content-based image retrieval, object tracking, robot navigation, automated surveillance, etc. are some that come to mind.

Object recognition is usually described as a high-dimensional classification problem where only a small number of training samples (compared to the data dimensionality) are available. Consequently, it is extremely difficult, if not impossible, to construct an efficient single classification rule. Ensemble learning is a method for constructing accurate classifiers from an ensemble of weak predictors or base classifiers. During the past decade, ensemble learning methods have been extensive studied for various disciplines. In this chapter, we review influential works along recent topics of ensemble learning approaches devised for recognizing and tracking objects.

11.1 Introduction

The notion of Ensemble Learning is to aggregate several predictions using multiple learners. Ensemble learning can be also intuitively defined as solving hard classification problems (such as computer vision problems) through breaking it down into smaller ones. Computer vision applications usually encounter high-dimensional feature vectors with only a small number of training samples (compared to the data dimensionality). As a result, it is extremely difficult, if not impossible, to construct an efficient single classification rule.

---

1   NICTA, PO Box 6020, St Lucia, QLD 4067, Australia The University of Queensland, School of ITEE, QLD 4072, Australia. E-mail: mehrtash.harandi@nicta.com.au.

2   The University of Sydney, School of Information Technologies, NSW 2006, Australia. E-mail: javid.taheri@sydney.edu.au.

3   NICTA, PO Box 6020, St Lucia, QLD 4067, Australia The University of Queensland, School of ITEE, QLD 4072, Australia. E-mail: lovell@itee.uq.edu.au.

4   Currently image tagging is the best and most practical approach to make images visible by search engines.

To solve such issues, ensemble learning techniques have become extremely popular over the last few years in computer vision applications. In fact, several studies even showed that ensemble of classifiers significantly outperform their single base counterparts [2].

As an instance, in the approximating task of a target function: $f(x)$: $\mathbb{R}^N \to \mathbb{R}$, theory of ensemble learning is used to approximate the target function using L distinct predictors (base or weak classifier) $c_k(x); k = 1..L$ in the form of $C(x) = g(c_1(\boldsymbol{x}), c_2(\boldsymbol{x}), ..., c_L(\boldsymbol{x}))$ (see Fig. 11.1. for a graphical interpretation).

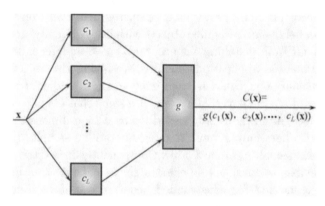

Fig. 11.1   Using a proper combination of individual base classifiers $\{c_1(\boldsymbol{x}),$ $c_2(\boldsymbol{x}), ..., c_L(\boldsymbol{x})\}$ a strong classifier $C(x) = g(c_1(\boldsymbol{x}), c_2(\boldsymbol{x}), ..., c_L(\boldsymbol{x}))$ can be designed to boost the performance.

This formalism arouses two basic questions:
1) How base classifiers can be generated?
2) How results of base classifiers can be combined together?

Ensemble is a set of weak, at the same time compatible, classifiers in which the base classifiers are different — it makes no sense to combine identical classifiers. Outputs of such classifiers are represented so that a combining classifier can use them as inputs.

Different classifiers can be generated by:

Different initializations: If the training is initialization dependent, different initializations may result in different classifiers.

Different classifiers: Different classifiers trained over the same feature set and training data can produce an ensemble. For example one might use support vector machines or decision trees to form an ensemble.

Different training sets: Different samples from the same training with or without replacement data may be used to generate the ensemble. A well-known example is bagging [3] where different training sets are generated by bootstrapping. It is also possible to form different sets by clustering, for example if each class is first split by cluster analysis, then different clusters can be used in different sets. This approach has been widely used for generating robust identification systems to pose variation, where each cluster contains images belong to a specific pose [4].

Different feature sets: In some applications objects may be represented in entirely different feature domains, e.g., in identification by speech and images [5]. Random subspace method [6] that generates random subsets out of a large feature set is also proven to be successful in computer vision [7, 8].

Basic methods to ensemble outputs of base classifiers are: Crisp output, soft output, and classification on classifiers' output.

• Crisp output: If only labels are available a majority vote [9, 10] or ranking [11] can be used.

• Soft output: if posterior probabilities are provided, linear combinations like averaging have been suggested in the literature [12, 9, 11]. If the classifier outputs are interpreted as fuzzy membership values, fuzzy rules [13 – 15], Dempster-Shafer techniques [16 – 19] can be considered.

• Classification on classifier outputs: It is possible to consider the output of classifiers in an ensemble as new features for another classifier [20, 21].

The first two categories, i.e., approaches based on labels and posterior probabilities have been considered more frequently in the literature. If the outputs of the base classifiers are just class-labels, the simplest way to combine classifiers is majority voting. In majority voting, votes for each class over the input classifiers are counted and the class with the maximum number of votes is selected as the winner. Another approach when crisp outputs are available is Borda count [11], a rank-based scheme for voting. For any class q, the Borda count is the sum of the number of classes ranked below q by each classifier. If $B_j(q)$ is the number of classes ranked below the class q by the jth classifier, then the Borda count for class q is set as $B(q) = \sum_{j=1}^{L} B_j(q)$, where L is the number of classifiers used. The final decision is given by selecting the class yielding the largest Borda count.

Although classifiers producing crisp class labels can perform well and could be applied to a variety of real-life problems, they provide the least amount of useful information for combination processes. If some measures of evidence like posterior probabilities, likelihood values, confidence measures, and distances are provided by the base classifiers, then fusion methods try to reduce the level of uncertainty [12].

In the sequel, a few applications of ensemble learning in computer vision, such as object detection, object recognition and tracking will be explained. These examples are intentionally chosen as they are classic problems in computer vision with various applications including surveillance, robotics, multimedia processing, and Human–computer Interaction (HCI).

11.2 Random Subspace Method

Random subspace [6] is a popular random sampling technique to enforce weak classifiers. In the random subspace method, a set of low dimensional subspaces is first generated by randomly sampling from the original feature vector (usually high dimensional); then, multiple classifiers are designed (in the random subspaces) and combined to form the final decision. Wang et al. first applied RSM to face recognition [7, 22, 23]. In random sampling Linear Discriminant Analysis (LDA) method, face images are first projected into the Principal Component Analysis (PCA) subspace; then, random sampling on the PCA subspace is performed in a more elaborate fashion [23]. Each random subspace comprises two parts and is spanned by $N_0 + N_1$ dimensions as follows:

1) Fixed part: N_0 features from the first N_0 PCA directions (associated with larger eigenvalues) are selected and used in all the subspaces constantly.

2) Random part: Randomly select N_1 directions from the remaining PCA directions.

After generating the PCA random subspaces, LDA is applied to form the Random LDA subspaces. The RS-LDA algorithm is illustrated in Table 11.1. The fusion rule could be the simple majority voting scheme or sum rule. It is mainly because LDA space usually has satisfactory accuracy. Fusion with the sum rule requires estimates of the likelihood values $P(x|c_i(x))$. A simple and efficient way for estimating likelihood values is the cosine distance between the centre of a class and the query in LDA subspaces. Mathematically, if m_i is the centre of class i and W_j is the transform applied to project a sample into the LDA random subspace j; then, the likelihood would be as follows:

$$\widehat{P}_i(x|c_k(x)) = \frac{1}{2}\left(1 + \frac{(W_j^\mathrm{T}x)^\mathrm{T}(W_j^\mathrm{T}m_i)}{\|W_j^\mathrm{T}x\|\|W_j^\mathrm{T}m_i\|}\right) \tag{11.1}$$

Similar ideas have been also applied to other feature spaces; e.g., Chawla et al. applied RSM on PCA space [24]. Considering a $n \times (n-1)$ dimensional eigenspace, where n is the number of images, Chawla et al. proposed to generate multiple random subspaces of size $n \times p$, L times, where p is the size of the randomly selected subspace. In order to match a query, the 1-nearest neighbor on each random subspace is computed and the most probable class is found by aggregating the obtained distances or class labels.

Table 11.1   Pseudo code for RS-LDA Algorithm

Input: Training set $\{X_1, X_2, ..., X_n\}$, where $\boldsymbol{X}_i \in R^D$, N_0 is the number of fixed features in random subspaces, N_1 is the number of random features in subspaces, and L is the number of random subspaces.

- Apply PCA to training set to generate eigenface space $\Phi = \{\varphi_1, \varphi_2, ..., \varphi_{n-1}\}$ as candidates for random subspaces.
- Generate L random subspaces $\{S_i\}_{i=1}^L$. Each random subspace S_i is spanned by $N_0 + N_1$. The first N_0 dimensions are fixed as the N_0 largest eigenfaces in Φ, i.e., $\{\varphi_1, \varphi_2, ..., \varphi_{N_0}\}$. The remaining N_1 dimensions are randomly selected from the other $n - N_0 - 1$ eigenfaces in Φ.
- Apply LDA to random subspaces $\{S_i\}_{i=1}^L$ to form L random LDA space.
- L classifiers $\{c_i(x)\}_{i=1}^L$ are constructed from the L random LDA subspaces.

Output: Combine all L classifiers $\{c_i(x)\}_{i=1}^L$ to form the strong classifier.

In Random Independent Subspace (RIS) proposed by Cheng et al., the RSM method is applied prior to Independent Component Analysis (ICA) [25]. ICA [26] is an extension of PCA and seeks to find mutually independent linear basis vectors, i.e., their cross-correlations as well as all their higher order moments are zero. Three sampling schemes have been deeply studied [25]. In the first scheme, random sampling was performed in the original feature space to reduce the dimensionality of feature space; in the second scheme, re-sampling is performed in whitened (whitening is a pre-process in ICA) feature space; in the third scheme, a two-level cascade re-sampling structure is adopted to re-sample both original feature space and the whitened feature space. The experimental results suggest that the third scheme can significantly improve the performance of ICA classifier.

It is generally believed that holistic representation of faces are prone to various factors like illumination [27] and expression changes [28, 29]. Beside the logical 3D solutions [30, 31], local face representation has been attracted a growing interest recently [32–35]. Recently Zhu et al. proposed Semi Random Subspace (Semi-RS) method for face recognition [36] in which sampling is applied on each local patch (obtained from partitioning the whole image). More specifically, a face image is deterministically split into several sub-images, then a set of base classifiers is constructed on different feature subsets that are randomly sampled from each sub-images set. All base classifiers are combined for the final decision using a combination rule. The Semi-RS algorithm is explained in Table 11.2.

Unlike RS-LDA, Semi-RS makes use of the face's local structure information so that it can achieve more robustness in recognizing faces with occlusion, illumination and expression changes. This can be interpreted by the concept of diversity in ensemble learning. It is widely accepted that classifiers to be combined should be diverse. Diversity (negative dependence, independence, orthogonality, and complementary) among the ensemble has been recognized as a key requirement for obtaining good results. A committee of independent classifiers is theoretically expected to improve upon the single best classifier when majority voting is used; for a dependent set of classifiers however, a classifier may be either better than the independent set or worse than the single worst member of the team [37].

Table 11.2   Pseudo code for Semi-RS Algorithm

Input: Training set $\{X_1, X_2, ..., X_n\}$, where $\boldsymbol{X}_i \in R^D$, n is the size of local patches $p \times q$ and L is the number of random sampling classifiers.

Partition each face image into several sub-images, collect all patches at the same position to create K patch image sets $\{T_1, T_2, ..., T_K\}$, where $T_i \in R^{p \times q}$.

Apply PCA to patch image sets T_i to obtain patch eigenspace $\{\varphi_1, \varphi_2, ..., \varphi_K\}$.

For each patch eigenspace φ_i generate L random sampling subspace $\{S_{i,1}, S_{i,2}, ..., S_{i,L}\}$.

Output: Combine all random subspaces to form the strong classifier.

Although random feature selection improves accuracy without affecting error rates, it is not always free of serious difficulties and drawbacks. The most evident drawback here is to guarantee that selected inputs for random selection of features have the necessary discriminant information. A result of such poor classifiers is usually damage to the ensemble [38].

From studies that attempt to give an answer to the question *"how we can construct the ensemble in RSM with less poor subspaces?"*, we will consider Directed Random Subspaces (DRS) next [8].

In DRS, selecting the lower dimensional subspace is not totally random. In fact, it is done based on a measure of generalization ability through a learning mechanism. The feature discrimination map $\boldsymbol{Q}$ for a classification problem defined in R^D is a two dimensional $D \times D$ table. Element (i, j) of this table represents the expected discrimination power of selecting feature j right after selecting feature i. The method is inspired by the feature selection method using a Reinforcement Learning (RL) framework for face recognition [39]. In summary, an intelligent agent repeatedly interacts with the R^D environment and picks a feature from R^D in each interaction. The environment responds to the intelligent agent by providing reward signals to proper selections. By exploring the environment, the intelligent agent learns what actions results in maximum rewards. Cell (a_{i-1}, a_i) of FDM is updated based on the current selected feature (a_i), the last selected feature (a_{i-1}) and the received reward (r), according to (2). Equation (2) is the Q-learning updating rule where $\alpha(0 < \alpha \leqslant 1)$ and $\gamma(0 \leqslant \gamma \leqslant 1)$ are the learning rate and the discount factor, respectively [40].

$$\boldsymbol{Q}(a_{i-1}, a_i) = \boldsymbol{Q}(a_{i-1}, a_i) + \alpha \left(r + \gamma \max_{l=1}^{D} \boldsymbol{Q}(a_i, a_l) - \boldsymbol{Q}(a_{i-1}, a_i) \right) \quad (11.2)$$

The obtained FDM is then used in the recall mode to generate directed random subspaces. Since each cell of FDM encodes a measure of discriminative power, cells with high discrimination powers can be appropriate candidates for generating directed subspaces (Fig. 11.2). Let Λ be the set of all cells in FDM which lie in a certain distance from the global maximum of FDM, i.e., $\Lambda = \{(i,j)|\boldsymbol{Q}(i,j) > \beta \max(\boldsymbol{Q}), 0 < \beta < 1\}$. In order to generate one directed random subspace, a sorted set of features must be derived first. This is accomplished by randomly selecting a starting point from Λ. Then, an agent traverses the FDM in D steps to generate the sorted set of features. In order to benefit from the randomness nature of RSM, the agent's policy in traversing FDM is epsilon greedy, i.e.,

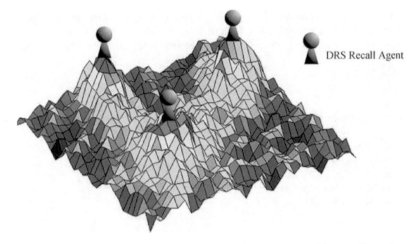

DRS Recall Agent

Fig. 11.2 A sample image of FDM and the agents used in recall mode.

1) It selects the best available feature $(a_l = \arg\max_j(\boldsymbol{Q}(a_{l-1}, a_j)|\boldsymbol{h}(j) = 0))$ with probability p.

2) It selects a random feature by probability $1\text{-}p$ from the available features.

The binary D dimensional vector $\boldsymbol{h}$ is used to avoid selecting a feature twice. An example of a feature discrimination map for a four dimensional feature space is shown in Fig. 11.3. The maximum value for this example is observed in the second row. Assume $\Lambda = \{(2,1),(4,2),(1,4)\}$ in which (4, 2) is randomly selected as the starting point to generate a sorted feature set. This results in $S = \{f_4\}$ in the first step. Here, the recall agent can either select the best feature in the fourth row with probability p, feature f_2, with discrimination power of 92.3 or randomly select a feature from the available feature set, $\{f_1, f_2, f_3\}$. Suppose that the best feature is selected, i.e., $S = \{f_4, f_2\}$. In this case, the recall agent is in the second row and the process continues until all features are selected. The pseudo code for generating a sorted set of feature is shown in Table 11.3.

Fig. 11.3 An example of a FDM in a four dimensional feature space.

Table 11.3   Pseudo code for generating a sorted set of features from FDM

Input: Training set $\{(\boldsymbol{X}_1, y_1), (\boldsymbol{X}_2, y_2), ..., (\boldsymbol{X}_n, y_n)\}$, Feature Discriminant Map $\boldsymbol{Q}$

Find the candidate set, Λ, from the feature discriminant map $\boldsymbol{Q}$ by thresholding.

Select a candidate randomly from Λ and let a_1 be the row index of the selected candidate.

$$\boldsymbol{h} = \left(\overbrace{0, 0, ...0}^{D} \right)$$

$\boldsymbol{h}(a_1) = 1$

repeat

Select a feature a_i by ε greedy policy, i.e., select the best feature $a_i = \arg\max_{j}(\boldsymbol{Q}(a_{i-1}, a_j)|\boldsymbol{h}(j) = 0)$ with probability p or select a feature randomly form available features with probability 1-p.

Update the selected feature vector, $\boldsymbol{h}(a_i) = 1$.

$S = \{S, a_i\}$

until all the features are selected.

Output: Sorted set of features, S.

The final stage of deriving a directed random subspace is to select a number of highest ranked features from the sorted feature set. This is simply accomplished by evaluating the recognition accuracy for D subsets $\{ \Psi_1, \Psi_2, ..., \Psi_D \}$ with leave one out scheme where $\Psi_i = \{mS_1, mS_2, ..., mS_i\}$ and mS_j is the jth member of sorted set of features S. The DRS correspond to a sorted set of features in the Ψ_i with maximum accuracy.

To conclude random subspace methods, a comparison between Eigenface, Fisherface, RS-LDA, and DRS has been performed on the BANCA database [41].

BANCA database contains image sets for 52 people (26 males and 26 females). For each person video recordings were made under various conditions (illumination, pose and camera variations), while the person was talking. Figure 11.4 shows some sample images from this database. In our experiment, all images were cropped to 46×56 pixels using manually labeled eye positions. Five images per each individual were selected randomly as the training data and the rest were used for testing. The random partitioning of image sets was repeated five times for each data set. The mean and standard deviation of recognition accuracy is reported in Table 11.4. Table 11.4 reveals that DRS algorithm outperforms the other studied methods with significant margin.

Fig. 11.4 Sample images of BANCA face database.

Table 11.4 Correct recognition rate and its standard deviation for the methods on the Banca database

Method	BANCA
Eigenface [42]	53.00%, σ = 4.11
Fisherface [43]	74.85%, σ = 4.02
Random subspace LDA[23]	76.08%, σ = 3.62
Directed Random Subspace [8]	81.39%, σ = 3.14

11.3 Boosting Method

Likewise other ensemble methods, in boosting one builds a classifier by combining several weak classifiers. The renowned AdaBoost algorithm [2] is a greedy method to combine a set of classifiers linearly. Boosting algorithms are designed based on a well-known combination theory, i.e., as long as the weak classifiers do better than chance, it is possible to boost their performance by combining them linearly [44]. In the sequel, we review how boosting algorithm can be utilized in designing robust classifier for various disciplines in computer vision like object detection, object recognition and.

11.3.1 Object detection/recognition

Viola and Jones in their seminal work showed how face detection can be considered as a binary classification problem in high-dimensional feature space [45]. For practical reasons, face detection must be performed real-time. To avoid complications in high-dimensional space, Viola and Jones proposed to use a version of boosting algorithm for feature selection. Feature selection aims to determine useful features after feature extraction based on appropriate rules. Using a proper feature selection, the dimensionality of the feature space can be significantly reduced where redundant features (irrelevant or noisy) are removed. Thus, it improves the data quality, and increases the accuracy of the resultant classification scheme [46]. Standard feature selection methods can be broadly categorized as filters and wrappers [47].

Filter methods usually use a criterion like cross-correlation, or information theoretic measures to estimate the relative importance of different features. More importantly, filters (e.g., those based on mutual information criteria) provide a generic selection of variables, not tuned for/by a given learning machine. As a result, filtering can be used as a pre-processing step to reduce space dimensionality and overcome overfitting. Filter methods have exhibited a number of drawbacks, for example some algorithms do not handle noise in data, and others require that the level of noise be roughly known a priori.

Wrappers are usually much slower than filters and work by directly evaluating the performance of a subset of features on a model trained on that subset. Wrappers can be considered as exhaustive search engines to investigate all possible feature subsets—usually equal to two to the power of the number of features. Although wrappers have the advantage of delivering the optimal subset for a given learning algorithm, they may become inefficient and time consuming at times.

Tieu and Viola adapted the AdaBoost algorithm for feature selection and used it for natural image retrieval [48]. They made the weak learner work in a single feature each time. Therefore, after T rounds of boosting, T features are selected along with their weak classifiers. The final strong classifier (hypothesis) is a linear combination of the T weak classifier where the weights are inversely proportional to their training errors. Tieu and Viola's version of AdaBoost considerably reduces the computational complexity of original Freund and Schapire's AdaBoost algorithm as in a D-dimensional problem; here, T comparisons are required instead of the original $T \times D$. In Tieu and Viola's version of AdaBoost for each feature, the weak learner determines the optimal threshold classification function, such that the minimum number of examples is misclassified. As a result, a weak classifier consists of a feature and a threshold. Tieu and Viola's version of AdaBoost is described by Table 11.5.

Viola and Jones later used this version of AdaBoost to select Haar-like features (Fig. 11.5) for face detection [45]. In their influential work, Viola and Jones showed that effective face detectors can be constructed with even two Haar-like features. The feature value of a Haar-like feature is calculated as the sum of pixels within rectangular regions, which are either positive (black regions) or negative (white regions). All these feature types can be efficiently computed using integral images [45]. An integral image, denoted as SI, sums up all the pixel values from the upper left up to the current position and is

Table 11.5   Pseudo code for Tieu and Viola's Adaboost Algorithm

Input: Training set $S = \{(\boldsymbol{X}_1, y_1), (\boldsymbol{X}_2, y_2), ..., (\boldsymbol{X}_n, y_n)\}$, where $\boldsymbol{X}_i \in R^D$ and $y_i \in \{0, 1\}$ for negative and positive samples, T is the number of iterations, i.e., selected features.

Initialize weights $w_i^1 = \begin{cases} \frac{1}{2m} & y = 0 \\ \frac{1}{2l} & y = 1 \end{cases}$, where the number of negative and positives samples are m and l respectively.

for $t = 1$ to T **do**

• Train one hypothesis h_j for each feature j using w_j^t and compute the error $e_j = Pr(h_j(\boldsymbol{X}_i) \neq y_i)$ over all samples.

• Choose the smallest error as $\varepsilon^t = \min_j \{e_j\}$ and corresponding hypothesis as h_t.

• Update the weights $w_i^{t+1} = \begin{cases} w_i^t \frac{\varepsilon^t}{1-\varepsilon^t} & \text{if } \boldsymbol{X}_i \text{ is classified correctly.} \\ w_i^t & \text{if } \boldsymbol{X}_i \text{ is classified correctly.} \end{cases}$

• Normalize the weights so that w_i^{t+1} becomes a distribution using $w_i^{t+1} \leftarrow \dfrac{w_i^{t+1}}{\sum\limits_{j=1}^{n} w_j^{t+1}}$.

end for

Output: The final hypothesis is $h(x) = \sum\limits_{t=1}^{T} \alpha_t h_t(x) \geqslant \dfrac{1}{2} \sum\limits_{t=1}^{T} \alpha_t$, where $\alpha_t = \ln\left(\dfrac{1-\varepsilon_t}{\varepsilon_t}\right)$.

Fig. 11.5   Haar-like features used by Viola and Jones for face detection.

defined as follows:

$$SI(x, y) = \sum_{i=1}^{x} \sum_{j=1}^{y} I(i, j).$$

The pre-calculation of an integral image for all pixels can be efficiently implemented in one pass over the image. Upon this calculation, any rectangular region can be computed by only three additions as shown in Fig. 11.6. As a result, exhaustive template matching can be achieved when scanning the whole image. Complex forms of Haar-like features have been introduced later. For example Lienhard et al. introduced an additional set of rotated features [49]. The work of Viola and Jones, paved the way for boosting in the area of computer vision. Many authors analyzed and used this approach with different extensions. Application of more complex feature types with boosting algorithm has been studied extensively. (e.g., [50, 51]).

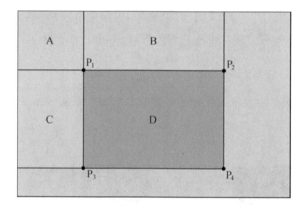

Fig. 11.6 Efficient calculation of the sum over a rectangular area. The value of the integral image at Position P_1 is the sum of the pixel values in region A. The sum over the area D can be calculated by $P_4 + P_1 - P_2 - P_3$.

11.3.2 Image-set Recognition

A novel methodology in comparing image-sets is the notion of distances between linear subspaces [52–57] Mutual Subspace Method (MSM) [52] was one of the earliest (and still competitive) approaches within this school of thought. In MSM the principal angle between two subspaces is considered as the similarity measure (see Fig. 11.7). Each principal angle carries some information for discrimination between the corresponding two subspaces. Later it was argued that one of the weaknesses of MSM approach is the use of only the first few principal angles [58]. For example Maeda et al. showed that the third principal angle is useful for discriminating between sets of images of a face and its photograph [59]. To circumvent the shortcoming of MSM, Kim et al. proposed to use the Adaboost to learn a weighting to combine the principal angles [58]. Here, in summary, weak classifiers are considered as a threshold over principal angles and Adaboost learns the weightings $\{w_i\}$ to generate the strong classifier as $M(\theta) = sign\left[\sum_{i=1}^{N} w_i M(\theta_i) - \frac{1}{2}\sum_{i=1}^{N} w_i\right]$, where N is the number of available principal angles.

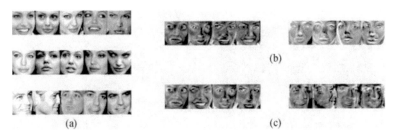

Fig. 11.7 Principal vectors in MSM: (a) Five samples of three typical face sets, the upper and middle images belong to the same person while the bottom belongs to a different individual. The first 4 pairs of principal vectors for a comparison of two linear subspaces corresponding to the same (b) and different individuals (c). In the former case, the most similar modes of pattern variation, represented by principal vectors, are very much alike in spite of different illumination, pose, and expression while the latter is quite different.

11.3.3 Object Tracking

A renowned approach in object tracking due to its low computational complexity and robustness to appearance changes is the Mean-Shift Tracking (MST) proposed by [60]. MST is a feature-based approach that primarily uses color-based object representation. MST and its extensions all assume that the histogram of the tracked object does not change much during tracking. This assumption can however be easily violated by the change in camera viewpoint and/or lighting conditions. These adverse situations call for online adaptation of the target model.

Considering tracking as a classification problem where a classifier is trained to distinguish an object from its background; a feature vector like color for every pixel in the reference image is constructed and a classifier is trained to separate pixels that belong to the object from pixels that belong to the background. Given a new video frame, the classifier is used to form a confidence map. The peak of the map is where the object is believed to be moved to (See Fig. 11.8 for a graphical illustration). Variation in the object and background over time was deemed for adapting the tracker accordingly. To alleviate this problem, ensemble tracking constantly updates a collection of weak classifiers to separate the foreground object from the background. The weak classifiers can be added or removed to reflect changes in object appearance or the background. As a result, an ensemble of classifiers is used to determine whether a pixel belongs to the object or not. A general algorithm for this purpose is

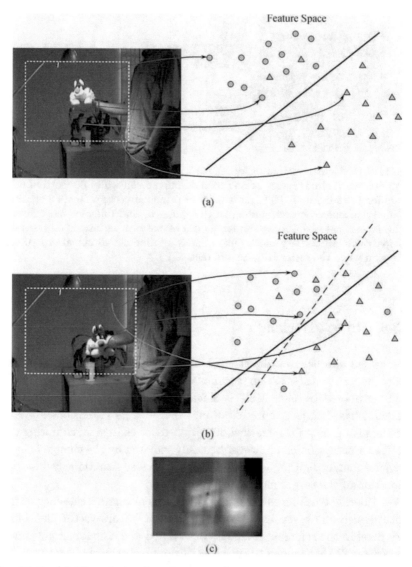

Fig. 11.8 (a) The pixels of an image at frame $t - 1$ are mapped to a feature space (circles for positive examples and triangles are negative examples). Pixels inside the red solid rectangle are assumed to belong to the object, pixels inside the white dashed rectangle and outside red solid rectangle are assumed to belong to the background. (b) At time t, all the pixels inside white dashed rectangle are classified using previously trained classifier, the result of this classification generate a confidence map (c) which is used to locate the position of the target.

given in Table 11.6 [61].

Table 11.6   General Ensemble Tracking

Input: n video frames $\{I_1, I_2, ..., I_n\}$ and the location of the target o_1 at first frame I_1.
Generate an ensemble by training L weak classifiers using o_1 and I_1.
For frame I_2 to I_n do
Test all pixels in current frame using the current strong classifier and generate a confidence map.
Locate the target by finding the maximum in the confidence map.
Use the extracted location to label the target and background in the current frame.
Keep K best weak classifiers.
Train new $L - K$ weak classifiers from the current frame and add them to the ensemble.
Output: Location of the target, $\{o_2, o_3, ..., o_n\}$ for frames $\{I_2, I_3, ..., I_n\}$.

Here, after initialization (generating the ensemble using the first frame and the location of the target), strong classifiers calculated using AdaBoost are used to classify pixels in the new frame. This results in producing a confidence map of the pixels, for example the classification margin can be used as the confidence measure. The peak of the map is the location of the object and can be efficiently extracted using mean shift algorithms. Once the detection for the new frame is completed, a new set of weak classifiers on the new frame are trained and added to the ensemble, and the process is repeated all over again.

11.3.4 Where to go from here

To conclude this chapter, we would like to bring the reader's attention to a very interesting topic of online-boosting. The basic idea of online-boosting is to modify the boosting algorithm in a way that it can operate on a single example and discards it after updating. This kind of learning can lead us to design systems that can adapt themselves to changes in their database (for example in a face recognition system, new persons may be added/deleted to/from the gallery without the need to train the system from the scratch). Also, when the number of training samples is large, online methods are more efficient in terms of computational costs. To reformulate this problem, assume a set of weak classifier $(h_1, h_2, ..., h_L)$ with their associated weights $\{\alpha_1, \alpha_2, ..., \alpha_L\}$ are given (which form the current strong classifier $H(x)$), the problem then becomes "how to update base classifiers and their weights whenever a new sample arrives?". On-line updating of the classifiers has a rich literature [62–64]. Updating the samples weight distribution is a crucial step because no knowledge about the difficulty of a sample is available a priori i.e., the algorithm does not know whether it has seen the sample before or not. In order to estimate the difficulty of a sample, Grabner et al.

propagate the sample through a set of weak classifiers. The difficulty of a sample is increased proportionally to the error of the weak classifier if the sample is still misclassified [65]. Online-boosting and its variances have recently been successfully applied in object detection and recognition [66 – 69], object tracking [65, 70 – 72] and image retrieval [73].

References

[1]   Ponce J, Hebert M, Schmid C et al (2007) Toward Category-Level Object Recognition. In: Lecture Notes in Computer Science. Springer, New York
[2]   Freund Y, Schapire R (1996) Experiments with a New Boosting Algorithm. In: International Conference on Machine Learning (ICML), pp 148 – 156
[3]   Breiman L (1996) Bagging Predictors. Journal of Machine Learning, 24(2): 123 – 140
[4]   Pentland A, Moghaddam B, Starner T (1994) View-based and Modular Eigenspaces for Face Recognition. In: IEEE Conference on Computer Vision and Pattern Recognition (CVPR'94), pp 84 – 91
[5]   Sanderson C, Paliwal K K (2004) Identity Verification Using Speech and Face Information. Digital Signal Processing, 14(5): 449 – 480
[6]   Ho T K (1998) The Random Subspace Method for Constructing Decision Forests. IEEE Trans Pattern Anal Mach Intell, 20(8): 832 – 844
[7]   Wang X, Tang X (2004) Random Sampling LDA for Face Recognition. In: IEEE Conference on Computer Vision and Pattern Recognition (CVPR'04), pp 259 – 265
[8]   Harandi M T, Ahmadabadi M N, Araabi B N et al (2010) Directed Random Subspace Method for Face Recognition. In: International Conference on Pattern Recognition (ICPR'10), 2010a. pp 2688 – 2691
[9]   Kuncheva L I (2002) A Theoretical Study on Six Classifier Fusion Strategies. IEEE Trans Pattern Anal Mach Intell, 24(2): 281 – 286
[10] Lam L, Suen C Y (1997) Application of Majority Voting to Pattern Recognition: An Analysis of its Behavior and Performance. IEEE Transactions on Systems, Man, and Cybernetics Part A:Systems and Humans, 27(5): 553 – 568
[11] Ho T K, Hull J J, Srihari S N (1994) Decision Combination in Multiple Classifier Systems. IEEE Trans Pattern Anal Mach Intell, 16(1): 66 – 75
[12] Kittler J, Hatef M, Duin R P W et al (1998) On Combining Classifiers. IEEE Trans Pattern Anal Mach Intell, 20(3): 226 – 239
[13] Cho S-B, Kim J H (1995) Combining Multiple Neural Networks by Fuzzy Integral for Robust Classification. IEEE Transactions on Systems, Man and Cybernetics, 25(2): 380 – 384
[14] Kuncheva L I (2003) "Fuzzy" Versus "Nonfuzzy" in Combining Classifiers Designed by Boosting. IEEE Trans on Fuzzy Systems, 11(6): 729 – 741
[15] Verikas A, Lipnickas A, Malmqvist K et al (1999) Soft Combination of Neural Classifiers: A Comparative Study. Pattern Recognition Letters, 20(4): 429 – 444
[16] Altincay H (2005) A Dempster-shafer Theoretic Framework for Boosting Based Ensemble Design. Pattern Analysis and Applications, 8(3): 287 – 302
[17] Bi Y, Guan J, Bell D (2008) The Combination of Multiple Classifiers Using an Evidential Reasoning Approach. Artificial Intelligence, 172(15): 1731 – 1751
[18] Al-Ani A, Deriche M (2002) A new Technique for Combining Multiple Classifiers Using the Dempster-Shafer Theory of Evidence. Journal of Artificial Intelligence Research, 17: 333 – 361
[19] Quost B, Denoux T, Masson M H (2007) Pairwise Classifier Combination Using Belief Functions. Pattern Recognition Letters, 28(5): 644 – 653
[20] Gama J, Brazdil P (2000) Cascade Generalization. Machine Learning, 41(3): 315 – 343
[21] Wolpert D H (1992) Stacked Generalization. Neural Networks, 5(2): 241 – 259
[22] Wang X, Tang X (2005) Subspace Analysis Using Random Mixture Models. In: IEEE Conference on Computer Vision and Pattern Recognition (CVPR'05), pp 574 – 580
[23] Wang X, Tang X (2006) Random Sampling for Subspace face Recognition. International Journal of Computer Vision, 70(1): 91 – 104

[24] Chawla N V, Bowye K W (2005) Random Subspaces and Subsampling for 2D face Recognition. In: IEEE Computer Society Conference on Computer Vision and Pattern Recognition (CVPR'05). IEEE Computer Society, pp 582–589

[25] Cheng J, Liu Q, Lu H et al (2006) Ensemble Learning for Independent Component Analysis. Pattern Recognition, 39(1): 81–88

[26] Hyvarinen A, Karhunen J, Oja E (2001) Independent Component Analysis. Wiley-Interscience

[27] Basri R, Jacobs D W (2003) Lambertian Reflectance and Linear Subspaces. IEEE Trans Pattern Anal Mach Intell, 25(2): 218–233

[28] Shih F Y, Chuang C F, Wang P S P (2008) Performance Comparisons of Facial Expression Recognition in JAFFE Database. International Journal of Pattern Recognition and Artificial Intelligence, 22(3): 445–459

[29] Luo Y, Gavrilova M L, Wang P S P (2008) Facial Metamorphosis Using Geometrical Methods for Biometric Applications. International Journal of Pattern Recognition and Artificial Intelligence, 22(3): 555–584

[30] Lee S W, Wang P S P, Yanushkevich S N (2008) Noniterative 3D face Reconstruction Based on Photometric Stereo. International Journal of Pattern Recognition and Artificial Intelligence, 22(3): 389–410

[31] Blanz V, Vetter T (2003) Face Recognition Based on Fitting a 3D Morphable Model. IEEE Trans Pattern Anal Mach Intell, 25(9): 1063–1074

[32] Ahonen T, Hadid A, Pietikainen M (2006) Face Description with Local Binary Patterns: Application to Face Recognition. IEEE Trans Pattern Anal Mach Intell, 28(12): 2037–2041

[33] Li S Z, Hou X W, Zhang H J et al (2001) Learning Spatially Localized, Parts-based Representation. In: IEEE Conference on Computer Vision and Pattern Recognition (CVPR'01), pp 207–212

[34] Schneiderman H, Kanade T (1998) Probabilistic Modeling of Local Appearance and Spatial Relationships for Object Recognition. In: IEEE Conference on Computer Vision and Pattern Recognition (CVPR'98), pp 45–51

[35] Zou J, Ji Q, Nagy G (2007) A Comparative Study of Local Matching Approach for Face Recognition. IEEE Trans on Image Processing, 16(10): 2617–2628

[36] Zhu Y, Liu J, Chen S (2009) Semi-random Subspace Method for face Recognition. Image and Vision Computing, 27(9): 1358–1370

[37] Kuncheva L I, Whitaker C J (2003) Measures of Diversity in Classifier Ensembles and Their Relationship with the Ensemble Accuracy. Machine Learning, 51(2): 181–207

[38] Garcia-Pedrajas N, Ortiz-Boyer D (2008) Boosting Random Subspace Method. Neural Networks, 21(9): 1344–1362

[39] Harandi M T, Nili Ahmadabadi M, Araabi B N (2009) Optimal Local Basis: A Reinforcement Learning Approach for face recognition. International Journal of Computer Vision, 81(2): 191–204

[40] Sutton R S, Barto A G (1998) Introduction to Reinforcement Learning. MIT Press, Massachusetts

[41] Bailly-Bailliére E, Bengio S, Bimbot F et al (2003) The BANCA Database and Evaluation Protocol. In: Kittler J, Nixon M (eds) Audio- and Video-Based Biometric Person Authentication, vol 2688. Lecture Notes in Computer Science. Springer, Heidelberg, pp 1057

[42] Turk M, Pentland A (1991) Eigenfaces for Recognition. J Cognitive Neuroscience, 3(1): 71–86

[43] Belhumeur P N, Hespanha J P, Kriegman D J (1997) Eigenfaces vs. Fisherfaces: Recognition Using Class Specific Linear Projection. IEEE Trans Pattern Anal Mach Intell, 19(7): 711–720

[44] Freund Y, Schapire R (1995) A Desicion-theoretic Generalization of on-line Learning and an Application to Boosting. In: Vitányi P (ed) Computational Learning Theory, vol 904. Lecture Notes in Computer Science. Springer, Heidelberg, pp 23–37

[45] Viola P, Jones M J (2004) Robust Real-Time Face Detection. International Journal of Computer Vision, 57(2): 137–154

[46] Blum A L, Langley P (1997) Selection of Relevant Features and Examples in Machine Learning. Artificial Intelligence, 97(1–2): 245–271

[47] Guyon I, Elisseeff A (2003) An introduction to Variable and Feature Selection. Journal of Machine Learning Research, 3: 1157–1182

[48] Tieu K, Viola P (2000) Boosting Image Retrieval. In: IEEE Conference on Computer Vision and Pattern Recognition (CVPR'00), pp 228–235

[49] Lienhart R, Maydt J (2002) An Extended set of Haar-like Features for Rapid Object

Detection. In: International Conference on Image Processing (ICIP'02). pp 900–903

[50]   Yang P, Shan S, Gao W et al (2004) Face Recognition Using Ada-Boosted Gabor Features. In: IEEE International Conference on Automatic Face and Gesture Recognition (FG'04), pp 356–361

[51]   Opelt A, Pinz A, Fussenegger M et al (2006) Generic Object Recognition With Boosting. IEEE Trans Pattern Anal Mach Intell, 28(3): 416–431

[52]   Yamaguchi O, Fukui K, Maeda K (1998) Face Recognition Using Temporal Image Sequence. In: International Conference on Face and Gesture Recognition (FG'98). pp 318–323

[53]   Hamm J, Lee D D (2008) Grassmann Discriminant Analysis: a Unifying View on Subspace-based Learning. In: International Conference on Machine Learning (ICML'08), Helsinki, Finland. ACM, pp 376–383

[54]   Wang T, Shi P (2009) Kernel Grassmannian Distances and Discriminant Analysis for face Recognition from image sets. Pattern Recog Lett, 30(13): 1161–1165

[55]   Kim T K, Kittler J, Cipolla R (2007a) Discriminative Learning and Recognition of Image Set Classes Using Canonical Correlations. IEEE Trans Pattern Anal Mach Intell, 29(6): 1005–1018

[56]   Wang R, Shan S, Chen X et al (2008) Manifold-manifold Distance with Application to Face Recognition Based on Image set. In: IEEE Conference on Computer Vision and Pattern Recognition (CVPR'08), pp 1–8

[57]   Harandi M T, Bigdeli A, Lovell B C (2010b) Image-set Face Recognition Based on Transductive Learning. In: International Conference on Image Processing (ICIP'10) Hong Kong, 2010, pp 2425–2428

[58]   Kim T K, Arandjelovic O, Cipolla R (2007b) Boosted Manifold Principal Angles For Image Set-based Recognition. Pattern Recognition, 40(9): 2475–2484

[59]   Maeda K I, Yamaguchi O, Fukui K (2004) Towards 3-dimensional Pattern Recognition. In: Fred A, Caelli T, Duin RPW, Campilho A, Ridder Dd (eds) Structural, Syntactic, and Statistical Pattern Recognition, vol 3138. Lecture Notes in Computer Science. Springer, Heidelberg, pp 1061–1068

[60]   Comaniciu D, Ramesh V, Meer P (2003) Kernel-Based Object Tracking. IEEE Trans Pattern Anal Mach Intell, 25(5): 564–575

[61]   Avidan S (2007) Ensemble tracking. IEEE Trans Pattern Anal Mach Intell, 29(2): 261–271

[62]   Bordes A, Ertekin S, Weston J et al (2005) Fast Kernel Classifiers With Online and Active Learning. Journal of Machine Learning Research, 6

[63]   Kivinen J, Smola A J, Williamson R C (2004) Online Learning with Kernels. IEEE Trans on Signal Processing, 52(8): 2165–2176

[64]   Li Y, Long P M (2002) The Relaxed Online Maximum Margin Algorithm. Machine Learning, 46(1–3): 361–387

[65]   Grabner H, Bischof H (2006) On-line Boosting and Vision. In: IEEE Conference on Computer Vision and Pattern Recognition (CVPR'06), 17–22 June 2006: 260–267

[66]   Nguyen T T, Binh N D, Bischof H (2008) An Active Boosting-based Learning Framework for Real-time Hand Detection. In: International Conference on Automatic Face Gesture Recognition (FG'08), pp 1–6

[67]   Chang W C, Cho C W (2010) Online Boosting for Vehicle Detection. IEEE Transactions on Systems, Man, and Cybernetics, Part B: Cybernetics, 40(3): 892–902

[68]   Huang C, Ai H, Yamashita T et al (2007) Incremental Learning of Boosted Face Detector. In: International Conference on Computer Vision (ICCV'07), pp 1–8

[69]   Masip D, Lapedriza Á, Vitrià J (2009) Boosted Online Learning for Face Recognition. IEEE Transactions on Systems, Man, and Cybernetics, Part B: Cybernetics, 39(2): 530–538

[70]   Grabner H, Leistner C, Bischof H (2008) Semi-supervised on-line Boosting for Robust Tracking. In: European Conference on Computer Vision, Marseille, France, 2008. Springer, pp 234–247

[71]   Binh N D, Nguyen T T, Bischof H (2008) On-line Boosting Learning for Hand Tracking and Recognition. In: International Conference on Image Processing, Computer Vision, and Pattern Recognition (IPCV'08), pp 345–351

[72]   Liu X, Yu T (2007) Gradient Feature Selection for Online Boosting. In: International Conference on Computer Vision (ICCV'07), pp 1–8

[73]   Jiang W, Er G, Dai Q et al (2006) Similarity-based Online Feature Selection in Content-based Image Retrieval. IEEE Trans on Image Processing, 15(3): 702–712

12 Depth Image Based Rendering

Michael Schmeing[1] and Xiaoyi Jiang[2]

Abstract   Depth Image Based Rendering is a technique to render new views from a video stream. The scene geometry is given by an additional depth stream which stores for each pixel its distance to the camera. This technique allows for a wide variety of applications including 2D-3D video conversion. This chapter gives an introduction to Depth Image Based Rendering and discusses some challenges including proposed solutions. The main focus lies on the disocclusion problem which is due to the fact that the scene is only observed from a single view which can lead to missing information in the generated virtual views. Other challenges include imperfect depth maps, ghosting, and the cardboard effect.

12.1 Introduction

In recent years, 3D television and 3D cinema have become a booming market. Many cinemas have meanwhile 3D-ready screens and 3D home entertainment systems become more and more affordable. With the growing number of player devices, there is also a growing demand for 3D content. Unfortunately, only some dozen 3D movies are made per year and 3D television channels are rare, which limits the content available for 3D devices. This unsatisfying situation could be overcome with a reliable 2D-3D conversion process, which would generate a 3D version of ordinary 2D movies or even of live 2D television on the fly.

The problem of converting a 2D movie into a 3D movie though is a very difficult one. This is due to the fact that this problem is heavily underdetermined. This means that the very same image that a 2D camera observes may be induced by a variety of 3D scenes. Figure 12.1 shows two examples. Figure 12.1 (a) shows one of the famous fun pictures of the Leaning Tower of Pisa. It could either show a giant touching the tower or a person standing way nearer to the camera than the tower. The picture itself gives no information about the true situation, only our common sense lets us tend towards the latter one. In Fig. 12.1 (b), the helicopter looks almost toylike but is in fact more than 30 meters long, six meters tall and can carry about 30 people.

1   University of Münster, Germany, Department of Mathematics and Computer Science.
    E-mail: m.schmeing@uni-muenster.de.
2   University of Münster, Germany, Department of Mathematics and Computer Science.
    E-mail: xjiang@uni-muenster.de.

(a)   (b)

Fig. 12.1 (a) The Leaning Tower of Pisa touched by a giant. (b) The helicopter is more than 30 meters long but appears way smaller than it is. Its real size cannot be deduced from the image.

Another example can be found in the movie *The Lord of the Rings: The Fellowship of the Ring*. Some of the characters, the *Hobbits*, are substantially smaller than others. Many of the scenes featuring both types of characters were shot utilizing *forced perspective*: The non-Hobbit characters were simply placed nearer to the camera than the Hobbits and the camera was appropriately adjusted so that the characters would appear to interact with each other. Employing a complex movable scene construction, this was even performed with a moving camera.

These examples show that it is a tricky task to infer 3D information from 2D images. In all those scenes, the information about the real setting cannot be inferred using only the video frames. Nevertheless, some images exhibit certain information, called *depth cues*, which can be interpreted by our brain to recover 3D information about the scene. These are, for example, *texture* [1, 2], *perspective* [3–5], *shading* [6, 7], *focus* [8–10], *motion* [11–15], and so on. There is a wide variety of approaches that try to infer 3D information about a scene from these cues. This can work very well, especially in the case of motion. On the other hand, these algorithms are very sensitive and fail if no cues are available.

Depth Image Based Rendering is a technique that is able to offer a different approach to 2D-3D video conversion. The human depth perception is based on the fact that our eyes capture the scene from two slightly different views. Our brain fuses them to generate the depth perception. The inter-eye distance of an average human is about 63mm, so two cameras horizontally mounted in that distance capture the scene in a way a human would do. Using larger or smaller inter-eye distances changes the depth perception and can be used as a creative effect. Since ordinary 2D video can only provide one video stream, 2D-3D conversion is basically about reconstructing a second view in addition to the original 2D video stream.

With Depth Image Based Rendering, we can generate the required second stream for (almost) arbitrary 2D video. Depth Image Based Rendering relies on additional geometric information about the scene. This information is given as a *depth map* of the scene. A depth map stores for each pixel of the video stream its distance to the camera, see Fig. 12.2 for an example. Using this depth information, a simple 3D model can be obtained from which virtual camera views are rendered.

(a)   (b)

Fig. 12.2 (a) A video frame. (b) The according depth map. The lighter the object appears in the depth map, the nearer it is to the camera.

Applications

The 2010 3D movie *Alice in Wonderland* was originally shot in 2D and converted to 3D afterwards. The original video stream was taken for one eye and a Depth Image Based Rendering technique was then used to generate a second video stream for the other one. The required depth stream was defined by artists, a process which caused a lot of manual work: More than 300 people worked half a year at this project [16].

Manual work is not the only way to obtain depth maps. Range scanners, like the ZCAM^{TM} system[3], can record the depth of a scene directly by measuring the reflections of either a laser or of infrared signals. *Shape from structured light* [17 – 19] obtains 3D information by projecting regular light patterns onto an object and analysing their deformations. A consumer example is the KINECT-controller for the XBOX360 from Microsoft[4]. Another way to obtain a depth map (or *disparity map*) is to infer it from one [20] or more video streams observing a scene using *stereo* techniques [21 – 23].

An application are autostereoscopic displays which are able to display up to 60 [24] views of a scene. Even native 3D movies offer only one 3D view (the left-eye- and right-eye-channel). Additional views have to be rendered by the display. This can be done by using Depth Image Based Rendering with the depth map infered from the two input video streams.

One primarily application of Depth Image Based Rendering is 3DTV [25 – 27]. Here it can help to replace the common 3D scene representation *video-*

---

3   http://pro.jvc.com/prof/attributes/features.jsp?model_id=MDL101309.
4   http://www.xbox.com/kinect/.

plus-video with a new one called *video-plus-depth*. Figure 12.3 shows a typical 3D video processing chain.

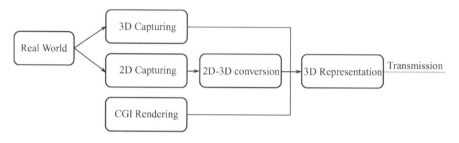

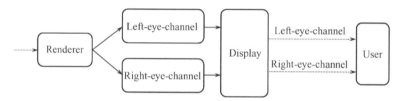

Fig. 12.3 A 3D processing chain.

Independently of its actual realization, each 3D scene representation must be able to provide a left and a right video channel, which are separately transmitted to the user's eyes by the display. The basic possible 3D representation is to directly deal with these two channels in the complete processing chain: The camera records two different views of the scene, which are transmitted to the user and projected by the display. We call this representation video-plus-video. This representation has the advantage that most, in particular new, 3D scene material is already in this format. But there are several limitations, which generate the need for another 3D representation.

The main limitation of video-plus-video is that the stereo geometry is fixed at the recording site, i.e., the *stereo basis* and the *plane of convergence*. The stereo basis is the horizontal distance of the two channels and corresponds with the strength of the 3D effect. In video-plus-video the amount of perceived depth cannot be changed by the user to his personal taste. This is a severe limitation since the perceived depth depends, for example, on the display size and even on the age of the user (since children have a smaller inter-eye distance than adults). Both factors are unknown at the recorder side and vary for different users. Controlling the convergence distance allows the user to decide which parts of the scene pop out of the scene and which parts appear behind the screen, see Section 12.2.1 for details. An additional disadvantage of video-plus-video is the comparatively high amount of data, which has to be stored and transmitted.

If we represent a 3D movie by one video stream and its according depth stream (*video-plus-depth*), we can render new virtual views, for example a new right-eye-channel, using Depth Image Based Rendering. The crucial difference to ordinary video-plus-video is that when rendering the new camera views, the user can select the stereo basis and the plane of convergence to his personal taste. An additional advantage of *video-plus-depth* is that the depth stream can be compressed efficiently [25].

In the following, we will explain Depth Image Based Rendering in detail and show some of the challenges that occur in this process as well as possible solutions for them. The remaining chapter is organized as follows. In the next section, we will explain the details about rendering new views from video-plus-depth. In Section 12.3 we describe the disocclusion problem which is one of the main challenges in Depth Image Based Rendering together with some proposed solutions. Section 12.4 will briefly cover some other challenges like dealing with *imperfect depth maps*, *ghosting* and the *cardboard effect*. We will conclude this chapter with Section 12.5.

12.2 Depth Image Based Rendering

Image Based Rendering and Modeling [28 – 31] deals with real world images and tries to extract some geometric information about the depicted scene to generate a model of the scene and to generate new virtual views employing this model. In [23], for example, 8 cameras are used to obtain 8 different views of a scene. Using these views, new virtual views can be rendered.

In *Depth Image Based Rendering* we only have one reference view of a scene. The geometric information is obtained from an according depth map which stores for each pixel of the reference image its distance to the camera. Figure 12.4 shows an image of the *Stanford Bunny*[5] and an according depth map. With the video and depth information, we can generate a new view of the bunny. The general idea consists of two steps. First, the reference view is projected into the 3D space according to the depth value of each pixel. This way, we obtain a point cloud consisting of pixels from the reference view. In a second step, the point cloud is projected onto a virtual camera, called *target view*. This process is illustrated in Fig. 12.5.

The process of computing the new view is called *3D warping* [26]. This process maps each pixel from the image plane of the reference view onto the image plane of the target view. So for each pixel with coordinates (u, v) in the reference view, we compute its new coordinates $(\tilde{u}, \tilde{v})$ in the image plane

---

5 The Stanford Bunny, the *Happy Buddha* and the *Dragon*, seen in the following, are from the *Stanford 3D Scanning Repository*, http://graphics.stanford.edu/data/3Dscanrep/.

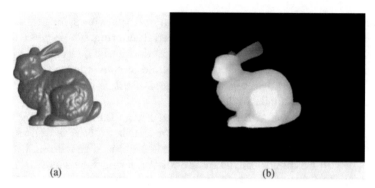

Fig. 12.4 (a) The Stanford Bunny. (b) The according depth map.

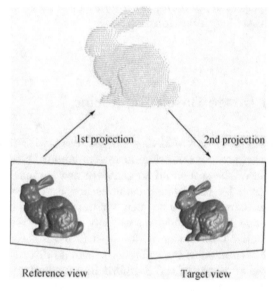

Fig. 12.5 The reference view is projected into the 3D space to obtain a 3D model of the scene, which basically is a point cloud consisting of the pixels of the reference view. From this model, the target view is obtained by projecting the point cloud onto the image plane of the target camera.

of the target view employing a transformation τ:

$$\tau : [1 \dots M] \times [1 \dots N] \longrightarrow \mathbb{R} \times \mathbb{R}, \tag{12.1}$$

$$\tau(u, v) = (\tilde{u}, \tilde{v}). \tag{12.2}$$

Before we examine the geometric details, we define some notation. A video I is a sequence of n frames $I_t, t \in [1 \dots n]$. Each frame is an image with M

rows, N columns and C color channels:

$$I = \{I_t : t \in [1..n]\}, \tag{12.3}$$
$$I_t = \{x_{t,p,c} \in [0\ldots255] : p \in [1\ldots M] \times [1\ldots N], c \in [1\ldots C]\}. \tag{12.4}$$

Common values for C are 1 (grayscale images) or 3 (RGB color images). In this chapter we always assume $C = 3$. In Depth Image Based Rendering, the scene geometry is given by a depth map D. The depth map stores for each pixel in video I its distance to the camera:

$$D = \{D_t : t \in [1\ldots n]\}, \tag{12.5}$$
$$D_t = \{d_{t,p} \in [0\ldots255] : p \in [0\ldots M] \times [0\ldots N]\}. \tag{12.6}$$

To shorten the notation, we may omit subscript t and c and denote a single pixel by x_p or $x(u,v)$. The according depth value is then denoted by d_p or $d(u,v)$, respectively. The depth frame D_t corresponding to frame I_t is sometimes referred to as depth map of I_t.

As can be seen in Eq. (12.6), the depth of the scene is not stored in absolute values but quantized in the range of $[0\ldots255]$. This has the advantage that we can deal with a depth map just like with a video, i.e., we can perform computer vision algorithms on it like on any other video or image. Therefore, if we want to obtain the absolute distance value Z_p of pixel x_p, we have to infer it from the depth map. This is usually done by linear interpolation [25, 26]:

$$Z_p = Z_{far} - d_p \frac{Z_{far} - Z_{near}}{255}, \tag{12.7}$$

where Z_{far} and Z_{near} denote the *far plane* and *near plane*, respectively, which represent the smallest and largest depth shown in the scene. In [32], another way to translate the values of the depth map into real depth values is proposed which is oriented towards human perception.

Geometry

In Depth Image Based Rendering we are given a video stream with associated depth. Our goal is to compute a new video stream that would be observed by a virtual camera, the *target view*, horizontally shifted with respect to the reference camera.

Remember that Depth Image Based Rendering uses 3D Warping to generate the target view. Warping is performed frame-wise. This means that for each pixel x_p in the reference view I_t we compute its new location $\tilde{p} = \tau(p)$ in the coordinate system of target view $\tilde{I}_t$ and copy its color value to this position.

In Fig. 12.6, two cameras C_1 and C_2 observe a point P. C_1 is the reference view and C_2 is the target view. C_2 is shifted to the right by an amount of b. In the image plane of camera C_2 the image of point P, denoted by P'', is shifted to the left with respect to its position in camera C_1, denoted by P'. So we know the position of P' in the reference view and try to compute the position of P'' in the target view. This new position is given by shifting P' to the left by a certain amount. This amount s_P is given by

$$s_P = b - |P'P''|. \tag{12.8}$$

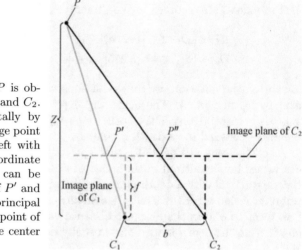

Fig. 12.6 A point P is observed by cameras C_1 and C_2. C_2 is shifted horizontally by the amount b. The image point P'' is shifted to the left with respect to P' in the coordinate system of C_2 (which can be seen by the distance of P' and P'' to their according principal points, i.e., the closest point of the image plane to the center of projection).

By Intercept Theorem, we have

$$\frac{|P'P''|}{b} = \frac{|PP'|}{|PC_1|} = \frac{Z - f}{Z}. \tag{12.9}$$

The depth Z of point P and can be inferred from the depth map as we have seen above in Eq.(12.7). Now we can compute s_P:

$$s_P = b - |P'P''| = b - b\frac{Z - f}{Z} = b\frac{f}{Z}. \tag{12.10}$$

Altogether, we are now able to render the view of a rightwards shifted camera. The target view $\tilde{I}_t$ is given by

$$\tilde{I}_t = \{x_{\tau(p)} : p \in [1 \ldots M] \times [1 \ldots N]\}. \tag{12.11}$$

So far, we have modeled a situation with two parallel cameras observing a scene. We now extend this concept to a setting of two cameras with *shifted sensors*. To understand why this is necessary, we have to understand how depth perception works when watching a movie on a screen.

3D movies are famous for embedding the audience into the center of action. This is mainly due to the fact that 3D movies can let objects appear to protrude from the screen into the audience. Figure 12.7 shows an observer with eyes at positions C_1 and C_2, respectively, observing a scene at the screen. The light rays observed by his right eye are shown with slashed lines and the light rays observed by his left eye are solid. P_2 appears to lay right on the screen. We see that if we want a point to appear in the screen plane, as it is the case in ordinary 2D video, the left- and the right-eye-channel must not exhibit any disparity. A point behind the screen (P_3) exhibits *positive parallax*, i.e., the right-eye-channel is shifted to the right with respect to the left-eye-channel. Finally, a point appearing in front of the screen (P_1) exhibits *negative parallax*, i.e., the right-eye-channel is shifted to the left with respect to the left-eye-channel.

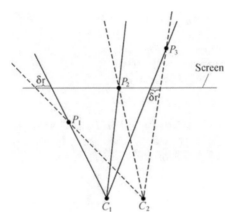

Fig. 12.7 There are three different areas where a point P can appear when watching a 3D video: before, right on and behind the screen. To make a point appear in front of the screen, the right-eye-channel must be projected on the left of the left-eye-channel, see point P_1. To make a point appear right on the screen, the right-eye- and left-eye-channel must not exhibit any parallax, see point P_2. This is the case in ordinary 2D video. To make a point appear behind the screen, the right-eye-channel must be projected on the right of the left-eye-channel, see point P_3.

In our model with parallel cameras, each pixel x_p in the right-eye-channel exhibits a shift to the left by a the (positive) amount of s_P. This means that the complete scene would appear in front of the screen. This is highly undesirable since a too excessive use of "pop-out" effects can cause eye strain and headache to the audience.

To overcome this situation, the camera model above can be extended to the so called *shift-sensor approach* [26], see Fig. 12.8. In this setting, we have two cameras C_1 and C_2 centered around the reference view O. They parallely observe the scene. In contrast to the previous setting, the image planes, i.e., in real cameras the CCD-sensors, are shifted inwards by an amount of h. This way, we establish a *plane of convergence* with distance Z_c. All objects of distance Z_c exhibit no parallax on the cameras, they are projected onto the same coordinates in each image. These objects will appear on the screen when watching the screen on a 3D monitor. All objects closer appear in front of the screen, and all objects further behind.

Fig. 12.8 The cameras C_1 and C_2 observe a scene horizontally shifted to the left and right, respectively, by the amount of b. Their image sensors are shifted inwards by the amount of h. Each point in distance Z_c is projected onto the same coordinate value in each camera's coordinate system, so objects in this distance exhibit no parallax. When C_1 is used as a left-eye-stream and C_2 as a right-eye-stream, each object in distance Z_c will appear right on the screen, see Fig. 12.7. Objects closer to the cameras will appear in front of the screen and object further will appear behind.

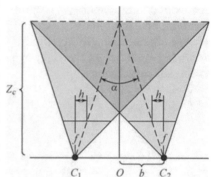

If we have set the values for Z_{near} and Z_{far} for a scene, we can choose which parts of the scene plane appear on the screen by choosing Z_c appropriately. This defines the needed sensor shift h: We have

$$\frac{Z_c}{b} = \frac{f}{h} \tag{12.12}$$

$$\Leftrightarrow \quad h = b\frac{f}{Z_c}. \tag{12.13}$$

So, in the right camera of the shift-sensor approach, a pixel gets shifted to the left by s_P and to the right by h. Analogously, in the left camera a pixel gets shifted by s_p to the right and by h to the left. With

$$r := -b\frac{f}{Z} + b\frac{f}{Z_c}, \tag{12.14}$$

we get the final transformation:

$$\tau(u, v) = (u, v + \delta r), \quad \begin{cases} \delta = 1, & \text{right camera,} \\ \delta = -1, & \text{left camera.} \end{cases} \tag{12.15}$$

Note that r is measured in mm, and has to be translated into pixels when implementing this formula. This is done by assuming an image sensor size for the virtual cameras (e.g., 36×24mm). The width K of the sensor and the width N of the image are then used to obtain an adjusted version r^* of r:

$$r^* := r\frac{N}{K}. \tag{12.16}$$

When rendering a target view using Eq. (12.15) it may happen that two pixels of the reference view are warped onto the same location in the target view. In this case, called *occlusion problem*, we have to decide which pixel should be visible in the target view. This problem is related to the disocclusion problem, see Section 3 which does not have simple solution like the occlusion problem: Since we know for each pixel its distance to the camera we simply can take the closer one. This works as a solution for the occlusion problem but introduces a large computational overhead since we have to keep a list of warped pixels for each location in the target view and sort this list according to the depth values. The work of [33] shows that this problem can be elegantly solved by processing the pixels of the reference view in the so-called *occlusion-compatible warp order* which only depends on the relative position of the reference view and the target view.

While there are 18 different processing orders for the general problem to warp one view onto another, in our case, though, where reference view and target view only differ by a horizontal shift, only two processing orders are needed, depending on whether rendering a rightwards, or leftwards shifted view, respectively: For a horizontally leftwards shifted view, the columns of the reference view have to be processed from the right towards the left image border. Accordingly, for a horizontally rightwards shifted view, the columns of the reference view have to be processed from right to left.

Now we have all necessary tools to render additional views from a video and its according depth map. In the following, we will describe some challenges that occur during this process. The main problem is called *Disocclusion Problem* and is caused by the fact that some areas in the target view might be occluded in the reference view by objects closer to the camera. The next section deals with approaches to fill these areas appropriately. We will focus in this chapter on the Disocclusion Problem. Section 4 briefly deals with other challenges, namely non-optimal depth maps, ghosting and the cardboard effect.

12.3 Disocclusions

Figure 12.9 shows a frame of the *granguardia* sequence[6] with according depth map and the result of Depth Image Based Rendering using Eq. (12.15) for rendering a rightwards shifted virtual view. We can see that in the new view, white areas get exposed, especially at object boundaries. Figure 12.10 shows how these blank areas occur.

Disocclusions occur when areas that are not visible in the reference view get disoccluded in the target view. This typically happens at the edge of foreground objects: When a camera moves rightwards, new areas on the right side of the foreground object come into view. There is no information

---

6   See [34], used with kind permission of A. Colombari and A. Fusiello.

available in the reference view on how these areas look like. In the target view, these areas remain blank which leads to white areas as can be seen in Fig. 12.9 (c) and (d).

Fig. 12.9 (a) A video frame of the granguardia sequence. (b) The according depth frame. (c) After 3D warping, blank areas at object boundaries and holes occur. (d) Magnification of (c).

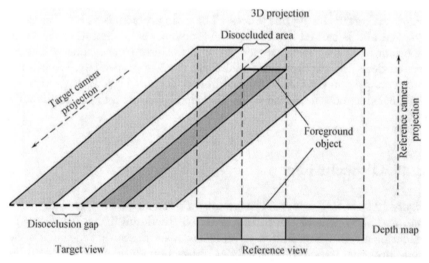

Fig. 12.10 In 3D warping, the reference view is projected into the 3D space according to the depth information stored in the depth map. This 3D model is then projected onto the image plane of the target view. In this process, areas that are not visible in the reference view get disoccluded in the targed view.

Figure 12.9 (d) shows a magnification of the new view. We can see that there are basically two kinds of blank areas: *Holes* and *disocclusions*. Holes appear in regions with only gradual depth changes whereas disocclusions appear on depth boundaries.

12.3.1 Holes

Holes appear in regions with smooth depth changes and are due to rounding errors. The transformation τ, which computes the new position for each image pixel x_p in the reference view, will in most cases have non-integer values. Since pixels can only be shifted by an integer-amount, rounding errors occur. See Table 12.1 for an example.

Table 12.1   Consider one column in image I_t with some color values x. The disparity computed by τ usually is not integer-valued. In the rounded disparity, which is used for the actual pixel-shifting, there is a discontinuity between the fourth and fifth column. $\tilde{I}_t$ exhibits a hole at this position. The last pixel gets shifted out of the view.

Image				Values			
Column	1	2	3	4	5	6	
I_t	x	x	x	x	x	x	
Disparity	0	0	$\frac{1}{3}$	$\frac{1}{3}$	$\frac{2}{3}$	1	
Rounded_Disparity	0	0	0	0	1	1	
$\tilde{I}_t$	x	x	x	x	–	x	(x)

Holes often occur when surfaces get enlarged in the virtual view. This is indicated in Fig. 12.5 where one can see that at the right side of the point cloud (which is observed by a rightwards shifted camera) the points are more sparsely distributed than at the left. Consequently, the target view exhibits holes at these areas. Another example is given in Fig. 12.11. Figure 12.11 (b) shows a right shifted view of the reference view, Fig. 12.11 (a). The left wall gets enlarged and therefore exhibits holes.

Dealing with holes, though, is quite simple. Holes occur in smooth regions of the depth map and are therefore *inside* of objects. This means that the texture surrounding the holes will look like what is supposed to be shown

(a)   (b)

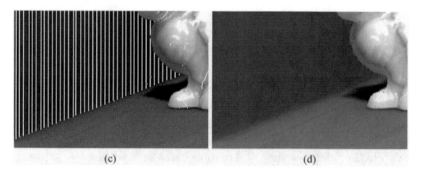

Fig. 12.11 (a) Reference view. (b) Target view. The left wall exhibits holes due to enlargement. (c) Magnification of hole region. (d) Holes filled by linear interpolation.

in the holes. Therefore, holes can be filled with simple linear interpolation of the horizontal neighbor pixels, see Fig. 12.11 (d) for an example.

12.3.2 Disocclusions

The main cause for blank areas in virtual views computed with Depth Image Based Rendering are *disocclusions*. They correspond with areas that are occluded by foreground objects in the reference view and get exposed (or *disoccluded*) in the target view. This happens when the virtual camera "looks behind" foreground objects and sees areas that were not visible in the reference view. The large white margin at the right side of the Buddha statue in Fig. 12.11 (b) is due to disocclusions. The area behind was not visible in Fig. 12.11 (a) and therefore we have to think how to fill in these regions. There is a wide variety of approaches to deal with this problem. They can be roughly classified into the following categories [35]: Post-processing, pre-processing, layered depth, and faithful rendering.

Post-processing

Post-processing techniques work directly on views that exhibit disocclusions. The basic idea is to fill disocclusions by infering the missing color values from surrounding pixels. In contrast to holes, disocclusions occur at object boundaries so that the inferred color values are a mixture of different textures. This can lead to blurry artifacts in these regions.

Simple post-processing methods include filling the disocclusions with constant color values, horizontal linear interpolation as well as extrapolating the background (i.e., the region with lower depth value) into the disoccluded area [36, 37]. Another way includes *image inpainting* [38]. In [35], the inpainting method of [39] is used for depth aware inpainting. All these methods have in common that they "guess" the actual missing color values.

Figure 12.12 gives an example. It uses the inpainting method of the

OpenCV[7]. Figure 12.12 (c) and (d) show enlarged parts of a virtual view generated with plain Depth Image Based Rendering and using inpainting, respectively. We can see that inpainting produces blurry edges which are not desirable since they disturb 3D perception when watched on a 3D screen.

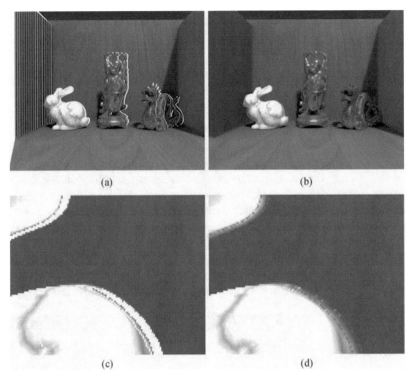

(a)

(b)

(c)

(d)

Fig. 12.12 (a) A virtual view generated with plain Depth Image Based Rendering. (b) A virtual view generated with Depth Image Based Rendering and additional inpainting to fill disocclusions. (c) Magnification of (a). At the object boundary disocclusions can be seen. (d) Magnification of (b). The disocclusions are filled but the edge is heavily blurred.

Pre-processing

The next category is called *pre-processing*. The main idea is to manipulate the depth map before the transformation in a way that disocclusions do not occur or get at least significantly diminished. Pre-processing approaches utilize the fact that disocclusions occur at object boundaries, i.e., at edges in the depth map. So reducing the edges in the depth map reduces the disocclusions. This is usually done by smoothing with a Gaussian filter $g(x, \sigma)$ [40]:

$$g(x, \sigma) = \frac{1}{\sqrt{2\pi}\sigma} e^{-\frac{x^2}{\sigma^2}}, \quad -\frac{w}{2} \leqslant x \leqslant \frac{w}{2}, \tag{12.17}$$

---

7   http://opencv.willowgarage.com/wiki/.

where σ is the standard deviation and w the filter's window size, which can be set to 3σ. For a depth pixel $d(u,v)$, its smoothed version $d^*(u,v)$ is given by

$$d^*(u,v) = \frac{\sum\limits_{i=-\frac{w}{2}}^{\frac{w}{2}}\left(\sum\limits_{j=-\frac{w}{2}}^{\frac{w}{2}} d(u-i,v-j)g(j,\sigma)g(i,\sigma)\right)}{\sum\limits_{i=-\frac{w}{2}}^{\frac{w}{2}}\left(\sum\limits_{j=-\frac{w}{2}}^{\frac{w}{2}} g(j,\sigma)g(i,\sigma)\right)}. \tag{12.18}$$

The value of σ controls the strength of the smoothing effect. Figure 12.13 illustrates the effect of of rendering with a smoothed depth map D^*: On the left, we can see one row of a video stream with its according depth map which exhibits a discontinuity. The lighter part of the depth map defines that the green part of the video stream is closer to the camera than the red part. When projecting the video stream into the 3D space (which is the first step in Depth Image Based Rendering), the resulting model exhibits a gap. A horizontally rightwards shifted camera can see behind the foreground object. This produces a blank area in the new view. A common approach is to smooth the depth map before rendering [25, 26] (see Fig. 12.13 on the right). Smoothing reduces the edges in the depth map so that dark areas near the edge become lighter and light areas near the edges become darker. This

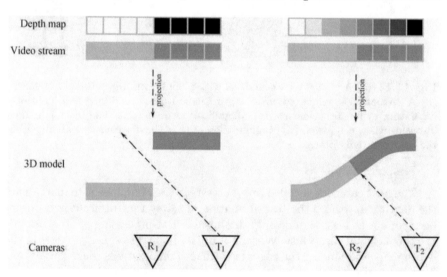

Fig. 12.13 The left side shows the rendering process using the original depth map. The harsh edge in the depth map causes a discontinuity in the model. The target view T_1 is not defined in this area. On the right side, the rendering process with a smoothed depth map is shown. The resulting model exhibits no discontinuity and in the target view T_2 no disocclusions appear.

results in a change of the scene geometry: The background object gets bent slightly to the front whereas the foreground object gets bent to the back. This way, the gap in the model vanishes or is at least reduced. A horizontally rightwards shifted camera sees less disocclusions.

Figure 12.14 shows different depth frames and their smoothed versions. Figure 12.15 depicts the rendered results: Column (a) shows Depth Image Based Rendering using just Eq. (12.15). Column (b) shows the results with a smoothed version of the depth map. We can see that smoothing reduces the disocclusions from a width of some pixel (e.g., the cube at the bottom) to a size of just one pixel. These remaining disocclusions can be eliminated by using one of the post-processing techniques, presented in the previous paragraph.

Fig. 12.14 First row: Sample depth maps. Second row: Symmetric smoothing with $\sigma = 30$. Third row: Asymmetric smoothing with $\sigma_1 = 10$ and $\sigma_2 = 90$.

Fig. 12.15 (a) Two virtual views rendered with plain Depth Image Based Rendering. (b) Two frames rendered with smoothing of the depth map. The parameter σ was set to 30.

There are some extensions of this method. In [36], for example, edge dependent filtering is used. In [41], the depth map is smoothed for the left and the right view independently. The most common smoothing method, though, is asymmetric smoothing [40]: The main motivation of this approach lies in one of the general drawbacks of smoothing. Since the depth map stores information about the scene geometry, each change in the depth map changes the scene geometry. Smoothing the depth map introduces geometric distortions in the scene. These can be seen in Fig. 12.16 (c). The top of the cube is distorted.

Asymmetric smoothing exploits some characteristics of human vision [40]. Vertical distortions are more pronounced for the human eye than horizontal ones. In asymmetric smoothing, the horizontal direction (which influences vertical edges) is smoothed less than the vertical direction. This can be done by introducing two different parameters σ_1 and σ_2, which control the strength of smoothing in horizontal and vertical direction, respectively. The depth pixel d^{**} of an asymmetrical smoothed depth map is given by

Fig. 12.16 (a) The reference view shows a cube laying on a plane. (b) The according depth map. (c) A rightwards shifted virtual view generated with a symmetrical smoothed depth map. Note the geometric distortions at the top of the cube. (d) A rightwards shifted virtual view generated with an asymmetrical smoothed depth map. The distortions are diminished. (e) Magnification of (c). (f) Magnification of (d).

$$
d^{**}(u,v) = \frac{\displaystyle\sum_{i=-\frac{w}{2}}^{\frac{w}{2}} \left(\sum_{j=-\frac{w}{2}}^{\frac{w}{2}} d(u-i,v-j)g(j,\sigma_1)g(i,\sigma_2) \right)}{\displaystyle\sum_{i=-\frac{w}{2}}^{\frac{w}{2}} \left(\sum_{j=-\frac{w}{2}}^{\frac{w}{2}} g(j,\sigma_1)g(i,\sigma_2) \right)}. \tag{12.19}
$$

The third row in Fig. 12.14 shows two depth maps after asymmetric smoothing. Figure 12 16 was rendered with $\sigma_1 = 10$ and $\sigma_2 = 90$. The top part of the cube does not exhibit a pronounced distortion any more.

Layered Depth

Layered depth [42, 23] does not use plain *video-plus-depth* but stores additional information about the scene, i.e., the disoccluded regions. For each pixel not one but several color values are stored, each corresponding to one plane in the scene. This way, information about disoccluded areas is available in the scene representation so that disocclusions can be filled.

This method delivers high quality images but it is challenging to store and process the layered depth images. In addition, layered scene information is not always available (e.g., when doing 2D-3D conversion). Since this method is not about filling disocclusions in video-plus-depth, we will not deal with it in detail.

Faithful Rendering

All the above mentioned methods have in common that they work framewise, i.e., they do not exploit the fact that I is a video of continuous frames I_t. The filling process is simply performed on each frame and its according depth frame independently.

Faithful rendering [43] is based on the observation that disoccluded areas may not be visible in the current frame but somewhere else in the video. This means that an area that is covered by a foreground object in the original view and gets exposed in the new virtual view may be visible not in the current frame but in some other frame. This is due to the fact that the foreground object moves and exposes the area behind in other frames. The key is to find the counterpart for a disoccluded area in the other frames and to use those color values to fill the disocclusion.

The basic idea of faithful rendering is to assume that the video I consists of a *foreground* and a *background*. The disocclusions that appear at foreground object boundaries are then filled with the corresponding background color values. This way, the disocclusions are filled with *true* color values, in contrast to the aforementioned methods which only can find *plausible* ones.

Faithful rendering works in a two-stage process, a pre-processing stage and the actual processing stage. The following listing summarizes the complete workflow for rendering a virtual view with faithful rendering.

- Pre-processing
    - The first step is to obtain a *background model* of video I. The background of I is assumed to be static, so the background model is simply an image B depicting the static parts of the scene, see Fig. 12.17 (e) for an example.
    - The virtual view $\tilde{B}$ of background B is computed.
    - A depth map X of the background image B is obtained, see Fig. 12.17 (f) for an example.

– For each frame I_t, a mask M_t is obtained, which identifies the foreground objects. Two example mask frames are shown in Fig. 12.17 (c)
  and (d).
- Rendering
  – For each frame I_t, a horizontally shifted virtual view $\tilde{I}_t$ is computed:
    The foreground objects defined by the mask M are set onto the new
    background view B according to their new positions, which are com-
    puted with Depth Image Based Rendering

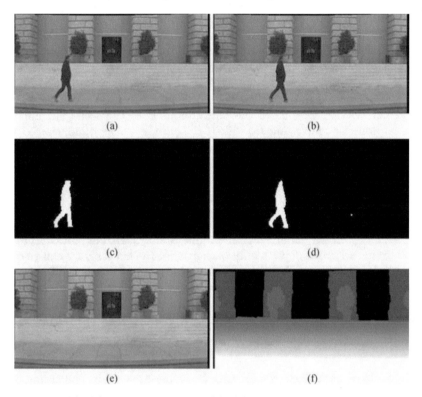

Fig. 12.17 (a), (b) Two video frames. (c), (d) The according foreground masks.
(e) The background of the scene. (f) A background depth map.

The details of each step are given in the following paragraphs.

Background Generation

First, the background image B is obtained. B corresponds with the static
parts of the scene. The background modeling step is crucial to this approach
since the background model will provide the color values that the disoccluded
areas are filled with. Simple background modeling methods are *median* and
mean:

$$B_{median}(p) := median(\{x_{p,t}, t \in [1 \ldots n]\}), \tag{12.20}$$

$$B_{mean}(p) := mean(\{x_{p,t}, t \in [1 \ldots n]\}). \tag{12.21}$$

The *median* method requires the background to be visible in more than 50% of the frames. The *mean* method incorporates each foreground pixel into the background model so that even minor occurrences of foreground objects lead to corrupt background images. In addition, the video needs to be shot with a static camera. In [43], the method of [34] is used. This method can restore the background of videos if each small region (i.e., square *patches*) of the background is visible in at least one frame. On the other hand, it requires the video I to be shot with a static camera and under constant illumination. Figure 17 (e) shows a background generated with this method. Every other method, which is able to obtain an image of the static parts of the scene, also suits this purpose.

Background Warping

A new virtual view $\tilde{B}$ of the background B of the scene is computed. This is done by employing Eq. (12.15). The new background view $\tilde{B}$ may exhibit disocclusions and holes. These have to be filled either by one of the aforementioned methods or manually. This is feasible since it has to be done just in this one frame for the complete video I.

Background Depth Map Generation

In the next step, a depth map X of the background B is obtained, see Fig. 12.17 (f) for an example. X is originally not available unless there is a video frame I_t which depicts only background. In this case, we can take the according depth frame D_t as background depth map. There are different ways to acquire X, like applying a background generation algorithm to the depth map D, defining the depth map manually or taking the time-wise minimum value:

$$X(p) = min(\{d_{p,t} : t \in [1 \ldots n]\}). \tag{12.22}$$

This works if each background pixel is visible at least in one frame. This assumption is also made by the background generation methods described above.

Foreground Extraction

In this step, a foreground mask $M = \{M_t : t \in [1 \ldots n]\}$ is obtained:

$$M_t = \{m_{t,p} \in [0,1]\},$$

$$m_{t,p} = \begin{cases} 0, & : x_{t,p} \text{ is background pixel}, \\ 1, & : x_{t,p} \text{ is foreground pixel}. \end{cases}$$

The mask M defines the exact location and shape of the foreground objects in the video. Foreground objects are regions where the current frame I_t differs from the background B, see Fig. 12.17 (c) and (d) for two examples.

The mask is computed by defining an energy function that assigns a score to a candidate mask [43]. The energy function gives a high score to masks that do not comply with the actual foreground objects, i.e., masks whose defined foreground is very similar to the background B. In addition, a mask is punished for exhibiting noise, i.e., foreground regions of only a few pixels size surrounded by background and vice versa.

The mask with the lowest score is accepted. Instead of computing the score for each of the 2^{MN} possible masks, the energy function is minimized using a graph-based approach, see [44, 45] for details. These methods can minimize certain energy functions despite the huge search space of 2^{MN} very efficiently. For a typical image size the execution time is only a few seconds, see [46] for details.

After the pre-processing stage, the actual rendering is done as follows:

Rendering

For a given video I, a horizontally shifted view $\tilde{I}_t$ is rendered as follows. The video is processed frame-wise. First, frame $\tilde{I}_t$ is initialized with $\tilde{B}$. Note that $\tilde{B}$ does not exhibit holes or disocclusions. After this, the foreground objects in I_t are identified using the foreground mask M_t. The foreground objects are then warped, i.e., their positions in the new view $\tilde{I}$ are computed. Their color values are copied to the respective positions in $\tilde{I}_t$.

$$\text{Initialization: } \tilde{I}_t := \tilde{B}, \tag{12.23}$$

$$\text{Foreground rendering: } \forall p, m_{t,p} = 1, \quad \tilde{x}_{t,\tau(p)} := x_{t,p}. \tag{12.24}$$

This way, for each frame only the foreground pixels are transformed. Since $\tilde{I}_t$ is initialized with the new background view $\tilde{B}$, $\tilde{I}_t$ exhibits no blank areas after the rendering. Areas that would have gotten disoccluded when using plain Depth Image Based Rendering are now filled with the true background color values. Since the background parts of frame I_t and the background image B only differ by noise, this does not corrupt $\tilde{I}_t$ but can rather help improve the image quality, see [43] for details.

Figure 12.18 shows the effect of faithful rendering. In Fig. 12.18 (a) we can see the setting of the scene: A foreground object (in this case a red Buddha statue) moves in front of a background (an array of dragons). This scene is observed by a camera whose view is depicted in Fig. 12.18 (b). Fig. 12.18 (c) shows the view of a horizontally leftwards shifted camera (ground truth). Note that the yellow dragon which is occluded in the reference view comes into sight in this view. Figure 12.18 (d) shows the view of a horizontally leftwards shifted *virtual* camera computed with plain Depth Image Based Rendering without handling disocclusions. In Fig. 12.18 (e) the disocclusions are eliminated by smoothing the depth map with a Gaussian filter. Note that the disocclusions indeed are gone but the blank areas are filled not with the correct values but by changing the geometry, especially of the foreground object. The yellow dragon cannot be seen. Figure 12.18 (f) shows the virtual view computed with faithful rendering. Note that the yellow dragon in

Fig. 12.18 (a) The setting of the scene: A camera observes a foreground object (the red Buddha statue) moving in front of a static background (the array of dragons). (b) The view of the camera (reference view). Note that the yellow dragon in the background is completely occluded by the foreground object. (c) The view of a leftwards shifted real camera (ground truth). Note that the yellow dragon in the background comes into sight. (d) The leftwards shifted view generated with Depth Image Based Rendering without handling disocclusions. (e) The leftwards shifted view generated with Depth Image Based Rendering and asymmetrical smoothing of the depth map. The disocclusions are eliminated but the yellow dragon is still not visible. (f) The leftwards shifted view generated with faithful rendering. The yellow dragon is correctly reconstructed.

the background is correctly recovered, compare with the ground truth shown in Fig. 12.18 (c). The holes inside of the foreground object can be filled with simple linear interpolation. Figure 12.19 shows a magnifications of Fig. 12.18.

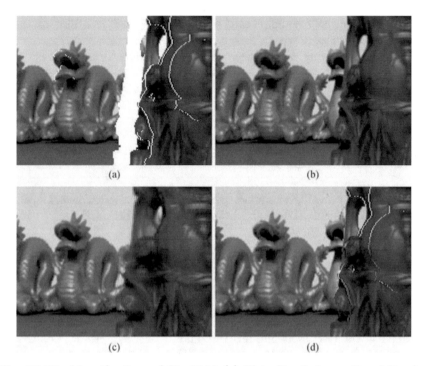

Fig. 12.19 Magnifications of Fig. 12.18 (a) Plain Depth Image Based Rendering (Fig. 12.18 (d)). (b) Ground truth left view (Fig. 12.18 (c)). (c) Rendered with a smoothed Depth Map (Fig. 12.18 (e)). (d) Rendered with faithful rendering (Fig. 12.18 (f)).

12.3.3 Comparison

We have seen different ways to fill disocclusions: Inpainting works as a post-processing technique directly on the image that exhibits the disocclusions. These are filled with color values that are inferred from the surrounding pixels. The disocclusions are diminished but the processed image exhibits artifacts at the positions where the disocclusions were filled. These artifacts are disturbing when watching the content in 3D.

Symmetric and asymmetric smoothing of the depth map with a Gaussian filter before warping falls into the category of pre-processing techniques. The smoothing changes the geometry of the scene in a way so that disocclusions do not appear or are diminished after warping. A subsequent interpolation step might be necessary to eliminate small holes. The disocclusions are eliminated but on the downside, geometric distortions are introduced. Asymmetric smoothing produces less visual distortion than symmetric smoothing but the effect remains visible. Like the pre-processing techniques, smoothing is not able to recover the true color values for a disoccluded area but just fills it with plausible ones.

Faithful rendering can recover the true color values. It also does not introduce any geometric distortions. However, it is applicable only on videos where a background model can be generated. Future work is necessary to expand the set of suitable videos.

12.4 Other Challenges

In this section we will give a short overview on other challenges in Depth Image Based Rendering. These are dealing with *imperfect depth maps, ghosting* and the *cardboard effect*.

12.4.1 Imperfect Depth Maps

Most of the depth maps shown in this chapter are rendered directly from 3D models via a 3D modeling program and are therefore exact, i.e., they do not exhibit noise or any other artifacts. These depth maps are very useful to demonstrate the principles of Depth Image Based Rendering. Real-world depth maps, though, may not be exact. The depth map can be corrupted by noise, depth estimation or coding of the depth map [40]. This can be alleviated by smoothing the depth map with a Gaussian filter. See the *ballet* [23] example in Fig. 12.20.

Misaligned video and depth streams can also lead to artifacts when rendering a new view, see Fig. 12.21. In Ref. [47], an edge based registration method is used to improve the spatial alignment of depth and video.

(a)　　　　　　　　　　　　(b)

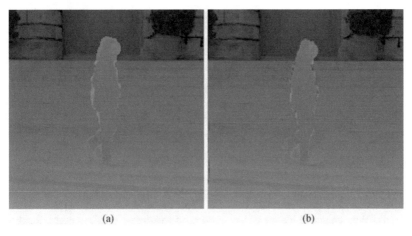

Fig. 12.20 (a) One frame of the ballet scene. (b) The according depth map generated with the method of [23]. Note that, for example, the floor does not exhibit a smooth depth gradient. (c) A virtual view generated with plain Depth Image Based Rendering. The floor exhibits heavy cracks. (d) A virtual view generated with a smoothed depth map. Note that the cracks in the floor are diminished.

Fig. 12.21 (a) Overlay of video and original depth map. (b) Overlay of video and registered depth map. The depth map fits better to the foreground object.

12.4.2 Ghosting

Ghosting refers to fine artifacts in a view generated with Depth Image Based Rendering. These artifacts are due to the fact that depth maps define a pixel to belong exactly to one object and therefore assume that each edge of an object has a width of zero. In a image the edge between two objects is usually not given by a harsh change of color values but by a gradient, see Fig. 12.22 (a). The color value of some edge pixels is biased from the foreground as well as from the background. A depth map, though, usually defines a pixel to belong to exactly one object, i.e., the depth of a pixel is not a mixture of the depth of a foreground object and a background object, see Fig. 12.22 (b) for an example.

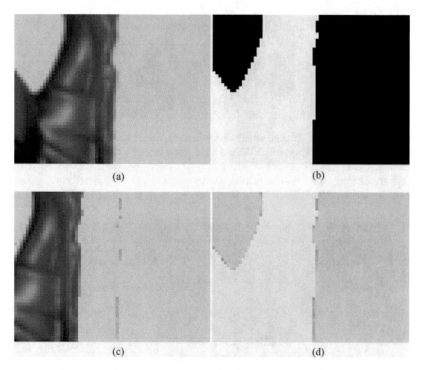

Fig. 12.22 (a) Magnified reference view. Note that the edges of the red foreground object are some pixels thick. (b) The according depth map. (c) The target view after Depth Image Based Rendering and disocclusions handling. The outer pixels of the foreground object were not shifted. (d) Overlay of video and depth map (green colored). We can see that the depth map cuts through the edge of the foreground object.

When using Depth Image Based Rendering for generating a new view, this leads to ghosting artifacts of the edges even after disocclusion handling, see Fig. 12.22 (c). This is due to the fact that the depth map cuts through the edge, making some edge pixels foreground and some background. In Fig. 12.22 (d) the depth of the foreground object (marked green to increase visibility) is overlayed with the video frame. We can see that the depth map does not cover some of the pixels whose color values are influenced by the foreground object.

This problem is similar to a problem in the domain of object segmentation. When segmenting objects from an image, a rigorous yes-no decision whether a pixel belongs to an object or not can lead to bad results especially in regions with fine detail like hair or fur. *Digital matting* [48 − 50]assumes each pixel of an image to be a convex combination of foreground and background:

$$x_p = \alpha_p F_p + (1 - \alpha_p) B_p \qquad (12.25)$$

with F and B being foreground and background, respectively, and α, the *alpha matte* describing the opacity of each object. The matte can be thought of as an extension of the mask M using, for example, 256 nuances instead of just two. The work of [23] successfully employs digital matting to enhance the quality of the virtual views. In [51], prior to warping, a labeling process is employed to avoid ghosting.

12.4.3 Cardboard Effect

In Eq. (12.6), we have defined a depth map in the following way:

$$D_t = \{d_{t,p} \in [0 \dots 255] : p \in [0 \dots M] \times [0 \dots N]\}. \qquad (12.26)$$

We can see that the depth is quantized, i.e., it can take 256 different values. This is not affected by the depth interpolation described in Eq. (12.7). This means that the scene model described by a depth map consists of 256 slides instead of a smooth "volume". This usually is sufficient to provide a good depth perception. Problems may arise, when an object is completely repre- sented by just one slide. This means that each pixel of an object has the same depth value. This, for example, can be the case for some automatic depth generation algorithms: In [52] the depth of an object is defined by the aver- age depth value of the ground it touches. This way, an object is assigned just one depth value. If such a depth map is used to compute a second view for 3D perception, the object appears flat when watched on a 3D scene. It seems to have as little volume as a cardboard. A real object would have different depth values throughout its area.

   A way to diminish the cardboard effect is to smoothen the depth map [40]. The smoothing affects the outer pixels of the object in a way that their depth values are converged to the background depth, i.e., the edges of the object get slightly bent to the background which provides the object with more volume when watched in 3D. The same effect is used to diminish disocclusions, see Fig. 12.13.

12.5 Conclusion

Depth Image Based Rendering is a technique to compute virtual views of a scene. It relies on a video stream and an associated depth stream which stores for each pixel in the video its distance to the camera. This representation, called *video-plus-depth*, has some advantages over the common *video-plus- video* format. It can be stored efficiently since the depth stream can be better compressed than a second video stream. It is more flexible since the second video stream is generated at the receiver side so that the user can tune the stereo geometry to his personal preferences. Depth Image Based Rendering also serves as a technique in a general 2D-3D video conversion process.

In this chapter we have described the basic ideas and concepts behind Depth Image Based Rendering. We have seen what challenges can occur and how to handle them. Disocclusions are one of the main challenges and are due to the fact that the single video stream in video-plus-depth cannot store all color information necessary to generate an arbitrary new view. We have presented some techniques to fill disoccluded areas like Gaussian filtering or faithful rendering.

References

[1]   Kanatani K I, Chou T C (1989) Shape From Texture: General Principle. Artificial Intelligence, 38(1): 1–48
[2]   Witkin A P (1981) Recovering Surface Shape and Orientation from Texture. Artificial Intelligence, 17(1–3): 17–45, August 1981
[3]   Brillault-O'Mahony B (1992) High level 3D Structures from a Single View. Image and Vision Computing, 10(7): 508–520
[4]   Chen Z (2004) Extracting Projective Invariant of 3D Line Set from a Single View. In International Conference on Image Processing, pp 2119–2122
[5]   Mulgaonkar P G, Shapiro L G, Haralick R M (1986) Shape from Perspective: A Rule-Based Approach. Graphical Models and Image Processing, 36(2/3): 298–320
[6]   Horn B K P (1970) Shape from Shading: A Method for Obtaining the Shape of a Smooth Opaque Object from One View. In MIT Artificial Intelligence Laboratory
[7]   Zhang R, Tsai P S, Cryer J E (1999) Shape from Shading: A Survey. IEEE Transactions on Pattern Analysis and Machine Intelligence, 21(8): 690–706
[8]   Asada N, Fujiwara H, Matsuyama T (1998) Edge and Depth from Focus. International Journal of Computer Vision, 26(2): 153–163
[9]   Grossmann P Depth from Focus (1987) Pysical Review Letters, 5(1): 63–69
[10]   Nayar S K, Nakagawa Y (1994) Shape from Focus. IEEE Transactions on Pattern Analysis and Machine Intelligence, 16(8): 824–831
[11]   Jin H L, Favaro P, Soatto S (2003) A Semi-direct Approach to Structure from Motion. The Visual Computer, 19(6): 377–394
[12]   Pollefeys M (2008) 3D Modeling of Real-world Objects, Scenes and Events from Videos. In: Proceedings of the True Vision - Capture, Transmission and Display of 3D Video (3DTV-CON), pp 5–6
[13]   Szeliski R (2010) Computer Vision: Algorithms and Applications. Springer, Heidelberg
[14]   Tomasi C, Kanade T (1992) Shape and Motion from Image Streams under Orthography: A Factorization Method. International Journal of Computer Vision, 9: 137–154
[15]   Ullman S (1979) The Interpretation of Structure from Motion. Proceedings of the Royal Society, B-203: 405–426
[16]   www.studiodaily.com. In-Three on the Workflow Behind 3D Conversion. http://opencv.willowgarage.com/wiki/. Accessed 19 October 2010
[17]   Guan C, Hassebrook L, Lau D (2003) Composite Structured Light Pattern for Three-dimensional Video. Opt. Express, 11(5): 406–417
[18]   Kim C, Park J, Yi J et al (2005) Structured Light Based Depth Edge Detection for Object Shape Recovery. In Proceedings of the 2005 IEEE Computer Society Conference on Computer Vision and Pattern Recognition
[19]   Scharstein D, Szeliski R S (2003) High-accuracy Stereo Depth Maps using Structured Light. In Computer Vision and Pattern Recognition, pp 195–202
[20]   Zhang G, Jia J, Wong T T (2008) Recovering Consistent Video Depth Maps via Bundle Optimization. In: Proceedings of IEEE Conference on Computer Vision and

Pattern Recognition CVPR 2008, pp 1 – 8
[21] Lee E K, Kang Y S, Jung Y K et al (2010) Three-dimensional Video Generation using Foreground Separation and Disocclusion Detection. In: Proceedings of Conference on the True Vision: Capture, Transmission and Display of 3D Video (3DTV-CON), pp 1 – 4
[22] Lucas B D, Kanade T (1981) An Iterative Image Registration Technique with an Application to Stereo Vision. In International Joint Conferences on Artificial Intelligence, pp 674 – 679
[23] Zitnick C L, Kang S B, Uyttendaele M et al (2004) High-quality Video View Interpolation using a Layered Representation. ACM Transactions on Graphics, 23(3): 600 – 608
[24] Nozick V, Saito H (2007) Online Multiple View Computation for Autostereoscopic Display. In PSIVT'07: Proceedings of the 2nd Pacific Rim Conference on Advances in Image and Video Technology, pp 399 – 412. Springer, Heidelberg
[25] Fehn C. Depth-Image-Based Rendering (DIBR), Compression, and Transmission for a new Approach on 3D-TV. In Stereoscopic Displays and Virtual Reality Systems XI, vol 5291: 93 – 104
[26] Fehn C, Barre R de la, Pastoor R S (2006) Interactive 3-DTV-Concepts and Key Technologies. Proceedings of the IEEE, 94: 524 – 538
[27] Fehn C, Kauff P, Op De Beeck M (2002) Sexton. An Evolutionary and Optimised Approach on 3D-TV. In: Proceedings of International Broadcast Conference, pp 357 – 365
[28] Burschka D, Hager G D, Dodds Z (2003) Recent Methods for Image-based Modeling and Rendering. In: Proceedings of IEEE Virtual Reality
[29] Chan S C, Shum H Y, Ng K T (2007) Image-Based Rendering and Synthesis. IEEE Signal Processing Magazine, 24(6): 22 – 33
[30] Lin Z, Shum H Y (2004) A Geometric Analysis of Light Field Rendering. International Journal of Computer Vision, 58(2): 121 – 138
[31] Shum H Y, Chan S C, Kang S B (2007) Image-Based Rendering. Springer, Heidelberg
[32] Fan Y C, Chi T C (2008) The Novel Non-Hole-Filling Approach of Depth Image Based Rendering. In Proceedings of the True Vision: Capture, Transmission and Display of 3D Video, pp 325 – 328
[33] McMillan L (1997) An Image-Based Approach to Three-Dimensional Computer Graphics. PhD thesis, University of North Carolina at Chapel Hill, Chapel Hill, NC, USA, 1997.
[34] Colombari A, Fusiello A (2010) Patch-Based Background Initialization in Heavily Cluttered Video. IEEE Transactions on Image Processing, 19(4): 926 – 933
[35] Luo K, Li D X, Feng Y M et al (2009) Depth-aided Inpainting for Disocclusion Restoration of Multi-view Images using Depth-Image-Based Rendering. Journal of Zhejiang University-Science A, 10: 1738 – 1749
[36] Lee S B, Ho Y S (2009) Discontinuity-adaptive Depth Map Filtering for 3D View Generation. In: Proceedings of the 2nd International Conference on Immersive Telecommunications, pp 1 – 6, Brussels, Belgium, 2009. ICST (Institute for Computer Sciences, Social-Informatics and Telecommunications Engineering).
[37] Vázquez C, Tam W J, Speranza F (2006) Stereoscopic Imaging: Filling Disoccluded Areas in Depth Image-Based Rendering. In Society of Photo-Optical Instrumentation Engineers (SPIE) Conference Series, vol 6392
[38] Tauber Z, Li Z N, Drew M S (2007) Review and Preview: Disocclusion by Inpainting for Image-Based Rendering. IEEE Transactions on Systems, Man, and Cybernetics, Part C: Applications and Reviews, 37(4): 527 – 540
[39] Criminisi A, Perez P, Toyama K (2004) Region Filling and Object Removal by Exemplar-based Image Inpainting. IEEE Transactions on Image Processing, 13(9): 1200 – 1212
[40] Zhang L, Tam W J (2005) Stereoscopic Image Generation Based on Depth Images for 3D TV. IEEE Transactions on Broadcasting, 51(2): 191 – 199
[41] Chen W Y, Chang Y L, Lin S F et al (2005) Efficient Depth Image Based Rendering with Edge Dependent Depth Filter and Interpolation. In: Proceedings of the 2005 IEEE International Conference on Multimedia and Expo, ICME 2005, July 6-9, 2005, Amsterdam, The Netherlands, pp 1314 – 1317
[42] Shade J, Gortler S, He L W (1998) Layered Depth Images. In: Proceedings of the 25th Annual Conference on Computer Graphics and Interactive Techniques, pp 231 – 242, New York, USA, 1998
[43] Schmeing M, Jiang X (2010) Depth Image Based Rendering: A Faithful Approach for

the Disocclusion Problem. In Proceedings of the True Vision - Capture, Transmission and Display of 3D Video (3DTV-CON), pp 1–4

[44]   Kolmogorov V, Zabin R (2004) What Energy Functions can be minimized via Graph Cuts? IEEE Transactions on Pattern Analysis and Machine Intelligence, 26(2): 147–159

[45]   Sheikh Y, Shah M (2005) Bayesian Modeling of Dynamic Scenes for Object Detection. IEEE Transactions on Pattern Analysis and Machine Intelligence, 27(11): 1778–1792

[46]   Schmeing M, Jiang X. (2009) Robust Background Subtraction for Depth Map Generation. In: Proceedings of 3D Stereo MEDIA, 2009

[47]   Fieseler M, Jiang X Discontinuity-based Registration of Depth-and Video Data in Depth Image Based Rendering. Signal, Image and Video Processing, 2010. (to appear).

[48]   Chuang Y Y, Curless B, Salesin D H et al (2001) A Bayesian Approach to Digital Matting. Computer Vision and Pattern Recognition, IEEE Computer Society Conference on, 2:264+

[49]   Levin A, Lischinski D, Weiss Y (2008) A Closed-Form Solution to Natural Image Matting. IEEE Transactions on Pattern Analysis and Machine Intelligence, 30(2): 228–242

[50]   Porter T, Duff T (1984) Compositing Digital Images. In: Proceedings of the 11th Annual Conference on Computer Graphics and Interactive Techniques, pp 253–259, New York, NY, USA, 1984

[51]   Do L, Zinger S, de With P H N (2010) Quality improving Techniques for Free-viewpoint DIBR. In Society of Photo-Optical Instrumentation Engineers (SPIE) Conference Series, vol 7524

[52]   Rothaus S, Rothaus K, Jiang X (2008) Synthesizing 3D Videos by a Motion-conditioned Background Mosaic. In Proceedings of the Internatinal Conference on Pattern Recognition ICPR 2008, 1: 4

Part III: Face Recognition and Forensics

13 Gender and Race Identification by Man and Machine

Usman Tariq[1], Yuxiao Hu[2] and Thomas S. Huang[1]

Abstract   This work details a comprehensive study on gender and race identification from different facial representations. The major contributions of this work are: comparison of human and machine performance, qualitative analysis of use of color for race identification, combining different facial views for gender identification and extensive human experiments for both gender and race recognition from four different facial representations.

13.1 Introduction

The scientific community has long been investigating the analysis of biometric information. This resulted into products for face recognition, iris identification, finger-print verification, etc. Apart from analysis, biometric synthesis has also captured interest. It refers to the generation of biometric information from a computer model of a person [1]. One example may be synthesizing expressions for the 3D-face model of a person. It is believed that this shall be a very fertile area of future research and development [1]. A comprehensive review of various topics on synthesis and analysis in biometrics can be found in Ref. [1].

The general public has been reluctant to accept biometrics because of privacy issues. However, because of the elevated security concerns, the balance of the debate is now shifting, as also pointed out in [2]. But computer vision, despite of its importance in security surveillance, faces major challenges. Though there has been a great deal of progress in face detection and recognition in the recent past, but a number of problems still remain unsolved. For instance, face recognition and detection is appreciably affected with large pose variations, illumination changes and occlusions [3].

In this chapter we present our work on two biometric "attributes" identification (gender and race). These are not just specific to one person alone, but are shared by groups of people. (We refer to these "attributes" as *"soft biometrics"*). These present another major challenge in face processing. They

1   Department of Electrical and Computer Engineering, Coordinated Science Laboratory and Beckman Institute for Advanced Science and Technology, University of Illinois at Urbana-Champaign, 405 N. Mathews Ave., Urbana, IL 61801, USA. E-mail: {utariq2, huang}@ifp.uiuc.edu.
2   Microsoft Corporation, 1 Microsoft Way, Redmond, WA 98052, USA. E-mail: Yuxiao. Hu@microsoft.com.

have an increasing number of applications as visual surveillance technologies and human-computer interaction (HCI) [4] evolves. For instance, Ref. [5] gives an interesting example of human-robot interaction. Gender identification can improve the user experience by modifying the robot's behavior accordingly.

Gender identification can also be helpful in face recognition, as this shall reduce the problem of comparing the face with half of the database (provided both males and females are equally probable in the database). Race identification shall reduce this problem even further (provided the database is multi-racial) [6]. These may also be helpful in designing a computer agent [4]. These may also aid shopkeepers, for example, to know the demographic distribution of the customers over a period of time [7]. It can also be used for targeted advertisements i.e., offering a person the options which may be more specific and useful to, say, a particular gender, etc.

In the proceeding sections, we shall first point towards the current references on the topic, and then we shall discuss novelty of our work followed by sections about computer and human experiments. Please note that the term race, in this paper, refers to a group of people who share similar facial features, which perceptually distinguish them from other groups (races).

13.2 Background

Humans are very accurate in deciding the gender of a face even when cues from hairstyle, makeup and facial hair are minimized [8]. The results from [9] show that both the shape and color are vital in deciding the sex and race from a face by humans. Bruce et al. in [8] presented that the "average" male face differs from an "average" female face in the 3D representation of the face (acquired by laser scanning), by having a more protuberant nose/brow and more prominent chin/jaw. Shape was also found significant in race decisions as shown in [9]. Davidenko in Ref. [10] reveals the presence of cues for humans for gender identification even in silhouetted face profiles. However, most of the existing work in the computer vision community uses frontal face images for gender or race identification. Some examples of the existing literature on gender identification may be found in Refs. [11 – 29].

Existing work for race identification may be found in Refs. [12, 16, 21, 30 – 32]. Color, although found useful in Ref. [9] (a psychology paper) for race identification, has not been experimented with earlier. Also except Gutta et al. [12], people have used two racial categories or three in Ref. [32], but for Gutta et al., different representation of the same subject may also appear in training. We used four racial categories and the testing and training sets were disjoint in our case. In addition, there has not been any effort to benchmark human performance on the same database, to the authors' best knowledge.

The significance of this work is neither to compete against the previous results from other researchers, nor to invent a new technique. The novelty of the work is to give a new research direction and to explore some ideas which were never investigated earlier. The contributions are, primarily, comparison of human and computer experiments, use of color for race identification and combining different views for gender identification; any of which was not experimented with in the past. Another contribution is empirically proving the presence of information in silhouetted face profiles with extensive human experiments. This work is an extension of the authors' work in Refs. [7, 33]. The human experiments in Ref. [33] are re-designed so that the stimuli (face sample presented to the subjects) are balanced in all the gender-racial categories and the subject pool is a lot more diverse and in twice the number to that in Ref. [33]. Also more views are experimented with than that in Ref. [33]. In addition to that, the stimuli used for silhouetted face profiles are different from the ones in Ref. [33] and are the same as were used for computer experiments.

In the following sections, we will present the experiments on silhouetted profile faces, followed by the experiments on frontal faces, then the result of combining the two classifiers, then finally human experiments and their results.

13.3 Silhouetted Profile Faces

This section describes the results of experimentation on silhouetted profile faces. We used shape context based matching [34]. We also experimented with chamfer matching; its results were marginally worse. Shape context based matching is a recent development and has been proven to outperform other shape matching techniques in situations similar to our application [34].

13.3.1 Database

The silhouetted face profiles in the database were generated from the 3D face models collected by Hu et al. in Ref. [35]. The age of subjects in the database for the experiments was primarily within the range $18-30$ years to avoid any age-related distortions on the silhouettes. Males whose beards and moustaches distorted their facial contours were also not included.

The database had 441 images. It was divided into the following four racial categories based upon the demographic information given by the participants:

1) Black (B): Africans or African Americans.

2) East and Southeast Asian (ESEA): Chinese, Japanese, Korean, Vietnamese, Filipino, Singaporean, etc.

3) South Asian (SA): Indian, Pakistani, Sri-Lankan, Bangladeshi, etc.

4) White (W): Caucasian, Middle Eastern.

The demographic distribution of the database is given in Table 13.1

(where M denotes male and F denotes female). Some example silhouettes and their (part of) extracted profile contours (later used for processing and referred to as profile edge) are shown in Fig. 13.1.

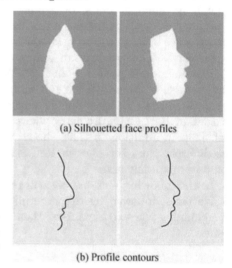

(a) Silhouetted face profiles

Fig. 13.1   Silhouetted face profiles from the database and extracted portion of respective profile contours used for processing.

(b) Profile contours

Table 13.1 Demographic distribution of the database for silhouetted profile view face experiments

B		W		ESEA		SA	
67		225		107		42	
M	F	M	F	M	F	M	F
32	35	118	107	55	52	25	17

13.3.2 Shape Context Based Matching

In this method, the shape of an object is represented by a discrete set of points sampled from its internal and external contours. For a point p_i on a shape having n sampled points, the coarse histogram h_i of the coordinates of the remaining $n - 1$ points, relative to p_i, is known as the shape context of p_i [7, 34]. Mathematically,

$$h_i(k) = \#\{q \neq p_i : (q - p_i) \in bin(k)\}. \tag{13.1}$$

The shape context is made more sensitive to nearby points to p_i, by using uniform bins in log-polar space. The cost of matching the point p_i on first shape P to the point q_j on the second shape Q, denoted by C_{ij}, is given by the chi-square test statistic [34],

$$C_{ij} \equiv C(p_i, q_j) = \frac{1}{2} \sum_{k=1}^{K} \frac{[h_i(k) - h_j(k)]^2}{h_i(k) + h_j(k)}, \tag{13.2}$$

where $h_i(k)$ and $h_j(k)$ denote the k-bin normalized histograms at p_i and q_j respectively.

If one is given the costs C_{ij} between all pairs of points $p_i q_j$ on the two shapes, the total cost of matching is given by [34]

$$H(\pi) = \sum_i C(p_i, q_{\pi(i)}), \tag{13.3}$$

where $H(\pi)$ is to be minimized subject to the constraint that matching is one-to-one; in other words, π is a permutation [34]. This is solved using Hungarian [43].

Then a plane transformation T, given by $T : R^2 \to R^2$, can be estimated using the correspondence between points. This may be an affine transformation. But here, the thin plate spline model (TPS) [44, 45] was used. TPS is commonly used for flexible coordinate transformation [34]. In its regularized form, it includes the affine model as a special case [34].

The TPS interpolant $f(x, y)$ minimizes the bending energy I_f and is given by [44]

$$f(x, y) = a_1 + a_x x + a_y y + \sum_{i=1}^{n} \omega_i U(||(x_i, y_i) - (x, y)||), \tag{13.4}$$

where (x_i, y_i) are the x and y co-ordinates of the point p_i and $i = 1, 2, ...n$; $U(r)$ is the kernel function, given by $U(r) = r^2 log r^2, U(0) = 0$; ω_i are weights; a_1, a_x and a_y are constants. Now for $f(x, y)$ to have square integrable derivatives it is required that [34, 44]

$$\sum_{i=1}^{n} \omega_i = 0, \ also \ \sum_{i=1}^{n} \omega_i x_i = \sum_{i=1}^{n} \omega_i y_i = 0. \tag{13.5}$$

Bending energy is given by [44]

$$I_f \propto \boldsymbol{\omega}^{\mathrm{T}} \boldsymbol{K} \boldsymbol{\omega} \equiv D_{be}, \tag{13.6}$$

where $K_{ij} = U(||(x_i, y_i) - (x_j, y_j)||)$. Regularization may be used to relax the exact interpolation requirement [34].

To model the coordinate transformation, two separate TPS functions were used [34]:

$$T(x, y) = (f_x(x, y), f_y(x, y)). \tag{13.7}$$

Now the shape context distance D_{sc} between the two shapes P and Q, is calculated as the symmetric sum of the shape context matching costs over best matching points [34], as following,

$$D_{sc}(P, Q) = \frac{1}{n} \sum_{p \in P} \underset{q \in Q}{\operatorname{argmin}} C(p, T(q)) + \frac{1}{m} \sum_{q \in Q} \underset{p \in P}{\operatorname{argmin}} C(p, T(q)). \tag{13.8}$$

Then finally, the shape distance, D_{sh}, between the two shapes may be calculated by a weighted sum of shape context distance D_{sc}, and D_{be} [34].

13.3.3 Experimental Setup

The database outlined in Section 13.3.1 was divided into a reference set and a testing set. The profile edges were extracted; their heights were normalized to 300 pixels and were sampled by 100 points. The sampling considerations can be found in [34]. It was found advantageous to have a certain minimum distance between the sampled points in [34].

For each test profile edge, shape distances were calculated with each of the reference profile edge, and classification was done using a kNN (k-nearest neighbors) classifier.

The reference database was made equally balanced in terms of each gender-racial category (male Black, female Black, male South Asian and so on), so that the prior probabilities may not affect the result of the nearest neighbor classification. The reference dataset had 128 images. First the remaining dataset was tested on the reference database. Then a replace-one approach was used to test the members of reference dataset, one-by-one, by replacement with an image of the same race and gender in the reference dataset. In this way, the approach was tested on a total of 441 images.

13.3.4 Results

The resulting accuracies achieved are summarized in Table 13.2. The class confusion matrices for the results in Table 13.2 are given for gender in Table 13.3 and for race in Table 13.4 and Table 13.5. For explanation of the

Table 13.2   Results achieved using shape context based matching

	Gender	Standard Error	**Race**	Standard Error
Accuracy	71.20%	± 2.16%	71.66%	± 2.15%

Table 13.3   Confusion matrix for gender for results with shape context based matching

Silhouetted Profile		Recognized As		Recognized As (%)			
				Female		Male	
		Female	Male	%	SE%	%	SE%
Ground Truth	Female	176	35	**83.41**	±2.56	16.59	±2.56
	Male	92	138	40	±3.23	**60**	±3.23

Table 13.4   Confusion matrix for race for results with shape context based matching

Silhouetted Profile		Recognized As			
		B	ESEA	SA	W
	B	39	24	0	4
Ground Truth	ESEA	7	86	6	8
	SA	0	13	14	15
	W	1	12	35	177

abbreviations used for races, please refer to Section 13.3.1. The standard error (SE) is given by $SE = \sqrt{p\,(1-p)/n}$.

Table 13.5 Confusion matrix for race for results with shape context based matching (percentages)

Silhouetted Profile		Recognized As B %	B SE%	ESEA %	ESEA SE%	SA %	SA SE%	W %	W SE%
Ground Truth	B	**58.2**	±6.0	35.8	±5.9	0.0	±0.0	6.0	±2.9
	ESEA	6.5	±2.3	**80.4**	±3.8	5.6	±2.2	7.5	±2.5
	SA	0.0	±0.0	31.0	±7.1	**33.3**	±7.3	35.7	±7.4
	W	0.4	±0.4	5.3	±1.5	15.6	±2.4	**78.7**	±2.7

13.4 Frontal Faces

This section presents the work on frontal view faces.

13.4.1 Database for Frontal Faces

The frontal faces in our database were generated from the same database which was used for generating silhouetted profile faces. The faces were aligned using manually labeled key points on the eye corners, nose-tip and mouth corners, in the least square sense. Some example images are shown in Fig. 13.2. The database had 400 images. The demographic distribution of the database is given in Table 13.6.

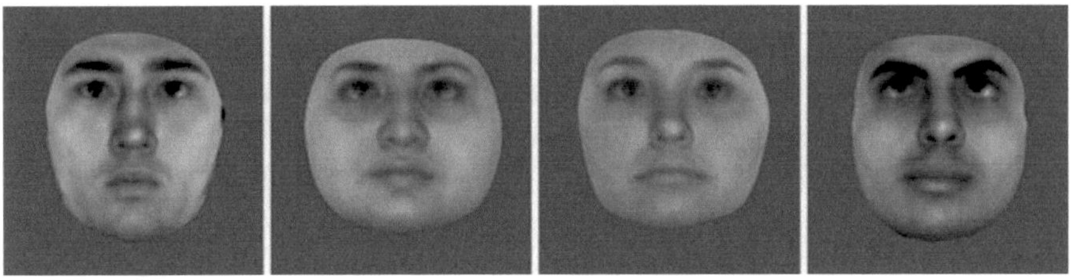

Fig. 13.2 Example frontal-view face images from the database.

Table 13.6 Demographic distribution of the database for frontal view face experiments

B		ESEA		SA		W	
64		110		29		197	
M	F	M	F	M	F	M	F
36	28	61	49	21	8	108	89

13.4.2 Experiments

Three sets of experiments were conducted. First gender identification was experimented with, by using gray scale images. Then qualitative effect of using color for race identification was investigated. And then the effect of number of training samples on the performance was also examined. The reason of using SVM as the classifier is that it may be considered state of the art; as was observed in a recent evaluation of techniques for gender identification [22]. Also PCA+SVM gave better results in our experiments than using SVM alone.

The YIQ color model was used, so that different combinations of grayscale and color components could be experimented with. Y corresponds to the luminance (gray-scale) while I and Q correspond to the chrominance (color) components. Another reason for using this color model was that the human eye is more sensitive to changes in the orange-blue (I) range than in the purple-green range (Q) [36].

For the first set of experiments, 5-fold cross-validation was used. Images were sub-sampled to 64×64 pixels. Then PCA was used for dimension reduction. Then different combinations of principal components of Y $(15, 30, \cdots, 120)$ were used for SVM [37] classification using linear and nonlinear SVMs (with rbf kernel). For the second set of experiments (race identification), 5-fold cross-validation was used as well. Images were again subsampled to 64×64 pixels. Here too, PCA was used for dimension reduction. Then different combinations of principal components of Y, I and Q were used for SVM classification. For instance, for Y; $15, 30, \cdots, 120$ principal components were used. And for each set of Y principal components, 16 different combinations of principal components of I and Q were experimented with. For example, for 15 principal components of Y; 0, 5, 10 and 15 principal components of I were used and for each of them; 0, 5, 10 and 15 components of Q were experimented with. Similarly, for 30 principal components of Y; 0, 10, 20 and 30 principal components of I were used, and for each of them; 0, 10, 20 and 30 components of Q were experimented with, and so on.

It may be argued that the comparison is not completely fair, as the feature vectors having features corresponding to color principal components, in addition to the features corresponding to a specific number of Y principal components, are in a higher dimension. But this does give a qualitative comparison, answering the question: Does the color really help?

Experimentation was done both with linear and non-linear SVMs using rbf kernel. This method of dimension reduction using PCA and then classification with SVM was experimented by Graf and Wichmann [11] for gender classification. They did experimentation on the original MPI head database [38].

For the third set of experiments, the effect of number of training samples on performance was examined (for one of the combinations of principal components in the experiment set 1 and 2). We experimented with training on 10%, 11%, 13%, 14%, 17%, 20%, 25%, 33%, 50%, 67%, 75%, 80%, 83%, 86%, 88%, 89% and 90% of data and tested on the rest. The process was randomized for 4 times and the results were averaged.

13.4.3 Results for Experiments with Frontal View Face Images

The best results for gender identification are summarized in Table 13.7 and those for race identification are shown in Table 13.8. The class confusion matrices for gender identification, corresponding to the results for linear kernel in Table 13.7, are given in Table 13.9. Similarly, the class confusion matrices for race identification, corresponding to the results for RBF kernel in Table 13.8, are given in Tables 13.10 and 13.11.

Table 13.7   Results for gender identification on frontal faces

Kernel Type	No. of principal components used for Y	Accuracy
RBF	$Y = 75$	88.25%
Linear	$Y = 105$	88.25%

Table 13.8   Results for race identification on frontal faces

Kernel Type	With/without color components	No. of principal components used for Y, I, Q	Accuracy
RBF	With color	Y=45, I=0, Q=15	93.25 %
	Without color	$Y = 30$	92.25 %
Linear	With color	Y=15, I=15, Q=15	91.75 %
	Without color	$Y = 30$	90.00 %

Table 13.9   Class confusion matrix for gender identification with linear kernel in Table 13.7

Frontal View		Recognized As		Recognized As (%)	
		Female	Male	Female	Male
Ground Truth	Female	148	26	85.06	14.94
	Male	21	205	9.29	90.71

Table 13.10   Class confusion matrix for race identification without using color for the RBF kernel in Table 13.8

Frontal View (grey-scale)		Recognized As %			
		B	ESEA	SA	W
Ground Truth	B	**92.19**	1.56	0.00	6.25
	ESEA	0.91	**93.64**	0.91	4.55
	SA	13.79	3.45	**65.52**	17.24
	W	1.02	2.03	1.52	**95.43**

Table 13.11 Class confusion matrix for race identification with the use of color for the RBF kernel in Table 13.8

Frontal View (color images)		Recognized As %			
		B	ESEA	SA	W
	B	**90.63**	1.56	1.56	6.25
Ground Truth	ESEA	0.00	**91.82**	1.82	6.36
	SA	17.24	6.90	**68.97**	6.90
	W	0.51	0.51	0.51	**98.48**

Other combinations of principal components for experimental setups 1 and 2 yielded similar performance. The results for race identification for both linear and non-linear SVMs are shown graphically in Fig. 13.3. The abscissa corresponds to the number of principal components of Y. And each bunch of bars corresponds to the 16 different combinations of color components used with a specific number of Y principal components, as outlined in the previous section, with the leftmost bar in each bunch representing the results with only the Y principal components used. The ordinate corresponds to the accuracy.

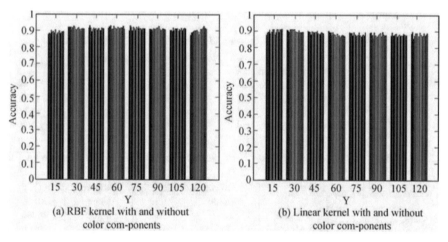

(a) RBF kernel with and without
color com-ponents

(b) Linear kernel with and without
color com-ponents

Fig. 13.3 Results for the race identification: (a) RBF kernel; (b) Linear kernel.

The results for the third set of experiments, as outlined in the previous section, are presented in Fig. 13.4.

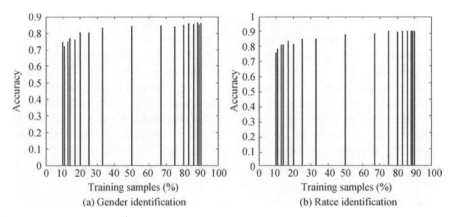

Fig. 13.4 Effect of number of training samples: (a) on gender identification; (b) on race identification.

13.5 Fusing the Frontal View and Silhouetted Profile View Classifiers

This section presents the attempt for fusing the frontal view and silhouetted profile view classifiers.

13.5.1 Database for Classifier Fusion

The number of faces common to the experiments for both the profile and frontal views was 357. Table 13.12 gives the distribution of the database.

Table 13.12 Demographic distribution of the database common to frontal view and silhouetted view face experiments

B		ESEA		SA		W	
56		93		28		180	
M	F	M	F	M	F	M	F
30	26	52	41	20	8	95	85

13.5.2 Results for Classifier Fusion

A careful analysis of the results corresponding to faces of both the views revealed that there was more room for improvement in gender identification classifiers than in race identification classifiers. For instance, out of the 42 mistakes that the frontal classifier made, 32 were classified correctly by the shape context base classifier.

For fusing the classifiers, a probability measure was required for both the profile and the frontal view classifiers. In the case of profile, the probability measure may be derived from the number of votes in kNN, and in the case of frontal view, SVM spits out a probability measure as well. A number of different methods were experimented with, including the product of the probabilities, weighted sum of the probabilities and training an SVM classifier on the probability outputs of both the classifiers. Of all these, the best performance was given by the product of the probabilities of both the classifiers. This correctly detected 319 images in total compared with 315 correct detections of frontal view alone, thus increasing the accuracy from 88.24% to 89.36%.

13.6 Human Experiments

These experiments were performed with the intent to benchmark the human performance on the database at hand for different face representations. Another purpose was to further investigate whether there is information present for race identification for humans in silhouetted profiles in addition to that of gender identification (information for gender identification was shown in [10]). This study was approved by the Institutional Review Board (IRB) at the University of Illinois at Urbana-Champaign.

13.6.1 Database and Experimental Design

The following sub-sections describe the database of the face samples used, demographics of the participants and the experimental design of the human subjective tests.

Face samples

The number of face samples selected for these experiments was 128. These were divided into 4 sample groups (SAM_G). Each group was completely balanced in terms of each gender-racial category; i.e., each group of 32 face samples had 4 face samples from male white racial category, 4 face samples from female white racial category and so on. The face samples were primarily within the age range of 18 – 30 years. Each group was shown either in silhouetted profile, 45 ° view, 2D frontal face image or 3D face model representation to one participant. In this manner each participant saw no two representations of the same face and saw 128 images in total (32 in silhouetted profile, 32 in 45 ° view, 32 in frontal view and 32 in 3D).

Participants

A total of 42 people participated in the study. Each participant was paid

$8/hr. The participant pool was reasonably diverse and had exposure to other racial groups. Their demographic distribution is given in Table 13.13.

Table 13.13   Demographic distribution of the participants

B		ESEA		SA		W	
2		12		16		12	
M	F	M	F	M	F	M	F
0	2	7	5	9	7	3	9

The participants were randomly divided into 4 subject groups (SUB_G). Each SUB_G was shown 4 different SAM_Gs, with 1 face representation for each sample group, in the manner shown in Table 13.14. Each subject group had at least 10 participants, so each image was labeled by at least 10 people.

Table 13.14   Task assignment of different subject groups

Subject Group	Silhouette	45° View	2D Frontal	3D
SUB_G#1	SAM_G#1	SAM_G#2	SAM_G#3	SAM_G#4
SUB_G#2	SAM_G#4	SAM_G#1	SAM_G#2	SAM_G#3
SUB_G#3	SAM_G#3	SAM_G#4	SAM_G#1	SAM_G#2
SUB_G#4	SAM_G#2	SAM_G#3	SAM_G#4	SAM_G#1

Further design considerations

The participants were requested to mark the perceived gender and race of a face sample in a labeling tool, shown in Fig. 13.5, in a single sitting. An effort was made to exclude males who had grown moustaches and beards in the face images. All the participants were first shown images with face in profile, then in near frontal (or 45° frontal), then frontal, and then in 3D with texture. There was no time limit in which the participants were required to label an image, they could take as long as they liked. When they were shown

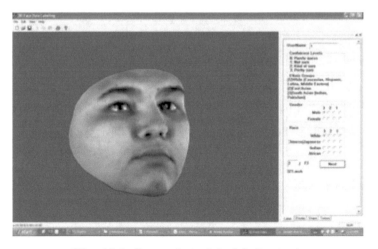

Fig. 13.5   Screen-shot of the labeling tool.

face samples in 3D, they could have rotated the face and zoomed in and out. Some example images shown to the participants are given in Fig. 13.6. It can be seen that the hair from head are concealed from the subjects.

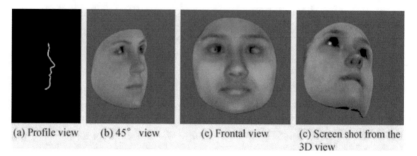

(a) Profile view   (b) 45° view   (c) Frontal view   (c) Screen shot from the 3D view

Fig. 13.6 Different face presentations shown to participants in the study.

The order of images within each SAM_G was randomized. As there were four images from each of the eight gender-racial categories, the probability that a participant saw more than 3 images from the same gender-racial category was indeed very low. This pretty well ensures that the participants do not get trained by viewing a series of images from the same gender-racial category.

13.6.2 Results

The performance of participants is summarized in Table 13.15. Also, the class confusion matrices for all the participants on different facial representations are given for gender in Table 13.16 and for race identification in Table 13.17.

Table 13.15   Summary of the performance of participants

Accuracy	Gender		Race	
	Mean	Std. Dev.	Mean	Std. Dev.
Profile view	59.82%	8.55%	47.77%	9.74%
45° view	88.10%	7.15%	88.76%	9.13%
Frontal view	86.97%	7.10%	83.33%	8.69%
3D view	88.84%	7.21%	87.28%	6.42%

Table 13.16   Confusion matrices for gender identification

		Recognized As (%)	
Profile Silhouette		F	M
Ground Truth	F	**47.8**	52.2
	M	28.1	**71.9**
45° Frontal Face Image		F	M
Ground Truth	F	**84.1**	15.9
	M	7.9	**92.1**

		Continued	
		Recognized As (%)	
2D Frontal Face Image		F	M
Ground Truth	F	**81.4**	18.6
	M	7.4	**92.6**
3D View		F	M
Ground Truth	F	**82.7**	17.3
	M	5.1	**94.9**

Table 13.17 Confusion matrices for race recognition

		Recognized As (%)			
Profile Silhouette		B	ESEA	SA	W
	B	**60.4**	19	9.2	11.3
Ground Truth	ESEA	22.9	**37.5**	17	22.6
	SA	10.7	13.4	**29.2**	46.7
	W	3.3	17.3	15.5	**64**
45° Frontal Face Image		B	ESEA	SA	W
	B	**87.5**	0.9	5.1	6.5
Ground Truth	ESEA	0.6	**93.2**	0.9	5.4
	SA	2.7	1.2	**81**	15.2
	W	0	3.3	3.3	**93.5**
2D Frontal Face Image		B	ESEA	SA	W
	B	**76.5**	2.7	14	6.8
Ground Truth	ESEA	0.9	**91.4**	2.4	5.4
	SA	5.7	1.5	**72.6**	20.2
	W	0	3.3	3.9	**92.9**
3D View		B	ESEA	SA	W
	B	**79.8**	3	10.1	7.1
Ground Truth	ESEA	0.3	**92.9**	3	3.9
	SA	3	2.1	**82.4**	12.5
	W	0	2.1	3.9	**94**

Table 13.18 gives the average performance of male and female subjects calculated separately on male and female face samples, while Table 13.19 gives the average performance of the subjects of different racial groups on facial samples of different races.

Table 13.18 Average performance of subjects calculated separately by gender

		Recognition accuracy (%) on	
Profile Silhouette		F	M
Gender of subjects	F	46.20	**72.55**
	M	49.67	**71.05**
45° Frontal Face Image		F	M
Gender of subjects	F	82.34	**91.30**
	M	86.18	**93.09**
2D Frontal Face Image		F	M
Gender of subjects	F	77.99	**91.30**
	M	85.53	**94.08**

Continued

3D View		Recognition accuracy (%) on	
		F	M
Gender of subjects	F	79.35	**95.11**
	M	86.84	**94.74**

Table 13.19 Average performance of subjects calculated separately by race

		Recognition accuracy (%) on			
Profile Silhouette		B	ESEA	SA	W
	B	56.25	50	25	37.5
Race of subjects	ESEA	54.17	48.96	19.79	**63.54**
	SA	**64.84**	33.59	31.25	60.94
	W	61.46	29.17	36.46	**72.92**
45° Frontal Face Image		B	ESEA	SA	W
	B	81.25	100	68.75	100
Race of subjects	ESEA	89.58	**93.75**	86.46	**92.71**
	SA	86.72	**95.31**	**89.06**	92.97
	W	87.5	88.54	66.67	**93.75**
2D Frontal Face Image		B	ESEA	SA	W
	B	68.75	100	56.25	68.75
Race of subjects	ESEA	76.04	**95.83**	71.88	**93.75**
	SA	78.13	**92.19**	**85.94**	92.19
	W	76.04	84.38	58.33	**96.88**
3D View		B	ESEA	SA	W
	B	81.25	100	75	87.5
Race of subjects	ESEA	71.88	**93.75**	77.08	**92.71**
	SA	82.81	**92.97**	**90.63**	94.53
	W	83.33	90.63	78.13	**95.83**

A comparison of human average performance with that of computer algorithms (shape context based matching for silhouette and PCA+SVM for frontal faces) is given for gender identification in Table 13.20 and for race identification in Table 13.21.

Table 13.20 Comparison of human average performance with computer algorithms for gender identification

	Humans average performance	Computer Algorithms
Silhouetted profile view	59.82%	**71.20%**
Frontal view	86.97%	**88.25%**

Table 13.21 Comparison of human average performance with computer algorithms for race identification

	Humans average performance	Computer Algorithms
Silhouetted profile view	47.77 %	**71.66 %**
Frontal view	83.33 %	**93.25 %**

13.7 Observations and Discussion

It may be observed from computer experiments, that females seem to be more reliably identified than males from silhouetted profile faces, evident by their accuracy of $(83.41\pm2.56)\%$ shown in Table 13.3. This is in contrast to the frontal views, in which generally the bias is towards males in performance, as noted in [16] and also in this work given by Table 13.9.

Also from the computer experiments with the profile view, East and South East Asians seems to be the most reliably identified racial category, evident by their accuracy of $(80.37\pm3.8)\%$ shown in Table 13.5. The Black racial category also shares a higher accuracy of true identification.

For the computer experiments with the frontal view, it may be observed from Fig. 13.3 that, in qualitative terms, color in general does help but the improvement in terms of accuracy is not much. In fact the results without using the color components are comparable to the results with the color components. Thus it is recommended not to use the chrominance information in practice because color is prone to changes in uncontrolled lighting conditions.

It may also be observed from Table 13.8 that non-linear SVMs gave only slightly better performance than linear SVMs for race identification. Also please note just a slight improvement in performance with the use of color components.

The class confusion matrices in Tables 13.10 and 13.11 reveal that, in qualitative terms, the performance of the Black and East and South Asian ethnic categories degrades with the use of color. Also please note the improvement in performance, in qualitative terms, of White and South Asian ethnic categories with the use of color information.

Also for the work with frontal view faces, it may be noted from Fig. 13.4 that accuracy initially increases with the increase in number of training samples but then plateaus, e.g., for number of training images greater than 70% of the database in case of race identification.

Classifier fusion does improve the performance but only slightly.

A closer look on the human experiments discloses some noticeable observations. Performance of the participants on the silhouetted face profiles was well above chance (which was 50% for gender identification and 25% for race identification) both for gender and racial identification, as seen in Table 13.15. This proves the presence of cues for humans for racial identification in silhouetted face profiles; more specifically in the "profile edge".

There was an evident bias towards males in the performance for gender identification for all the views, as seen in Table 13.16. This is not due to the own-gender bias as shown in Table 13.18; both male and female participants performed better at identifying male faces than female faces for all views.

White seems to be the most reliably identified racial category, followed by East and South East Asians, as seen in Table 13.17. Participants seemed to have performed the best on near frontal views (45° view) and 3D, as shown in Table 13.15, particularly for race identification. The better performance on near frontal and 3D may be attributed to the presence of geometric information in these views, which may have helped the participants to make correct decisions.

The authors were unable to find a reference in literature where the focus is to experiment with the ability of subjects to only identify the race of the face samples. Levin, in [39], experiments with one such scenario during the search of the faces of a race, but those experiments were time constrained. Also they used two races, white and black, and their subject pool was also not very diverse. In general, the papers report face-recognition performance, rather than that of race/gender identification. An asymmetric own-race bias has been reported in a number of studies in recognition of faces [10, 40–42]. For instance, Caucasian subjects often show better recognition performance with Caucasian faces compared to African American or Asian faces. However, African American or Asian subjects perform just as well with both types of faces. In our study, which was targeted towards race identification (please see Table 13.19 and please ignore the values corresponding to the subjects from the Black group as they were scarce in number); the trend was different across views and across the racial group of subjects. Participants from the East and South East Asian group performed poorly on South Asian faces for profile views. In the other three views they performed almost equally well for both East and South East Asian faces and White faces. However, their performance on the other racial groups was worse. Also, the participants from South Asian group performed poorly both on the South Asian faces and East and South East Asian faces for the profile view. In other views, however, their best performance was in general with East and South East Asian faces. The participants from White group showed consistently better performance with White faces.

Computer algorithms which were used in this work outperformed the average human performance, evident from Table 13.20 and Table 13.21. The performance is particularly better in case of race identification.

13.8 Concluding Remarks

This chapter presents a comprehensive study of gender and race identification by both man and machine from different facial views. However, because of the limited number of images the results may be preliminary. But this chapter presents new research directions for gender and race identification. Fusing cues from different views for the purpose have been experimented with. The qualitative importance of color has been evaluated for race identification. The results are benchmarked by the human experiments which also reveal some other interesting observations.

Acknowledgements

This chapter has had the benefit of some very useful suggestions from Prof. Neal J. Cohen on human experiments and from Prof. David Forsyth. In this project, the Matlab implementation of Hungarian by Niclas Borlin was used. This research is supported in part by U. S. Government VACE Program.

References

[1] Yanushkevich S, Gavrilova M, Wang P et al (eds) (2007) Image Pattern Recognition, Synthesis and Analysis in Biometrics. World Scientific Publishing Company, Singapore

[2] Yanushkevich S, Hurley D, Wang P (eds) (2008) Special Issue: Pattern Recognition and Artificial Intelligence in Biometrics, International Journal of Pattern Recognition and Artificial Intelligence, 22(3): 367–369

[3] Grgic M, Shan S, Lukac R et al (eds) (2009) Special Issue: Facial Image Processing and Analysis, International Journal of Pattern Recognition and Artificial Intelligence, 23(3): 355–358

[4] Pratt J, Hauser K, Ugray Z et al (2007) Looking at Human-computer Interface Design: Effects of Ethnicity in Computer Agents. Interacting with Computers, 19(4): 512–523

[5] Chang M S, Chou J H (2010) A Robust And Friendly Human-Robot Interface System Based On Natural Human Gestures. International Journal of Pattern Recognition and Artificial Intelligence, 24(6): 847–866

[6] Veropoulos K, Bebis G, Webster M (2005) Investigating the Impact of Face Categorization on Recognition Performance. In: Proceedings of First International Symposium in Advances in Visual Computing ISVC 2005, LCNS, vol 3804: 207–218

[7] Tariq U, Hu Y, Huang T (2009) Gender and Ethnicity Identification from Silhouetted Face Profiles. In: 2009 IEEE International Conference on Image Processing (ICIP), Cairo, Egypt

[8] Bruce V, Burton A M, Hanna E et al (1993) Sex Discrimination: How Do We Tell the Difference Between Male and Female Faces? Perception, 22: 131–152

[9] Hill H, Bruce V, Akamatsu S (1995) Perceiving the Sex and Race of Faces: The Role of Shape and color. In: Proceedings of the Royal Society-Biological Sciences (Series B), vol 261: 367–373

[10] Davidenko N (2007) Silhouetted Face Profiles: A new Methodology for Face Perception Research. Journal of Vision, 7: 1–17

[11] Golomb B A, Lawrence D T, Sejnowski T J (1990) Sexnet: a Neural Network Identifies Sex from Human Faces. In: Lipmann R P (ed) Advances in Neural Information Processing Systems (NIPS), San Mateo, vol 3: 572–577

[12] Gutta S, Huang J R J, Jonathon P et al (2000) Mixture of Experts for Classification of Gender, Ethnic Origin, and Pose of Human Faces. IEEE Trans on Neural Networks, 11: 948–960

[13] O'Toole J, Abdi H, Deffenbacher K et al (1993) Low Dimensional Representation of Faces in Higher Dimensions of the Face Space. Journal of the Optical Society of America, 10: 405–410

[14] Moghaddam B, Yang M (2002) Learning Gender with Support Faces. IEEE Trans on Pattern Analysis and Machine Intelligence, 24: 707–711

[15] Graf A B A, Wichmann F A (2002) Gender Classification of Human Faces. In: Biologically Motivated Computer Vision 2002, LNCS, vol 2525: 491–500

[16] Shakhnarovich G, Viola P A et al (2002) A Unified Learning Framework for Real Time Face Detection and Classification. In: IEEE International Conference on Automatic Face & Gesture Recognition (FG), pp 14–21

[17] Sun Z, Bebis G, Yuan X et al (2002) Genetic Feature Subset Selection for Gender Classification: a Comparison Study. In: IEEE Workshop on Application of Computer Vision, pp 165–170

[18] Walawalkar L, Yeasin M, Narasimhamurthy A et al (2003) Support Vector Learning for Gender Classification Using Audio and Visual Cues. International Journal on

Pattern Recognition and Artificial Intelligence, 17: 417–439

[19]   Jain A, Huang J (2004) Integrating Independent Component Analysis and Linear Discriminant Analysis for Gender Classification. In: IEEE International Conference on Automatic Face & Gesture Recognition (FG), pp 159–163

[20]   Shan C, Gong S, McOwan P (2008) Fusing Gait and Face Cues for Human Gender Recognition. Neurocomputing, 71: 1931–1938

[21]   Lu X G, Chen H, Jain A K (2006) Multimodal Facial Gender and Ethnicity Identification. In: ICB 2006, pp 554–561

[22]   Raisamo E (2008) Evaluation of Gender Classification Methods with Automatically Detected and Aligned Faces. IEEE Trans. Pattern Analysis and Machine Intelligence, 30(3): 541–547

[23]   Rowley S (2007) Boosting Sex Identification Performance. Proc of International Journal of Computer Vision, 71(1): 111–119

[24]   Wang X (2008) Improving Generalization for Gender Classification. In: International Conference on Image Processing, pp 1656–1659

[25]   Ueki K, Komatsu H, Imaizumi S et al (2004) A Method of Gender Classification by Integrating Facial, Hairstyle, and Clothing Images. In: Proc International Conference on Pattern Recognition, pp 446–449

[26]   Costen N, Brown M, Akamatsu S (2004) Sparse Models for Gender Classification. In: Proc. of International Conference on Automatic Face and Gesture Recognition, pp 201–206

[27]   Kim H, Kim D, Ghahramani Z et al (2006) Appearance-based Gender Classification with Gaussian Processes. Pattern Recognition Letters, 27(6): 618–626

[28]   Samal A, Subramani V, Marx D (2007) Analysis of Sexual Dimorphism in Human Face. Journal of Visual Communication and Image Representation, 18(6): 453–463

[29]   O'Toole A, Vetter T, Troje N F et al (1997) Sex Classification is Better with Three-dimensional Head Structure Than with Image Intensity Information. Perception, 26: 75–84

[30]   O'Toole A, Abidi H, Deffenbacher et al (1991) Classifying Faces by Race and sex Using an Autoassociative Memory Trained for Recognition. In: Hammond K (ed) Proc, 13th Annu Conf Cognitive Sci Soc, Lawrence Erlbaum, NJ

[31]   Lu X, Jain A K (2004) Ethnicity Identification From face Images. In: Biometric Technology for Human Identification. Proceedings of SPIE: The International Society for Optical Engineering, 540: 114–123

[32]   Hosoi S, Takikawa E, Kawade M (2004) Ethnicity Estimation with Facial Images. In: IEEE Conf on Automatic Face and Gesture Recognition, pp 195–200

[33]   Hu Y, Fu Y, Tariq U et al (2010) Subjective Experiments on Gender and Ethnicity. In: The 16th International Conference on MultiMedia Modeling (MMM). Chongqing

[34]   Belongie S, Malik J, Puzicha J et al (2002) Shape Matching and Object Recognition Using Shape Contexts. IEEE Trans on Pattern Analysis and Machine Intelligence, 22: 509–522

[35]   Hu Y, Zhang Z, Xu X et al (2007) Building Large Scale 3D Face Database for Face Analysis. In: MCAM 2007, Weihai, China, pp 343–350

[36]   Buchsbaum, Walter H (1975) Color TV Servicing, 3rd edn. Prentice Hall, NJ

[37]   Chang C C, Lin C In: LIBSVM: A Library for Support Vector Machines. Available at: http://www.csie.ntu.edu.tw/cjlin/libsvm. Accessed 1 January, 2011

[38]   Blanz V, Vetter T (1999) A Morphable Model for the Synthesis of 3D Faces. In: Proc. Siggraph99, Los Angeles, pp 187–194

[39]   Levin D (2000) Race as a Visual Feature: Using Visual Search and Perceptual Discrimination Tasks to Understand Face Categories and the Cross-race Recognition Deficit. Journal of Experimental Psychology: General, 129: 559–574

[40]   Bothwell R K, Brigham J C, Malpass R S (1989) Cross-racial identification. Personality and Social Psychology Bulletin, 15: 19–25

[41]   Meissner C A, Brigham J C (2001) Thirty Years of Investigating the Own-Race Bias in Memory for Faces: A Meta-analytic Review. Psychology, Public Policy, and Law, 7: 3–35

[42]   Sporer S L (2001) Recognizing Faces of Other Ethnic Groups: An Integration of Theories. Psychology, Public Policy, and Law, 7: 36–97

[43]   Papadimitriou C, Stieglitz K (1982) Combinatorial Optimization: Algorithms and Complexity. Prentice Hall, NJ

[44]   Bookstein F L (1989) Principal Warps-Thin-Plate Splines and the Decomposition of Deformations. IEEE Trans on Pattern Analysis and Machine Intelligence, 11: 567–585

[45]   Meinguet J (1979) Multivariate Interpolation at Arbitrary Points Made Simple. Journal of Applied Math and Physics (ZAMP), 30: 292 – 304

14 Common Vector Based Face Recognition Algorithm

Ying Wen[1], Yue Lu[2], Pengfei Shi[3] and Patrick S. P. Wang[4]

Abstract   The common vector represents the common invariant properties of the respective class, which was proposed and originally introduced
for isolated word recognition problems. Inspired by the idea of the common
vector, researchers proposed the discriminant common vector approach and
the common faces approach for face recognition. In this chapter, we study
an approach for face recognition based on the difference vector plus the kernel PCA (Principal component analysis). Difference vector is the difference
between the original image and the common vector which is obtained by the
images processed by the Gram-Schmidt orthogonalization and represents the
common invariant properties of the class. The optimal feature vectors are
obtained by KPCA (kernel-based PCA) procedure for the difference vectors.
Recognition result is derived from finding the minimum distance between the
testing difference feature vectors and the training difference feature vectors.
A comparative study among them in face recognition (or generally in the SSS
problem) was carried out. To test and evaluate these approach performance,
a series of experiments are performed on five face databases: ORL, YALE,
FERET, AR and JAFFE face databases and the experimental results show
that the common vector based face recognition algorithm is encouraging.

14.1 Introduction

Face recognition by machine has been started since 1970s, due to military,
commercial, and law enforcement applications, currently becoming an active
and important research area, especially after 9.11 tragedy, for security purposes, including personal identification, verification and preventing terrorists
from entering the region. Face recognition can be defined as the identification
of individual's from images of their faces by using a stored database of faces
labeled with people's identities. This task is complex and can be decomposed
into the smaller steps of detection of faces in a cluttered background, local-

1   Department of Psychiatry, Columbia University, New York, NY 10032, USA. E-mail:
    yw2365@columbia.edu.
2   Department of Computer Science and Technology, East China Normal University,
    Shanghai 200062, China. E-mail: ylu@cs.ecnu.edu.cn.
3   Institute of Image Processing and Pattern Recognition, Shanghai Jiao Tong Univer-
    sity, Shanghai, 200030, China. E-mail: pfshi@sjtu.edu.cn.
4   College of Computer Science, Northeastern University, Boston, MA, 02115, USA.
    E-mail: patwang@ieee.org.

ization of these faces followed by extraction of features from the face regions, and, finally, recognition and verification [1]. It is a difficult problem as there are numerous factors such as 3D pose, facial expression, hair style, make up, mustache, beard, wearing glasses etc., which affect the appearance of an individual's facial features. In addition to these varying factors, lighting, shadowing, background, scale changes, and topological variations etc. make this task even more challenging. Additional problematic conditions include noise, occlusion, and many other possible factors. Furthermore, a more challenging class of application imagery includes real-time detection and recognition of faces in surveillance video images, which presents additional constraints in terms of speed and processing requirements [2]. These issues make face recognition a difficult task.

Many methods have been proposed for face recognition within the last two decades [2, 3]. Linear and nonlinear discriminant algorithms such as Eigenface method, LDA (Linear Discriminant Analysis) method [4], improved LDA method [5], two-dimensional PCA (Principal Component Analysis) [6] and Kernel PCA method [20] were proposed. Among these methods, appearance based approaches operate directly on images or appearances of face objects, and process the images as two-dimensional holistic patterns. In these approaches, a wh-dimensional image with w by h pixels is represented by a vector in a high dimensional space. Therefore, each facial image corresponds to a point in this space. This space is called the sample space or the image space, and its dimension typically is very high [7]. However, since face images have similar structure, the image vectors are correlated, and any image in the sample space can be represented in a lower-dimensional subspace without losing a significant amount of information. The Eigenface method has been proposed for finding such a lower-dimensional subspace [8]. The key idea behind the Eigenface method is to find the best set of projection directions in the sample space that will maximize the total scatter across all images such that

$$J_{\mathrm{PCA}} = argmax_{\boldsymbol{W}} |\boldsymbol{W}^{\mathrm{T}} \boldsymbol{S}_{\mathrm{T}} \boldsymbol{W}|$$

is maximized. Here, $\boldsymbol{S}_{\mathrm{T}}$ is the total scatter matrix of the training set samples, and $\boldsymbol{W}$ is the matrix whose columns are the orthonormal projection vectors. The projection directions are also called the eigenfaces. Any face image in the sample space can be approximated by a linear combination of the significant eigenfaces. The sum of the eigenvalues that correspond to the eigenfaces not used in reconstruction gives the mean square error of reconstruction. This method is an unsupervised technique since it does not consider the classes within the training set data.

The Linear Discriminant Analysis (LDA) method is proposed in [9] and [10].

This method overcomes the limitations of the Eigenface method by applying the Fisher's Linear Discriminant criterion. This criterion tries to maximize the ratio

$$J_{\mathrm{FLD}} = argmax_{\boldsymbol{W}} \frac{|\boldsymbol{W}^{\mathrm{T}} \boldsymbol{S}_{\mathrm{B}} \boldsymbol{W}|}{|\boldsymbol{W}^{\mathrm{T}} \boldsymbol{S}_{\mathrm{W}} \boldsymbol{W}|},$$

where $\boldsymbol{S}_{\mathrm{B}}$ is the between-class scatter matrix, and $\boldsymbol{S}_{\mathrm{W}}$ is the within-class scatter matrix. Thus, by applying this method, we find the projection directions that on the one hand maximize the Euclidean distance between the face images of different classes and on the other hand minimize the distance between the face images of the same class. This ratio is maximized when the column vectors of the projection matrix $\boldsymbol{W}$ are the eigenvectors of $\boldsymbol{S}_{\mathrm{W}} \boldsymbol{S}_{\mathrm{B}}$. In face recognition tasks, this method cannot be applied directly since the dimension of the sample space is typically larger than the number of samples in the training set. As a consequence, $\boldsymbol{S}_{\mathrm{W}}$ is singular in this case. This problem is also known as the "small sample size problem" [11].

Due to the high dimensionality of face image, the dimension reduction technique is important to face recognition. Principal component analysis (PCA) is a well known method for dimension reduction. By calculating the eigenvectors of the covariance matrix of the original inputs, PCA linearly transforms a high-dimensional input vector into a low-dimensional ones whose components are uncorrelated [12, 13]. Although PCA has many advantages, it also has many shortcomings, such as its sensitiveness to noise, and its limitation to data description. To eliminate these shortcomings, many methods have been proposed to improve PCA algorithm. These methods mainly focus on two aspects. On the one hand, the sensitiveness of the traditional PCA to noise seriously affects its precision. Many scholars analyzed the robustness of PCA in detail, and presented many improved PCA algorithms [14–16]. On the other hand, for the traditional PCA, principal components are determined exclusively by the second-order statistics of the input data, which can only describe the data with smooth Gaussian distribution. In order to describe the data with Non-Gaussian distribution, many researchers such as Karhunen et al. introduced an appropriate nonlinear process into the traditional PCA and developed a nonlinear PCA [17] according to the distribution of the input samples. In these improved PCA algorithms, kernel based PCA (KPCA) [15] proposed by Scholkopf et al. is a state-of-the-art one as a nonlinear PCA algorithm. KPCA utilizes kernel function to gain the random high-order correlation between input variants, and finds the principal components needed through the inner production between input data. KPCA cannot only successfully describe the data with Gaussian distribution, but also can commendably describe the data with Non-Gaussian distribution. More and more people are interested in this field and have carried out some relevant researches [19, 20].

Recently, the common vector was proposed and originally introduced for isolated word recognition problems in the case where the number of samples in each class was less than or equal to the dimensionality of the sample space [21, 22]. The common vector presents the common properties of a training set. Inspired by this idea, Hakan Cevikalp et al. [23] proposed an approach of the discriminative common vectors (DCV) for face recognition, which uses the subspace methods and the Gram Schmidt orthogonalization procedure to obtain the discriminative common vectors. Afterwards, He et al. [24] also presented an algorithm called kernel DCV by combining DCV with kernel method (KDCV). In addition, they also proposed a common face approach for face recognition (CVP) [25]. The recognition result of CVP is derived by finding the minimum distance between a face image and the common vectors. And then, they also extend kernel method to CVP (KCVP).

Furthermore, in addition to those references up to 2005, there are also many other methods in the literature regarding face analysis, and recognition and classifications, since 2006 till 2010. For examples: D. Zhang et al. [32] and X.G. You et al. [37, 39] proposed methods using non-tensor wavelets for analyzing face shapes and structures. S.W. Lee, P.S.P. Wang et al. [33] discussed 3D face reconstruction based on photometric stereo facial images. F. Shih et al. [34, 35] examined some new ideas of colored fscila images and tested on JEFFE facial databases with rather satisfactory results. Y. Luo et al. [36] proposed a method of facial metamorphosis using geometrical methods for biometric applications to face analysis and recognition. L.C. Zhang et al. [38] discussed face analysis and face recognition using scale invariant feature transform and support vector machine. P.S.P. Wang [40] shows importance to understand basic definitions of "Similarity Distance" and shows some significant concerns on the measurement for biometric analysis and applications to facial image analysis and classifications. K. Soundar et al. [41] discussed new ideas of preserving global and local features as well as a combined approach for recognizing face images. Y. Zheng et al. [42] proposed a new method of 3D modeling of faces from near infrared images using statistical learning, which is different from that of [33] in that it uses different camera tools to get different structural images, i.e., infrared facial images, which can capture inner information between skins and bones (skeletons). Such information is helpful for characterizing facial structures necessary to distinguish from each other, and cannot be obtained by normal means such as conventional digital cameras. M.Q. JING et al. [43] proposed a novel method for horizontal eye line detection under various environments, which is an inherently important and essential portion for classifying face. Finally, last but not least, in M. Grgic [44], there is a special issue solely devoted for facial image analysis and recognition at IJPRAI 2009, v23, which contains 12 excellent chapters regarding all aspects of face analysis and recognition. All these constitute an invaluable source for inspiring of our method here using common vector based algorithm for face recognition.

In this section, we study a face recognition method based on the difference vectors plus KPCA approach. The common vector of each class is gotten by the procedure of the Gram-Schmidt orthogonalization, while the difference vector is derived by the difference between the original image and the common vector. The optimal feature vector is obtained by KPCA procedure of difference vectors. Recognition result is derived from finding the minimum distance of the feature vectors. Experiments on ORL, YALE, FERET (The Facial Recognition Technology), AR (Purdue University) and JAFFE face databases indicate that the common vector based face recognition is encouraging.

14.2 Algorithm Description

Common vector was proposed and originally introduced for isolated word recognition problems in the case where the number of samples in each class was less than or equal to the dimensionality of the sample space. The common vector presents the common properties of a training set.

14.2.1 Common Vector

Assume that there are C classes, and each class contains N samples. Let $x_j^i, i = 1, 2, \cdots, C, j = 1, 2, \cdots, N$ be a n-dimensional column vector which denotes the j_{th} sample from the i_{th} class. There are a total of $M = NC$ samples in the training set. Construct B^i of the i_{th} class $(i = 1, 2, \cdots, C)$ whose columns span a difference subspace for the i_{th} class is defined as follows:

$$B^i = [b_1^i, b_2^i, \cdots, b_{N-1}^i], \tag{14.1}$$

where $b_k^i = x_{k+1}^i - x_1^i, k = 1, 2, \cdots, N-1$, x_1^i is called reference vector which can be randomly selected from the i_{th} class, and the first sample is selected in this section.

By performing Gram-Schmidt orthogonalization procedure, the orthonormal vector set $y_1^i, y_2^i, \cdots, y_{N-1}^i$ which spans the difference subspace $L(B^i)$ is obtained. Then a sample x_k^i randomly selected from the i_{th} class is projected on the orthonormal vector $y_k^i (k = 1, 2, \cdots, N-1)$, and the summation of the projection is computed as follows:

$$s^i = <x_k^i, y_1^i> y_1^i + <x_k^i, y_2^i> y_2^i + \cdots + <x_k^i, y_{N-1}^i> y_{N-1}^i, \tag{14.2}$$

where $< x, y >= x_1 y_1 + x_2 y_2 + \cdots + x_n y_n$ denotes the scalar product of $x = (x_1, x_2, \cdots, x_n) \in R^n$ and $y = (y_1, y_2, \cdots, y_n) \in R^n$.

Then common vector x_{common}^i of the i_{th} face class is derived as follows:

$$x_{\text{common}}^i = x_k^i - s^i. \tag{14.3}$$

It was proved that the common vector x_{common}^i is unique and independent of the randomly selected sample x_k^i (Theorem 1). Therefore, the common vector x_{common}^i represents the common invariant properties of the i_{th} face class. As the above statement, we obtain C common vectors. The above fact is proven in the following theorem.

Common vector of the i_{th} class $x^i_{common} = x^i_k - s^i (k = 1, \cdots, N)$ (in Eq.(14.3)) is independent from a selected sample x^i_k, i.e.,

$$x^i_{common-k} = x^i_{common-h}, k, h = 1, \cdots, N. \tag{14.4}$$

Proof. (the definition of b_k and B in the section) Since $b_k \in B$ and $\{y^i_1, \cdots, y^i_{N-1}\}$ is a basis of B, there exists constants $K_1, \cdots, K_{N-1}$ such that

$$b_k = K_1 y^i_1 + \cdots + K_{N-1} y^i_{N-1}.$$

If the above equation is scalarly multiplied by each vector $\{y^i_1, \cdots, y^i_{N-1}\}$ respectively and using $< y^i_h, y^i_k > = \delta_{kh} = \{1, \text{ if } k = h; 0, \text{ if } k \neq h\}$, we obtain $K_h = < b_k, y^i_h > (h = 1, \cdots, N-1)$. Thus, the equalities can be written:

$$b_k = < b_k, y^i_1 > y^i_1 + \cdots + < b_k, y^i_{N-1} > y^i_{N-1}$$

or

$$x^i_k - x_1 = < x^i_k - x_1, y^i_1 > y^i_1 + \cdots + < x^i_k - x_1, y^i_{N-1} > y^i_{N-1}$$

The following equality is written:

$$x^i_k - [< x^i_k, y^i_1 > y^i_1 + \cdots + < x^i_k, y^i_{N-1} > y^i_{N-1}] =$$

$$x^i_1 - [< x^i_1, y^i_1 > y^i_1 + \cdots + < x^i_1, y^i_{N-1} > y^i_{N-1}], \quad k = 1, \cdots, N \tag{14.5}$$

According to the above notation, it is clear that

$$x^i_{common-k} = x^i_{common-1}, \quad \forall k = 1, \cdots, N.$$

Hence,

$$x^i_{common-k} = x^i_{common-h}, \quad k, h = 1, \cdots, N.$$

14.2.2 Principal Component Analysis

Principal component analysis is a popular technique for feature extraction and dimensionality reduction. Suppose that $X = \{x_p; p = 1, 2, \cdots, M\}$ is a set of centered observations of an n-dimensional zero-mean variable. Let

$$\sum_{p=1}^{M} x_p = 0. \tag{14.6}$$

The covariance matrix of the variable can be estimated as follows:

$$C_x = \frac{1}{M} \sum_{p=1}^{M} x_p x_p^T. \tag{14.7}$$

PCA aims at making the covariance matrix C_x in Eq.(14.5) be diagonal. It leads to an eigenvalue problem:

$$\lambda u = C_x u, \tag{14.8}$$

where λ is the eigenvalues of C_x and u is the corresponding eigenvectors.

PCA linearly transforms x into a new one z:

$$z = u^T x. \tag{14.9}$$

The new components are called principal components. By using only the first several eigenvectors sorted in descending order of the eigenvalues, the number of principal components in z can be reduced. So PCA has the dimensional reduction characteristic and the principal components are uncorrelated.

14.2.3 Relation between Common Vector and PCA

The face image can be regarded as a summation of common vectors of the i_{th} face class which represents the common invariant properties of the i_{th} face class, while a difference vector represents the specific properties of a face image. However, what does the difference vector imply?

The eigenvalues of the i_{th} class covariance matrix C_x are nonnegative and they can be written in decreasing order: $z = u^T x$. Let $u_1, u_2, \cdots, u_n$ be the orthonormal eigenvectors corresponding to these eigenvalues. The first $(m-1)$ eigenvectors of the covariance matrix correspond to the nonzero eigenvalues.

Let $KerC_x$ be the space of all eigenvectors corresponding to the zero eigenvalues of C_x, and $B^\perp$ be the orthogonal complement of the difference subspace B. Further, since the space B is $(m-1)$ dimensional, the space $B^\perp$ is $(n-m+1)$ dimensional [29].

$$KerC_x = \{x \in R^{n \times 1} : C_x x = 0\},$$

$$B^\perp = \{x \in R^{n \times 1} : \;\; < x, b >= 0 \;\forall b \in B\}.$$

Here, we prove some theorems Refs. [22, 30].

$$KerC_x = B^\perp \tag{14.10}$$

Proof. (a) If any $x \in B^\perp$, then $x \perp B$.

From the following scalar multiplication must be zero:

$$< x_i - x_j, x >= 0, i, j = 1, 2, \cdots, m$$

or

$$< x_i - x_{\text{ave}}, x >= 0, i = 1, 2, \cdots, m.$$

Therefore

$$C_x x = \sum_{i=1}^{m} [(x_i - x_{\text{ave}})(x_i - x_{\text{ave}})^{\mathrm{T}} x] = \sum_{i=1}^{m} [(x_i - x_{\text{ave}}) < x_i - x_{\text{ave}}, x >] = 0 \tag{14.11}$$

or $x \in KerC_x$. Since the element $x \in B^{\perp}$ was arbitrary, then $B^{\perp} \subseteq KerC_x$ is true.

(b) If $a_i = x_i - x_{\text{ave}}$ for $i = 1, 2, \cdots, m$, and if the matrix A is defined as $A = (a_1, \cdots, a_m)$, then the covariance matrix C can be written as

$$X = AA^{\mathrm{T}}.$$

Let u_i be one of the eigenvectors corresponding to a zero eigenvalue. Thus

$$Cu_i = 0,$$

$$AA^{\mathrm{T}} u_i = 0.$$

By premultiplying the above equality with u_i^{T}, we get

$$u_i^{\mathrm{T}} AA^{\mathrm{T}} u_i = 0$$

or

$$||A^{\mathrm{T}} u_i|| = 0.$$

From here

$$a_1^{\mathrm{T}} u_i = a_2^{\mathrm{T}} u_i = \cdots = a_m^{\mathrm{T}} u_i = 0.$$

It can be easily shown that the space spanned by $a_1, a_2, \cdots, a_m$ has $(m = 1)$ dimension. On the other hand, the dimension of B is also $(m = 1)$ and since $a_i \in B$ then according to the above equation we obtain $u_i \perp B$ or equivalently $u_i \in B$.

Thus all the eigenvectors u_i corresponding to zero eigenvalues must belong to $B^{\perp}$. Since the set of the eigenvectors u_i corresponding to zero eigenvalues is the basis of $KerC_x$, then, $KerC_x \subseteq B^{\perp}$.

From the (a) and (b), one concludes that

$$KerC_x = B^{\perp}.$$

Because x_{common} is orthogonal to any vector in the difference subspace. Let $(x_i - x_j) \in B$, so

$$< x_i - x_j, x_{\text{common}} >= 0. \tag{14.12}$$

Proof. Since $< x_i - x_j > \in B$, then the following can be written:

$$x_i - x_j = < x_i - x_j, y_1 > y_1 + \cdots + < x_i - x_j, y_{m-1} > y_{m-1}.$$

The definition of the common vector states that

$$x_{\text{common}} = x_j - < x_j, y_1 > y_1 - \cdots - < x_j, y_{m-1} > y_{m-1}.$$

The scalar multiplication of the last two equality gives theorem 3.

In the following theorem, it will be shown that the common vector satisfies the eigenvalue and eigenvector equation for the covariance matrix for the zero eigenvalues.

$$C_x x_{\text{common}} = 0. \tag{14.13}$$

Proof. From Theorem 3, the following can be written:

$$< ((x_1 + x_2 + \cdots + x_m)/m - x_1),\ x_{\text{common}} >= 0.$$

Since the first term in the parenthesis is the average of all $x_i, i = 1, 2, \cdots, m$, the following can be written:

$$< (x_1 - x_{\text{ave}}),\ x_{\text{common}} >= 0.$$

Similarly $< (x_i - x_{\text{ave}}),\ x_{\text{common}} >= 0$ for $i = 2, 3, \cdots, m$. From the above equations, it follows that theorem 4 holds.

The first $(m - 1)$ eigenvectors of the covariance matrix correspond to the nonzero eigenvalues. Now it will be shown that B is the span of the eigenvectors corresponding to the nonzero eigenvalues.

$$B^\perp = \text{span}[u_m, u_{m+1}, \cdots, u_n] \tag{14.14}$$

or

$$B = \text{span}[u_1, u_2, \cdots, u_{m-1}]. \tag{14.15}$$

Here $[u_1, u_2, \cdots, u_{m-1}]$ are the eigenvectors corresponding to the nonzero eigenvalues of C_x.

Proof. For any $x \in \text{span}\ [u_1, u_2, \cdots, u_{m-1}]$,

$$x \perp B^\perp = \text{span}[u_1, u_2, \cdots, u_{m-1}].$$

Therefore, $x \in (B^\perp)^\perp$ or $x \in B$, since B is finite dimensional and therefore $B = (B^\perp)^\perp$. Thus, span $[u_1, u_2, \cdots, u_{m-1}] \subseteq B$.

Since the span $[u_1, u_2, \cdots, u_{m-1}]$ and B have the same $(m-1)$ dimensions; from this, we get theorem 5.

Based on the theorems $(2-5)$, any feature vector x can be written as

$$\begin{aligned} x =&< x, u_1 > u_1 + \cdots + < x, u_{m-1} > u_{m-1} + \\ & < x, u_m > u_m + \cdots + < x, u_n > u_n \end{aligned} \tag{14.16}$$

or

$$x = x^* + x^\perp, \tag{14.17}$$

where

$$x^* =< x, u_1 > u_1 + \cdots + < x, u_{m-1} > u_{m-1} x^* \in B \tag{14.18}$$

and

$$x^\perp = x_{\text{common}} =< x, u_m > u_m + \cdots + < x, u_n > u_n x^\perp \in B^\perp \tag{14.19}$$

So, for any feature vector x, the common vector x_{common} can also be written as

$$x_{common} = x - x^*. \tag{14.20}$$

As seen from the above derivations, the common vector can be determined by using the eigenvectors corresponding to the zero or nonzero eigenvalues of the covariance matrix C_x. But it should be noted that common vector does not include information in the directions corresponding to the nonzero eigenvalues. The common vector is orthogonal to all the eigenvectors that correspond to the nonzero eigenvalues of the covariance matrix. The common vectors is unique for its class and contains all the common invariant features of its own class.

For a i_{th} class sample, the difference vector is the difference between the sample and common vector of the i_{th} class, which is defined by

$$x_{diff}^i = x_{sample}^i - x_{common}^i. \tag{14.21}$$

As we know, the common vector that corresponds to the zero eigenvalues of the covariance matrix contains the invariant features of the class, while the difference vector x_{diff}^i reserves all the components of a feature vector that are along the eigenvectors corresponding to the nonzero eigenvalues of the covariance matrix. So x_{diff}^i contains more details of the within-individual variations. In our method, the difference vectors replacing the original face images are used for face recognition.

14.2.4 Kernel Principal Component Analysis (KPCA)

KPCA is an approach to generalize linear PCA into nonlinear case using the kernel method. The idea of KPCA is to firstly map the original input vectors x_p into a high-dimensional feature space $\Phi(x_p)$ and then to calculate the linear PCA in $\Phi(x_p)$. By mapping x_p into $\Phi(x_p)$ whose dimension is assumed to be larger than the number of training samples M,

$$\Phi : x \to \Phi(x) \tag{14.22}$$

Assume that the mapped observations are centered, i.e,

$$\sum_{p=1}^{M} \Phi(x_p) = 0 \tag{14.23}$$

Then the covariance matrix in the feature space is

$$\Sigma = \frac{1}{M} \sum_{p=1}^{M} \Phi(x_p) \Phi(x_p)^{\mathrm{T}} \tag{14.24}$$

The corresponding eigenvalue problem is

$$\lambda\mu = \Sigma\mu \tag{14.25}$$

Eq. (14.25) can be transformed to the eigenvalue problem of kernel matrix:

$$\delta_p \alpha_p = K\alpha_p, \quad p = 1, \cdots, M, \tag{14.26}$$

where K is the $M \times M$ kernel matrix, δ_p is one of the eigenvalue of K and α_p is the corresponding eigenvector of K. Finally, the principal components for x_p are calculated by [26]

$$Z_p = \sum_{p=1}^{M} \frac{\delta_p}{\sqrt{\delta_p}} K(x_p, x). \tag{14.27}$$

By the mapping Φ, we assume that an original nonlinear problem in the input space can be transformed to a linear problem in the high dimensional feature space. However, it is impossible to compute the matrix K directly without carrying out the mapping Φ. Fortunately, for certain mapping Φ and corresponding feature spaces, there is a highly effective trick for computing the dot product $(\Phi(x) \cdot \Phi(y))$ in feature spaces using kernel functions. If the mapping Φ satisfies the Mercer's condition, the dot product can be replaced by a kernel function as follows:

$$K(x, y) = (\Phi(x) \cdot \Phi(y)), \tag{14.28}$$

which allows us to compute the value of the dot product in the high dimensional feature space without having to carry out the mapping Φ explicitly. The following polynomial kernel is a commonly used kernel function:

$$K(x, y) = (x \cdot y)^l, \tag{14.29}$$

where l is any positive integer.

In general, Gaussian kernel gives good performance when the optimal parameter is used. However, the optimal parameter selection is difficult. It is reported that normalized polynomial kernel gives the comparable performance with Gaussian kernel [27]. In addition, parameter dependency of normalized polynomial kernel is low. Therefore, we use normalized polynomial kernel as the kernel function [27, 28]. Normalized polynomial kernel is defined as

$$K(x, y) = \frac{(1 + x^T y)^l}{\sqrt{(1 + x^T x)^l (1 + y^T y)^l}}. \tag{14.30}$$

In the following experiments, parameter l is set to 3 by preliminary experiment.

14.2.5 Classification Method

In our method, difference vector of each sample is taken as the input. The optimal feature vectors are obtained by KPCA procedure on difference vectors. For a training difference vector $A_j^i = x_j^i - x_{common}^i$, let

$$Z_{train}^i = \frac{1}{N} \sum_{j=0}^{N} Z_j^i = \frac{1}{N} \sum_{j=0}^{N} \sum_{p=1}^{d} \frac{\alpha_p}{\sqrt{\delta_p}} K(A_j^i, x_{diff}^p),$$

$$i = 1, 2, \cdots, C; j = 1, 2, \cdots, N; p = 1, 2, \cdots, d, \tag{14.31}$$

where, x_{diff}^p is a training sample difference vectors and Z_j^i is a feature vector obtained from the j_{th} difference vector of the i_{th} class processed by KPCA. And Z_{train}^i is the average feature vector of the i_{th} class.

For a test difference vector $A_{test}^i = x_{test} - x_{common}^i$, its feature vector is also gotten by:

$$Z_{test}^i = \sum_{p=1}^{d} \frac{\alpha_p}{\sqrt{\delta_p}} K(A_{test}^i, x_{diff}^p), i = 1, 2, \cdots, C; p = 1, 2, \cdots, d. \tag{14.32}$$

Suppose that there are C classes. So we get C average feature vectors Z_{train}^i of training samples and C feature vectors Z_{test}^i of a testing sample. The classifier is adopted as follow:

$$D^r(x) = \min\{D^i(x)\} = \min\{\|Z_{test}^i - Z_{train}^i\|^2\}, i = 1, 2, \cdots, C. \tag{14.33}$$

If there is a minimum dissimilarity among C dissimilarities $D^i(x)$, it indicates that the class with the minimum dissimilarity is the recognition result. Then the testing sample belongs to the r class.

In this section, a method for face recognition based on difference vector plus KPCA is proposed (we call the method KDVP here after). If PCA is taken as feature extraction technique, the method is defined as DVP. KDVP algorithm performs the following computation:

Step 1 Construct difference subspace B^i of class i ($i = 1, 2, \cdots, C$). Apply the Gram-Schmidt orthogonalization to obtain an orthonormal basis for B^i. Choose a sample from each class and project it onto the orthogonal complement of B^i to obtain common vector for each class. By these procedures, we get C common vectors and obtain all training and testing difference vectors.

Step 2 Construct difference vector kernel matrix and center the kernel matrix K. Computer the nonzero eigenvalues and corresponding eigenvectors of K and the coefficients in terms of the corresponding eigenvector and eigenvalue of kernel matrix. Use these eigenvectors to form the projection vectors which are used to obtain feature vectors.

Step 3 Implement the recognition work according to the classification criteria.

14.3 Two Methods Based on Common Vector

In this section, the discriminant common vector (DCV) and the common faces method (CVP) are, respectively, described.

14.3.1 Discriminant Common Vectors (DCV)

Before describing DCV, we first introduce the idea of common vectors from which DCV is originated. The idea of common vectors is originally introduced for isolated word recognition problems [21, 22] in the case where the number of samples in each class is less than or equal to the dimensionality of sample space. These approaches extract the common properties of classes in the training set by eliminating the differences of the samples in each class. A common vector for each individual class is obtained by removing all the features that are in the range space of the scatter matrix of its own class and then the obtained common vectors are used for recognition.

To solve the small sample size problem, the DCV method utilizes the idea of common vector. However, instead of using a given class's own scatter matrix, it uses the within-class scatter matrix of all the classes to obtain the common vectors. The major characteristic of the DCV method is that its projection matrix P_{DCV} resides in the null space of the within-class scatter matrix. Consequently, P_{DCV} concentrates the samples from the same class to a unique discriminant common vector and the Fisher's linear discriminant criterion achieves a maximum (infinite in fact). In [23], the authors gave two theoretically identical ways for implementing the DCV method, i.e., one by eigen-decomposition and the other by the difference subspace and the Gram-Schmidt orthogonalization procedure. Here, we introduce these two methods for the discriminant common vectors.

1) Discriminant common vectors based on eigen-decomposition:

Step 1   Compute the nonzero eigenvalues and corresponding eigenvectors of S_{w}. Here, we design $S_w = AA^{\mathrm{T}}$, where A is defined:

$$A = [x_1^\ell - \mu_1 \cdots x_N^\ell - \mu_1, x_1^\ell - \mu_2, \cdots, x_N^C - \mu_C], \tag{14.34}$$

where, μ is the mean of the class. Set $U = [u_1, \cdots, u_r]$, where r is the rank of S_{w}. U is an orthonormal matrix whose column vectors span the range space of the within-class matrix.

Step 2   Choose any sample from each class and project it to the null space of the within-class matrix through the following equation

$$x_{\mathrm{common}}^i = x_j^i - UU^{\mathrm{T}}x_j^i, j = 1, \cdots, N, i = 1, \cdots, C, \tag{14.35}$$

where x_{common}^i is a common vector of the i_{th} class and is independent of index j.

Step 3   Compute the eigenvectors W of S_{common}:

$$S_{\text{common}} = (x^i_{\text{common}} - \mu_{\text{common}})(x^i_{\text{common}} - \mu_{\text{common}})^{\text{T}}. \qquad (14.36)$$

There are at most $C - 1$ eigenvectors that that correspond to the nonzero eigenvalues.

Step 4   Use these eigenvectors to form the projection matrix $P_{\text{DCV}} = [w_1, \cdots, w_{C-1}]$ to extract discriminant common vectors.

$$\Omega^i_{\text{train}} = P^{\text{T}}_{\text{DCV}} x^i_m, m = 1, \cdots, N, \ i = 1, \cdots, C, \qquad (14.37)$$

$$\Omega_{\text{test}} = P^{\text{T}}_{\text{DCV}} x_{\text{test}}. \qquad (14.38)$$

2) Discriminant common vectors based on the difference subspace and the Gram-Schmidt:

Step 1   Calculate the range space of the within-class matrix, which is identical to the range space of the difference subspace H_{w}. Here, H_{w} is defined as

$$H_{\text{w}} = [b^1_1, \cdots, b^1_{N_1-1}, b^2_1, \cdots, b^c_{N_c-1}], \qquad (14.39)$$

where

$$b^i_j = x^i_j - x^i_{N_i}, i = 1, 2, \cdots, C, j = 1, 2, \cdots, N_i - 1$$

is the j_{th} difference vector of the i_{th} class. Apply the Gram-Schmidt orthogonalization procedure to H_{w}, and get

$$H_{\text{w}} = UV_1 \qquad (14.40)$$

then U is an orthonormal matrix whose column vectors span the range space of the within-class matrix.

Step 2   Choose any sample from each class (typically, the last sample of the i_{th} class $x^i_{N_i}$) and project it to the null space of the within-class matrix through the following equation

$$x^i_{\text{common}} = x^i_j - UU^{\text{T}} x^i_j = x^i_{N_i} - UU^{\text{T}} x^i_{N_i}, \qquad (14.41)$$

where x^i_{common} is a common vector of the i_{th} class and is independent of index j.

Step 3   Form the matrix B_{common}, where

$$B_{\text{common}} = [b^1_{\text{common}}, b^2_{\text{common}}, \cdots, b^{C-1}_{\text{common}}] \qquad (14.42)$$

and

$$b^i_{\text{common}} = x^i_{\text{common}} - x^C_{\text{common}}, i = 1, 2, \cdots, C - 1. \qquad (14.43)$$

Apply the Gram-Schmidt orthogonalization procedure to B_{common}, and get

$$B_{\text{common}} = P_{\text{DCV}} V_2 \qquad (14.44)$$

then P_{DCV} is the projection matrix calculated by DCV.

14.3.2 Common Faces Method (CVP)

Based on the common vector, He et al. [24] proposed the common faces method for face recognition. According to common vector theory (Eqs. (14.1–14.3)), the common vector x^i_{common} of i_{th} face class is obtained. It was proved that the common vector x^i_{common} is unique and independent of the randomly selected sample. Thus the common vector x^i_{common} can be used to represent the common invariant properties of i_{th} face class. The face image x^i_m in training set is then regarded as a summation of common vector x^i_{common} of i_{th} face class which represents the common invariant properties of i_{th} face class, and a difference vector $x^i_{m,\text{diff}}$ which represents the specific properties of the face image x^i_m due to the specific pose and expression variations in this face image as follows:

$$s^i_m = s^i_{\text{common}} + x^i_{m,\text{diff}}. \tag{14.45}$$

Obviously, the common vector of each face class is useful information for classification purpose, and the difference vector should be removed from face image to eliminate the within-individual variations which can be regarded as noise effects and may deteriorate classification performance.

For query face x, the vector $x^i_{\text{remaining}}$ called remaining vector of i_{th} face class is derived as

$$x^i_{\text{remaining}} = x - (< x, y^i_1 > y^i_1 + < x, y^i_2 > y^i_2 + \cdots + < x, y^i_{N_i-1} > y^i_{N_i-1}). \tag{14.46}$$

It was shown in Refs. [21, 22] that the remaining vector is usually closer to common vector of its own face class than to the common vector of other face class, and therefore in recognition stage, the query face can be assigned to class by finding the minimum distance between the remaining vector and the common vectors.

The angle between the remaining vector of the query face and the common vector of each face class is used as distance criteria as follows:

$$c = \min_i \left(\arccos \frac{< x^i_{\text{remaining}}, x^i_{\text{common}} >}{||x^i_{\text{remaining}}|| ||x^i_{\text{common}}||} \right), i = 1, 2, \cdots, C. \tag{14.47}$$

The common face method based on CVP is essentially classification method. The query face can be directly classified, and there is no need to extract the facial features in advance.

In addition, He et al. further extend the CVP to kernel CVP (KCVP) by using kernel method in which the sample x is nonlinearly mapped to a high-dimensional feature space. The implicit nonlinear mapping is unknown and the computation is done by computing the inner product in feature space with a kernel function $k(x, y) = \varphi(x)^{\mathrm{T}} \varphi(y)$ (called kernel trick) [18].

14.4 Experiments and Results

In Section 14.2 and 14.3, we have presented our theoretical arguments. And now, we carry out experiments on four well-known face datasets: ORL, YALE, FERET and AR in order to experimentally evaluate the recognition performance of these face recognition approaches. All experiments are carried out on a PC machine with P4 2.8 GHz CPU and 512 MB RAM memory under Matlab 7 platform.

14.4.1 Experiments on ORL Face Database

In this group of experiments, the Olivetti-Oracle Research Lab (ORL) face database is adopted (http://www.uk.research.att.com/facedatabase.html) and used to test the performance of face recognition algorithms under the condition of minor variation of scaling and rotation. ORL face database contains 112×92 sized 400 frontal faces: 10 tightly, cropped images of 40 individuals with variation in pose, illumination, facial expression (open/closed eyes, smiling/not smiling) and facial details (glasses/no glasses). Figure 14.1 shows 20 images of two persons in this dataset. All the images were taken against a dark homogeneous background with the subjects in an upright frontal position, and with tolerance for some tilting and rotation of up to about 20.

Fig. 14.1 Twenty images from two persons in the ORL face database.

In this experiment, we randomly select five face images from each face class. Figure 14.2 shows the original face images on ORL database (row 1) and common vector corresponding to their respective faces (row 2). The experiments verify that common vector possesses the common invariant properties of the class. It also indicates that common vector is independent of the randomly selected sample. Figure 14.3 shows that four classes face images and their corresponding difference vectors. It can be seen from Fig. 14.3 that the difference face images represent the specific properties of the respective face images due to the variants such as face appearance, pose and expression variants. In this section, we use difference face images for face recognition.

In the following experiments, the normalized polynomial kernel of KDVP

Fig. 14.2 Face images on ORL database (row 1). Common vectors corresponding to their respective faces (row 2).

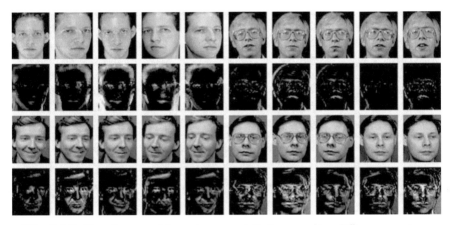

Fig. 14.3 Face images of four classes and their respective difference vectors.

is adopted and the parameter l is equal to 3. We employ two groups to test the recognition performances of our proposed approach. First, we experimentally evaluate our proposed approach with PCA and LDA approaches which employ the technique for dimensionality reduction. And then, we experimentally evaluate our proposed approach with CVP, KCVP, DCV, KDCV approaches which employ the common vector approach.

Experiments of the approaches with lower-dimension technique

In the experiments, we split the whole database into two parts. One part is used for training and the other part for testing. The samples are randomly selected in each face class for training and the remaining for testing. In the first experiment, we randomly select five images of each class as the training set and the rest as the testing set. This experiment is repeated 20 times. For PCA method the most significant eigenvectors were chosen such that corresponding eigenvalues contain 95 percent of the total energy [10]. The average recognition rates of different algorithms are, respectively, summarized in Table 14.1. It also gives the comparisons of four approaches on the recognition rate, the corresponding dimension of feature vector and the running time. We see from Table 14.1 that the recognition rate of KDVP is better than that of PCA, LDA, and DVP, but not in terms of the running time.

Table 14.1 Performance comparison of four methods on ORL database

Method	Average recognition rate (%)	Dimension	Running time (ms)
PCA	93.32	80	21.78
LDA	94.55	39	10.05
DVP	95.55	100	26.84
KDVP	96.05	80	22.32

Now, let us design a series of experiments to compare PCA, LDA, DVP, and KDVP methods under conditions where the sample size is varied. Here, four tests are performed with a varying number of training samples. We randomly select some images from each class to construct the training set and the remaining images as the testing set. To ensure the sufficient training samples, at least 4 samples are used. All of experiments are repeated 20 times. Table 14.2 shows the comparison of the average error rates of different approaches on ORL database for varying number of training samples. The error rate of the proposed approach is the minimum in all tests. As the number of training samples increases, the error rate decreases obviously.

Table 14.2 Average error rate (%) comparison of different approaches on ORL database

Method/Number of Training samples	4	5	6	7
PCA	8.05	5.68	4.78	3.25
LDA	7.23	5.45	4.03	3.04
DVP	7.34	4.45	3.97	2.21
KDVP	6.55	3.95	3.48	2.08

Experiments of the approaches with common vector

We design a series of experiments to compare the performance of the DVP, KDVP, CVP, KCVP, DCV, and KDCV methods under conditions where the training sample size is varied. We randomly select $3, 5, 7$ samples from each class for the training, and the remaining samples of each class for the testing. The experiments are also repeated 20 times. The averaging recognition rates of these experiments are listed in Table 14.3. From the table, we also see that the recognition rate of KDVP is comparable to that of CVP, KCVP, DCV,

Table 14.3 Average recognition rate (%) comparison of six methods on ORL database

Method/Number of Training samples	3	5	7
CVP[24]	89.21	93.6	96.67
KCVP[24]	90.39	95.50	97.75
DCV[22]	90.70	96.11	97.77
KDCV[22]	90.54	96.32	97.65
DVP	88.45	95.55	97.79
KDVP	89.15	96.05	97.92

KDCV, and DVP. Actually, from the experimental results, the recognition performances of DCV and KDVP are comparable and similar. In this section, we focus on the verification which is the difference vector good for face recognition, while it is undeniable that DVP is also indeed a good method. Although different feature extraction, the good performances of DVP and KDVP are really obtained.

14.4.2 Experiments on YALE Database

YALE database (http://cvc.yale.edu/projects/yalefaces/yalefaces.html) consists of images from 15 different people, using 11 images from each person, for a total of 165 images. The images contain variations with the following facial expressions or configurations: center-light, with glasses, happy, left-light, without glasses, normal, right-light, sad, sleepy, surprised, and wink. We preprocessed these images by aligning and scaling them so that the distances between the eyes were the same for all images and also ensuring that the eyes occurred in the same coordinates of the image. The resulting image was then cropped. The final image size was 100×80. We considered this database in order to evaluate the performance of methods under the condition when facial expression and lighting conditions are changed. Some images from YALE face database are shown in Fig 14.4.

Fig. 14.4 First low: Some original images on YALE database. Second low: normalized images.

As previously mentioned, we also repeat the experiment 20 times in this experiments. Table 14.4 shows the average recognition rate obtained by different methods with varying number of training samples and reveals that KDVP is comparable to PCA and DVP methods. The recognition rate of PCA, DVP, and KDVP with varying dimension of feature vectors for five training samples is given in Fig. 14.5. It can be easily ascertained from Fig. 14.5 that KDVP, with reduced feature vector, obtains the same or even good recognition rate when compared with other methods.

Table 14.4 Average recognition rate (%) comparison of different approaches on YALE database

Method/Number of Training samples	2	4	6	8
PCA	76.67	89.67	94.00	99.11
DVP	80.22	92.33	94.56	99.22
KDVP	82.19	93.12	95.13	99.44

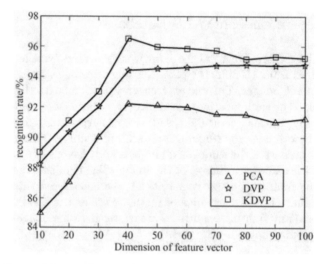

Fig. 14.5 Recognition performance of different approaches with varying dimension of feature vectors on YALE database.

14.4.3 Experiments on FERET Database

We analyze the performance of the KDVP method using FERET database. Comparative results are also given for other methods such as KPCA, PCA and LDA. FERET (The Facial Recognition Technology) database is a standard test set for face recognition technologies [31]. 450 frontal face images corresponding to 150 subjects are extracted from the database for the experiments. The face images are acquired under varying illumination conditions and facial expressions. The following procedures are applied to normalize the face images prior to the experiments:

1) The centers of the eyes of each image are manually marked, each image is then rotated and scaled to align the centers of the eyes.

2) Each face image is cropped to the size of 128 × 128 to extract facial region.

Figure 14.6 shows some images from FERET database. The images are captured at different photo sessions so that they display different illumination and facial expressions. To test the algorithms, two images of each subject are randomly chosen for the training, while the remaining one is used for the testing. The experiments are repeated 3 times. The performance of the proposed method is compared with PCA, KPCA and LDA. The average results of the experiments are shown in Table 14.5. It can be observed that the proposed method has improved the performance of PCA and LDA significantly.

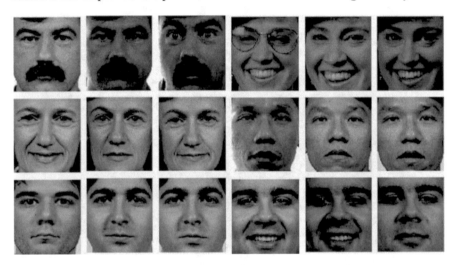

Fig. 14.6 Normalized face images on FERET face database.

Table 14.5 Recognition rate (%) comparison of different approaches on FERET database

Method	PCA	KPCA	LDA	DVP	KDVP
Recognition rate (%)	42.3	48.6	69.7	70.6	71.9

14.4.4 Experiments on AR Database

AR database was created by Aleix Martinez and Robert Benavente in the Computer Vision Center (CVC) at the U.A.B. It contains over 4,000 color images corresponding to 126 people's faces (70 men and 56 women). Images feature frontal view faces with different facial expressions, illumination conditions, and occlusions (sun glasses and scarf). No restrictions on wear (clothes, glasses, etc.), make-up, hair style, etc. were imposed to participants. We select 1,300 images of 50 males and 50 females, and each person has 13 images to test our method. The original images are of 768 by 576 pixels, and then normalized to 128×128. Figure 14.7 shows 13 images from one person in AR dataset and the normalized images. We randomly select $5, 6, 7, 8$ samples from each class for the training, and the rest for the testing. The processes are repeated 5 times, and the results are shown in Table 14.6.

Fig. 14.7 The 13 images from one person in AR face database and their respective normalized images.

Table 14.6 Average recognition rate (%) comparison of different approaches on AR database

Method/number of the training sample	PCA	LDA	CVP	DCV
5	65.57	75.14	84.62	85.47
6	73.39	78.23	87.53	89.46
7	78.38	85.48	90.75	92.56
8	80.21	89.57	92.49	93.78

Table 14.6 reveals that common vector based face recognition methods are comparable to PCA and LDA methods in terms of recognition rate and the number of the training samples. From the table, it can be easily ascertained, with increased training number, the proposed method obtains good recognition rate when compared with other methods.

14.4.5 Experiments on JAFFE Database

The image database we use in this experiment is the JAFFE (Japanese Female Facial Expression) database. The database contains ten Japanese females. There are seven different facial expressions, such as neutral, happy, angry, disgust, fear, sad and surprise. Each female has two to four examples for each expression. Totally, there are 213 grayscale facial expression images in this database. Each image is of size 256×256. Figure 14.8 shows two expressers comprising seven different facial expressions from the JAFFE database.

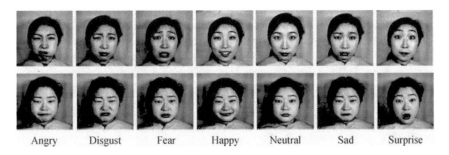

| Angry | Disgust | Fear | Happy | Neutral | Sad | Surprise |

Fig. 14.8 Samples of two expressers containing seven different facial expressions.

We apply cross-validation strategy and leave-one-out strategy to perform comparisons with other existing methods. For the cross validation strategy, we randomly divide the database into ten segments in terms of different facial expressions. Each time, we train nine out of the ten segments and test the remaining segment. We perform the same procedure of training and testing repeatedly for 20 times. At last, we average all the 20 recognition rates to obtain the final performance of the method. For the leave-one-out strategy, each time we only test one image in each class, and the remaining images are used for training.

Table 14.7 shows the performance comparisons in JAFFE database. Comparative results are also given for other methods such as PCA, independent component analysis (ICA), LDA and two dimensional LDA (2DLDA). From Table 14.7, we observe that no matter which strategy is used, the common vector based method outperforms the others.

Table 14.7   The performance comparisons (%) in JAFFE database

Feature Extraction Methods	Recognition Rates	
	Cross Validation	Leave-One-Out
PCA	83.21	85.43
ICA	85.17	86.28
LDA	86.36	87.67
2DLDA	88.47	88.89
CVP	88.35	89.14
DCV	89.56	90.22

14.5 Conclusion and Future Research

The common vector formulated in this section is in accordance with previous derivations for the insufficient data. Therefore, the common vector is in the direction of the eigenvectors that belong to the smallest (including zero) eigenvalues of the within-class scatter matrix of a class. Difference vector is gotten by the difference between the original image and common vector of its class. Based on this idea, an efficient face recognition approach based on difference vector plus KPCA is studied. The recognition result is obtained by finding the minimum distance between the difference feature vectors of the test image and the training image. Paper reviewed the discriminant common vector approach and the common faces approach and a comparative study among them in face recognition was carried out. Experiments on ORL, YALE, FERET, AR and JAFFE database indicate that the common vector based face recognition is encouraging.

For future research, we would be glad to continue to refine our current method, and apply it to more specifically to personal identification and verification for the purpose of national and social security. Also we believe our method paves solid ground for further study and research for facial emotional image analysis, classification, recognition and visualization, and can be tested on more facial image databases.

References

[1] Zhao W, Chellappa R, Krishnaswamy A (1998) Discriminant Analysis of Principal Components for Face Recognition. In: Proceedings of 3rd IEEE International Conference of Automatic Face and Gesture Recognition, pp 336–341
[2] Chellappa R, Wilson C L, Sirohey S (1995) Human and Machine Recognition of Faces: A Survey. Proceedings of IEEE, 83: 705–740
[3] Wen Y, Zhou Z et al (2010) An Improved Locally Linear Embedding for Sparse Data Sets. IEEE 17th International Conference on Image Processing, pp 26–29
[4] Wen Y, Shi P F (2008) An Approach to Numeral Recognition based on Improved LDA and Bhattacharyya Distance. The Third International Symposium on Communications, Control and Signal Processing, pp 309–311
[5] Chen L F, Liao H et al (2000) A New LDA-based Face Recognition System Which Can Solve the Small Sample Size Problem. Pattern Recognition, 33: 1713–1726
[6] Yang J, Zhang D, Frangi A F et al (2004) Two-dimensional PCA: a new approach to appearance based face representation and recognition. IEEE Trans on Pattern Analysis and Machine Intelligence, 26(1): 131–137
[7] Turk M (2001) A Random Walk Through Eigenspace. IEICE Transaction on Information Systems, E84-D, 12: 1586–1695
[8] Turk M, Pentland A (1991) Eigenfaces for Recognition. Journal of Cognitive Neuroscience, 3(1): 71–86
[9] Belhumeur P N, Hespanha J P, Kriegman D J (1997) Eigenfaces vs Fisherfaces: Recognition Using Class Specific Linear Projection. IEEE Trans on Pattern Analysis

and Machine Intelligence, 19(7): 711 – 720
[10] Swets D L, Weng J (1996) Using Discriminant Eigenfeatures for Image Retrieval. IEEE Transaction on Pattern Analysis and Machine Intelligence, 18(8): 831 – 836
[11] Fukunaga K (1990) Introduction to Statistical Pattern Recognition, 2nd Edn. pp 31 – 34, 39 – 40, 220 – 221. Academic Press, New York
[12] Chatterjee C, Kang Z, Roychowdhury V (2000) Algorithms for Accelerated Convergence of Adaptive PCA. IEEE Trans on Neural Networks, 11(2): 338 – 355
[13] Draper B A, Baek K, Bartlett M S et al (2003) Recognizing faces with PCA and ICA. Computer Vision and Image Understanding, 91: 115 – 137
[14] Partridge M, Jabri M (2000) Robust Principal Component Analysis. Proceedings of the 2000 IEEE Signal Processing Society Workshop, vol 1: 289 – 298
[15] Shan Y, Zheng B et al (2000) Robust Recursive Least Squares Learning Algorithm for Principal Component Analysis. IEEE Transaction on Neural Networks, 11(1): 215 – 221
[16] Wen Y, Shi P F (2007) Image PCA: A New Approach For face Recognition. 2007 IEEE International Conference on Acoustics, Speech, and Signal Processing, pp 1241 – 1244
[17] Karhunen J, Joutsensalo J (1995) Generalizations of Principal Component Analysis, Optimization Problems, and Neural Networks. Neural Neworks, 8 (4): 549 – 562
[18] Scholkopf B, Smola A, Muller K R (1998) Nonlinear Component Analysis as a Kernel Eigenvalue Problem. Neural Computer, 10: 1299 – 1319
[19] Wen Y, Lu Y, Shi P F (2007) Handwritten Bangla Numeral Recognition System and its Applicaiton to Postal Automation. Pattern Recognition, 40: 99 – 107
[20] Wen Y, Shi P F (2008) A Novel Classifier for Handwritten Numeral Recognition. 2008 IEEE International Conference on Acoustics, Speech, and Signal Processing, pp 1321 – 1324
[21] Gulmezoglu M B, Dzhafarov V, Keskin M et al (1999) A Novel Approach to Isolated Word Recognition. IEEE Trans on Speech and Audio Processing, 7(6): 620 – 628
[22] Gulmezoglu M B, Dzhafarov V, Barkana A (2001) The Common Vector Approach and Its Relation to Principal Component Analysis. IEEE Trans on Speech and Audio Processing, 9(6)
[23] Cevikalp H, Wilkes M, Barkana A (2005) Discriminative Common Vectors for Face Recognition. IEEE Transaction on Pattern Analysis and Machine Intelligence, 27(1): 4 – 13
[24] He Y, Zhao L, Zou C (2005) Kernel Discriminative Common Vectors for Face Recognition. Proceedings of Fourth International Conference on Machine Learning and Cybernetics, Guangzhou, pp 18 – 21
[25] He Y, Zhao L, Zou C (2006) Face Recognition Using Common Faces Method. Pattern Recognition, 39: 2218 – 2222
[26] Cao L J, Chua K S, Chong W K et al (2003) A Comparison of PCA, KPCA, and ICA for Dimensionality Reduction in Support Vector Machine. Neuro computing, 55: 321 – 336
[27] Debnath R, Takahashi H (2004) Kernel Selection for the Support Vector Machine. IEICE Transaction on Information and System, E87-D(12): 2903 – 2904
[28] Taylor J S, Cristianini N (2004) Kernel Methods for Pattern Analysis. Cambridge University Press, London
[29] Hoffman K, Kunze R (1961) Linear Algebra. Prentice Hall, New Jersey
[30] Edwards C H, Penney D E (1988) Elementary Linear Algebra. Prentice Hall, NJ
[31] Phillips P, Moon H et al (2000) The Feret Evaluation Methodology for Face Recognition Algorithms. IEEE Transaction on Pattern Analysis and Machine Intelligence, 22(10): 1090 – 1104
[32] Zhang D, You X G et al (2009) Facial Biometrics Using Non-tensor Product Wavelet and 2D Discriminant Technique. International Journal of Pattern Recognition and Artificial Intelligence, 23(3): 521 – 543
[33] Lee S W, P Wang P S et al (2008) Noniterative 3D Face Reconstruction Based On Photometric Stereo. International Journal of Pattern Recognition and Artificial Intelligence, 22(3): 389 – 410
[34] Shih F Y, Chuang C F, Wang P S P (2008) Performance Comparisons Of Facial Expression Recognition In Jaffe Database. International Journal of Pattern Recognition and Artificial Intelligence, 22(3): 445 – 459
[35] Shih F, Cheng S X, Chuang C F, Wang P S P (2008) Extracting Faces And Facial Features From Color Images. International Journal of Pattern Recognition and Artificial Intelligence, 22(3): 515 – 534

[36]   Luo Y, Gavrilova M L, Wang P S P (2008) Facial Metamorphosis Using Geometrical Methods for Biometric Applications. International Journal of Pattern Recognition and Artificial Intelligence, 22(3): 555–584

[37]   You X, Chen Q, Wang P S P et al (2007) Nontensor-Product-Wavelet-Based Facial Feature Representation. In: Image Pattern Recognition-Synthesis and Analysis in Biometrics, WSP, pp 207–224

[38]   Zhang L C, Chen J W, Yue Lu et al (2008) Face Recognition Using Scale Invariant Feature Transform and Support Vector Machine. Int. Conference for Young Computer Scientists, China, pp 1766–1770

[39]   You X G, Zhang D, Chen Q et al (2006) Face Representation By Using Non-tensor Product Wavelets. Proc. ICPR2006, pp 503–506, Hong Kong, China

[40]   Wang P S P (2007) Some Concerns on the Measurement for Biometric Analysis and Applications. In: Pattern Recognition: Synthesis and Analysis in Biometrics, WSP, pp 321–337

[41]   Soundar K R, Murugesan K (2010) Preserving Global and Local Features – A Combined Approach for Recognizing Face Images, 24: 39–53

[42]   Zeng Y et al (2010) 3D Modeling of Faces from Near Infrared Images using Statistical Learning. International Journal of Pattern Recognition and Artificial Intelligence, 24: 55–71

[43]   Jing M Q, Chen L H (2010) A Novel Method for Horizontal Eye Line Detection under Various Environments. International Journal of Pattern Recognition and Artificial Intelligence, 24: 475–498

[44]   Grgic M, Shan S, Lukac R et al (2009) Special issues on Face Recognition. International Journal of Pattern Recognition and Artificial Intelligence, WSP, 23(3): 355–358

15 A Look at Eye Detection for Unconstrained Environments

Brian C. Heflin[1], Walter J. Scheirer[1], Anderson Rocha[2] and Terrance E. Boult[1]

Abstract   Eye detection is a well studied problem for the *constrained face recognition problem*, where we find controlled distances, lighting, and limited pose variation. A far more difficult scenario for eye detection is the *unconstrained face recognition problem*, where we do not have any control over the environment or the subject. In this chapter, we take a look at two different approaches for eye detection under difficult acquisition circumstances, including low-light, distance, pose variation, and blur. A machine learning approach and several correlation filter approaches, including our own adaptive variant, are compared. We present experimental results for a variety of controlled data sets (derived from FERET and CMU PIE) that have been re-imaged under the difficult conditions of interest with an EMCCD based acquisition system, as well as on a realistic surveillance oriented set (SCface). The results of our experiments show that our detection approaches are extremely accurate under all tested conditions, and significantly improve detection accuracy compared to a leading commercial detector. This unique evaluation brings us one step closer to a better solution for the unconstrained face recognition problem.

15.1 Introduction

Eye detection is a necessary processing step for many face recognition algorithms. For some of these algorithms, the eye coordinates are required for proper geometric normalization before recognition. For others, the eyes serve as reference points to locate other significant features on the face, such as the nose and mouth. The eyes, containing significant discriminative information, can even be used by themselves as features for recognition. Eye detection is a well studied problem for the *constrained face recognition problem*, where we find controlled distances, lighting, and limited pose variation. A far more difficult scenario for eye detection is the *unconstrained face recognition problem*, where we do not have any control over the environment or the subject.

---

1   Vision and Security Technology Lab, University of Colorado at Colorado Springs, Colorado 80918, USA. E-mail: bheflin@uccs.edu.
2   Institute of Computing, University of Campinas (Unicamp), Campinas, Brazil. E-mail: anderson.rocha@ic.unicamp.br.

In this chapter, we will take a look at eye detection for the latter, which encompasses problems of flexible authentication, surveillance, and intelligence collection.

A multitude of problems affect the acquisition of face imagery in unconstrained environments, with major problems related to lighting, distance, motion and pose. Existing work on lighting [1, 2] has focused on algorithmic issues (specifically, normalization), and not the direct impact of acquisition. Under difficult acquisition circumstances, normalization is not enough to produce the best possible recognition results — considerations must be made for image intensification, thermal imagery and electron multiplication. Long distances between the subject and acquisition system present a host of problems, including high f-numbers from very long focal lengths, which significantly reduces the amount of light reaching the sensor, and a smaller amount of pixels on the faces, as a function of distance and sensor resolution. Further, the interplay between motion blur and optics exasperates the lighting problems, as we require faster shutter speeds to compensate for the subjects movement during exposure, which again limits the amount of light reaching the sensor. In general, we'll have to face some level of motion blur in order for the sensor to collect enough light. Pose variation, as is well known, impacts the nature of facial features required for recognition, inducing partial occlusion and orientation variation, which might differ significantly from what a feature detector expects.

Both lighting and distance should influence sensor choice, where nonstandard technologies can mitigate some of the problem discussed above. For instance, EMCCD sensors have emerged as an attractive solution for low-light surveillance (where low-light is both conditional, and induced by long-range optics), because they preserve a great bit of detail on the face and can use traditional imagery for the gallery (as opposed to midwave-IR sensors). This makes them very attractive for biometric application as well. However, the noise induced by the cooling of the sensor also presents new challenges for facial feature detection and recognition. In this chapter, for the reasons cited above, we use the EMCCD to acquire our difficult imagery under a variety of different conditions, and apply several different eye detectors on the acquired images.

In Section 15.2 we take a brief survey of the existing literature related to difficult detection and recognition problems, as well as the pattern recognition works relevant to the detection techniques discussed in this chapter. In Section 15.3 we introduce a machine learning based approach to feature detection for difficult scenarios, with background on the learning and feature approach used. In Section 15.4 we introduce the correlation filter approach for feature detection, including a new adaptive variant. Our experimental protocol is defined in Section 15.5, followed by a thorough series of experiments to evaluate the detection approaches. Finally, in Section 15.6, we make some concluding remarks on our examination of algorithms for difficult feature detection.

15.2 Related Work

On the algorithm front, we find only a few references directly related to difficult facial feature detection and recognition. Super-resolution and deblurring were considered in Ref. [3] as techniques to enhance images degraded by long distance acquisition (50 – 300 m). That work goes further to show recognition performance improvement for images processed with those techniques compared to the original images. The test data set for outdoor conditions is taken as sequential video under daylight conditions; the super-resolution process considers direct sequences of detected faces from the captured frames. The problem with this approach is that under truly difficult conditions, as opposed to the very controlled settings of Ref. [3] (full frontal imagery, with a constant inter-ocular distance), it is likely that a collection of detected faces in a direct temporal sequence will not be possible, thus reducing the potential of such algorithms. Real-time techniques to recover facial images degraded by motion and atmospheric blur were explored in Ref. [4]. The experiments of Ref. [4] with standard data sets and live data captured at 100 m showed how even moderate amounts of motion and atmospheric blur can effectively cripple a facial recognition system. The work of Refs. [5, 4] is more along the lines of what is explored in this paper, including a thorough discussion of the underlying issues that impact algorithm design, as well as an explanation of how to perform realistic controlled experiments under difficult conditions, and algorithmic issues such as predicting when a recognition algorithm is failing in order to enhance recognition performance.

In the more general pattern recognition literature, we do find several learning techniques applied to standard data sets for eye detection. Many different learning techniques have been shown to be quite effective for the eye detection problem. The work of Ref. [6] is most closely related to the learning technique presented in this work in a feature sense, with PCA features derived from the eyes used as input to a neural network learning system. Using a data set of 240 images of 40 different full frontal faces, this technique is shown to be as accurate as several other popular eye detection algorithms. Ref. [7] uses color information and wavelet features together with a new efficient Support Vector Machine (eSVM) to locate eyes. The eSVM, based on the idea of minimizing the maximum margin of misclassified samples, is defined on fewer support vectors than the standard SVM, which can achieve faster detection speed and comparable or ever higher detection accuracy [7]. The method of Ref. [7] consists of two steps. In the first step selects possible eye candidate regions using a color distribution analysis in YcbCr color space. The second validation step consists of applying 2D Haar wavelets to the image for multi-scale image representations followed by PCA for dimensionality reduction and using the eSVM to detect the center of the eye. Ref. [8] uses normalized eye images projected onto weighted eigenspace terrain features as features for an SVM learning system. Ref. [9] uses a recursive non-parametric discriminant feature as input to an AdaBoost learning system.

For recognition, a very large volume of work exists for correlation,but we find some important work on feature detection as well. Correlation filters [10, 11] are a family of approaches that are tolerant to variations in pose and expression, making them quite attractive for detection and recognition problems. Excellent face recognition results have been reported for the PIE data set [12] and the FRGC data set [13]. For the specific problem of eye detection, [14] first demonstrated the feasibility of correlation filters, while [15] introduced a more sophisticated class of filters that are more insensitive to over-fitting during training, more flexible towards training data selection, and more robust to structured backgrounds. All of these approaches have been tested on standard well-known data sets, and not the more difficult imagery we consider in this chapter. We discuss correlation in detail in Section 15.4.

Of course we should reduce the impact of difficult conditions using better sensors and optics, which is why we choose to use EMCCD sensors to allow faster shutter speeds. For the optics, one possibility gaining attention is the use of advanced Adaptive Optics (AO) models [16], which have proved effective for astronomy, though most do not apply to biometric systems. Astronomy has natural and easily added artificial "point sources" for guiding the AO process. Secondly, astronomical imaging is vertical, which changes the type and spatial character of distortions. More significantly, they have near point sources for guides, allowing for specialized algorithms for estimation of the distortions. Horizontal terrestrial atmospheric turbulence is much larger and spatially more complex making it much more difficult to address. To date, no published papers discuss an effective AO system for outdoor biometrics. While companies such as AOptix[1] have made interesting claims, public demonstrations to date have been stationary targets indoor at less than 20 m, where there is no atmospherics and minimal motion blur .

A critical limiting question for adaptive optics is the assumption of wave-front distortion and measurement. For visible and NIR light, the isoplanatic angle is about 2 arc seconds (0.00027 degrees or motion of about 0.08 mm at 50 m). Motion outside the isoplanatic angle violates the wave-front model needed for AO correction [17]. An AO system may be able to compensate for micro-motion on stationary targets, where a wave-front isoplanatic compensation AO correction approach would be a first-order isoplanatic approximation to small motions, but it's unclear how it could apply to walking motions that are not well modeled as a wave-front error.

15.3 Machine Learning Approach

The core concept of our machine learning approach for detection is to use a sliding window search for the object feature, using image features extracted from the window and applying a classifier to those features. For different

---

3   http://www.aoptix.com/

difficult environments we can learn different classifiers. We first review the learning and image features used.

15.3.1 Learning Techniques

Supervised learning is a machine learning approach that aims to estimate a classification function f from a *training data set*. Such a training data set consists of pairs of input values X and its desired outcomes Y [18]. Observed values in X are denoted by x_i, i.e., x_i is the ith observation in X. Often, x is as simple as a sequence of numbers that represent some observed features. The number of variables or features in each $x \in X$ is p. Therefore, X is formed by N input examples (vectors) and each input example is composed by p features or variables.

The commonest output of the function f is a label (class indicator) of the input object under analysis. The learning task is to predict the function outcome of any valid input object after having seen a sufficient number of training examples.

In the literature, there are many different approaches for supervised learning such as Linear Discriminant Analysis, Support Vector Machines (SVMs), Classification Trees, and Neural Networks. We focus on an SVM-based solution.

15.3.2 PCA Features

Principle Components Analysis [19], that battle-worn method of statistics, is well suited to the dimensionality reduction of image data. Mathematically defined, PCA is an orthogonal linear transformation, which after transforming data leaves the greatest variance by any projection of data on the first coordinate (the *principal component*), and each subsequent level of variance on the following coordinates. For a data matrix X^T, after mean subtraction, the PCA transformation is given as

$$Y^T = X^T W = V \Sigma, \tag{15.1}$$

where $V \Sigma W^T$ is the singular value decomposition of X^T. In essence, for feature detection, PCA provides a series of coefficients that become a feature vector for machine learning. Varying numbers of coefficients can be retained, depending on the energy level that provides the best detection resolution.

15.3.3 PCA + Learning Algorithm

A learning based feature detection approach allows us to learn over features gathered in the appropriate scenarios in which a recognition system will operate, including illumination, distance, pose, and weather (Fig. 15.1). By projecting a set of candidate pixels against a pre-computed PCA subspace for a particular condition, and classifying the resulting coefficients using a machine learning system yields an extremely powerful detection approach. The basic algorithm, depicted in Fig. 15.2, begins with the results of the Viola-Jones face detector [20], implemented to return a face region that is symmetrical. With the assumption of symmetry, the face can be separated into feature regions, which will be scanned by a sliding window of a pre-defined size $w \times h$. Each positive marginal distance returned by an SVM classifier is compared against a saved maximum, with new maximums and corresponding x and y coordinates being saved. When all valid window positions are exhausted, the maximum marginal value indicates the candidate feature coordinate with the highest confidence. While for this work we are only interested in the eyes, we do note that the generic nature of the proposed approach allows for the detection of any defined feature.

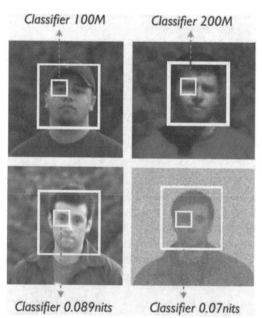

Fig. 15.1 The approach: build classifiers for different conditions, such as distance and illumination. While general enough for any feature that can be represented by a window of pixels, the eye is shown here, and in subsequent figures, as an example.

The speed of such a sliding window approach is of concern. If we assume the window slides one pixel at a time, a 50×45 window (a window size suitable for eye detection on faces 160 pixels across) in an 80×60 potential feature region, 496 locations must be scanned. One approach to enhancing speed is through the use of multiple resolutions of feature regions. Figure 15.3 depicts this, with the full feature region scaled down by 1/4 as the lowest resolution region considered by the detector. The best positive window (if any) then determines a point to center around for the second (1/2 resolution)

Fig. 15.2 The basic algorithm is straightforward. First, a feature region is isolated (using pre-set coordinates) from the face region returned by the face detector (separate from the feature detection). Next, using a pre-defined sliding window over the feature region, candidate pixels are collected. PCA feature generation is then performed using the pixels in the window. Finally, the PCA coefficients are treated as feature vectors for an SVM learning system, which produces the positive or negative detection result.

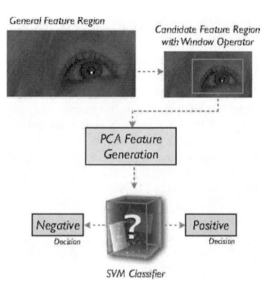

scale's search, with a more limited bounding box defined around this point for the search. The process then repeats again for the highest resolution. Presuming a strong classifier, the positive windows will cluster tightly around the correct eye region. A further enhancement to the algorithm is to determine the best positive window for the first row of the feature region where positive detections have occurred. From the x coordinate of this best window x_{best}, the scanning procedure can reduce the search space to $(x_{best} + c) - (x_{best} - c) + 1$ windows per row of the feature region, where c is some pixel constant set to ensure flexibility for the search region. c pixels will be searched on both the left and right sides of x_{best}. This approach does come with a drawback-the space requirement for PCA subspaces and SVM classifiers increases by number of features × number of scales.

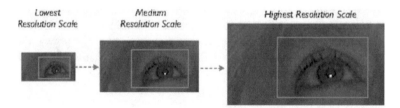

Fig. 15.3 The speed of sliding window approaches is always a concern. To increase computational performance, a multi-resolution approach can be used to reduce the area that must be scanned. While reducing time, this does increase the space requirement for PCA subspaces and SVM classifiers (number of features × number of scales).

15.4 Correlation Filter Approach

Correlation filters as considered in this work consist of Minimum Average Correlation Energy (MACE) filters [10], Unconstrained Minimum Average Correlation Energy (UMACE) filters [11], and our own Adaptive Average Correlation Energy (AACE) filters. These approaches produce a single correlation filter for a set of training images. For feature detection, these techniques produce a sharp correlation peak after filtering in the positive case, from which the correct coordinates for the feature can be derived (an example of this is shown in Fig. 15.6). The variations among MACE, UMACE, and AACE are described below.

15.4.1 MACE Filter for Feature Detection

Synthesis of the Minimum Average Correlation Energy (MACE) filter began with cropping out 40×32 regions from our training data with the eye centered at coordinates (21,19). Figure 15.4 (a) shows an example cropped eye from one of our training images.

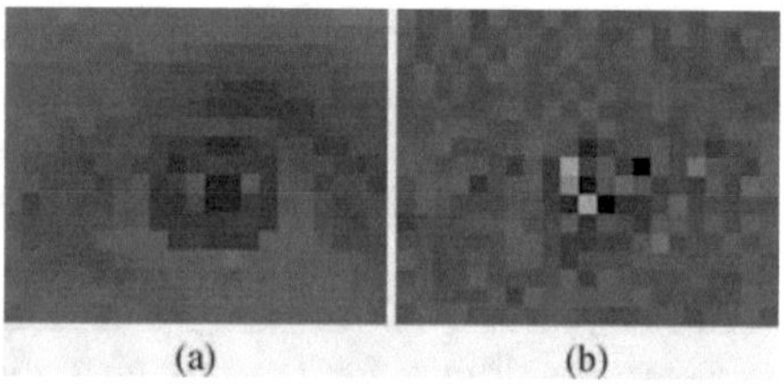

Fig. 15.4   (a) Example cropped eye for MACE filter training. (b) Impulse response from MACE filter.

(a)   (b)

The MACE filter specifies a single correlation value per input image, which is the value that should be returned when the filter is centered upon the training image. Unfortunately when more than $4-6$ training images are used this leads to over fitting of the training data and decreases accuracy in eye detection. After cropping the eye region, it is transformed to the frequency domain using a 2D Fourier transform. Next the average of the power spectrum of all of the training images is obtained. Then MACE filter is synthesized using the following formula:

$$h = D^{-1}X(X'D^{-1}X)^{-1}u, \qquad (15.2)$$

where D is the average power spectrum of the N training images, X is a matrix containing the 2D Fourier transform of the N training images, and u is the desired filter output. Separate MACE filters were designed for both the left and right eyes. The impulse response of the MACE filter for experiments shown in Figs. 15.15 and 15.16 is shown in Fig. 15.4 (b).

To incorporate a motion blur estimate into the MACE filter, an additional convolution operation must be executed prior to eye detection which can be performed on at run time on a per image basis. Finally, after the normalized cross correlation operation is performed the global maximum or peak location is chosen as the detected eye location in the original image with the appropriate offsets.

15.4.2 UMACE and AACE Filters for Feature Detection

Synthesis of our Adaptive Average Correlation Energy (AACE) filter is based on a UMACE filter. We start the filter design by cropping out regions of size 64×64 for the training data, with the eye centered at coordinates $(32, 32)$. After the eyes are cropped, each cropped eye region is transformed to the frequency domain using a 2D Fourier transform. Next, the average training images and the average of the power spectrum is calculated. The base UMACE filter for our AACE filter is synthesized using the following formula:

$$h = D^{-1}m, \tag{15.3}$$

where D is the average power spectrum of the N training images, and m is the 2D Fourier transform of the average training image.

Separate filters were designed for both the left and right eyes. The UMACE filter is stored in its frequency domain representation to eliminate another 2D Fourier transform before the correlation operation is performed. Since we are performing the correlation operation in the frequency domain the UMACE filter had to be preprocessed by a Hamming window to help reduce the edge effects and impact of high frequency noise that is prevalent in the spectrum of low-light EMCCD imagery. Our experiments showed that windowing both the filter and input image decreased the accuracy of the UMACE eye detector. Since the UMACE filter is trained off line it was chosen as the input that was preprocessed by the Hamming window. One advantage of the UMACE filter over the MACE filter is that over-fitting of the training data is avoided by averaging the training images. Furthermore, we found that training data taken under ideal lighting conditions performed well for difficult detection scenarios when combined with an effective lighting normalization algorithm (discussed in Section 15.5.4.1). This allows us to build an extremely robust filter that can operate in a wide array of lighting conditions, instead of requiring different training data for different lighting levels, as was the case with the machine learning based detector.

Furthermore, our motion blur estimate or point spread function (PSF) can be convolved into UMACE filter using only a point wise multiply of the motion blur Optical Transfer Function (OTF) and the UMACE filter. The resulting filter is what we call our Adaptive Average Correlation Energy (AACE) filter. The concept of the AACE filter is to take the UMACE filter, trained on good data, and adapt it, per image, for the environmental degradations using estimates of blur and noise. The AACE filter is synthesized using the following formula:

$$h = (D^{-1}m) \otimes BlurOTF. \tag{15.4}$$

Fig. 15.5 shows the impulse response of a unblurred and motion blurrred AACE filter.

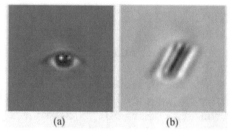

Fig. 15.5 Impulse response of AACE filter: Unblurred (a); Motion Blurred (b).

(a)   (b)

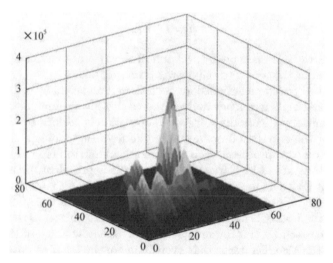

Fig. 15.6 Example Correlation Output with the Detected Eye Centered at Coordinates (40, 36).

Finally, after the correlation operation is performed the global maximum or peak location is chosen as the detected eye location in the original image with the appropriate offsets. Figure 15.6 shows an example correlation output with the detected eye centered at coordinates (40, 36).

15.5 Experiments

Generating statistically significant datasets for difficult acquisition circumstances is a laborious and time consuming process. Capturing real world variables such as atmospheric turbulence, specific lighting conditions, and othe real world scenarios exacerbate the problem further. A specialized experimental setup called "photo-head" introduced by [5] showed using quality guided-synthetic data was a feasible evaluation technique for face recognition algorithm development.

15.5.1 The Photo-head Testing Protocol

In the setup described in that work, two cameras were placed 94 ft and 182 ft from a weather-proof LCD panel in an outdoor setting. The FERET data set was·displayed on the panel at various points throughout the day, where it was re-imaged by the cameras over the course of several years. This unique re-imaging model is well suited to biometric experiments, as we can control for distance, lighting and pose, as well as capture statistically meaningful samples in a timely fashion. Further, it allows for reproducible experiments by use of standard data sets that are re-imaged.

In our own setup, instead of imaging an LCD panel, we used a Mitsubishi PK10 LCD pocket projector, which has a resolution of 800×600 pixels and outputs 25 ANSI Lumens, to project images onto a blank screen. The experimental apparatus was contained in a sealed room, where lighting could be directly controlled via the application of polarization filters to the projector. The camera used for acquisition was a SI-1M30-EM low-light EMCCD unit from FLIR Systems. At its core, this camera utilizes the TI TX285SPD-B0 EMCCD sensor, with a declared resolution of $1\,004 \times 1\,002$ (the effective resolution is actually $1\,008 \times 1\,010$) pixels. To simulate distance, all collected faces were roughly 160 pixels in width (from our own work in long distance acquisition, this is typical of what we would find at 100M with the appropriate optics). Photo-head images can be seen in Fig. 15.9.

In order to assess and adjust the light levels of the photo-head imagery, we directly measure the light leaving the projected face in the direction of the sensor-its *luminance*. The candela per square meter $\left(\dfrac{cd}{m^2}\right)$ is the SI unit of luminance; nit is a common non-SI name also used for this unit (and used throughout the rest of this paper). Luminance is valuable because it describes the "brightness" of the face and does not vary with distance. For our experiments, luminance is the better measure to assess how well a particular target can be viewed - what is most important for biometric acquisition. More details on this issue of light and face acquisition can be found in [21].

15.5.2 Evaluation of Machine Learning Approach

In order to assess the viability of the detector described in Section 15.3.3, a series experiments under very difficult conditions were devised. First, using the photo-head methodology of Section 15.5.1, a subset of the CMU PIE [22] dataset was re-imaged in a controlled (face sizes at approximately the same width as what we would collect at 100M), dark indoor setting (0.043 – 0.017 face nits). Defined feature points are the eyes, with a window size of 45 × 35 pixels. For SVM training, the base positive set consisted of 250 images × (8 1-pixel offsets from the ground-truth + ground-truth point), for each feature. The base negative set consisted of 250 images × 9 pre-defined negative regions around the ground-truth, for each feature. The testing set consisted of 150 images per feature. The actual data used to train the PCA subspaces and SVM classifiers varies by feature, and was determined experimentally based on performance. For the left eye, 1 000 training samples were provided for subspace training, and for the right eye, 1 200 samples were provided. The experiments presented in this section are tailored to assess accuracy of the base technique, and are performed at the highest resolution possible, with the window sliding 1 pixel at a time.

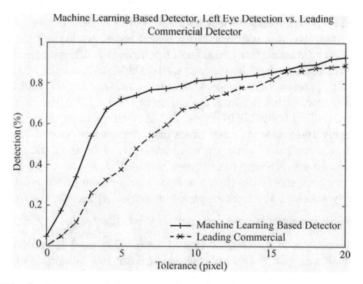

Fig. 15.7 Performance of the proposed machine learning based detector against a leading commercial detector for the left eye under dark conditions. The machine learning based detector clearly outperforms the commercial detector.

The results for eye detection are shown in Figs. 15.7 and 15.8. On each plot, the *x* axis represents the pixel tolerance as a function of distance from the ground-truth for detection, and the *y* axis represents the detection percentage at each tolerance. The proposed detection approach shows excellent

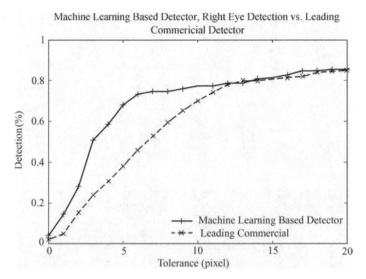

Fig. 15.8 Performance of the proposed machine learning based detector against a leading commercial detector for the right eye under dark conditions. Results are similar to the left eye in Fig. 15.7.

performance for the photo-head imagery. For comparison, the performance of a leading commercial detector (chosen for its inclusion in a face recognition suite that scored at or near the top of every test in FRVT 2006), is also plotted. The proposed detection approach clearly outperforms it till both approaches start to converge after the pixel tolerance of 10. Examples of the detected feature points from the eye comparison experiment are shown in Fig. 15.9.

Even more extreme conditions were of interest for this research. Another photo-head set was collected based on the FERET [23] dataset between $0.0108 - 0.002$ nits. For an eye feature (left is shown here), a window of 50×45 pixels was defined. The Gallery subset was used for training, with a subspace of 1,100 training samples, and a classifier composed of 4,200 training samples (with additional images generated using the perturbation protocol above). For testing, all of the FAFC subset was submitted to the detector. The results of this experiment are shown in Fig. 15.10; the commercial detector is once again used for comparison. From the plot, we can see the commercial detector failing nearly outright at these very difficult conditions, while the proposed detector performs rather well.

Blur is another difficult scenario that we have looked at. For this set of experiments, we produced a subset of images from the FERET dataset (including the ba, bj, and bk subsets) for three different uniform linear motion models of blur: blur length of 15 pixels, at an angle of 122 degrees; blur length of 17 pixels, at an angle of 59 degrees; blur length of 20 pixels, at an angle of 52 degrees. Sample images from each of these sets are shown in Fig. 15.11. The classifier for detection was trained using 2,000 base images

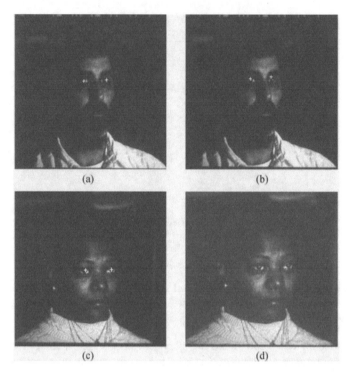

Fig. 15.9 Qualitative results for the proposed machine learning based detector (left) and a leading commercial detector (right) for comparison. The commercial detector is not able to find any eyes in image (d).

of the left eye (split evenly between positive and negative training samples), derived from 112 base images (again, additional images were generated using the perturbation protocol above) at the blur length of 20 pixels, at an angle of 52 degrees. The subspace was trained on 1 000 positive images, with the same blur model. The testing set consisted of 150 images, with the left eye as the feature target for each of the three blur models. The results for this experiment are shown in Fig. 15.12. From this figure, we can see that the machine learning based approach has a slight advantage over the commercial detector for the blur length of 20 pixels — the blur model it was trained with. For testing with the other blur models, performance is acceptable, but drops noticeably. Thus, we conclude that incorrect blur estimations can negatively impact this detection approach.

Reduced resolution imagery (face sizes $\leqslant 90 \times 90$ pixels), is another difficult scenario that we have explored. The performance of most face recognition algorithms degrades substantially whenever the input images are of low resolution or size, which is often the case whenever the images are taken by a surveillance camera in an uncontrolled setting, since these algorithms were designed and developed with high or average quality images at close ranges $\leqslant$

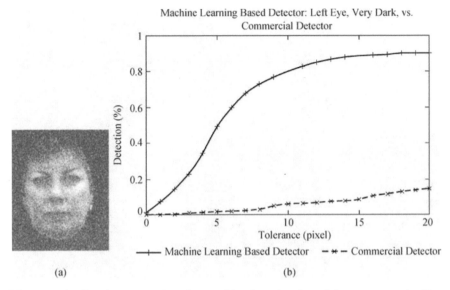

(a)   (b)

Fig. 15.10 Results comparing the machine learning based detector to a leading commercial detector, for the left eye, with very dark imagery (0.0108 – 0.002 nits). A sample image (a) is provided to signify the difficulty of this test (histogram equalized here to show "detail"). The commercial detector fails regularly under these conditions.

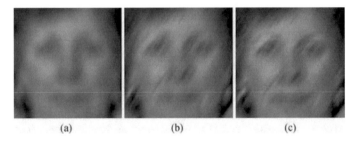

(a)   (b)   (c)

Fig. 15.11 Examples of blurred imagery for three different blur models used for experimentation. (a) Blur length of 15 pixels, at an angle of 122 degrees. (b) Blur length of 17 pixels, at an angle of 59 degrees. (c) Blur length of 20 pixels, at an angle of 52 degrees.

3 meters. Recent work from the face recognition community is addressing the issue of recognizing subjects from low quality or reduced resolution images [24 – 26]. However, accurate eye detection is still vital to provide optimal performance when using these reduced resolution face recognition algorithms. This set of experiments was designed to examine how our machine learning based detector performs on the same dataset at full resolution and at a reduced resolution; down sampled by 2× in each direction.

For this set of experiments, we again used a subset of images from the FERET dataset (including the ba, bj, and bk subsets) at the full and re-

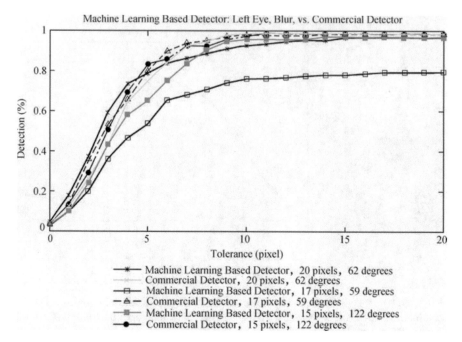

Machine Learning Based Detector: Left Eye, Blur, vs. Commercial Detector

——*—— Machine Learning Based Detector, 20 pixels, 62 degrees
———— Commercial Detector, 20 pixels, 62 degrees
——ⴲ—— Machine Learning Based Detector, 17 pixels, 59 degrees
— △ — Commercial Detector, 17 pixels, 59 degrees
——■—— Machine Learning Based Detector, 15 pixels, 122 degrees
——●—— Commercial Detector, 15 pixels, 122 degrees

Fig. 15.12 Results comparing the machine learning based detector to a leading commercial detector, for the left eye, with three varying degrees of blur. The detector was trained using the blur length of 20 pixels at an angle of 52 degrees.

duced resolution, where images were down sampled by 2 × in each direction. Sample images from each of these sets are shown in Fig. 15.13. The classifier for eye detection was trained using 2 000 base images of the left eye (split evenly between positive and negative training samples), derived from 200 base images (additional images were generated using the perturbation protocol). The subspace was trained with the 1 000 positive images. The testing set consisted of 150 images, with the left eye as the feature target for each of the models. The results for this experiment are shown in Fig. 15.14. From this figure, we can see that the machine learning based approach has a slight

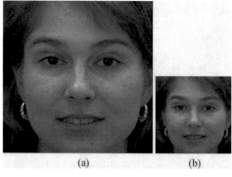

Fig. 15.13 Examples of full and reduced resolution imagery used for experimentation. (a) Sample image full resolution 176 × 176. (b) Sample image reduced resolution 89 × 89.

(a)   (b)

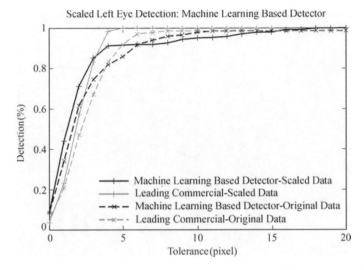

Scaled Left Eye Detection: Machine Learning Based Detector

Fig. 15.14 Results comparing the machine learning based detector to a leading commercial detector, for the left eye, for full and reduced resolution images. The machine learning based detector outperforms the commercial detector for pixel tolerances < 5 (following this both detectors converge).

advantage over the commercial detector for pixel tolerances < 5 (following this both detectors converge).

15.5.3 Evaluation of Correlation Approach

The experiments performed for the correlation approach are identical to the ones we performed for the machine learning approach, with the following training details. The MACE filters used in Figs. 15.15 and 15.16 were trained with 6 eye images, while the MACE filter for Fig. 15.17 used 4 training images (these values were determined experimentally, and yield the best performance). For the experiments of Figs. 15.15 and 15.16, the AACE filter was synthesized with 266 images, for the experiment of Fig. 15.17, the filter was synthesized with 588 images. For the AACE filter used in the experiment of Fig. 15.18, the filter was synthesized with 1 500 images, incorporating the exact same blur model as the machine learning experiments into the convolution operator. Furthermore, the training data for the experiments in Figs. 15.15, 15.16, and 15.17 used images taken under ideal lighting conditions. For the AACE filter used in the experiment of Fig. 15.19, the filter was synthesized with the same 1 500 images for the motion blur experiment though no PSF model was incorporated into the AACE filter.

Comparing the AACE approach to the machine learning approach, the correlation filter detector shows a significant performance gain over the learning based detector on blurry imagery (Fig. 15.12 vs. Figure 15.18). What can

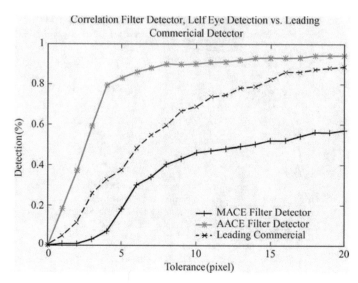

Fig. 15.15 Performance of the correlation filter detectors against a leading commercial detector for the left eye under dark conditions.

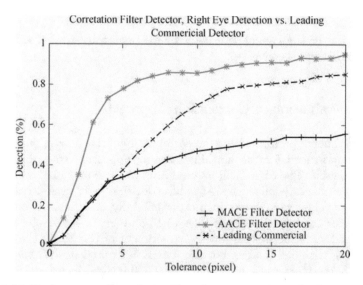

Fig. 15.16 Performance of correlation filter detectors against a leading commercial detector for the right eye under dark conditions.

also be seen from our experiments is a stronger tolerance for incorrect blur estimation, with the blur length of 17 pixels, 59 degrees performing just as well as the training blur model; this was not the case with the machine learning based detector. In all other experiments, the AACE filter detector produced a modest performance gain over the machine learning based detector. The

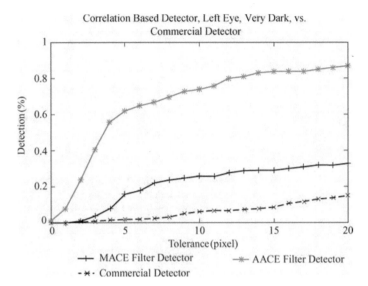

Fig. 15.17 Results for the correlation filter detectors for the left eye, with very dark imagery (0.0108 – 0.002 nits.)

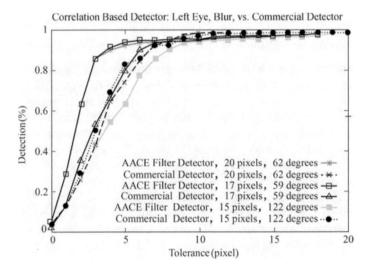

Fig. 15.18 Results comparing the AACE correlation filter detector to a leading commercial detector, for the left eye, with three varying degrees of blur. The filters were trained using the blur length of 20 pixels, at an angle of 52 degrees.

performance of MACE was poor for all test instances that it was applied to.

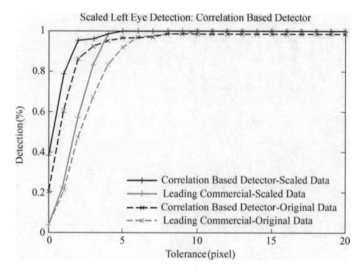

Fig. 15.19 Results comparing the AACE correlation filter detector to a leading commercial detector, for the left eye, for full and reduced resolution images.

15.5.4 Methods to Improve Correlation Approach

Lighting Normalization

In addition to using multiple AACE eye filter models for different lighting situations, we decided to implement and test a lighting normalization algorithm to see whether it would increase the accuracy of the eye detector. A key motive for using lighting normalization in conjunction with our correlation eye detector came from some of our daytime experiments where the faces had extreme shadows and gradients on them. These shadows and gradients on the face were causing the eye detector to improperly identify the position of the eye as shown below in Fig. 15.20.

Our lighting normalization algorithm is presented below. We are currently using a modified version of the Self-Quotient illumination (SQI) lighting normalization algorithm. Self-Quotient illumination (SQI) normalization is based on the work of [4]. The SQI image is formed by dividing the original face image $f(x, y)$ with the original image convolved with a Gaussian function that acts as a smoothing kernel function $S(x, y)$.

$$Q(x, y) = \frac{f(x, y)}{S(x, y)} = \frac{f(x, y)}{G(x, y) \otimes f(x, y)}. \tag{15.5}$$

The subsequent task of the lighting normalization method is to normalize $Q(x, y)$ to have pixel intensity between 0 and 1, and to increase the contrast of the image by applying linear transformation function.

$$Q'(x, y) = \frac{Q(x, y) - Q_{min}}{Q_{max} - Q_{min}}, \tag{15.6}$$

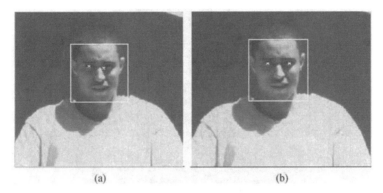

(a)   (b)

Fig. 15.20 (a) Output of eye detector without lighting normalization; the right eye position is not properly identified. (b) Output of eye detector with lighting normalization; the right eye position is properly identified.

$$Q_{norm}(x,y) = 1 - e^{-\frac{Q'(x,y)}{E(Q'(x,y))}}, \tag{15.7}$$

where Q_{max} and Q_{min} are maximum and minimum values of Q respectively, and $E(.)$ is a mean value. Therefore, Q_{norm} is a normalized Gaussian quotient image and is used as an image for eye detection as shown below in Fig. 15.21 and 15.22.

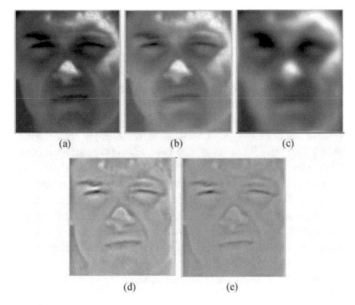

(a)   (b)   (c)

(d)   (e)

Fig. 15.21 Lighting normalization algorithm with example daytime image. (a) Original image. (b) Gamma correction of image. (c) Smoothed image. (d) Quotient image. (e) Normalized quotient image.

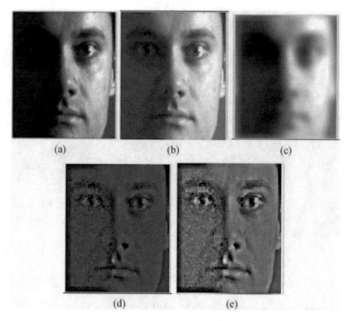

Fig. 15.22 Lighting normalization algorithm with example low-light image. (a) Original image. (b) Gamma correction of image. (c) Smoothed image. (d) Quotient image. (e) Normalized quotient image.

Eye Location Perturbations

A known problem with correlation based eye detectors is that they will also show a high response to eyebrows, nostrils, dark rimmed glasses, and strong lighting such as glare from eye glasses and return these points as the coordinates of the eye. Through our analysis of the problem we have discovered that when an invalid location has the highest correlation peak value, a second or third correlation peak with a value slightly less than the highest peak is usually the true location of the eye. Therefore, our eye detection algorithm has been modified to search for multiple correlation peaks on each side of the face and then determine which correlation peak is the true location of the eye. Once the initial correlation output is returned it is thresholded at 80% of the maximum value to eliminate all but the salient structures in the correlation output. A unique label is then assigned to each structure using connected component labeling [27]. The location of the maximum peak within each label is then located and returned as a possible eye location. This process is repeated for both sides of the face.

Our ultimate goal is to determine the location of the left and right eye and then send the input image and the eye locations to a geometric normalization algorithm. However, we are taking a different approach by sending all of the initial eye locations to the geometric normalization algorithm and then determining the "best" geometrically normalized image from all of the normalized images. Geometric normalization is a vital step in our face recognition pipeline since it reduces the variation between gallery and probe images. The geometrically normalized image is of uniform size and if the input eye coordinates are correct the output image will contain a face chip with uniform orientation. All of the geometrically normalized images are compared against an "average" face using normalized cross-correlation. Our "average" face was formed by first geometrically normalizing and then averaging all of the faces from the FERET data set [23]. Normalized cross-correlation is only performed on a small region around the center of the image. The left and right (x, y) eye coordinates corresponding to the image with the highest similarity are returned as the true eye coordinates. Additionally, since we have already performed geometrically normalization this step does not need to be performed again in our pipeline. A summary of the new algorithm is shown in Fig. 15.25. Only the top two eye coordinates were considered on each side of the face.

Evaluation of Eye Location Perturbations

To evaluate the performance of the correlation approach using the eye location perturbation algorithm presented in Figs. 15.23 – 15.25 we performed an experiment using 128 images from the SCface – Surveillance Cameras Face Database [25]. The images in the SCface database are taken from various surveillance cameras with uncontrolled lighting and the images are of various quality and resolution. Example images from our test set are shown in Fig. 15.26. For the correlation filter used in the experiment of Fig. 15.27, the filter was synthesized with the same 1 500 images from the motion blur experiment with no PSF model being incorporated into the AACE filter. The lighting normalization algorithm presented in Section 15.5.4 was used on the

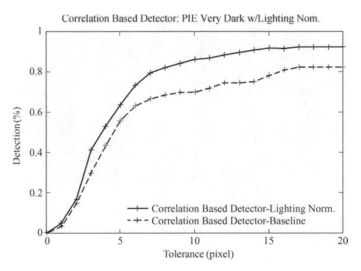

Fig. 15.23 Results for the correlation filter detector for the left eye, with very dark imagery (0.043 – 0.017 nits) with and without lighting normalization.

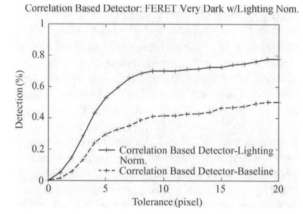

Fig. 15.24 Results for the correlation filter detector for the left eye, with very dark imagery (0.0108 − 0.002 nits) with and without lighting normalization.

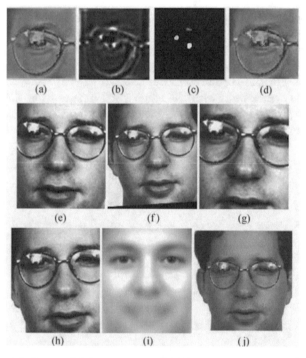

Fig. 15.25 (a) Cropped left eye area. (b) Correlation output. (c) Connected components image derived from thresholded correlation output. (d) Cropped left eye area with top two initial eye locations returned. (e−h) Image perturbations using top two initial left and right eye locations. (i) "Average Face". (j) Final eye coordinates returned based on top score using perturbation algorithm.

images prior to eye detection. Only two (x, y) eye coordinates were considered on each side of the face for this experiment. The results for eye detection are shown in Fig. 15.27. The proposed detection approach shows a moderate performance gain for the difficult imagery.

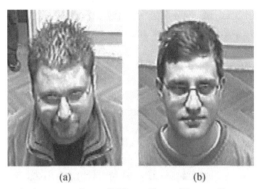

Fig. 15.26 Example imagery from SCface–Surveillance Cameras Face Database.

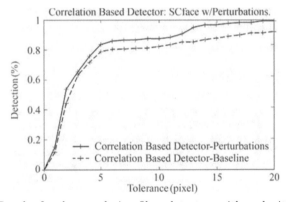

Fig. 15.27 Results for the correlation filter detectors with and without using eye location perturbation algorithm.

15.6 Conclusions

As face recognition moves forward, difficult imagery becomes a primary concern. But before we can even attempt face recognition, we often need to perform some necessary pre-processing steps, including geometric normalization and facial feature localization, with the eyes providing the necessary reference points. Thus, in this paper, we have concentrated on the eye detection problem for unconstrained environments. First, we introduced an EMCCD approach for low-light acquisition, and subsequently described an experimental protocol for simulating low-light conditions, distance, pose variation and motion blur. Next, we described two different detection algorithms: a novel machine learning based algorithm and a novel adaptive correlation filter based algorithm. Finally, using the data generated by our testing protocol, we performed a thorough series of experiments incorporating the aforementioned conditions. Both approaches show significant performance improvement over a leading commercial eye detector.

Comparing both approaches, our new AACE correlation filter detector shows a significant performance gain over the learning based detector on blurry imagery, and a moderate performance gain on low-light imagery. Our lighting normalization results showed that we could build a AACE correlation filter that can operate in a wide array of lighting conditions, instead of requiring different training data for different lighting levels. The perturbation algorithm showed that we could use multiple eye estimates to ultimately help select the real eye locations. Based on the presented results, we conclude that both approaches are suitable for the problem at hand–the choice of one as a solution can be based upon implementation requirements. As far as we know, this is the first study of feature detection under a multitude of difficult acquisition circumstances, and its results give us confidence for tackling the next steps for unconstrained face recognition.

Acknowledgements

This chapter was supported by ONR STTR Biometrics in the Maritime Domain (Award Number N00014-07-M-0421), ONR MURI (Award Number N00014-08-1-0638), São Paulo Research Foundation, FAPESP (Award Number 2010/05647-4) and Unicamp's PAPDIC program (Award Number 34/010). Portions of the research in this paper use the SCface database of facial images. Credit is hereby given to the University of Zagreb, Faculty of Electrical Engineering and Computing for providing the database of facial images.

References

[1] Phillips P, Vardi Y (1996) Efficient Illumination Normalization of Facial Images. Pattern Recogn Lett, 17(8): 921 – 927
[2] Chen T, Yin W, Zhou X et al (2006) Total Variation Models for Variable Lighting face Recognition. IEEE Trans Pattern Anal Mach Intell, 28(9): 1519 – 1524
[3] Yao Y, Abidi B, Kalka N et al (2008) Improving long Range and High Magnification Face Recognition: Database Acquisition, Evaluation, and Enhancement. Comput Vis Image Underst, 111(2): 111 – 125
[4] Heflin B, Parks B, Scheirer W et al (2010) Single Image Deblurring for a Real-time Face Recognition System. Paper Presented at the 36th Annual Conference of the IEEE Industrial Electronics Society, Glendale, AZ, 7 – 10 November 2010
[5] Boult T, Scheirer W, Woodworth R (2008) FAAD: Face at a Distance. Paper Presented at the SPIE Defense and Security Symposium, Orlando, April 2008
[6] Leite B, Pereira E, Gomes H et al (2007) A Learning-based Eye Detector Coupled with Eye Candidate Filtering and PCA Features. In: Proceedings of the XX Brazilian

Symposium on Computer Graphics and Image Processing, Washington DC

[7] Shuo C, Liu C (2010) Eye Detection Using Color Information and a New Efficient SVM. Paper Presented at the IEEE 4th International Conference on Biometrics: Theory, Applications and Systems, Washington DC, 27–29 September 2010

[8] Wang P, Green M, Ji Q et al (2005) Automatic eye Detection and its Validation. In: Proceedings of the 2005 IEEE Computer Society Conference on Computer Vision and Pattern Recognition-workshops, Washington DC

[9] Jin L, Yuan X, Satoh S et al (2006) A Hybrid Classifier for Precise and Robust eye Detection. In: Proceedings of the 18th International Conference on Pattern Recognition, Washington DC

[10] Mahalanobis A, Kumar B, Casasent D (1987) Minimum Average Correlation Energy Filters. Appl Opt, 26: 3633–3640

[11] Savvides M, Kumar B (2003) Efficient Design of Advanced Correlation Filters for Robust Distortion-tolerant Face Recognition. In: Proceedings of the IEEE Conference on Advanced Video and Signal Based Surveillance, Washington DC

[12] Savvides M, Kumar B, Khosla P (2004) "Corefaces": Robust Shift Invariant PCA Based Correlation Filter for Illumination Tolerant Face Recognition. In: Proceedings of the 2004 IEEE Computer Society Conference on Computer Vision and Pattern Recognition, Washington DC

[13] Savvides M, Abiantun R, Heo J et al (2006) Partial & Holistic Face Recognition on FRGC-II Data Using Support Vector Machine. In: Proceedings of the 2006 Conference on Computer Vision and Pattern Recognition Workshop, Washington DC

[14] Brunelli R, Poggio T (1995) Template Matching: Matched Spatial Filters and Beyond. Pattern Recogn, 30: 751–768

[15] Bolme D, Draper B, Beveridge J (2009) Average of Synthetic Exact Filters. In: Proceedings of the 2009 IEEE Conference on Computer Vision and Pattern Recognition, Miami, 20–26 June 2009

[16] Tyson R (2000) Introduction to Adaptive Optics. SPIE Publications, Bellingham

[17] Carroll J, Gray D, Roorda A et al (2005) Recent Advances in Retinal Imaging with Adaptive Optics. Opt Photon News, 16(1): 36–42

[18] Bishop C (2006) Pattern Recognition and Machine Learning (Information Science and Statistics). Springer, New York

[19] Smith L (2002) A Tutorial on Principal Components Analysis Introduction. Statistics, 51: 52

[20] Viola P, Jones M (2004) Robust Real-time Face Detection. Int J Comput Vision, 57(2): 137–154

[21] Boult T, Scheirer W (2009) Long-range Facial Image Acquisition and Quality. In: Handbook of Remote Biometrics, Springer, London

[22] Sim T, Baker S, Bsat M (2002) The CMU Pose, Illumination, and Expression (PIE) Database. In: Proceedings of the 5th IEEE International Conference on Automatic Face and Gesture Recognition, Washington DC

[23] Phillips P, Moonb H, Rizvi S et al (2000) The FERET Evaluation Methodology for Face-recognition Algorithms. IEEE Trans Pattern Anal Mach Intell, 22(10): 1090–1104

[24] Sapkota A, Parks B, Scheirer W et al (2010) Face-Grab: Face Recognition with General Region Assigned to Binary Operator. In: Proceedings of the 2010 Conference on Computer Vision and Pattern Recognition Workshop on Biometrics, San Francisco, 13–18 June 2010

[25] Grgic M, Delac K, Grgic S (2009) SCface: A Surveillance Cameras Face Database. Multimedia Tools and Applications, Multimed Tools Appl (2011)51: 863–879. Springer

[26] Iyer V, Kirkbride S, Parks B et al (2010) A Taxonomy of Face Models for System Evaluation. In: Proceedings of the 2010 IEEE Workshop on Analysis and Modeling of Faces and Gestures, San Francisco

[27] Shapiro L, Stockman G (2002) Computer Vision. Prentice Hall, NJ

16 Kernel Methods for Facial Image Preprocessing

Weishi Zheng[1], Jianhuang Lai[2], Xiaohua Xie[3], Yan Liang[4],
Pong C. Yuen[5] and Yaoxian Zou[6]

Abstract Kernel methods have been widely used for classification in pattern recognition due to the kernel trick that enables algorithms to perform nonlinear transformation of data for learning a better nonlinear decision boundary. Despite extensive development of kernel methods for classification problems, it is still a rather more open topic to use kernel methods for nonlinear image preprocessing. Kernel methods, especially kernel principal component analysis (KPCA), are able to model nonlinear variations of data, so as to alleviate high-order noisy variations more effectively. In this chapter, we introduce recent advanced developments of using KPCA for nonlinear image preprocessing. Particularly, we concentrate on the pre-image learning problem in KPCA as well as other kernel methods, which is a vital step and also a seriously ill-posed estimation problem for applying kernel methods to image preprocessing. A general framework for pre-image learning in kernel methods is presented and a weakly supervised pre-image learning method is detailed. Extensive applications for face image preprocessing, including facial expression normalization, face image denoising, face image occlusion recovery, illumination normalization, and face hallucination are demonstrated.

16.1 Introduction

Kernel methods are recently developed techniques in machine learning and widely applied to pattern recognition and computer vision. Kernel methods are developed based on the theory of Reproducing Kernel Hilbert Space (RKHS) [1]. The essential idea is to utilize the kernel trick [1] supported by Mercer kernels to extend linear methods so as to develop corresponding nonlinear methods. For any Mercer kernel κ, the kernel trick implies that a nonlinear function φ is induced by κ such that $\kappa(\boldsymbol{x}, \boldsymbol{y}) = <\varphi(\boldsymbol{x}), \varphi(\boldsymbol{y})>$,

1   Sun Yat-sen University, Guangzhou. E-mail: wszheng@ieee.org.
2   Sun Yat-sen University, Guangzhou China. E-mail: stsljh@mail.sysu.edu.cn (correspondence author).
3   Sun Yat-sen University, Guangzhou China, and Shenzhen Institutes of Advanced Technology, Chinese Academy of Sciences, Shengzhen China. E-mail: sysuxiexh@gmail.com.
4   Sun Yat-sen University, Guangzhou China. E-mail: liangyan_709@163.com.
5   Hong Kong Baptist University, Hong Kong. E-mail: pcyuen@comp.hkbu.edu.hk.
6   Sun Yat-sen University, Guangzhou China.

where $\boldsymbol{x}$, $\boldsymbol{y}$ are samples in the data space $\boldsymbol{X}$. The implicitly defined nonlinear mapping φ always maps any input data to a higher (even infinite) dimensional reproducing kernel Hilbert space (RKHS) $\boldsymbol{H}_\kappa$ associated to the kernel function κ, if the most popular kernels such as radial basis function (RBF) and polynomial kernels are used. Since data are always not linearly separated and cannot be linearly represented in the original space, the kernel trick technique gives great benefits that pattern recognition and computer vision problems can be considered in a higher dimensional space, where linear techniques can be more plausibly applied to solve the nonlinear problem. Typical kernel methods for pattern recognition include kernel svm (KSVM) [1], kernel linear discriminant analysis (KLDA) [2], kernel principal component analysis (KPCA) [3] and etc.

Face image analysis is also benefited by using kernel methods. Two typical examples are KLDA and KPCA. From the recognition point of view, linear discriminant analysis (LDA) is a widely used technique that learns a discriminant subspace where face images of the same person collapse and face images of different people scatter, and KLDA extends LDA so that the linear separation assumption is alleviated and a more suitable low-rank discriminant subspace can be found. So far, a lot of variants of KLDA and more other kernel methods such as Kernel ICA [4] and Kernel CCA [5] have extensively been developed and investigated for the classification problem.

From the image preprocessing point of view, the effect and usefulness of kernel methods are still not extensively investigated, with the exception of using KPCA [6–9]. KPCA can significantly improve the ability of principal component analysis (PCA) in normalizing face images with large and nonlinear variations such as occlusion, corruption, illumination, and expression as shown in Ref. [9]. This is because by learning the major variations of data in the feature space, KPCA is able to eliminate these variations more effectively in the input image in the feature space induced by kernel function, i.e., in a nonlinear way. However, several challenges and open problems are still remained, for examples the pre-image learning problem in KPCA.

This chapter introduces some recent advanced developments in solving the pre-image problem in KPCA and their applications for face image analysis. In particular, we present a general framework for pre-image learning in kernel methods and detail a weakly supervised pre-image learning method. In order to start our topic, we first briefly review the methodology of KPCA [1] in Section 16.2. After that, we introduce the pre-image problem in Section 16.3 and extensively demonstrate its application for face image analysis in Section 16.4, including facial expression normalization, face image denoising and occlusion recovery, face image illumination normalization and face hallucination. Finally, we summarize our story in Section 16.5.

16.2 Kernel PCA

KPCA performs linear PCA in the feature space $\mathcal{H}_\kappa$ in order to learn the maximum variations of data and can be regarded as a non-linear description of data. Suppose $\{x_1, \cdots, x_N\}$, $x_i \in \mathcal{X}$ are N training samples. Let

$$\Psi = [x_1, \cdots, x_N],$$
$$\Psi_\varphi = [\varphi(x_1), \cdots, \varphi(x_N)],$$
$$\mu^\varphi = \frac{1}{N} \sum_{i=1}^{N} \varphi(x_i) = \frac{1}{N} \Psi_\varphi e_N,$$

where $e_N = (1, \cdots, 1)^\mathrm{T} \in \Re^N$. Denote $O_t^\varphi = \frac{1}{\sqrt{N}}(\varphi(x_1) - \mu^\varphi, \cdots, \varphi(x_N) - \mu^\varphi)$, then KPCA is to solve the following eigenvalue function:

$$S_t^\varphi U^\varphi = U^\varphi \Lambda^\varphi. \tag{16.1}$$

where $S_t^\varphi = O_t^\varphi O_t^{\varphi^\mathrm{T}}$, $U^\varphi = (u_1^\varphi, \cdots, u_q^\varphi)$, $\Lambda^\varphi = diag(\lambda_1^\varphi, \cdots, \lambda_q^\varphi)$, $\lambda_1^\varphi \geqslant \cdots \geqslant \lambda_q^\varphi > 0$. Noting that $O_t^\varphi = N^{-0.5}(\Psi_\varphi - \mu^\varphi e_N^\mathrm{T}) = N^{-0.5}\Psi_\varphi(I_N - N^{-1}e_N e_N^\mathrm{T})$ and $(I_N - N^{-1}e_N e_N^\mathrm{T})(I_N - N^{-1}e_N e_N^\mathrm{T})^\mathrm{T} = (I_N - N^{-1}e_N e_N^\mathrm{T})$. We have

$$S_t^\varphi = O_t^\varphi O_t^{\varphi^\mathrm{T}} = N^{-1}\Psi_\varphi(I_N - N^{-1}e_N e_N^\mathrm{T})\Psi_\varphi^\mathrm{T}. \tag{16.2}$$

Therefore, for each principal component u_i^φ, there exists a coefficient vector $\alpha_i = (\alpha_1^i, \ldots, \alpha_N^i)^\mathrm{T}$ [1] so that

$$u_i^\varphi = \sum_{j=1}^{N} \alpha_j^i \frac{1}{\sqrt{N}}(\varphi(x_j) - \mu^\varphi) = O_t^\varphi \alpha_i = \Psi_\varphi p_i, \tag{16.3}$$

where $p_i = N^{-0.5}(I_N - N^{-1}e_N e_N^\mathrm{T})\alpha_i$. So $U^\varphi = \Psi_\varphi P$, $P = (p_1, \cdots, p_q)$. Denote $U_{q_0}^\varphi = \Psi_\varphi P_{q_0}$ as the matrix consists of the q_0 principal components corresponding to the largest q_0 eigenvalue, $P_{q_0} = (p_1, \cdots, p_{q_0})$. Then, for a given sample x, the projection of $\varphi(x)$ in the principal component subspace spanned by matrix $U_{q_0}^\varphi$ is $P_\kappa \varphi(x)$:

$$P_\kappa \varphi(x) = U_{q_0}^\varphi U_{q_0}^{\varphi^\mathrm{T}}(\varphi(x) - \mu^\varphi) + \mu^\varphi$$
$$= \Psi_\varphi P_{q_0} P_{q_0}^\mathrm{T} \Psi_\varphi^\mathrm{T}(\varphi(x) - N^{-1}\Psi_\varphi e_N) + N^{-1}\Psi_\varphi e_N$$
$$= \Psi_\varphi \gamma^x, \tag{16.4}$$

where $\gamma^x = (\gamma_1^x, \cdots, \gamma_N^x)^\mathrm{T} = P_{q_0} P_{q_0}^\mathrm{T} \Psi_\varphi^\mathrm{T} \varphi(x) - N^{-1} P_{q_0} P_{q_0}^\mathrm{T} K e_N + N^{-1} e_N$, and the kernel matrix is $K = \Psi_\varphi^\mathrm{T} \Psi_\varphi$.

16.3 Kernel Methods for Nonlinear Image Preprocessing

Principal Component Analysis (PCA) has been widely applied for data pre-processing. However, it cannot process the image data as well as many other kinds of data very well, because real-world data are always generated by a nonlinear system and the nonlinearity hidden in data is hard to be explored by the linear PCA. Recently, as a well-known nonlinear extension, kernel principal component analysis (KPCA) has been used for nonlinear data pre-processing. Encouraging results have been seen in image denoising [6–9], image compression [8], image super-resolution [22] etc.

In order to realize the nonlinear data preprocessing using KPCA, the pre-image learning in KPCA is an important step. To specify the pre-image problem, we demonstrate KPCA + Pre-image Learning for data preprocessing in Fig. 16.1. Any input pattern $x(\in X)$ is first mapped to $\varphi(x)$ in the feature space H_κ. Then, $\varphi(x)$ is projected onto the kernel principal component subspace in H_κ and the projection is denoted by $P_\kappa\varphi(x)$. Finally, a pre-image $\hat{x}$ is found in the input data space X such that

$$\varphi(\hat{x}) = P_\kappa\varphi(x). \tag{16.5}$$

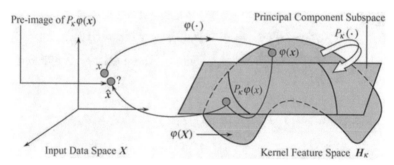

Fig. 16.1 Illustration of the process "KPCA+Pre-image learning".

16.3.1 The Pre-image Learning in KPCA

In the nonlinear image preprocessing using KPCA, the pre-image learning at the final stage is the central part. It is, however, hard (always impossible) to find an exact pre-image entirely satisfying $\varphi(\hat{x}) = P_\kappa\varphi(x)$ [9]. To address this problem, there are at least four major approaches to solve this problem. These include (1) Mika et al.'s least squares method [6], (2) Kwok and Tsang's distance constraint based method [7], (3) Bakır et al.'s pre-image map [8], and (4) Zheng et al.'s penalized pre-image learning method [9]. In addition, Dambreville et al. proposed a simplified Mika's method in order to avoid iteration [10] and Aria et al. utilized the Nyström extension to improve any existing pre-image learning methods [11].

More specifically, suppose there are N training data $\boldsymbol{x}_1, \cdots, \boldsymbol{x}_N \in \boldsymbol{X}$ and Let $\boldsymbol{\Psi} = [\boldsymbol{x}_1, \cdots, \boldsymbol{x}_N]$. The Mika's method learns a pre-image value by optimizing the following least squares criterion [6]:

$$\hat{\boldsymbol{x}} = \arg\min_{\boldsymbol{x}'} G(\boldsymbol{x}'), \quad G(\boldsymbol{x}') = \|\varphi(\boldsymbol{x}') - P_\kappa \varphi(\boldsymbol{x})\|^2. \tag{16.6}$$

For RBF kernel $\kappa(\boldsymbol{x}, \boldsymbol{x}') = \exp\left(-\dfrac{\|\boldsymbol{x} - \boldsymbol{x}'\|^2}{c}\right)$, where $\boldsymbol{x}, \boldsymbol{x}' \in \mathcal{X}$ and $c > 0$, the pre-image can be estimated by the following iterative algorithm

$$\hat{\boldsymbol{x}}_{t+1} = \frac{\displaystyle\sum_{i=1}^{N} \gamma_i^{\boldsymbol{x}} \exp(-c^{-1}\|\hat{\boldsymbol{x}}_t - \boldsymbol{x}_i\|^2)\boldsymbol{x}_i}{\displaystyle\sum_{j=1}^{N} \gamma_j^{\boldsymbol{x}} \exp(-c^{-1}\|\hat{\boldsymbol{x}}_t - \boldsymbol{x}_j\|^2)}. \tag{16.7}$$

The Kwok and Tsang's method imposes the distance values $d_1^\varphi, \cdots, d_s^\varphi$ between the feature point $P_\kappa \varphi(\boldsymbol{x})$ and its neighboring s data points $\varphi(\tilde{\boldsymbol{x}}_1), \cdots,$ $\varphi(\tilde{\boldsymbol{x}}_s)$ as constraint for inferring the pre-image value [7]. Let $\boldsymbol{\Psi}_{s,\boldsymbol{x}} = [\tilde{\boldsymbol{x}}_1, \cdots, \tilde{\boldsymbol{x}}_s]$, and $\boldsymbol{H}_s = \boldsymbol{I}_s - s^{-1}\boldsymbol{e}_s\boldsymbol{e}_s^{\mathrm{T}}$, where $\boldsymbol{I}_s$ is the $s \times s$ identity matrix and $\boldsymbol{e}_s = (1, \cdots, 1)^{\mathrm{T}} \in \Re^s$. Also, let the SVD decomposition $\boldsymbol{\Psi}_{s,\boldsymbol{x}}\boldsymbol{H}_s$ be $\boldsymbol{\Psi}_{s,\boldsymbol{x}}\boldsymbol{H}_s = \boldsymbol{U}\boldsymbol{\Lambda}\boldsymbol{V}^{\mathrm{T}}$. Then, the pre-image $\hat{\boldsymbol{x}}$ is estimated by [7]

$$\hat{\boldsymbol{x}} = \boldsymbol{U}\hat{\boldsymbol{z}} + \bar{\boldsymbol{x}}, \tag{16.8}$$

where $\hat{\boldsymbol{z}} = -0.5 \cdot (\boldsymbol{Z}_s\boldsymbol{Z}_s^{\mathrm{T}})^{-1}\boldsymbol{Z}_s(\boldsymbol{d}_s^2 - \boldsymbol{d}_{s,0}^2)$, $\boldsymbol{Z}_s = [\boldsymbol{z}_1, \cdots, \boldsymbol{z}_s] = \boldsymbol{\Lambda}\boldsymbol{V}^{\mathrm{T}}$, $\boldsymbol{d}_s^2 = (d_1^2, \cdots, d_s^2)^{\mathrm{T}}$, $\boldsymbol{d}_{s,0}^2 = (\|\boldsymbol{z}_1\|^2, \cdots, \|\boldsymbol{z}_s\|^2)^{\mathrm{T}}$, and $\bar{\boldsymbol{x}} = s^{-1}\displaystyle\sum_{j=1}^{s} \tilde{\boldsymbol{x}}_j$.

Bak\i r et al. addressed the problem in a different route. They aimed to learn a kernel regression based mapping for estimating the pre-image of $P_\kappa \varphi(\boldsymbol{x})$ [8], that is

$$\Gamma(P_\kappa \varphi(\boldsymbol{x})) = [\Gamma_1(P_\kappa \varphi(\boldsymbol{x})), \cdots, \Gamma_n(P_\kappa \varphi(\boldsymbol{x}))]^{\mathrm{T}},$$

where $\Gamma_j(P_\kappa \varphi(\boldsymbol{x})) = \displaystyle\sum_{i'=1}^{N} \beta_{i'}^j \kappa'(P_\kappa \varphi(\boldsymbol{x}), P_\kappa \varphi(\boldsymbol{x}_{i'}))$ and κ' is another Mercer kernel function different from κ. Let $\boldsymbol{\beta}^j = (\beta_1^j, \cdots, \beta_N^j)^{\mathrm{T}}$, by optimizing the following criterion:

$$\Gamma = \arg\min_{\Gamma} \sum_{i=1}^{N} \|\boldsymbol{x}_i - \Gamma(P_\kappa \varphi(\boldsymbol{x}_i))\|^2 + \lambda \sum_{j=1}^{n} \|\boldsymbol{\beta}^j\|^2, \quad \lambda \geqslant 0 \tag{16.9}$$

finding the pre-image is equal to finding the optimal $\boldsymbol{B}$ where $\boldsymbol{B} = [\beta^1, \cdots, \beta^n]$ as follows:

$$\boldsymbol{B} = (\boldsymbol{K}'^{\mathrm{T}} \boldsymbol{K}' + \lambda \cdot \boldsymbol{I}_N)^{-1} \boldsymbol{K}'^{\mathrm{T}} \boldsymbol{\Psi}^{\mathrm{T}}. \tag{16.10}$$

where $\boldsymbol{K}' = (\kappa'(P_\kappa \varphi(\boldsymbol{x}_i)), \kappa'(P_\kappa \varphi(\boldsymbol{x}_j)))$. Hence, $\hat{\boldsymbol{x}} = \Gamma(P_\kappa \varphi(\boldsymbol{x})) = (\boldsymbol{\kappa}'_{\boldsymbol{x}} \boldsymbol{B})^{\mathrm{T}}$, where $\boldsymbol{\kappa}'_{\boldsymbol{x}} = (\kappa'(P_\kappa \varphi(\boldsymbol{x}), P_\kappa \varphi(\boldsymbol{x}_1)), \cdots, \kappa'(P_\kappa \varphi(\boldsymbol{x}), P_\kappa \varphi(\boldsymbol{x}_N)))$.

So far, it has been verified that all the above methods can be formulated under the following general framework [9]:

$$\hat{\boldsymbol{x}} = \boldsymbol{\Psi} \boldsymbol{w}_{\boldsymbol{x}} = \sum_{i=1}^{N} w_i^{\boldsymbol{x}} \boldsymbol{x}_i,$$

where

$$\begin{cases} \boldsymbol{w}_{\boldsymbol{x}} = \underset{\boldsymbol{w} = (w_1, \cdots, w_N)^{\mathrm{T}} \in \mathfrak{R}^N}{\arg \min} G(\boldsymbol{w} \,|\, P_\kappa \varphi(\boldsymbol{x}), \{\boldsymbol{x}_i\}), \\ s.t. \ \boldsymbol{w}_{\boldsymbol{x}} \ \text{satisfies some constraints.} \end{cases} \tag{16.11}$$

Recently, based on the above framework, Zheng et al. have developed an advanced algorithm called the penalized pre-image learning (P²L) [9]. The P²L can utilize weakly supervised information to alleviate the ill-posed estimation in pre-image learning and will be detailed in the following sections.

16.3.2 Pre-image Learning Using Weakly Supervised Penalty

Given the projection of the feature vector $\varphi(\boldsymbol{x})$ onto the kernel principal component subspace $P_\kappa \varphi(\boldsymbol{x})$, we define the s nearest neighborhood set $U_s(\boldsymbol{x}) = \{\varphi(\tilde{\boldsymbol{x}}_1), \cdots, \varphi(\tilde{\boldsymbol{x}}_s)\}$ of $\varphi(\boldsymbol{x})$ and let $\boldsymbol{\Psi}_{s,\varphi(\boldsymbol{x})} = [\varphi(\tilde{\boldsymbol{x}}_1), \cdots, \varphi(\tilde{\boldsymbol{x}}_s)]$, where $\tilde{\boldsymbol{x}}_i \in U(\boldsymbol{x}, s)$, $i = 1, \cdots, s$ and $\varphi(\tilde{\boldsymbol{x}}_j) \neq \varphi(\tilde{\boldsymbol{x}}_{j'})$ for any $j \neq j$. We specifically consider the following criterion

$$\hat{\boldsymbol{x}} = \boldsymbol{\Psi}_{s,\boldsymbol{x}} \boldsymbol{w}_{\boldsymbol{x}} = \sum_{i=1}^{s} w_i^{\boldsymbol{x}} \tilde{\boldsymbol{x}}_i,$$

where

$$\begin{cases} \boldsymbol{w}_{\boldsymbol{x}} = (w_1^{\boldsymbol{x}}, \cdots, w_s^{\boldsymbol{x}})^{\mathrm{T}} = \underset{\boldsymbol{w} = (w_1, \cdots, w_s)^{\mathrm{T}} \in \mathfrak{R}^s}{\arg \min} \left\| \boldsymbol{\Psi}_{s,\varphi(\boldsymbol{x})} \boldsymbol{w} - P_\kappa \varphi(\boldsymbol{x}) \right\|^2 \\ s.t. \ \sum_{i=1}^{s} w_i^{\boldsymbol{x}} = 1 \ \& \ w_i^{\boldsymbol{x}} \geqslant 0, \quad i = 1, \cdots, s. \end{cases}$$

$$\tag{16.12}$$

where the convexity constraint is imposed in order to ensure the learned pre-image to be well-defined, i.e., the entries of $\hat{\boldsymbol{x}}$ are not out of range. For example, the range of entry is always set to be $0-1$ or $0-255$ for 8-bit image. Moreover, in this model the idea of locally linear embedding (LLE) [12] is absorbed, which preserves local information through the local least square reconstruction weights. This is motivated by the perspective that the dimension of feature space $\boldsymbol{H}_\kappa$ is always much higher than that of input data space $\boldsymbol{X}$ when using popular kernels such as RBF. Thus, finding the pre-image of $P_\kappa \varphi(\boldsymbol{x})$ could be viewed as finding an embedding point for $P_\kappa \varphi(\boldsymbol{x})$ in $\boldsymbol{X}$.

In order to guide the pre-image learning using prior knowledge, the above criterion for learning $\boldsymbol{w_x}$ can be further penalized by integrating a penalty function $F(\boldsymbol{w})$ into the optimization function as follows:

$$\hat{\boldsymbol{x}} = \boldsymbol{\Psi}_{s,\boldsymbol{x}} \boldsymbol{w_x} = \sum_{i=1}^{s} w_i^{\boldsymbol{x}} \tilde{\boldsymbol{x}}_i,$$

where

$$
\begin{cases}
\boldsymbol{w_x} = \underset{\boldsymbol{w}=(w_1,\cdots,w_s)^{\mathrm{T}} \in \Re^s}{\arg\min} \quad \boldsymbol{w}^{\mathrm{T}} \boldsymbol{\Psi}_{s,\varphi(\boldsymbol{x})}^{\mathrm{T}} \boldsymbol{\Psi}_{s,\varphi(\boldsymbol{x})} \boldsymbol{w} \\
\qquad -2(\boldsymbol{\Psi}_\varphi \boldsymbol{\gamma}^{\boldsymbol{x}})^{\mathrm{T}} \boldsymbol{\Psi}_{s,\varphi(\boldsymbol{x})} \boldsymbol{w} + \lambda \cdot F(\boldsymbol{w}), \\
s.t. \sum_{i=1}^{s} w_i^{\boldsymbol{x}} = 1, \quad w_i^{\boldsymbol{x}} \geqslant 0, \quad i = 1, \cdots, s,
\end{cases}
\tag{16.13}
$$

where $\lambda \geqslant 0$ and the term $(\boldsymbol{\Psi}_\varphi \boldsymbol{\gamma}^{\boldsymbol{x}})^{\mathrm{T}} \boldsymbol{\Psi}_\varphi \boldsymbol{\gamma}^{\boldsymbol{x}}$ which is independent of $\boldsymbol{w}$ in the first square part of the objective function is excluded.

To incorporate any prior positive and negative information which benefits for the target applications, the weakly supervised prior knowledge is modeled here. The weakly supervised knowledge means only positive class information and negative class information are available and the exact class labels of samples are indeed unknown. For pre-image learning, positive class is defined as the sample set which the pre-image is expected close to and the negative class is defined as the sample set which the pre-image is expected far away from.

Denote the positive class by $C^+ = \{z_1^+, \cdots, z_{N_+}^+\}$ and the negative class by $C^- = \{z_1^-, \cdots, z_{N_-}^-\}$, where z_i^+ and z_j^- are samples in the positive class and negative class respectively. For penalization, the learned pre-image is expected close to the local positive class information and far away from the local negative class information. Therefore, the weakly supervised penalty is introduced as follows:

$$
\begin{aligned}
F(\boldsymbol{w}) &= \eta^+ |H_\theta^+(\boldsymbol{x})|^{-1} \Big[\sum_{z^+ \in H_\theta^+(\boldsymbol{x})} \| \boldsymbol{\Psi}_{s,\boldsymbol{x}} \boldsymbol{w} - z^+ \|^2 \Big] \\
&\quad -\eta^- |H_\theta^-(\boldsymbol{x})|^{-1} \Big[\sum_{z^- \in H_\theta^-(\boldsymbol{x})} \| \boldsymbol{\Psi}_{s,\boldsymbol{x}} \boldsymbol{w} - z^- \|^2 \Big] \\
&= \eta^+ |H_\theta^+(\boldsymbol{x})|^{-1} \Big[\sum_{z^+ \in H_\theta^+(\boldsymbol{x})} \boldsymbol{w}^{\mathrm{T}} \boldsymbol{\Psi}_{s,\boldsymbol{x}}^{\mathrm{T}} \boldsymbol{\Psi}_{s,\boldsymbol{x}} \boldsymbol{w} - 2z^{+\mathrm{T}} \boldsymbol{\Psi}_{s,\boldsymbol{x}} \boldsymbol{w} + z^{+\mathrm{T}} z^+ \Big] \\
&\quad -\eta^- |H_\theta^-(\boldsymbol{x})|^{-1} \Big[\sum_{z^- \in H_\theta^-(\boldsymbol{x})} \boldsymbol{w}^{\mathrm{T}} \boldsymbol{\Psi}_{s,\boldsymbol{x}}^{\mathrm{T}} \boldsymbol{\Psi}_{s,\boldsymbol{x}} \boldsymbol{w} - 2z^{-\mathrm{T}} \boldsymbol{\Psi}_{s,\boldsymbol{x}} \boldsymbol{w} + z^{-\mathrm{T}} z^- \Big] \\
&= (\eta^+ - \eta^-) \boldsymbol{w}^{\mathrm{T}} \boldsymbol{\Psi}_{s,\boldsymbol{x}}^{\mathrm{T}} \boldsymbol{\Psi}_{s,\boldsymbol{x}} \boldsymbol{w}
\end{aligned}
$$

$$
\begin{aligned}
&-(2\eta^{+}|H_{\theta}^{+}(\boldsymbol{x})|^{-1}\sum_{\boldsymbol{z}^{+}\in H_{\theta}^{+}(\boldsymbol{x})}\boldsymbol{z}^{+\mathrm{T}}\\
&-2\eta^{-}|H_{\theta}^{-}(\boldsymbol{x})|^{-1}\sum_{\boldsymbol{z}^{-}\in H_{\theta}^{-}(\boldsymbol{x})}\boldsymbol{z}^{-\mathrm{T}})\,\boldsymbol{\varPsi}_{s,\boldsymbol{x}}\boldsymbol{w}\\
&+(\eta^{+}|H_{\theta}^{+}(\boldsymbol{x})|^{-1}\sum_{\boldsymbol{z}^{+}\in H_{\theta}^{+}(\boldsymbol{x})}\boldsymbol{z}^{+\mathrm{T}}\boldsymbol{z}^{+}\\
&-\eta^{-}|H_{\theta}^{-}(\boldsymbol{x})|^{-1}\sum_{\boldsymbol{z}^{-}\in H_{\theta}^{-}(\boldsymbol{x})}\boldsymbol{z}^{-\mathrm{T}}\boldsymbol{z}^{-}),
\end{aligned}
\tag{16.14}
$$

where $H_{\theta}^{+}(\boldsymbol{x})=\{\boldsymbol{z}^{+}|\boldsymbol{z}^{+}\in C^{+},\quad \varphi(\boldsymbol{z}^{+})$ is one of the θ^{+} samples of $P_{\kappa}\varphi(\boldsymbol{x})\}$, $H_{\theta}^{-}(\boldsymbol{x})=\{\boldsymbol{z}^{-}|\boldsymbol{z}^{-}\in C^{-},\quad \varphi(\boldsymbol{z}^{-})$, is one of the θ^{+} samples of $P_{\kappa}\varphi(\boldsymbol{x})\}$, $\eta^{+}\geqslant 0$, $\eta^{-}\geqslant 0$, $|H_{\theta}^{+}(\boldsymbol{x})|$ and $|H_{\theta}^{-}(\boldsymbol{x})|$ are the cardinalities of $H_{\theta}^{+}(\boldsymbol{x})$ and $H_{\theta}^{-}(\boldsymbol{x})$ respectively.

Define $\boldsymbol{z}_{\boldsymbol{x}}^{\eta^{+},\eta^{-},\theta}=2\eta^{+}|H_{\theta}^{+}(\boldsymbol{x})|^{-1}\sum_{\boldsymbol{z}^{+}\in H_{\theta}^{+}(\boldsymbol{x})}\boldsymbol{z}^{+\mathrm{T}}-2\eta^{-}|H_{\theta}^{-}(\boldsymbol{x})|^{-1}\times$

$\sum_{\boldsymbol{z}^{-}\in H_{\theta}^{-}(\boldsymbol{x})}\boldsymbol{z}^{-\mathrm{T}}$. Then by removing any terms independent of $\boldsymbol{w}$, the pre-image learning using weakly supervised penalty is formulated as follow:

$$
\boldsymbol{w}_{\boldsymbol{x}}=\underset{\boldsymbol{w}=(w_{1},\cdots,w_{s})^{\mathrm{T}}}{\arg\min}\left\{
\begin{array}{c}
\boldsymbol{w}^{\mathrm{T}}(\boldsymbol{\varPsi}_{s,\varphi(\boldsymbol{x})}^{\mathrm{T}}\boldsymbol{\varPsi}_{s,\varphi(\boldsymbol{x})}+(\eta^{+}-\eta^{-})\boldsymbol{\varPsi}_{s,\boldsymbol{x}}^{\mathrm{T}}\boldsymbol{\varPsi}_{s,\boldsymbol{x}})\boldsymbol{w}\\
-[2(\boldsymbol{\varPsi}_{\varphi}\boldsymbol{\gamma}^{\boldsymbol{x}})^{\mathrm{T}}\boldsymbol{\varPsi}_{s,\varphi(\boldsymbol{x})}+\boldsymbol{z}_{\boldsymbol{x}}^{\eta^{+},\eta^{-},\theta}\boldsymbol{\varPsi}_{s,\boldsymbol{x}}]\boldsymbol{w}
\end{array}
\right\}
$$

$$
s.t.\ \sum_{i=1}^{s}w_{i}^{\boldsymbol{x}}=1,\quad w_{i}^{\boldsymbol{x}}\geqslant 0,\quad i=1,\cdots,s,
\tag{16.15}
$$

where we let $\eta^{+}\leftarrow\lambda\eta^{+}$ and $\eta^{-}\leftarrow\lambda\eta^{-}$.

Note that, when the convexity constraint is not considered, P^{2}L becomes the regularized locality preserving method for pre-image learning [16.13]. It has been shown in Ref. [9] that P^{2}L in such a special case also works more effectively as compared to other pre-image learning method and has approximate performance with the complete P^{2}L if the weakly supervised penalty is not used.

Finally, the insight of the penalized model is discussed as follows. Define a function $\Theta_{s,\lambda,F}(P_{\kappa}\varphi(\boldsymbol{x}))=\sum_{i=1}^{s}w_{i}^{\boldsymbol{x}}\varphi(\tilde{\boldsymbol{x}}_{i})$, where $w_{i}^{\boldsymbol{x}}$ are learned by Criterion (16.13). Since $\sum_{i=1}^{s}w_{i}^{\boldsymbol{x}}=1$, we can estimate the following conditional probability $P(\varphi(\tilde{\boldsymbol{x}}_{i})|\Theta_{s,\lambda,F}(P_{\kappa}\varphi(\boldsymbol{x})))=w_{i}^{\boldsymbol{x}}$, $i=1,\cdots,s$.

The $P(\varphi(\tilde{\boldsymbol{x}}_{i})|\Theta_{s,\lambda,F}(P_{\kappa}\varphi(\boldsymbol{x})))$ is called the penalized probability relationship between $P_{\kappa}\varphi(\boldsymbol{x})$ and its neighbor $\varphi(\tilde{\boldsymbol{x}}_{i})$, as $\Theta_{s,\lambda,F}(P_{\kappa}\varphi(\boldsymbol{x}))$ can be viewed as a penalized approximation value to $P_{\kappa}\varphi(\boldsymbol{x})$ given penalty function F and parameters s and λ. Also, as the pre-image $\hat{\boldsymbol{x}}$ is found by additive combination of s neighboring pre-images of $\varphi(\tilde{\boldsymbol{x}}_i)$, so we can also estimate the conditional probability $P(\tilde{\boldsymbol{x}}_i|\hat{\boldsymbol{x}}, s, \lambda, F)$ as follows:

$$P(\tilde{\boldsymbol{x}}_i|\hat{\boldsymbol{x}}, s, \lambda, F) = w_i^x = P(\varphi(\tilde{\boldsymbol{x}}_i)|\Theta_{s,\lambda,F}(P_\kappa \varphi(\boldsymbol{x}))). \tag{16.16}$$

To this end, the penalized probability relationship between $P_\kappa \varphi(\boldsymbol{x})$ and its neighbors $\varphi(\tilde{\boldsymbol{x}}_i)$ in the feature space is preserved by the learned pre-image in the input data space. Moreover, if it is assumed that the prior probabilities $P(\varphi(\tilde{\boldsymbol{x}}_i))$ and $P(\tilde{\boldsymbol{x}}_i)$ are the same, i.e., $P(\varphi(\tilde{\boldsymbol{x}}_i)) = P(\tilde{\boldsymbol{x}}_i) = \dfrac{1}{N}$, where N is the sample size, we can further have the following relationship:

$$\begin{aligned}
SI&(\varphi(\tilde{\boldsymbol{x}}_i), \Theta_{s,\lambda,F}(P_\kappa \varphi(\boldsymbol{x}))\,|s, \lambda, F) \\
&= log\,\frac{P(\varphi(\tilde{\boldsymbol{x}}_i), \Theta_{s,\lambda,F}(P_\kappa \varphi(\boldsymbol{x}))\,|s, \lambda, F)}{P(\varphi(\tilde{\boldsymbol{x}}_i))P(\Theta_{s,\lambda,F}(P_\kappa \varphi(\boldsymbol{x}))\,|s, \lambda, F)} \\
&= log\,\frac{P(\varphi(\tilde{\boldsymbol{x}}_i)\,|\Theta_{s,\lambda,F}(P_\kappa \varphi(\boldsymbol{x})))}{P(\varphi(\tilde{\boldsymbol{x}}_i))} \\
&= log(Nw_i^x) \\
&= log\,\frac{P(\tilde{\boldsymbol{x}}_i\,|\hat{\boldsymbol{x}}, s, \lambda, F)P(\hat{\boldsymbol{x}}\,|s, \lambda, F)}{P(\tilde{\boldsymbol{x}}_i)P(\hat{\boldsymbol{x}}\,|s, \lambda, F)} \\
&= SI(\tilde{\boldsymbol{x}}_i, \hat{\boldsymbol{x}}\,|s, \lambda, F),
\end{aligned} \tag{16.17}$$

where $P(\varphi(\tilde{\boldsymbol{x}}_i)\,|\Theta_{s,\lambda,F}(P_\kappa \varphi(\boldsymbol{x})), s, \lambda, F) = P(\varphi(\tilde{\boldsymbol{x}}_i)\,|\Theta_{s,\lambda,F}(P_\kappa \varphi(\boldsymbol{x})))$, $P(\varphi(\tilde{\boldsymbol{x}}_i)\,|s, \lambda, F) = P(\varphi(\tilde{\boldsymbol{x}}_i))$ and $P(\tilde{\boldsymbol{x}}_i\,|s, \lambda, F) = P(\tilde{\boldsymbol{x}}_i)$. We call $SI(\varphi(\tilde{\boldsymbol{x}}_i)$, $\Theta_{s,\lambda,F}(P_\kappa \varphi(\boldsymbol{x}))\,|s, \lambda, F)$ the penalized pointwise conditional mutual information between $P_\kappa \varphi(\boldsymbol{x})$ and its neighbor $\varphi(\tilde{\boldsymbol{x}}_i)$. This indicates that the penalized pointwise mutual information between $P_\kappa \varphi(\boldsymbol{x})$ and $\varphi(\tilde{\boldsymbol{x}}_i)$ in the feature space is preserved as $SI(\tilde{\boldsymbol{x}}_i, \hat{\boldsymbol{x}}\,|s, \lambda, F)$ the pointwise conditional mutual information between the pre-images $\hat{\boldsymbol{x}}$ and $\tilde{\boldsymbol{x}}_i$ in the input data space.

16.4 Face Image Preprocessing Using KPCA

In this section, we demonstrate the performance of KPCA+Pre-image Learning by applying it to five different face image preprocessing applications, namely three standard evaluations on facial expression normalization, image denoising and occlusion recovery, and two more challenging applications including illumination normalization and face image hallucination.

For all applications, all images were linearly stretched to full range of pixel values of $[0, 1]$. We demonstrate the three standard applications based on two kernel functions, namely a RBF kernel $\kappa(\boldsymbol{x}_i, \boldsymbol{x}_j) = \exp(-||\boldsymbol{x}_i - \boldsymbol{x}_j||^2/c)$, $c = 10^{-4}$ and a polynomial kernel $\kappa(\boldsymbol{x}_i, \boldsymbol{x}_j) = (< \boldsymbol{x}_i, \boldsymbol{x}_j > + \theta)^d$, $\theta = 0.1$ and $d = 0.5$, and only use the RBF kernel for the other two more challenging applications. In all applications, after KPCA was learned and a KPCA subspace was determined by retaining the largest kernel principal components that were selected to preserve 95% energy. For learning the pre-image, the penalized pre-image learning (P²L) algorithm [9] and its variant [13, 9] was used.

16.4.1 Expression Normalization

Figure 16.2 shows the facial expression normalization results as compared to the PCA method. The experiment was conducted on the CKFE database [14]. CKFE includes image sequences of different people, and each sequence describes the variations from natural expression to a particular facial expression. All face images were manually cropped from the original images. Totally, there are 8795 images of different types of expressions from 97 people, including surprise, fear, joy, sadness, disgust, and anger.

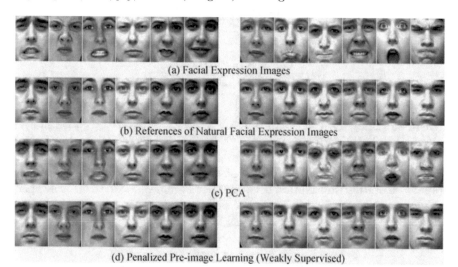

(a) Facial Expression Images

(b) References of Natural Facial Expression Images

(c) PCA

(d) Penalized Pre-image Learning (Weakly Supervised)

Fig. 16.2 Facial expression normalization using KPCA with pre-image learning. In the last row, the left six images are results based on RBF kernel and the right six images are results based on polynomial kernel. Five neighbors are used in P²L. The importance weight λ in P²L is 10^{-4}.

In this application, the first three images from each sequence of facial expression were used to train the KPCA model, since they are nearly natural facial expression. The rest images in the sequence except the last image were used for testing. So there were 1461 training images and 6847 testing images. The last image of each facial expression image sequence was treated as the negative sample for modelling the weakly supervised penalty. Note that in our experiments, for modelling the weakly supervised penalty, we only incorporate negative information. Totally, there were 487 negative class samples. For the P²L, we set $s = 10$, $\eta^+ = 0$ and $\eta^- = 10^{-4}$ in Eq. (16.15) for modelling the weakly supervised penalty.

As shown by using pre-image learning, much better visual results are obtained, particularly around the eyes and mouth in an image. The quantified results in terms of MSE can be found in Table 16.1. As there is no unique ground truth natural facial expression for each person in CKFE, but a set of training images of nearly natural facial expression is available. The mean square error (MSE) value of a normalized facial expression image was then obtained by computing the minimum MSE between it and all training images of the same person. As shown, a much lower MSE is obtained by using KPCA+Pre-image Learning.

Table 16.1 MSE: PCA vs. KPCA + Pre-image Learning

Methods		Expression Normalization	Denoising	Occlusion Recovery
PCA		48.76	53.381	78.924
P^2L	RBF	11.417	36.536	28.506
	Polyn.	14.013	39.696	26.909

16.4.2 Image Denoising and Occlusion Recovery

The YALEB database [15] was selected for demonstrating these two applications. YALEB consists of 10 people with 9 different poses. For each pose, there are 65 face images undergoing various illuminations. Face images of each pose in YALEB are divided into 5 subsets according to the light-source directions. The light-source directions of images in subset 1 are frontal or nearly frontal. In the application, each image was aligned and resized to 60×80.

The applications were conducted on subset 1 of YALEB database, of which images were captured under normal or nearly normal illumination condition. In that subset, for each pose, there are 7 images for each person. The training set was established by randomly selecting 6 images from each pose for each person in the subset 1 to train a KPCA model, and the rest 1 image of each pose from each person was treated as the clean testing image. For denoising, two testing noisy images were generated by adding noise on each clean testing image. Note that only Gaussian noise with zero mean and random variance between 0 and 1 was used. For occlusion recovery, each occluded image was generated by placing a rectangle black patch onto each clean testing image at a random coordinate. The width and height of the patch were random, and they varied from 15 to 40 in pixels. Hence, totally 1800 testing noisy images were processed over the 10 run experiment.

The penalized pre-image learning model using weakly supervised penalty was again used with the same setting as for expression normalization. For modelling the weakly supervised penalty, for denoising, two noisy images were generated on each image in the training set using Gaussian noise with zero mean and random variance, and for occlusion recovery two occluded face images were generated with rectangle black patch randomly placed in the image. So there were also 1080 negative class samples for training for both applications in each run.

Figure 16.3 and 16.4 show the denoising and occlusion recovery results

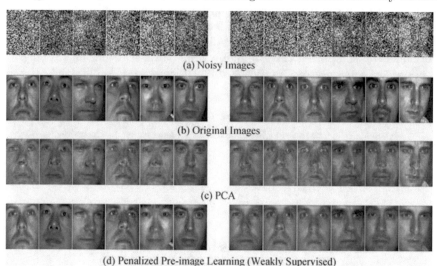

(a) Noisy Images

(b) Original Images

(c) PCA

(d) Penalized Pre-image Learning (Weakly Supervised)

Fig. 16.3 Facial image denoising using KPCA with pre-image learning. In the last row, the left six images are results based on RBF kernel and the right six images are results based on polynomial kernel. Five neighbors are used in P^2L. The importance weight λ in P^2L is 10^{-4}.

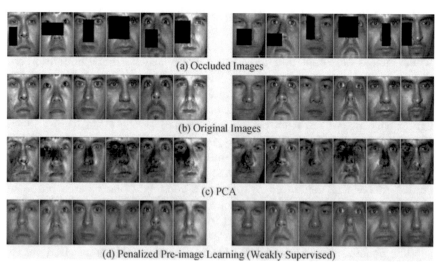

(a) Occluded Images

(b) Original Images

(c) PCA

(d) Penalized Pre-image Learning (Weakly Supervised)

Fig. 16.4 Facial expression normalization using KPCA with pre-image learning. In the last row, the left six images are results based on RBF kernel and the right six images are results based on polynomial kernel. Five neighbors are used in P^2L. The importance weight λ in P^2L is 10^{-4}.

respectively, and the MSE comparison can also be found in Table 16.1. As shown, similar to the comparison results for expression normalization, the KPCA+Pre-image Learning does obtain significant improvement no matter in terms of the image quality of reconstructed images or the MSE values.

16.4.3 Illumination Normalization

The illumination normalization here is to restore a frontal illuminated face image based on a single input face image [16].

Based on the multi-linear representation of face images [17], the set of face images from all objects, under the same illumination condition, can be approximated by a linear subspace. Accordingly, a frontal illuminated face image can be reconstructed based on an input face image by using the KPCA+Pre-image technology.

Unfortunately, the P^2L algorithm cannot be directly used in this case, since it requires many samples for each person. Furthermore, due to the individual characteristics, other pre-image algorithms always confront the "generalization problem", that is, it cannot accurately reconstruct an image of a face that is outside the training set. To address this problem, the KPCA+Pre-image technology would be incorporated into the small- and large-scale features based illumination-normalization framework [18]. In details, the KPCA+Pre-image based method consists of the following four steps:

1) Performing image decomposition.

The input image I is first decomposed into the small-scale features ρ and large-scale features S by using the Logarithmic Total Variation (LTV) model [19]. Based on the Lambertian reflectance model, Chen et al. [19] proposed the large- and small-scale features model as follows:

$$I(x, y) = \rho(x, y)S(x, y), \qquad (16.18)$$

where I is the intensity, the small-scale features ρ contain the small intrinsic structures of a face, and the large-scale features S contains the extrinsic illumination, shadows cast by bigger objects, and the large intrinsic facial structures. Generally, ρ is illumination invariant and S is affected by illumination variations. To obtain the decomposition in Eq. (16.18). The LTV model is utilized. LTV has good capabilities of edge-preserving [19], so that it can extract illumination-invariant features as well. The decomposition can be formulated as follows:

$$f = \log(I), \qquad (16.19)$$

$$\hat{u} = arg \min_{u} \left\{ \int |\nabla u| + \lambda \|f - u\|_{L^1} \right\}, \qquad (16.20)$$

$$\hat{v} = f - \hat{u}, \qquad (16.21)$$

$$\rho \approx exp(\hat{v}), S \approx exp(\hat{u}), \qquad (16.22)$$

where $\int |\nabla u|$ is the total variation of u, and λ is a scalar threshold. Generally, $\lambda = 0.4$ is a suitable value for the image of resolution size of 100×100. The interior-point Second-Order Cone Program (SOCP) algorithm [19] or the Parametric Maximum Flow (PMF) algorithm [20] can be used for the computation of the TV model. The PMF algorithm produces almost the same result as the SOCP algorithm, but its computation is much faster and is almost in real time.

2) Conducting Smooth operation on the small-scale features.

In cast shadows, the small-scale features ρ may not be correctly recovered and some "light spots" may appear in the estimated result. For obtaining a better visual result, a threshold-average filtering is performed on ρ. Suppose (x_0, y_0) is the center of the convolution region, then the average filter kernel will convolute if only if $\rho(x_0, y_0) > \theta$, where θ is the threshold. Here, a 10×10 filtering mask was used, and the θ was chosen individually for each ρ, such that the values of 99% pixels were smaller than θ, and the values of the remaining 1% pixels were larger than θ. In other words, this threshold-average filtering replaces the largest 1% value of ρ by the local average value. This threshold smoothing does not change the facial structures almost.

3) Performing illumination normalization on the large-scale features by using KPCA+Pre-image technology

First, the large-scale features, which are decomposed from the frontal-and nearly frontal- illuminated face images, are used to learn a KPCA space where 95% of the energy is maintained. Then, for an input image I, its large-scale features S are projected onto that KPCA subspace to eliminate the illumination variations. Finally, a normalized large-scale features S_{norm} are obtained by learning the pre-image of that projection.

4) Reconstructing normalized image.

Similar to Eq. (16.18). The frontal illuminated face image, I_{norm}, is finally produced by a combination of the normalized large-scale features S_{norm} and the smoothed small-scale features ρ_{norm}:

$$I_{norm}(x, y) = \rho_{norm}(x, y) S_{norm}(x, y). \qquad (16.23)$$

A block diagram of the demonstrated method and the result of each step are illustrated in Fig. 16.5, where the procedure of using KPCA+Pre-image technology for normalizing the large-scale features is highlighted by the red (dotted) box. In this method, the large-scale features can be normalized by KPCA, while the facial details (small-scale features) can be well preserved, making KPCA a feasible approach to restoring the frontal illuminated face image from a single input image. The demonstrated KPCA+Pre-image method does not require any shape modeling or lighting estimation. As a holistic reconstruction, KPCA+Pre-image technology incurs less local distortion.

Figure 16.6 illustrates the results of illumination normalization on face images. For comparison, the results of the following algorithms are also shown: Non-Point Light Quotient Image relighting (NPL-QI) [21], and the

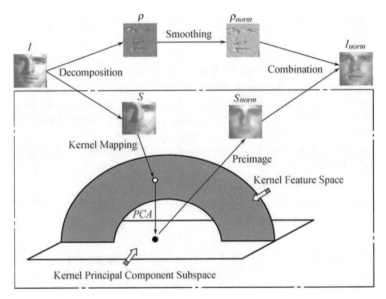

Fig. 16.5 Diagram of the KPCA+Pre-image based method for illumination normalization. The process of KPCA+Pre-image is performing inside the blue (dotted) box.

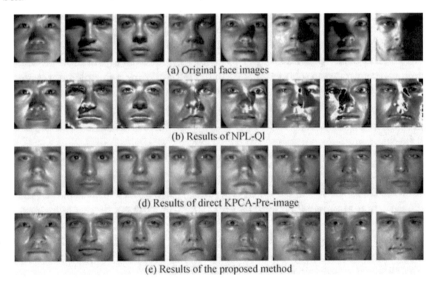

Fig. 16.6 Reconstruction of frontal illuminated face images by using different methods. The first row are the original face images, the second are the processed results of NPL-QI, the third row are the results of direct KPCA+Pre-image, the fourth row illustrates the results of KPCA+Pre-image used in the small- and large-scale framework.

direct KPCA+Pre-image approach. By using KPCA+Pre-image algorithm, five frontal and nearly frontal illuminated images for each person and their large-scale features were respectively used for learning the kernel matrix and the KPCA subspace. All images of the processed object were excluded from the training set. It shows that our demonstrated KPCA+Pre-image methodology preserves the intrinsic facial features well. Compared to directly applying KPCA+Pre-image technology on original images, the performance can be better at preprocessing a face image that is outside the training set. Note that by using direct KPCA+Pre-image approach, the identities of the processed faces are always changed.

16.4.4 Face Hallucination

Face hallucination, which is one of the major challenges in face image processing, is to recover the high-resolution image from its low-resolution one. In this section, we demonstrate that KPCA+Pre-image learning can be used to generate good results for face hallucination [22].

For face hallucination, the KPCA here is used to learn the relationship between the high low resolution face images in kernel feature space, and the pre-image learning technology is applied to find a high resolution face image in a high resolution face image space. More specifically, let $\Im^\Delta, \Delta \in \{l, h\}$ be the image data space, where $\Im^l$ denotes low-resolution image space, $\Im^h$ denotes the corresponding high-resolution image space, and low-resolution images are supposed to be from the smoothed and down-sampled corresponding high-resolution counterparts. In this section, we use "Δ" to tell different resolution samples, kernels, features or other information related to low- and high- resolution images. According to formulation (4), for a given sample x^Δ, we have

$$P_\kappa^\Delta \varphi^\Delta(x^\Delta) = \Psi_\varphi^\Delta \gamma^{x^\Delta}. \tag{16.24}$$

The weight vector γ^{x^Δ} is called the intrinsic feature of x^Δ in its corresponding kernel principal subspace. To this end, the following model is developed for face hallucination:

Model: The low- and high- resolution images approximately have the same intrinsic features in their corresponding kernel principal subspaces, if suitable kernels are selected for them respectively.

To interpret this model, the following theorem is introduced.

Theorem 16.1   Suppose the high-resolution kernel principal components have $U_{q_0}^{\varphi h} = \Psi_\varphi^h P_{q_0}^h$ and the low-resolution kernel principal components have $U_{q_0}^{\varphi l} = \Psi_\varphi^l P_{q_0}^l$. If $\kappa(x_i^h, x_j^h) = \kappa(x_i^l, x_j^l)$ for any $x_i^\Delta, x_j^\Delta \in \Im^\Delta$, then we have $P_{q_0}^h = P_{q_0}^l$.

Proof of the theorem can be found in A 16.1 According to Theorem 16.1 and formulation (16.4) under the assumption $\kappa(x_i^h, x_j^h) = \kappa(x_i^l, x_j^l)$, we can say $\gamma^{x^h} = \gamma^{x^l}$.

As mentioned in the beginning of this chapter, we apply RBF kernel for face hallucination. In practice, the RBF kernel well generates two approximate kernels for low-resolution and high-resolution data respectively. For visualization, we show two-dimensional results in Fig. 16.7.

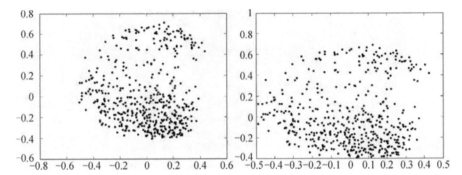

Fig. 16.7 KPCA projections of the low-resolution (left) and high-resolution (right) samples.

It can be seen that the topologies of the projections of different resolution training data onto their corresponding kernel subspaces are approximately the same, which indicates that different resolution images can approximately have the same intrinsic feature in their corresponding kernel subspaces when RBF kernels are applied.

Based on the model 1, a KPCA-based face hallucination method is introduced as follows. For an input low-resolution image, the weights γ^{x^l} are learnt by Eq. (16.24). Then, the projection of high-resolution image onto high-resolution kernel principal component subspace can be reconstructed using the same weights, namely $P_\kappa^h \varphi^h(x^h) = \Psi_\varphi^h \gamma^{x^h}$, where $\gamma^{x^h} = \gamma^{x^l}$. As $P_\kappa^h \varphi^h(x^h)$ is a feature point in kernel feature space rather than a real image, its pre-image, which is the expected high-resolution image x^h in the high-resolution image space, is finally estimated. The diagram of the algorithm is shown in Fig. 16.8.

Always, single global image based hallucination is not enough to achieve satisfactory results. Residue compensation [23 – 25] is widely used to compensate the detailed information lacked in global image. Among them, neighbor embedding for residue compensation [25] is selected to perform the residue compensation. The basic idea of this method is to consider each residue image as a patch matrix composed of overlapped square patches. Similar to the idea of Locally Linear Embedding (LLE), for a given low-resolution residue patch, it can be reconstructed by a combination of k-nearest neighboring low-resolution training residue patches. Then, the high-resolution residue patch can be synthesized by replacing the low-resolution training residue patches with high-resolution training residue patches while retaining the same combi-

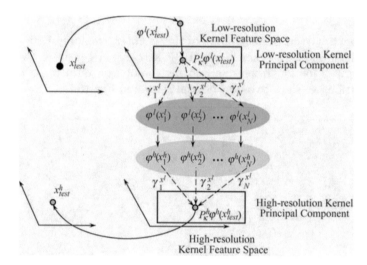

Fig. 16.8 Face hallucination using KPCA.

nation weights. After that, the overlapped parts of the high-resolution residue patches are postprocessed by averaging. Finally, the high-resolution residue face is obtained by integrating these high-resolution residue patches. By using this method, we will add the inferred high-resolution residue face to the global face which is obtained by our model.

Figure 16.9 illustrates the comparative experimental results among the KPCA-based method, bilinear interpolation and PCA-based method [26] on CAS-PEAL-R1 face database [27], which consists of 99 594 images of 1 040 individuals. In this experiment, a subset of CAS-PEAL is established by selecting face images of all people under good illumination condition. All face images are simply aligned according to the eyes and mouth. The high resolution face images are with size 96×128. By smoothing and down-sampling of high-resolution images, the corresponding low-resolution images are obtained with size 24×32. The results are similar to the original high-resolution face

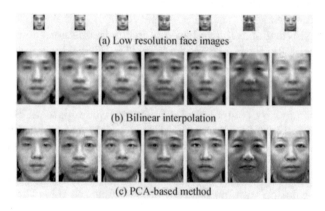

(a) Low resolution face images

(b) Bilinear interpolation

(c) PCA-based method

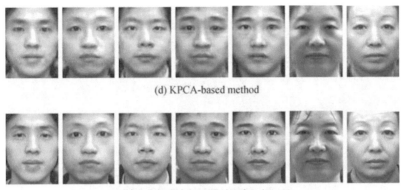

(d) KPCA-based method

(e) Original high resolution face images

Fig. 16.9 Face hallucination using different methods.

and detailed facial features can be recovered much better.

16.5 Summary

Recent developments of using kernel methods, especially KPCA, for face image analysis are introduced. Especially, a general framework for pre-image learning in kernel methods and the penalized pre-image learning in KPCA are presented. With these new developments, it shows that KPCA is effective in performing nonlinear image preprocessing such as expression normalization, illumination normalization, occlusion recovery and image denoising so as to obtain significantly better image quality after preprocessing. By using KPCA plus pre-image learning, the benefit for face image preprocessing including facial expression normalization, face image denoising and occlusion recovery, face image illumination normalization and face hallucination are shown.

Research on kernel methods for pattern recognition and computer vision is still attractive nowadays. Compared to the classification problem, it is still not widely investigated how much and how well kernel methods can perform for nonlinear image preprocessing and more open problems are still there. Nonetheless, the demonstrations shown here have shown the perspective of kernel methods in this route.

Acknowledgements

This project was supported by the NSFC-GuangDong (U0835005), NSFC (60803083), and the 985 project in Sun Yat-sen University with grant no. 35000-3181305.

Appendix

Proof of Theorem 16.1

Let $K^{\Delta} = \Psi_{\varphi}^{\Delta^{\mathrm{T}}} \Psi_{\varphi}^{\Delta}$, $H = I_N - N^{-1}e_N e_N^{\mathrm{T}}$, and $D^{\Delta} = O_t^{\varphi\Delta^{\mathrm{T}}} O_t^{\varphi\Delta}$. We can first solve the eigenvalue problem below:

$$D^{\Delta} V^{\Delta} = V^{\Delta} \Lambda^{\Delta} \tag{A16.1}$$

According to Eqs. (16.1) and (16.3), we have

$$U_{q_0}^{\varphi\Delta} = \Psi_{\varphi}^{\Delta} P_{q_0}^{\Delta} = \Psi_{\varphi}^{\Delta}(N^{-1/2} H V^{\Delta} \Lambda^{\Delta^{-1/2}}) \tag{A16.2}$$

It can be seen that V^{Δ} and Λ^{Δ} are only determined by K^{Δ}, and H is a constant, so $P_{q_0}^{\Delta}$ is only determined by K^{Δ}. Therefore if $\kappa(x_i^h, x_j^h) = \kappa(x_i^l, x_j^l)$, which means $K^l = K^h$, we have $P_{q_0}^h = P_{q_0}^l$. This proves Theorem 16.1.

References

[1] Schölkopf B, Smola A J (2002) Learning with Kernels. MIT Press, Cambridge
[2] Mika S, Ratsch G, Weston J et al (1999) Fisher Discriminant Analysis with Kernels. Proceedings of the IEEE Signal Processing Society Workshop, pp 41–48
[3] Schölkopf B, Smola A, Müller K (1998) Advances in Neural Information Processing Systems Neural Computation, 10(5): 1299–1319
[4] Yang J, Gao X, Zhang D et al (2005) Kernel Ica: An Alternative Formulation and its Application to Face Recognition. Pattern Recognition, 38(10): 1784–1787
[5] Zheng W, Zhou X, Zou C et al (2006) Facial Expression Recognition Using Kernel Canonical Correlation Analysis. IEEE Transactions on Neural Networks, 17(1): 233–238
[6] Mika S, Schölkopf B, Smola A et al (1999) In Kernel Pca and De-noising in Feature Spaces. Advances in Neural Information Processing Systems, 11: 536–524
[7] Kwok J T, Tsang I W (2004) The Pre-image Problem in Kernel Methods. IEEE Transacations on Neural Networks, 15(6): 1517–1525
[8] Bakır G H, Weston J, Schölkopf B (2004) Learning to Find Pre-images. Advances in Neural Information Processing Systems, 16: 449–456
[9] Zheng W S, Lai J, Yuen P C (2010) Penalized Preimage Learning in Kernel Principal Component Analysis. IEEE Transactions on Neural Networks, 21(4): 551–570
[10] Dambreville S, Rathi Y, Tannenbaum A (2006) Statistical Shape Analysis Using Kernel Pca. IS&T/SPIE Symposium on Electronic Imaging
[11] Arias P, Randall G, Sapiro G (2007) Connecting The Out-of-Sample and Pre-image Problems in Kernel Methods. IEEE Conference on Computer Vision and Pattern Recognition, October 2007, Rio de Janeiro, Brazil
[12] Tenenbaum J B, Silvam V D, Langford J C (2000) A Global Geometric Framework for Nonlinear Dimensionality Reduction. Science, 290(4): 2319–2323
[13] Zheng W S, Lai J (2006) Regularized Locality Preserving Learning of Pre-image Problem in Kernel Principal Component Analysis. ICPR, 2: 456–459
[14] Tian Y l, Kanade T, Cohn J F (2001) Recognizing Action Units for Facial Expression Analysis. IEEE Transactions on Pattern Analysis and Machine Intelligence, 23(2): 97–115
[15] Sim T, Baker S, Bsat M (2003) The Cmu Pose, Illumination, and Expression Database. IEEE Trans on Pattern Anal Mach Intell, 25(12): 1615–,1619
[16] Xie X, Zheng W S, Lai J et al (2010) Normalization of Face Illumination Based on Large-and Small-scale Features. IEEE Trans on Image Processing (Accepted)

[17] Vasilescu M, Terzopoulos D (2002) Multilinear Analysis of Image Ensembles: Ten-sorfaces. In: European Conference on Computer Vision

[18] Xie X, Zheng W S, Lai J et al (2008) Face Illumination Normalization on Large and Small Scale Features. In: IEEE Computer Society Conference on Computer Vision and Pattern Recognition, In Proceedings of ECCV, 1: 447 – 460

[19] Chen T, Yin W, Zhou X et al (2006) Total Variation Models for Variable Lighting Face Recognition. IEEE Transactions on Pattern Analysis and Machine Intelligence, 28(9): 1519 – 1524

[20] Goldfarb D, Yin W (2009) Parametric Maximum Flow Algorithms for Fast Total Variation Minimization. SIAM Journal on Scientific Computing, 31(5): 3712 – 3743

[21] Wang H, Li S, Wang Y (2004) Generalized Quotient Image. In: Proceedings IEEE International Conference on Computer Vision and Pattern Recognition, pp 819 – 824

[22] Liang Y, Lai J H, Zou Y X et al (2009) Face Hallucination Through Kpca. In: Proceedings of CISP '09 2nd edn International Congress on Image and Signal Processing (CISP), pp 1 – 5

[23] Liu C, Shum H, Zhang C S (2001) A Two-step Approach to Hallucinating Faces: Global Parametric Model and Local Nonparametric Model. IEEE Transactions on Computer Vision and Pattern Recognition, pp 192 – 198

[24] Liu W, Lin D, Tang X (2005) Neighbor Combination and Transformation for Hallucinating Faces. In: IEEE International Conference on Multimedia and Expo, p 4

[25] Zhuang Y, Zhang J, Wu F (2007) Hallucinating Face: Lph Super-resolution and Neighbor Reconstruction for Residue Compensation. Pattern Recognition, 40(11): 3178 – 3194

[26] Wang H, Li S Z, Wang Y (2004) Generalized Quotient Image. In Proceedings of IEEE International Conference on Computer Vision and Pattern Recognition, 27 June – 2 July, 2004, Washington DC

[27] Gao W, Cao B, Shan S et al (2008) The Cas-peal Large-scale Chinese Face Database and Baseline Evaluations. IEEE Transacations on System Man and Cybernetics, 38(1): 149 – 161

17 Fingerprint Identification — Ideas, Influences, and Trends of New Age

Sangita Bharkad[1] and Manesh Kokare[1]

Abstract   Contrary to popular belief, despite decades of research in fingerprints, reliable fingerprint recognition from large database is an open problem. Extracting features out of poor quality prints is the most challenging problem faced in this area. For that we need effective and efficient fingerprint matching algorithms that meet user requirements, to identify similarity. This paper gives a brief survey of current fingerprint matching methods and technical achievement in this area. The survey includes a large number of papers covering the research aspects of system design and applications of fingerprint matching, image feature representation and extraction. Furthermore future research directions are suggested.

17.1 Introduction

The term biometric comes from the Greek words bios (life) and metrikos (measure). Biometric technologies are becoming the foundation of an extensive array of highly secure identification and personal verification solutions. It is well known that humans intuitively use some body characteristics such as face, gait or voice to recognize each other. Since, today, a wide variety of applications require reliable verification schemes to confirm the identity of an individual, recognizing humans based on their body characteristics became more and more interesting in emerging technology applications. This technology acts as a front end to a system that requires precise identification before it can be accessed or used.

17.1.1 Motivation

The proliferation of information access terminals and the growing use of applications (such as E-commerce, e-voting and e-banking) involving the transfer of personal data make it essential to provide reliable systems that are user-friendly and generally acceptable. With conventional identity verification systems for access control, such as passports and identity cards, passwords and

---

1   S.G.G.S. Institute of Engg. and Technology, Vishnupuri, Nanded, Maharashtra, India.
    E-mails: sbharkad@yahoo.co.in, mbkokare@yahoo.com.

secret codes can easily be falsified. Biometrics seems to provide a way of overcoming some of these systems drawbacks, by basing verification on aspects that are specific to each individual. For a long time the use of biometrics has remained limited to policing applications, but in view of its potential advantages, this technology is now being considered for many other tasks. Commercial applications have thus been developed, based on facial features, fingerprints, iris, retinal scans, hand, and finger geometry; or behavioral, the traits idiosyncratic of the individual, such as voice print, gait, signature, and key stroking with distinct levels of merits.

These biometrics identifiers suffer with limitations which are overcome by using unique pattern of ridges and valleys in fingerprint image to each person or fingerprint. The two basic ideas scientists believe about permanence and individuality of fingerprints are:

1) Fingerprints never change. Small ridges form on a person's hands and feet before they are born and do not change for as long as the person lives.

2) No two fingerprints are alike. The ridges on the hands and feet of all persons have three characteristics (ridge endings, bifurcations and dots) which appear in combinations that are never repeated on the hands or feet of any two persons. A ridge ending is simply the end of a ridge. A bifurcation is a Y-shaped split of one ridge into two. A dot is a very short ridge that looks like a "dot".

The fingerprint is a unique organ which has more discriminators than any other biometric feature currently in use. Fingerprints remain unchanged throughout a lifetime. Fingerprints of even identical twins are different. Maltoni et al. [1] have observed that the maximum global similarity is present in the monozygotic (identical) twins, which is the closest genetic relationship. However, fingerprints of identical twins have different micro details which can be used for identification purposes. The overall ridge structure, singular points of fingerprint image are shown in Fig. 17.1.

Figure 17.2 shows the possible organization of various facets of the fingerprint recognition system. Figure 17.3 shows the referred journal publications based on the fingerprint matching over the last 20 years. After rigorous survey of last 20 years publications we found that the following challenges are still present in the fingerprint recognition system. Firstly, accuracy is one of the main parameter of the fingerprint recognition system. Hence, it is required to propose a novel fingerprint recognition system with minimum Equal Error Rate (ERR), possibly zero. The most of the research work is based on minutiae points (Ridge ending and ridge bifurcation) of the fingerprint image which does not work well for poor quality images. Researchers explored non-minutiae representations of fingerprints considering the fingerprint images as oriented textures that combine both the global and the local information present in a fingerprint. Gabor filter based features provide global (core point and delta point) as well as local (ridge ending and ridge bifurcation) information of the fingerprint image for accurate recognition of the fingerprint image. Secondly, coping with nonlinear distortion in the fingerprint matching is a

Fig. 17.1 Minutiæ (Ridge ending and Ridge bifurcation) and singular points (Core and Delta) of fingerprint image.

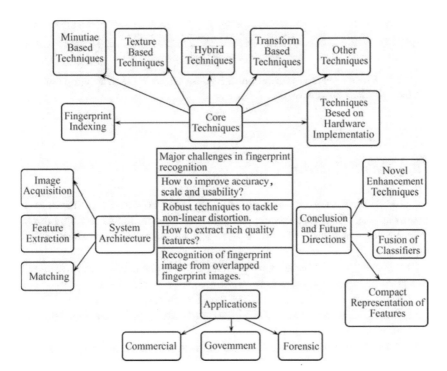

Fig. 17.2 Various facets of the fingerprint recognition system.

challenging task. Generally a fingerprint image is non-linearly deformed by torsion and traction when a fingerprint is pressed on the sensor. This nonlinear deformation creates spurious minutiae. Thirdly, speed requirement is also the main problem to be attacked. As the query image needs to be compared with each and every image in the database to find a match of the fingerprint image, it requires huge time. This needs to be reduced possibly to milliseconds. Finally, the recognition of the fingerprint image from overlapped fingerprint images.

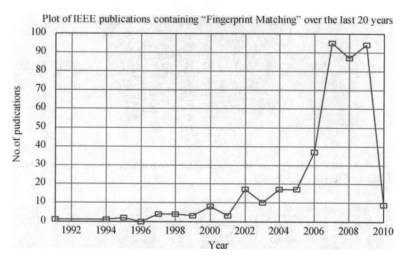

Fig. 17.3 Plot of referred journal publications containing the "Fingerprint Matching" over the last 20 years.

17.1.2 Main Contribution:

In this section, we comprehensively survey, analyze current progress and future prospects of the fingerprint recognition. It contributes in following way:

1) Various facets of the fingerprint identification as a research field are proposed.

2) Selected publications over the last 20 years are briefly elaborated which gives the core ideas about the research progress of this field.

3) Key contributions of initial phase of research are discussed.

4) Core issues of current techniques and future prospectus of the fingerprint recognition are overviewed.

Rest of the section is organized as follows. We discuss the system architecture and applications of the fingerprint matching in 17.2. 17.3 gives gradual progress of preliminary phase in research and development of the fingerprint matching. In 17.4, we review various core features extraction techniques to facilitate the fast search in large database. Conclusion and future research directions are presented in 17.5 to enhance the overall response of the fingerprint matching system.

17.2 System Architecture and Applications of Fingerprint Matching

A typical fingerprint verification process is shown in Fig. 17.4, which works in two phases: namely fingerprint enrollment phase and fingerprint matching phase. In the fingerprint enrollment phase, a sensor captures the fingerprint image from which the various features of the fingerprint image are extracted, processed, and stored as a "master template." In the fingerprint identification phase, the above process repeats, resulting in the generation of a "live template." The two fingerprint templates are matched to determine a similarity score of the two fingerprints. Matching section finds the similarity score, which can be found out using different similarity metrics. Decision is taken from the matching score of two-fingerprint image. Block diagram of the fingerprint matching system is shown in Fig. 17.4 which consists of following blocks like image acquisition system, input image database, feature extraction, feature database, feature matching and decision, etc.

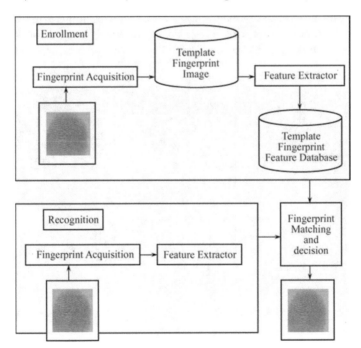

Fig. 17.4 System Architecture.

17.2.1 Image Acquisition and Image Database

The fingerprint images may be acquired either by an offline or an online process. The fingerprint images acquired by the offline process are known as the "inked" fingerprints while the images acquired by the online process are known as the "live-scan" fingerprints. In offline process, ink is applied on the fingerprint and then fingerprint impressions are taken on the paper, that is then scanned by a standard grayscale scanner. A live-scan fingerprint is obtained directly from the finger without the intermediate use of the paper. Typically, live-scan sensors capture a series of dab fingerprints when a finger is pressed on the sensor surface. Latent fingerprints are formed when the fingers leave a thin layer of sweat and grease on the surfaces that they touch due to the presence of sweat pores in our fingertips. The commercially available live-scan sensors are based on several different technologies. The optical fingerprint sensor from Digital Biometrics Inc. (model FC21RS1) is based on the "optical total internal reflection" technology (http://www.digitalbiometrics.com). The Thompson-CFS chip-based sensor works on the thermal sensing of temperature difference across the ridges and valleys (http://www.tcs.thomson-csf.com/FChome.htm). The Veridicom (www.veridicon.com) and the Siemens (www.siemens.com) sensors are based on differential capacitance. These acquired fingerprint images are stored in template fingerprint image database which is used for feature extraction. The fingerprint images acquired by using various sensors are shown in Fig. 17.5.

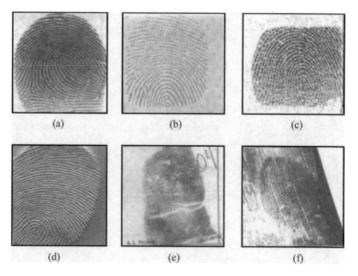

Fig. 17.5 Fingerprint images captured from (a) live scan optical Sensor; (b) live scan capacitive sensor; (c) live scan piezoelectric sensor; (d) live scan thermal sensor; (e) an offline inked impression and (f) a latent fingerprint.

17.2.2 Feature Extractor and Feature Database

A complete image cannot be used or processed every time as the memory required by it may be large. This reduces the processing speed. Hence only prominent features are extracted from each image and an image feature database is formed. Also pixel by pixel comparison with the query image may consume most of the time of the processor and a shift by one pixel value may lead to complete mismatch and pseudo results. To overcome above lacunae feature extraction is essential. The fingerprint recognition technology looks at the unique characteristics of fingerprints, minutiae points and singular points of the fingerprint image. Various feature extraction techniques with distinct features are given below.

1) In minutiae based techniques feature vector is formed by $x - y$ position and orientation of minutiae points (ridge ending and ridge bifurcation) given by following equation.

$$P = ((x_1, y_1, \theta_1)^{\mathrm{T}}, \cdots, (x_m, y_m, \theta_m)^{\mathrm{T}}), \tag{17.1}$$
$$Q = ((x_1, y_1, \theta_1)^{\mathrm{T}}, \cdots, (x_m, y_m, \theta_m)^{\mathrm{T}}), \tag{17.2}$$

where P is the feature vector of database image and Q is the feature vector of query image.

2) In texture based feature extraction techniques, Gabor filter or Discrete wavelet transform is used to extract ridge pattern of the fingerprint image. Some statistical features like mean, variance, standard deviation, and average absolute deviation are computed from ridge pattern of the fingerprint image to form feature vectors. The mean, standard deviation and average absolute deviation are given by following equations.

$$Mean = \frac{1}{M} \sum_{i=1}^{M} P_i, \tag{17.3}$$

$$std = \left(\frac{1}{M} \sum_{i=1}^{M} (P_i - \bar{P})^2 \right)^{\frac{1}{2}}, \tag{17.4}$$

$$avg.abs.std = \left(\frac{1}{M} \sum_{i=1}^{M} |P_i - \bar{P}| \right), \tag{17.5}$$

where P_i is a feature vector and $\bar{P}$ is the mean of the feature vector P and variance is square of standard deviation.

3) In hybrid fingerprint matching techniques the feature vector is formed by combining minutiae features and texture based features.

4) In transform based techniques, various image transform like discrete cosine transform (DCT), fast Fourier transform (FFT), hough transform and correlation of sub blocks fingerprint image are computed. Energy and statistical features of transformed image are computed to form the feature vector.

Energy is given by following equation.

$$E_j = \sum_{i=1}^{m} (P_{ji})^2, \tag{17.6}$$

where E_j is the energy of jth sub block of transformed imaged and P_j is jth sub block of transformed image.

The feature extraction unit computes the features of template image as well as input image. Features of template image are stored in template feature database. Features of input image are used for comparison with features of template images. The input image should be one among the database images otherwise there is no match between input image and template database image.

17.2.3 Fingerprint Matching and Decision

Given two (test and database) representations, the matching module determines whether the prints are impressions of the same finger. The matching phase typically defines a metric of the similarity between two fingerprint representations. The similarity metrics used for comparison of features of test image and reference image are given below.

1) Euclidean Metric

$$d_E(P, Q) = \sqrt{\sum_{i=1}^{M} (P_i - Q_i)^2}, \tag{17.7}$$

where P and Q are M dimensional feature vectors of test image and reference image respectively.

2) Mahalanobis Metric

$$d_{Maha}(P, Q) = \sqrt{\sum_{i=1}^{M} |P_i - Q_i|}. \tag{17.8}$$

3) Canberra Metric

$$d_{Can}(P, Q) = \sqrt{\sum_{i=1}^{M} \frac{|P_i - Q_i|}{|P_i| + |Q_i|}}. \tag{17.9}$$

4) City block

$$d_{City}(P, Q) = \sum_{i=1}^{M} |P_i - Q_i|. \tag{17.10}$$

The test image will be more similar to the database images if the distance is smaller. The distance of the test image is computed with the database images using above similarity metrics. Smallest distance of the database image with the test image gives identity of the test image.

17.2.4 Applications

The fingerprint is a rapidly evolving technology that has been widely used in forensics, such as criminal identification and prison security. The fingerprint identification is also under serious consideration for adoption in a broad range of civilian applications. E-commerce and e-banking are two of the most important application areas due to the rapid progress in electronic transactions. Following are the few commercial, government and forensic applications of the fingerprint recognition.

Table 17.1 Applications of fingerprint recognition system

Commercial	Government	Forensic
Computer Network Logon, Electronic Data Security, E-Commerce, Internet Access, ATM, Credit Card, Physical Access Control, Cellular Phones Personal Digital Assistant, Medical Records, Distance Leaning, etc.	National ID card, Correctional Facilities, Driver's License, Social Security, Welfare Disbursement, Border Control, Passport Control, etc.	Corpse Identification Criminal Investigation, Terrorist Identification, Parenthood determination, Missing Children, etc.

17.3 The Early Years

The fingerprint analysis, also known in the US as dactylography, is the science of using fingerprints to identify a person. Fingerprints are the most commonly used biometric and have been used for identification since the 1890s [2]. In South American countries a system devised by Juan Vucetich in 1892 is widely used [2]. These manual classification systems are, however, being replaced by other techniques which are more suitable for large scale electronic storage and analysis. The fingerprint identification is well established and a mature science. It has also been extensively tested in various legal systems and is accepted as an international standard for identification. Developments in early years are elaborated in this section.

17.3.1 Development in Nineteenth Century

During the period of nineteenth century research progress on fingerprints as a biometric identifier matching was rather slow. Therefore this period can be considered as preliminary phase in research and development of the fingerprint recognition. The feature extraction is a key stage in any biometric identification system. Earlier research techniques are based on only minutiae features of the fingerprint image. Few earlier research techniques based on rigorous study of characteristics of the fingerprint image called minutiae and their extraction are briefed here.

In 1684, English plant morphologist Grew published a paper on the ridge, furrow, and pore structure in fingerprints, which is believed to be the first scientific paper on fingerprints [3]. In 1788, Mayer made a detailed description of the anatomical formations of fingerprints [3]. Starting from 1809, Bewick began to use his fingerprint as his trademark, which is believed to be one of the most important contribution in the early scientific study of the fingerprint identification [3]. Purkinje proposed the first fingerprint classification scheme in 1823 which classifies the fingerprints into nine categories according to the ridge configurations. Herschel [3] used fingerprints on legal contracts in Bengal [3]. Fauld and Hershel, in 1880, first time scientifically suggested the individuality and uniqueness of fingerprints. This discovery established the foundation of the modern fingerprint identification. In the late 19th century, Sir Francis Galton conducted an extensive study of fingerprints and identified the minutiae features of the fingerprint as demonstrated in [3].

Development in minutiae based the feature extraction of the fingerprint image is initiated in nineteenth century and gradually progressed in twentieth century which is elaborated in the next section.

17.3.2 Development in Twentieth Century

In twentieth century, peoples were started working on preprocessing, enhancement, minutiae based feature extraction and minutiae feature matching for the partition of the fingerprint database and fingerprint classification. Henry [4] gave the classification and uses of fingerprints. Newham's [5] biometric report gives brief illustration of the initial phase of the modern fingerprint techniques. Since then, a number of researchers have invested huge amounts of effort in studying unique features of fingerprints. Various preprocessing and enhancement, ridge detection, minutiae extraction and fingerprint classification algorithms are elaborated in following sections.

Development in Ridge Features

Fingerprint ridge features play important role in the minutiae extraction. Cummins and Midlo [6] have given a detailed description of the anatomical formations of fingerprints made by Mayer in which number of fingerprint ridge characteristics was identified. Chang [7] found out the singularity points in fingerprint image using texture analysis method. Mehtre et al. [8] proposed the segmentation of fingerprint images using the directional image. To extract the ridges of the fingerprint image, Verma et al. [9] introduced new technique for edge detection in fingerprints. O'Gorman and Nickerson [10] designad filter for good ridge separation, continuity, and background noise reduction. This can be done by determining the local ridge orientations throughout the image, smoothing this orientation image, pixel-by-pixel image enhancement by application of oriented, matched filter masks, and post-processing to reduce background and boundary noise. The orientation of ridges is an important feature to distinguish the two fingerprint images. Ratha et al. [11] have discussed flow orientation based texture extraction as a new feature over minutiae.

Chen and Kuo [12] developed the ridge pattern based on a structural model of minutiae. They used the non-ink fingerprint input device called optical prism to capture the fingerprint images. This physical fingerprint image is converted into the video signal using video cameras. The analog video signal is then digitized via a frame grabber and stored as an array of grey values. Here the print images obtained have a better image quality than those captured from the finger impressions with ink. A set of standard image preprocessing techniques consisting of binarization, thinning and tracing is applied to the print image. As a consequence, a ridge image is obtained. By definition, the beginning or ending of a ridge is identified as a ridge ending and the fork point of three ridge segments is identified as a bifurcation. They are the two primitive minutiae. In practice, due to imperfect image processing, a ridge ending may appear as a bifurcation and vice versa. However, this algorithm fails to extract minutiae from whorl fingerprint image. In poor quality images it is very difficult to locate the minutiae points which results in the generation of spurious minutiae.

Preprocessing and Enhancement

Remarkable progress is made on models of the structure of fingerprints, techniques for the acquisition of prints, and the development of commercial automated fingerprint recognition systems. Despite these advances, there remain considerable opportunities for improvement. The speed of retrieval, and the ability to recognize partial or distorted prints are prominent among those areas that require improvement. Hrechak and Mchugh [13] developed the structural fingerprint matching system based on local structural relations among features, and an associated automated recognition system which addresses the limitations of existing fingerprint models. After that Sherlock et al. [14] presented a fingerprint enhancement algorithm to minimize the artifacts in poor quality fingerprint image. Watson and Wilson [15] proposed standard fingerprint database NIST (National Institute of Standards and Technology) which highlights the challenges in the fingerprint matching. Srinivasan and Murthy [16] used directional histograms of the directional image of a fingerprint for the detection of singular points. Coetzee and Botha [17] proposed the preprocessing algorithms for noise removal and binarization which performs well on low quality images. They used three different classifiers, neural network, linear classifier and nearest neighbor. Hung [18] presented the scheme for enhancing and purifying fingerprint images. The purification of characteristics is based on both foreground and background characteristics; that is, ridge and valley minutiae. This duality allows a fast and effective purification for ridge breaks and gives good results for distorted and damaged fingerprint images. Kamei and Mizoguchi [19] discussed an image filter based on Fourier transform for the fingerprint enhancement. Filters are designed in the Fourier domain with the consideration given to the frequency and direction features of fingerprint ridges. There are two filters, a frequency filter and a directional filter, and this separability of filters increases the process speed of image enhancement. This image enhancement method is based on the energy minimization principle and helps to minimize the spurious minutiae points from low quality fingerprint images. Wahab et al. [20] have discussed the preprocessing techniques like histogram equalization, modification of directional codes, dynamic threshold and ridgelines thinning, but these preprocessing techniques are time consuming.

Minutiae Extraction

Asai et al. [21] developed a fingerprint identification system based on the rigid deformation model for minutiae. Malleswara [22] proposed the fingerprint feature extraction technique to verify whether a minutia is created as a result of noise or imperfect image processing. Sclove [23] prove that the occurrence of fingerprint characteristics is a two dimensional process. Isenor and Zaky [24], Moayer and Fu [25] introduced graph based the fingerprint matching. The algorithm is intended to be insensitive to imperfections introduced during the fingerprint registration, such as noise, distortion and displacement. A fingerprint is represented in the form of a graph whose nodes correspond to ridges in the print. Edges of the graph connect nodes that represent neighboring or intersecting ridges. Hence the graph structure captures the topological relationships within the fingerprint. The algorithm has been implemented and tested using a library of real-life fingerprint images. Kawagoe and Tojo [26] proposed a technique to classify the fingerprint minutiae pattern. Stoney and Thornton [27] gave critical analysis of quantitative fingerprint individuality models. Asai et al. [28] proposed automated fingerprint identification feature-matching process. Tomko [29] gave method and apparatus for fingerprint verification. Kravchinsky et al. [30] and Wilson et al. [31] developed a minutiae classification algorithm based on the neural network, which is trained to classify the minutiae points, ridge ending and ridge bifurcation. Sherlock and Monro [32] developed a model for interpreting fingerprint topology. Khanna and Shen [33] proposed benchmarks and evaluation methods for the automatic fingerprint matching system. The benchmarks are based on standard NIST special database 4. FPGA based implementation of the fingerprint matching algorithm for real time applications proposed by Ratha et al. [34]. The benchmarks provide an open standard to compare fingerprint pattern matching systems but the benchmark does not measure scanning performance. Sankar and Mitra [35] used multilayer perceptron for the classification of noisy fingerprint patterns. Algorithm is based on fuzzy geometrical features, texture-based and directional features of image. O'Gorman [36] describes overview of fingerprint verification methods and technologies. Farina et al. [37] proposed the minutiae extraction algorithm from binary image which works well in dirty areas and on the background. Jain et al. [38] integrated various fingerprint matching algorithms to improve the performance of a fingerprint verification system.

Fingerprint Classification

Clerici [39] introduced a new classification scheme based on a set of concentric circles. Rao and Black [40] proposed global structures of ridge patterns to partition the fingerprint database. Karu and Jain [41] proposed the fingerprint classification method to classify the fingerprint into five categories: arch, tented arch, left loop, right loop and whorl. The algorithm extracts singular points (cores and deltas) in a fingerprint image and performs classification based on the number and locations of the detected singular points. The classifier is invariant to rotation, translation and small amounts of scale changes. This type of classification helps to increase the matching speed of the system. Lumini et al. [42] introduced the continuous verses exclusive classification for the fingerprint retrieval in a huge database. Traditional approaches adopt exclusive classification of fingerprints; the paper shows that a continuous classification can improve the performance of fingerprint retrieval tasks significantly. This approach is based on the extraction of numerical vectors from the directional images of the fingerprints; the retrieval is thus performed in a multidimensional space by using similarity criteria. It provides significant improvements in matching performance in large database.

These earlier research techniques are not suitable for tackling the challenges in fingerprint matching. Hence although in this earlier research phase many people worked to tackle the problems arrived in the fingerprint matching and to enhance the progress of the fingerprint recognition system, but still some research areas like feature extraction, enhancement, speed of retrieval are need to be improved. Innovations of the new age to enhance the above research areas are covered extensively in our survey in next section.

17.4 Recent Feature Extraction Techniques — Addressing Core Problem

The feature extraction plays an important role in the fingerprint matching system to support for the efficient and fast matching from large databases. Significant features must be first extracted from image data. The local features (ridge ending, ridge bifurcation, crossover, island and pore) and global features (core point and delta point) of fingerprints are defined as:

1) Ridge Ending   The point where a ridge ends abruptly.

2) Ridge Bifurcation   The point where a ridge forks or diverges in to branch ridges.

3) Crossover   Two ridges which cross each other.

4) Island   Ridges slightly longer than dots, occupying space between two temporarily divergent ridges.

5) Pore   Small dot.

6) Core   The maximum curvature point, these can be at the most two in a given fingerprint type as demonstrated in [1].

7) Delta   The triangular portion formed due to the flow, at the most there can be two deltas in a given fingerprint type.

8) Minutiae   Ridge ending, bifurcation, crossover, island and pore are called as minutiae points of fingerprint image. A good quality fingerprint contains such 60 to 80 points.

Local features like crossover, dots are not visible in Fig. 17.1. Global (Core point and delta point) and local features (ridge ending, ridge bifurcation, crossover, island and pore) of the fingerprint image are clearly visible in Fig. 17.6. Advanced features like loops, islands can be formed by combination of all above minutiae points. Following are the various feature extraction techniques addressing core problems in the fingerprint matching.

Fig. 17.6 Global and local features of fingerprint image.

17.4.1 Minutiae Based Features

Early days research on the fingerprint matching is based on minutiae points of the fingerprint image. Set of minutiae points (ridge ending and ridge bifurcation) of the fingerprint image is used for matching. Matching score is calculated using Euclidean distance between the query image and database image.

Preprocessing

To match the query image and database image, the number of minutiae points should be same. However, in case of noisy image, spurious minutiae points are created. In that case there is more possibility of false recognition of the query image. To improve the quality of the noisy fingerprint image, some preprocessing techniques are used. Yuanbao et al. [43] proposed fast preprocessing algorithm to increase the speed and accuracy of the fingerprint identification process. This algorithm has the fine enhancement effect to the polluted fingerprint images, and enhances the ridges and valleys of a fingerprint image.

Thinning is the part of pre-processing techniques. To improve the thinning effect, Hongbin et al. [44] proposed a hit-miss transform which is based on mathematical morphology. Hashad et al. [45] also proposed the morphological fingerprint enhancement algorithm. Saleh et al. [46] introduced a fast thinning algorithm works directly on the gray-scale image not the binarized one, where the binarization of the fingerprints causes many spurious minutiae and also removes many important features. Zhang and Zhang [47] had done the rigorous study on the key technologies of the fingerprint pretreatment process in order to improve the quality of image and processing speed. Numerous works have been done in preprocessing image but still the extraction of best set of features from noisy images is a big challenge in front of researchers.

Preprocessing gives the fine improvement in the quality of images. Enhancement algorithms play a key role to improve the poor quality images. Hong et al. [48] proposed an enhancement algorithm based on a frequency domain approach. However, this enhancement approach suffers from few drawbacks which are given below.

1) They are computationally expensive.

2) Their underlying model of oriented fingerprint ridges based on the sinusoid grayscale profile which creates feature artifacts and adversely affects performance of the system.

3) They are not easily tunable.

To deal with the above drawbacks a hierarchical partition-based least square filter bank is proposed by Ghosal et al. [49], which are computationally inexpensive and easily tunable. To overcome the limitations of preprocessing technique, Tico and Kuosmanen [50] and Prabhakar et al. [51] proposed the post processing and a feedback path for the feature extraction stage for improving the quality of minutiae. Wilson [52] demonstrates that Fourier transform matching and neural networks can be used to match fingerprints which have too low image quality to be matched using minutia-based methods. Khalil et al. [53] proposed a statistical descriptor to improve matching accuracy by overcoming the shortcomings of the poor image quality.

Willis and Myers [54] developed a robust algorithm allowing good recognition of low-quality fingerprints with inexpensive hardware. This algorithm simultaneously smooths and enhances the poor quality images derived from a database of imperfect prints. Features are extracted from the enhanced images using a number of approaches including a novel wedge ring overlay minutia detector that is particularly robust to imperfections.

Rusyn et al. [55] carried out the fingerprint image enhancement by adaptive filtering, median filtering, thresholding, and histogram transformation. Enhancement result of this method is shown in Fig. 17.7. However, the enhanced fingerprint image shown in Fig. 17.6 is unable to show the unique ridge pattern of fingerprint images which is used to distinguish between two fingerprints. It is a very difficult task to extract minutiae features from such image.

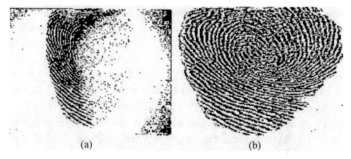

(a)   (b)

Fig. 17.7 (a) Input Image; (b) enhanced image.

Wang et al. [56] introduced the Log-Gabor filter to overcome limitations of the traditional Gabor filter and promote the fingerprint enhancement performance. Hsieh et al. [57] proposed an effective algorithm of fingerprint image enhancement, which can much improve the clarity and continuity of ridge structures based on the multiresolution analysis of global texture and local orientation by the wavelet transform. False minutiae also produced due to geometric transformations of fingerprint images. Bazen and Gerez [58], Almansa and Cohen [59] proposed a thin-plate spline model as a more accurate model to reduce false minutiae. Uz et al. [60] proposed the minutiae-based template synthesis and matching by combining minutiae information from multiple impressions of the same finger in order to increase coverage area, restore missing minutiae, and eliminate spurious ones.

Algorithms to Tackle the Non-linear Distortion

Prominent ridge frequency and ridge direction decides the number of genuine minutiae points. The second reason of false minutiae is non-linear distortion in fingerprint images. Coping with nonlinear distortion in the fingerprint matching is a challenging task. Because of non-linear distortion fingerprint images shown Fig. 17.8 (a) and (b) are impressions of same finger but they look different. The fingerprint images shown Fig. 17.8 (c) and (d) are of different fingers but they look similar. Generally a fingerprint image is non-linearly deformed by torsion and traction when a fingerprint is pressed on the sensor.

This non-linear deformation changes both position and orientation of minutiae and decreases the reliability of minutiae. Lee et al. [61], He et al. [62] and Ross et al. [63] tried to minimize non-linear distortion using local alignment of minutiae. A feature called Local Relative Location Error Descriptor (LRLED) is proposed by Tong et al. [64] to overcome non-linear distortion. Cao et al. [65] introduced the novel features to deal with nonlinear distortion in fingerprints. To match minutiae points of deformed fingerprint images, graph-based fingerprint representation with a fuzzy feature map and a normalized fuzzy similarity measure algorithm is proposed by Chen and Tian [66].

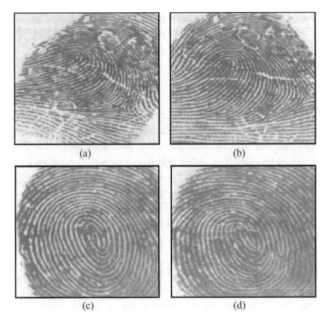

Fig. 17.8 Fingerprint images (a) and (b) look different but are impressions of same finger.

Chen et al. [67] used a local triangle feature set to match the deformed fingerprints. Ratha and Pandit [68] introduced a graph-based representation and a heuristic matching algorithm which allows for anomalies like missing/spurious minutiae, elastic distortion, rotation and translation of fingerprint images. Ito et al. [69] used the phase components in two dimensional discrete Fourier transforms of fingerprint images to achieve the highly robust fingerprint recognition for low-quality fingerprints.

A fingerprint image can be enhanced by eliminating the LL4 sub-band component of a hierarchical Discrete Wavelet transform (DWT) as discussed in Yang's et al. [70]. The misalignment of fingerprint images is dramatically reduced by maximizing mutual information between features extracted from orientation fields of template and input fingerprint images. Lui et al. [71]

proposed algorithm which gives great emphasis on mutual information. A method that is simple and effective to do the fingerprint image alignment is discussed in Luo et al. [72] and Zhu et al. [73].

Minutiae Extraction

After preprocessing and enhancement, the next significant step is the feature extraction. The binary tree-based minutiae matching is carried out by Jain et al. [74]. Bhowmick and Bhattacharya [75], Dixit et al. [76] proposed the minutiae-based efficient data structure technique called kdtree. The neural network based on new approach of minutiae extraction is introduced by Josef et al. [77]. The minutiae-based matching algorithm does not give a satisfactory matching rate because only ridge ending and ridge bifurcation do not give enough information for matching. The ridge orientation plays an important role to increase the recognition rate. Mei et al. [78] proposed the gradient-based method to compute the orientation in fingerprint images. Ram et al. [79] proposed a novel method for the fingerprint ridge orientation modelling using Legendre polynomials. Tico and Pauli [80], Gu et al. [81], and Qi and Wang [82] elaborated the concept of the orientation-based minutiae descriptor. To enhance the matching rate, some non-minutiae points are integrated with minutiae points. This provides additional non-minutiae information for calculating a more reliable degree of similarity between fingerprint impressions and also improves the matching rate considerably. The minutiae with energy as a feature can improve the recognition rate up to some extent as discussed by Nagaty [83]. Feng et al. [84] proposed a novel fingerprint matching algorithm which establishes both the ridge correspondences and the minutia correspondences between two fingerprints to improve the performance.

Matching of latent fingerprints is one of the difficult tasks in the fingerprint matching because the challenges involved in latent print matching are quite different from plain or rolled fingerprint matching. Poor quality of friction ridge impressions, small finger area and large non-linear distortion are some of the main difficulties in latent fingerprint matching. The fingerprints acquired by this method are often blurred, incomplete, degraded, and their spatial definition is not clear, that results in false minutiae. To tackle these problems, Ko [85] analyzed some spatial nonlinear filters and frequency domain filters using Fast Fourier Transform which gives prominent features of fingerprint images. Jain et al. [86] proposed the system for matching the latent images which is based on additional features like orientation field and quality map.

Ray et al. [87] obtained sweat pores as additional descriminary information from gray scale fingerprint images and proved that ridge ending and ridge bifurcation information are not sufficient for large databases. Jain et al. [88] captured level1 (pattern), level2 (minutiae points) and level3 (pores and ridge shape) features of fingerprint images using the wavelet transform and Gabor filter. Shi et al. [89] proposed the novel feature extraction algorithm based on minutiae and global statistical features. Aguilar et al. [90] used the FFT and Gabor filter for the enhancement and detection of features from fingerprint images. The impact of age group on the number minutiae points and the performance of minutiae-based algorithms on different age groups are elaborated by Modi et al. [91].

Jain et al. [92], Chen and Jain [93], Vatsa et al. [94], Vatsa et al. [95], and Zhao et al. [96] extract the micro details of fingerprint images to boost the recognition rate which are summarized in Fig. 17.9.

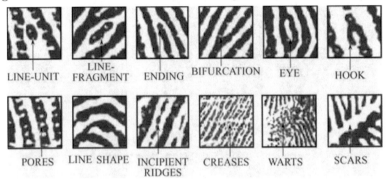

Fig. 17.9 Types of minutiae points.

17.4.2 Texture Based Features

Most of the research work is based on minutiae points of fingerprints. The widely used minutiae-based representation does not utilize a significant component of the rich discriminatory information available in fingerprints. This limitation of minutiae-based algorithm is overcome by using the texture information of fingerprint images. The texture information contains local information (ridge ending and ridge bifurcation) as well as global information (core point and delta point) of fingerprint images. The Gabor filter has optimally combined the spatial and frequency domains.

To capture the texture information of fingerprint images, Gabor filter with eight orientations is proposed by Jain et al. [97, 98] as shown in Fig. 17.10. The eight directional-sensitive filters capture most of the global ridge directionality information as well as the local ridge characteristics present in a fingerprint. A fingerprint convolved with 0° oriented filter accentuates those ridges which are parallel to the x axis and smoothes the ridges in the other directions. Filters tuned to other directions work in a similar way. Figure shows the system architecture proposed by the Jain et al. [97, 98]. The transfer function of even Gabor filter is shown as follows:

$$G(x, y; f, \theta) = \exp\left\{\frac{1}{2} - \left[\frac{x'^2}{\sigma_{x'}^2} + \frac{y'^2}{\sigma_{y'}^2}\right]\right\} \cos(2\pi f x'), \qquad (17.11)$$

$$x' = x \sin\theta + y \cos\theta,$$
$$y' = x \cos\theta - y \sin\theta, \qquad\qquad (17.12)$$

where f is the frequency of the sinusoidal plane wave along the direction θ from the x axis, and $\delta_{x'}$ and $\delta_{y'}$ are the space constants of the Gaussian envelope along x' and y' axes, respectively.

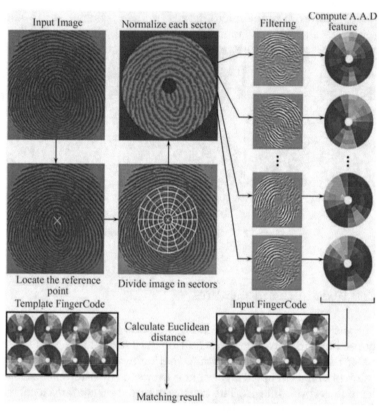

Fig. 17.10 Filtered Images after convolution with eight Gabor Filters of orientations (0, 22.5, 45, 67.5, 90, 112.5, 135, 157.5 degrees).

Patil et al. [99] used four orientations of Gabor filters for the extraction of fingerprint features from a gray scale fingerprint image cropped in the size of 128×128 pixels using its core point as the center. These algorithms proved that they give better performance over minutiae-based algorithms.

Horton et al. [100] combined both even and odd features of fingerprint images using the complex Gabor filters which did not show any improvement in the results obtained from the even Gabor filter but it increases the computational cost. Munir and Javed [101] used sixteen orientations of Gabor filters for feature extraction and proved that the increase in number of orientations of Gabor filters also increases the overall matching rate. Because if we increase the number of orientations of Gabor filters, the computational complexity also increases which increases the overall matching time. Sixteen Gabor filtered images with different sixteen orientations are shown in Fig. 17.11.

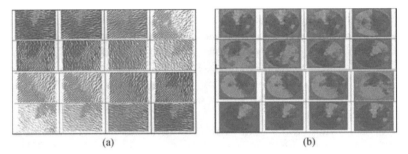

(a)   (b)

Fig. 17.11 (a) Filtered Images after convolution with sixteen orientations of Gabor Filters (0, 11.25, 22.5, 33.75, 45, 56.25, 67.5, 78.75, 90, 101.25, 112.5, 123.75, 135, 146.25, 157.5, 168.75 degrees). (b) Corresponding Feature vectors of sixteen filtered images.

The first major limitation of Gabor filter-based algorithms is computational complexity because the Gabor filter is not orthogonal. Second limitation of Gabor filter-based algorithm is that they suffer the problem of exact location of reference point called core point. Above discussed algorithms fails to locate exact location of reference point in noisy and distorted images. Sha et al. [102] introduced a novel method for finding exact location of reference point. Huang and Aviyente [103], and Khan and Javed [104] used wavelet-based analysis which provides rich discriminory texture structure of the fingerprint verification and this method is not based on reference point location. Haar wavelet with minimum square error is used by Mokji and Abu-Bakar [105]. To capture global transformation between two fingerprint images, a genetic algorithm is used by Tan and Bhannu [106]. Though the wavelet-based algorithms are computationally efficient since wavelet transform is orthogonal transform, it does not provide quality features.

To reduce the computational complexity, Dadgostar et al. [107] proposed feature extraction based on the Gabor filters and recursive fisher linear discriminate. Cheng and Xin-Ming [108] used the wavelet and Gabor based features for the fingerprint matching.

This shows that the accuracy with a low computational complexity is a big challenge in the area of the fingerprint matching.

17.4.3 Minutiae and Texture-based Features

The combination of minutiae and texture information of fingerprint images performs well over minutiae and texture. The minutiae and texture with both local and global information of fingerprint images are explored by Jain et al.

[109], Ross et al. [110], Youssif et al. [111], Feng [112], Nanni and Lumini [113]. They proved that it gives better result against only local and global information. Gu et al. [114] combines the global structure (orientation field) and the local cue (minutiae) of fingerprint images. They showed that the global orientation field is beneficial to the alignment of the fingerprints which are either incomplete or poor-qualities. Minutiae and square finger code gives good results for large databases proved by Sha et al. [115]. To capture local and global information of fingerprints Hidden Morkov model-based algorithm is discussed in Guo [116]. The pretreatment process of this algorithm is simple and fast as it skipped the processes of thinning ridge images and selecting minutiae.

Wen et al. [117] and Nikam et al. [118] demonstrated that by integrating a number of features like core point, Gabor, wavelet, standard deviation and Fourier Mellin transform improves matching rate significantly. However, at the same time it increases the overall matching time of the system. Four level decomposition using wavelet is shown in Fig. 17.12. Detail and approximate coefficients of decomposition using wavelet are given by Eqs. (17.12) and (17.13) respectively.

$$L_j(j_0, m, n) = \frac{1}{\sqrt{MN}} \sum_{x=0}^{M-1} \sum_{y=0}^{N-1} f(x, y)\varphi_{j0,m,n}(x, y), \qquad (17.13)$$

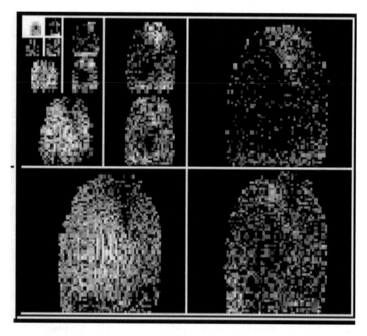

Fig. 17.12 Four level decomposition using wavelet.

$$D_j(j_0, m, n) = \frac{1}{\sqrt{MN}} \sum_{x=0}^{M-1} \sum_{y=0}^{N-1} f(x, y) \psi_{j0,m,n}(x, y), \qquad (17.14)$$

where L_j is detail coefficients and D_j is approximate coefficients, φ and ψ are scaling and wavelet functions.

Despite the fact that the integration of minutiae and texture features increases the matching rate significantly, at the same time it increases the computational complexity. Thus these algorithms are not good for real time applications. There is a desperate need of the feature extraction techniques which will tackle this problem.

17.4.4 Transform Based Features

The term image transform refers to a class of unitary matrices used for representing images. Just as one dimensional signal can be represented by an orthogonal series of basis functions, an image can also be expanded in terms discrete set of basis arrays called basis images. Images are expanded in terms of a discrete set of basis arrays called basis images [119]. Energy conservation, energy compaction and decorrelation are the important properties of these transforms. Various image transforms with statistical features can be used for the fingerprint matching. The discrete cosine transform is considered a good choice of image transform as it offers energy compaction efficiency and its decomposition leads to the fast implementation. The two dimensional DCT of an $M \times N$ image $f(x, y)$ is a separable transform defined as

$$C(u, v) = \alpha(u)\alpha(v) \sum_{x=0}^{M-1} \sum_{y=0}^{N-1} f(x, y) \cos\left(\frac{(2x+1)u\pi}{2M}\right) \cos\left(\frac{(2y+1)v\pi}{2N}\right),$$
$$(17.15)$$

where $0 \leqslant u \leqslant M - 1, 0 \leqslant v \leqslant N - 1$. The DCT is a real, orthogonal, fast and separable transform. It has excellent energy compaction for highly correlated data. The discrete cosine transform coefficients with some statistical features are used by Tachaphetpiboon and Amornraksa [120, 121] to enhance recognition rate and processing time.

The Fourier Transform is an important image processing tool which is used to decompose an image into its sine and cosine components. The output of the transformation represents the image in the Fourier or frequency domain, while the input image is the spatial domain equivalent. In the Fourier domain image, each point represents a particular frequency contained in the spatial domain image. It can be used for feature extraction of fingerprint images. The two dimensional discrete Fourier transform (DFT) of an $M \times N$ image $f(x, y)$ is a separable transform defined as

$$F(u, v) = \frac{1}{MN} \sum_{x=0}^{M-1} \sum_{y=0}^{N-1} f(x, y) e^{-j2\pi(ux/M + vy/N)}. \tag{17.16}$$

Inverse Fourier transform of $F(u, v)$ is given by

$$f(x, y) = \frac{1}{MN} \sum_{u=0}^{M-1} \sum_{v=0}^{N-1} F(u, v) e^{j2\pi(ux/M + vy/N)}, \tag{17.17}$$

where $0 \leqslant x, u \leqslant M - 1, 0 \leqslant y, v \leqslant N - 1$. Tachaphetpiboon and Amornraksa [122] proposed a FFT-(Fast Fourier Transform) based features for the fingerprint matching. Ridge features can be used to align and match fingerprints in the fingerprint matching technique. Straight lines that approximate each fingerprint ridge are separately extracted using the Hough transform. Qi et al. [123], Marana and Jain [124] proposed a Hough transform-based ridge pattern extraction algorithm. All detected Hough space peaks are used to estimate the rigid transformation parameters between the query and database fingerprint images. After the alignment, a matching score is computed from a matrix of ridge alignments. Some genuine fingerprint pairs, which were rejected by the minutia-based matcher, are correctly aligned and accepted by the ridge-based Hough transform matcher. The correlation is also one kind of transform which gives the degree of similarity between two images. This feature of correlation is used by Seow et al. [125] for image-based fingerprint verification to find out correspondence between two fingerprint images instead of minutiae points. A fingerprint-matching algorithm that combines the minutia-based matching method with the correlation-based matching method is discussed in Jiang et al. [126]. These traditional correlation-based methods are not so efficient. Fourier-Mellin transform and phase only correlation method are introduced by Zhang et al. [127], Ouyang et al. [128] respectively which give better result over the traditional correlation-based methods. Lindoso et al. [129], and Chen and Gao [130] used cross correlation to capture minute information like level3 features of fingerprints which works superior over minutiae and phase correlation.

Image projections with radon transform for the fingerprint matching is discussed in Haddad et al. [131]. A new multiresolution analysis tool called digital curvelet transform for the fingerprint matching is proposed by Mandal et al. [132]. Nikam and Agarwal [133] proposed the ridge-based feature extraction algorithm for fake fingerprint detection which works better than wavelet-based methods. Wavelets are very effective in representing objects with isolated point singularities, but failed to represent line singularities. The ridgelet transform allows representing singularities along lines in a more efficient way than wavelets.

Various image transforms give the compact representation of a signal but still there is a big challenge to extract the quality feature set.

17.4.5 Other Features

Other features like the integration of facial features with fingerprint images, agent computing-based fingerprint matching, video and audio fingerprints can be used for person identification. Hong and Jain [134], Patra and Das [135], Khan and Zhang [136], and Marcialis et al. [137] developed a prototype biometric system which integrates faces and fingerprints. The concepts of video-fingerprints and audio-fingerprints, based on normalized spectral subband and gradient orientations is proposed by Seo et al. [138], Lee and Yoo [139], Lee and yoo [140]. Tan and Bhanu [141] proposed a fingerprint-matching approach based on genetic algorithms (GA), which tries to find the optimal transformation between two different fingerprints. Girgis et al. [142] developed a genetic algorithm to improve the deformed ridges and complex distortions in the fingerprint verification system. Yang and Park [143] used the tessellated invariant features to improve the matching accuracy and processing speed by overcoming the demerits of previous methods over poor-quality images.

The effect of controlled acquisition on features of fingerprint images is analyzed by Ratha and Bolle [144]. An agent computing-based novel approach for a parallel fingerprint matching system is discussed by Nagaty and Hattab [145]. Analysis and comparison of eight different fingerprint matching methods are given by Duan et al. [146]. Bossen et al. [147] used a frequency-domain optical coherence tomography (FD-OCT) system to capture a 3D image of a finger and the information of the internal fingerprint. They proved that OCT imaging of internal fingerprints can be used for the accurate and reliable fingerprint recognition. Wang and Hu [148] proposed the Global Ridge Orientation Modeling to tackle the problem of partial fingerprint identification. Jain et al. [149] describe an automated fingerprint recognition system and identify key challenges and research opportunities in this field.

Though integration of multiple biometric patterns increases the accuracy, but the limitation is that it will also increase the memory requirement, computational complexity and overall matching time.

17.4.6 Features Used to Reduce Matching Time

To reduce the overall matching time, the whole database can be classified into different classes. A query image is matched with that particular class instead of matching whole database. This section gives the brief explanation of various algorithms for the fingerprint indexing, computational complexity and hardware implementation for real time applications.

Fingerprint Indexing

The fingerprint indexing is a method of classifying fingerprint images into six major classes of fingerprints. These six classes are twin loop, whorl, right loop, left loop, arch and tented arch which are shown in Fig. 17.13. Based on the global information (i.e., number of core points and delta points), fingerprints are classified into above major six classes.

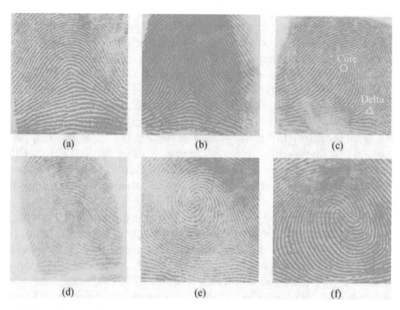

Fig. 17.13 Fingerprints and a fingerprint classification schema of six categories: (a) arch; (b) tented arch; (c) right loop; (d) left loop; (e) whorl; (f) twin loop; critical points in a fingerprint, called core and delta, are marked on one of these images (c).

Yasha and Xiaoping [150] proposed an image sampling method for fast core point detection in a fingerprint image. Robust fingerprint indexing can be done using the minutiae neighborhood structure and Delaunay triangle [151]. Chikkerur and Ratha [152] proposed a method for computing core and delta points of fingerprint images and proved that it performs better over minutiae. Sen et al. [153] proposed directional field-based Poincare index to give good indexing for the fingerprint classification. Effective fingerprint classifications can be achieved using neural network and fuzzy artmap proposed by Daghar et al. [154]. Nilsson and Bigun [155] use complex filters to detect the singular points in fingerprint images. Jain et al. [156] used Gabor filter-based texture information for fingerprint indexing. However, this algorithm suffers from the requirement that the region of interest is correctly located, requiring the accurate detection of center point in fingerprint images. A new reference point called focal point is proposed by Rerkrai and Areekul [157] for fingerprint indexing of large database. However, the stability of the focal point for a fingerprint depends on the convergence condition with constraints; i.e., limited effective area and limited effective cross-points. Based on the focal point or core point, a fast matching algorithm is discussed in Zhang and Wang [158]. The exclusive and continuous classification of fingerprint images reduces the considerable matching time proposed by Sha and Tang [159]. Ahmadyfard and Nosrati [160] extracted the horizontal, vertical and diagonal information of fingerprints and used for accurate detection of singular points. Park et al. [161] presented a new fingerprint singular-point detection method that is type-distinguishable and applicable to various fingerprint images regardless of their resolutions. This method detects singular points by analyzing the shapes of the local directional fields of a fingerprint image. Areekul and Boonchaiseree [162], Biswas et al. [163], Mohammadi and Farajzadeh [164], Li et al. [165], Jirachaweng et al. [166] used the orientation and curvature information of fingerprints for fast indexing-based core point.

Though more number of peoples has proposed their work for fingerprint indexing still it is a challenging task to locate the core and delta points in poor quality fingerprint images.

Computational Complexity

Both computational complexity and matching time are important factors for real-time application. The less complexity and matching time prove the efficiency of algorithm. These two factors can be reduced by using all images into one larger image but it lowers the matching rate discussed in Ramoser et al. [167]. Germain et al. [168] used geometric hashing and proves suitable for one-to-many matching of fingerprints on large-scale databases. To tackle the above problem of complexity and matching time, a synergetic-based algorithm is proposed by Guo et al. [169] for matching and verification. Driche [170] used the compression technique to reduce matching time without affecting on the minutiae points. Two methods for direction and density images are presented in Malalur et al. [171] for fast efficient and robust verification. Instead of finding features of whole fingerprint image, features of subimages can increase the speed of verification introduced in Chan et al. [172].

The computational complexity is one of the major factors which decides the overall efficiency of the matching system. Thus it is required to develop the algorithms which extract best set of features in a compact form. This reduces the feature extraction time, feature matching time and memory requirement of the system.

Hardware Implementation of Fingerprint Algorithms

A complete parallel fingerprint matching system is not possible with software implementation of algorithms. A smart card-based ultra low memory fingerprint matching algorithm is developed by Pan et al. [173] and it is implemented on 32 bit smart card for real-time application. A fast and efficient memory approach is proposed in Allah [174] to reduce memory requirement for storing image buffer and to reduce the image processing time for implementation of fingerprint matching algorithms on smart card. A computationally efficient fingerprint matching algorithm on smart card implementation is proposed by Govan and Buggy [175] and Yahaya et al. [176]. Yang et al. [177] developed a thumb pod which executes minutiae extraction algorithm in lesser time compared to the traditional hardware implementation.

For improving the performance of fingerprint-based authentication, and the hardware/software enhancement that includes a generic set of custom instruction extensions to an embedded processor's instruction set architecture, a memory-aware software re-design and fixed-point arithmetic are proposed by Gupta et al. [178], Fons et al. [179]. It shows that the custom instruction set extensions proposed in this work are generic enough to speed up many fingerprint matching algorithms and even other geometric algorithms. The hardware software co-design of a computational platform is responsible for matching two fingerprint minutiae sets. A novel system concept is suggested by making use of reconfigurable architecture in Fons et al. [180]. Lindoso et al. [181] proposed zero mean normalized cross correlation using vertex-4 FPGA for online application. A Gabor, filter-based finger code with smart card can be used for online application discussed in Yeswanth and Sakthi [182].

Lopez and Canto [183] implemented the minutiae extraction fingerprint algorithm on Spartan-3 FPGA which is most suitable for portable devices and for the low-cost consumer market. Militello et al. [184] prototyped the whole fingerprint authentication system based on singularity points on the Celoxica RC203E board, equipped with a Xilinx Virtex II FPGA. This is the first authentication system based on singularity points. Lorrentz et al. [185] explored the biometric identification and verification of human subjects via fingerprints utilizing an adaptive FPGA-based weightless neural network. The hardware implementation of fingerprint matching algorithms for huge database is a big challenge in this research area.

17.5 Conclusion and Future Directions

Most of the recent efforts in the fingerprint matching system have focused on either minutiae points or texture of fingerprint images. But only minutiae points or texture information of fingerprint images give limited knowledge to the fingerprint system to take correct matching decision. Due to this matching efficiency is affected in few percent, very few fingerprint-matching algorithms are based on both local and global information of fingerprint images. These hybrid algorithms give better efficiency but consume more time for the feature extraction, matching process and require large memory. Hence it needs to be developed a system that would deal with these limitations of hybrid algorithms.

The enhancement plays a key role in the quality feature extraction. Existing enhancement techniques are not competent enough to tackle the nonlinear distortion present due to traction and torsion when fingerprints placed on the sensor. Fingerprint images from same fingerprints shows large variability in different impressions. Main factors responsible for intraclass variations are displacement, rotation, partial overlap, non-linear distortion, variable pressure, skin condition, noise and feature extraction errors. Innovative enhancement techniques are needed to be investigated to improve the quality of fingerprint images and to extract quality features.

However most of the current techniques work effectively on small databases, which cannot produce satisfactory results for large databases. So the ways of improving matching results for large databases are urgently needed. Speed is also an important parameter to be focused. In case of large database, the time required to establish the identity of test images is more. For real-time application it should be less. Therefore, the new rich information features with compact representation need to be found to reduce the overall matching time. The final complexity is the recognition of overlapped fingerprint images. This problem is the novel challenge in the research area of the fingerprint recognition. Some pioneering and effective work should be needed in this direction. In this chapter, we argue that for a fingerprint matching system to be successful we need to develop approaches robust to poor quality images, rotation invariant and also produce good results for huge databases.

References

[1]   Maltoni D, Maio D, Jain A K (2003) Handbook of Fingerprint Recognition. Springer, New York
[2]   Roberts C (2006) Biometrics Technologies – Fingerprints. Publisher name and location are missing, pp 1 – 23
[3]   Lee H C, Gaensslen R E (1991) Advances in Fingerprint Technology. Elsevier, Science Publisher, New York
[4]   Henry E R (1900) Classification and Uses of Fingerprints. Routledge, pp 54 – 58 London
[5]   Newham E (1995) The Biometric Report. SJB Services, New York
[6]   Cummins H, Midlo C (1961) Finger Prints, Palms and Soles. Dover Publications Inc., New York
[7]   Chang T (1980) Texture Analysis of Digitized Fingerprints for Singularity Detection. In: Proceedings of the 5th International Conference on Pattern Recognition, pp 478 – 480
[8]   Mehtre B M, Murthy N N, Kapoor S et al (1987) Segmentation of Fingerprint Images Using the Directional Image. In: Proceedings on Pattern Recognition, 20: 425 – 429
[9]   Verma V K, Majumdar A K, Chatterjee B (1987) Edge Detection in Fingerprints. Trans on Pattern Recog, 20(5): 513 – 523
[10] O'Gorman, Nickerson L J (1989) An Approach to Fingerprint Filter Design. Pattern Recognition, 22(1): 29 – 38
[11] Ratha N K, Chen S, Jain A K (1995) Adaptive Flow Orientation Based Texture Extraction in Fingerprint Images. IEEE Proc on Pattern Recog, 28: 1657 – 1672
[12] Chen Z, Kuo C H (1991) A Toplogy-based Matching Algorithm for Fingerprint Authentication. IEEE Proc on Image Processing, pp 84 – 87
[13] Hrechak A K, Mchugh J A (1990) Automated Fingerprint Recognition Using Structural Matching. Trans on Pattern Recognition, 23(8): 893 – 904
[14] Sherlock B G, Monro D M, Millard K (1992) Algorithm for Enhancing Fingerprint Images. Proc Electron Lett, 28: 1720 – 1721
[15] Watson C I, Wilson C L (1992) NIST Special Database 4, Fingerprint Database. National Institute of Standards and Technology

[16] Srinivasan V S, Murthy N N (1992) Detection of Singular Points in Fingerprint Images. Pattern Recognition, 25(2): 139 – 153
[17] Coetzee L, Botha E C (1993) Fingerprint Recognition in Low Quality Images. Pattern Recognition, 26(10): 1441 – 1460
[18] Hung D D (1993) Enhancement and Feature Purification of Fingerprint Images. Pattern Recognition, 26(11): 1661 – 1671
[19] Kamei T, Mizoguchi H (1995) Image Filter Design for Fingerprint Enhancement. Proc IEEE on Computer Vision, pp 109 – 114
[20] Wahab A, Chin S H, Ta n E C (1998) A Novel Approach to Automated Fingerprint Matching. Proc IEEE on Vision Image Signal Proc, 145: 160 – 166
[21] Asai k, Hoshino Y, Yamashita N et al (1975) Fingerprint Identification System. Second USA-Japan Computer Conference, pp 30 – 35
[22] Malleswara T C (1976) Feature Extraction for Fingerprint Classification. Trans on Pattern Recognition, vol 8: 181 – 192
[23] Sclove S L (1979) The Occurrence of Fingerprint Characteristics as a Two Dimensional Process. Journal of American Statistical Association, 74(367): 588 – 595
[24] Isenor D K, Zaky S G (1986) Fingerprint Identification Using Graph Matching. Trans on Pattern Recognition, 19(2): 113 – 122
[25] Moayer B, Fu K S (1986) A Tree System Approach for Fingerprint Pattern Recognition. IEEE Trans on Pattern Analysis and Machine Intelligence, vol 8: 376 – 387
[26] Kawagoe M, Tojo A (1984) Fingerprint Pattern Classification. IEEE Trans on Pattern Recognition, 17: 295 – 303
[27] Stoney D A, Thornton J I (1986) A Critical Analysis of Quantitative Fingerprint Individuality Models. Journal of Forensic Sciences, 31(4): 1187 – 1216
[28] Asai K, Hoshino Y, Kiji K (1989) Automated Fingerprint Identification Feature-matching Process. IEICE Trans, J72-D-II (5): 733 – 740
[29] Tomko G J (1989) Method and Apparatus for Fingerprint Verification. US Patent 4876725
[30] Kravchinsky L V, Kuznetsov S O, Nuidel I V et al (1993) Application of Neural Networks for Analyzing and Encoding of Fingerprints. Neurocomputing, 4 (1 – 2): 65 – 74
[31] Wilson C L, Candela G T, Watson C I (1993) Neural Network Fingerprint Classification. Trans on Artificial Neural Networks, 1, 203 – 228
[32] Sherlock B G, Monro D M (1993) A Model for Interpreting Fingerprint Topology. IEEE Trans on Pattern Recognition, 26, 1047 – 1055
[33] Khanna R, Shen W (1994) Automated Fingerprint Identification System (AFIS) Benchmarking Using the National Institute of Standards and Technology (NIST) Special Database 4. In: IEEE Proc, pp 188 – 194
[34] Ratha N K, Jain A K, Rover D T (1995) An FPGA-based Point Pattern Matching Processor with Application to Fingerprint Matching. In: Proc. IEEE on Computer Architectures for Machine Perception, 18: 394 – 401
[35] Sankar K P, Mitra S (1996) Noisy Fingerprint Classification Using Multilayer Perceptron with Fuzzy Geometrical and Textural Features. Fuzzy Sets and Systems, 80(2): 121 – 132
[36] O'Gorman L (1998) An Overview of Fingerprint Verification Technologies. Information Security Technical Report, 3(1): 21 – 32
[37] Farina A, Kovács-Vajna Z M, Leone A (1999) Fingerprint Minutiae Extraction from Skeletonized Binary Images. Pattern Recognition, 32(5): 877 – 889
[38] Jain A K, Prabhakar S, Chen S (1999) Combining Multiple Matchers for a High Security Fingerprint Verification System. Pattern Recognition Letters, 20(11 – 13): 1371 – 1379
[39] Clerici R (1969) Fingerprints: A New Classification Scheme. Nature, 224: 779 – 780
[40] Rao K, Black K (1980) Type Classification of Fingerprints: A Syntactic Approach. IEEE Trans on Pattern Analysis and Machine Intelligence, PAMI-2 (3): 223 – 231
[41] Karu K, Jain A K (1996) Fingerprint Classification. Pattern Recognition, 29(3): 389 – 404
[42] Lumini A, Maio D, Maltoni D (1997) Continuous Versus Exclusive Classification for Fingerprint Retrieval. Pattern Recognition Letters, 18(10): 1027 – 1034
[43] YuanBao W, GuiMing H, Xiong Z et al (2007) A Fast Fingerprint Identification Pre-Processing Algorithm. In: Proc of the First IEEE International Conf on Bioinformatics and Biomedical Engineering (ICBBE-07), pp 596 – 598
[44] Hongbin P, Junali C, Yashe Z (2007) Fingerprint Thinning Algorithm Based on Mathematical Morphology. In: Proc of the Eighth IEEE International Conf on

Electronic Measurement and Instruments (ICEMI-07), 2: 618–621

[45] Hashad F G, Halim T M, Diab S M et al (2009) A Hybrid Algorithm for Fingerprint Enhancement. In: Proc of the IEEE Conf on Computer Engineering & Systems (ICCES-09), pp 57–62

[46] Saleh A M, Bahaa Eldin A M, Wahdan A.-M A (2009) A Modified Thinning Algorithm for Fingerprint Identification Systems. In: Proc of the IEEE Conf on Computer Engineering & Systems (ICCES-09), pp 371–376

[47] Zhang Q, Zhang X (2010) Research of Key Algorithm in the Technology of Fingerprint Identification. In: Proc of the IEEE Conf on Computer Modeling and Simulation (ICCMS-10), Vol. 4: 282–284

[48] Hong L, Wan Y, Jain A K (1998) Fingerprint Image Enhancement: Algorithm and Evaluation. IEEE Trans on Pattern Analysis and Machine Intelligence, 20: 777–789

[49] Goshal S, Ratha N K, Udupa R et al (2000) Hierarchical Partitioned Least Squares Filter Bank for Fingerprint Enhancement. In: Proc IEEE on Pattern Recognition, 3: 334–337

[50] Tico M, Kuosmanen P (2000) Algorithm for Fingerprint Image Postprocessing. In: Proc IEEE on Signals, Systems and Computer, 2: 1735–1739

[51] Prabhakar S, Jain A K, Wang J et al (2000) Minutia Verification and Classification for Fingerprint Matching. In: Proc IEEE on Pattern Recognition, 1: 25–29

[52] Wilson C L, Watson C I, Paek E G (2000) Effect of Resolution and Image Quality on Combined Optical and Neural Network Fingerprint Matching. Pattern Recognition, 33(2): 317–331

[53] Khalil M S, Mohamad D, Khan M K et al (2010) Fingerprint Verification Using Statistical Descriptors. Digital Signal Processing, 20(4): 1264–1273

[54] Willis A J, Myers L (2001) A Cost-effective Fingerprint Recognition System for Use with Low-quality Prints and Damaged Fingertips. Pattern Recognition, 34(2): 255–270

[55] Rusyn B, Prudyus I, Ostap V (2001) Fingerprint Image Enhancement Algorithm. In: Proc IEEE Designing and Applications of CAD Systems in Microelectronics, pp 193–194

[56] Wang W, Li J, Huang F et al (2008) Design and Implementation of Log-Gabor Filter in Fingerprint Image Enhancement. Pattern Recognition Letters 29(3): 301–308

[57] Hsieh C, Lai E, Wang Y (2003) An Effective Algorithm for Fingerprint Image Enhancement Based on Wavelet Transform. Pattern Recognition, 36 (2): 303–312

[58] Bazen A M, Gerez S H (2003) Fingerprint Matching by Thin-plate Spline Modelling of Elastic Deformations. Pattern Recognition, 36(8): 1859–1867

[59] Almansa A, Cohen L (2000) Fingerprint Image Matching by Minimization of a Thin-plate Energy Using a Two-step Algorithm with Auxiliary Variables. In: Proc IEEE on Applications of Computer Vision, 2: 35–40

[60] Uz T, Bebis G, Erol A et al (2009) Minutiae-based Template Synthesis and Matching for Fingerprint Authentication. Computer Vision and Image Understanding, 113(9): 979–992

[61] Lee D, Choi K, Kim J (2002) A Robust Fingerprint Matching Algorithm Using Local Alignment. In: Proc IEEE on Pattern Recog. 3 pp 803–806

[62] He Y, Tian J, Luo X et al (2003) Image Enhancement and Minutiae Matching in Fingerprint Verification. Pattern Recognition Letters, 24(9–10): 1349–1360

[63] Ross A, Dass S, Jain A (2005) A Deformable Model for Fingerprint Matching. Pattern Recognition, 38(1): 95–103

[64] Tong X, Liu S, Huang J et al (2008) Local Relative Location Error Descriptor-based Fingerprint Minutiae Matching. Pattern Recognition Letters, 29(3): 286–294

[65] Cao C, Yang X, Tao X et al (2010) Combining Features for Distorted Fingerprint Matching. Journal of Network and Computer Applications, 33(3): 258–267

[66] Chen X, Tain J (2006) A new Algorithm for Distorted Fingerprints Matching Based on Normalized Fuzzy Similarity Measure. IEEE Trans on Image Proc, 15: 767–776

[67] Chen X, Tain J, Yang X et al (2006) An Algorithm for Distorted Fingerprint Matching Based on Local Triangle Feature set. IEEE Transacations on Information Forensics and Security, 1: 169–177

[68] Ratha N K, Pandit V D (2000) Fingerprint Authentication Using Local Structural Similarity. Proc IEEE on Applications of Computer Vision, pp 29–34

[69] Ito K, Morita A, Aoki T et al (2005) A Fingerprint Recognition Algorithm Using Phase-based Image Matching for Low-quality Fingerprints. In: Proc IEEE on Image Processing, 2: 33–38

[70] Yang j, Shin J, Min B et al (2006) Fingerprint Matching Using Invariant Moment

Finger-code and Learning Vector Quantization Neural Network. In: Proc IEEE on Computational Intelligence and Security, 1: 735–738

[71] Lui L, Jiang T, Yang J et al (2006) Fingerprint Registration by Maximization of Mutual Information. IEEE Trans on Image Processing, 15: 1101–1110

[72] Luo X, Tian J, Wu Y (2000) A Minutia Matching Algorithm in Fingerprint Verification. In: Proc IEEE on Pattern Recognition, 4: 833–836

[73] Zhu E Yin J, Zhang G (2005) Fingerprint Matching Based on Global Alignment of Multiple Reference Minutiae. Pattern Recognition, 38(10): 1685–1694

[74] Jain M D, Nalin P S, Prakash C et al (2006) Binary Tree Based Linear Time Fingerprint Matching. Proc IEEE on Image Procssing, pp 309–312

[75] Bhowmick P, Bhattacharya B (2004) Approximate Fingerprint Matching Using Kd-tree. Proc IEEE on Pattern Recognition, 1: 544–547

[76] Dixit V, Singh D, Raj P et al (2008) kd-tree based fingerprint identification system. proc of the IEEE conf on Anti-counterfeiting, Security and Identification (ASID-08), pp 5–10

[77] Josef S B, Mikael N, Jorgen N et al (2006) Neural Network Based Minutiae Extraction From Skeletonized Fingerprints. Proc IEEE on TENCON, pp 1–4

[78] Mei Y, Sun H, Xia D (2009) A Gradient-based Combined Method for the Computation of Fingerprints' Orientation Field. Image and Vision Computing, 27(8): 1169–1177

[79] Ram S, Bischof H, Birchbauer J (2010) Modelling Fingerprint Ridge Orientation Using Legendre polynomials. Pattern Recognition, 43(1): 342–357.

[80] Tico M, Pauli K (2003) Fingerprint Matching Using an Orientation-based Minutia Descriptor. IEEE Trans on Pattern Analysis and Machine Intelligence, 25: 1009–1014

[81] Gu J, Zhou J, Zhang D (2004) A Combination Model for Orientation Field of Fingerprints. Pattern Recognition, 37(3): 543–553

[82] Qi J, Wang Y (2005) A Robust Fingerprint Matching Method. Pattern Recognition, 38(10): 1665–1671

[83] Nagaty K A, Hattab E (2004) An Approach to a Fingerprints Multi-agent Parallel Matching System. Proc IEEE on Systems, Man and cybernetica, 4: 4750–4756

[84] Feng J, Ouyang Z, Cai A (2006) Fingerprint Matching Using Ridges. Pattern Recognition, 39(11): 2131–2140

[85] Ko T (2002) Fingerprint Enhancement by Spectral Analysis Techniques. Proc IEEE on Image and Pattern Recognition Workshop, pp 7695–1863

[86] Jain A K, Jianjiang Feng, Nagar A et al (2008) On Matching Latent Fingerprints. Proc of the IEEE Workshop on Computer Vision and Pattern Recognition (CVPRW-08), pp 1–8

[87] Ray M, Meenen P, Adhami R (2005) A Novel Approach to Fingerprint Pore Extraction. Proc IEEE on system Theory, pp 282–286

[88] Jain A, Chen Y, Demirkus M (2006) Pores and Ridges: Fingerprint Matching Using Level 3 Features. Proc IEEE on Pattern Recognition, 4: 477–480

[89] Shi P, Tian J, Su Q et al (2007) A Novel Fingerprint Matching Algorithm Based on Minutiae and Global Statistical Features. In: Proc. IEEE conf. on Biometrics: Theory, Applications, and Systems (BTAS-07), pp 1–6

[90] Aguilar G, Sanchez G, Toscano K et al (2007) Fingerprint Recognition. Proc of the IEEE conf. on Internet Monitoring and Protection (ICIMP-07), pp 32–32

[91] Modi S K, Elliott S J, Whetsone J et al (2007) Impact of Age Groups on Fingerprint Recognition Performance. Proc of the IEEE Workshop on Automatic Identification Advanced Technologies, pp 19–23

[92] Jain A, Chen Y, Demirkus M (2007) Pores and Ridges: High-Resolution Fingerprint Matching Using Level 3 Features. IEEE Trans on Pattern Analysis and Machine Intelligence, 29: 15–27

[93] Chen Y, Jain A K (2008) Dots and Incipients: Extended Features for Partial Fingerprint Matching. Proc of the IEEE conf on Biometrics Symposium, pp 1–6

[94] Vatsa M, Singh R, Noore A et al (2008) Quality Induced Fingerprint Identification using Extended Feature Set. Proc of the IEEE Conf on Biometrics: Theory, Applications and Systems (BTAS-2008), pp 1–6

[95] Vatsa M, Singh R, Noore A et al (2009) Quality-augmented Fusion of Level-2 and Level-3 Fingerprint Information Using DSm Theory. International Journal of Approximate Reasoning, 50(1): 51–61

[96] Zhao Q, Zhang Z, Zhang L et al (2010) Adaptive Fingerprint Pore Modeling and Extraction. Pattern Recognition, 43(8): 2833–2844

[97]   Jain A K, Prabhakar S, Hong L et al (1999) Finger Code: A filterbank for Finger-
       print Representation and Matching. Proc IEEE On Computer Vision and Pattern
       Recognition, vol 2: 187 – 193
[98]   Jain A K, Prabhakar S, Pankanti S (2000) Filterbank-based Fingerprint Matching.
       IEEE Trans on Image Proc 9, 846 – 859
[99]   Patil P M, Suralkar R S, Abhyankar H K (2005) Fingerprint Verification Based on
       Fixed Length Square Finger Code. Proc IEEE on Tools with Artificial Intelligence
       (ICTAI'05), 14: 657 – 662
[100] Horton M, Meenen P, Adhami R et al (2002) The Costs and Benefits of Using
       Complex 2 D Gabor Filter in a Filter Based Fingerprint Matching System. Proc
       IEEE on System Theory, pp 171 – 175
[101] Munir M U, Javed M Y (2005) Fingerprint Matching Using Ridge Patterns. IEEE
       Proc on Inform and Commun Technology, pp 116 – 120
[102] Sha L, Zhao F, Tang X (2003) Improved Fingercode for Filterbank-based Finger-
       print Matching. Proc IEEE on Image Processing, 2: 895 – 898
[103] Huang K, Aviyente S (2004) Choosing Best Basis in Wavelet Packets for Fingerprint
       Matching. Proc IEEE on Image Processing, 2: 1249 – 1252
[104] Khan N Y, Javed M Y (2007) Efficient Fingerprint Matching Technique Using
       Wavelet Based Features. Proc IEEE on Digital Image Computing Techniques and
       Applications, 3: 253 – 259
[105] Mokji M, Abu-Bakar S A R (2004) Fingerprint Matching Based on Directional
       Image Constructed Using Expanded Harr Wavelet Transform. Proc IEEE, pp 149 –
       152
[106] Tan X, Bhanu B (2002) Fingerprint Verification Using Genetic Algorithms. Proc
       of the sixth IEEE Workshop on Applicayions of Computer Vision (WACV02), 39:
       465 – 477
[107] Dadgostar M, Tabrizi P R, Fatemizadeh E et al (2009) Feature Extraction Using
       Gabor-Filter and Recursive Fisher Linear Discriminant with Application in Finger-
       print Identification. Proc of the Advances in Pattern Recognition (ICAPR-090), pp
       217 – 220
[108] Cheng X, Xin-Ming C (2009) An Algorithm for Fingerprint Identification Based
       on Wavelet Transform and Gabor Feature. Proc of the IEEE Conf on Genetic and
       Evolutionary Computing (WGEC-09), pp 827 – 830
[109] Jain A, Ross A, Prabhakar S (2001) Fingerprint Matching Using Minutiae and
       Texture Features. Proc IEEE on Image Processing, 2: 282 – 285
[110] Ross A, Jain A, Reisman J (2002) A Hybrid Fingerprint Matching. Proc IEEE on
       Pattern Recognition, 36: 795 – 798
[111] Youssif A A A, Chowdhary M U, Ray S et al (2007) Fingerprint Recognition Sys-
       tem Using Hybrid Matching Techniques. Proc IEEE on Computer and Information
       Science, 11: 234 – 240
[112] Feng J (2008) Combining Minutiae Descriptors for Fingerprint Matching. Pattern
       Recognition, 41(1): 342 – 352
[113] Nanni L, Lumini A (2008) Local Binary Patterns for a Hybrid Fingerprint Matcher.
       Pattern Recognition, 41(11): 3461 – 3466
[114] Gu J, Zhou J, Tang X (2006) Fingerprint Recognition by Combining Global Struc-
       ture and Local Cues. IEEE Trans on Image Processing, 15: 1951 – 1964
[115] Sha L, Zhao F, Tang X (2005) Fingerprint Matching Using Minutiae and
       Interpolation-based Square Tessellation Fingercode. Proc IEEE on Image Process-
       ing, 2: 41 – 44
[116] Guo H (2005) A Hidden Markov Model Fingerprint Matching Approach. Proc IEEE
       on Machine Learning and Cybernetics, 8: 5505 – 5509
[117] Wen M, Liang Y, Pan Q et al (2006) Integration of Multiple Fingerprint Matching
       Algorithms. Proc IEEE on Machine Learning and Cybernetics, Dalian, pp 3186 –
       3189
[118] Nikam S B, Goel P, Tapadar R et al (2007) Combining Gabor Local Texture Pattern
       and Wavelet Global Features for Fingerprint Matching. Proc IEEE on Computa-
       tional Intelligence and Multimedia Applications, 2: 409 – 416
[119] Jain A (1989) Fundamentals of Digital Image Processing, Prentice-Hall Englewood
       Cliffs
[120] Tachaphetpiboon S, Amornraksa T (2005) A Fingerprint Matching Method Using
       DCT Features. Proc IEEE on Communications and Information Technology, 1:
       446 – 449
[121] Tachaphetpiboon S, Amornraksa T (2007) Fingerprint Features Extraction Using

Curve-scanned DCT Coefficients. Proc IEEE on Communications, 18: 33–36

[122] Tachaphetpiboont S, Amornraksa T (2006) Applying FFT Features for Fingerprint Matching. Proc IEEE on Wireless Pervasive Computing, 16: 1–5

[123] Qi J, Shi Z, Zhao X et al (2004) A Novel Fingerprint Matching Method Based on the Hough Transform Without Quantization of the Hough Space. Proc IEEE on Image and Graphics, pp 262–265

[124] Marana A N, Jain A K (2005) Ridge-based Fingerprint Matching Using Hough Transform. Proc IEEE on Computer Graphics and Image Processing, pp 112–119

[125] Seow B C, Yeoh S K, Lai S L et al (2002) Image Based Fingerprint Verification. Proc IEEE on Research and Development, pp 58–61

[126] Jiang L, Sergery T, Venu G (2007) Verifying Fingerprint Match by Local Correlation Methods. Proc IEEE on Biometrics: Theory, Applications and Systems, pp 1–5

[127] Zhang J, Ou Z, Wei H (2006) Fingerprint Matching Using Phase-only Correlation and Fourier-Mellin Transforms. Proc IEEE on Intelligent Systems Design and Applications, 2: 379–383

[128] Ouyang Z, Feng J, Su F et al (2006) Fingerprint Matching with Rotation-descriptor Texture Features. Proc IEEE on Pattern Recognition (ICPR'06), 4: 417–420

[129] Lindoso A, Entrena L, Liu-Jimenez J et al (2007) Increasing Security with Correlation-based Fingerprint Matching. Proc IEEE on Security Technology, 8: 37–43

[130] Chen W, Gao Y (2007) A minutiae-based Fingerprint Matching Algorithm Using Phase Correlation. Proc IEEE on Digital Image Computing Techniques and Applications, pp 233–238

[131] Haddad Z, Beghdadi A, Serir A et al (2008) Fingerprint Identification using Radon Transform. Proc of the IEEE workshop on Image Processing Theory, Tools and Applications (IPTA-08), pp 1–7

[132] Mandal T, Wu Q M J (2008) A Small Scale Fingerprint Matching Scheme Using Digital Curvelet Transform. Proc of the IEEE Conf on Systems, Man and Cybernetics (SMC-08), pp 1534–1538

[133] Nikam S B, Agarwal S (2009) Ridgelet-based Fake Fingerprint Detection. Neurocomputing, 72(10–12): 2491–2506

[134] Hong L, Jain A (1998) Integrating Faces and Fingerprints for Personal Identification. IEEE Trans on Pattern Analysis and Machine Intelligence, 20: 1295–1307

[135] Patra A, Das S (2008) Enhancing Decision Combination of Face and Fingerprint by Exploitation of Individual Classifier Space: An Approach to Multi-modal Biometry. Pattern Recognition, 41(7): 2298–2308

[136] Khan M K, Zhang J (2008) Multimodal Face and Fingerprint Biometrics Authentication on Space-limited Tokens. Neurocomputing, 71(31–15): 3026–3031

[137] Marcialis G L, Roli F, Didaci L (2009) Personal Identity Verification by Serial Fusion of Fingerprint and Face Matchers. Pattern Recognition, 42(11): 2807–2817

[138] Seo J S, Jin M, Lee S (2006) Audio Fingerprinting Based on Normalized Spectral Subband Moments. IEEE Signal Processing Letters, 13: 209–212

[139] Lee S, Yoo C D (2006) Video Fingerprinting Based on Centroids of Gradient Orientations. Proc IEEE on Acoustics, Speech and Signal Processing, 2: 401–404

[140] Lee S, Yoo C D (2008) Robust Video Fingerprinting for Content-based Video Identification. IEEE Trans on Circuits and Systems for Video Technology, 18: 983–988

[141] Tan X, Bhanu B (2006) Fingerprint Matching by Genetic Algorithms. Pattern Recognition, 39(3): 465–477

[142] Girgis M R, Sewisy A A, Mansour R F (2009) A Robust Method for Partial Deformed Fingerprints Verification Using Genetic Algorithm. Expert Systems with Applications, 36(2): 2008–2016

[143] Yang J, Park D S (2008) A Fingerprint Verification Algorithm Using Tessellated Invariant Moment Features. Neurocomputing, 71(10–12): 1939–1946

[144] Ratha N K, Bolle R M (1998) Effect of Controlled Image Acquisition on Fingerprint Matching. Proc IEEE on Pattern recognition, 2: 1659–1661

[145] Nagaty K A, Hattab E (2004) An Approach to a Fingerprints Multi-agent Parallel Matching System. Proc IEEE on Systems, Man and cybernetica, 4: 4750–4756

[146] Duan J, Dixon S L, Lowrie J F et al (2010) Analysis and Comparison of 2D Fingerprints: Insights into Database Screening Performance Using Eight Fingerprint Methods. Journal of Molecular Graphics and Modelling, 29(2): 157–170

[147] Bossen A, Lehmann R, Meier C (2010) Internal Fingerprint Identification With Optical Coherence Tomography. Proc of the IEEE Photonics Technology Letters, 4(7): 507–509

[148]   Wang Y, Hu J (2010) Global Ridge Orientation Modeling for Partial Fingerprint Identification. Proc of the IEEE Trans Pattern Analysis and Machine Intelligence, Issue 99: 1−1
[149]   Jain A K, Jianjiang F, Nandakumar K (2010) Fingerprint Matching. Proc of the IEEE Conf on Computer, 44(2): 36−44
[150]   Yasha C, Xiaoping W (2007) Core-Point Location Method Using Image Sampling Techniques. Proc of the IEEE Conf on Bioinformatics and Biomedical Engineering (ICBBE-07), pp 1019−1021
[151]   Liang X, Bishnu A, Asano T, (2007) A Robust Fingerprint Indexing Scheme Using Minutia Neighborhood Structure and Low-Order Delaunay Triangles. Proc of the IEEE Trans on Information Forensics and Security, 2(4): 721−733
[152]   Chikkerur S, Ratha N (2005) Impact of Singular Point Detection on Fingerprint Matching Performance. Proc IEEE on Automatic Identification Advanced Technologies, pp 207−212
[153]   Sen W, Zhang W W, Wang Y S (2002) Fingerprint Classification by Directional Fields. Proc IEEE on Multimodal Interfaces, pp 395−398
[154]   Dagher I, Helwe W, Yassine F (2002) Fingerprint Recognition Using Fuzzy ARTMAP Neural Network Architecture. Proc IEEE on Microelectronics, 11: 157−160
[155]   Nilsson K, Bigun J (2003) Localization of Corresponding Points in Fingerprints by Complex Filtering. Pattern Recognition Letters, 24(13): 2135−2144
[156]   Jain A K, Prabhakar S, Hong L (1999) A Multichannel Approach to Fingerprint Classification. IEEE Trans on Pattern Analysis and Machine Intelligence, 21: 348−359
[157]   Rerkrai K, Areekul V (2000) A New Reference Point for Fingerprint Recognition. Proc IEEE on Image Processing, 2: 499−502
[158]   Zhang W, Wang Y (2002) Core Based Structure Matching Algorithm of Fingerprint Verification. Proc IEEE on Pattern Recognition, 1: 70−74
[159]   Sha L, Tang X (2004) Combining Exclusive and Continuous Fingerprint Classification. Proc IEEE on Image Proc, 2, pp 1245−1248
[160]   Ahmadyfard A, Nosrati M S (2007) A Novel Approach for Fingerprint Singular Points Detection Using 2D-Wavelet. Proc of the IEEE Conf on Computer Systems and Applications, pp 688−691
[161]   Park C, Lee J, Smith M et al (2006) Singular Point Detection by Shape Analysis of Directional Fields in Fingerprints. Pattern Recognition, 39(5): 839−855
[162]   Areekul V, Boonchaiseree N (2008) Fast Focal Point Localization Algorithm for Fingerprint Registration. Proc of the IEEE Conf on Industrial Electronics and Applications (ICIEA-08), pp 2089−2094
[163]   Biswas S, Ratha N K, Aggarwal G et al (2008) Exploring Ridge Curvature for Fingerprint Indexing. Proc of the IEEE Conf on Biometrics: Theory, Applications and Systems (BTAS-08), pp 1−6
[164]   Mohammadi S, Farajzadeh A (2009) Fingerprint Reference Point Detection Using Orientation Field and Curvature Measurements. Proc of the IEEE Conf on Intelligent Computing and Intelligent Systems (ICIS-09), 4: 25−29
[165]   Li J, Yau W, Wang H (2008) Combining Singular Points and Orientation Image Information for Fingerprint Classification. Pattern Recognition, 41(1): 353−366
[166]   Jirachaweng S, Hou Z, Yau W et al (2011) Residual Orientation Modeling for Fingerprint Enhancement and Singular Point Detection. Pattern Recognition, 44(2): pp 431−442
[167]   Ramoser H, Wavhmann B, Bischof H (2002) Efficient Alignment of Fingerprint Images. Proc IEEE on Pattern Recognition, 3: 748−751
[168]   Germain R S, Califano A, Colville S (1997) Fingerprint Matching Using Transformation Parameter Clustering. Proc IEEE on Computational Science & Engineering, 4: 42−49
[169]   Guo J, Dong H, Chen D et al (2003) Research on Synergetic Fingerprint Classification and Matching. Proc IEEE on Machine Learning and Cybernencs, 5: 3068−3071
[170]   Deriche M (2001) An Algorithm for Reducing the Effect of Compression/Decompression Techniques on Fingerprint Minutiae. Proc IEEE on Intelligent Information System, pp 243−246
[171]   Malalur S S, Manry M T, Narasimha P L (2004) A Pseudo-spectral Fusion Approach to Fingerprint Matching. Proc IEEE on Signals, Systems and Computers, 1: 572−576
[172]   Chan K C, Moon Y S, Cheng P S (2004) Fast Fingerprint Verification Using Subregions of Fingerprint Images. IEEE Trans on Circuits and Systems for Video Tech-

nology, 14: 95 – 101

[173] Pan S B, Moon D, Gil Y et al (2003) An Ultra-low Memory Fingerprint Matching Algorithm and its Implementation on a 32-bit Smart Card. Proc IEEE on Consumer Electronics, 49: 453 – 459

[174] Allah M M A (2005) A Fast and Memory Efficient Approach for Fingerprint Authentication System. Proc IEEE on Advanced Video and Signal Based surveillance, 15: 259 – 263

[175] Govan M, Buggy T (2007) A Computationally Efficient Fingerprint Matching Algorithm for Implementation on Smartcards. Proc IEEE on Signals, Systems and Computers, pp 1 – 6

[176] Yahaya Y H, Isa M, Aziz M I (2009) Fingerprint Biometrics Authentication on Smart Card. Proc of the IEEE Conf on Computer and Electrical Engineering (ICCEE-09), 2: 671 – 673

[177] Yang S, Sakiyama K, Verbauwhede I M (2003) A Compact and Efficient Fingerprint Verification System for Secure Embedded Devices. Proc IEEE On Signals Systems and Computers, 2: 2058 – 2062

[178] Gupta P, Ravi S, Raghunathan A et al (2005) Efficient Fingerprint-based User Authentication for Embedded Systems. Proc IEEE On Design Automation, pp 244 – 247

[179] Fons M, Fons F, Canto E et al (2007) Design of a Hardware Accelerator for Fingerprint Alignment. Proc IEEE on Field Programmable Logic and Applications, pp 485 – 488

[180] Fons M, Fons F, Canto (2006) Design of an Embedded Fingerprint Matcher System. Proc IEEE on Consumer Electronics, pp 1 – 6

[181] Lindoso A, Entrena L, López-Ongil C et al (2005) Correlation-based Fingerprint Matching Using FPGAs. Proc IEEE on Field Programmable Technology, pp 87 – 94

[182] Yeswanth K P, Sakthi G T (2005) Integration of Smart Card and Gabor Filter Method Based Fingerprint Matching for Faster Verification. Proc IEEE on INDICON, pp 526 – 529

[183] Lopez M, Canto E (2008) FPGA Implementation of a Minutiae Extraction Fingerprint Algorithm. Proc of the IEEE Conf on Industrial Electronics (ISIE-08), pp 1920 – 1925.

[184] Militello C, Conti V, Sorbello F et al (2008) A Novel Embedded Fingerprints Authentication System Based on Singularity Points. Proc of the IEEE Conf on Complex, Intelligent and Software Intensive Systems (CISIS-08), pp 72 – 78

[185] Lorrentz P, Howells W G J, McDonald-Maier K D (2009) A Fingerprint Identification System Using Adaptive FPGA-Based Enhanced Probabilistic Convergent Network. Proc of the IEEE Conf on Adaptive Hardware and Systems (AHS-09), pp 204 – 211

18 Subspaces Versus Submanifolds — A Comparative Study of Face Recognition

Hong Huang[1]

Abstract   Automatic face recognition is a challenging problem in the biometrics area, where the dimension of the sample space is typically larger than the number of samples in the training set and consequently the so-called small sample size problem exists. Recently, neuroscientists emphasized the manifold ways of perception, and showed the face images may reside on a nonlinear submanifold hidden in the image space. Many manifold learning methods, such as Isometric feature mapping, Locally Linear Embedding, and Locally Linear Coordination are proposed. These methods achieved the submanifold by collectively analyzing the overlapped local neighborhoods and all claimed to be superior to such subspace methods as Eigenfaces and Fisherfaces in terms of classification accuracy. However, in literature, no systematic comparative study for face recognition is performed among them. In this paper, we carry out a comparative study among them in face recognition, and this study considers theoretical aspects as well as simulations performed using CMU PIE and FERET face databases.

18.1 Introduction

Research in face recognition is motivated not only by posed by the fundamental challenges this recognition problem but also by numerous practical applications where human identification is needed [1, 2]. Face recognition, as one of the primary biometric technologies, becomes more and more important owing to rapid advancement in technologies such as digital cameras, the Internet and mobile devices, and increasing demands on security. Face recognition has several advantages over other biometric technologies: it is natural, nonintrusive, and easy to use [3 – 5].

Although progress in face recognition has been encouraging, the task has also turned out to be a difficult endeavor, especially for unconstrained tasks where viewpoint, illumination, expression, occlusion, accessories, and so on vary considerably [6]. Face recognition has become one of the most challenging problems in the field of computer vision and pattern recognition. Numerous methods have been proposed for face recognition over the past few decades [7 – 12]. In general, these methods can be divided into two groups: geometric

---

1 Key Lab. on Opto-electronic Technique and Systems, Ministry of Education, Chongqing University, 400044/Chongqing, China. E-mail: hhuang.cqu@gmail.com.

feature-based method and appearance-based method [13].

For decades, geometric feature-based methods have used properties and relations (e.g., distances and angles) between facial features such as eyes, mouth, nose, and chin to perform recognition. Despite their economical representation and insensitivity to small variations in illumination and viewpoint, feature-based methods are quite sensitive to the feature extraction and measurement process. It has been argued that existing techniques for the extraction and measurement of facial features are not reliable enough. It has also been claimed that methods for face recognition based on finding local image features and inferring identity by the geometric relations of these features are often ineffective.

The most popular algorithms are appearance-based approaches. They differ from feature-based techniques in that their low-dimensional representation is, in a least-squares sense, faithful to the original image. One primary advantage of appearance-based methods is that they do not need to create representations or models for face images. Since for a given face image, its model is now implicitly defined in the face image itself. When using appearance-based methods, we usually represent an image by a vector in a high dimensional space. Here the feature vector used for classification is a linear projection of the face image into a lower-dimensional linear subspace. In extreme cases, the feature vector is chosen as the entire image, with each element in the feature vector taken from a pixel in the image. Although such an appearance-based representation is a simple form, the corresponding dimensionality is too large to realize robust and fast recognition, and is typically lager than the number of samples in the training set which leads to the so-called small sample size problem (SSS) [14 − 16].

A common way to resolve this problem is to use dimensionality reduction methods. The dimensionality reduction algorithms for face recognition can be divided into two groups: subspace-based algorithm and submanifold-based algorithm [17]. Two of the most popular methods for this purpose are Principal Component Analysis (PCA) [7] and Linear Discriminant Analysis (LDA) [8], which assume that the samples lie on a linear embedded subspace and aim at preserving the global Euclidean structure of the image space. However, a lot of researches have shown that facial images possibly lie on a submanifold [18 − 23, 25]. When using PCA and LDA for dimensionality reduction, they will fail to discover the intrinsic manifold structure of the image space [24, 26 − 30]. In the last ten years, a number of manifold learning methods have been proposed to discover the nonlinear structure of the manifold by investigating the geometric structures of samples, such as Locally Linear Embedding(LLE) [31], Isometric feature mapping(Isomap) [32, 33] and Locally Linear Coordination(LLC) [47]. These methods are appropriate for representation of nonlinear data, and they all claimed to be superior to such methods as PCA and LDA in terms of classification accuracy. Some comparisons are preformed in Refs. [34 − 37, 60]. However, all the comparisons are very limited in scope with respect to the number of methods and tasks that are addressed, there is no systematic comparative study for face recognition among these algorithms in literature. The purpose of this paper is to present a systematic comparative study of subspace and submanifold algorithms. It performs both theoretical analysis and simulations of methods for face recognition and gets some sight from such comparative study.

The remainder of the paper is structured as follows. In Section 18.2, we introduce the definitions and notation used in the later sections. Section 18.2 discusses briefly two subspace methods for face recognition. Subsequently, Section 18.4 reviews the manifold learning methods. Then, in Section 18.5 we describes the experimental results based on some open databases and discussions. Section 18.6 summarizes the paper and indicates the main interests for future work.

18.2 Notation and Definitions

Throughout this paper, the following mathematical notations are used:

- The high-dimensional input points will be referred to as $x_1, x_2, \ldots, x_n$. At times it will be convenient to work with these points as a single matrix X, where the ith row of X is x_i.
- The low-dimensional representations that the dimensionality reduction algorithms find will be referred to as $y_1, y_2, \ldots, y_n$. Y is the matrix of these points.
- n is the number of points in the input.
- D is the dimensionality of the input (i.e., $x_i \in \Re^D$).
- d is the dimensionality of the manifold that the input is assumed to lie on and, accordingly, the dimensionality of the output (i.e., $y_i \in \Re^d$).
- k is the number of nearest neighbors used by a specific algorithm.
- $knn(i)$ denotes the set of the k-nearest neighbors of x_i.
- The (eigenvector, eigenvalue) pairs are ordered by the eigenvalues. That is, if the (eigenvector, eigenvalue) pairs are (v_i, λ_i), for $i = 1, \ldots, n$, then $\lambda_1 \geqslant \lambda_2 \geqslant \ldots \geqslant \lambda_n$. We refer to $v_1, \ldots v_d$ as the top d eigenvectors, and $v_{n-d+1}, \ldots v_n$ as the bottom d eigenvectors.
- Σ is a covariance matrix.
- μ is the mean of all samples.
- S_t is the total scatter matrix.
- S_b is the between-class matrix.
- S_w is the within-class matrix.
- G denotes the adjacency graph.
- ψ represents a non-linear mapping from an input space to a high dimensional implicit output space.
- V represents a transformation matrix from a high dimensional implicit input space to a low dimensional embedding space.
- K denotes a Mercer kernel function.

- I is the identity matrix.
- $\ell_i \in \{1,\ldots,c\}$ denotes the class label of x_i.

Let us quickly review some basic terminology from geometry and topology in order to crystallize this notion of subspace and submanifold. We begin with a description of the dimensionality reduction problem.

The generic dimensionality reduction problem is as follows. Given n data points $\{x_i, x_2, \ldots, x_n\} \in \Re^D$ sampled from one underlying manifold $\mathscr{M}$. The goal of dimension reduction is to map $(X \in \Re^D) \mapsto (Y \in \Re^d)$ where $d \ll D$ by using the information of examples. Linear techniques consist of replacing the original data X by a matrix of the form

$$Y = V^{\mathrm{T}} X, \ where \ V \in \Re^{m \times d} \tag{18.1}$$

Thus, each vector x_i is replaced by $y_i = V^{\mathrm{T}} x_i$, a member of the d-dimensional space $\Re^d$. If V is a unitary matrix, then Y represents the orthogonal projection of X into the embedding space Y.

Definition 18.1   A Euclidean subspace (or subspace of $\Re^n$) is a set of vectors (Fig. 18.1) that is closed under addition and scalar multiplication. Geometrically, a subspace is a flat in n-dimensional Euclidean space that passes through the origin.

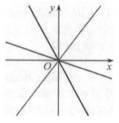

Fig. 18.1 Three one-dimensional subspaces of $\Re^2$.

Definition 18.2   A homeomorphism is a continuous function whose inverse is also a continuous function.

Definition 18.3   A d-dimensional manifold $\mathscr{M}$ is set that is locally homeomorphic with $\Re^d$. That is, for each $x \in \mathscr{M}$, there is an open neighborhood around x, N_x, and a homeomorphism f : $N_x \longrightarrow \Re^d$. These neighborhoods are referred to as coordinate patches, and the map is referred to as a coordinate chart. The image of the coordinate charts is referred to as the parameter space (Fig. 18.2).

Definition 18.4   A submanifold of a manifold $\mathscr{M}$ is a subset S which itself has the structure of a manifold, and for which the inclusion map $S \longrightarrow \mathscr{M}$ satisfies certain properties.

Definition 18.5   A tangent space of a manifold is a concept which facilitates the generalization of vectors from affine spaces to general manifolds, since in the latter case one cannot simply subtract two points to obtain a vector pointing from one to the other.

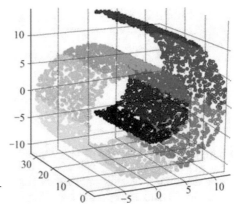

Fig. 18.2 A two-dimensional manifold embedded in three-dimensions.

Manifolds are well-studied in mathematics, but we will be interested only in the case where $\mathcal{M}$ is a subset of $\Re^d$, where D is typically much larger than d. In other words, the manifold will lie in a high dimensional space ($\Re^D$), but will be homeomorphic with a low-dimensional space ($\Re^d$, with $d \ll D$).

With this background in mind, we proceed to the main topic of this paper.

18.3 Brief Review of Subspace-Based Face Recognition Algorithms

In this section, we provide a brief review of two main subspace algorithms: (1) Eigenfaces (PCA) [7] and (2) Fisherfaces(PCA + LDA) [8]. Since last decade, subspace-based methods, originated from Turk's Eigenfaces based on the PCA and improved by Belhumeur's Fisherfaces based on LDA, have dominated the two algorithms in face recognition for good performance and computational feasibility.

18.3.1 PCA

Principal component analysis (PCA) is a popular dimensionality reduction technique, seeking an orthonormal set of principal axes, that is to say, a set of subspace basic vectors correspond to the maximum variance direction in the original image space. In other words, the objective function for PCA is

$$J_F(\boldsymbol{V}) = \arg\max |\boldsymbol{V}^{\mathrm{T}} \boldsymbol{S}_t \boldsymbol{V}|, \tag{18.2}$$

where $\boldsymbol{S}_T$ is the *total scatter* matrix as

$$
\begin{aligned}
\boldsymbol{S}_t &= \frac{1}{n} \sum_{i=1}^{n} (\boldsymbol{x}_i - \mu)(\boldsymbol{x}_i - \mu^{\mathrm{T}}) \\
&= \frac{1}{n} \boldsymbol{X} \left(\boldsymbol{I} - \frac{1}{n} \boldsymbol{e} \boldsymbol{e}^{\mathrm{T}} \right) \boldsymbol{X}^{\mathrm{T}} \\
&= \boldsymbol{X} \boldsymbol{G} \boldsymbol{X}^{\mathrm{T}},
\end{aligned} \tag{18.3}
$$

where $\mu = \dfrac{1}{n} \sum_{i=1}^{n} \boldsymbol{x}_i$, $\boldsymbol{G} = \dfrac{1}{n} \boldsymbol{I} - \dfrac{1}{n^2} \boldsymbol{e} \boldsymbol{e}^{\mathrm{T}}$, and $\boldsymbol{e} = (1, \dots, 1)^{\mathrm{T}}$.

A drawback of this method is that the scatter being maximized is due not only to the between-class scatter that is useful for classification, but also to the within-class scatter that, for classification purposes, is unwanted information.

18.3.2   LDA

A popular supervised dimensionality reduction technique is linear discriminant analysis (LDA). LDA seeks to find a projection axis such that the Fisher criterion (i.e., the ratio of the *between-class scatter* to the *within-class scatter*) is maximized after the projection of samples. The *between-class* matrix $\boldsymbol{S}_b$ and *within-class scatter* matrix $\boldsymbol{S}_w$ are defined by

$$
\boldsymbol{S}_b = \frac{1}{n} \sum_{i=1}^{c} n_i (\mu_i - \mu)(\mu_i - \mu)^{\mathrm{T}}, \tag{18.4}
$$

$$
\boldsymbol{S}_w = \frac{1}{n} \sum_{i=1}^{c} \sum_{j=1}^{\ell_i} (x_{ij} - \mu_i)(x_{ij} - \mu_i)^{\mathrm{T}}. \tag{18.5}
$$

where x_{ij} denotes the j-th training sample in class i, n_i is the number of training samples in class i, μ_i is the mean of the training samples in class i, and μ is the mean of all samples.

It is easy to show that $\boldsymbol{S}_b$ and $\boldsymbol{S}_w$ are both non-negative definite matrix and satisfy $\boldsymbol{S}_t = \boldsymbol{S}_w + \boldsymbol{S}_b$. The Fisher criterion is defined by

$$
J_F(\boldsymbol{V}) = \arg\max \frac{trace(\boldsymbol{v}^{\mathrm{T}} \boldsymbol{S}_b \boldsymbol{v})}{trace(\boldsymbol{v}^{\mathrm{T}} \boldsymbol{S}_w \boldsymbol{v})}. \tag{18.6}
$$

The stationary points of $J_F(\boldsymbol{v})$ are the generalized eigenvectors $\boldsymbol{v}_1, \boldsymbol{v}_2, \dots, \boldsymbol{v}_d$ of $\boldsymbol{S}_b \boldsymbol{v} = \lambda \boldsymbol{S}_w \boldsymbol{v}$ corresponding to the d largest eigenvalues.

The two algorithms both try to transform a given set of face images into a smaller set of basis images by using matrix decomposition techniques. The unsupervised Eigenfaces intends to maximize the covariance, while the supervised Fisherfaces intends to maximize the discriminability.

18.4 Submanifold-Based Algorithms for Face Recognition

In section 3, we have discussed the two main subspace methods for face recognition, which are established and well understood. However, when using PCA and LDA for dimensionality reduction, they may fail to discover the intrinsic manifold structure of the image space. So, Many manifold learning algorithms have been proposed more recently. In this section, we will discuss some submanifold methods for face recognition. Manifold learning algorithms can be subdivided into four main types:

1) Algorithms that attempt to preserve global properties of the original data in the low-dimensional representation;

2) Algorithms that attempt to preserve local properties of the original data in the low-dimensional representation;

3) Algorithms that perform global alignment of a mixture of local models;

4) Algorithms that extend classical manifold learning algorithms to take prior information into account or relate the algorithms to Mercer kernel machines, and extend new data points to the low dimensional embedding space. The first three types introduce manifold learning methods in those basic forms, so we introduce three useful generalizations as the fourth type of manifold learning methods:

1) The supervised manifold learning methods;

2) The kernel manifold learning methods;

3) Out-of-sample extensions.

Figure 18.3 shows a taxonomy of manifold learning algorithms.

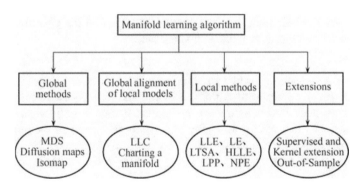

Fig. 18.3 Taxonomy of manifold learning algorithms.

18.4.1 Global Algorithms

Global manifold learning algorithms attempt to preserve global properties of data. This subsection presents three global algorithms for face recognition:

(1) Multidimensional scaling (MDS) [38], (2) Isomap, (3) Diffusion maps (DM) [39].

Multidimensional Scaling

MDS represents a collection of nonlinear methods that map the high dimensional data representation to a low-dimensional representation while retaining the pairwise distances between the data points as much as possible. The quality of the mapping is expressed in the stress function, a measure of the error between the pairwise distances in the low-dimensional and that in the high-dimensional representation of data.

Two important examples of stress functions are the raw stress function and the Sammon cost function. The raw stress function is defined by

$$J(Y) = \sum_{ij} (\|x_i - x_j\| - \|y_i - y_j\|)^2 \tag{18.7}$$

in which $\|x_i - x_j\|$ is the Euclidean distance between the high dimensional data points x_i and x_j and $\|y_i - y_j\|$ is the Euclidean distance between the low-dimensional data points y_i and y_j.

The Sammon cost function is given by

$$J(Y) = \frac{1}{\Sigma_{ij}\|x_i - x_j\|} \sum_{ij} \frac{(\|x_i - x_j\| - \|y_i - y_j\|)^2}{\|x_i - x_j\|}. \tag{18.8}$$

The minimization of the stress function can be performed using various methods, such as the eigen decomposition of a pairwise dissimilarity matrix, the conjugate gradient method, or a pseudo-Newton method [38, 40].

MDS is relatively simple to implement, very useful for visualization, and able to uncover hidden structure in the data.

Isomap

Isomap generalizes MDS to submanifolds. It is based on replacing Euclidean distance by an approximation of the geodesic distance on the manifold. The geodesic distance between two points p and q is defined as the length of the shortest path from p to q. Suppose there is a path in a connected domain between points p_1 and p_n, i.e., p_i and p_{i+1} are connected neighbors and p_i belongs to the domain for all i. The path length $l(P)$ is defined as

$$l_p = \sum_{i=1}^{n-1} d_N(p_i, p_{i+1}), \tag{18.9}$$

where the sum of the neighbor distances d_N between adjacent points in the path.

The Isomap algorithm obtains the normalized matrix $\tilde{M}$ from which the embedding is derived by transforming the raw pairwise distances matrix as follows:

(1) compute the matrix $M_{ij} = l_p^2(x_i, x_j)$ of squared geodesic distances with respect to the data $\boldsymbol{X}$; (2) apply this matrix to the distance-to-dot-product transformation as follows

$$\tilde{M}_{ij} = -\frac{1}{2}\left(M_{ij} - \frac{1}{n}S_i - \frac{1}{n}S_j + \frac{1}{n^2}S_iS_j\right). \qquad (18.10)$$

Note that S_i be the *ith* row sum of the affinity matrix M

$$S_i = \sum_j M_{ij}. \qquad (18.11)$$

Practically, Isomap goes through the following steps.

Algorithm Outline: Isomap

Input
    1) Training samples $\{\boldsymbol{x}_1, \ldots, \boldsymbol{x}_n\} \in \Re^D$.
    2) Dimensionality of embedding space $d(1 \leqslant d \leqslant D)$.
Output
    The d-dimensional embedding coordinates $\boldsymbol{Y}$ for the input points $\boldsymbol{X}$.
Algorithm steps
    Step 1: Construct neighborhood graph: Define the graph $\boldsymbol{G}$ over all data points by connecting points $\boldsymbol{x}_i$ and $\boldsymbol{x}_j$ if (as measured by $d(\boldsymbol{x}_i, \boldsymbol{x}_j)$) they are closer than ϵ, or if $\boldsymbol{x}_i$ is one of the k nearest neighbors of $\boldsymbol{x}_j$. Set edge lengths equal to $d(\boldsymbol{x}_i, \boldsymbol{x}_j)$.
    Step 2: Compute shortest paths as Eq. (18.9).
    Step 3: Construct d-dimensional embedding: Let λ_k be the k-th eigenvalue (in decreasing order) of the matrix $\tilde{M}_{ij}$, and v_k^i be the i-th component of the k-th eigenvector. Then set the k-th component of the d-dimensional coordinate vector $\boldsymbol{y}_i$ equal to $\sqrt{\lambda_k}v_k^i$.

Isomap was successfully applied to tasks such as face recognition, wood inspection, visualization of biomedical data, and head pose estimation.

Diffusion Maps

Diffusion maps (DM) framework originates from the field of dynamical systems. DM is based on defining a Markov random walk on the graph of data. By performing the random walk for a number of time steps, a measure for the proximity of data points is obtained. Using this measure, the so-called diffusion distance is defined. In the low-dimensional representation of data, the pairwise diffusion distances are retained as well as possible. In the diffusion maps framework, a graph of data is constructed first. The weights of the edges in the graph are computed using the Gaussian kernel function, leading to a matrix W with a formula as

$$W_{ij} = \mathrm{e}^{-\frac{\|x_i - x_j\|^2}{2\sigma^2}}, \qquad (18.12)$$

where σ indicates the variance of the Gaussian. Subsequently, normalization of the matrix W is performed in such a way that its rows add up to 1. In this way, a matrix P is formed with a formula as

$$P_{ij} = \frac{w_{ij}}{\Sigma_k w_{ik}}. \qquad (18.13)$$

The matrix P represents the probability of a transition from one data point to another data point in a single time step. Using the random walk forward probabilities p_{ij}, the diffusion distance is defined by

$$\phi_{(x_i, x_j)} = \sum_k \frac{(p_{ik} - p_{jk})^2}{\psi_0(x_k)}. \tag{18.14}$$

In the equation, $\psi_0(x_i)$ is a term that attributes more weight to parts of the graph with high density. In the low dimensional representation of data Y, diffusion maps attempt to retain the diffusion distances. Using spectral theory on the random walk, it can be shown that the low-dimensional representation Y which retains the diffusion distances is formed by the d nontrivial principal eigenvectors of the eigen problem $\boldsymbol{PY} = \lambda \boldsymbol{Y}$.

The low-dimensional representation $\boldsymbol{Y}$ is given by the next d principal eigenvectors. In the low-dimensional representation, the eigenvectors are normalized by their corresponding eigenvalues. Hence, the low-dimensional data representation is given by

$$\boldsymbol{Y} = \begin{bmatrix} \boldsymbol{y}_1 \\ \boldsymbol{y}_2 \\ \vdots \\ \boldsymbol{y}_d \end{bmatrix} = \begin{bmatrix} \lambda_1 \boldsymbol{v}_1^{\mathrm{T}} \\ \lambda_2 \boldsymbol{v}_2^{\mathrm{T}} \\ \vdots \\ \lambda_d \boldsymbol{v}_d^{\mathrm{T}} \end{bmatrix}. \tag{18.15}$$

The DM algorithm is summarized below.

Algorithm Outline: Diffusion maps

Input
   1) Training samples $\{\boldsymbol{x}_1, \ldots, \boldsymbol{x}_n\} \in \Re^D$.
   2) Dimensionality of embedding space d ($1 \leqslant d \leqslant D$).
Output
   The d-dimensional embedding coordinates $\boldsymbol{Y}$ for the input points $\boldsymbol{X}$.
Algorithm steps
   Step 1: Define a kernel, $k(x, y)$ and create a kernel matrix, $\boldsymbol{K}$, such that $\boldsymbol{K}_{i,j} = k(\boldsymbol{x}_i, \boldsymbol{x}_j)$.
   Step 2: Create the diffusion matrix by normalizing the rows of the kernel matrix.
   Step 3: Calculate the eigenvectors of the diffusion matrix.
   Step 4: Map to the d-dimensional diffusion space at time t, using the d dominant eigenvectors and values.

18.4.2   Local Algorithms

Local algorithms for manifold learning attempt to preserve the local geometry of data; essentially, they seek to map nearby points on the manifold to nearby points in the low-dimensional representation. These methods include LLE [31], Laplacian Eigenmaps (LE) [41], Locality preserving projection (LPP)

[42], Local Tangent Space Analysis (LTSA) [43], Hessian LLE (HLLE) [44] and Neighborhood Preserving Embedding (NPE) [45], etc. Note that the dimensionality of the actual low-dimensional representation by Hessian LLE and LTSA is not higher than the value of k, our comparative review does not cover the two algorithms.

LLE

LLE algorithm looks for an embedding that preserves the local geometry in the neighborhood of each data point. Provided that sufficient data are available by sampling well from the manifold, the goal of LLE is to find a low-dimensional embedding of X by mapping the D-dimensional data into a single global coordinate system in $\Re^d$.

The LLE algorithm consists of three main steps as follows.

Step 1   Finding sets of points constituting local patches in a high-dimensional space. The first step of the LLE algorithm is to find the neighborhood of each data point $\boldsymbol{x}_i$, $i \in 1, 2, \ldots, n$. This can be done either by identifying a fixed number of nearest neighbors k per data point in terms of Euclidean distances or by choosing all points within a ball of fixed radius.

Step 2   Assigning weights to pairs of neighboring points. this step is to assign a weight to every pair of neighboring points. These weights form a weight matrix $\boldsymbol{W}$, each element of which (w_{ij}) characterizes a degree of closeness of two particular points $(\boldsymbol{x}_i$ and $\boldsymbol{x}_j)$.

Let $\boldsymbol{x}_i$ and $\boldsymbol{x}_j$ be neighbors, the weight values w_{ij} are defined as contributions to the reconstruction of the given point from its nearest neighbors. In other words, the following minimization task must be solved:

$$\varepsilon(\boldsymbol{W}) = \sum_{i=1}^n \left\| \boldsymbol{x}_i - \sum_j w_{ij} \boldsymbol{x}_j \right\| \tag{18.16}$$

subject to constraints $w_{ij} = 0$, if $\boldsymbol{x}_i$ and $\boldsymbol{x}_j$ are not neighbors, and $\sum_{i=1}^n w_{ij} = 1$.

Step 3   Computing the low-dimensional embedding. The goal of the LLE algorithm is to preserve a local linear structure of a high-dimensional space as accurately as possible in a low-dimensional space. In particular, the weights w_{ij} that reconstruct point $\boldsymbol{x}_i$ in D-dimensional space from its neighbors should reconstruct its projected manifold coordinates in d-dimensional space. Hence, the weights w_{ij} are kept fixed and embedded coordinates $\boldsymbol{y}_i$ are sought by minimizing the following cost function:

$$J_F(\boldsymbol{Y}) = argmin \sum_{i=1}^n \left\| \boldsymbol{y}_i - \sum_j w_{ij} \boldsymbol{y}_j \right\|. \tag{18.17}$$

To make the problem well-posed, the following constraints are imposed:

$$\frac{1}{n} \sum_{i=1}^n \boldsymbol{y}_i \boldsymbol{y}_i^{\mathrm{T}} = \boldsymbol{I}_{d \times d}, \tag{18.18}$$

$$\frac{1}{n}\sum_{i=1}^{n}\boldsymbol{y}_i = \boldsymbol{0}_{d\times d}, \tag{18.19}$$

where $\boldsymbol{0}_{d\times d}$ is a zero-vector of length d.

The embedded coordinates have to be normalized unit covariance as in Eq. (18.18) in order to remove the rotational degree of freedom and to fix the scale. Eq. (18.19) removes the translation degree of freedom by requiring the outputs to be centered at the origin. As a result, a unique solution is obtained.

To find the embedding coordinates minimizing Eq. (18.17), which are composed in the matrix $\boldsymbol{Y}$ of size $d \times n$, under the constraints given in Eqs. (18.18) and (18.19), a new matrix $\boldsymbol{M}$ is constructed, based on the weight matrix $\boldsymbol{W}$:

$$\boldsymbol{M} = (\boldsymbol{I} - \boldsymbol{W})^{\mathrm{T}}(\boldsymbol{I} - \boldsymbol{W}). \tag{18.20}$$

The cost matrix $\boldsymbol{M}$ is sparse, symmetrical, and positive semidefinite. LLE then computes the bottom $(d+1)$ eigenvectors of $\boldsymbol{M}$, associated with the $(d+1)$ smallest eigenvalues. The first eigenvector (composed of 1's) whose eigenvalue is close to zero is excluded. The remaining d eigenvectors yield the final embedding $\boldsymbol{Y}$.

The LLE algorithm can be summarized as follows:

Algorithm Outline: LLE

Input
1) Training samples $\{\boldsymbol{x}_1, \ldots, \boldsymbol{x}_n\} \in \Re^D$.
2) Dimensionality of embedding space d ($1 \leqslant d \leqslant D$).
Output
The d-dimensional embedding coordinates $\boldsymbol{Y}$ for the input points $\boldsymbol{X}$.
Algorithm steps
Step 1: Finding k nearest neighbors for each data point $\boldsymbol{x}_i$, $i \in 1, 2, \ldots, n$.
Step 2: Calculating the weight matrix $\boldsymbol{W}$ of weights between pairs of neighbors.
Step 3: Constructing the cost matrix $\boldsymbol{M}$ (based on $\boldsymbol{W}$) and computing its bottom $(d+1)$ eigenvectors.

Laplacian Eigenmaps

The Laplacian Eigenmaps method is an application of spectral graph theory. Given a set of n multivariate observations embedded as vectors $\boldsymbol{X} \in \Re^D$, a weighted graph $\boldsymbol{G}$ is built over the endpoints of these vectors. It consists of n nodes, one for each point and a set of edges connecting neighboring points. Consider the problem of mapping the weighted graph $\boldsymbol{G}$ to a line so that connected points stay as close as possible. If two points are close enough, then there is an edge between them. Let $\boldsymbol{Y} \in \Re^d, d \ll D$ be such a map. A reasonable criterion for choosing a "good" map is to minimize the following objective function

$$\min \sum_{ij}(\boldsymbol{y}_i - \boldsymbol{y}_j)^2 W_{ij} \tag{18.21}$$

where $\boldsymbol{W}$ is the weight matrix defined as Eq. (18.12).

The objective function with our choice of weights $\boldsymbol{W}_{ij}$ incurs a heavy penalty if neighboring points $\boldsymbol{x}_i$ and $\boldsymbol{x}_j$ are mapped far apart. Therefore, minimizing it is an attempt to ensure that if $\boldsymbol{x}_i$ and $\boldsymbol{x}_j$ are close then $\boldsymbol{y}_i$ and $\boldsymbol{y}_j$ are close as well. It turns out that the minimization problem reduces to finding

$$\begin{array}{c} \arg\min \ trace(\boldsymbol{Y}^{\mathrm{T}}\boldsymbol{L}\boldsymbol{Y}), \\ subject\ to\ \boldsymbol{Y}^{\mathrm{T}}\boldsymbol{D}\boldsymbol{Y} = \boldsymbol{I}, \end{array} \qquad (18.22)$$

where $\boldsymbol{L} = \boldsymbol{D} - \boldsymbol{W}$ is the Laplacian matrix; $\boldsymbol{D}$ a diagonal matrix with $D_{ii} = \sum_j W_{ij}$. The Laplacian matrix $\boldsymbol{L}$ is symmetric positive semidefinite, which can be regarded as an operator on function s defines on vertices of G.

The algorithmic procedure is formally stated below:

Algorithm Outline: Laplacian Eigenmaps

Input

    1) Training samples $\{\boldsymbol{x}_1, \ldots, \boldsymbol{x}_n\} \in \Re^D$.

    2) Dimensionality of embedding space d $(1 \leqslant d \leqslant D)$.

Output

    The d-dimensional embedding coordinates $\boldsymbol{Y}$ for the input points $\boldsymbol{X}$.

Algorithm steps

    Step 1: Constructing the adjacency graph. Nodes $\boldsymbol{x}_i$ and $\boldsymbol{x}_j$ are connected by an edge if $\boldsymbol{x}_i$ and $\boldsymbol{x}_j$ are "close". There are two variations: (a) ε neighborhoods. Nodes $\boldsymbol{x}_i$ and $\boldsymbol{x}_j$ are connected by an edge if $\|\boldsymbol{x}_i - \boldsymbol{x}_j\|^2 < \varepsilon$. (b) k nearest neighbors. Nodes $\boldsymbol{x}_i$ and $\boldsymbol{x}_j$ are connected by an edge if $\boldsymbol{x}_i$ is among k nearest neighbors of $\boldsymbol{x}_j$ or $\boldsymbol{x}_j$ is among k nearest neighbors of $\boldsymbol{x}_i$.

    In Fig. 18.4, we show an example of an adjacency graph on the Swiss-roll dataset using five nearest neighbors.

    Step 2: Compute the weighted graph $\boldsymbol{W}$ according to Eq. (18.12).

    Step 3: Assume the graph $\boldsymbol{G}$, constructed above, is connected component. Compute eigenvalues and eigenvectors for the generalized eigenvector problem:

$$\boldsymbol{L}\boldsymbol{Y} = \lambda\boldsymbol{Y} \qquad (18.23)$$

    Let $\boldsymbol{y}_1, \ldots, \boldsymbol{y}_{n-1}$ be the solutions of Eq. (18.23) ordered according to their eigenvalues with $\boldsymbol{y}_0$ having the smallest eigenvalue (in fact 0). The image of $\boldsymbol{x}_i$ under the embedding into the lower dimensional space $\Re^d$ is given by $(\boldsymbol{y}_1(i), \ldots, \boldsymbol{y}_m(i))$.

LPP

LPP [42] is unsupervised and performs a linear transformation. It models the manifold structure by constructing an adjacency graph (see Fig. 18.4), which is a graph expressing local nearness of data. This is highly desirable for face recognition, since it is significantly less computationally expensive and more importantly, it is defined in all points and not just in the training points as Isomap and Laplacian Eigenmaps.

It constructs a weighted graph $G = (\nu, \varepsilon, \boldsymbol{W})$, where ν is the set of all points; ε is the set of edges connecting the points; $\boldsymbol{W}$ is a similarity matrix with weights characterizing the likelihood of two points. The criterion

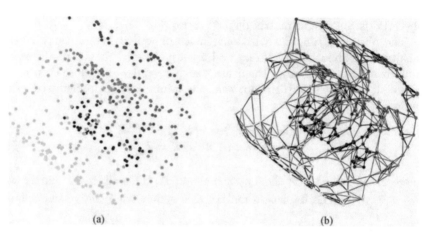

(a)   (b)

Fig. 18.4 An example of constructing the adjacency graph. (a) Swiss-roll dataset ($n = 400$), (b) the adjacency graph using five nearest neighbors.

function of LPP is as follows:

$$\min \sum_{ij} \|\boldsymbol{y}_i - \boldsymbol{y}_j\|^2 W_{ij}, \tag{18.24}$$

where the weights can be defined in different ways as well. Two common choices are weights of the heat kernel $W_{ij} = \exp\left(-\dfrac{\|\boldsymbol{x}_i - \boldsymbol{x}_j\|^2}{t}\right)$ and constant weights ($W_{ij} = 1$ if i and j are adjacent; otherwise, $W_{ij} = 0$). Note that the entries of $\boldsymbol{W}$ are non-negative and that $\boldsymbol{W}$ is sparse and symmetric.

Follow some algebraic steps, we see that

$$\frac{1}{2} \sum_{ij} \|\boldsymbol{y}_i - \boldsymbol{y}_j\|^2 W_{ij} \tag{18.25}$$

$$= \frac{1}{2} \sum_{ij} (\boldsymbol{V}^{\mathrm{T}}\boldsymbol{x}_i - \boldsymbol{V}^{\mathrm{T}}\boldsymbol{x}_j)(\boldsymbol{V}^{\mathrm{T}}\boldsymbol{x}_i - \boldsymbol{V}^{\mathrm{T}}\boldsymbol{x}_j)W_{ij}$$
$$= trace(\boldsymbol{V}^{\mathrm{T}}\boldsymbol{X}\boldsymbol{D}\boldsymbol{X}^{\mathrm{T}}\boldsymbol{V}) - trace(\boldsymbol{V}^{\mathrm{T}}\boldsymbol{X}\boldsymbol{W}\boldsymbol{X}^{\mathrm{T}}\boldsymbol{V})$$
$$= trace(\boldsymbol{V}^{\mathrm{T}}\boldsymbol{X}(\boldsymbol{D} - \boldsymbol{W})\boldsymbol{X}^{\mathrm{T}}\boldsymbol{V})$$
$$= trace(\boldsymbol{V}^{\mathrm{T}}\boldsymbol{X}\boldsymbol{L}\boldsymbol{X}^{\mathrm{T}}\boldsymbol{V}), \tag{18.26}$$

where $\boldsymbol{D}$ is a diagonal matrix with $D_{ii} = \displaystyle\sum_j W_{ij}$; $\boldsymbol{L} = \boldsymbol{D} - \boldsymbol{W}$ is the Laplacian matrix.

In order to remove the arbitrary scaling factor in the embedding, LPP imposes a constraint as follows:

$$\boldsymbol{Y}\boldsymbol{D}\boldsymbol{Y}^{\mathrm{T}} = \boldsymbol{I} \Rightarrow \boldsymbol{V}^{\mathrm{T}}\boldsymbol{X}\boldsymbol{D}\boldsymbol{X}^{\mathrm{T}}\boldsymbol{V} = \boldsymbol{I}. \tag{18.27}$$

This constraint sets the mapping (embedding) scale and makes the vertices with high similarities to be mapped nearer to the origin. Finally, the minimization problem is reduced to

$$\arg\min \ trace(\boldsymbol{V}^{\mathrm{T}}\boldsymbol{X}\boldsymbol{L}\boldsymbol{X}^{\mathrm{T}}\boldsymbol{V}),$$
$$subject \ to \ \boldsymbol{V}^{\mathrm{T}}\boldsymbol{X}\boldsymbol{D}\boldsymbol{X}^{\mathrm{T}}\boldsymbol{V} = \boldsymbol{I}. \tag{18.28}$$

The transformation matrix $\boldsymbol{V}$ that minimizes the objective function can be obtained by solving the generalized eigenvalue problem:

$$\boldsymbol{X}\boldsymbol{L}\boldsymbol{X}^{\mathrm{T}}\boldsymbol{v} = \lambda\boldsymbol{X}\boldsymbol{D}\boldsymbol{X}^{\mathrm{T}}\boldsymbol{v}. \tag{18.29}$$

That is, LPP seeks a transformation matrix $\boldsymbol{V}$ such that nearby data pairs in the original space $\Re^m$ are kept close in the embedding space $\Re^d$. Thus, LPP tends to preserve the local structure of the data.

He and Niyogi proposed a Laplacianfaces method [46], which is an optimal linear approximation to Laplacian Beltrami operator on the face manifold, and very flexible in connection with both PCA/LDA versus clustering/classification.

NPE

NPE [45] is a linear approximation to the LLE algorithm, and it aims at preserving the local manifold structure. Here, by "local structure" mean that each data point can be represented as a linear combination of its neighbors. In many cases, the data points might reside on a nonlinear submanifold, but it might be reasonable to assume that each local neighborhood is linear. Thus, the local geometry of these patches can be characterized by linear coefficients that reconstruct each data point from its neighbors. Reconstruction errors are measured by the cost function (Eq. (18.16)).

Consider the problem of mapping the original data points to a line so that each data point on the line can be represented as a linear combination of its neighbors with the coefficients $\boldsymbol{W}_{ij}$. A reasonable criterion for choosing a "good" map is to minimize the cost function (Eq. (18.17)).

Following some algebraic formulations, the cost function can be reduced to

$$\begin{aligned} J_F(\boldsymbol{V}) &= \sum_{ij} \|\boldsymbol{y}_i - W_{ij}\boldsymbol{y}_j\|^2 \\ &= \boldsymbol{Y}^{\mathrm{T}}(\boldsymbol{I} - \boldsymbol{W})^{\mathrm{T}}(\boldsymbol{I} - \boldsymbol{W})\boldsymbol{Y} \\ &= \boldsymbol{V}^{\mathrm{T}}\boldsymbol{X}(\boldsymbol{I} - \boldsymbol{W})^{\mathrm{T}}(\boldsymbol{I} - \boldsymbol{W})\boldsymbol{X}^{\mathrm{T}}\boldsymbol{V} \\ &= \boldsymbol{V}^{\mathrm{T}}\boldsymbol{X}\boldsymbol{M}\boldsymbol{X}^{\mathrm{T}}\boldsymbol{V}, \end{aligned} \tag{18.30}$$

where $\boldsymbol{M} = (\boldsymbol{I} - \boldsymbol{W})^{\mathrm{T}}(\boldsymbol{I} - \boldsymbol{W})$. Clearly, the matrix $\boldsymbol{X}\boldsymbol{M}\boldsymbol{X}^{\mathrm{T}}$ is symmetric and semi-positive definite. In order to remove an arbitrary scaling factor in the projection, NEP impose a constraint as follows:

$$\boldsymbol{Y}^{\mathrm{T}}\boldsymbol{Y} = \boldsymbol{I} \mapsto \boldsymbol{V}^{\mathrm{T}}\boldsymbol{X}\boldsymbol{X}^{\mathrm{T}}\boldsymbol{V} = \boldsymbol{I}. \tag{18.31}$$

Finally, the minimization problem reduces to finding:

$$\arg\min \ \boldsymbol{V}^{\mathrm{T}}\boldsymbol{X}\boldsymbol{M}\boldsymbol{X}^{\mathrm{T}}\boldsymbol{V},$$
$$subject \ to \ \boldsymbol{V}^{\mathrm{T}}\boldsymbol{X}\boldsymbol{X}^{\mathrm{T}}\boldsymbol{V} = \boldsymbol{I}. \tag{18.32}$$

The transformation vector v that minimizes the objective function is given by the minimum eigenvalue solution to the following generalized eigenvector problem:

$$XMX^{\mathrm{T}}v = \lambda XX^{\mathrm{T}}v \qquad (18.33)$$

The algorithmic procedure is stated as follows:

Algorithm Outline: NPE

Input
    1) Training samples $\{x_1, \ldots, x_n\} \in \Re^D$.
    2) Dimensionality of embedding space d $(1 \leqslant d \leqslant D)$.
Output
    1) The d-dimensional embedding coordinates Y for the input points X.
    2) the $D \times d$ transformation matrix V.
Algorithm steps
    Step 1: Constructing an adjacency graph G as Fig. 18.4.
    Step 2: In this step, we compute the weights on the edges. Let W denote the weight matrix with W_{ij} having the weight of the edge from node i to node j, and 0 if there is no such edge. The weights on the edges can be computed by minimizing the following objective function (Eq. (18.16)).
    Step 3: Computing the projections: In this step, we compute the linear projections. Let $V = v_0, v_1, \ldots, v_{d-1}$ be the solutions of Eq. (18.23) ordered according to their eigenvalues, and the embedding is as follows: $Y = V^{\mathrm{T}}X$.

NPE has similar neighborhood preserving properties as LLE, but the main disadvantage of LLE is that, it is defined only on the training samples, and there are no natural maps of the testing sample. Instead, NPE is defined everywhere.

18.4.3 Global Alignment of Local Models

The algorithms are automatic alignment procedures, which map the disparate internal representations learned by several local, dimensionality reduction methods into a single, coherent global coordinate system for the original data space. They can be applied to any set of methods, each of them produces a low-dimensional local representation of a high dimensional input. These algorithms compute a number of local models and perform a global alignment of these local models. This subsection presents two algorithms for manifold learning: 1) LLC [47], 2) Charting a Manifold [17].

LLC

Locally Linear Coordination (LLC) computes a number of locally linear models and subsequently performs a global alignment of the local models. The process of LLC is described in Fig. 18.5.

Two key ideas motivate the method. First, to use a convex cost function whose unique minimum is attained at the desired global coordinates. Second, to restrict the global coordinates y_n depending on the data x_n only through

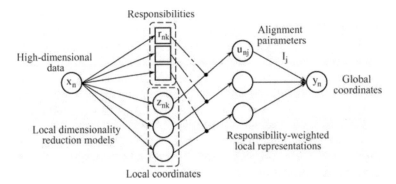

Fig. 18.5 Obtaining global coordinates from data via responsibility-weighted local coordinates.

the local representations z_{nk} and responsibilities r_{nk}, thereby leveraging the structure of the mixture model to regularize and reduce the effective size of the optimization problem.

The LLC algorithm can be summarized as follows:

Algorithm Outline: LLC

Input
    1) Training samples $\{\boldsymbol{x}_1, \ldots, \boldsymbol{x}_n\} \in \Re^D$.
    2) Dimensionality of embedding space d $(1 \leqslant d \leqslant D)$.
Output
    The d-dimensional embedding coordinates $\boldsymbol{Y}$ for the input points $\boldsymbol{X}$.
Algorithm steps
    Step 1:   Using data $\boldsymbol{X}$ compute local linear reconstruction weights $\boldsymbol{W}_{ij}$ using Eq. (18.16).
    Step 2:   Train or receive a pre-trained mixture of local dimensionality reducers. Apply this mixture to $\boldsymbol{X}$, obtaining a local representation $\boldsymbol{z}_{nk}$ and responsibility $\boldsymbol{r}_{nk}$ for each submodel $\boldsymbol{k}$ and each data point $\boldsymbol{x}_i$.
    Step 3:   Form the matrix $\boldsymbol{U}$ with $u_{nj} = r_{nk}\boldsymbol{z}_{nk}^i$ and calculate $\boldsymbol{A}$ and $\boldsymbol{B}$ as follows:

$$\boldsymbol{A} = \boldsymbol{U}^{\mathrm{T}}(\boldsymbol{I} - \boldsymbol{W})^{\mathrm{T}}(\boldsymbol{I} - \boldsymbol{W})U, \tag{18.34}$$

$$\boldsymbol{B} = \frac{1}{n}\boldsymbol{U}^{\mathrm{T}}U. \tag{18.35}$$

    Step 4:   Find the eigenvectors corresponding to the smallest $(d + 1)$ eigenvalues of the generalized eigenvalue system $\boldsymbol{Av} = \lambda \boldsymbol{Bv}$.
    Step 5:   Let $\boldsymbol{L}$ be a matrix with columns formed by the 2nd to $(d + 1)$ smallest eigenvectors. Return the j-th row of $\boldsymbol{L}$ as alignment weight l_k^i.
    Step 6:   Compute the global manifold coordinates as $\boldsymbol{Y} = \boldsymbol{U}\boldsymbol{L}$.

LLC emphasizes a powerful but often overlooked interpretation of local mixture models. Rather than considering the output of such systems to be a single quantity, such as a density estimate or an expert-weighted regression, it is possible to view them as networks which convert high-dimensional inputs into a vector of internal coordinates from each submodel, accompanied by responsibilities.

Charting a Manifold

The charting algorithm casts manifold learning as a density estimation problem. In particular, charting first fits a mixture of Gaussian densities to the data, and then the local coordinates implied by each Gaussian's covariance into a single, global system. The density model underlying charting naturally provides a function mapping all coordinates to the high-dimensional manifold, rather than just an embedding of the given data.

The charting a manifold method retains LLE's basic three-step structure, including the attractive property that the optimal solution to each stage may be computed in a closed form:

Algorithm Outline: Charting a Manifold

Input

    1) Training samples $\{\boldsymbol{x}_1, \ldots, \boldsymbol{x}_n\} \in \Re^D$.

    2) Dimensionality of embedding space d ($1 \leqslant d \leqslant D$).

Output

    The d-dimensional embedding coordinates $\boldsymbol{Y}$ for the input points $\boldsymbol{X}$.

Algorithm steps

    Step 1: Soft nearest neighbor assignment: For each $\boldsymbol{x}_i$, assign a weight W_{ij} to each $\boldsymbol{x}_j$, $j \neq i$, according to a Gaussian kernel centered at $\boldsymbol{x}_i$ as in Eq. (18.12).

    Step 2: Fit Gaussian mixture model: Let $N(\boldsymbol{x}; \mu; \Lambda)$ denote a Gaussian density with mean μ and covariance Λ, evaluated at the point $\boldsymbol{x}$. In charting, we model the high dimensional data space by an n-component Gaussian mixture, where the component means are set to the observed data points $\boldsymbol{x}_i$:

$$p(\boldsymbol{x}|\Lambda) = \frac{1}{n} \sum_{i=1}^{n} N(\boldsymbol{x}; \boldsymbol{x}_i, \Lambda_i). \tag{18.36}$$

    Step 3: Connect local charts: Let $U_k = [u_{k1}, \ldots, u_{kn}]$ denote the projection of the n data points into the q dimensional subspace spanned by the q dominant eigenvectors of Λ_k. For each chart, we would like to determine a low dimensional affine projection $G_k \in \Re^{q+1}$ which maps these points into the global coordinate frame. We couple these projections by requiring them to agree on data points for which they share responsibility, as encoded by the following objective:

$$\Phi_{chart}(G) = \sum_{k \neq j} \sum_{i=1}^{n} p_k(\boldsymbol{x}_i) p_j(\boldsymbol{x}_i) \|G_k[\begin{smallmatrix} u_{ki} \\ 1 \end{smallmatrix}] - G_j[\begin{smallmatrix} u_{ji} \\ 1 \end{smallmatrix}]\|_F^2, \tag{18.37}$$

where $Q \doteq \Sigma_{j \neq k}(u_j - u_k)p_j p_k$. Similarly to LLE, the optimal embedding may be obtained by finding the bottom eigenvectors of an $n \times n$ symmetric matrix.

From the above analysis, we can see that it is presumed in the algorithm that the data lies on or near a low dimensional manifold embedded in the high dimensional space, and that there exists a 1-to-1 smooth nonlinear transform between the manifold and the high dimensional space.

18.4.4 Extensions

This subsection presents three extensions for manifold learning algorithms: (1) supervised extension, (2) Kernel extension, (3) Out-of-sample extension.

Supervised Extension

Many classical algorithms, such as LLE, LPP and Isomap are all unsupervised learning algorithms, that is, they assume no prior information on the input data. In general, prior knowledge can be expressed in diverse forms, such as class labels, pairwise constraints or other prior information.

In this paper, we focus on prior knowledge in form of class labels. In many real-world applications, supervised learning settings are required [19, 23, 28]. The extent to which manifold learning algorithms support supervised learning is of importance for such applications. This subsection presents three algorithms for manifold learning: (1) Supervised Isomap (S-Isomap) [48], (2) Supervised LPP (S-LPP) [49], (3) Marginal Fisher Analysis (MFA) [50], (4) Local Discriminant Embedding (LDE) [51].

1) S-Isomap   Gen et al. proposed a method called supervised Isomap (S-Isomap) [48], which utilizes class information to guide the procedure of nonlinear dimensionality reduction.

To make the Isomap algorithm more robust for both visualization and classification, a more sophisticated method is proposed for S-Isomap. The S-Isomap method defines the dissimilarity between two points $\boldsymbol{x}_i$ and $\boldsymbol{x}_j$ as

$$
D(\boldsymbol{x}_i, \boldsymbol{x}_j) = \begin{cases} \sqrt{1 - \exp\left(-\dfrac{d^2(\boldsymbol{x}_i, \boldsymbol{x}_j)}{\beta}\right)}, & \text{if } l_i = l_j, \\[3ex] \sqrt{\exp\left(\dfrac{d^2(\boldsymbol{x}_i, \boldsymbol{x}_j)}{\beta}\right) - \alpha}, & \text{if } l_i \neq l_j, \end{cases} \tag{18.38}
$$

where $d(\boldsymbol{x}_i, \boldsymbol{x}_j)$ denotes the Euclidean distance between $\boldsymbol{x}_i$ and $\boldsymbol{x}_j$. Since the Euclidean distance $d(\boldsymbol{x}_i, \boldsymbol{x}_j)$ is in the exponent, the parameter is used to prevent $D(\boldsymbol{x}_i, \boldsymbol{x}_j)$ to increase too fast when $d(\boldsymbol{x}_i, \boldsymbol{x}_j)$ is relatively large. Thus the value of β should depend on the "density" of the data set. Usually, β is set to be the average Euclidean distance between all pairs of data points. The parameter α gives a certain chance to the points in different classes to be "more similar", i.e., to have a smaller value of dissimilarity, than those in the same class. A typical plot of $D(\boldsymbol{x}_i, \boldsymbol{x}_j)$ as a function of $d^2(\boldsymbol{x}_i, \boldsymbol{x}_j)/\beta$ is shown in Fig. 18.6.

In Fig. 18.6, the dissimilarity between two points is equal to or larger than 1 if their class labels are different and is less than 1 if otherwise. Thus the inter-class dissimilarity is definitely larger than the intra-class dissimilarity, which is a very good property for classification.

Since $D(\boldsymbol{x}_i, \boldsymbol{x}_j)$ integrates the class information, so it can be used in the procedure of S-Isomap to address the robustness problem in visualization and classification.

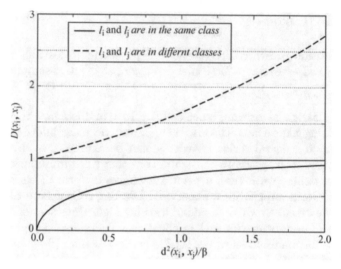

Fig. 18.6 Typical plot of $D(\boldsymbol{x}_i, \boldsymbol{x}_j)$ as a function of $d^2(\boldsymbol{x}_i, \boldsymbol{x}_j)/\beta$ $(\alpha = 1)$.

2) SLPP   Being unsupervised, the original LPP does not utilize class information of each point to be projected, so supervised LPP(SLPP) [49] has been proposed to improve the original LPP for classification, and claimed to have good performance.

SLPP employs prior information about the original data set to perform learning the feature subspace. The essence of SLPP algorithm is how to choose the similarity measure matrix $\boldsymbol{W}$ in Eq. (18.24). In LPP, $\boldsymbol{W}$ is only related to the neighborhood or the nearest neighbors. In other words, the selection of $\boldsymbol{W}$ is independent of class information through the samples in neighborhood may belong to the same class. Due to the locality preserving property of LPP, the points corresponding to samples would be close to each other in the reduced space if they are in neighborhood, through they may belong to different classes. This will result in an unfavorable situation in pattern analysis, especially in classification problem.

To make use of class membership relation of each point, the weighted matrix $\boldsymbol{W}$ is defined as follows:

$$\boldsymbol{W}_{ij} = \begin{cases} \exp\left(-\dfrac{\| x_i - x_j \|^2}{t}\right), & \text{if } x_i \text{ and } x_j \text{ are from the same class,} \\ 0, & \text{otherwise.} \end{cases}$$

(18.39)

The other steps for SLPP are the same as LPP.

3) MFA   Marginal fisher analysis (MFA) [50] develops a new criteria to characterize intra-class compactness and inter-class separability, in which, the within-class compactness is represented as the sum of distances between each sample and its intra-class neighbors, and the between-class separability is characterized as the sum of distances between each point and its inter-class

neighboring points.

MFA designs an intrinsic graph that characterizes the intraclass compactness and another penalty graph which characterizes the interclass separability, both shown in the intrinsic graph G_c illustrates the intraclass point adjacency relationship, and each sample is connected to its k_1-nearest neighbors of the same class. The penalty graph G_p illustrates the interclass marginal point adjacency relationship and the marginal point pairs of different classes are connected.

In MFA, the within-class scatter matrix is defined as

$$
\begin{aligned}
\tilde{S}_w &= \sum_i \sum_{i \in N_{k1}^+(j) \, or \, j \in N_{k1}^+(i)} \\
&= trace\left\{ \boldsymbol{V}^{\mathrm{T}} \sum_{ij} ((\boldsymbol{x}_i - \boldsymbol{x}_j) \boldsymbol{W}_{ij}^c (\boldsymbol{x}_i - \boldsymbol{x}_j)^{\mathrm{T}}) \boldsymbol{V} \right\} \\
&= trace\{ \boldsymbol{V}^{\mathrm{T}} (2\boldsymbol{X}\boldsymbol{D}^c\boldsymbol{X}^{\mathrm{T}} - 2\boldsymbol{X}\boldsymbol{W}^c\boldsymbol{X}^{\mathrm{T}}) \boldsymbol{V} \} \\
&= 2trace\{ \boldsymbol{V}^{\mathrm{T}} \boldsymbol{X}(\boldsymbol{D}^c - \boldsymbol{W}^c)\boldsymbol{X}^{\mathrm{T}} \boldsymbol{V} \},
\end{aligned} \tag{18.40}
$$

where $\boldsymbol{W}^c$ is defined as follows:

$$
\boldsymbol{W}_{ij}^c = \begin{cases} 1, & \text{if } \boldsymbol{x}_i \in N_{k1}^+(j) \text{ or } \boldsymbol{x}_j \in N_{k1}^+(i), \\ 0, & \text{otherwise,} \end{cases} \tag{18.41}
$$

here, $N_{k1}^+(i)$ indicates the index set of the k_1 nearest neighbors of the sample $\boldsymbol{x}_i$ in the same class.

Interclass separability is characterized by a penalty graph with the term

$$
\begin{aligned}
\tilde{S}_p &= \sum_i \sum_{(\boldsymbol{x}_i,\boldsymbol{x}_j) \in P_{k2}(c_i) \, or \, (\boldsymbol{x}_i,\boldsymbol{x}_j) \in P_{k2}(c_j)} \\
&= trace\left\{ \boldsymbol{V}^{\mathrm{T}} \sum_{ij} ((\boldsymbol{x}_i - \boldsymbol{x}_j) \boldsymbol{W}_{ij}^p (\boldsymbol{x}_i - \boldsymbol{x}_j)^{\mathrm{T}}) \boldsymbol{V} \right\} \\
&= trace\{ \boldsymbol{V}^{\mathrm{T}} (2\boldsymbol{X}\boldsymbol{D}^p\boldsymbol{X}^{\mathrm{T}} - 2\boldsymbol{X}\boldsymbol{W}^p\boldsymbol{X}^{\mathrm{T}}) \boldsymbol{V} \} \\
&= 2trace\{ \boldsymbol{V}^{\mathrm{T}} \boldsymbol{X}(\boldsymbol{D}^p - \boldsymbol{W}^p)\boldsymbol{X}^{\mathrm{T}} \boldsymbol{V} \},
\end{aligned} \tag{18.42}
$$

where $\boldsymbol{W}^p$ is defined as follows:

$$
\boldsymbol{W}_{ij}^p = \begin{cases} 1, & \text{if } (\boldsymbol{x}_i, \boldsymbol{x}_j) \in \boldsymbol{P}_{k2}(c_i) \text{ or } (\boldsymbol{x}_i, \boldsymbol{x}_j) \in \boldsymbol{P}_{k2}(c_j), \\ 0 & \text{otherwise,} \end{cases} \tag{18.43}
$$

here, $P_{k2}(c)$ is a set of data pairs that are the k_2 nearest pairs among the set $\{(\boldsymbol{x}_i, \boldsymbol{x}_j), l_i = l_j\}$.

With $\boldsymbol{S}_c$ and $\boldsymbol{S}_p$, the objection function of MFA is defined as

$$
J_F(\boldsymbol{V}) = \arg\min \frac{\boldsymbol{V}^{\mathrm{T}} \boldsymbol{X}(\boldsymbol{D}^c - \boldsymbol{W}^c)\boldsymbol{X}^{\mathrm{T}} \boldsymbol{V}}{\boldsymbol{V}^{\mathrm{T}} \boldsymbol{X}(\boldsymbol{D}^p - \boldsymbol{W}^p)\boldsymbol{X}^{\mathrm{T}} \boldsymbol{V}}. \tag{18.44}
$$

The algorithmic procedure of Marginal Fisher Analysis algorithm is formally stated as follows:

Algorithm Outline: MFA

Input
  1) Training samples $\{(x_1, \ell_1), \ldots, (x_n, \ell_n)\} \in \Re^D$.
  2) Dimensionality of embedding space d ($1 \leqslant d \leqslant D$).
Output
  1) The d-dimensional embedding coordinates Y for the input points X.
  2) the $D \times d$ transformation matrix V.
Algorithm steps
  Step 1:   PCA projection. We first project the data set into the PCA subspace by retaining $n - n_c$ dimensions or a certain energy. Let V_{PCA} denote the transformation matrix of PCA.
  Step 2:   Constructing the intraclass compactness graph G_c and interclass separability graph G_p.
  Step 3:   Computing the intraclass compactness weight W_c and interclass separability weight W_p.
  Step 4:   Complete the embedding. Find the generalized eigenvectors $v_1, v_2, \ldots, v_d$ that correspond to the d smallest eigenvalues in

$$X(D^c - W^c)X^T v = X(D^p - W^p)X^T v. \tag{18.45}$$

  Step 5:   Output the final linear projection direction as $V^* = V_{PCA}V_{MFA}$.

4) LDE   The LDE method [51] explores the local relations between neighboring data points. Since the class information is given, it is reasonable to require that, after embedding, the neighbor relations can better reflect the class relations. That is, in the low-dimensional embedding subspace, we want to keep neighboring points close if they have the same label, whereas prevent points of other classes from entering the neighborhood. With these two aspects of consideration, we arrive at the following constrained optimization problem:

$$\arg\max J_F(V) = \sum_{ij} \|V^T x_i - V^T x_j\|^2 W'_{ij},$$
$$subject\ to\ \sum_{ij} \|V^T x_i - V^T x_j\|^2 W_{ij} = 1. \tag{18.46}$$

The optimization formulation essentially uses the class and neighbor information through the two affinity matrices W (for $l_i = l_j$ in neighborhoods, computed as Eq. (18.38)) and W' (for $l_i \neq l_j$ in neighborhoods). W' can be computed as follows:

$$W'_{ij} = \begin{cases} \exp\left(-\dfrac{\|x_i - x_j\|^2}{t}\right), & \text{if } x_i \text{ and } x_j \text{ are from different class,} \\ 0, & \text{otherwise.} \end{cases} \tag{18.47}$$

Follow some algebraic steps, we see that

$$J_F(V) = \sum_{ij} \|V^T x_i - V^T x_j\|^2 W'_{ij}$$

$$= trace\left\{\boldsymbol{V}^{\mathrm{T}}\sum_{ij}((\boldsymbol{x}_i - \boldsymbol{x}_j)\boldsymbol{W}'_{ij}(\boldsymbol{x}_i - \boldsymbol{x}_j)^{\mathrm{T}})\boldsymbol{V}\right\}$$

$$= trace\{\boldsymbol{V}^{\mathrm{T}}(2\boldsymbol{X}\boldsymbol{D}'\boldsymbol{X}^{\mathrm{T}} - 2\boldsymbol{X}\boldsymbol{W}'\boldsymbol{X}^{\mathrm{T}})\boldsymbol{V}\}$$

$$= 2trace\{\boldsymbol{V}^{\mathrm{T}}\boldsymbol{X}(\boldsymbol{D}' - \boldsymbol{W}')\boldsymbol{X}^{\mathrm{T}}\boldsymbol{V}\}, \tag{18.48}$$

where $\boldsymbol{D}'$ is a diagonal matrix with $D'_{ii} = \sum_j W'_{ij}$.

The objective function and the constraint in Eq. (18.45) can be reformulated as

$$\begin{aligned}\arg\max J_F(V) &= 2trace\{\boldsymbol{V}^{\mathrm{T}}\boldsymbol{X}(\boldsymbol{D}' - \boldsymbol{W}')\boldsymbol{X}^{\mathrm{T}}\boldsymbol{V}\},\\ subject\ to\ \ &2trace\{\boldsymbol{V}^{\mathrm{T}}\boldsymbol{X}(\boldsymbol{D} - \boldsymbol{W})\boldsymbol{X}^{\mathrm{T}}\boldsymbol{V}\} = 1.\end{aligned} \tag{18.49}$$

Thus, the columns of an optimal V are the generalized eigenvectors corresponding to the d largest eigenvalues in

$$\boldsymbol{X}(\boldsymbol{D}' - \boldsymbol{W}')\boldsymbol{X}^{\mathrm{T}}\boldsymbol{v} = \boldsymbol{X}(\boldsymbol{D} - \boldsymbol{W})\boldsymbol{X}^{\mathrm{T}}\boldsymbol{v}. \tag{18.50}$$

Kernel Extension

Not uncommonly, one needs to consider nonlinear transformations of data in order to apply learning algorithms. One efficient method for doing this is via a kernel that computes a similarity measure between any two data points. Kernel-based learning algorithms attempt to embed the data into a Hilbert space, and search for linear relations in such a space. Performed implicitly, the embedding specifies the inner product between each pair of points instead of giving their coordinates explicitly. The advantages of kernel methods come from the fact that often the inner product in the embedding space can be computed much more easily than the coordinates of the points themselves.

Suppose we have a dataset $\boldsymbol{X}$ with n samples, and a map $\psi : \boldsymbol{X} \mapsto \boldsymbol{F}$, where $\boldsymbol{F}$ is an embedding space. The kernel matrix is defined as $\boldsymbol{K} = (k(\boldsymbol{x}_i, \boldsymbol{x}_j))^n_{i,j=1}$, and $\boldsymbol{K}_{ij} = k(\boldsymbol{x}_i, \boldsymbol{x}_j) = \langle\psi(\boldsymbol{x}_i), \psi(\boldsymbol{x}_i)\rangle$. In this way, $\boldsymbol{K}$ completely determines the relative positions between those points in the embedding space.

Proposition: Every positive definite and symmetric matrix is a kernel matrix, that is, an inner product matrix in some embedding space. Conversely, every kernel matrix is symmetric and positive definite.

Mercer kernels have been used quite successfully for learning in Support Vector Machines (SVM) and in kernel principal components analysis (KPCA) [52, 53, 55]. The manifold learning algorithms can be generalized for nonlinear manifolds by employing the kernel trick, so that the generalization property naturally emerges. In this subsection, we provide a brief review of two main kernel manifold learning methods: kernel Isomap (K-Isomap) [54] and kernel LPP (K-LPP) [42].

1) Kernel Isomap   In Ref. [54], the authors presented a robust kernel Isomap (K-Isomap) method, and claimed to have solved two critical issues

that were not considered in Isomap, which are: (1) generalization property (projection property); (2) topological stability.

The basic idea in Isomap is to use geodesic distances on a neighborhood graph in the framework of the classical scaling, in order to incorporate with the manifold structure, instead of subspace. The sum of edge weights along the shortest path between two nodes, is assigned as geodesic distance. The top n eigenvectors of the geodesic distance matrix, represent the coordinates in the n-dimensional Euclidean space.

Following the connection between the classical scaling and PCA, metric MDS can be interpreted as kernel PCA (KPCA) [55]. In a similar manner, the geodesic distance matrix in Isomap, can be viewed as a kernel matrix. The doubly centered geodesic distance matrix $\boldsymbol{K}$ in Isomap is of the form

$$\boldsymbol{K} = -\frac{1}{2}\boldsymbol{H}\boldsymbol{D}^2\boldsymbol{H}, \tag{18.51}$$

where $\boldsymbol{D}^2 = [\boldsymbol{D}_{ij}^2]$ means the element-wise square of the geodesic distance matrix , $\boldsymbol{H}$ is the centering matrix, given by $\boldsymbol{H} = \boldsymbol{I} - \frac{1}{n}\boldsymbol{e}_n\boldsymbol{e}_n^{\mathrm{T}}$ for $\boldsymbol{e}_n = [1,\ldots,1]^{\mathrm{T}} \in \Re^n$.

Algorithm Outline: Kernel Isomap

Input

    1) Training samples $\{\boldsymbol{x}_1,\ldots,\boldsymbol{x}_n\} \in \Re^D$.

    2) Dimensionality of embedding space d $(1 \leqslant d \leqslant D)$.

Output

    The d-dimensional embedding coordinates $\boldsymbol{Y}$ for the input points $\boldsymbol{X}$.

Algorithm steps

    Step 1:   Identify k nearest neighbors (or ε-ball neighborhood) of each input data point and construct a neighborhood graph where edge lengths between points in a neighborhood are set as their Euclidean distances.

    Step 2:   Compute geodesic distances, D_{ij}, that are associated with the sum of edge weights along shortest paths between all pairs of points.

    Step 3:   Construct a matrix $\boldsymbol{K}$ according to Eq. (18.50).

    Step 4:   Compute the largest eigenvalue, c^*, of the matrix

$$\begin{bmatrix} 0 & 2\boldsymbol{K}(\boldsymbol{D}^2) \\ -I & -4\boldsymbol{K}(\boldsymbol{D}) \end{bmatrix} \tag{18.52}$$

and construct a Mercer kernel matrix $\tilde{\boldsymbol{K}} = \tilde{\boldsymbol{K}}(\tilde{\boldsymbol{D}}^2)$ that is of the form

$$\tilde{\boldsymbol{K}} = \boldsymbol{K}(\boldsymbol{D}^2) + 2c\boldsymbol{K}(\boldsymbol{D}) + \frac{1}{2}c^2\boldsymbol{H}, \tag{18.53}$$

where $\tilde{\boldsymbol{K}}$ is guaranteed to be positive semidefinite for $c \geqslant c^*$.

    Step 5:   Compute top d eigenvectors of $\tilde{\boldsymbol{K}}$, which leads to the eigenvector matrix $\boldsymbol{V} \in \Re^{n \times d}$ and the eigenvalue matrix $\Phi \in \Re^{d \times d}$.

    Step 6:   The coordinates of the n points in the d-dimensional Euclidean space are given by the column vectors of $\boldsymbol{Y} = \Phi^{\frac{1}{2}}\boldsymbol{V}^{\mathrm{T}}$.

The kernel Isomap method finds an implicit mapping which places n points in a low-dimensional space. In contrast to Isomap, K-Isomap can project test data points onto a low-dimensional space, as well, through a kernel trick. K-Isomap mainly exploits the additive constant problem, the goal of which is to find an appropriate constant to be added to all dissimilarities (or distances), apart from the self-dissimilarities, that makes the matrix $\boldsymbol{K}$ to be positive semidefinite, which leads to $\tilde{\boldsymbol{K}}$. In fact, the additive constant problem was extensively studied in the context of MDS and recently in embedding. The matrix $\tilde{\boldsymbol{K}}$ induced by a constant-shifting method, has a Euclidean representation and becomes a Mercer kernel matrix. The kernel Isomap algorithm is summarized below.

2) Kernel LPP   He et al. proposed a method called Kernel LPP (K-LPP) [42], which attempts to extend LPP algorithm to high dimension feature space.

Suppose that the Euclidean space $\Re^D$ is mapped to a Hilbert space $\boldsymbol{F}$ through a nonlinear mapping function ψ. Now, the eigenvector problem in the Hilbert space can be written as follows:

$$[\psi(\boldsymbol{X})\boldsymbol{L}\psi^{\mathrm{T}}(\boldsymbol{X})]\boldsymbol{v} = \lambda[\psi(\boldsymbol{X})\boldsymbol{D}\psi^{\mathrm{T}}(\boldsymbol{X})]\boldsymbol{v}. \tag{18.54}$$

To generalize LPP to the nonlinear case, we formulate it in a way that uses dot product exclusively. Therefore, we consider an expression of dot product on the Hilbert space $\boldsymbol{F}$ given by the following kernel function:

$$\boldsymbol{K}(\boldsymbol{x}_i, \boldsymbol{x}_j) = (\psi(\boldsymbol{x}_i) \cdot \psi(\boldsymbol{x}_j)) = \psi^{\mathrm{T}}(\boldsymbol{x}_i)\psi(\boldsymbol{x}_j). \tag{18.55}$$

Because the eigenvectors of Eq. (18.53) are linear combinations of $\psi(\boldsymbol{x}_1)$, $\psi(\boldsymbol{x}_2), \ldots, \psi(\boldsymbol{x}_n)$, there exist coefficients $\alpha_i, i = 1, 2, \ldots, n$ such that

$$\boldsymbol{v} = \sum_{i=1}^{n} \alpha_i\psi(\boldsymbol{x}_i) = \psi(\boldsymbol{X})\alpha, \tag{18.56}$$

where $\alpha = [\alpha_1, \alpha_2, \ldots, \alpha_n]^{\mathrm{T}}$.

By simple algebra formulation, we can finally obtain the following eigenvector problem:

$$\boldsymbol{K}\boldsymbol{L}\boldsymbol{K}\alpha = \lambda\boldsymbol{K}\boldsymbol{D}\boldsymbol{K}\alpha. \tag{18.57}$$

Let the column vectors $\alpha^1, \alpha^2, \ldots, \alpha^n$ be the solutions of Eq. (18.56). For a test point $\boldsymbol{x}$, we compute projections onto the eigenvectors $\boldsymbol{v}^k$ according to

$$(\boldsymbol{v}^k \cdot \psi(\boldsymbol{x})) = \sum_{i=1}^{n} \alpha_i^k(\psi(\boldsymbol{x}_i) \cdot \psi(\boldsymbol{x}_j)) = \sum_{i=1}^{n} \alpha_i^k \boldsymbol{K}(\boldsymbol{x}_i, \boldsymbol{x}_j), \tag{18.58}$$

where α_i^k is the ith element of the vector α^k. For the original training points, the maps can be obtained by $\boldsymbol{y} = \boldsymbol{K}\alpha$, where the ith element of $\boldsymbol{y}$ is the one-dimensional representation of $\boldsymbol{x}_i$. Furthermore, Eq. (18.56) can be reduced to

$$\boldsymbol{L}\boldsymbol{Y} = \lambda\boldsymbol{D}\boldsymbol{Y} \tag{18.59}$$

which is identical to the eigenvalue problem of Laplacian Eigenmaps. This shows that Kernel LPP yields the same results as Laplacian Eigenmaps on the training points.

Out-of-Sample Extension

Out-of-sample extension attempts to embed a new data point in the low dimensional space. Many manifold learning algorithms based on an eigen decomposition provide either an embedding or a clustering only for given training points, with no straightforward extension for out-of-sample examples short of recomputing eigenvectors. Those methods differ from the subspace methods, in PCA and LDA, the out-of-sample extension is computed by multiplying the new data point with the linear mapping matrix. The method for the out-of-sample extensions of Isomap, LLE, and Laplacian Eigenmaps has been presented in which the algorithms are redefined as kernel methods [53]. In Refs. [55–57], in order to obtain the new embedded coordinates, the intuition of manifold learning is used, i.e., many nonlinear manifolds can be considered as locally linear. This linearity is used to build a linear relation between high and low-dimensional points belonging to a particular neighborhood of data. There are two possibilities for linear generalization :

LG1: Let us put the *knn(i)* of x_{n+1} and the corresponding embedded points into the matrices: $X^{n+1} = (x_{n+1}^1, x_{n+1}^2, \ldots, x_{n+1}^k)$ and $Y^{n+1} = (y_{n+1}^1, y_{n+1}^2, \ldots, y_{n+1}^k)$. By taking into consideration the assumption that the manifold is locally linear, the following equation is approximately true: $Y^{n+1} = ZX^{n+1}$, where Z is an unknown linear transformation matrix of size $d \times D$, which can be straightforwardly determined as $Z = Y^{n+1}(X^{n+1})^{-1}$, where $(X^{n+1})^{-1}$ is pseudoinverse matrix of X^{n+1}.

LG2: To find y_{n+1}, first, the k nearest neighbors of x_{n+1} are detected among points in the high-dimensional space: $x_i \in X, i = 1, \ldots, n$. Then, the linear weights, w_{n+1}, that best reconstruct x_{n+1} from its neighbors, are computed using Eq. (18.16) with the sum-to-one constraint: $\Sigma_{j=1} w_{n+1j}$. Finally, the new output Y^{n+1} is found: $Y^{n+1} = \Sigma_{j=1} w_{n+1j} y_j$, where the sum is over the y_i's corresponding to the k nearest neighbors of x_{n+1}.

However, the out-of-sample extensions have to be obtained using the above estimation algorithms, which decreases the performance of classifier in face recognition.

18.5 Experiments Results and Analysis

In this section, we implement experiments on CMU PIE and FERET databases to evaluate the subspace and submanifold algorithms for face recognition. Firstly, we select the procedure parameters with cross-validation method, i.e., k for k nearest neighbor measure, kernel parameters. Secondly, we evaluate the performance of different algorithms on recognition accuracy.

For simplicity, all neighborhood graphs for manifold learning algorithms are constructed based on the k-nearest-neighbor criterion. As discussed above, while the neighborhood graphs for unsupervised methods are constructed without incorporating any label information, the neighborhood graphs for supervised algorithms are constructed based on the same k-nearest-neighbor criterion by incorporating label information. For those methods (Fisherfaces, LPP and NPE) which need PCA to dimensionality reduction firstly, we decide the number of principal components by retaining 95% energy. As some manifold learning algorithms (i.e., Isomap, LLE) are defined only on given training points, with no straightforward mapping matrix for out-of-sample examples. We adopt an estimation method for out-of-sample extension in **LG1**, which can be applied to all nonlinear dimensionality reduction methods. Note that, for Fisherfaces, there are at most c-1 nonzero generalized eigenvalues and, so, an upper bound on the dimension of the reduced space is c-1, where c is the number of individuals [8].

In short, the recognition process has three steps. First, we calculate the feature vectors from the training set of face images; then the new face image to be identified is projected into the face subspace or submanifold spanned by different methods; finally, the new face image is identified by a classifier. For its simplicity, the nearest neighbor classifier using Euclidean metric was employed.

18.5.1 Experiments on CMU PIE Face Database

The CMU PIE face database [59] contains 68 subjects with 41 368 face images as a whole. The face images were captured by 13 synchronized cameras and 21 flashes, under varying pose, illumination, and expression. For simplicity, we use the dataset collected by He [46]. Each face image is cropped to 32×32 sizes and one individual holds 170 images. Figure 18.7 shows some of the faces with pose, illumination and expression variations in the database.

Fig. 18.7 The sample cropped face images of two individuals from PIE database.

To compare the recognition performance with different training set sizes, r images per subject are randomly selected for training and the rest for testing. We firstly use a case to compare the performance of different methods on randomly selected training set and the classification results based on the nearest neighbor classifier are reported in Fig. 18.8.

From Fig. 18.8 (a) – (c), we can find that the performances of global algorithms do not outperform subspace algorithms on CMU PIE database as well as the algorithms of Global alignment of local models, despite their ability to learn the structure of complex nonlinear manifolds. Fisherfaces method

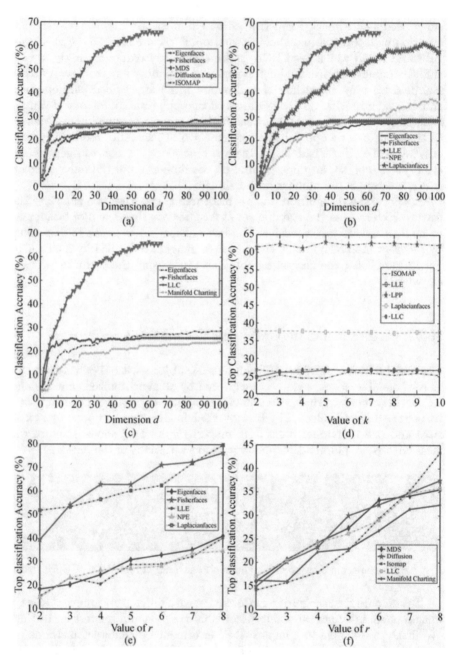

Fig. 18.8 Comparative study based on the CMU PIE database. (a)–(c) Classification accuracy (%) vs. dimensionality d, with $k = 5$, $r = 5$. (d) Top classification accuracy (%) vs. neighborhood size k, with $r = 5$; (e)–(f) Top classification accuracy (%) vs. training set size r, with $k = 5$.

outperforms the others because it considers the most discriminative vector in the low dimensional space, while Laplacianfaces achieve the best performance among the unsupervised algorithms. Almost all the local algorithms outperform Eigenfaces. We also find that the recognition curve of MDS and Isomap is similar. As shown in Fig. 18.8 (d), the most algorithms achieve the highest recognition accuracy with $k = 5$.

We further repeat 10 times to get the average values of the top classification accuracy of supervised and unsupervised manifold learning under different dimension d, with $k = 5$, $r = 5$. The result is reported in Table 18.1 and Table 18.2. We also test the average recognition performance of the kernel algorithms. The results of Polynomial kernel $k(x, y) = (x \cdot y)^d$ and Gaussian kernel $k(x, y) = \exp(-\|x - y\|^2/2\sigma^2)(\sigma > 0)$ with the different kernel parameters are show in Table 18.3.

Table 18.1 Comparative study under unsupervised algorithms based on the CMU PIE database

Algorithms	Top Classification Accuracy (%)	Max Dim.(Reduced)
Eigenfaces	28.82±2.10	103(100)
MDS	26.76±2.08	103(66)
Diffusion Maps	26.76±2.09	103(45)
ISOMAP	24.41±1.97	103(59)
LLE	28.53±1.92	103(79)
NPE	37.65±1.80	103(95)
Laplacianfaces	61.18±1.76	103(91)
LLC	25.59±2.26	103(101)
Manifold Charting	24.12±2.14	103(103)

Table 18.2 Comparative study under supervised algorithms based on the CMU PIE database

Algorithms	Top Classification Accuracy (%)	Max Dim.(Reduced)
Fisherfaces	68.24±1.58	67(63)
S-Isomap	31.76±1.74	103(16)
S-LPP	74.12±1.26	103(110)
MFA	75.71±1.60	103(55)
LDE	85.29±1.15	103(47)

Table 18.3 Comparative study under kernel manifold learning methods based on the CMU PIE database with different kernel parameters

Algorithms	Classification Accuracy (%) (Reduced Dim.)					
	Polynomial kernel			Gaussian kernel		
	$d = 1$	$d = 2$	$d = 3$	$\sigma = 1$	$\sigma = 10$	$\sigma = 100$
KPCA	20.39±3.15	22.94±2.78	19.94±2.94	32.12±2.50	25.59±2.63	25.88±2.70
	(98)	(93)	(97)	(82)	(82)	(99)
KDA	35.29±2.12	39.76±1.96	32.23±2.17	71.76±1.90	52.65±1.97	52.35±2.01
	(66)	(59)	(61)	(15)	(65)	(36)
K-LPP	25.12±1.16	28.29±1.07	24.73±1.23	62.94±1.02	39.41±1.35	31.47±1.40
	(78)	(75)	(74)	(95)	(18)	(25)
K-Isoamp	11.47±2.30	11.47±2.30	11.47±2.30	11.47±2.30	11.47±2.30	11.47±2.30
	(25)	(25)	(25)	(25)	(25)	(25)

As results shown in Table 18.1, linearization extension of manifold learning algorithms, NPE and Laplacianfaces, achieve higher recognition rate than nonlinear algorithms. From Table 18.2, we can see that LDE algorithm outperform other supervised algorithms over a wide range of dimensionality choices. As results shown in Table 18.3, most kernel algorithms achieve the highest recognition accuracy on the CMU PIE dataset with Gaussian kernel with $\sigma = 1$.

18.5.2 Experiments on FERET Face Database

The FERET face database [60] was collected in 15 sessions between August 1993 and July 1996. The database contains 1564 sets of images for a total of 14 126 images that includes 1199 individuals and 365 duplicate sets of images, which has become the de facto benchmark for evaluating face recognition algorithms. We randomly select 150 subjects from the FERET database with 10 different gray-scale images for each subject. In all the experiments, all images are cropped and rectified according to the manually located eye positions supplied with the FERET data. Then, the facial areas were cropped into the final images for matching. The size of each cropped image in all the experiments is 46×56 pixels, with 256 gray levels per pixel, which can be encoded in a 2576-dimensional vectorized representation. We lastly normalize them to be zero-mean and unit-variance vectors. Figure 18.9 shows an example of the original face image and the cropped image.

Fig. 18.9 The original face image and the cropped image from FERET database.

In face recognition applications, the standard situation (i.e., a realistic scenario) is that just 1, 2, or 3 face images are available for each individual in the face databases. We firstly use a case to compare the performance of different methods. To compare the recognition performance in realistic scenarios, a random subset with two images per individual (hence, 300 images in total) was taken with labels to form the training set. The rest of database was considered to be the testing. The recognition results are shown in Fig. 18.10.

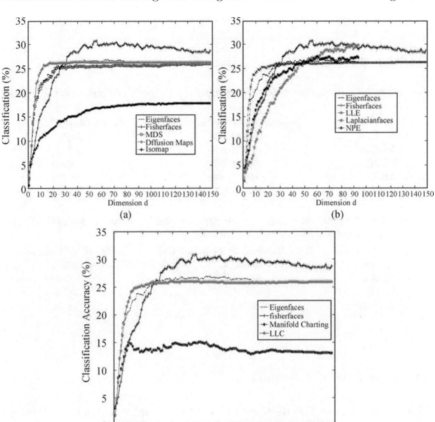

Fig. 18.10 Comparative study based on the FERET database. (a) – (c) Classification accuracy (%) vs. dimensionality d, with $k = 5$, $r = 2$.

We further repeat 10 times to get the average values of the best recognition rate of each algorithm. The results are reported in Tables 18.4 – 18.6.

Table 18.4 Comparative study under unsupervised algorithms based on the FERET database

Algorithms	Top Classification Accuracy (%)	Max Dim.(Reduced)
Eigenfaces	28.92±3.52	98(63)
MDS	26.08±3.17	148(126)
Diffusion Maps	26.42±3.05	148(37)
ISOMAP	17.83±2.78	148(119)
LLE	26.33±2.23	148(85)

Algorithms	Top Classification Accuracy (%)	Continued Max Dim.(Reduced)
NPE	26.58±2.09	98(60)
Laplacianfaces	28.00±1.94	98(88)
LLC	26.08±3.15	148(39)
Manifold Charting	13.08±3.36	148(58)

Table 18.5 Comparative study under supervised algorithms based on the FERET database

Algorithms	Top Classification Accuracy (%)	Max Dim.(Reduced)
Fisherfaces	28.83±2.34	98(56)
S-Isomap	27.67±2.59	148(28)
S-LPP	29.25±2.01	98(96)
MFA	30.00±1.98	98(17)
LDE	35.75±1.73	98(30)

Table 18.6 Comparative study under kernel manifold learning methods based on the FERET database with Gaussian kernel($\sigma=1$)

Algorithms	Top Classification Accuracy (%)	Max Dim.(Reduced)
KPCA	30.83±2.70	(98)88
KDA	32.17±2.56	(148)47
K-LPP	28.83±2.07	(98)98
K-Isoamp	21.08±2.48	(98)64

18.5.3 Analysis of Results

The experiments on two databases have been carried out. On each database, we use the recognition rate obtained in one case and the highest average classification accuracy to compare different face recognition algorithms. We try to keep the impartiality of each algorithm.

By analyzing the YaleB database results (Fig. 18.8 and Tables 18.1 – 18.3), the following can be concluded.

1) Comparing to the global manifold learning methods, the local manifold learning methods consistently perform better than Eigenfaces (Fig. 18.8 and Table 18.1). Especially, Laplacianfaces and NPE significantly outperform Eigenfaces. The local algorithms encode more discriminating information in the low-dimensional face submanifold by preserving the local structure which is more important than the global structure for classification, especially when nearest-neighbor-like classifiers are used. Moreover, the local method takes advantage of more training samples, which is important to the real-world face recognition systems.

2) Fisherfaces generally has higher recognition rates than unsupervised methods over a wide range of dimension (Fig. 18.8 and Table 18.2). However, it models only the global structure of data. Lack of neighborhood properties makes the Fisherfaces method less appealing to the supervised local mani-

fold learning methods such as MFA and LDE, especially when the nearest neighbor criterion is used for classification. The performance of our implementation of supervised methods is quite similar to the originally reported results in Refs. [50 – 51].

3) The methods of global alignment of local models do not have a relative good performance, and the main drawback of these algorithms is that the construction of a mixture of local models is performed using an EM algorithm. EM algorithms suffer from local minima and are very sensitive to outliers [57].

4) Regarding the kernel methods (Table 18.3), they improve the classification performance of linear algorithms for face recognition. However, the performance of kernel algorithms depending on the selection of a proper kernel function, and the algorithm achieves the different performance with the different parameters. KDA gives better results than other methods, because KDA uses class membership information when projecting each point to a low-dimensional space, good class separation suggests that KDA are well suitable for face recognition [34, 35].

5) From the viewpoint of dimensionality reduction, most of the methods based on manifold learning can achieve the presetting object more quickly. While some methods such as PCA are not so effective.

As previously explained, one simulation was performed with the FERET-database, one using 150 classes and two disjoint training sets and eight disjoint target sets (Fig. 18.10 and Tables 18.4 – 18.6). By analyzing these experimental results, the following can be concluded.

1) The experimental results reveal that the local manifold learning methods do not consistently outperform the global methods as they do in CMU PIE database (Fig. 18.10 and Table 18.4). The poor performance of local methods based on manifold learning is due to the following weaknesses. Firstly, the local manifold learning methods want a guarantee that the manifold will be sampled sufficiently densely everywhere. However, the FERET database contains many classes and few images per class (2 – 3), it is almost impossible to learn the manifold structures using only local manifold learning. Secondly, the intrinsic dimensionality of data is much higher and the data have complex structures with high noise, so that the data points lie along a manifold only very roughly, and it is difficult for local manifold learning methods to feature a manifold structure.

2) LDE outperforms all other supervised methods over a wide range of dimensionality choices (Table 18.5), which get a set of most discriminative vectors in discriminative submanifold of the matrices L and L', while L describes the similarity in the same class and L' encodes the dissimilarity information between different classes. Then it exploits the information in the class labels by means of the Fisher criterion, so it can get the promising performance among all the algorithms.

3) Regarding the subspace-based methods (Fig. 18.10), the performance of Fisherfaces method is better than Eigenfaces method when the number of images per class is 2. The reason for this is that the random training-

subset selection would strongly affect the face recognition results [11]. However, when the average best recognition rate is employed (Tables 18.4 and 18.5), Eigenfaces gives better results than Fisherfaces. These results are in accordance with the results in Refs. [11, 35], which concludes that Eigenfaces might outperform Fisherfaces when the number of samples per class is small.

4) KDA gives better results than KPCA (Table 18.6), the results in Table 18.5 are similar to the reported results in Ref. [35] using euclidian distances as similarity measures. By analyzing the number of employed dimensions, it can be seen that KDA employs a much lower numbers of dimensions than other methods. KPCA and KDA give better results than K-LPP and K-ISOMAP, so the kernelized subspace methods have a better generalization ability over the kernelized manifold learning methods.

5) From the experimental results, we observe there is no single method that outperforms all other methods for all dimensions. Most of manifold learning methods gives better results than eigenfaces when the number of dimension is small, and eigenfaces gives better results with the increasing of dimension. Manifold learning methods can effectively learn the inner structure embedded in the high-dimensional image space. By discovering the face manifold structure, these algorithms can identify the person with different expressions, poses, and lighting conditions more quickly. Eigenfaces method is based on the statistical representation of a random variable, which cannot discover the intrinsic manifold structure of the image space, and then it is not so effective as manifold learning methods. Moreover, MFA and LDE can obtain the best recognition rate with very small dimensions. The reason is that MFA and LDE consider both label information and manifold structure to address classification problems, thus they can extract more discriminant information for the small sample size problem.

18.6 Conclusion

The central claim of this paper is to present a systematic comparative study between subspace-based methods and submanifold-based methods for face recognition. The study considered standard, kernel and supervised methods. The manifold leaning algorithm for dimensionality reduction shows a lot of promise and has achieved some successes. When it comes to face recognition, from the results obtained using CMU PIE database and FERET database, we can make the following observations.

1) Submanifold-based methods are not yet capable of outperforming subspace-based methods very effectively, especially in the large face image databases with only one or two training samples available for each subject.

2) Local manifold learning methods give better results in CMU PIE database, and local methods do not consistently outperform the global methods in FERET database. Local methods tend to characterize the local geom-

etry of manifolds accurately, but break down at the global level. Moreover, Local methods require that the data are sampled dense. For global methods, the local topology of points is often corrupted in the embedding process. The drawback of local methods and global methods could be overcome by preserving the local neighborhood information and the global structure simultaneously.

3) LDE and MFA outperform both Fisherfaces and S-LPP in most cases, which demonstrates the importance of utilizing label information and manifold structure, as well as characterize the separability of different classes with the margin criterion. However, in many practical applications, unlabeled training examples are readily available but labeled ones are fairly expensive to obtain, an approach that might resolve this problem is semi-supervised learning.

Future work should focus on overcoming the limitations of manifold learning methods. Most importantly, how to preserve local and global manifold structure in face images and the sensitivity to noise/outlier has to be solved. To address these problems, manifold learning algorithms should be designed in such a way that they can share information about the local and global structure of the manifold, coming from different regions of space, and prior information on the input data. Local and global structure preserving algorithms and supervised/semi-supervised manifold learning algorithms form a good step towards the aim, as the results of our experiments on face databases suggest.

Acknowledgements

This chapter is partially supported by the Natural Science Foundation Project (No. CSTC2009BB2195) and the Key Science Projects (No. CSTC2009AB2231) of Chongqing Science and Technology Commission, Chongqing, China. The authors would like to thank Carnegie Mellon University and National Institute of Standards and Technology for providing the face databases; and L.J.P. van der Maaten of Maastricht University for providing Matlab@Dimensionality Reduction Toolbox.

References

[1] Brunelli R, Poggio T (1993) Face Recognition: Features Versus Templates. IEEE Trans Pattern Anal Mach Intell, 15(10): 1042 – 1052
[2] Kirby M, Sirovich L (1990) Application of the Karhunen-Loeve Procedure for the Characterization of Human Faces. IEEE Trans Pattern Anal Mach Intell, 12(1): 103 – 108
[3] Sinha P, Balas B, Ostrovsky Y et al (2006) Face Recognition by Humans: 19 Results All Computer Vision Researchers Should Know About. Proc the IEEE, November 2006, 94 (11): 1948 – 1962
[4] Gross R, Baker S, Matthews I et al (2004) Face Recognition Across Pose and Illumination. In: Li S Z, Jain A K(eds) Handbook of Face Recognition, Springer, Heidelberg

[5]   Gross R, Shi J, Cohn J et al (2001) Face Recognition? The Current State of the Art in Face Recognition, Technical Report, Robotics Institute, Carnegie Mellon University, Pittsburgh, PA, USA

[6]   Torres L (2004) Is There any Hope for Face Recognition? Proc of the 5th International Workshop on Image Analysis for Multimedia Interactive Services, WIAMIS 2004, Lisboa, Portugal

[7]   Turk M, Pentland (1991) Eigenfaces for Recognition. Journal of Cognitive Neuroscience, 3(1): 71–86

[8]   Belhumeur P N, Hespanha J P (1997) Eigenfaces vs. Fisherfaces: Recognition Using Class Specific Linear Projection. IEEE Trans Pattern Anal Mach Intell, 19(7): 711–720

[9]   Wiskott L, Fellous J M, Kruger N et al (1997) Face Recognition by Elastic Bunch Graph Matching. IEEE Trans Pattern Anal Mach Intell, 19(7): 775–779

[10]   Batur A U and Hayes M H (2001) Linear Subspace for Illumination Robust Face Recognition. Proc IEEE Int Conf Computer Vision and Pattern Recognition, pp 296–301

[11]   Martinez A M and Kak A C (2001) PCA Versus LDA. IEEE Trans Pattern Anal Mach Intell, 23(2): 228–233

[12]   Zhao D L , Lin Z C, Tang X O (2007) Laplacian PCA and its Applications. 11th IEEE International Conference on Computer Vision, pp 2012–2019

[13]   Li S Z, Jain A K (eds) (2004) Handbook of Face Recognition. Springer, Heidelberg

[14]   Wang H H, Zhou Y, Ge X L et al (2007) Subspace Evolution Analysis for Face Representation and Recognition. Patt Recogn, 40: 335–338

[15]   Howland P, Wang J L, Park H (2006) Solving the Small Sample Size Problem in Face Recognition Using Generalized Discriminant Analysis. Patt Recogn, 39: 277–287

[16]   Zhang F (2006) Nonlinear Feature Extraction and Dimension Reduction by Polygonal Principal Curves. Int J PRAI, 20(1): 63–78

[17]   Brand M (2004) From Subspaces to Submanifolds. Proc 15th British Machine Vision Conference. British Machine Vision Association, London

[18]   Yang M H (2002) Extended Isomap for Pattern Classification. Proc 18th National Conf on Artificial Intelligence (AAAI 2002), pp 224–229

[19]   Wu Y M, Chan K L, Wang L (2004) Face Recognition Based on Discriminative Manifold Learning. Proc 17th Int Conf on Pattern Recognition, 4: 171–174

[20]   Zhang J P, Li S Z, Wang J (2004) Nearest Manifold Approach for Face Recognition. Proc 6th IEEE Int Conf on Automatic Face and Gesture Recognition, pp 223–228

[21]   Mekuz N, Bauckhage C, Tsotsos J K (2005) Face Recognition with Weighted Locally Linear Embedding. Proc 2th Canadian Conf on Computer and Robot Vision, pp 290–296

[22]   Geng X, Zhan D C, Zhou Z H (2005) Supervised Nonlinear Dimensionality Reduction for Visualization and Classification. IEEE Trans Syst Man and Cybern Part B-Cybernetics, 35: 1098–1107

[23]   Chen H J and Wei W (2006) Geodesic Gabriel Graph Based Supervised Nonlinear Manifold Learning. Intelligent Computing in Signal Processing and Pattern Recognition, 345: 882–888

[24]   Dickens M P, Smith W A, Jing W et al (2007) Face Recognition Using Principal Geodesic Analysis and Manifold Learning. Proc 3th Iberian Conf on Pattern Recognition and Image Analysis, pp 426–434

[25]   David L D, Carrie G (2002) When Does Isomap Recover the Natural Parametrization of Families of Articulated Images? Technical report, Department of Statistics, Stanford University

[26]   Guang D and Dit-Yan Y (2006) Tensor Embedding Methods. Proc 21th National Conf on Artificial Intelligence (AAAI-2006), pp 330–335

[27]   Wand H X, Zheng W M, Hu Z L et al (2007) Local and Weighted Maximum Margin Discriminant Analysis. Proc IEEE Conf on Computer Vision and Pattern Recognition(CVPR'07), pp 532–539

[28]   Li X L, Lin S, Yan S C et al (2007) Discriminant Locally Linear Embedding with High-order Tensor Data. IEEE Trans Syst Man and Cyberne Part B-Cybernetics, 38: 342–352

[29]   Liu X M, Yin J W, Feng Z L et al (2007) Orthogonal Neighborhood Preserving Embedding for Face Recognition. Proc Ieee Int Conf on Image Processing, pp 133–136

[30]   Yang J, Zhang D, Yang J Y et al (2007) Globally Maximizing, Locally Minimizing: Unsupervised Discriminant Projection with Applications to Face and Palm Biomet-

rics. IEEE Trans Pattern Anal Mach Intell, 29: 650 – 664

[31] Roweis S T and Saul L K (2000) Nonlinear Dimensionality Reduction by Locally Linear Embedding, Science, 290(5500): 2323 – 2326

[32] Tenenbaum J B (1998) Mapping a Manifold of Perceptual Observations. In Advances in Neural Information Processing Systems. MIT Press, Cambridge

[33] Tenenbaum J B, Silva V de, Langford J C (2000) A Global Geometric Framework for Nonlinear Dimensionality Reduction. Science, 290(5500): 2319 – 2323

[34] Navarrete P, Ruiz-del-Solar J (2002) Analysis and Comparison of Eigenspace-based Face Recognition Approaches. Int J PRAI, 16(2): 817 – 830

[35] Ruiz-del-Solar J, Navarrete P (2005) Eigenspace-based Face Recognition: A Comparative Study of Different Approaches. IEEE Trans Syst Man and Cybern: Part C, 35(3): 315 – 325

[36] Zhao W, Chellappa R, Phillips P J et al (2003) Face Recognition: A Literature Survey. ACM Comput Surv, 35(4): 399 – 458

[37] Batagelj, Borut, Solina (2006) Face Recognition in Different Subspaces-A Comparative Study. Proc ICEIS'06 (Paphos, Cyprus), pp 71 – 80

[38] Cox T F, Cox M A A (2001) Multidimensional Scaling, 2nd edn. Chapman and Hall/CRC

[39] Lafon S, Lee A B (2006) Diffusion Maps and Coarse-graining: A Unified Framework for Dimensionality Reduction, Graph Partitioning, and Dataset Parameterization. IEEE Trans Pattern Anal Mach Intell, 28(9): 1393 – 1403

[40] Saltenis V (2005) Constrained Optimization of the Stress Function for Multidimensional Scaling. In Lecture Notes on Computer Science. Verlag, Berlin, Germany, vol 3991: 704 – 711

[41] Belkin M, Niyogi P (2001) Laplacian Eigenmaps and Spectral Techniques for Embedding and Clustering. In Advances in Neural Information Processing Systems(NIPS), pp 585 – 591

[42] He X and Niyogi P (2004) Locality Preserving Projections. In Advances in Neural Information Processing Systems, vol 16

[43] Zhang Z, Zha H (2004) Principal Manifolds And Nonlinear Dimensionality Reduction via Local Tangent Space Alignment. SIAM Journal of Scientific Computing, 26(1): 313 – 338

[44] Donoho D L and Grimes C (2005) Hessian Eigenmaps: New Locally Linear Embedding Techniques for High-dimensional Data. Proc of the National Academy of Sciences, 102(21): 7426 – 7431

[45] He X, Deng C, Yan S et al (2005) Neighborhood Preserving Embedding. Proc 10th IEEE Int Conf on Computer Vision (ICCV'05), pp 1208 – 1213

[46] He X, Yan S, Hu Y et al (2005) Face Recognition Using Laplacianfaces. IEEE Trans Pattern Anal Mach Intell, 27(3): 328 – 340

[47] Teh Y W, Roweis S T (2002) Automatic Alignment of Hidden Representations. In Advances in Neural Information Processing Systems, 15: 841 – 848

[48] Geng X, Zhan D C, Zhou Z H (2005) Supervised Nonlinear Dimensionality Reduction for Visualization and Classification. IEEE Trans Syst Man and Cybern: Part B-Cybernetics, 35(5): 1098 – 1107

[49] Lu K, He X F (2005) Image Retrieval Based on Incremental Subspace Learning. Patt Recogn, 38(11): 2047 – 2054

[50] Xu D, Yan S C, Tao D C et al (2007) Marginal Fisher Analysis and its Variants for Human Gait Recognition and Content-based Image Retrieval. IEEE Trans Image Processing, 162: 811 – 2821

[51] Hwann-Tzong C, Huang-Wei C, Tyng-Luh L (2005) Local Discriminant Embedding and its Variants. Proc IEEE Computer Society Conference on Computer Vision and Pattern Recognition, vol 2: 846 – 53

[52] Navarrete P and Ruiz-del-Solar J (2003) Kernel-based Face Recognition by a Reformulation of Kernel Machines. Advances in Soft Computing-Engineering, Design and Manufacturing, Springer Engineering Series, pp 183 – 196

[53] Bengio Y, Delalleau O, Le Roux N (2004) Learning Eigenfunctions Links Spectral Embedding and Kernel PCA. Neural Computation, 16(10): 2197 – 2219

[54] Choi H, Choi S (2007) Robust Kernel Isomap. Patt Recogn, 40(3): 853 – 862

[55] Bengio Y, Paiement J F, Vincent P et al (2004) Out-of-sample Extensions for LLE, Isomap, MDS, eigenmaps, and spectral clustering. In Advances in Neural Information Processing Systems, 16: 177 – 184

[56] Li H, Teng L, and Chen W et al (2005) Supervised Learning on Local Tangent Space. In Lecture Notes on Computer Science. Springer Berlin, vol 3496: 546 – 551

[57]   Kouropteva O, Okun O, Hadid A et al (2002) Beyond Locally Linear Embedding Algorithm. Technical Report MVG-01-2002, University of Oulu

[58]   Saul L, Roweis S (2003) Think Globally, Fit Locally: Unsupervised Learning of Nonlinear Manifolds. Journal of Machine Learning Research, 4: 119 – 155

[59]   Sim T, Baker S, Bsat M (2003) The CMU Database. IEEE Trans Pattern Anal Mach Intell, 25: 1615 – 1618

[60]   Phillips P J, Moon H, Rizvi S et al (2000) The FERET Evaluation Methodology for Face-Recognition Algorithms. IEEE Trans Pattern Anal Mach Intell, 22(10): 1090 – 1104

19 Linear and Nonlinear Feature Extraction Approaches for Face Recognition

Wensheng Chen[1], Pong C. Yuen[2], Bin Fang[3] and
Patrick S.P. Wang[4]

Abstract   This chapter introduces recent progress and existing challenges in the area of face recognition (FR). Main problems encountered in FR such as curse of dimensionality problem, nonlinear problem, and small sample size problem are discussed. To deal with these limitations, some classical methods and state of the art schemes, including linear feature and non-linear feature extraction methodologies, are proposed and recommended for face recognition research community.

19.1 Introduction

Face recognition is an important biometric authentication technology and has become one of the most active research areas in computer vision and pattern recognition ([1 – 40]). FR has more and more applications in our real world, such as security card or identity card verification, passport verification, access control for computer, facility access, advanced video surveillance, etc. A typical face recognition system involves four steps: facial image acquiring, face detection/segmentation, face feature extraction, and face identification. Based on different face feature extraction criteria, many FR algorithms have been developed during the last decade. Two very good reviews on FR can be found in [41, 42]. Face recognition approaches can be divided into two main categories, namely geometric feature-based and appearance-based [1]. The geometric feature-based approach performs recognition based on the shape and the location of facial components (such as eyes, eyebrows, nose, mouth). The facial features are extracted to form a face geometric feature vector. The appearance-based approach relies on the global facial features, which are extracted to generate a whole facial feature vector for face classification. This chapter focuses on the appearance-based approaches.

1   College of Mathematics and Computational Science, Shenzhen University, Shenzhen 518060, China. E-mail: chenws@szu.edu.cn.

2   Department of Computer Science, Hong Kong Baptist University, Hong Kong, China. E-mail: pcyuen@comp.hkbu.edu.hk.

3   College of Computer Science, Chongqing University, Chongqing 400044, China. E-mail: fb@cqu.edu.cn.

4   College of Computer and Information Science, Northeastern University, Boston, MA 02115, USA. E-mail: pwang@ccs.neu.edu.

It is known that the dimension of facial pattern vector obtained by vectorizing a facial image is very high. For example, if the resolution of a facial image is 112×92, then the size of facial vector attains 10 304. This leads to the so-called curse of dimensionality problem. Various dimensionality reduction technologies have been presented to tackle this problem. Two famous appeared-based methods for dimensionality reduction are principal component analysis (PCA) [2] and linear discriminant analysis (LDA) [3]. PCA method, proposed by Turk et al. in 1991, is an unsupervised linear method. It has an alternative name called eigenface method in face recognition. All eigenfaces form a group of basis images which can generate the original face feature space. PCA selects eigenfaces corresponding to large eigenvalues of covariance matrix to account for most distributions. LDA method, known as Fisherface in FR, was developed by Belhumeur et al. in 1997. LDA aims to find an optimal projection such that the ratio of between-class distance against within-class distance achieves maximum in the LDA-transformed lower-dimensional feature space. Therefore, from the classification point of view, LDA should surpass PCA in pattern recognition tasks. Figure 19.1 shows the projection directions of PCA and LDA.

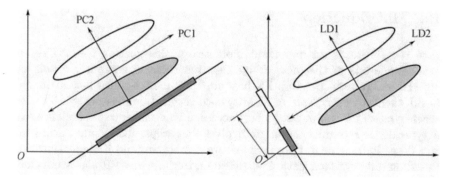

Fig. 19.1 PCA versus LDA.

LDA is theoretically sound and various LDA-based face recognition approaches have been reported in many literatures [3–17, 20–24, 28–30, 44]. However, LDA suffers from the small sample size (3S) problem. This problem always occurs when the total number of training samples is smaller than the dimension of feature vector. Under this situation, the eigen-system in LDA method becomes singular and direct solving its eigenvalue problem is infeasible. To overcome 3S problem, a number of research works [3–17] have been proposed. These methods, in applying to face recognition, includes two categories, namely subspace approach and regularization approach.

Two-stage subspace approach is to reduce the dimension of the original feature vector so that the eigen-system is non-singular in the mapped feature space, and therefore LDA can be applied. PCA is a popular and typical technique employed in this approach. The method, which employs

PCA for dimensionality reduction before applying LDA for FR, is known as Fisherface. The Fisherface approach is straightforward but suffers from two limitations. First, selection of the principal components for LDA is still an unsolved problem. Many researchers have demonstrated that the eigenvectors corresponding to the largest eigenvalues are not the best for recognition from the discriminant analysis point of view. As far as we know, existing methods select the largest $N - k$ eigenvalues for recognition, where N is the total number of training samples and k is the number of classes. Second, removing the small eigenvalues (including zero eigenvalues) in within scatter matrix S_w means that the null space of S_w is discarded. It has been shown that the null space of S_w contains useful discriminant information [4]. In turn, the performance will be degraded.

Another LDA based subspace approach is to find the projection matrix W^* from a restricted subspace in within-class scatter matrix S_w, between-class scatter matrix S_b or total scatter matrix $S_t(= S_b + S_w)$ via simultaneous diagonalized technique. Comparing with the two-stage approach, the subspace approach gives a more rigorous theoretical justification. The main reason is that the transformation can be found in the null space of S_w, in which the Fisher index will be maximum if the W^* is outside the null space of S_b. Moreover, another advantage of subspace approach is that no selection of principal components is required. However, this approach suffers from one limitation in solving the 3S problem. The transformation W^* is not unique. The Fisher index is maximum as long as the W^* is found in the null space of S_w and out of the null space of S_b.

It can be seen that subspace approaches find the projection directions only in the restricted subspace(s), instead of entire feature space. So it may not obtain the global optimal solution. To find the solution in full feature space, regularization approaches [11, 12] are proposed. Dai et al. [11] presented three parameters RDA method to solve 3S problem. Although this method is executed in the full sample space, it is difficult to determine three optimal parameters because of the computational complexity. To overcome the complexity problem, Chen et al. [12] further proposed a single parameter RDA algorithm. It greatly reduces the computational complexity from $O(n^3)$ to $O(n)$.

However, PCA, LDA-based approaches all are linear feature extraction methods and thus hard to solve nonlinear problem. It is well-known that the distributions of face image variations under different pose and illumination are complex and nonlinear. Therefore, like other appearance-based methods, the performance of LDA-based method will degrade under pose and illumination variations. To overcome this drawback, kernel method is employed. The basic idea is to apply a nonlinear mapping $\Phi : x \in R^d \rightarrow \Phi(x) \in F$ to the input data vector $x \in R^d$ and then to perform the LDA on the mapped feature space F. This method is the so-called Kernel Discriminant Analysis (KDA) [42, 44]. The feature space F could be considered as a linearized space. However, its dimensionality could be arbitrarily large and possibly infinite.

Direct applying LDA method to feature space is impossible. Fortunately, the exact mapping function Φ does not need to be used directly. The feature space can become implicit by using kernel trick, in which the inner products $\langle \Phi(x_i), \Phi(x_j) \rangle$ in F can be replaced with a kernel function $K(x_i, x_j)$, i.e., $K(x_i, x_j) = \langle \Phi(x_i), \Phi(x_j) \rangle$, where $x_i, x_j \in R^d$ are input pattern vectors. So the nonlinear mapping Φ can be performed implicitly in input space R^d.

Owing to the nonlinear property of KDA and its strong theoretical background, it has been applied in face recognition to solve the pose and illumination problems since 2000 [19–31]. It is found that kernel-based methods [19–21] outperform linear methods and independent component analysis (ICA) [45]. Liu et al. [23] proposed to improve kernel discriminant analysis using cosine kernel and performed a good experimental analysis on the KDA and showed that KDA, in general, gives better performance than that of KPCA [46]. Lu et al. proposed [24] to solve the pose problem using kernel direct-LDA (KDDA). They experimentally showed that five individuals with different poses could be linearly separated in feature space while these images are overlapped in input space. Moreover, the performance of KDDA outperforms that of KPCA and general discriminant analysis (GDA) [25].

Recently, some researches on KDA are to construct the optimal Mercer kernel functions and automatically choose kernel parameters. For kernel construction technique, literature [26] utilizes discriminant Gram matrix and interpolatory strategy to construct a data-dependent Mercer kernel function, the performance of which greatly outperforms RBF kernel function. Based on Fourier criterion, literature [27] has mathematically shown that by using the cubic spline function, wavelet kernel can be constructed and applied for the KDA on face recognition. In the area of kernel parameter estimation, Chen and Yuen et al. have developed gradient descent-based algorithms [28, 29] to automatically select the best set of RBF kernel and regularization parameters for the KRDA algorithm. On the other hand, they developed algorithm to find the optimal parameters by optimizing the Fisher criterion function in their proposed kernel subspace LDA algorithm [30]. Along this direction, they have further enhanced the algorithm based on the concept of stability. The marginal maximization criterion is adopted and an eigenvalue stability bounded margin maximization algorithm [31] was developed to find the optimal RBF kernel parameters.

The rest of this chapter is organized as follows. Section 19.2 describes existing linear feature learning methods for FR. Section 19.3 reports the nonlinear feature extraction approaches based on kernel machine learning. Finally, Section 19.4 draws the conclusions.

19.2 Linear Feature Extraction Methods

This section gives some linear feature extraction methodologies for face recognition. Before going into theory analysis, let's define some symbols to be used in this section.

19.2.1 Some Definitions

Let d be the dimensionality of original sample feature space and k be the number of sample classes, the total original sample $C = \{C_1, C_2, \ldots, C_k\}$, the jth class C_j contains N_j samples, i.e., $C_j = \{x_1^{(j)}, x_2^{(j)}, \ldots, x_{N_j}^{(j)}\}, j = 1, 2, \ldots, k$. Let N be the total number of original training samples, $\bar{x}_j = \frac{1}{N_j} \sum_{x \in C_j} x$ be the mean of the sample class C_j and $\bar{x} = \frac{1}{N} \sum_{j=1}^{k} \sum_{x \in C_j} x$ be the global mean of the total original sample C. In discriminant analysis, three scatter matrices, namely within-class, between-class and total scatter matrices, are defined respectively as follows:

$$S_w = \frac{1}{N} \sum_{j=1}^{k} \sum_{x \in C_j} (x - \bar{x}_j)(x - \bar{x}_j)^{\mathrm{T}} = \Phi_w \Phi_w^{\mathrm{T}}, \tag{19.1}$$

$$S_b = \frac{1}{N} \sum_{j=1}^{k} N_j(\bar{x}_j - \bar{x})(\bar{x}_j - \bar{x})^{\mathrm{T}} = \Phi_b \Phi_b^{\mathrm{T}},$$

$$S_t = S_w + S_b = \frac{1}{N} \sum_{j=1}^{k} \sum_{x \in X_j} (x - \bar{x})(x - \bar{x})^{\mathrm{T}} = \Phi_t \Phi_t^{\mathrm{T}},$$

where $\Phi_w, \Phi_t \in R^{d \times N}$ and $\Phi_b \in R^{d \times k}$.

19.2.2 PCA

Principal component analysis (PCA), also known as Eigenface method, is a popular statistic linear method for dimensionality reduction in face recognition. PCA is based on Karhunen-Loeve transform. The principle of PCA is as follows.

Let S_t be the covariance matrix (total scatter matrix) and perform eigenvalue decomposition on S_t:

$$S_t = U_t \Lambda U_t^{\mathrm{T}},$$

where $U_t = [u_1, u_2, \ldots, u_d] \in R^{d \times d}$ is a orthonormal matrix and Λ is a diagonal matrix with its diagonal entries in decreasing order, namely $\Lambda = diag\{\lambda_1, \ldots, \lambda_\tau, 0, \ldots, 0\}$ and $\lambda_1 \geqslant \ldots \geqslant \lambda_\tau > 0$. The basis vectors $u_i(i = 1, 2, \ldots, d)$ are called eigenfaces. Several original facial images and the first six eigenface images from ORL face database are plotted in Fig. 19.2.

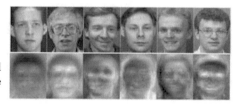

Fig. 19.2 Original faces (top) and Eigenfaces (down) from ORL face database.

Every image $x \in R^d$ can be recovered by the eigenfaces. The PCA projection matrix W_{PCA} is selected as $W_{PCA} = [u_1, u_2, \ldots, u_m] \in R^{d \times m}$, where m can be determined by

$$\frac{\sum_{i=1}^{m} \lambda_i}{\sum_{i=1}^{\tau} \lambda_i} \geqslant k,$$

where k can be chosen as 95% or 98%.

In geometric sense, PCA chooses the first few principal components to account for most distributions. Therefore, PCA cannot exploit all of the feature classification information and then does not lead to good performance.

19.2.3 Classical LDA

The aim of linear (Fisher) discriminant analysis is to find an optimal projection matrix W^*, from original feature space R^d to a transformed lower-dimensional feature space R^m, such that the ratio of between-class distance against within-class distance achieves maximum. i.e.,

$$W^* : R^d \to R^m, \ d > m, \quad W^* = \arg\max_w J(W),$$

where $J(W)$ is the Fisher criterion function defined by

$$J(W) = \frac{tr(W^T S_b W)}{tr(W^T S_w W)},$$

where matrix $W \in R^{d \times m}$ and $tr(\cdot)$ denotes the trace of a matrix.

The above problem is equivalent to solving the following eigenvalue problem:

$$(S_w)^{-1} S_b W = W \Lambda,$$

where Λ is a $d \times d$ diagonal eigenvalue matrix.

But LDA fails to work when 3S problem or single training sample problem occurred. To tackle 3S problem, subspace methods and regularized methods will be introduced in the following subsections.

19.2.4 Subspace Method — Fisherface

Fisherface method [3] is a two-stage method to overcome 3S problem in FR. It applies an intermediate dimensionality reduction stage using PCA on original feature space such that the within-class scatter matrix S_w is invertible and then the traditional LDA can be performed in the PCA-transformed feature space. In detail, it selects proper principal exponents to form a PCA projection matrix W_{PCA} such that $\tilde{S}_w = W_{PCA}^{\mathrm{T}} S_w W_{PCA}$ is full rank. Then the second projection matrix W_{LDA} can be solved from the following eigensystem, where $\tilde{S}_b = W_{PCA}^{\mathrm{T}} S_b W_{PCA}$:

$$(\tilde{S}_w)^{-1}\tilde{S}_b W = W\Lambda.$$

The final LDA projection matrix is $W^* = W_{PCA}W_{LDA} \in R^{d\times(k-1)}$. The column vectors of matrix W^* are called Fisherfaces. Several original facial images and Fisherface images from FERET face dataset are shown in Fig. 19.3.

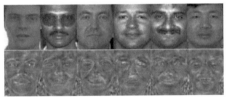

Fig. 19.3 Original faces (top) and Fisherfaces (down) from FERET face database.

As we know, the null space of S_w contains many useful discriminant information for PR. Therefore, discarding all the null space of S_w to guarantee that S_w is a nonsingular matrix will lose many important discriminant information.

19.2.5 Subspace Method – Direct LDA

Yu et al. [5] proposed a the so-called direct LDA (DLDA) method to deal with 3S problem. The idea of DLPP is first to remove the null space of between-class scatter matrix S_b and then keep the null space of within-class scatter matrix S_w. It obtains the direct LDA projection matrix W_{DLDA} by diagonalizing both S_b and S_w simultaneously, namely

$$W_{DLDA}S_b W_{DLDA}^{\mathrm{T}} = I, \quad W_{DLDA}S_w W_{DLDA}^{\mathrm{T}} = \Lambda,$$

where I is an identity matrix and Λ is a diagonal matrix. Its diagonal entries are the zero or small eigenvalues of S_w.

The first six feature basis images of DLDA from ORL face database are shown in Fig. 19.4. The top row of Fig. 19.2 is some original facial images from different individuals.

Fig. 19.4 DLDA feature basis images from ORL face database.

Since Rank $(S_b) \leqslant$ Rank (S_w), first discarding the null space of S_b leads to discarding partial or entire null space of S_w as well. It will decrease DLDA classification power. But DLDA method has lower computational complexity.

19.2.6 Regularized LDA

Other than subspace approaches, regularization technique is also commonly used to overcome the 3S issue of LDA. It searches the projection directions in the full feature space. This subsection introduces a two-step single parameter regularization discriminant analysis (2SRDA) method [14] in face recognition.

Regularization Strategy

To deal with 3S problem, 2SRDA method exploits two-stage regularization strategy, namely rank lifting step and three-to-one regularized step. Below is the details on 2SRDA method.

Rank lifting scheme

In the rank lifting stage, it needs the following theorem.

Theorem 19.1   If vector set $\bigcup\limits_{i=1}^{k}\{\alpha_j^{(i)}|j = 1, 2, \ldots, n_i\}$ is a linearly independent set and $m_i = \dfrac{1}{n_i}\sum\limits_{j=1}^{n_i}\alpha_j^{(i)}, i = 1, 2, \ldots, k$, then for any constant $t(t \neq 1)$, the following vector set

$$\bigcup_{i=1}^{k}\{\alpha_1^{(i)} - m_i, \alpha_2^{(i)} - m_i, \ldots, \alpha_{n_i-1}^{(i)} - m_i, \alpha_{n_i}^{(i)} - t \cdot m_i\}$$

is a linearly independent set as well.

Proof.   Assume there exists a group of constants $\bigcup\limits_{i=1}^{k}\{k_j^{(i)}|j = 1, 2, \ldots, n_i\}$, such that

$$\sum_{i=1}^{k}\sum_{j=1}^{n_i}k_j^{(i)}\left(\alpha_j^{(i)} - t_j^{(i)} \cdot m_i\right) = 0,$$

where $t_j^{(i)} = \begin{cases} 1, & 1 \leqslant j \leqslant n_i - 1 \\ t, & j = n_i \end{cases}$. It yields that

$$\sum_{i=1}^{k} \sum_{j=1}^{n_i} k_j^{(i)} \left(\alpha_j^{(i)} - \frac{t_j^{(i)}}{n_i} \sum_{l=1}^{n_i} \alpha_l^{(i)} \right) = 0.$$

Therefore,

$$\sum_{i=1}^{k} \sum_{l=1}^{n_i} \left(k_l^{(i)} - \frac{1}{n_i} \sum_{j=1}^{n_i} t_j^{(i)} \cdot k_j^{(i)} \right) \alpha_l^{(i)} = 0.$$

Since the vector set $\bigcup_{i=1}^{k} \{\alpha_j^{(i)} | j = 1, 2, \ldots, n_i\}$ is a linearly independent set, we have

$$k_l^{(i)} - \frac{1}{n_i} \sum_{j=1}^{n_i} t_j^{(i)} \cdot k_j^{(i)} = 0, \quad i = 1, 2, \ldots, k, \quad l = 1, 2, \ldots, n_i.$$

For each fixed i $(i = 1, 2, \ldots, k)$, $(k_1^{(i)}, k_2^{(i)}, \ldots, k_{n_i}^{(i)})$ is the solution of the following homogeneous system:

$$n_i \cdot k_l^{(i)} - \sum_{j=1}^{n_i} t_j^{(i)} \cdot k_j^{(i)} = 0, \quad l = 1, 2, \ldots, n_i.$$

The coefficient matrix associated with the above system is a $n_i \times n_i$ matrix as follows,

$$A_i = \begin{pmatrix} n_i - 1 & -1 & -1 & \cdots & -1 & -t \\ -1 & n_i - 1 & -1 & \cdots & -1 & -t \\ \vdots & \vdots & \vdots & \cdots & \vdots & \vdots \\ -1 & -1 & -1 & \cdots & n_i - 1 & -t \\ -1 & -1 & -1 & \cdots & -1 & n_i - t \end{pmatrix}_{n_i \times n_i}.$$

Since $\det(A_i) = (1 - t) \cdot n_i^{n_i - 1} \neq 0$ $(t \neq 1)$, it implies that A_i is a nonsingular matrix, so we have

$$k_j^{(i)} = 0, \quad i = 1, 2, \ldots, k, \quad l = 1, 2, \ldots, n_i.$$

This concludes the theorem immediately.

Assume the total training samples $\bigcup_{i=1}^{k} \{x_j^{(i)} | j = 1, 2, \ldots, N_i\}$ is a linearly independent vector set. For the matrix Φ_w defined in (1), we have

$$\Phi_w = \frac{1}{\sqrt{N}} [x_1^{(1)} - \bar{x}_1| \ldots |x_{N_1}^{(1)} - \bar{x}_1|, \ldots, |x_1^{(k)} - \bar{x}_k| \ldots |x_{N_k}^{(k)} - \bar{x}_k],$$

and then define matrix $\tilde{\Phi}_w$ by

$$\tilde{\Phi}_w = \frac{1}{\sqrt{N}}[x_1^{(1)} - \bar{x}_1| \ldots |x_{N_1}^{(1)} - \bar{x}_1 + t \cdot \bar{x}_1|, \ldots, |x_1^{(k)} - \bar{x}_k| \ldots |x_{N_k}^{(k)} - \bar{x}_k + t \cdot \bar{x}_k],$$
(19.2)

where $t \neq 0$. By theorem 1, we know that the matrix $\tilde{\Phi}_w$ is a full rank matrix. The semi-regularized within-class scatter matrix $\tilde{S}_w(= \tilde{\Phi}_w \tilde{\Phi}_w^T)$ can be obtained. If $\tilde{S}_w$ is nondegenerate, the classical LDA method can be applied directly. Otherwise, the second regularization technique proposed in [12] will be applied to $\tilde{S}_w$ and the final invertible within-class scatter matrix $\tilde{S}_w^R$ is determined.

Moreover, it can be derived from Eqs. (19.1) and (19.2) that

$$\tilde{S}_w = S_w + \frac{t}{N}S_1 + \frac{t^2}{N}S_2,$$
(19.3)

where $S_1 = M\alpha^T + \alpha M^T - 2MM^T, S_2 = MM^T \in R^{d \times d}, \alpha = [x_{N_1}^{(1)}, x_{N_2}^{(2)}, \ldots, x_{N_k}^{(k)}], M = [\bar{x}_1, \bar{x}_2, \ldots, \bar{x}_k] \in R^{d \times k}$.

Three-to-one regularized step

Performing eigenvalue decomposition on the semi-regularized within-class scatter matrix $\tilde{S}_w$, we obtain

$$\tilde{S}_w = U_w \Lambda_w U_w^T,$$
(19.4)

where $\Lambda_w = \text{diag}\{\lambda_1, \ldots, \lambda_N, 0, \ldots, 0\} \in R^{d \times d}$ with $\lambda_1 \geqslant \ldots \geqslant \lambda_N > 0$.

Based on Eq. (19.4), the three-parameter family regularization matrices $S_w^{\alpha\beta\gamma}$ can be difined as follows,

$$S_w^{\alpha\beta\gamma} = U_w \hat{\Lambda}_w U_w^T,$$

where $\hat{\Lambda}_w$ is a diagonal matrix with its diagonal elements ξ_i $(i = 1, 2, \ldots d)$ defined by

$$\xi_i = \begin{cases} (\alpha\lambda_i + \beta)/M, & i = 1, 2, \ldots, N, \\ \gamma, & i = N+1, \ldots, d, \end{cases}$$
(19.5)

and M is a normalization constant given by

$$M = \frac{\alpha tr(\tilde{S}_w) + N\beta}{tr(\tilde{S}_w) - (d-N)\gamma},$$

where $\alpha \geqslant 1$, $\beta \geqslant 0$, $\gamma > 0$ and $\alpha\lambda_N + \beta \geqslant M\gamma$.

By the three-to-one technique proposed in literature [12], the three-parameter regularized matrix $S_w^{\alpha\beta\gamma}$ can be reduced to one parameter regularized within-class scatter matrix $\tilde{S}_w^R$ by using the following formulae:

$$\alpha = t+1, \quad \beta = \frac{tr(\tilde{S}_w)}{N(d-N)} \cdot t, \quad \gamma = N \cdot \beta, \quad (0 < t \leqslant 1)$$
(19.6)

In addition, it is easy to show that the regularized within-class scatter matrix $\tilde{S}_w^R$ approaches to original within-class scatter matrix S_w as the regularization parameter t tends to zero. Namely, we have the following approximation theorem.

Theorem 19.2 The regularized within-class scatter matrix $\tilde{S}_w^R$ approaches to original within-class scatter matrix S_w as the regularization parameter t tends to zero. i.e.,

$$\tilde{S}_w^R \to S_w, \quad as\ t \to 0.$$

Proof. For the Frobenius norm, it derives from Eq. (19.3) that

$$\|\tilde{S}_w^R - S_w\|_F \leqslant \|\tilde{S}_w^R - \tilde{S}_w\|_F + \|\tilde{S}_w - S_w\|_F$$

$$= \|U_w(\hat{\Lambda}_w - \Lambda_w)U_w^{\mathrm{T}}\|_F + \|\frac{t}{N}S_1 + \frac{t^2}{N}S_2\|_F$$

$$= \sum_{i=1}^{N}(\xi_i - \lambda_i)^2 + \sum_{i=N+1}^{d} \xi_i^2 + \frac{t}{N}\|S_1 + t \cdot S_2\|_F. \quad (19.7)$$

From Eqs. (19.5) and (19.6), we know that

$$\alpha \to 1, \quad \beta \to 0, \quad \gamma \to 0, \quad M \to 1, \quad as\ \ t \to 0.$$

It yields that

$$\xi_i \to \lambda_i\ (i = 1, 2, \ldots, N)\ \text{and}\ \xi_i \to 0\ (i = N + 1, \ldots, d),\ \text{as}\ \ t \to 0.$$

Therefore, Inequality (19.7) shows that

$$\|\tilde{S}_w^R - S_w\|_F \to 0, \quad \text{as}\ \ t \to 0.$$

This concludes the theorem.

2SRDA Algorithm Design

Based on above analysis, this subsection will give a detail flow of 2SRDA algorithm.

After finishing regularization stage, the regularized Fisher discriminant analysis method is to solve the the following eigenvalue system:

$$(\tilde{S}_w^R)^{-1}S_b W = W\Lambda, \quad (19.8)$$

where Λ is a diagonal eigenvalue matrix with its diagonal elements in decreasing order and W is an eigenvector matrix.

Let $\hat{S}_b = YS_bY^{\mathrm{T}}, Y = \hat{\Lambda}_w^{-1/2}U_w^{\mathrm{T}}$ and $V = \hat{\Lambda}_w^{1/2}U_w^{\mathrm{T}}W$, then Eq. (19.8) is equivalent to solving the following symmetric eigen-system:

$$\hat{S}_b V = V\Lambda. \quad (19.9)$$

If the solution V is solved from Eq. (19.9), the solution of Eq. (19.8) can be obtained by

$$W = U_w \widehat{\Lambda}_w^{-1/2} V.$$

Therefore, the 2SRDA algorithm is developed as follows.

Rank Lifting Stage: $(S_w \rightarrow \tilde{S}_w)$

Step 1 Set regularized parameter $t = 0.01$, calculate matrix $\tilde{\Phi}_w$ defined in Eq. (19.2), then Rank $(\tilde{\Phi}_w) = N$.

Step 2 Perform eigenvalue decomposition on $N \times N$ matrix $\tilde{\Phi}_w^T \tilde{\Phi}_w \overset{\text{evd}}{=} V \Lambda V^T$, where $V = [v_1, \ldots, v_N] \in R^{N \times N}$ is an orthonormal matrix, $\Lambda = \text{diag}\{\lambda_1, \ldots, \lambda_N\} \in R^{N \times N}$ with $\lambda_1 > \ldots > \lambda_N > 0$. Calculate $\tilde{v}_i = \tilde{\Phi}_w v_i / \|\tilde{\Phi}_w v_i\| \in R^{d \times 1}$ $(i = 1, \ldots, N)$ and construct an orthonormal matrix $\tilde{U} = [\tilde{v}_1, \ldots, \tilde{v}_N, \tilde{v}_{N+1}, \ldots, \tilde{v}_d] \in R^{d \times d}$, where vectors $\tilde{v}_{N+1}, \ldots, \tilde{v}_d$ are just subject to orthonormal constraint.

Final regularized Stage: $(\tilde{S}_w \rightarrow \tilde{S}_w^R)$

Step 3 Calculate $\alpha = t + 1, \beta = \dfrac{\text{tr}(\tilde{S}_w)}{N(d - N)} \cdot t, \gamma = N \cdot \beta$,

$$M = \frac{\alpha \, \text{tr}(\tilde{S}_w) + N\beta}{\text{tr}(\tilde{S}_w) - (d - N)\gamma}, \quad \xi_i = \begin{cases} (\alpha\lambda_i + \beta)/M, & i = 1, \ldots, N \\ \gamma, & i = N + 1, \ldots, d \end{cases},$$

and let $\widehat{\Lambda}_w = \text{diag}\{\xi_1, \ldots, \xi_d\}$.

Find RDA projection Matrix.

Step 4 Compute $Y = \widehat{\Lambda}_w^{-1} \tilde{U}^T \Phi_B \in R^{d \times k}$ and perform eigenvalue decomposition on $k \times k$ matrix $Y^T Y \overset{\text{evd}}{=} V_Y \Lambda_Y V_Y^T$, where $V_Y = [y_1, \ldots, y_k] \in R^{k \times k}$ is an orthonormal matrix. Calculate $\tilde{y}_i = Y y_i / \|Y y_i\| \in R^{d \times 1}$ $(i = 1, \ldots, k-1)$ and construct an $d \times (k - 1)$ matrix $\tilde{Y}_{k-1} = [\tilde{y}_1 | \tilde{y}_2 | \ldots | \tilde{y}_{k-1}]$.

Step 5 The optimal projection matrix $W_{2SRDA} = \tilde{U} \widehat{\Lambda}_w^{-1} \tilde{Y}_{k-1}$.

Performance Evaluations

The experimental results are reported in this subsection to evaluate the effectiveness of the 2SRDA method. A comparison with some existing LDA-based methods for solving the small sample size problem, namely Fisherface [3], Direct LDA [5] and Huang et al. method [9], is also presented. In the regularization procedure, the value of regularization parameter is given as $t = 0.01$.

Two popular and public available databases, namely ORL database and FERET database, are selected for the evaluation.

Face Database – ORL

There are 40 persons involved in ORL database and each person consists of 10 images with different facial expressions, small variations in scales and orientations. Image variations of one person in the database are shown in Fig. 19.5.

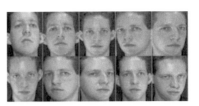

Fig. 19.5 Images of one person from ORL database.

Face Database – FERET

For FERET database, we select 72 people, 6 images for each individual. The six images are extracted from 4 different sets, namely Fa, Fb, Fc, and duplicate. Fa and Fb are sets of images taken with the same camera at the same day but with different facial expressions. Fc is a set of images taken with different camera at the same day. Duplicate is a set of images taken around $6-12$ months after the day taking the Fa and Fb photos. Details of the characteristics of each set can be found in [18]. All images are aligned by the centers of eyes and mouth and then normalized with resolution 92×112. This resolution is the same as that in ORL database. Images from one individual are shown in Fig. 19.6.

Fig. 19.6 Images of one person from FERET database.

Performance on ORL Database

The experimental setting on ORL database is as follows. We randomly selected n images from each individual for training while the rest $(10 - n)$ images are for testing. The experiments are repeated 50 times and the average accuracy is then calculated. The average accuracies are recorded and tabulated in Table 19.1. It can be seen from Table 19.1 that the recognition accuracy of the 2SRDA method increases from 79.72% with 2 training images to 97.75% with 9 training images.

Table 19.1 Performance comparison on ORL face database

TN	Fisherface [3]	DLDA [5]	Huang [9]	2SRDA [14]
2	61.28%	78.13%	79.63%	79.72%
3	73.36%	83.54%	87.84%	86.50%
4	67.22%	89.33%	91.62%	91.50%
5	88.30%	91.50%	92.78%	93.95%
6	90.88%	91.56%	94.27%	96.37%
7	93.67%	93.08%	94.18%	95.58%
8	96.25%	94.13%	94.50%	96.63%
9	96.00%	95.25%	95.40%	97.75%

In order to compare the performance of the 2SRDA method with existing methods, the same experiments are conducted using Fisherface method [3] and Direct LDA method [5] and Huang et al. method [9]. The results are recorded and tabulated in Table 19.1. It can be seen that when 2 images are used for training, the accuracies for Fisherface, Direct LDA and Huang et al. methods are 61.28%,78.13%, and 79.63% respectively. When the number of training images is equal to 9, the accuracy for Fisherface is increased to 96.00% while Direct LDA and Huang et al. methos are 95.25% and 95.40% respectively. The results show that 2SRDA method gives the best performance for all cases on ORL database except that Huang et al. method is slightly better than 2SRDA method for training numbers 3 and 4.

Performance on FERET Database

The experimental setting for the FERET database is similar with that of ORL database. As the number of images for each individual is 6, the number of training images is ranged from 2 to 5. The average accuracies are recorded and tabulated in Table 19.2. It can be seen from Table 19.2 that the recognition accuracy of 2SRDA method increases from 81.49% with 2 training images to 94.58% with 5 training images.

Table 19.2   Performance comparison on FERET face database

TN	2	3	4	5
Fisherface	62.60%	78.81%	86.33%	91.92%
DLDA	71.25%	77.53%	82.92%	89.00%
Huang	73.04%	84.28%	88.64%	90.71%
2SRDA	81.49%	88.75%	91.74%	94.58%

The same experiments are implemented by using Fisherface method, Direct LDA method and Huang et al. method with the same lists of training and testing images. The results are recorded and tabulated in Table 19.2. It can be seen that when 2 images are used for training, the accuracies for Fisherface, Direct LDA and Huang et al. methods are 62.60%, 71.25% and 73.04% respectively. When the number of training images is equal to 5, the accuracy for Fisherface increases to 91.92% while Direct LDA and Huang et al. methods are 89.00% and 90.71% respectively. Experiment results show on FERET dataset that the 2SRDA method gives the best performance for all cases.

Computational Complexity

As the major difference on complexity of different methods is on the training process, the computational complexity on the training processes in different methods is discussed in this part.

The computational cost of Fisherface includes solving one N by N eigensystem, $O(N^3)$, and two $(N - k) \times (N - k)$ eigensystem, $O((N - k)^3)$. In Direct LDA, the computational cost involves solving one k by k eigensystem, $O(k^3)$ and one $m \times m(m \leqslant k - 1)$ eigensystem, $O(m^3)$. In Huang's method, the computational cost consists of three parts. The first part is to solve one N by N eigensystem, $O(N^3)$. The rest parts are to solve one $(N-1)$ by $(N-1)$ eigensystem, $O((N-1)^3)$ and one $(k-1) \times (k-1)$ eigensystem, $O((k-1)^3)$. The computational cost of 2SRDA method consists of two parts, which involves solving one N by N eigensystem, $O(N^3)$ and one $k \times k$ eigensystem, $O(k^3)$. The computational complexity of the 2SRDA method and other methods are summarized in Table 19.3.

Table 19.3 Computational complexity

	Computing $S_w(\varPhi_w), S_b(\varPhi_b), S_t(\varPhi_t)$	Solving eigensystem
Fisherface	$\varPhi_w, \varPhi_b, \varPhi_t$	One $N \times N$
		Two $(N-k) \times (N-k)$
DLDA	$\varPhi_w, \varPhi_b$	One $k \times k$
		One $m \times m$ $(m \leqslant k-1)$
Huang' method	$S_w, S_b, \varPhi_t$	One $N \times N$, One $(N-1) \times (N-1)$
		One $(k-1) \times (k-1)$
2SRDA	$\varPhi_w, \varPhi_b$	One $N \times N$
		One $k \times k$

In a word, the 2SRDA method not only gives better performance, but also has good computational efficiency.

19.3 Non-Linear Feature Extraction Methods

The variations of pose and illumination are the great challenges in face recognition. It causes that the distribution of facial images is very complicated in the input feature space. Under this situation, linear classifiers will not give good performance. In order to enhance the nonlinear feature extraction ability, kernel method is employed. The basic idea is to transform the input feature space into a high dimensional feature space with a nonlinear mapping $\varPhi : x \in R^d \to \varPhi(x) \in F$ such that the mapped feature space F is linearly separable, and then linear approaches can be conducted. Actually, in all kernel based algorithms, the nonlinear mapping $\varPhi$ only appears in the inner product $\langle \varPhi(x_i), \varPhi(x_j) \rangle$ which can be replaced by a Mercer kernel function $K(x_i, x_j)$, i.e., $K(x_i, x_j) = \langle \varPhi(x_i), \varPhi(x_j) \rangle$. The theoretical justification of kernel method is based on Mercer kernel theory [47]. This section will discuss Mercer kernel construction, kernel parameter selection and their applications to face recognition.

19.3.1 Notations

Let the dimensionality of original sample feature space be d and the number of sample classes be C, the total original sample $X = \{X_1, X_2, \ldots, X_C\}$, the jth class X_j contains N_j samples, namely $X_j = \{x_1^j, x_2^j, \ldots, x_{N_j}^j\}$, $j = 1, 2, \ldots, C$, N be the total number of original training samples, and then $N = \sum_{j=1}^{C} N_j$.

If $\Phi : x \in R^d \to \Phi(x) \in F$ is the kernel nonlinear mapping, where F is the mapped feature space and df is the dimension of F, the total mapped sample set and the jth mapped class are given by $\Phi(X) = \{\Phi(X_1), \Phi(X_2), \ldots, \Phi(X_C)\}$ and $\Phi(X_j) = \{\Phi(x_1^j), \Phi(x_2^j), \ldots, \Phi(x_{N_j}^j)\}$ respectively. The mean of the mapped sample class $\Phi(X_j)$ and the global mean of the total mapped sample $\Phi(X)$ are given by $m_j = \dfrac{1}{N_j} \sum_{x \in X_j} \Phi(x)$ and $m = \dfrac{1}{N} \sum_{j=1}^{C} \sum_{x \in X_j} \Phi(x)$ respectively. The within-class scatter matrix S_w^{Φ} in F and between-class scatter matrix S_b^{Φ} in F are defined as

$$S_w^{\Phi} = \frac{1}{N} \sum_{j=1}^{C} \sum_{x \in X_j} (\Phi(x) - m_j)(\Phi(x) - m_j)^{\mathrm{T}} = \Phi_w \Phi_w^{\mathrm{T}},$$

$$S_b^{\Phi} = \frac{1}{N} \sum_{j=1}^{C} N_j (m_j - m)(m_j - m)^{\mathrm{T}} = \Phi_b \Phi_b^{\mathrm{T}},$$

where $\Phi_w = [\Phi_{wj}^i]_{i=1,\ldots,C, j=1,\ldots,N_i}$ is a df by N matrix and $\Phi_b = [\Phi_b^i]_{i=1,\ldots,C}$ is a df by C matrix.

19.3.2   Mercer Kernel Theory

The Mercer condition is essential to kernel design, as it is the key requirement for a unique global optimal solution to the kernel-based classifiers based on convex optimization [48].

    Theorem 19.3 (Mercer) [47]   Suppose χ is compact, $K(x, y) : \chi \times \chi \to R$ is symmetric continuous function, s.t. $T_K f = \int K(\cdot, x) f(x)\, dx$ is a positive semi-definite operator: $T_K \geqslant 0$, i.e.,

$$\iint_{\chi^2} K(x, y) f(x) f(y) dx dy \geqslant 0, \ \ for\ all\ f \in L^2(\chi), \tag{19.10}$$

then there exists an orthonormal feature basis of eigen-functions $\psi_j(x) \in L^2(\chi)$ associated with nonzero eigenvalues λ_j, such that:
    1) $\{\lambda_j\}_{j \in N} \in l^1$;
    2) $\psi_j(x) \in L^\infty(\chi)$ and $\sup_j \|\psi_j\|_{L^\infty} < \infty$;
    3) $K(x, y) = \sum_{j \in N} \lambda_j \psi_j(x) \psi_j(y)$, where the series converges absolutely and uniformly for almost all $(x, y) \in \chi \times \chi$.

Positive semi-definite condition (19.10) is called Mercer condition. Strictly speaking, $K(x, y)$ is a Mercer kernel if and only if it satisfies the Mercer condition. The condition (10) can be replaced by a discrete form as stated in Lemma 19.1.

Lemma 19.1 [49] If $K(x, y)$ is a symmetric function defined on $R^d \times R^d$, and for any finite data set $\{y_1, \ldots, y_m\} \subset R^d$, it always yields a symmetric and positive semi-definite matrix $\boldsymbol{K} = (k_{ij})_{m \times m}$, where $k_{ij} = k(y_i, y_j)$, $i, j = 1, 2, \ldots, m$, then function $K(x, y)$ is a Mercer kernel function.

19.3.3 Mercer Kernel Constructions

Gram matrix, also called kernel matrix, is a symmetric and positive semi-definite matrix and plays very important role in kernel based machine learning. The question is what kind of kernel function is good for kernel based classifier? It is natural to hope that the similarities determined by the kernel are higher among within-class samples and lower among between-class samples. However, the Gram matrices, which are computed by the commonly used kernels such as RBF/polynomial kernels on the training data, are full matrices. It means that the between-class data possibly have higher similarity and this leads to degrading the performance of kernel based learning methods. So, it is reasonable to think that such a kernel is a better kernel, if its Gram matrix generated from the the training data is a block diagonal matrix. This subsection will present a methodology to construct high performance Mercer kernel with interpolatory technique [26].

Cholesky Decomposition

Denote RBF kernel $K_{RBF}(x, y)$ by $K_{RBF}(x, y) = \exp\left(-\dfrac{\|x - y\|^2}{t}\right)$ with $t > 0$ and Define matrices $\boldsymbol{K}_i = (k_{jk}^i)_{N_i \times N_i} \in R^{N_i \times N_i}$, where $k_{jk}^i = K_{RBF}(x_j^i, x_k^i)$, $i = 1, 2, \ldots, C$. So, $\boldsymbol{K}_i(i = 1, 2, \ldots, C)$ all are symmetric and positive semi-definite matrices. If let

$$\boldsymbol{K} = \mathrm{diag}\{\boldsymbol{K}_1, \ldots, \boldsymbol{K}_C\} \in R^{N \times N}, \tag{19.11}$$

then $\boldsymbol{K}$ is a symmetric and positive semi-definite matrix as well.

Let matrix $\boldsymbol{K}$ be the matrix define by Eq. (19.11). Since submatrices $\boldsymbol{K}_i(i = 1, 2, \ldots, C)$ are generated by RBF kernel and thus are symmetric and positive semi-definite matrix.

By performing Cholesky decomposition on matrix $\boldsymbol{K}_i$, it yields that $\boldsymbol{K}_i = U_i^T U_i \in R^{N_i \times N_i}$, where U_i is a unique $N_i \times N_i$ upper triangular matrix. Denote that $U = \mathrm{diag}\{U_1, U_2, \ldots, U_C\} \in R^{N \times N}$, then U is also an upper triangular matrix. The Cholesky decomposition of matrix $\boldsymbol{K}$ can be written as $\boldsymbol{K} = U^T U \in R^{N \times N}$. We rewrite matrix U as $U = [u_1^1, \ldots, u_{N_1}^1 | u_1^2, \ldots, u_{N_2}^2 | \cdots$

$|u_1^C, \ldots, u_{N_C}^C]$, where $u_j^i \in R^N$ is the $\left(j + \sum_{k=1}^{i-1} N_k\right)$ column vector. Define nonlinear feature mapping $\varPhi$ on the training data set X as

$$\varPhi(x_j^i) = u_j^i, \tag{19.12}$$

where $j = 1, 2, \ldots, N_i$ and $i = 1, 2, \ldots, C$.

Interpolatory Strategy

By using interpolatory technique, this subsection will extend the nonlinear mapping $\varPhi(x)$ (see Eq. (19.12)), which is just well-defined on training sample set, to the whole input space. To this end, we define N Lagrange interpolatory basis functions $L_j^i(x)$ as

$$L_j^i(x) = \frac{\displaystyle\prod_{(u,v)\neq(i,j)} \|x - x_v^u\|_2}{\displaystyle\prod_{(u,v)\neq(i,j)} \|x_j^i - x_v^u\|_2}, \quad x \in R^d.$$

Apparently, above interpolatory basis functions satisfy the following property

$$L_j^i(x_v^u) = \begin{cases} 1, (u, v) = (i, j) \\ 0, (u, v) \neq (i, j) \end{cases}, \text{ for all } x_v^u \in X.$$

Therefore, the nonlinear mapping $\varPhi(x)$ can be extended to the whole input feature space R^d as follows:

$$\varPhi(x) = \sum_{i=1}^{C} \sum_{j=1}^{N_i} L_j^i(x) u_j^i. \tag{19.13}$$

Interpolatory Mercer Kernel Construction

Based on the nonlinear feature mapping defined in Eq. (19.13), we can construct the kernel function on $R^d \times R^d$ as follows:

$$K(x, y) = \langle \varphi(x), \varphi(y) \rangle$$

$$= \left\{ \sum_{i=1}^{C} \sum_{j=1}^{N_i} L_j^i(x) u_j^i \right\}^{\mathrm{T}} \cdot \left\{ \sum_{u=1}^{C} \sum_{v=1}^{N_u} L_v^u(y) u_v^u \right\}. \tag{19.14}$$

Obviously, a function $K(x, y)$ is a symmetric function. The following Theorem 19.4 demonstrates that above $K(x, y)$ is indeed a Mercer kernel function.

Theorem 19.4   Function $K(x, y)$ defined by Eq. (19.14) is a Mercer kernel function.

Proof. It just needs to show that $K(x, y)$ is a positive semi-definite function. To this end, we first denote a column vector $\boldsymbol{L}(x) \in R^N$ as follows:

$$\boldsymbol{L}(x) = [L_1^1(x), \ldots, L_{N_1}^1(x)|, \ldots, |L_1^C(x), \ldots, L_{N_C}^C(x)]^{\mathrm{T}},$$

then the function $K(x, y)$ can be written as

$$K(x, y) = \boldsymbol{L}(x)^{\mathrm{T}} \boldsymbol{K} \boldsymbol{L}(y).$$

For any finite training data set $\{x_l | l = 1, 2, \ldots, n\} \subset R^d$, the Gram matrix $\boldsymbol{G}$ generated by the kernel function $K(x, y)$ on this n training data set is $\boldsymbol{G} = [K(x_l, x_m)]_{n \times n}$, where $K(x_l, x_m) = \boldsymbol{L}(x_l)^{\mathrm{T}} \boldsymbol{K} \boldsymbol{L}(x_m), l, m = 1, 2, \ldots n$. Let $\boldsymbol{L}_n = [\boldsymbol{L}(x_l), \boldsymbol{L}(x_2), \ldots, \boldsymbol{L}(x_n)]_{N \times n}$, the Gram matrix $\boldsymbol{G}$ can be written as $\boldsymbol{G} = \boldsymbol{L}_n^{\mathrm{T}} \boldsymbol{K} \boldsymbol{L}_n$. Thereby, $\boldsymbol{G}$ is a symmetric matrix. As $\boldsymbol{K}$ is a positive semi-definite matrix, for all $\theta \in R^n$, we have

$$\theta^{\mathrm{T}} \boldsymbol{G} \theta = \theta^{\mathrm{T}} \boldsymbol{L}_n^{\mathrm{T}} \boldsymbol{K} \boldsymbol{L}_n \theta = (\boldsymbol{L}_n \theta)^{\mathrm{T}} \boldsymbol{K}(\boldsymbol{L}_n \theta) \geqslant 0.$$

It means that Gram matrix $\boldsymbol{G}$ is a positive semi-definite matrix. Hence by lemma 1, we know that $K(x, y)$ is a Mercer kernel.

It is not difficult to verify that the Gram matrix $\boldsymbol{G}_X$, which is generated by IM kernel Eq. (19.14) on the training data set X, is exactly the block diagonal positive semi-definite matrix $\boldsymbol{K}$. This indicates that the similarities among between-class data are zeros, while the similarities among within-class data are greater than zeros. Therefore, IM kernel is good for measuring the similarity between two samples and will enhance the the classification power of Kernel based machine learning approaches.

IM Kernel Based KDDA Algorithm

Based on analysis in above sections, IM-KDDA algorithm is designed as follows.

Step 1 Construct symmetric and positive semi-definite matrix

$$\boldsymbol{K} = \mathrm{diag}\{\boldsymbol{K}_1, \ldots, \boldsymbol{K}_C\} \in R^{N \times N},$$

where $\boldsymbol{K}_i = [K_{RBF}(x_j^i, x_k^i)]_{N_i \times N_i}, x_j^i, x_k^i \in X_i$, and $K_{RBF}(x_j^i, x_k^i) = \exp\left(\dfrac{-\|x_j^i - x_k^i\|^2}{t}\right)$.

Step 2 Let $\boldsymbol{L}(x) = [L_j^i(x)] \in R^{N \times 1}$, where $L_j^i(x)$ are the Lagrange intepolatory basis functions defined by

$$L_j^i(x) = \frac{\prod\limits_{(u,v) \neq (i,j)} \|x - x_v^u\|_2}{\prod\limits_{(u,v) \neq (i,j)} \|x_j^i - x_v^u\|_2}, \quad x_v^u \in X_u, x_j^i \in X_i.$$

Step 3 The interpolatory Mercer kernel is constructed as $K(x, y) = \boldsymbol{L}^{\mathrm{T}}(x) \boldsymbol{K} \boldsymbol{L}^{\mathrm{T}}(x)$.

Step 4 KDDA [24] with IM kernel is performed for face recognition.

Remark   In the above algorithm, if the value of some interpolatory basis function exceeds a given large threshold, then its value is set to zero.

Kernel Performance Comparisons

Two databases, namely FERET and CMU PIE databases, are selected to evaluate the performance of constructed IM kernel for kernel direct linear discriminant analysis algorithm.

Face Image Datasets

For FERET database, we select 120 people, 6 images for each individual. Face image variations in FERET database include pose, illumination, facial expression and aging. Images from one individual are shown in Fig. 19.7.

CMU PIE face database, includes totally 68 people. There are 13 pose variations ranged from full right profile image to full left profile image and 43 different lighting conditions, 21 flashes with ambient light on or off. In the experiments, for each people, we select 56 images including 13 poses with neutral expression and 43 different lighting conditions in frontal view. Several images of one people are shown in Fig. 19.7.

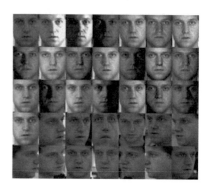

Fig. 19.7 Some images of one person on CMU PIE database.

Results on FERET Dataset

This section reports the results of IM-KDDA method on FERET database. We randomly select n ($n=2$ to 5) images from each people for training , while the rest $(6 - n)$ images of each individual are selected for testing. The experiments are repeated 10 times and the average accuracies are recorded in Table 19.4. It can be seen that the recognition rate of KDDA with IM kernel increases from 73.06% with training number 2 to 92.00% with training number 5, while the recognition accuracy of KDDA with RBF kernel increases from 69.13% with training number 2 to 91.50% with training number 5 respectively.

Comparing with KDDA with RBF kernel, KDDA with IM kernel gives around 2.81% entire mean accuracy improvement.

Table 19.4 Average accuracy of rank 1 versus Training Number (TN) on FERET database

TN	2	3	4	5
RBF-KDDA	69.13%	80.89%	89.17%	91.50%
IM-KDDA	73.06%	84.08%	90.33%	92.00%

Results on CMU PIE Dataset

The experimental setting on the CMU PIE database is similar with that of FERET database. As the number of images for each individual is 56, the number of training images is ranged from 5 to 10. The experiments are repeated 10 times and the average accuracy of KDDA with IM kernel is then calculated. The average accuracies are recorded and tabulated in the last row of Table 19.5. It can be seen from Table 19.5 that the recognition accuracy of IM-KDDA method increases from 86.03% with 5 training images to 94.26% with 10 training images.

Table 19.5 Average accuracy (%) of rank 1 versus Training Number on CMU PIE database.

TN	5	6	7	8	9	10
RBF-KDDA	67.51	68.11	70.79	72.34	72.74	72.91
IM-KDDA	86.03	89.15	90.78	92.16	88.16	94.26

The same experiments are implemented by using KDDA with RBF kernel function. The results are also recorded and tabulated in Table 19.5. It can be seen that when 5 images are used for training, the accuracy for KDDA with RBF kernel is 67.51%. When the number of training images is equal to 10, the accuracy for RBF kernel based KDDA increases to 72.91%. Comparing with RBF based KDDA method, KDDA with IM kernel gives around 19.36% entire average accuracy improvement.

In the 10 repeated experiments with training number 9, we found that the abnormal situations occurred in 2 times running, namely the value of some interpolatory basis function exceeds a given large threshold and probably attains infinite. So, we set its value to zero in practice. The 10 times mean accuracy with training number 9 is 88.16%. If excluding 2 abnormal cases, the mean accuracy of the rest 8 times running improves to 93.84%. The results are tabulated in Table 19.6. Comparing with RBF based KDDA method, KDDA with IM kernel gives around 20.30% entire mean accuracy improvement. It can be seen that IM kernel based KDDA approach gives the best performance for all cases.

Table 19.6 Average accuracy (%) on CMU PIE dataset, excluding 2 abnormal cases of training number 9.

TN	5	6	7	8	9	10
RBF-KDDA	67.51	68.11	70.79	72.34	72.74	72.91
IM-KDDA	86.03	89.15	90.78	92.16	93.84	94.26

19.3.4 Automatic Kernel Parameter Selections

This subsection will introduce a single parameter regularization KDA method (1PKRDA) [29] for face recognition. 1PKRDA consists of two parameters, namely the regularization parameter and kernel parameter. More interesting is that it automatically determines the optimal kernel parameter in RBF kernel and regularized parameter in within-class scatter matrix simultaneously based on the conjugate gradient method (CGM). Details are discussed as follows.

Regularized Kernel Discriminant Analysis

By the theory of linear algebra, it is easy to show that: $rank(S_w^\Phi) \leqslant \min(df, N - C)$ and $rank(S_b^\Phi) \leqslant \min(df, C - 1)$. The Fisher index $J_\Phi(w)$ in the feature space F is defined as

$$J_\Phi(w) = \frac{w^T S_b^\Phi w}{w^T S_w^\Phi w}, \tag{19.15}$$

where $w \in F$. LDA is used to find a projection in the feature space F that minimizes within-class distance and simultaneously maximizes between-class distance.

According to Mercer kernel function theory [50], any solutions $w \in F$ belong to the span of all training patterns in F. Hence there exists a group of constants $\{\tilde{w}_k^l\}_{1 \leqslant l \leqslant C, 1 \leqslant k \leqslant N_l}$ such that

$$w = \sum_{l=1}^{C} \sum_{k=1}^{N_l} \tilde{w}_k^l \, \Phi(x_k^l), \tag{19.16}$$

Substituting Eq. (19.16) into Eq. (19.15), the Fisher criterion function in the mapped feature space F can be written as follows,

$$J_\Phi(\tilde{w}) = \frac{\tilde{w}^T P_\Phi \tilde{w}}{\tilde{w}^T Q_\Phi \tilde{w}},$$

where $\tilde{w} = (\tilde{w}_k^l)_{1 \leqslant l \leqslant C, 1 \leqslant k \leqslant N_l} \in R^N, Q_\Phi = \frac{1}{N} Q_1 Q_1^T$ and $Q_1 = \tilde{K} \cdot (I - D_{C \times C} \cdot 1_{C \times C})$, $D_{C \times C}$ is a C by C block diagonal matrix, its jth $(1 \leqslant j \leqslant C)$ block is a diagonal matrix with diagonal elements all equal to $\frac{1}{N_j}$ and $1_{C \times C}$ also is a C by C block diagonal matrix, the jth block is a $N_j \times N_j$ diagonal matrix with terms all equal to 1.

$P_\Phi = \tilde{P}_1 \tilde{P}_1^T$, where $\tilde{P}_1 = \tilde{K} \cdot (1_C \cdot D_1 - 1_{N \times C} \cdot D_2) \cdot D_3$ and 1_C is a C by C block diagonal matrix, the jth block is a $N_j \times 1$ matrix with terms equal to 1. $1_{N \times C}$ is a N by C matrix with all terms equal to 1. $D_1 = diag\left(\frac{1}{N_1}, \ldots, \frac{1}{N_C}\right), D_2 = diag\left(\frac{1}{N}, \ldots, \frac{1}{N}\right)_{C \times C}, D_3 = diag\left(\sqrt{\frac{N_1}{N}}, \right.$

$$\sqrt{\frac{N_2}{N}}, \ldots, \sqrt{\frac{N_C}{N}}\Big)_{C \times C}, \text{ the kernel matrix } \tilde{K} = \big(K(x_j^i, x_k^l)\big)_{\substack{i,j=1,2,\ldots,C; \\ l,k=1,2,\ldots,C}}, \text{ and}$$
$$K(x_j^i, x_k^l) = \Phi(x_j^i)^{\mathrm{T}} \cdot \Phi(x_k^l).$$

After finding the solution $\tilde{w}^* = \arg\max_{\tilde{w}} J_\Phi(\tilde{w})$, the optimal projection vector can be obtained as $w^* = \sum_{l=1}^{C} \sum_{k=1}^{N_l} \tilde{w}_k^l \Phi(x_k^l)$. For any testing sample vector $x \in R^d$, the projection of x onto $w^* \in F$ is given by

$$(w^*)^{\mathrm{T}} \Phi(x) = \left(\sum_{l=1}^{C} \sum_{k=1}^{N_l} \tilde{w}_k^l \Phi(x_k^l)\right)^{\mathrm{T}} \Phi(x) = \sum_{l=1}^{C} \sum_{k=1}^{N_l} \tilde{w}_k^l K(x_k^l, x).$$

The problem $\tilde{w}^* = \arg\max_{\tilde{w}} J_\Phi(\tilde{w})$ is equivalent to the following eigenvalue problem:

$$(Q_\Phi^{-1} P_\Phi) W = W\Lambda,$$

where Λ is a diagonal eigenvalue matrix with its diagonal elements in decreasing order and W is an eigenvector matrix.

However, when 3S problem occurs, the traditional LDA method cannot be used directly. In turn, it proposes to regularize the matrix Q_Φ.

In designing the regularized matrix Q_Φ^R for the singular matrix Q_Φ, the criteria as suggested by Krzanowski etc [51] are used in this algorithm.

Assume $Q_\Phi = U_Q \Lambda_Q U_Q^{\mathrm{T}}$ is the eigenvalue decomposition of matrix Q_Φ. We define the two-parameter family regularization $Q_\Phi^{\alpha\beta}$ for Q_Φ as $Q_\Phi^{\alpha\beta} = U_Q \widehat{\Lambda}_Q U_Q^{\mathrm{T}}$, where $\widehat{\Lambda}_Q$ is a diagonal matrix with its diagonal elements ξ_i $(i = 1, 2, \ldots, d)$ given by

$$\xi_i = \begin{cases} (\lambda_i + \alpha)/M, & i = 1, 2, \ldots, \tau, \\ \beta, & i = \tau + 1, \ldots, N, \end{cases} \tag{19.17}$$

where M is a normalization constant and is given by

$$M = \frac{tr(Q_\Phi) + \tau\alpha}{tr(Q_\Phi) - (N - \tau)\beta},$$

with $\alpha \geqslant 0, \beta > 0$ and $(\xi_\tau + \alpha)/M - \beta \geqslant 0$. It is easy to verify that the regularized matrix $Q_\Phi^{\alpha\beta}$ satisfies all the criteria listed in [51].

Formulating Single Parameter Regularization

This subsection derives the single parameter formulation from the above defined two parameters regularization.

Denote $G = diag(I_\tau, 0) \in R^{N \times N}, \bar{G} = I_N - G, a = w^{\mathrm{T}} Q_\Phi w, b = w^{\mathrm{T}} U_Q G U_Q^{\mathrm{T}} w, c = w^{\mathrm{T}} U_Q \bar{G} U_Q^{\mathrm{T}} w, e = b + c = w^{\mathrm{T}} w$. Then the regularized Fisher index $J_\Phi^{\alpha\beta}(w) = \dfrac{w^{\mathrm{T}} P_\Phi w}{w^{\mathrm{T}} Q_\Phi^{\alpha\beta} w}, w \in R^N$ can be written as

$$J_\Phi^{\alpha\beta}(w) = \frac{(tr(Q_\Phi) + \tau\alpha)\, w^{\mathrm{T}} P_\Phi w}{(tr(Q_\Phi) - (N - \tau)\beta)\,(a + b\alpha) + c\beta\,(tr(Q_\Phi) + \tau\alpha)}.$$

Two optimal parameters α, β can be determined by solving equations $\nabla_{\alpha\beta} J_{\Phi}^{\alpha\beta}(w) = 0$, where ∇ is a gradient operator, as follows,

$$\alpha = \frac{tr(Q_{\Phi})c - (N - \tau)a}{bN - e\tau} \text{ and } \beta = \frac{tr(Q_{\Phi})}{N - \tau}.$$

On the other hand, we hope that above two parameters α, β can be reduced to one parameter t and when t tends to zero, the regularized matrix tends to the original matrix, i.e., $\alpha(t) \to 0$ and $\beta(t) \to 0$ as $t \to 0$. So it can slightly modify the above formula as

$$\alpha(t) = \left| \frac{tr(Q_{\Phi})c - (N - \tau)a}{bN - e\tau} \right| \cdot t, \beta(t) = \frac{tr(Q_{\Phi})}{N - \tau} \cdot t, \quad (0 < t < 1). \quad (19.18)$$

1PKRDA Algorithm

Based on above analysis, the 1PKRDA algorithm is designed as follows.

Step 1 Give a initial value $\Theta = (\theta, t)$ and $w \in R^N$, calculate matrices Q_{Φ} and $P_{\Phi} \in R^{N \times N}$.

Step 2 Perform eigenvalue decomposition $Q_{\Phi} = U_Q \Lambda_Q U_Q^T$, where

$$\Lambda_Q = diag(\lambda_1, \ldots, \lambda_{\tau}, 0, \ldots, 0) \in R^{N \times N}, \lambda_1 > \lambda_2 > \ldots > \lambda_{\tau} > 0.$$

Step 3 Calculate α, β defined in (18) and ξ_i $(i = 1, 2, \ldots, N)$ defined in Eq. (19.17).

Step 4 Let $Y = \hat{\Lambda}_Q^{-1/2} U_Q^T$ and $\hat{P}_{\Phi} = Y P_{\Phi} Y^T$, where $\hat{\Lambda}_Q = diag(\xi_1, \ldots, \xi_N)$, then conduct eigenvalue decomposition $\hat{P}_{\Phi} = V \Lambda_P V^T$, where Λ_P is a diagonal eigenvalue matrix with its diagonal elements in decreasing order and V is an orthogonal eigenvector matrix. Rewrite $V = (v_1, \ldots, v_{C-1}, \ldots, v_N)$ and let $V_{C-1} = (v_1, v_2, \ldots, v_{C-1}) \in R^{N \times (C-1)}$.

Step 5 Calculate matrix $\tilde{W} = U_Q \hat{\Lambda}_Q^{-1/2} V_{C-1} = [\tilde{w}_1, \tilde{w}_2, \ldots, \tilde{w}_{C-1}]$, where $\tilde{w}_j = (\tilde{w}_{jk}^l)_{1 \leqslant l \leqslant C, 1 \leqslant k \leqslant N_j} \in R^N, 1 \leqslant j \leqslant C - 1$, and let $w_j = \sum_{l=1}^{C} \sum_{k=1}^{N_l} \tilde{w}_{jk}^l \Phi(x_k^l)$. Therefore, the optimal projection matrix $W_{1PKRDA} = [w_1, w_2, \ldots, w_{C-1}]$.

Optimal Parameter Determination

In this subsection, the conjugate gradient method (CGM) will be exploited to determine the optimal parameters. The detail CGM algorithm is given as follows.

1) Give initial value $\Theta_1 = (\theta_1, t_1)$ and $w_0 \in R^N$, calculate matrices $Q_{\Phi}^{(1)}, P_{\Phi}^{(1)}$, via 1PKRDA algorithm to get w_1.

2) Compute searching direction: $S_1 = \nabla J(\Theta_1, w_1)$, let $\hat{S}_1 = S_1 / \|S_1\|$.

3) For $k \geqslant 1, \Theta_{k+1} = \Theta_k + \rho_k \cdot \hat{S}_k$, where $\hat{S}_k = S_k / \|S_k\|$, where $S_k = \nabla J(\Theta_k, w_k) + v_{k-1} \cdot S_{k-1}$ and

$$v_{k-1} = \|\nabla J(\Theta_k, w_k)\|^2 / \|\nabla J(\Theta_{k-1}, w_{k-1})\|^2.$$

4) Calculate matrices $Q_\Phi^{(k+1)}$ and $P_\Phi^{(k+1)}$, via 1PKRDA algorithm to obtain w_{k+1}. If $J(\Theta_{k+1}, w_{k+1}) < J(\Theta_k, w_k)$, then go to step 3 to search the next points.

5) The CGM iterative procedure of conjugate gradient method will terminate while $t < 0$ or $(\xi_\tau + \alpha)/M - \beta < 0$.

19.3.4.1   Experimental Results

To test 1PKRDA method, an in-depth investigation of the influence on performance of pose and illumination variations is performed on YaleB database in this subsection. we select Gaussian RBF kernel as $K(x, y) = \exp(-2^{-1}\theta^2\|x - y\|^2)$.

The YaleB database contains 5850 source images of 10 subjects each seen under 585 viewing conditions (9 poses × 65 illumination conditions). In the experiments, it only uses images under 45 illumination conditions and these images has been divided into four subsets according to the angle the light source direction makes with the camera axis [52]. Some images from one individual are shown in Fig. 19.8.

Fig. 19.8 Images of one person from *Yale B* database.

Fixed Pose with Illumination Variations

This part will investigate the influence of illumination variations upon performance of LDA-based face recognition algorithm. Results of fixing the pose and testing the illumination variations are shown in Fig. 19.9 (a) – (b).

For each pose, we randomly select 2 images from each subset for training ($2 \times 4 = 8$ images for training per individual), and all the other images from the 4 subsets are selected for testing (37 images for testing per individual). The experiments are repeated 10 times and the average accuracies of rank 1 are shown in the Fig. 19.9 (a). The mean accuracies of rank 1 to rank 3 of all poses are plotted in Fig. 19.9 (b). In the CGM iterative procedure, the initial values of parameters are given as: $t = 0.005, \theta = 0.045$, the step length $\rho = 0.000\,125$ and $w_0 = ones(N, 1) \in R^N$.

From the results shown in Fig. 19.9 (a), we can see that, the performance of 1PKRDA method outperforms other five methods under all illumination variations except that Kernel Direct LDA is slightly better than 1PKRDA method in pose 3 and K1PRFD [28] is a little better than 1PKRDA method in cases of pose 1, 4, 6, and 7.

From the results shown in Fig. 19.9 (b), it can been seen that the recognition accuracy of 1PKRDA method [29] increases from 88.92% (rank1) to

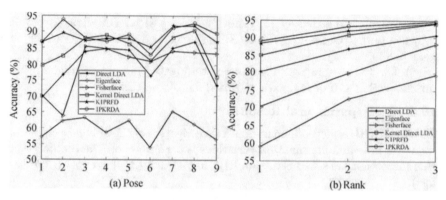

Fig. 19.9 For each pose, randomly select 2 images from each subset for training, and all the other images from the 4 subsets are selected for testing.

94.49% (rank3). The recognition accuracies of Eigenface [2], Fisherface [3], direct LDA [5], Kernel direct LDA [24] and K1PRFD [28] increase from 58.98%, 70.42%, 80.39%, 84.86% and 88.37% (rank1) to 79.43%, 88.2%, 90.63%, 94.11% and 94.17% (rank3) respectively. The results show that the 1PKRDA method surpasses other five methods as well.

Both Pose and Illumination Variations

This part will select the training samples including pose and illumination variations. The initial values of parameters are given as: $t = 0.005, \theta = 0.045$, the step length $\rho = 0.001\,25$ and $w_0 = ones(N, 1) \in R^N$. The experimental setting are as follows.

For each pose, it selects 2 images from each illumination subset of 4 subsets in all. This is to say that 720 images (10 persons × 9 poses × 4 subsets × 2 images) are randomly selected for training. Then the rest images, say 3330 images (10 persons × 9 poses × 37 images), are for testing. The experiments are repeated 10 times and the average rank1 to rank 3 accuracies are recorded and shown in Fig. 19.10. The results show that the recognition accuracy of 1PKRDA method increases from 90.90% with rank 1 to 96.13% with rank 3. The results show that the 1PKRDA method outperforms the other five methods except that K1PRFD method [44] is very close to 1PKRDA method.

Finally, the experiments are conducted one time to demonstrate the CGM iterative procedure. For fixed pose 2 with illumination variation case, the CGM starts from the initial values $\theta_0 = 0.029\,5, t_0 = 0.003\,9$, step length $= 0.001\,25$ and $w_0 = ones(N, 1) \in R^N$. The CGM iterative procedure does not terminate at the iterative number 7 until the regularized parameter $t_7 = -0.0001 < 0$. The results show that the rank1 accuracy increases from 86.49% with $\theta_1 = 0.029, t_1 = 0.002\,7$ to 90.27% with the final optimal parameter values $\theta_6 = 0.037\,3, t_6 = 0.000\,4$. The regularized parameter θ and the kernel parameter t versus Rank1 accuracy are recorded and plotted in Fig. 19.11 (a) and (b), respectively.

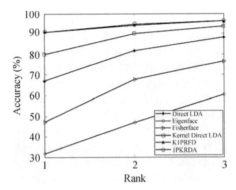

Fig. 19.10 Performance evaluation on pose and illumination.

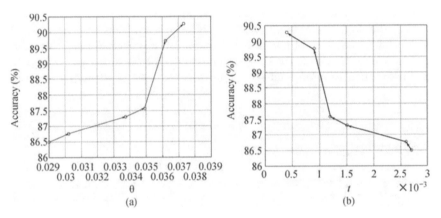

Fig. 19.11 Initial values $\theta_0 = 0.029\ 5$, $t_0 = 0.003\ 9$, step length$=0.001\ 25$ and $w_0 = ones(N, 1) \in R^N$. (a) The regularized parameter θ. (b) The kernel parameter t versus Rank1 accuracy.

19.4 Conclusions

This chapter discussed challenge issues encountered in face recognition. Some traditional and state of the art linear and nonlinear feature extraction algorithms are introduced to face recognition community.

Over the past decade, a number of appearance-based linear feature extraction approaches has been developed and successfully applied to face recognition. Among them, LDA is a commonly used statistic linear method, which has more discriminant power. However, LDA usually suffers from 3S problem in face recognition. To tackle 3S problem, subspace-based and regularized-based approaches has been presented and reported in many literatures ([3 – 17]). It may be still worthwhile to give an in-depth investigation on this problem.

Another problem in FR is the so-called nonlinear problem caused by pose and illumination variations and so on. In such a case, the performance of linear-based approaches will degrade. Kernel method, based on Mercer kernel theory, is a promising nonlinear feature extraction technology and a number of kernel discriminant analysis (KDA) approaches have been proposed to deal with nonlinear problem [19 – 31]. In spite of kernel method achieving better performance, most existing KDA-based research works on face recognition focused on the experimental results with limited theoretical justifications on the kernel selection and kernel parameters estimation. On the other hand, most of the researchers pointed out that different kernels and kernel parameters affect the performance of kernel based classifiers significantly. In addition, there also exists the computational complexity problem for kernel machine learning methods. Therefore, the future research directions on KDA based FR should include the optimal kernel design, automatically kernel parameter determination and low computational complexity etc.

In summary, the existing problems of face recognition has gained significant attention and encouraging results has been reported in many literatures during the past decade. More efforts should still be made to improve the performance of FR algorithms/systems under uncontrolled environment conditions.

Acknowledgements

This work was partially supported by NSF of China Grant (60873168, 60873092) and Science & Technology Planning Project of Shenzhen City (JC200903130300A, JYC200903250175A) and the Opening Project of Guangdong Province Key Laboratory of Computational Science of Sun Yat-Sen University. We would like to thank Olivetti Research Laboratory in Cambridge UK for providing the ORL face database, Amy Research Laboratory for the FERET face database and Yale University for contribution of the Yale Group B database.

References

[1]   Brunelli R, Poggio T (1993) Face Recognition: Feature Versus Templates. IEEE T PAMI, 15(10): 1042 – 1052
[2]   Turk M and Pentland A (1991) Eigenfaces For Recognition. J Cog Neuroscience, 3(1): 71 – 86
[3]   Belhumeur P N, Hespanha J P, Kriegman D J (1997) Eigenfaces vs. Fisherfaces: Recognition Using Class Specific Linear Projection. IEEE T PAMI, 19(7): 711 – 720
[4]   Chen L, Liao H, Ko M et al (2000) A New LDA-based Face Recognition System, Which Can Solve the Small Sample Size Problem. Patt Recog, 33(10): 1713 – 1726
[5]   Yu H and Yang J (2001) A Direct LDA Algorithm for High-dimensional Data-with Application to Face Recognition. Patt Recog, 34(10): 2067 – 2070
[6]   Jin Z, Yang J Y, Hu Z S et al (2001) Face Recognition Based on the Uncorrelated Discriminant Transform. Patt Recog, 34(7): 1405 – 1416
[7]   Zhao W, Chellappa R, Phillips P J (1999) Subspace Linear Discriminant Analysis For Face Recognition. Technical Report CAR-TR-914, CS-TR-4009, University of Maryland at College Park, USA

[8]   Yang J, Yang J Y, Zhang D (2002) What's Wrong With Fisher Criterion. Patt Recog, 35(11): 2665 – 2668

[9]   Huang R, Liu Q, Lu H et al (2002) Solving Small Sample Size Problem of LDA. Proceeding of International Conference in Pattern Recognition, 3: 29 – 32

[10] Lu J, Plataniotis K N, Venetsanopoulos A N (2003) Face Recognition Using LDA Based Algorithms. IEEE T Neural Networks, 14(1): 195 – 200

[11] Dai D Q, Yuen P C (2003) Regularized Discriminant Analysis and its Application on Face Recognition. Patt Recog, 36(3): 845 – 847

[12] Chen W S, Yuen P C, Huang J (2005) A New Regularized Linear Discriminant Analysis Methods to Solve Small Sample Size Problems. Int J Patt Recog & Artif Intel, 19(7): 917 – 936

[13] Huang J, Yuen P C, Chen W S et al (2005) Component-based Subspacec Linear Discriminant Analysis Method for Recognition of Face Images With one Training Sample. Opt Eng, 44(5): 057002

[14] Chen W S, Yuen P C, Huang J et al (2006) Two-Step Single Parameter Regularization Fisher Discriminant Method for Face Recognition. Int J Patt Recog & Artif Intel, 20(2): 189 – 208

[15] Zuo W, Zhang D, Yang J et al (2006) BDPCA plus LDA: a Novel Fast Feature Extraction Technique for Face Recognition. IEEE T SMC-B, 36(4): 946 – 953

[16] Wang H Y, Wang Z F, Xiao Y L et al (2007) PCA Plus F-LDA: A New Approach to Face Recognition. Int J Patt Recog & Artif Intel, 21(6): 1059 – 1068

[17] Kyperountas M, Tefas A, Pitas I (2007) Weighted Piecewise LDA for Solving the Small Sample Size Problem in Face Verification. IEEE T Neural Networks, 18(2): 506 – 519

[18] Phillips P J, Moon H, Rizvi S A et al (2000) The FERET Evaluation Methodology for Face Recognition Algorithms. IEEE T PAMI, 22(10): 1090 – 1104

[19] Yang M H (2001) Face Recognition Using Kernel Methods. Advances in Neural Information Processing Systems. MIT Press, 13: 960 – 966

[20] Yang M H (2002) Kernel Eigenfaces vs Kernel Fisherfaces: Face Recognition Using Kernel Methods. Proceedings of the Fifth IEEE International Conference on Automatic Face and Gesture Recognition (FG2002): 215 – 220

[21] Gupta H, Agrawal A K, Pruthi T et al (2002) An Experiment Evaluation of Linear and Kernel-based Methods for Face Recognition. Proceedings of IEEE Workshop on Applications on Computer Vision, December 2002

[22] Yang J, Frangi A F, Yang J Y et al (2005) KPCA plus LDA: A Complete Kernel Fisher Discriminant Framework for Feature Extraction and Recognition. IEEE T PAMI, 27(2): 230 – 244

[23] Liu Q S, L, Lu H Q, Ma S D (2004) Improving Kernel Fisher Discriminant Analysis for Face Recognition. IEEE T Circ Syst Vid, 14(1): 42 – 49

[24] Lu J, Plataniotis K N, Ventsanopoulos A N (2003) Face Recognition Using Kernel Direct Discriminant Analysis. IEEE T Neural Network, 14: 117 – 126

[25] Baudat G, Anouar F (2000) Generalized Discriminant Analysis Using a Kernel Approach. Neural Comput, 12(10): 2385 – 2404

[26] Chen W S, Yuen P C (2009) Interpolatory Mercer Kernel Construction for Kernel Direct LDA on Face Recognition. IEEE International Conference on Acoustics, Speech and Signal Processing(ICASSP2009), pp 857 – 860

[27] Chen W S, Yuen P C, Huang J et al (2006) Wavelet Kernel Construction in Kernel Discriminant Analysis for Face Recognition. Proceedings of the 2006 Conference on Computer Vision and Pattern Recognition Workshop (CVPRW'06), pp 47 – 47

[28] Chen W S, Yuen P C, Huang J et al (2005) Kernel Machine-based One-parameter Regularized Fisher Discriminant Method for Face Recognition. IEEE Transactions on System, Man and Cybernetics, Part B 35, 659 – 669

[29] Chen W S, Yuen P C, Huang J et al (2005) A Novel One-parameter Regularized Kernel Fisher Discriminant Method for Face Recognition. Lecture Notes in Computer Science, Springer, 353: 67 – 74

[30] Huang J, Yuen P C, Chen W S et al (2004) Kernel Subspace LDA with Optimized Kernel Parameters on Face Recognition. Proceeding of The Sixth International Conference on Automatic Face and Gesture Recognition (FG04), pp 327 – 332

[31] Huang J, Yuen P C, Chen W S et al (2007) Choosing Parameters of Kernel Subspace-lda for Recognition of Face Images Under Pose and Illumination Variations. IEEE T SMC-B, 37(4): 847 – 862

[32] You X G, Chen Q H, Wang P S P et al (2007) Nontensor-product-wavelet-based Facial Feature Representation. In: Yanushkevich S, Gavrilova M, Wang P S P (eds)

Image Pattern Recognition: Synthesis and Analysis in Biometrics, pp 207 – 224, World Scientific, Singapore

[33] You X G, Zhang D, Chen Q H et al (2006) Face Representation By Using Non-tensor Product Wavelets. Proc ICPR2006, Hong Kong, China, pp 503 – 506

[34] He Z, You X G, Tang Y Y et al (2006) Texture Image Retrieval Using Novel Nonseparable Filter Banks Based on Centrally Symmetric Matrices. Proc ICPR2006, Hong Kong, China, pp 161 – 164

[35] Lee S W, Wang P S P, Yanushkevich S et al (2008) Noniterative 3D Face Recognition Based on Photometric Stereo. Int J Patt Recog & Artif Intell, 22(3): 389 – 410

[36] Shih F Y, Chuang C F, Wang P S P (2008) Performance Comparisons of Facial Expression Recognition in JAFFE Database. Int J Patt Recog & Artif Intell, 22(3): 445 – 459

[37] Shih F Y, Cheng S X, Chuang C F et al (2008) Extracting Faces and Facial Features from Color Images. Int J Patt Recog & Artif Intell, 22(3): 515 – 534

[38] Luo Y, Gavrilova M L, Wang P S P (2008) Facial Metamorphosis Using Geometrical Methods for Biometric Applications. Int J Patt Recog & Artif Intell, 22(3): 555 – 584

[39] Chen X, Zhang J S (2010) Maximum Variance Difference Based Embedding Approach for Facial Feature Extraction. Int J Patt Recog, Artif Intell, 24(7): 1047 – 1060

[40] Grgic M, Shan S, Lukac R et al (2009) Facial Image Processing and Analysis. Int J Patt Recog & Artif Intell, 23(3): 355 – 358

[41] Chellappa R, Wilson C, Stacks et al (1995) Human and Machine Recognition of Faces: A Survey. IEEE J PROC, 83(5): 705 – 740

[42] Zhao W, Chellappa R, Rosenfeld A et al (2003) Face Recognition: A Literature Survey. ACM Computing Survey, 35: 399 – 458

[43] Mika S, Rätsch G, Müller KR (2001) A Mathematical Programming Approach to the Kernel Fisher Algorithm. Adv in Neural Inf Proc Syst, 13: 591 – 597

[44] Mika, S, Rätsch G, Weston J et al (1999) Fisher Discriminant Analysis with Kernels. Neural Networks for Signal Processing IX: 41 – 48

[45] Pierre Comon (1994) Independent Component Analysis: A New Concept. Elsevier, 36(3): 287 – 314

[46] Schölkopf B, Smola A, Müller KR (1998) Nonlinear Component Analysis as A Kernel Eigenvalue Problem. Neural Comput, 10: 1299 – 1319

[47] Mercer J (1909) Functions of Positive and Negative Type and Their Connection with the Theory of Integral Equations. Philos Trans Roy Soc London, A 209: 415 – 446

[48] Shawe-Taylor J, Cristianini N (2004) Kernel Methods for Pattern Analysis. Cambridge University Press, Cambridge

[49] Scholkopf B, Smola AJ (2002) Learning with Kernels-support Vector Machine, Regularization, Optimization, Beyond. MIT Press, Cambridge

[50] Saitoh S (1988) Theory of Reproducing Kernels and its Applications. Harlow England: Longman Scientific Technical, L

[51] Krzanowski W J, Jonathan P, McCarthy W V et al (1995) Discriminant Analysis With Singular Covariance Matrices: Methods and Applications to Spectroscopic data. Applied Statistics, 44: 101 – 115

[52] Georghiades A S, Belhumeur P N, Kriegman D J (2001) From Few to Many: Illumination Cone Models for Face Recognition Under Variable Lighting and Pose. IEEE PAMI, 23(6): 643 – 660

20 Facial Occlusion Reconstruction Using Direct Combined Model

Ching-Ting Tu[1,2] and Jenn-Jier James Lien[1]

Abstract   In real life, facial images are invariably occluded to a certain extent; therefore, the performance of automatic face recognition, facial expression analysis, and facial pose estimation schemes is inevitably degraded. Consequently, it is necessary to develop means to recover the occluded region(s) of the facial image such that the performance of these applications can be improved. An automatic occluded face recovery system based upon a novel learning algorithm called the direct combined model (DCM) approach is presented. The system comprises two basic DCM modules, namely a shape recovery module and a texture recovery module. Each module directly models the facial shapes (or facial textures) of occluded and non-occluded regions by DCM in order to extract the significant components of their correlations. When only the facial shape or texture information of non-occluded region is available, the optimal shape and texture of the occluded region can be recovered via the DCM transformation. To enhance the quality of the recovered results, the shape recovery module is rendered robust to facial feature point labeling errors by suppressing the effects of biased noises. Furthermore, the texture recovery module recovers the texture of the occluded facial image by synthesizing the global texture image and the local detailed texture image. The experimental results demonstrate that compared to existing facial recovery systems, the recovered results obtained using the proposed DCM-based scheme are quantitatively closer to the ground truth.

20.1 Introduction

The performance of automatic face recognition, facial expression analysis, face modeling, facial aging estimation or facial pose estimation schemes (e.g., Ref. [1–11]) is largely dependent upon the amount of information available in the input facial images. However, in real life, facial images are invariably occluded to a greater or lesser extent, and hence the performance of such schemes is inevitably degraded. It is necessary to develop the means to recover the occluded region(s) of the facial image such that the performance of these

---

1   Robotics Laboratory, Department of Computer Science and Information Engineering, Cheng Kung University, Tainan, Taiwan, China. E-mails: {vida, jjlien}@csie.ncku.edu. tw.

2   Department of Computer Science and Information Engineering, Tamkang University, Taiwan, China

applications can be improved.

Saito et al. [12] proposed a method for removing eyeglasses and reconstructing the facial image by applying principal component analysis (PCA) to eigenspaces having no eyeglass information. Similarly, Park et al. [13] removed eyeglasses from facial images by repainting the pixels in the occluded region of the image with the grayvalues of the corresponding region of the mean facial image prior to the PCA reconstruction process. However, in both studies, the reconstruction process was performed based upon eigenspaces derived from the entire facial image rather than from the occluded and non-occluded regions, respectively. As a result, the two schemes are capable only of reconstructing facial images with long and thin occluded regions, e.g., occlusion by a pair of eyeglasses. If the major facial features, e.g., the eyes or the nose, are occluded, the schemes yield highly unpredictable and unrealistic reconstruction results. Furthermore, the reconstructed images tend to be notably blurred since both schemes use the Gaussian-distributed PCA process to model the facial images, whereas such images typically have a non-Gaussian distribution. To resolve this problem, the facial reconstruction systems presented in Refs. [14 – 16] separated each facial image into its facial shape and facial texture, respectively, utilizing the face models introduced in Refs. [17 – 19]. In contrast to the iterative facial reconstruction process presented in [16], Hwang et al. [14, 15] proposed a non-iterative process for reconstructing the occluded region of an input face using facial shape and facial texture models. Each model consisted of one eigenspace and one sub-eigenspace, with the former containing the whole facial shape or texture information and the latter containing only the shape or texture information of the non-occluded region. In the proposed approach, the shape or texture information of the non-occluded region was reconstructed via a linear combination of the sub-eigenspace and the corresponding weight vector. The whole facial image was then reconstructed by applying the same weight vector to the whole-face eigenspace. However, the significant characters of the two eigenspaces are different, and thus inherent variances between two different subjects may be suppressed if the same weight vectors are applied to both.

In contrast to the methods described above, apply a Gaussian distributed PCA process, the patch-based non-parametric sampling methods presented in Refs. [20 – 22] synthesize facial images based upon local detailed features. In the psychological evaluations performed in Ref. [23], it was shown that facial features are correlated rather than independent. However, the localized characteristic of patch-based approaches results in a loss of information describing the overall geometric relationships between the individual facial features.

This study proposes a learning-based facial occlusion reconstruction system comprising two DCM modules, namely a shape reconstruction module and a texture reconstruction module. Adopting a similar approach to that used in Ref. [18], the proposed system normalizes the texture image by warping the facial image to the mean-shape coordinates. The DCM approach used in the two modules facilitates the direct analysis of the geometric and grayvalue correlations of the occluded and non-occluded regions of the face by coupling the shape and texture of the two regions within single shape and texture eigenspaces, respectively. Given the shape or texture of the non-occluded region of the face, the DCM modules enable the optimal shape or texture of the occluded region to be reconstructed even though the two regions of the face are modeled within a single eigenspace. In practice, the quality of the reconstructed facial shape is adversely effected by errors in the facial feature positions when labeling the features in the non-occluded region of the face. Consequently, the shape reconstruction module developed in this study is specifically designed to tolerate such misalignments by taking account of these noise sources. Furthermore, the quality of the texture reconstruction results is enhanced by synthesizing the global texture image, i.e., a smooth texture image containing the global geometric facial structure, and a local detailed texture image.

Previous Work

Developing low-dimensional linear models for high dimensional data representation is an important task in several computer vision applications. In particular, PCA is a popular and widely used technique. There are several variations of PCA developed for practical problems in computer vision applications: for example, the eigen-transformation method is proposed in [24] for facial sketch synthesis, the coupled PCA proposed is proposed in [25] for face hallucination, and in work [15], the least-square minimization is embedded in a morphable face model (a PCA process) for reconstruction of partially damaged face images. In these works, it is necessary to exploit correlations between two related classes, i.e., the pair of facial sketch and facial image, the pair of low-resolution and high-resolution images, or the pair of non-occluded and occluded textures in order to estimate one class from the other, i.e., the facial sketch from its input image, a high-resolution image from the given low resolution image, and a recovered texture of occluded region from the texture of non-occluded region. There are two general approaches for estimating one class from the other class; the first one (e.g., Refs. [8, 19, 26, 27]) separately learns two spaces for each class and further derives a regression matrix for the inter-correlation between them. However, each space does dimension reduction separately without considering their correlation; the significant features of their correlation are tended to be lost. Accordingly, Lin and Tang [28] proposed a learning method to construct two spaces for each class separately while maximizing their correlation. An alternative approach is to directly combine the two classes in one space for capturing their correlation (e.g., [18, 26, 29, 30]). When estimating one class from the other, the common approach is to project the input data (corresponding one class) into the corresponding subspace to obtain weights for synthesizing the corresponding result of the other class. However, each subspace is no longer orthogonal under such combination formulation. Significant features of the input data will disappear when projected into its subspace.

In this study, we propose a novel DCM algorithm that directly connects two related classes in one combined eigenspace in order to maximize their covariance. Subsequently, a robust DCM transform is derived directly based on the learned combined model. Such DCM transformation inherits the major properties of the combined space, i.e., minimizing pair-wise reconstruction error and extracting significant features of their correlation for the estimation tasks. For reconstruction of occluded faces, DCM is applied for modeling two types of pairs, i.e., the pair of facial shape of occluded and of non-occluded region and the pair of facial texture of these two regions. Such two types of DCMs are then used to carry out two major estimation tasks of the proposed reconstruction framework, inferring the facial shape/facial texture of the occluded regions from that of non-occluded regions.

20.2 Direct Combined Model Algorithm

The training process and testing process of DCM are introduced in the following subsections. During the training process, DCM is created to model the combined spaces which the significant features of two related classes will be extracted. In the testing process, only one class is observable while the other one is unknown. DCM transformation is derived for estimating the corresponding result of the observable class. Such transformation is based on the combined model, several advantages will be discussed in the following.

20.2.1 Combined Space Modeling

The DCM algorithm assumes the existence of two related classes, i.e., $\boldsymbol{X}$ and $\boldsymbol{Y}$. Given an observable (or known) vector $\boldsymbol{x} \in \boldsymbol{X}$, such as the shape or pixel grayvalues of the non-occluded facial region, the objective of the DCM modules developed in this study is to estimate (i.e., recover) the corresponding unobservable (or unknown) vector $\boldsymbol{y} \in \boldsymbol{Y}$, i.e., the shape or pixel grayvalues of the occluded region, based on a dataset with the structure $\{\boldsymbol{x}_i, \boldsymbol{y}_i\}_i^p$. In this dataset, the $\boldsymbol{x}$ feature vector ($\boldsymbol{x} \in \Re^m$) and the $\boldsymbol{y}$ feature vector ($\boldsymbol{y} \in \Re^n$) contain m and n elements, respectively; p is the total number of $(\boldsymbol{x}, \boldsymbol{y})$ feature pairs; and i is the ith $(\boldsymbol{x}, \boldsymbol{y})$ training pair of classes $\boldsymbol{X}$ and $\boldsymbol{Y}$.

As described in the following, the training dataset is fitted to a single joint Gaussian distribution using the PCA method. Let each training sample be represented by the pair $(\boldsymbol{x}, \boldsymbol{y})$. The training dataset can then be represented as an $(m + n)$ by p unbiased matrix, $(\boldsymbol{X}^\mathrm{T} \quad \boldsymbol{Y}^\mathrm{T})^\mathrm{T}$, in which each column corresponds to an unbiased, concatenated sample vector $((\boldsymbol{x} - \bar{\boldsymbol{x}})^\mathrm{T} (\boldsymbol{y} - \bar{\boldsymbol{y}})^\mathrm{T})^\mathrm{T}$, where $(\bar{\boldsymbol{x}}^\mathrm{T}, \bar{\boldsymbol{y}}^\mathrm{T})^\mathrm{T}$ is the mean vector of all p training $(\boldsymbol{x}, \boldsymbol{y})$ pairs within the dataset. Applying the singular value decomposition (SVD) process, the covariance matrix of the coupled training dataset can be expressed as

$$
\begin{bmatrix} X \\ Y \end{bmatrix} \begin{bmatrix} X \\ Y \end{bmatrix}^{\mathrm{T}} = \begin{bmatrix} U_X & U_\Delta \\ U_Y & \end{bmatrix} \begin{bmatrix} \Sigma_K^2 & 0 \\ 0 & \Sigma_\Delta^2 \end{bmatrix} \begin{bmatrix} U_X & U_\Delta \\ U_Y & \end{bmatrix}^{\mathrm{T}}
$$
$$
= \begin{bmatrix} U_X \Sigma_K^2 U_X^{\mathrm{T}} & U_X \Sigma_K^2 U_Y^{\mathrm{T}} \\ U_Y \Sigma_K^2 U_X^{\mathrm{T}} & U_Y \Sigma_K^2 U_Y^{\mathrm{T}} \end{bmatrix} + \left(U_\Delta \quad \Sigma_\Delta^2 U_\Delta^{\mathrm{T}} \right),
\tag{20.1}
$$

where U and Σ represent the eigenvector matrix and the eigenvalue matrix, respectively. According to basic PCA principles, the linear combination of the first $K(K \ll (m+n))$ eigenvectors, $[U_X^{\mathrm{T}} U_Y^{\mathrm{T}}]^{\mathrm{T}}$, sorted in decreasing order based on their corresponding eigenvalues, sufficiently represents all of the significant variances in the training dataset, and thus the remaining eigenvectors, U_Δ, can be discarded. Accordingly, the K-dimensional combined eigenspace, i.e., the DCM $[U_X^{\mathrm{T}} U_Y^{\mathrm{T}}]^{\mathrm{T}}$, can synthesize a new feature pair $(\hat{x}, \hat{y})$ by applying a process of linear combination using the corresponding K-dimensional weight vector, w, i.e.,

$$
\begin{bmatrix} x \\ y \end{bmatrix} \approx \begin{bmatrix} \hat{x} \\ \hat{y} \end{bmatrix} = \begin{bmatrix} U_X \\ U_Y \end{bmatrix} w + \begin{bmatrix} \bar{x} \\ \bar{y} \end{bmatrix},
\tag{20.2}
$$

where

$$
w = \begin{bmatrix} U_X \\ U_Y \end{bmatrix}^{\mathrm{T}} \left(\begin{bmatrix} x \\ y \end{bmatrix} - \begin{bmatrix} \bar{x} \\ \bar{y} \end{bmatrix} \right).
\tag{20.3}
$$

As shown, w is defined in terms of a set of parameters of the combined model. Furthermore, from Eq. (20.2), it is clear that specifying different element values in w will yield different coupled pair vectors $(\hat{x}, \hat{y})$.

20.2.2 Combined Model-Based Transformation

In the DCM reconstruction procedures performed in the reconstruction of occluded facial image, vector x (i.e., the grayvalues of non-occluded region in the facial texture image) is observable (or known) while the coupled vector y (i.e., the reconstructed grayvalues of occluded region in the facial texture image) is unobservable (or unknown). In general, a missing dimensionality problem arises when the combined input vector $[(x - \bar{x})^{\mathrm{T}} (y - \bar{y})^{\mathrm{T}}]^{\mathrm{T}}$ is projected onto the combined eigenspace $[U_X^{\mathrm{T}} U_Y^{\mathrm{T}}]^{\mathrm{T}}$. However, by exploiting a fundamental property of PCA, namely that a set of orthogonal eigenvectors can be found by minimizing the mean-square-error (MSE) between the input data and the corresponding reconstruction results, the scheme proposed in this study resolves this problem by using a minimum mean-square-error (MMSE) criterion.

Given x the MMSE estimator of y which satisfies the MMSE criterion has the form

$$
\begin{aligned}
\hat{y}(\boldsymbol{x}) &= \arg\min_{y} \iint (\boldsymbol{y} - \hat{\boldsymbol{y}})^2 P(\boldsymbol{x}, \boldsymbol{y}) \mathrm{d}\boldsymbol{x} \mathrm{d}\boldsymbol{y} \\
&= \arg\min_{y} \int P(\boldsymbol{x}) \int (\boldsymbol{y} - \hat{\boldsymbol{y}}(\boldsymbol{x}))^2 P_{\boldsymbol{Y}|\boldsymbol{X}}(\boldsymbol{Y}|\boldsymbol{X}) \mathrm{d}\boldsymbol{Y} \mathrm{d}\boldsymbol{X} \\
&= E[\boldsymbol{Y}|\boldsymbol{X} = \boldsymbol{x}] \\
&= \int_{\boldsymbol{Y}} \boldsymbol{y} P_{\boldsymbol{Y}|\boldsymbol{X}}(\boldsymbol{y}|\boldsymbol{x}) \mathrm{d}\boldsymbol{Y},
\end{aligned}
\tag{20.4}
$$

where $\hat{\boldsymbol{y}}$ is the estimated vector of $\boldsymbol{y}$; $P(\boldsymbol{x}, \boldsymbol{y})$ is the joint probability of the observable feature vector x and the unobservable feature vector $\boldsymbol{y}$; $\hat{\boldsymbol{y}}(\boldsymbol{x})$ is the estimated vector $\boldsymbol{y}$ for a given $\boldsymbol{x}$; $P_{\boldsymbol{Y}|\boldsymbol{X}}(\boldsymbol{y}|\boldsymbol{x})$ is the posterior probability; and $E[\boldsymbol{Y}|\boldsymbol{X} = \boldsymbol{x}]$ is the mean-square-error estimate of $\boldsymbol{Y}$ for a given $\boldsymbol{X} = \boldsymbol{x}$. According to the law of conditional probability, the MSE criterion, $\iint (\boldsymbol{y} - \hat{\boldsymbol{y}})^2 P(\boldsymbol{x}, \boldsymbol{y}) \mathrm{d}\boldsymbol{X} \mathrm{d}\boldsymbol{Y}$, can be rewritten in the form $\int P(\boldsymbol{x}) \int (\boldsymbol{y} - \hat{\boldsymbol{y}}(\boldsymbol{x}))^2 P_{\boldsymbol{Y}|\boldsymbol{X}}(\boldsymbol{y}|\boldsymbol{x}) \mathrm{d}\boldsymbol{Y} \mathrm{d}\boldsymbol{X}$. It can be seen that the inner integral, $\int (\boldsymbol{y} - \hat{\boldsymbol{y}}(\boldsymbol{x}))^2 P_{\boldsymbol{Y}|\boldsymbol{X}}(\boldsymbol{y}|\boldsymbol{x}) \mathrm{d}\boldsymbol{Y}$, is weighted by a positive value, $P(\boldsymbol{x})$. Therefore, minimizing this MSE criterion with respect to $\boldsymbol{y}$ is equivalent to minimizing the inner integral for each value of $\boldsymbol{x}$. In other words, the estimated vector $\hat{\boldsymbol{y}}(\boldsymbol{x})$ for a given vector $\boldsymbol{x}$ can be obtained by setting the derivative of the inner integral with respect to $\boldsymbol{y}$ equal to zero. Thus, the MMSE estimator of $\boldsymbol{y}$, i.e., Eq. (20.4), can be considered as the expected value or vector of the posterior probability of $\boldsymbol{y}$ for a given observation $\boldsymbol{x}$.

Applying the PCA process given in Eq. (20.1) under the assumption that the joint distribution of $\boldsymbol{X}$ and $\boldsymbol{Y}$ is a single joint Gaussian distribution, and utilizing the conditional expected value of vector $\boldsymbol{Y}$ obtained for a given vector $\boldsymbol{X}$, the MMSE estimator can be derived as

$$
\hat{y}(\boldsymbol{x}) = \bar{\boldsymbol{y}} + C_{\boldsymbol{YX}} C_{\boldsymbol{XX}}^{-1} (\boldsymbol{x} - \bar{\boldsymbol{x}}),
\tag{20.5}
$$

where $C_{\boldsymbol{YX}}$ is the cross-covariance matrix of $\boldsymbol{Y}$ and $\boldsymbol{X}$ and $C_{\boldsymbol{XX}}^{-1}$ represents the inverse covariance matrix of $\boldsymbol{X}$. Further, the conditional covariance matrix of such joint Gaussian is

$$
\sum_{\boldsymbol{Y}|\boldsymbol{X}=\boldsymbol{x}} = C_{\boldsymbol{YY}} - C_{\boldsymbol{YX}} C_{\boldsymbol{XX}}^{-1} C_{\boldsymbol{YX}}^{\mathrm{T}}.
\tag{20.6}
$$

Applying Eq. (20.1), the cross-covariance matrix, $C_{\boldsymbol{YX}}$, can be decomposed into $U_{\boldsymbol{Y}} \Sigma_K U_{\boldsymbol{X}}^{\mathrm{T}}$. Similarly, the covariance matrix, $C_{\boldsymbol{XX}}$, can be decomposed into $U_{\boldsymbol{X}} \Sigma_K U_{\boldsymbol{X}}^{\mathrm{T}}$, where $C_{\boldsymbol{XX}}$ is a square, invertible matrix. Integrating these decomposed terms with the MMSE estimator in Eq. (20.5) and then applying the Penrose condition [31], the MMSE estimator can be rewritten in the form

$$
\begin{aligned}
\hat{\boldsymbol{y}}(\boldsymbol{x}) &= C_{\boldsymbol{YX}} C_{\boldsymbol{XX}}^{-1} (\boldsymbol{x} - \bar{\boldsymbol{x}}) \\
&= (U_{\boldsymbol{Y}} \Sigma_K^{-1} U_{\boldsymbol{X}}^{\mathrm{T}})(U_{\boldsymbol{X}} \Sigma_K^{-1} U_{\boldsymbol{X}}^{\mathrm{T}})^{-1}(\boldsymbol{x} - \bar{\boldsymbol{x}}) \\
&= (U_{\boldsymbol{Y}} \Sigma_K^{-1} U_{\boldsymbol{X}}^{\mathrm{T}})(U_{\boldsymbol{X}}^{\mathrm{T}})^{-1}(\Sigma_K^{-1})^{-1} U_{\boldsymbol{X}}^{-1}(\boldsymbol{x} - \bar{\boldsymbol{x}}) \\
&= \bar{\boldsymbol{y}} + U_{\boldsymbol{Y}} U_{\boldsymbol{X}}^{\dagger}(\boldsymbol{x} - \bar{\boldsymbol{x}}) \hat{\boldsymbol{y}}(\boldsymbol{x}) = C_{\boldsymbol{YX}} C_{\boldsymbol{XX}}^{-1}(\boldsymbol{x} - \bar{\boldsymbol{x}}) \\
&= (U_{\boldsymbol{Y}} \Sigma_K^{-1} U_{\boldsymbol{X}}^{\mathrm{T}})(U_{\boldsymbol{X}} \Sigma_K^{-1} U_{\boldsymbol{X}}^{\mathrm{T}})^{-1}(\boldsymbol{x} - \bar{\boldsymbol{x}}) \\
&= (U_{\boldsymbol{Y}} \Sigma_K^{-1} U_{\boldsymbol{X}}^{\mathrm{T}})(U_{\boldsymbol{X}}^{\mathrm{T}})^{-1}(\Sigma_K^{-1})^{-1} U_{\boldsymbol{X}}^{-1}(\boldsymbol{x} - \bar{\boldsymbol{x}}) \\
&= \bar{\boldsymbol{y}} + U_{\boldsymbol{Y}} U_{\boldsymbol{X}}^{\dagger}(\boldsymbol{x} - \bar{\boldsymbol{x}}),
\end{aligned}
\tag{20.7}
$$

where $U_{\boldsymbol{X}}^{\dagger}$ is the right inverse matrix of $U_{\boldsymbol{X}}$, i.e., $U_{\boldsymbol{X}} U_{\boldsymbol{X}}^{\dagger} = I$, and is non-invertible. Further, the covariance of Gaussian distribution, i.e., Eq. (20.6), becomes

$$
\begin{aligned}
\Sigma_{\boldsymbol{Y}|\boldsymbol{X}=x} &= C_{\boldsymbol{YY}} - C_{\boldsymbol{YX}} C_{\boldsymbol{XX}}^{-1} C_{\boldsymbol{YX}}^{\mathrm{T}} \\
&= (U_{\boldsymbol{Y}} \Sigma_k^{-1} U_{\boldsymbol{Y}}^{\mathrm{T}}) - (U_{\boldsymbol{Y}} \Sigma_K^{-1} U_{\boldsymbol{X}}^{\mathrm{T}})(U_{\boldsymbol{X}} \Sigma_K^{-1} U_{\boldsymbol{X}}^{\mathrm{T}})^{-1}(U_{\boldsymbol{Y}} \Sigma_K^{-1} U_{\boldsymbol{X}}^{\mathrm{T}})^{\mathrm{T}} \\
&= (U_{\boldsymbol{Y}} \Sigma_k^{-1} U_{\boldsymbol{Y}}^{\mathrm{T}}) - U_{\boldsymbol{Y}} \Sigma_K^2 U_{\boldsymbol{X}}^{\mathrm{T}}(U_{\boldsymbol{X}}^{\mathrm{T}})^{-1}(\Sigma_K^2)^{-1}(U_{\boldsymbol{X}})^{-1} U_{\boldsymbol{X}}(\Sigma_K^2)^{\mathrm{T}} U_{\boldsymbol{Y}}^{\mathrm{T}} \\
&= (U_{\boldsymbol{Y}} \Sigma_k^{-1} U_{\boldsymbol{Y}}^{\mathrm{T}}) - U_{\boldsymbol{Y}} \Sigma_k^{-1} U_{\boldsymbol{Y}}^{\mathrm{T}} = 0,
\end{aligned}
\tag{20.8}
$$

which implies that, under the joint assumption, the transformation in Eq. (20.7) is the exact solution of the estimated $\boldsymbol{Y}$ given $\boldsymbol{X} = \boldsymbol{x}$.

Many previous studies (e.g., Refs [5, 13, 30]) have commented on the occurrence of the non-invertible problem during the inference process. In these studies, this problem was resolved by approximating the inverse of the non-invertible matrix indirectly using the formulation $U_{\boldsymbol{X}}^{\dagger} = \arg\min(I - U_{\boldsymbol{X}} U_{\boldsymbol{X}}^{\dagger})$, and will introduce an additional estimation error during this optimization procedure. Accordingly, the current study adopts an alternative approach (described in the following) to derive the inverse matrix $U_{\boldsymbol{X}}^{\dagger}$ directly in order to improve the accuracy of the synthesized results. Since $U_{\boldsymbol{X}}^{\mathrm{T}} I = U_{\boldsymbol{X}}^{\mathrm{T}}$ and $U_{\boldsymbol{X}}^{\dagger}$ is the right inverse matrix of $U_{\boldsymbol{X}}$, then

$$
U_{\boldsymbol{X}}^{\mathrm{T}}(U_{\boldsymbol{X}} U_{\boldsymbol{X}}^{\dagger}) = U_{\boldsymbol{X}}^{\mathrm{T}}.
\tag{20.9}
$$

In addition, the combined model, $[U_{\boldsymbol{X}}^{\mathrm{T}} U_{\boldsymbol{Y}}^{\mathrm{T}}]^{\mathrm{T}}$, is an orthonormal matrix, i.e., $U_{\boldsymbol{X}}^{\mathrm{T}} U_{\boldsymbol{X}} + U_{\boldsymbol{Y}}^{\mathrm{T}} U_{\boldsymbol{Y}} = I$ (or $U_{\boldsymbol{X}}^{\mathrm{T}} U_{\boldsymbol{X}} = I - U_{\boldsymbol{Y}}^{\mathrm{T}} U_{\boldsymbol{Y}}$), and thus Eq. (20.9) becomes

$$
(I - U_{\boldsymbol{Y}}^{\mathrm{T}} U_{\boldsymbol{Y}}) U_{\boldsymbol{X}}^{\dagger} = U_{\boldsymbol{X}}^{\mathrm{T}}.
\tag{20.10}
$$

Multiplying both sides by $U_{\boldsymbol{Y}}$ and then applying the associative rule, Eq. (20.10) becomes

$$
(I - U_{\boldsymbol{Y}} U_{\boldsymbol{Y}}^{\mathrm{T}}) U_{\boldsymbol{Y}} U_{\boldsymbol{X}}^{\dagger} = U_{\boldsymbol{Y}} U_{\boldsymbol{X}}^{\mathrm{T}}.
\tag{20.11}
$$

Here, if the square matrix $(I - U_{\boldsymbol{Y}} U_{\boldsymbol{Y}}^{\mathrm{T}})$ is invertible, then the $U_{\boldsymbol{Y}} U_{\boldsymbol{X}}^{\dagger}$ term in Eq. (20.7) can be expanded as

$$
(I - U_{\boldsymbol{Y}} U_{\boldsymbol{Y}}^{\mathrm{T}}) U_{\boldsymbol{Y}} U_{\boldsymbol{X}}^{\dagger} = U_{\boldsymbol{Y}} U_{\boldsymbol{X}}^{\mathrm{T}}.
\tag{20.12}
$$

The following check is then performed to verify whether or not the term $(I - U_Y U_Y^T)$ is invertible. Since $U_X^T U_X + U_Y^T U_Y = I$, if the SVD process is applied to the two $K \times K$ matrixes $U_X^T U_X$ and $U_Y^T U_Y$, respectively, $U_X^T U_X$ and $U_Y^T U_Y$ share a common eigenvector matrix V [11] and $U_X^T U_X + U_Y^T U_Y = I$ can be reformulated as

$$V \Sigma_Y^2 V^T + V \Sigma_X^2 V^T = V I V^T. \tag{20.13}$$

Furthermore, since X and Y are correlated, all the diagonal elements of the two $K \times K$ matrixes Σ_Y^2 and Σ_X^2 are greater than zero, but less than one. Consequently, the n eigenvalues of the $n \times n$ matrix $U_Y U_Y^T$ can be separated into two parts, namely: 1) the first K eigenvalues of the $n \times n$ matrix $U_Y U_Y^T$, which are identical to the K non-zero eigenvalues of the $K \times K$ matrix $U_Y^T U_Y$: $1 > \lambda_1 \geqslant \cdots \geqslant \lambda_K > 0$, where $\Sigma_Y^2 = \text{diag}(\lambda_1, \cdots, \lambda_K)$; and 2) the remaining $n - K$ eigenvalues of the $n \times n$ matrix $U_Y U_Y^T$, which are all zeros, i.e., $\lambda_{K+1} = \cdots = \lambda_n = 0$. Therefore, all n eigenvalues of the $n \times n$ matrix $I - U_Y U_Y^T$ are positive, i.e., $0 < \lambda_1' \leqslant \cdots \leqslant \lambda_K' < 1$ and $\lambda_{K+1}' = \cdots = \lambda_n' = 1$, where $\lambda_i' = 1 - \lambda_i$ for $i = 1, \cdots, n$. In other words, the $n \times n$ matrix $I - U_Y U_Y^T$ is non-singular (or invertible). Hence, it can be concluded that Eq. (20.12) does indeed exist. Therefore, replacing the term $U_Y U_X^\dagger$ in Eq. (20.7) with the expression given in Eq. (20.12), Eq. (20.7) can be rewritten as

$$\hat{y}(x) = \bar{y} + (I - U_Y U_Y^T)^{-1} U_Y U_X^T (x - \bar{x}), \tag{20.14}$$

which is the DCM transformation. Note that the DCM algorithm assumes the coupled classes X and Y to be statistically correlated since otherwise $U_Y U_X^T$ becomes a zero matrix.

20.2.3 Procedure of Direct Combined Model Learning

The DCM algorithm applied in the three modules of the facial feature location and sketch synthesis system proposed in this study can be summarized as follows:

1) Training process (based on Eq. (20.1)):

Construct the DCM (or eigenspace) $[U_X^T U_Y^T]^T$ by applying the SVD process to the covariance matrix of the combined unbiased matrix $[X^T Y^T]^T$.

2) Synthesizing process (based on the DCM transformation (20.14)):

(1) Project the input unbiased observation vector $x - \bar{x}$ onto the K-dimensional eigenvector matrix U_X to obtain the K-dimensional projection weight vector w'.

(2) Generate the initial synthesized result of the unknown vector y (a non-normalized vector) by multiplying the weight vector w' by the eigenvector matrix U_Y.

(3) Normalize the initial synthesized result of the unknown vector $\boldsymbol{y}$ by multiplying the result obtained in Step (2) by the inverse of the residual covariance matrix, $(I - U_{\boldsymbol{Y}} U_{\boldsymbol{Y}}^{\mathrm{T}})^{-1}$.

(4) Estimate the optimized result $\hat{\boldsymbol{y}}$ of the normalized synthesized vector by adding the mean vector $\bar{\boldsymbol{y}}$ to the result obtained in Step (3) in order to bias (or shift) the normalized synthesized result.

In summary, the PCA method provides a powerful approach for analyzing or extracting the significant features of a single class within a training dataset, and then synthesizing or reconstructing various results of this class based on a linear combination of these significant features. However, PCA is unsuitable for the processing of two related but different classes since the analysis results obtained from PCA are sensitive to differences in the units or variance, respectively, of the two classes. The combined AAM model [18] resolves the unit difference problem between the grayvalue model and the shape model by utilizing a normalization process. That is, the shape model is normalized by the ratio of the standard deviation of the intensity (i.e., the grayvalue) to the standard deviation of the shape before the combined eigenspace is built. By contrast, the DCM synthesizing process proposed in this study normalizes the concatenated feature vectors during the inference process, i.e., $(I - U_{\boldsymbol{Y}} U_{\boldsymbol{Y}}^{\mathrm{T}})^{-1}$ in Step (3) above. As described in the following, the DCM synthesizing process also resolves the problem of variance or scatter differences between the two different classes. In Eq. (20.5), the normalization term, $C_{\boldsymbol{XX}}^{-1}$, performs a similar role to that of the normalization process performed in the combined AAM model [18]. Moreover, the $U_{\boldsymbol{X}}^{\dagger}$ term in Eq. (20.7) obtained by applying the combined eigenspace to Eq. (20.5) can be considered as an enhanced normalization term since the significant features of the $\boldsymbol{X}$-covariance $C_{\boldsymbol{XX}}$ are extracted by translating them from the $\boldsymbol{X}$ space to the eigenspace. However, in Eq. (20.7), $U_{\boldsymbol{X}}^{\dagger}$, which is the right inverse matrix of $U_{\boldsymbol{X}}$, i.e., $U_{\boldsymbol{X}} U_{\boldsymbol{X}}^{+} = I$, is non-invertible. Therefore, by following the procedures described after Eq. (20.7) in the previous section, it is possible to derive the synthesizing function of the DCM algorithm, i.e., Eq. (20.14). This equation is capable of resolving the problem of variance or scale differences between different classes or feature vectors. That is, the inverse of the residual covariance matrix, i.e., $(I - U_{\boldsymbol{Y}} U_{\boldsymbol{Y}}^{\mathrm{T}})^{-1}$, utilized in Step (3) of the DCM algorithm can be considered as a normalization process for the initial synthesized vector $\boldsymbol{y}$ obtained in Step (2).

20.3 Reconstruction System

The proposed reconstruction system for partially-occluded facial images is based on a joint Gaussian distribution assumption. However, in practice, the distribution of facial images actually has the form of a complicated manifold in a high-dimensional space, and thus it is inappropriate to model this distribution using a Gaussian distribution model. To resolve this problem, the current system separates the facial shape and texture of each image, rendering both facial properties more suitable for modeling using a Gaussian approach.

As shown in Fig. 20.1, the proposed facial occlusion reconstruction system comprises two separate DCM modules, namely the shape reconstruction module and the texture reconstruction module. In the training process, the facial feature points of each facial image in the training set are manually labeled to generate the corresponding facial shape, S, and the mean facial shape coordinates, $\overline{S}$, are then derived. Thereafter, each facial texture image with facial shape coordinates S is warped to the mean facial shape $\overline{S}$ using a texture-warping transformation function W [18] to generate the corresponding normalized texture image T. The resulting facial shapes $\{S\}$ and canonical textures $\{T\}$ of the training images are then used in the shape and texture modules, respectively, as described in the following.

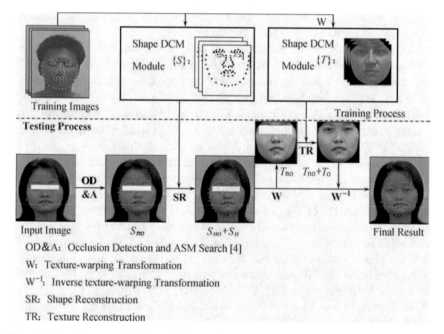

Fig. 20.1 Framework of partially-occluded facial image reconstruction system comprising shape reconstruction DCM module and texture reconstruction DCM module.

In the shape reconstruction module, the occluded region of the input image and the facial shape of the non-occluded region, i.e., S_{no}, are detected automatically using the method prescribed in Ref. [32]. The facial shape of the occluded region, i.e., S_o, is then reconstructed by the shape reconstruction DCM algorithm by applying a process of DCM transformation to the facial shape of the non-occluded region to give the complete facial shape, S. Meanwhile, in the texture reconstruction module, the input texture image of the non-occluded region is warped from its original shape coordinates S to the mean shape coordinates $\overline{S}$ using the transformation function W to generate the corresponding normalized texture image of the non-occluded region, i.e., T_{no}. The canonical facial texture of the occluded region, T_o, is then reconstructed from T_{no} using the texture reconstruction DCM algorithm. Finally, the complete canonical facial texture T (i.e., $T_{no} + T_o$) is warped from the mean facial shape coordinates $\overline{S}$ back to the original facial shape coordinates S in order to generate the final reconstruction result. The reconstruction process illustrated in Fig. 20.1 presents the particular case in which both eyes are occluded. However, due to the combined model approach, the reconstruction system developed in this study is capable of reconstructing frontal-view facial images containing occluded regions in other facial features, such as the nose and the mouth, without modeling an additional combined model.

20.3.1 Robustness of DCM Shape Reconstruction Module to Facial Feature Point Labeling Errors

As shown in Fig. 20.2, the DCM shape reconstruction module comprises a training process and a testing (or reconstruction) process. In the training process, a K-dimensional shape eigenspace is constructed based upon a total of p manually-labeled facial shapes S. The performance of the facial shape reconstruction module is highly dependent on the accuracy with which the individual facial feature points in the non-occluded region of the face image are identified. To improve the robustness of the shape reconstruction module, each training facial shape S is added by q number of random biased noises to generate a total of q biased facial shapes S'. The biased noise is randomly generated and is bounded by $\sigma_{cn}^2 (\Sigma_K^2 - \sigma_{cn}^2 I)^{-1}$ in accordance with the recommendations of the subspace sensitivity analyses presented in Ref. [11], where Σ_K is the matrix of the first K eigenvalues and σ_{cn}^2 is the norm of the covariance matrix of the expected residual vector based on the p training shape vectors S. Note that this residual vector is defined as the distance between the input facial shape and the corresponding reconstructed shape obtained using the K-dimensional shape eigenspace. A new facial shape eigenspace is then constructed based on the total of $p \times q$ facial shapes S'.

During the testing process, once the non-occluded and occluded regions of the input image have been detected and separated, the new facial shape eigenspace is rearranged according to the combined eigenspace formula of the DCM algorithm in Eq. (20.1), where the non-occluded part, $\boldsymbol{X}$, should be in the upper rows of the combined model, while the occluded part, $\boldsymbol{Y}$, should be in the lower rows. Importantly, rearranging the eigenspace has no effect on the reconstruction result since exchanging any two row vectors in the combined

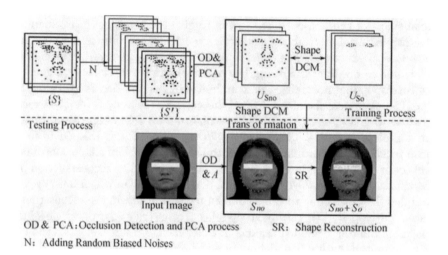

OD & PCA: Occlusion Detection and PCA process   SR: Shape Reconstruction

N: Adding Random Biased Noises

Fig. 20.2 Workflow of DCM shape reconstruction module.

eigenspace changes only their relative position in the eigenspace, i.e., the values of their elements are unchanged. Finally, the rearranged combined eigenspace is used to reconstruct the shape of the occluded region S_o by replacing x in Eq. (20.14) with S_{no}:

$$\hat{S}_o(S_{no}) = \bar{S}_o + (I - U_{S_o}U_{S_o}^{\mathrm{T}})^{-1}U_{S_o}U_{S_{no}}^{\mathrm{T}}(S_{no} - \bar{S}_{no}). \qquad (20.15)$$

20.3.2 Recovery of Global Structure and Local Detailed Texture Components Using DCM Texture Reconstruction Module

As shown in Fig. 20.3, the texture of an input image is reconstructed by integrating the global texture DCM and the local detailed texture DCM. The global texture image, i.e., T^g, is a smooth texture image containing the global geometric facial structure, while the local detailed texture image, i.e., T^l, represents the difference between T^g and the ground-truth texture image T, and contains the subtle details of the facial texture. The objective function of the DCM texture reconstruction module can be formulated as

$$\begin{aligned} T_o^g &= \arg\max P(T_{no}^g|T_o^g, \theta)P(T_o^g, \theta), \\ T_o^l &= \arg\max P(T_{no}^l|T_o^l, \theta)P(T_o^l, \theta), \end{aligned} \qquad (20.16)$$

where T_{no}^g and T_{no}^l are the global and local detailed texture components of T_{no}, respectively, and T_o^g and T_o^l are the global and local detailed texture components of T_o, respectively.

In the training process, the texture image training dataset, $\{T\}$, is used to construct a K-dimensional global texture eigenspace. The local detailed texture images of the training dataset $\{T^l\}$ are then calculated and used to

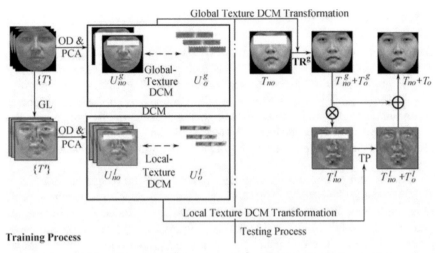

Fig. 20.3 Workflow of DCM texture reconstruction module.

construct the local detailed texture eigenspace. In this texture module, the local detailed texture image is derived by calculating the difference between its texture image, T, and the corresponding global texture image, T^g. In the testing process, the texture of the occluded region $T_o = T_o^g + T_o^l$ is inferred via the following procedure:

1) According to the occluded region and the non-occluded region of the input texture image, the global eigenspace and the local detailed eigenspace are rearranged using the combined eigenspace formula of DCM by SVD.

2) The global texture of the occluded region, i.e., T_o^g, is reconstructed using the global texture DCM from T_{no} by

$$\hat{T}_o^g(T_{no}) = \bar{T}_o^g + (I - U_{T_o^g} U_{T_o^g}^{\mathrm{T}})^{-1} U_{T_o^g} U_{T_{no}^g}^{\mathrm{T}} (T_{no} - \bar{T}_{no}^g). \qquad (20.17)$$

3) The image T', which contains the texture of the non-occluded region T_{no} and the current reconstruction result T_o^g, is projected onto the K-dimensional global texture eigenspace, and the corresponding projection weight is then used to reconstruct T''. The local detailed texture components of the non-occluded region are then extracted by calculating $T_{no}^l = T'' - T'$.

4) The local detailed texture of the occluded region T_o^l is reconstructed using the local detailed DCM from T_{no}^l by

$$\hat{T}_o^l(T_{no}^l) = \bar{T}_o^l + (I - U_{T_o^l} U_{T_o^l}^{\mathrm{T}})^{-1} U_{T_o^l} U_{T_{no}^l}^{\mathrm{T}} (T_{no}^l - \bar{T}_{no}^l). \qquad (20.18)$$

5) The final texture result is obtained by synthesizing T_{no} with the reconstruction results T_o^g and T_o, i.e., $T = T_{no} + T_o^g + T_o^l$.

20.4 Experimental Results

The performance of the proposed reconstruction system was evaluated by performing a series of experimental trials using training and testing databases comprising 205 and 60 facial images, respectively. The images were acquired using a commercial digital camera at different times, in various indoor environments. Eighty-four facial feature points were manually labeled on each training and testing facial image to represent the ground truth of the facial shape. Specific facial feature regions of the testing images were then occluded manually.

Figure 20.4 presents representative examples of the reconstruction results obtained using the proposed method for input images with a variety of occlusion conditions. Figure 20.4 (a) and (b) show the occluded facial images and the original facial images, respectively. Figure 20.4 (c) presents the reconstruction results obtained using the shape and global texture DCMs. Meanwhile, Fig. 20.4 (d) presents the reconstruction results obtained when the texture is reconstructed using not only the global texture DCM, but also the local detailed texture DSM. Comparing the images presented in Fig. 20.4 (d) with those presented in Fig. 20.4 (b), it can be seen that the use of the two tex-

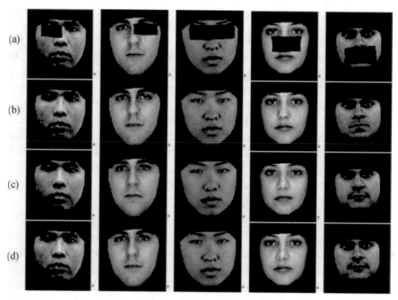

Fig. 20.4 Reconstruction results obtained using DCM method: (a) Occluded facial images. (b) Original facial images. (c) Reconstructed facial images using global texture DCM only. (d) Final reconstruction results using both global texture DCM and local texture DCM.

ture DCMs yields a highly accurate reconstruction of the original facial image. Table 20.1 presents the average reconstructed shape and texture errors computed over all the images in the testing database. In general, the results show

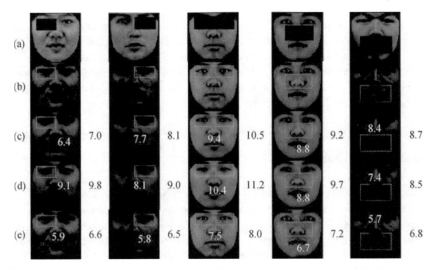

Fig. 20.5 Reconstruction results: (a) Occluded facial images. (b) Original facial images. (c) Reconstructed texture images using method presented in Ref. [13]. (d) Reconstructed texture images using method presented in Ref. [15]. (e) Reconstructed texture images using current DCM method. The digits within the images represent the average grayvalue evaluation error of the corresponding pixels in the original non-occluded image, while the digits in the columns next to these images represent the average grayvalue error over all of the images in the test database. Note that each facial texture image has a size of 100×100 pixels.

that the magnitudes of both errors increase as the level of occlusion increases or as the geometrical complexity of the occluded facial feature increases. Figure 20.5 compares the reconstruction results obtained using the proposed DCM method with those obtained using the occlusion recovery schemes presented in Refs. [13, 14], respectively. The data presented within the reconstructed images indicate the average difference between the grayvalues

Table 20.1 Average and standard deviation of facial shape and facial texture reconstruction errors for images in testing database with different levels of occlusion. Note that the occlusion rate data indicate the ratio of the occluded area to the non-occluded area in the facial image.

Facial features	Ave. Error (Pixel/Grayvalues)		Std. Dev. (Pixel/Grayvalues)		Occlusion Rate
	Shape	Texture	Shape	Texture	
Left Eye	1.2	6.6	1.1	1.7	10%
Right Eye	1.3	6.5	1.0	1.8	10%
Both Eye	1.4	8.0	1.7	3.6	24%
Nose	1.0	7.2	1.4	3.0	16%
Mouth	1.6	6.8	1.5	3.2	20%

Std. Dev: Standard deviation of errors.   Ave: Average

of the pixels in the restored region of the reconstructed image and the grayvalues of the corresponding pixels in the original non-occluded image, while the data in the columns next to these images indicate the average grayvalue error of the restored pixels in the occluded region computed over all 60 texture images within the test database. Overall, the results demonstrate that the images reconstructed by the current DCM-based method are closer to the original un-occluded facial images than those obtained using the schemes presented in Ref. [15] or Ref. [13].

20.5 Conclusions

This study has presented a facial occlusion reconstruction framework for facial shape and texture reconstructions through two DCM-based modules, namely a shape reconstruction module and a texture reconstruction module, respectively. The DCM algorithm is proposed to model the relation between the shapes/textures of the occluded and non-occluded regions in the facial image by a single combined eigenspace; thereby such DCM preserves the significant feature characteristics of their correlation. Given the shape/texture of the non-occluded region, the system optimally reconstructs the corresponding shape/texture of the occluded region without losing the significant characteristics of the original combined eigenspace. Our results show that facial appearances recovered by our system closely resemble the original ground truth facial appearance. The enhanced reconstruction performance of our framework is due to its robustness against facial feature misalignments and its ability to recover both the global facial structure and the local detailed facial textures. Overall, our results show that our framework is a promising way to improve the performance of existing automatic face recognition, facial expression recognition, and facial pose estimation applications.

References

[1]   Amin M A, Yan H (2009) An Empirical Study on The Characteristics of Gabor Representations for Face Recognition. International Journal of Pattern Recognition and Artificial Intelligence (IJPRAI), 23: 401–431
[2]   De La Hunty M, Asthana A and Goecke R (2010) Linear Facial Expression Transfer with Active Appearance Models. In: Proceedings of Intl Conf Pattern Recognition, pp 3789–3792
[3]   Fang H, Grant P W and Chen M (2010) Discriminant Feature Manifold for Facial Aging Estimation. Proceedings of Intl Conf Pattern Recognition, pp 593–596
[4]   Huang C, Ding X and Fang C (2010) Head Pose Estimation Based on Random Forests for Multiclass Classification. In: Proceedings of Intl Conf Pattern Recognition, pp 934–937
[5]   Jain A K, Flynn P and Ross A A (2007) Face Recognition. Hand Book of Biometrics, pp 43–70, Springer New York
[6]   Lee S W, Wang P S P, Yanushkevich S N et al (2008) Noniterative 3D Face Reconstruction Based on Photometric Stereo. International Journal of Pattern Recognition and Artificial Intelligence (IJPRAI), 2: 389–410
[7]   Rudovic O, Patras I and Pantic M (2010) Regression-based Multi-view Facial Expression Recognition. In: Proceedings of Intl Conf Pattern Recognition, pp 4121–4124

[8]   Tsagkatakis G and Savakis A (2010) Manifold Modeling with Learned Distance in Random Projection Space for Face Recognition. In: Proceedings of Intl Conf Pattern Recognition, pp 653–656

[9]   Xiong P F, Huang L and Liu C P (2010) Initialization and Pose Alignment in Active Shape Model. In Proceedings of Int'l Conf Pattern Recognition, pp 3971–3974

[10] Yang W, Sun C and Zhang L (2010) Face Recognition Using a Multi-manifold Discriminant Analysis Method. In: Proceedings of Intl Conf Pattern Recognition, pp 527–530

[11] Zheng Y, Li S Z, Chang J et al (2010) 3D Modeling of Faces from Near Infrared Images Using Statistical Learning. In: International Journal of Pattern Recognition and Artificial Intelligence (IJPRAI), 24: 55–71

[12] Saito Y, Kenmochi Y and Kotani K (1999) Estimation of Eyeglassless Facial Images Using Principal Component Analysis. In: Proceedings of ICIP, 4: 197–201

[13] Park J S, Oh Y H, Ahn S C et al (2005) Glasses Removal from Facial Image Using Recursive PCA Reconstruction. In IEEE Trans on PAMI, 27(5): 805–811

[14] Hwang B W, Blanz V, Vetter T et al (2000) Face Reconstruction from a Small Number of Feature Points. In: Proceedings of International Conference on Pattern Recognition, 2: 838–841

[15] Hwang B W and Lee S W (2003) Reconstruction of Partially Damaged Face Images Based on a Morphable Face Model, IEEE Transactions on Pattern Analysis and Machine Intelligence, 25(3): 365–372

[16] Jones M J and Poggio T (1998) Multidimensional Morphable Models: A Framework for Representing and Matching Object Classes. IJCV 29(2): 107–131

[17] Blanz V, Romdhani S and Vetter T (2002) Face Identification across Different Poses and Illuminations with a 3D Morphable Model. In Proceedings of the IEEE International Conference on Automatic Face and Gesture Recognition, pp 202–207

[18] Cootes T F and Taylor C J (2000) Statistical Models of Appearance for Computer Vision, Technical Report, University of Manchester

[19] Vetter T and Poggio T (1997) Linear Object Classes and Image Synthesis from a Single Example Image. IEEE Transactions on Pattern Analysis and Machine Intelligence, 19(7): 733–742

[20] Chen H, Xu Y Q, Shum H Y et al (2001) Example-Based Facial Sketch Generation with Non-Parametric Sampling, Proceedings of the IEEE International Conference on Computer Vision, pp 433–438

[21] Freeman W T, Pasztor E C and Carmichael O T (2000), Learning low-level vision, International Journal of Computer Vision, 40(1): 25–47

[22] Liu C, Shum H Y and Zhang C S (2001) A Two-Step Approach to Hallucinating Faces: Global Parametric Model and Local Nonparametric Model. IEEE International Conference on Computer Vision and Pattern Recognition, 1: 192–198

[23] Mo Z, Lewis J P and Neumann U (2004) Face Inpainting with Local Linear Representations, Proceedings of British Machine Vision Conference, 1: 347–356

[24] Tang X and Wang X (2003) Face Sketch Synthesis and Recognition. Proceedings of the IEEE International Conference on Computer Vision, 12: 687–694

[25] Liu W, Lin D and Tang X (2005) Hallucinating Faces: TensorPatch Super-Resolution and Coupled Residue Compensation. IEEE International Conference on Computer Vision and Pattern Recognition, 12: 478–484

[26] Donner R, Reiter M, Langs G et al (2006) Fast Active Appearance Model Search Using Canonical Correlation Analysis. IEEE Transactions on Pattern Analysis and Machine Intelligence, 28(10): 1690–1694

[27] Rosales R, Athitsos V, Sigal L et al (2001) 3D Hand Pose Estimation Using Specialized Mappings, ICCV, 378–387

[28] Lin D and Tang X (2009) Quality-Driven Face Occlusion Detection and Recovery, ICCV, pp 1–7

[29] Covell M (1996) Eigen-Points: Control-Point Location Using Principal Component Analysis, Proceedings of the IEEE International Conference on Automatic Face and Gesture Recognition, pp 122–127

[30] Wu C, Liu C, Shum H Y et al (2004) Automatic Eyeglasses Removal from Face Images. IEEE Transactions on Pattern Analysis and Machine Intelligence, 26(3): 322–336

[31] Golub G H and Van Loan C F (1996) Matrix Computations, 3rd edn. Johns Hopkins University Press, Baltimore, MD

[32] Lanitis A (2004) Person Identification from Heavily Occluded Face Images. ACM Symposium on Applied Computing (SAC), pp 5–9

[33]   De la Torre F and Black M J (2001) Dynamic Coupled Component Analysis. IEEE International Conference on Computer Vision and Pattern Recognition, 2: 643–650

21 Generative Models and Probability Evaluation for Forensic Evidence

Sargur N. Srihari[1] and Chang Su[2]

Abstract Generative approaches to pattern recognition and machine learning involve two aparts: first describing the underlying probability distributions and then using such models to compute probabilities or make classificatory decisions. We consider generative models for forensic evidence where the goal is to describe the distributions using graphical models and to use such models to compute probabilistic metrics for measuring the degree of individuality of a forensic modality or of a piece of evidence. The metrics are defined as variations of the probability of random correspondence (PRC) when evidence consists of a set of measurements and correspondence is within a tolerance. Three metrics are defined, the first two of which concern the modality and the third concerns evidence: 1) PRC of two samples, 2) PRC among a random set of n samples (nPRC), and 3) PRC between a specific sample among n others (specific nPRC). Computation of these probabilities are described using graphical models which makes all the variables explicit. The metrics are evaluated for several cases—some of which are illustrative (birthdays and heights) and others concern fingerprints. For birthdays, which are discrete-valued and exact, assuming uniformly distributed birthdays, while nPRC rapidly approaches unity with higher n (which is the well-known birthday paradox), specific nPRC grows much less rapidly. For human heights, which are continuous-valued scalars, a quantization is needed and results are gender-specific: assuming Gaussian distributed heights, the PRC for males is higher than for females due to lower variance. Two forms of fingerprint representation are considered: ridge flow and minutiae. Gaussian mixtures are used to model location and orientation of minutiae. With parameters estimated from standard databases (using expectation maximization), fingerprint PRCs are determined for given numbers of available and matching minutiae (which correspond to quantization), for the case of 36 available minutiae where 24 of them match, the PRC is 4.0×10^{-34}. The methodology put forward should be value to establish the relative value of different types of forensic evidence in courtroom scenarios.

---

1 The State University of New York at Buffalo, Buffalo, New York 14260. E-mail: srihari@cedar.buffalo.edu.

2 The State University of New York at Buffalo, Buffalo, New York 14260. E-mail: changsu@cedar.buffalo.edu.

21.1 Introduction

In forensics there are several modalities for identifying an object or individual from evidence. Examples of modality classes (and specific instances of modalities) are impression evidence (latent prints, handwriting, shoe-prints), trace evidence (paint, hair, fiber), biological evidence (DNA, blood type), etc. It is useful to establish the strength of a given modality or of a given piece of evidence within a given modality. Not least of the reasons for such analysis being court rulings that require a scientific basis for evidence presented [1].

The terms class characterization and individualization are commonly used in forensics. In addition terminology from the biometric domain, such as verification and identification are also present. Thus it is useful to first define these terms. Class-characterization is the narrowing down of the evidence into a sub-class within the forensic modality, e.g., ethnicity. Individualization is sometimes defined as the exclusion of all other sources for the given evidence. Verification is the determination of whether a given evidence is from a given source and is a binary decision. Identification is the determination of the best match of the evidence given a finite set of sources for that evidence. Finally individuality of a forensic modality or of a piece of forensic evidence is the degree of distinctiveness of that type of evidence in a population.

In many modalities, particularly those based on visual patterns, the set of measurements made from the evidence have an inherent variability even for the same object or individual. Since evidence in many forensic modalities can be characterized by quantitative measurements [2] a natural metric for individuality is a probability or a probability distribution. Such a metric can be evaluated, for a particular modality, measurement (or feature vector) or or a given sample of data, by using either one of the classical approaches of machine learning: discriminative and generative [3, 4]. In both approaches, representative samples of evidence are used to construct models during a training phase.

In the discriminative approach to measuring individuality, samples are directly used to construct either a two-class or a multi-class classifier [5]. One such approach is that based on determining a similarity (or kernel) function s whose value is high when the input and a template have the same origin and low when they are not. Such a method is used, for instance, in automated fingerprint identification systems that determine the degree of *match* between two fingerprints [6]. By thresholding the value of s into binary classes: *same* and *different*, an accuracy measure can be estimated from known training samples, e.g., average probability of error or risk. The accuracy estimate itself provides a measure of individuality of measurement x. Since the nature of the testing set is crucial, data from cohort groups such as twins are often used to determine error rates [7, 8].

Generative models are statistical models that represent the distribution of x. They are referred to as being generative in that given the distribution, samples can be generated from them. In these models, a distribution of x

is learnt through a training data set. Samples can be generated from this distribution to determine the probability of random correspondence. The training set used is immaterial as long as it is representative of the entire population.

Generative models contrast with discriminative models, in that a generative model is a full probability model of all variables, whereas a discriminative model provides a model only of the target variable(s) conditioned on the observed variables. A generative model can be used, for example, to simulate (i.e., generate) values of any variable in the model, whereas a discriminative model allows only sampling of the target variables conditioned on the observed quantities. If the observed data are truly sampled from the generative model, then fitting the parameters of the generative model to maximize the data likelihood is a common method.

Both approaches have their advantage and limitations. In the generative approach, the problems with realistic modeling of all the parameters of the measurement may become unsurmountable. But a good model would be a method of gaining insights into the fundamental accuracy bounds of the measurement; more typically, these models may be relatively less effective in predicting performance. In the discriminative approach, definition of decision thresholds based on minimizing risk is an issue [9]. Discriminative methods lead to higher performing automated systems, but do not readily lead to fundamental understanding of underlying issues [10].

This paper only focuses on generative models and defines probabilistic metrics that are useful for a given modality or given evidence within a modality (Section 21.2). These metrics are applied to several modalities: birthdays (Section 21.3), human heights of males and females (Section 21.4), and fingerprints using ridge flow and minutiae (Section 21.5). The modalities discussed provide a continuum in terms of problem complexity: birthdays are represented as discrete numbers with uniform distribution, heights are continuous-valued Gaussian distributed scalars and fingerprints involve multivariate continuous features and scores provided by fingerprint matching algorithms.

21.2 Generative Models of Individuality

The goal is to specify a model for randomly generating observed data from which the relevant metrics can be computed. Given the model values for the probability of two randomly chosen samples having the same value within some tolerance can be computed. The process involves the following steps:

Step 1   Consider a generative model (proposal) for measurement x;
Step 2   Formulate a method for estimating parameters of the model;
Step 3   Evaluate the parameters from a data set;
Step 4   Use the model to evaluate relevant individuality metrics.

The first four steps are to determine the distribution of the data. The final step involves computing the probability of match or correspondence within some tolerance between two samples. Tolerance specification depends on the modality and type of measurement, e.g., in the case of a continuous scalar it can be specified as $\pm\varepsilon$.

As measures of individuality, three probabilities can be defined: PRC (Probability of Random Correspondence), nPRC (Probability of Random Correspondence given n samples) and Specific nPRC (Probability of Random Correspondence of a specific x among n samples). These definitions are further described below.

1) PRC: probability that two randomly chosen samples have the same measured value x within specified tolerance ε.

2) nPRC: the probability that among a set of n samples, some pair have the same value x, within specified tolerance, where $n \geqslant 2$. Since there are $\binom{n}{2}$ pairs involved this probability is higher than PRC. Note that when $n = 2$, PRC=nPRC.

3) Specific nPRC: the probability that in a set of n samples, a specific one with value x coincides with another sample, within specified tolerance. Since we are trying to match a specific value x, this probability depends on the probability of x and is generally smaller than PRC. The exact relationship with respect to PRC depends on the distribution of x.

We note that the first two measures, PRC and nPRC characterize the forensic modality as described by a set of measurements and furthermore, the second is a function of the first. The third measure *specific nPRC* characterizes specific evidence, e.g., a specimen. It is dependent on the specific value as well as the distribution of the measurement.

Graphical Models

Since probability distributions and their computation play a central role in this discussion, it is useful to represent them using graphical models [4]. Directed graphical models for PRC, nPRC and specific nPRC are shown in Fig. 21.1. Each node represents a random variable (or a group of random

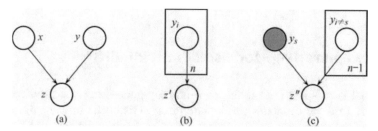

(a)   (b)   (c)

Fig. 21.1 Graphical models for computing coincidence probabilities. The models correspond to: (a) PRC, the probability of two samples having the same value. (b) nPRC, the probability of at least two samples among n having the same value. (c) Specific nPRC, the probability that a specific sample is among n. Here x, y, y_s, and y_i are feature vectors with identical distribution. We are interested in the distributions of z and z' which express probabilities of match/non-match. Note that y_s is shaded indicating that its value is observed.

variables), and the links express conditional probabilistic relationships between them. The variables are: $\boldsymbol{x}$ is a random variable corresponding to the
feature vector, $\boldsymbol{y}$ is another random variable with the same distribution, $\{\boldsymbol{y}_i\}$
where $i = [1, .., n]$ represents a set of n random variables using the plate
representation and $\boldsymbol{y}_s$ where $s \in [1, .., n]$ is a random variable among $\{\boldsymbol{y}_i\}$.

A binary-valued random variable z in Fig. 21.1 (a) indicates that if $\boldsymbol{x}$ is
the same as $\boldsymbol{y}$ within a tolerance ε, in which case $z = 1$ and $z = 0$ otherwise.
This leads to the following definition

$$p(z|\boldsymbol{x}, \boldsymbol{y}) = \begin{cases} p(z = 1|\boldsymbol{x}, \boldsymbol{y}) = 1 \text{ and } p(z = 0|\boldsymbol{x}, \boldsymbol{y}) = 0, & \text{if } \boldsymbol{x} = \boldsymbol{y}, \\ p(z = 1|\boldsymbol{x}, \boldsymbol{y}) = 0 \text{ and } p(z = 0|\boldsymbol{x}, \boldsymbol{y}) = 1, & \text{if } \boldsymbol{x} \neq \boldsymbol{y}. \end{cases} \quad (21.1)$$

By marginalizing the joint distribution as

$$p(z) = \sum_{\boldsymbol{x}} \sum_{\boldsymbol{y}} p(\boldsymbol{x}, \boldsymbol{y}, z), \quad (21.2)$$

applying Bayes rule to write

$$p(z) = \sum_{\boldsymbol{x}} \sum_{\boldsymbol{y}} p(z|\boldsymbol{x}, \boldsymbol{y}) p(\boldsymbol{x}, \boldsymbol{y}), \quad (21.3)$$

and noting the independence of $\boldsymbol{x}$ and $\boldsymbol{y}$

$$p(z) = \sum_{\boldsymbol{x}} \sum_{\boldsymbol{y}} p(z|\boldsymbol{x}, \boldsymbol{y}) p(\boldsymbol{x}) p(\boldsymbol{y}), \quad (21.4)$$

we obtain the PRC as $\rho \equiv p(z = 1)$.

Next we consider the case of nPRC where there are n identically dis-
tributed random variables $\boldsymbol{Y} = [\boldsymbol{y}_1, \boldsymbol{y}_2, \ldots, \boldsymbol{y}_n]$. This is shown using the plate
representation in Fig. 21.1 (b), where z' is the state indicating if in a set of
n random variables at least one value $\boldsymbol{y}_i$ is the same as another value $\boldsymbol{y}_j$ in
which case $z' = 1$ and $z' = 0$ otherwise. This leads to the following definition

$$p(z'|\boldsymbol{Y}) = \begin{cases} p(z' = 1|\boldsymbol{Y}) = 1 \text{ and } p(z' = 0|\boldsymbol{Y}) = 0, \\ \quad \text{if } \exists \boldsymbol{y}_i, \boldsymbol{y}_j \in \boldsymbol{Y} \ni \boldsymbol{y}_i = \boldsymbol{y}_j, \\ p(z' = 1|\boldsymbol{Y}) = 0 \text{ and } p(z' = 0|\boldsymbol{Y}) = 1, \\ \quad \text{if } \forall \boldsymbol{y}_i, \boldsymbol{y}_j \in \boldsymbol{Y} \ni \boldsymbol{y}_i \neq \boldsymbol{y}_j. \end{cases} \quad (21.5)$$

The marginal distribution of z' is obtained as

$$p(z') = \sum_{\boldsymbol{Y}} p(z'|\boldsymbol{Y}) p(\boldsymbol{Y}). \quad (21.6)$$

The specific instance of $p(z' = 1)$, which is the nPRC, can be written as

$$p(z' = 1) = 1 - p(z' = 0) = 1 - \left[1 - \sum_{\boldsymbol{x}} \sum_{\boldsymbol{y}} p(z = 1|\boldsymbol{x}, \boldsymbol{y}) p(\boldsymbol{x}) p(\boldsymbol{y}) \right]^{\frac{n(n-1)}{2}}.$$
$$(21.7)$$

Denoting the nPRC value as $\rho[n] \equiv p(z' = 1)$ and using Eq. (21.4), we have a relationship between PRC and nPRC as

$$\rho[n] = 1 - (1 - \rho)^{\frac{n(n-1)}{2}}. \qquad (21.8)$$

Finally we consider the case where the value of $\boldsymbol{y}_s \in \boldsymbol{Y}$ is known as represented by the shaded node in Fig. 21.1 (c). Given a specific value $\boldsymbol{y}_s$, we define z'' as

$$p(z''|\boldsymbol{y}_s, \boldsymbol{Y}) = \begin{cases} p(z'' = 1|\boldsymbol{y}_s, \boldsymbol{Y}) = 1 \text{ and } p(z'' = 0|\boldsymbol{y}_s, \boldsymbol{Y}) = 0, \\ \qquad \text{if } \exists \boldsymbol{y}_i \in \boldsymbol{Y} \ni \boldsymbol{y}_i = \boldsymbol{y}_s, \\ p(z'' = 1|\boldsymbol{y}_s, \boldsymbol{Y}) = 0 \text{ and } p(z'' = 0|\boldsymbol{y}_s, \boldsymbol{Y}) = 1, \\ \qquad \text{if } \forall \boldsymbol{y}_i \in \boldsymbol{Y} \ni \boldsymbol{y}_i \neq \boldsymbol{y}_s, \end{cases} \qquad (21.9)$$

the specific nPRC is then given by the marginal probability

$$p(z'' = 1|\boldsymbol{y}_s) = \sum_{\boldsymbol{Y'}} p(z'' = 1, \boldsymbol{Y'}|\boldsymbol{y}_s) = \sum_{\boldsymbol{Y'}} p(z'' = 1|\boldsymbol{y}_s, \boldsymbol{Y'}) p(\boldsymbol{Y'}), \quad (21.10)$$

where $\boldsymbol{Y'} = [\boldsymbol{y}_1, \ldots, \boldsymbol{y}_{s-1}, \boldsymbol{y}_{s+1}, \ldots, \boldsymbol{y}_n]$ and $p(\boldsymbol{Y'})$ is the joint probability of the $n - 1$ individuals. Since nPRC is dependent on the specific value $\boldsymbol{y} = \boldsymbol{y}_s$, it is useful to compute the expected value of nPRC, where the expectation is computed with respect to the distribution of $\boldsymbol{y}$.

The rest of this paper concerns the application of the developed methods to three cases: birthdays, human heights and fingerprints represented by minutiae and ridges. In the case of birthdays, which are discrete-valued, the tolerance is zero, i.e., requiring exact match. In the case of heights, which are continuous-valued scalars, a tolerance is specified in terms of the height differences that are considered to be the same. In the case of fingerprints, the measurement is a variable-length vector with each element represented as a triple, and tolerance is specified within the fingerprint matching algorithm.

21.3 Application to Birthdays

Birthdays are a standard case for the probability of coincidences [11]. As is typical, we disregard leap years and assume that the 365 possible birthdays are equally likely. We use a uniform density to model the birthday distribution and the probability that a person has any specific birthday is 1/365. As this case is for illustrative purpose only we will forego the goodness of fit test against a database of birthdays. Also we only consider the case of exact matches rather than almost matches, e.g., birthdays coincide within one day or k days.

The PRC, nPRC and specific nPRC are determined as follows.

21.3.1 PRC

PRC is the probability that any two persons have the same birthday. This is given by Eq. (21.4) and the definition in Eq. (21.1)

$$p_\varepsilon = p(z = 1) = \sum_{\boldsymbol{x}=1}^{365} \sum_{\boldsymbol{y}=1}^{365} p(z = 1|\boldsymbol{x}, \boldsymbol{y}) \left(\frac{1}{365}\right) \left(\frac{2\varepsilon + 1}{365}\right)$$

$$= \sum_{1}^{365} \left(\frac{2\varepsilon + 1}{365^2}\right) = \frac{2\varepsilon + 1}{365}, \tag{21.11}$$

where ε is the tolerance for birthdays. In the special case when $\varepsilon = 0$, the case of exactly matching birthdays, we have $p_\varepsilon = 1/365 = 0.002\,7$. Note that when $\varepsilon = 1$ the two birthdays can differ in three ways.

21.3.2 nPRC

The value of nPRC for birthdays is the probability that in a group of n persons, some pair of them have the same birthday. To compute it, it is easier to first calculate the probability $\bar{p}(n)$ that all n persons have different birthdays. The nPRC is complementary to $\bar{p}(n)$ and is given by

$$p(n) = 1 - \bar{p}(n) = 1 - (1 - p_\varepsilon)^{\frac{n(n-1)}{2}}, \tag{21.12}$$

where p_ε is the PRC value. Thus nPRC increases very rapidly with n, in fact exponentially with an exponent which is $O(n^2)$.

We can see how nPRC increases monotonically with n in Fig. 21.2. As an example consider the case of 40 randomly chosen individuals, or $n = 40$. We see the probability that two individuals have the same birthday in this group is 0.9. This is the well-known birthday paradox [11] which has been cited as a reason against claiming individualization in forensic evidence [12]. We shall see that there is a counter-paradox when we consider specific nPRCs.

21.3.3 Specific nPRC

To evaluate the probability of at least one individual sharing a birthday with a given individual among n individuals, the specific nPRCs are calculated. The event of at least one of the n persons having the same birthday with the given individual is complementary to all the n people having the different birthday with the specific people. Assuming that $p(b)$ is the probability of an individual has the birthday b, the specific nPRC is $p(b, n)$ which is given by

$$p(b, n) = 1 - \bar{p}(b, n) = 1 - (1 - p(b))^{n-1}, \tag{21.13}$$

where $\bar{p}(b, n)$ is the probability that no other person has birthday within ε days of b.

The expected value of specific nPRC is given by

$$E[p(b, n)] = \sum_{b=1}^{365} p(b)p(b, n) = \sum_{b=1}^{365} p(b)[1 - (1 - p(b))^{n-1}]. \quad (21.14)$$

Assuming a uniform distribution of birthdays $p(b) = 1/365$ and allowing for birthdays to match when they are ε of each other, the expected value of specific nPRC is $1 - (1 - p_\varepsilon)^{n-1}$. This increases exponentially as in Eq. (21.12) but with an exponent which is $O(n)$.

Figure 21.2 shows the expected value of specific nPRC for some values of n. We see that given a specific birthday, the probability of a second individual having that same birthday among 40 individuals is about 0.1, which is significantly smaller than the nPRC value which is 0.9. Even with $n = 1000$, which in nearly three times the number of days in a year, there is more than 6% probability that there is no other person with a specific birthday, say June 1st. In view of the birthday paradox, this is an interesting counter-paradox!

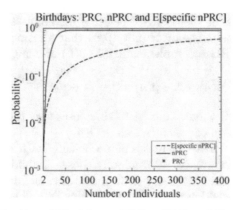

Fig. 21.2 Individuality of Birthdays: PRC, nPRC and Expectation of specific nPRC.

The expectation of specific nPRC is also plotted in Fig. 21.2 where it is seen that specific nPRC grows much slower than nPRC. In the case of birthdays the following inequality holds:

$$\text{PRC} \leqslant \text{E[Specific } n\text{PRC]} \leqslant n\text{PRC}$$

with equality holding when $n = 2$.

21.4 Application to Human Heights

We next consider a simple human measurement, height, which is continuous-valued and for which data is abundantly available. We use distributions of

heights for males and females in the United States as produced by the CDC
[13]. The Gaussian density is a reasonable model to fit the distribution. The
Gaussian probability density functions for heights (in inches) for males and
females aged 20 years and over is given in Fig. 21.3. The parameters for
these distributions are as follows: mean $\mu_f = 63.8$, $\mu_m = 69.3$ and standard
deviation $\sigma_f = 4.2$, $\sigma_m = 3.3$. Note that the standard deviation for females
in much larger than for males. Again we forego the goodness of fit for the
case of heights as it is also an illustrative case.

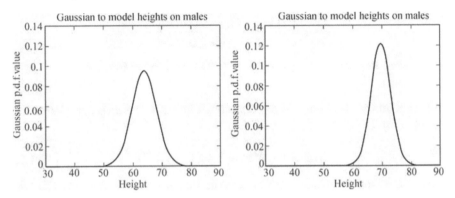

Fig. 21.3 Gaussian density models of human height: (a) female and (b) male,
using US-CDC data. Females have higher variance.

We derive a value for the probability of random correspondence (PRC)
for heights in terms of the known Gaussian distributions as follows. The
probability of two individuals having the same height with some tolerance
$\pm\varepsilon$, can be written as

$$p_\varepsilon = \int_{-\infty}^{\infty} \int_{y-\varepsilon}^{y+\varepsilon} P(x|\mu, \sigma) P(y|\mu, \sigma) \mathrm{d}x \mathrm{d}y, \qquad (21.15)$$

where

$$P(h|\mu, \sigma) \sim \mathcal{N}(\mu, \sigma) = \frac{1}{\sqrt{2\pi}\sigma} e^{-\frac{(h-\mu)^2}{2\sigma^2}}.$$

Eq. (21.15) can be numerically evaluated for values of μ and σ corresponding to the male and female height distributions. Using a tolerance of $\varepsilon = 0.1$
inches we can calculate the PRCs. The PRC for female height with a mean
height of 63.8 inches and standard deviation of 4.2 inches is $p_\varepsilon = 0.013\,4$
and the PRC for male height with a mean height of 69.3 inches and standard deviation of 3.3 inches is $p_\varepsilon = 0.017\,3$ (tolerance of 0.1 inch). The PRC
increases with ε as shown in Fig. 21.4.

The value of nPRC can be computed from the PRC value. This is the
probability that in a randomly chosen group some pair will have the same
height. The continuous variable can be handled by the implicit discreteness
due to the tolerance ε. It is easier to first calculate the probability $\bar{p}(n)$ that

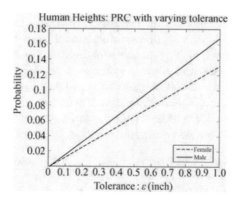

Fig. 21.4 Human Heights: PRC with varying tolerance ε.

all n heights are different from each other. The nPRC value is complementary to $\bar{p}(n)$ and given by

$$p(n) = 1 - \bar{p}(n) = 1 - (1 - p_\varepsilon)^{\frac{n(n-1)}{2}}, \qquad (21.16)$$

where p_ε is the PRC value. Values of nPRC with different n are shown in the second column of Tables 21.1 and 21.2.

Table 21.1 Female Height Probabilities: nPRC, Specific nPRC (57 inches) and Expectation of Specific nPRC

n	PRC $p(n)$	Specific nPRC ($h = 57$ inches) $p(h, n)$	Expectation of Specific nPRC $E[p(h,n)]$
2	0.0134	0.0051	0.0134
5	0.1262	0.0203	0.0525
10	0.4551	0.0452	0.1137
15	0.7574	0.0694	0.1703
20	0.9229	0.0930	0.2227
40	0.99995	0.1815	0.3965
80	1	0.3335	0.6214
160	1	0.5581	0.8233
320	1	0.8057	0.9316
800	1	0.9835	0.9784
1600	1	0.9997	0.9904

Since the variance of heights of females is higher than for males, nPRC grows slower for females than for males. If $n = 40$ it is almost certain that there will be two persons with the same height within a tolerance of 0.1 inch. As we shall see these probabilities will be much smaller in the case of specific heights.

To evaluate the probability that at least one among n individuals has the same height as a given individual, the specific nPRCs are estimated. To compute it, it is easier to first calculate the probability $\bar{p}(h, n)$ that all n heights are different from the given height h. The event of at least one of the

Table 21.2 Male Height Probabilities: nPRC, specific nPRC (68 inches) and expectation of specific nPRC

n	PRC $p(n)$	Specific nPRC ($h = 68$ inches) $p(h, n)$	Expectation of Specific nPRC $E[p(h, n)]$
2	0.0173	0.0224	0.0173
5	0.1601	0.0865	0.0664
10	0.5440	0.1842	0.1423
15	0.8400	0.2715	0.2111
20	0.9637	0.3494	0.2734
40	0.99999997	0.5862	0.4704
80	1	0.8326	0.6991
160	1	0.9726	0.8723
320	1	0.9993	0.9316
800	1	0.9999992	0.9838
1600	1	0.9999999999995	0.9927

n persons having the same height with the given individual is complementary to all n heights being different. Assuming that $p(h)$ is the probability of an individual at a height of h, the specific PRC $p(h, n)$ is given by

$$p(h, n) = 1 - \bar{p}(h, n) = 1 - (1 - p(h))^{n-1}, \qquad (21.17)$$

where $p(h) = \int_{h-\varepsilon}^{h+\varepsilon} P(l|\mu, \sigma) dl$, $P(l|\mu, \sigma)$ is the generative model for height.

The expected value of specific nPRC is given by

$$E[p(h, n)] = \int_0^\infty p(h, n) p(h) dh. \qquad (21.18)$$

The third and fourth columns of Tables 21.1 and 21.2 shows specific nPRCs and expectation of specific nPRC for some values of n. For example, among 40 randomly chosen females, the probability that one of them has height $h = 57$ inches (or 4ft 9in) is 0.18 and the expectation that one of them has same height as a given one is 0.39. Similarly, among 40 randomly chosen males, the probability that one of them has height $h = 68$ inches (or 6ft 8in) is 0.58 and the expectation that one of them has same height as a given one is 0.47. These probabilities are much smaller than the corresponding values for nPRC, which are close to certainty.

The values will depend specific heights h. PRC, nPRC, expected value of specific nPRC and specific nPRC of $5'9''$ and $5'2''$ on male height over varying number of individuals n are shown in Fig. 21.5. The value of nPRC increases rapidly with n. The rate of increase of specific nPRC depends on the specific height — with a more likely height ($5'9''$) having a higher value than a less likely value ($5'2''$), i.e., an unusual height will have a low specific nPRC. The expected value has an intermediate value.

In the case of human heights the following inequality holds:
$$\text{PRC} \leqslant \text{E[Specific } n\text{PRC]} \leqslant n\text{PRC}$$
with equality holding when $n = 2$.

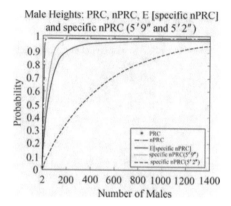

Fig. 21.5 Probabilities of male heights coinciding within 0.1 inch: PRC, nPRC, Expectation of specific nPRC and Specific nPRC of $5'9''$ and $5'2''$.

21.5 Application to Fingerprints

We consider next the evaluation of the relevant probabilities in a commonly used forensic modality — that of fingerprints. Their use in human identification has been based on two premises, that, (1) they do not change with time and (2) they are unique for each individual. While in the past, identification based on latent fingerprints had been accepted by courts, more recently their use has been questioned under the basis that the premises stated above have not been objectively tested and error rates not been scientifically established [14]. Though the first premise has been accepted, the second one on individuality is being challenged.

Fingerprint individuality studies date back to the late 1800s. More than twenty models have been proposed to establish the improbability of two random people (or fingers) have the same fingerprint [5]. These models can be classified into five different categories: grid-based [16, 17], ridge-based [17], fixed probability [18], relative measurement [19, 20] and generative [21–24]. All models try to quantify the uniqueness property, e.g., the probability of false correspondence. A match here does not necessarily mean an exact match but a match within given tolerance levels.

Features for representing fingerprints are classified into three types [25]. Level 1 features provide class-characterization of fingerprints based on ridge flow. They are divided into five primary classes: whorl, left loop, right loop, arch and tent (Fig. 21.6). Some of the primary classes have secondary classes resulting in a total of eight class types.

Level 2 features, which are more useful for identification, are also known as minutiae. Fingerprints such as those shown in Fig. 21.6 are first aligned. This is done manually where core points are identified and then the image is centered. The minutiae correspond to ridge endings and ridge bifurcations. Automatic fingerprint matching algorithms use minutiae as the salient fea-

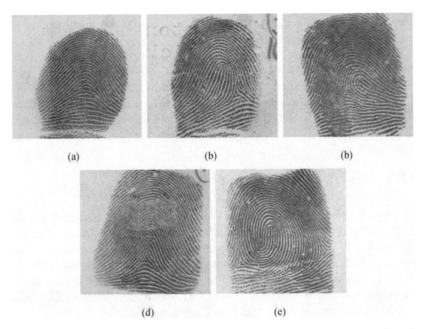

Fig. 21.6 Examples of five main types of ridge flow in fingerprints, referred to as Level 1 features: (a) arch, (b) left loop, (c) right loop, (d) tented arch, and (e) whorl. From NIST Special Database 4.

tures, e.g., [6], since they are stable and are reliably extracted. A minutia is represented by its location and direction; direction is determined by the ridge ending at the location (Fig. 21.7). The type of minutiae (either bifurcation or ending) is not distinguished since this information is not as reliable as the

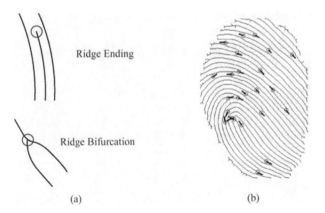

Fig. 21.7 Representation of fingerprints using minutiae: (a) locations of ridge endings and ridge bifurcations are indicated by circles, and (b) minutiae directions are indicated by line segments in a skeletonized fingerprint image.

information on location and direction. Level 3 features, such as pores and scars are ancillary features.

Our goal is to model the distribution of fingerprints based on features. The distributions considered are those based on ridge flow types (Section 21.5.1) and minutiae (Section 21.5.2). Methods for estimating distribution parameters are discussed in Section 21.5.3. PRCs calculation algorithms are proposed in Section 21.5.4. Section 21.5.5 shows the experiments and results.

21.5.1 Distribution of Ridge Flow Type

A simple distribution of the Level 1 ridge flow types is obtained by counting the relative frequency of each of the primary and secondary types in a fingerprint database. In one such evaluation [8] loops account for 64% of the fingers, with the secondary types being: 30% left loops, 27% right loops and 7% double loops. Arches account for 18% of the primary types, with the seondary types being: plain arches (13%) and tented arches (5%). Whorls account for the remainder of the Level 1 types (19%).

Level 1 features are clearly broad class characteristics which are useful for exclusion of individual fingers but not by themselves useful for the tasks of verification, identification and individualization.

Assuming that fingerprints are distinguished by 6 secondary types, the PRC for ridge flow types is calculated by Eq. 21.3. Given above type frequencies, we have PRC value $p_\varepsilon = 0.2233$. The nPRC and specific nPRC can be caculated by Eqs. 21.8 and 21.9. Tables 21.3 and 21.4 show the nPRC and Specific nPRC with different n.

Table 21.3   Ridge Flow Types: nPRC

n	2	3	4	5	6	7	8
nPRC	0.2233	0.5314	0.7805	0.9201	0.9774	1	1

Table 21.4   Ridge Flow Types: Specific nPRC belong to class LL = left loop, RL = right loop, DL = double loop, PA = plain arch, TA = tented arch and W = whorl

n	Specific nPRC					
	LL	RL	DL	PA	TA	W
2	0.3000	0.2700	0.0700	0.1300	0.0500	0.1900
4	0.6570	0.6110	0.1956	0.3415	0.1426	0.4686
6	0.8319	0.7927	0.3043	0.5016	0.2262	0.6513
8	0.9176	0.8895	0.3983	0.6227	0.3017	0.7712
10	0.9596	0.9411	0.4796	0.7145	0.3698	0.8499
12	0.9802	0.9686	0.5499	0.7839	0.4312	0.9015
14	0.9903	0.9833	0.6107	0.8364	0.4867	0.9354
16	0.9953	0.9911	0.6633	0.8762	0.5367	0.9576
18	0.9977	0.9953	0.7088	0.9063	0.5819	0.9722
20	0.9989	0.9975	0.7481	0.9291	0.6226	0.9818

21.5.2   Distribution of Minutiae

Each minutia is represented as $x = (s, \theta)$ where $s = (x_1, x_2)$ is its location and θ its direction. The distribution of minutiae location conditioned on ridge flow is shown in Fig. 21.8 where there were 400 fingerprints of each type. In the model we develop the combined distribution over all types is used (Fig. 8 (f)).

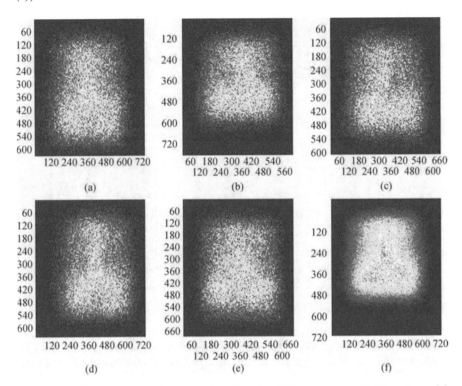

Fig. 21.8   Distribution of minutia location for different types of ridge flow: (a) arch, (b) left loop, (c) right loop, (d) tent, (e) whorl, and (f) all types combined.

Since minutia location has a multimodal distribution, a mixture of K Gaussians is a natural approach. For the data set considered a value of $K = 3$ provided a good fit, as validated by a goodness of fit test. Values of $K = 4, 5$ do not fit the data as well. A Gaussian mixture with $k = 3$ is shown in Fig. 21.9.

Since minutiae orientation is a periodic variable, it is modeled by a *circular normal* or von Mises distribution which itself is derived from the Gaussian [4, 26]. Such a model is better than mixtures of hyper-geometric and binomial distributions [21, 22].

Such a model for minutiae distributions involves a random variable z that represents the particular mixture component from which the minutia is

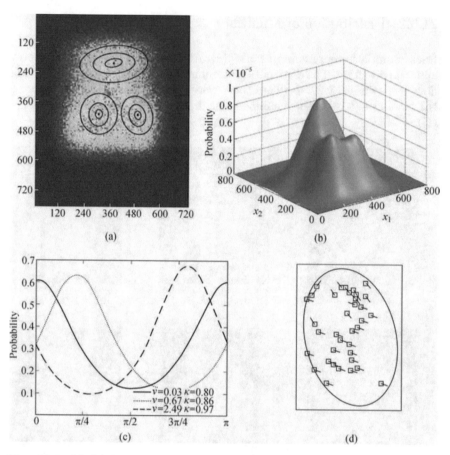

Fig. 21.9 Model for minutiae distribution using Gaussian mixture for location and von Mises for direction: (a) Gaussian mixture model for minutia location with three components, (b) three-dimensional plot of mixture model, (c) von Mises distributions of minutiae orientation for each of the three components, where the green curve corresponds to the upper cluster, blue the lower left cluster and red the lower right cluster, and (d) sample generated from model.

drawn. In this model both minutiae location and orientation depend on the component they belong to. Minutiae location and orientation are conditionally independent given the component. This is represented by

$$p(\boldsymbol{x}|z) = p(s, \theta|z) = p(s|z)p(\theta|z), \qquad (21.19)$$

whose graphical model is shown in Fig. 21.10 from which we have the joint distribution

$$p(\boldsymbol{x}, z) = p(z)p(\boldsymbol{x}|z). \qquad (21.20)$$

Marginalizing over the components, we have the distribution of minutiae

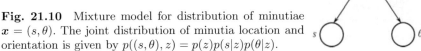

Fig. 21.10 Mixture model for distribution of minutiae $\boldsymbol{x} = (s, \theta)$. The joint distribution of minutia location and orientation is given by $p((s, \theta), z) = p(z)p(s|z)p(\theta|z)$.

as

$$p(\boldsymbol{x}) = \sum_z p(z)p(\boldsymbol{x}|z). \qquad (21.21)$$

Substituting Eq. (21.19) in Eq. (21.21) we have

$$p(\boldsymbol{x}) = \sum_z p(z)p(s|z)p(\theta|z). \qquad (21.22)$$

Since minutiae location within each component is Gaussian and minutiae orientation within each component is von Mises we can write

$$p(\boldsymbol{x}|\Theta) = \sum_{k=1}^{K} \pi_k \cdot \mathcal{N}(s|\mu_k, \Sigma_k) \cdot \mathcal{V}(\theta|\nu_k, \kappa_k), \qquad (21.23)$$

where K is the number of mixture components, π_k are non-negative component weights that sum to one, $\mathcal{N}(s|\mu_k, \Sigma_k)$ is the bivariate Gaussian probability density function of minutia with mean μ_k and covariance matrix Σ_k, $\mathcal{V}(\theta|\nu_k, \kappa_k)$ is the von Mises probability density function of minutiae orientation with mean angle ν_k and precision (inverse variance) κ_k, and $\Theta = \{\pi_k, \mu_k, \Sigma_k, \nu_k, \kappa_k, \rho_k\}$ where $k = 1, 2, \ldots, K$ is the set of all parameters of the k Gaussian and von Mises distributions.

Rather than using the standard form of the von Mises distribution for the range $[0, 2\pi]$, since minutiae orientations are represented as being in the range $[0, \pi)$, we use the alternate form [26] as follows:

$$\mathcal{V}(\theta|\nu_k, \kappa_k, \rho_k) = \rho_k \upsilon(\theta) \cdot I\{0 \leqslant \theta < \pi\}$$
$$+ (1 - \rho_k)\upsilon(\theta - \pi) \cdot I\{\pi \leqslant \theta < 2\pi\}, \qquad (21.24)$$

where $I\{A\}$ is the indicator function of the condition A,

$$\upsilon(\theta) \equiv \upsilon(\theta|\nu_k, \kappa_k) = \frac{2}{I_0(\kappa_i)} exp[\kappa_i \cos 2(\theta - \nu_k)], \qquad (21.25)$$

minutiae arising from the kth component have directions that are either θ or $\theta + \pi$ and the probabilities associated with these two occurrences are ρ_k and $1 - \rho_k$ respectively.

Since fingerprint ridges flow smoothly with very slow direction changes, direction of neighboring minutiae are strongly correlated, i.e., minutiae that are spatially close tend to have similar directions with each other. However, minutiae in different regions of a fingerprint tend to be associated with different region-specific minutiae directions thereby demonstrating independence [27, 28]. The model allows ridge orientations to be different at different regions (different regions can be denoted by different components) while it makes sure that nearby minutiae have similar orientations (as nearby minutiae will belong to the same component).

Since several minutiae are observed in a finger, a joint distribution model is needed. The minutiae are assumed to be independent of each other, with each minutiae, consisting of an (s, θ) pair, being distributed according to a mixture of component densities. This is discussed further when we consider parameter estimation in Section 21.5.3.

21.5.3 Parameter Estimation

We now develop an equivalent formulation of the mixture distribution given in Eq. (21.23) by involving an explicit latent variable. This will allow us to formulate the problem of parameter estimation in terms of the expectation maximization (EM) algorithm.

We define the joint distribution $p(\boldsymbol{x}, \boldsymbol{z})$ in terms of a marginal distribution $p(\boldsymbol{z})$ and a conditional distribution $p(\boldsymbol{x}|\boldsymbol{z})$, corresponding to the graphical model in Fig. 21.11 (a).

Given that the total number of minutiae observed in a finger is D, a joint distribution model is needed. The D minutiae are assumed to be independent of each other, with each minutiae, consisting of an $\boldsymbol{x}(s, \theta)$ pair, being distributed according to a mixture of component densities. This is shown in Fig. 21.11 (b).

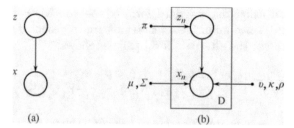

(a)   (b)

Fig. 21.11 Graphical models representing mixture for: (a) Single minutia whose distribution is expressed as $p(\boldsymbol{x}, \boldsymbol{z}) = p(\boldsymbol{x})p(\boldsymbol{x}|\boldsymbol{z})$. (b) Set of D identically distributed minutiae with corresponding latent points $\boldsymbol{z}_n$, where $n = 1, .., D$.

The K-dimensional random variable $\boldsymbol{z}$ has a 1-of-K representation in which a particular element z_k is equal to 1 and all other elements are equal to 0, we can write

$$p(\boldsymbol{z}) = \prod_{k=1}^{K} \pi_k^{z_k}. \tag{21.26}$$

Similarly the conditional distribution of $\boldsymbol{x}$ given a particular value for $\boldsymbol{z}$ is given by

$$p(\boldsymbol{x}|z_k = 1) = \mathcal{N}(\boldsymbol{x}|\mu_k, \Sigma_k) \cdot \mathcal{V}(\theta|\nu_k, \kappa_k, \rho_k), \qquad (21.27)$$

which can also be written in the form

$$p(\boldsymbol{x}|\boldsymbol{z}) = \prod_{k=1}^{K} \mathcal{N}(\boldsymbol{x}|\mu_k, \Sigma_k)^{z_k} \cdot \mathcal{V}(\theta|\nu_k, \kappa_k, \rho_k)^{z_k}. \qquad (21.28)$$

The joint distribution is given by $p(\boldsymbol{z})p(\boldsymbol{x}|\boldsymbol{z})$, and the marginal distribution of $\boldsymbol{x}$ is obtained by summing the joint distribution over all possible states of $\boldsymbol{z}$ to give

$$p(\boldsymbol{x}) = \sum_{\boldsymbol{z}} p(\boldsymbol{z})p(\boldsymbol{x}|\boldsymbol{z}) = \sum_{k=1}^{K} \pi_k \cdot \mathcal{N}(s|\mu_k, \Sigma_k) \cdot \mathcal{V}(\theta|\nu_k, \kappa_k, \rho_k), \qquad (21.29)$$

where we have made use of Eqs. (21.27) and (21.29). Thus the marginal distribution of $\boldsymbol{x}$ is a mixture of the form (24). If we have several observed minutiae $\boldsymbol{x}_1, \boldsymbol{x}_2, \ldots, \boldsymbol{x}_D$ then, because we have represented the marginal distribution in the form $p(\boldsymbol{x}) = \sum_{\boldsymbol{z}} p(\boldsymbol{z})p(\boldsymbol{x}|\boldsymbol{z})$, it follows that for every observed minutia $\boldsymbol{x}_n$, there is a corresponding latent variable $\boldsymbol{z}_n$.

We are now able to work with the joint distribution $p(\boldsymbol{x}, \boldsymbol{z})$ instead of the marginal distribution $p(\boldsymbol{x})$. To estimate the unknown parameters using the maximum likelihood approach, we use the EM algorithm. The number of components K for the mixture model was found after validation using k-means clustering.

E-Step Using γ_{dk} is to denote the responsibility of component k for minutiae $\boldsymbol{x}_d$, its value can be found using Bayes's theorem

$$\gamma_{dk} \equiv p(z_k = 1|\boldsymbol{x}_d) = \frac{p(z_k = 1)p(\boldsymbol{x}_d|z_k = 1)}{\sum\limits_{k=1}^{K} p(z_k = 1)p(\boldsymbol{x}_d|z_k = 1)}$$

$$= \frac{\pi_k \mathcal{N}(s_d|\mu_k, \sigma_k)\mathcal{V}(\theta_d|\nu_k, \kappa_k, \rho_k)}{\sum\limits_{k=1}^{K} \pi_k \mathcal{N}(s_d|\mu_k, \sigma_k)\mathcal{V}(\theta_d|\nu_k, \kappa_k, \rho_k)}. \qquad (21.30)$$

M-Step The estimates of the Gaussian distribution parameters π_k, μ_{mk}, and Σ_{mk} at the $(n+1)$th iteration, are given by

$$\pi_k^{(n+1)} = \frac{1}{D} \sum_{d=1}^{D} \gamma_{dk}^{(n)}, \qquad (21.31)$$

$$\mu_{mk}^{(n+1)} = \frac{\sum\limits_{d=1}^{D} \gamma_{dk}^{(n)} s_m}{\sum\limits_{d=1}^{D} \gamma_{dk}^{(n)}}, \tag{21.32}$$

$$\Sigma_{mk}^{(n+1)} = \frac{\sum\limits_{d=1}^{D} \gamma_{dk}^{(n)} (s_m - \mu_{mk}^{(n+1)})(s_m - \mu_{mk}^{(n+1)})^{\mathrm{T}}}{\sum\limits_{d=1}^{D} \gamma_{dk}^{(n)}}. \tag{21.33}$$

The parameters for orientation distributions are obtained using expectation maximization for the von Mises distribution [29]. The estimates of ν_{mk} and κ_{mk} at the $(n+1)$th iteration are given by

$$\nu_{mk}^{(n+1)} = \frac{1}{2} \tan^{-1} \left(\frac{\sum\limits_{d=1}^{D} \gamma_{dk}^{(n)} \sin 2\psi_d}{\sum\limits_{d=1}^{D} \gamma_{dk}^{(n)} \cos 2\psi_d} \right), \tag{21.34}$$

$$\frac{I_0'(\kappa_{mk}^{(n+1)})}{I_0(\kappa_{mk}^{(n+1)})} = \frac{\sum\limits_{d=1}^{D} r_{dk}^{(n)} \cos 2(\psi_d - \nu_k^{(n+1)})}{\sum\limits_{d=1}^{D} r_{dk}^{(n)}}. \tag{21.35}$$

The solution for Eq. (21.35), which involves Bessel functions, obtained using an iterative method gives the estimate for κ_{mk}. The estimate of ρ_{mk} is then obtained as

$$\rho_{mk}^{(n+1)} = \frac{\sum\limits_{d=1}^{D} I\{c_d^{(n+1)} = k, \theta_d \in [0, \pi)\}}{\sum\limits_{d=1}^{D} I\{c_d^{(n+1)} = k\}}, \tag{21.36}$$

where $c_d^{(n+1)} = \mathrm{argmax}_k \gamma_{dk}^{(n+1)}$ is the component label for the observation d at the $(n+1)$th iteration, ψ_j is the orientation of the minutiae m_j.

21.5.4 Evaluation of PRCs

To compute the PRCs, we first define correspondence, or match, between two minutiae. Let $x_a = (s_a, \theta_a)$ and $x_b = (s_b, \theta_b)$ be a pair of minutiae. The minutiae are said to correspond if for tolerance $\varepsilon = [\varepsilon_s, \varepsilon_\theta]$,

$$|s_a - s_b| \leqslant \varepsilon_s \text{ and } |\theta_a - \theta_b| \leqslant \varepsilon_\theta \tag{21.37}$$

where $|s_a - s_b|$, the Euclidean distance between the minutiae location $s_a = (x_{a1}, x_{a2})$ and $s_b = (x_{b1}, x_{b2})$, is given by

$$|s_a - s_b| = \sqrt{(x_{a1} - x_{b1})^2 + (x_{a2} - x_{b2})^2}. \tag{21.38}$$

Then, the probability that a random minutia $\boldsymbol{x}_a$ would match a random minutia $\boldsymbol{x}_b$ is given by

$$p_\varepsilon(\boldsymbol{x}) = p(|\boldsymbol{x}_a - \boldsymbol{x}_b| \leqslant \varepsilon | \Theta)$$
$$= \int_{\boldsymbol{x}_a} \int_{|\boldsymbol{x}_a - \boldsymbol{x}_b| \leqslant \varepsilon} p(\boldsymbol{x}_a | \Theta) p(\boldsymbol{x}_b | \Theta) \mathrm{d}\boldsymbol{x}_a \mathrm{d}\boldsymbol{x}_b, \tag{21.39}$$

where Θ is the set of parameters describing the distribution of the minutiae location and direction.

Finally, the PRC, or the probability of matching at least $\hat{m}$ pairs of minutiae within ε between two randomly chosen fingerprint f_1 and f_2 is calculated as

$$p_\varepsilon(\hat{m}, m_1, m_2) = \binom{m_1}{\hat{m}}\binom{m_2}{\hat{m}} \hat{m}! \cdot p_\varepsilon(\boldsymbol{x})^{\hat{m}} (1 - p_\varepsilon(\boldsymbol{x}))^{(m_1 - \hat{m}) \cdot (m_2 - \hat{m})}, \tag{21.40}$$

where m_1 and m_2 are numbers of minutiae in fingerprints f_1 and f_2, $p_\varepsilon(\boldsymbol{x})^{\hat{m}}$ is the probability of matching $\hat{m}$ specific pairs of minutiae between f_1 and f_2, $(1 - p_\varepsilon(\boldsymbol{x}))^{(m_1 - \hat{m}) \cdot (m_2 - \hat{m})}$ is the probability that none of minutiae pair would match between the rest of minutiae in f_1 and f_2 and $\binom{m_1}{\hat{m}}\binom{m_2}{\hat{m}}\hat{m}!$ is the number of different match sets that can be paired up.

Given n fingerprints and assuming that the number of minutiae in a fingerprint m can be modeled by the distribution $p(m)$, the general PRCs $p(n)$ is given by

$$p(n) = 1 - \bar{p}(n) = 1 - (1 - p_\varepsilon)^{\frac{n(n-1)}{2}}, \tag{21.41}$$

where p_ε is the probability of matching two random fingerprint from n fingerprints. If we set the tolerance in terms of number of matching minutiae to $\hat{m}$, p_ε is calculated by

$$p_\varepsilon = \sum_{m_1' \in M_1} \sum_{m_2' \in M_2} p(m_1') p(m_2') p_\varepsilon(\hat{m}, m_1', m_2'), \tag{21.42}$$

where M_1 and M_2 contain all possible numbers of minutiae in one fingerprint among n fingerprints, and $p_\varepsilon(\hat{m}, m_1', m_2')$ can be calculated by Eq. 21.40.

Given a specific fingerprint f, the specific nPRCs can be computed by

$$p(f, n) = 1 - (1 - p(f))^{n-1}, \tag{21.43}$$

where $p(f)$ is the probability that $\hat{m}$ pairs of minutiae are matched between the given fingerprint f and a randomly chosen fingerprint from n fingerprints.

$$p(f) = \sum_{m' \in M} p(m') \binom{m'}{\hat{m}} p(f_i), \qquad (21.44)$$

where M contains all possible numbers of minutiae in one fingerprint among n fingerprints, $p(m')$ is the probability of a figerprint having m' minutiae in n fingerprints, m_f is the number of minutiae in the given fingerprint f, minutiae set $f_i = (\boldsymbol{x}_{i1}, \boldsymbol{x}_{i2}, ..., \boldsymbol{x}_{i\hat{m}})$ is the subset of the minutiae set of given fingerprint and $p(f_i)$ is the joint probability of minutiae set f_i based on learned generative model.

21.5.5 Evaluation with Fingerprint Databases

Parameters of the fingerprint generative model introduced in Sections 21.5.2 was evaluated using the NIST fingerprint database. The NIST fingerprint database, NIST Special Database 4, contains 8-bit gray scale images of randomly selected fingerprints. Each print is 512×512 pixels with 32 rows of white space at the bottom of the print. The entire database contains fingerprints taken from 2 000 different fingers with 2 impression of the same finger. Thus, there are a total of 2 000 fingerprints using which the model has been developed. The fingerprints are classified into one of five categories (left loop, whirl, right loop, tented arch, and arch) with an equal number of prints from each class (400). The number of components K for the mixture model was found after validation using k-means clustering.

Values of PRC p_ε are calculated using the formula introduced in Section 21.5.4. The tolerance is set at $\varepsilon_s = 10$ pixels and $\varepsilon_\theta = \pi/8$. For comparison, the empirical PRC $\hat{p}_\varepsilon(\boldsymbol{x})$ was calculated with the same tolerance. To compute $\hat{p}_\varepsilon(\boldsymbol{x})$, the empirical probabilities of matching a minutiae pair between imposter fingerprints are calculated first by

$$\hat{p}_\varepsilon(\boldsymbol{x}) = \frac{1}{I} \sum_{i=1}^{I} \frac{\hat{m}_i}{m_i \times m_i'}, \qquad (21.45)$$

where I is the number of the imposter fingerprints pairs, $\hat{m}_i$ is the number of matched minutiae pairs and m_i and m_i' are the numbers of minutiae pairs in each of the two fingerprints. Then, the empirical PRC $\hat{p}_\varepsilon$ can be calculated by Eq. 21.41.

Both the theoretical and empirical PRCs are given in Table 21.5. The PRCs are calculated through varying number of minutiae in two randomly chosen fingerprint f_1 and f_2 and the number of matches between them. We can see that more minutiae the template and input fingerprint have, higher the PRC is. In experiments conducted on the NIST 4 dataset, the PRC values obtain here are smaller than the results in [22, 24]. The differences mainly result from use of differnt ways to evaluate PRC from gererative models which is described in Section 21.5.4. Different matching tolerance, which $p_\varepsilon(\boldsymbol{x})$ depends on may cause the differences also. Note that the theoritical PRC based on our model are close and have the same trend to empirical PRC. The consistency between the theoretical probilities and empirical probilities shows the validation of our generative model. The PRCs for the different m_1 and m_2 with 6, 26, 56, and 76 matches are shown in Fig. 21.12. It is obvious to note that, when $\hat{m}$ decreases or m_1 and m_2 increase, the probability of matching two random fingerprints is more.

Table 21.5   PRC for different fingerprint matches with varying m_1(number of minutiae in fingerprint f_1), m_2 (number of minutiae in fingerprint f_2) and $\hat{m}$ (number of matched minutiae). p_ε is the theoretical PRC for the general population and $\hat{p}_\varepsilon$ is the empirical PRC for NIST 4 dataset

m_1	m_2	$\hat{m}$	p_ε	$\hat{p}_\varepsilon$
		6	6.3×10^{-9}	1.0×10^{-8}
16	16	12	3.4×10^{-23}	1.0×10^{-22}
		16	1.2×10^{-37}	5.4×10^{-37}
		6	2.1×10^{-4}	3.4×10^{-4}
		12	1.1×10^{-11}	3.1×10^{-11}
36	36	24	4.0×10^{-34}	3.6×10^{-33}
		36	3.6×10^{-72}	9.9×10^{-71}
		6	1.8×10^{-2}	2.6×10^{-2}
		12	7.8×10^{-7}	2.1×10^{-6}
		24	2.6×10^{-21}	2.2×10^{-20}
56	56	36	1.7×10^{-42}	4.5×10^{-41}
		48	5.0×10^{-72}	4.1×10^{-70}
		56	1.1×10^{-101}	2.0×10^{-99}
		6	1.4×10^{-1}	1.8×10^{-1}
		12	5.0×10^{-4}	1.1×10^{-3}
		24	5.6×10^{-14}	4.2×10^{-13}
76	76	36	4.4×10^{-29}	1.1×10^{-27}
		48	3.6×10^{-49}	2.8×10^{-47}
		60	3.6×10^{-75}	8.8×10^{-73}
		76	4.9×10^{-128}	5.4×10^{-125}

Based on the PRC value, nPRC can be computed. Table 21.6 shows the nPRCs through varying numbers of matched minutiae pairs $\hat{m}$ in different numbers of fingerprints.

The specific nPRCs are also computed by Eq. (21.44) and given by Table 21.7. Here three fingerprints are chosen as query prints and they are shown in Fig. 21.13. The first one is a full print in good quality, the second one is a full print in low quality and the third one is a partial print. The specific nPRCs are calculated through varying number of minutiae in each template fingerprint (m) and the number of matches ($\hat{m}$), assuming that the number of fingerprints in template database (n) is 100 000. The numbers of minutiae in

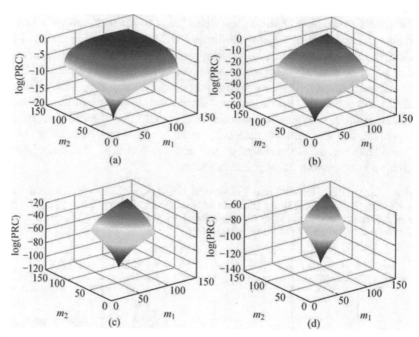

Fig. 21.12 PRCs with different number of the matched minutiae for (a) $\hat{m} = 6$, (b) $\hat{m} = 26$, (c) $\hat{m} = 56$, and (d) $\hat{m} = 76$.

3 given query fingerprint m_f are 71, 42, and 18. In 100 000 randomly chosen fingerprints there is only $5.137\,1 \times 10^{-39}$ probability that one of them have 12 matched minutiae with the fingerprint F_1.

Table 21.6 Fingerprint Probabilities: nPRCs with varying n and $\hat{m}$

Number of fingerprints n	Number of matches $\hat{m}$	$p(n)$
	16	1
	36	$4.173\,1 \times 10^{-5}$
1000	56	$1.721\,9 \times 10^{-18}$
	76	$4.854\,6 \times 10^{-38}$
	96	$4.236\,9 \times 10^{-64}$
	16	1
	36	$3.414\,6 \times 10^{-1}$
100 000	56	$1.723\,6 \times 10^{-14}$
	76	$4.859\,4 \times 10^{-34}$
	96	$4.241\,1 \times 10^{-60}$
	16	1
	36	1
10 000 000	56	$1.723\,6 \times 10^{-10}$
	76	$4.859\,4 \times 10^{-30}$
	96	$4.241\,1 \times 10^{-56}$

Fig. 21.13 Three specific fingerprints (from the same finger) used to calculate probabilities: (a) good quality full print F_1, (b) low quality full print F_2, and (c) partial print F_3.

Table 21.7 Fingerprint Probabilities: Specific nPRCs for fingerprints in Fig. 21.13 with $n = 100\,000$ fingerprints

Fingerprint f	Number of minutiae m_f	Number of matches $\hat{m}$	$p(n, f)$
F_1	71	55	8.5815×10^{-276}
		45	1.8829×10^{-211}
		35	2.0781×10^{-153}
		25	3.0360×10^{-102}
		12	5.1371×10^{-39}
		6	6.5876×10^{-15}
F_2	42	40	5.1783×10^{-212}
		35	2.4791×10^{-183}
		25	8.4136×10^{-115}
		12	1.6059×10^{-43}
		6	4.9787×10^{-16}
F_3	18	18	4.0256×10^{-100}
		12	3.4509×10^{-58}
		6	1.4068×10^{-22}

21.6 Summary

While forensic evidence of many modalities have long been used in the judicial system, e.g., impression evidence (latent prints, handwriting, shoe-prints), trace evidence (paint flakes, pollen, fibers, glass, hair), etc., characterizing their accuracy in identification is still needed to provide a scientific basis. The degree of individuality of a forensic modality can be established quantitatively by using either discriminative or generative approaches. In the former an error rate is determined after a suitable classifier is constructed. In the latter a probability distribution is determined from which different types of probabilities of random correspondence (PRC) are evaluated.

Generative models of individuality attempt to model the distribution of features and then use the models to determine the probability of random correspondence. We have proposed such models of individuality for birthdays, heights and fingerprints. Individuality is evaluated in terms of three probability measures: probability of random correspondence (PRC) between two individuals, general probability of random correspondence (nPRC) between two individuals among a group of n individuals and specific probability of random correspondence (specific nPRC) which is the probability of matching a given individual among n individuals.

The generative model for birthdays is based on assuming a uniform distribution where each date is assumed to be equally likely. The PRC for birthdays is 0.002 7. For 40 individuals, the nPRC is 0.891 2 and specific nPRC (any date) is 0.101 5. The nPRC value is the well-known birthday paradox where with $n = 23$ the value reaches a high value of 0.5.

The heights of female and male individuals are modeled by Gaussian distribution, where the parameters are learned from the health statistic data. The PRCs for female and male heights are 0.013 4 and 0.017 3. In a group of 40 people, the nPRCs for females and males are 0.999 95 and 0.999 999 97 and specific nPRCs for females (57 inches) and male(68 inches) are 0.181 5 and 0.586 2.

Models for fingerprint individuality have been proposed for ridge flows and minutiae. A mixture distribution was proposed to model minutiae information. The new generative model is compared by implementation and experiments with the empirical results on the NIST 4 dataset. The PRC obtained for a fingerprint template and input with 36 minutiae each with 12 matching minutiae is 1.1×10^{-11}. This probablity is very close to the empirical result which is 3.1×10^{-11}. nPRC and specific nPRC of fingerprints are computed also. Considering the case of 100 000 fingerprints, the nPRC with minutiae information where 56 minutiae pairs are matched is $1.723 6 \times 10^{-14}$. Given a specific fingerprint with 71 minutiae, the specific nPRC with minutiae information where 55 out of 71 minutiae are matched is $8.581 5 \times 10^{-276}$. The proposed generative model offers a reasonable and accurate fingerprint representation. The results provide a much stronger argument for the individuality of fingerprints in forensics than previous generative models.

Acknowledgements

This work was supported by a grant from the Department of Justice, Office of Justice Programs Grant NIJ 2005-DD-BX-K012. The opinions expressed are those of the authors and not of the DOJ.

References

[1]   Aitken C, Taroni F (2004) Statistics and the Evaluation of Evidence for Forensic Scientists. Wiley, New York

[2]   Ashbaugh D R (1999) Quantitative-Qualitative Friction Ridge Analysis: An Introduction to Basic and Advanced Ridgeology. CRC Press, Boca Raton

[3]   Banerjee A, Dhillon I, Ghosh J et al (2003) Generative Model-based Clustering of Directional Data. In Proceedings of the Ninth ACM SIGKDD International Conference on Knowledge Discovery and Data Mining

[4]   Bishop C (2006) Pattern Recognition and Machine Learning. Springer, New York

[5]   Bolle R M (2003) Guide to Biometrics. LLC, Springer, New York

[6]   Champod C, Margot P (1996) Computer Assisted Analysis of Minutiae Occurrences on Fingerprints. In: Almog J, Spinger E (eds) Proc International Symposium on Fingerprint Detection and Identification, p 305. Israel National Police, Jerusalem

[7]   Chen J, Moon Y s (2007) A Minutiae-based Fingerprint Individuality Model. Computer Vision and Pattern Recognition, 2007, CVPR '07

[8]   Dass S, Zhu Y, Jain A K (2005) Statistical Models for Assessing the Individuality of Fingerprints. Fourth IEEE Workshop on Automatic Identification Advanced Technologies, pp 3 − 9

[9]   Diaconia P, Mosteller F (1989) Methods for Studying Coincidences. J Am Stat Association, 84: 853 − 857

[10] Duda R O, Hart P E, Stork D G (2001) Pattern Classification, 2nd edn. Wiley, New York

[11] Fang G, Srihari S N, Srinivasan et al (2007) Use of ridge points in partial fingerprint matching. In: Biometric Technology for Human Identification IV, pp 65390D1 − 65390D9, SPIE

[12] Galton F (1892) Finger Prints. McMillan, London

[13] Henry E (1900) Classification and Uses of FingerPrints. Routledge & Sons London

[14] Hsu R V, Martin B (2008) An Analysis of Minutiae Neighborhood Probabilities. Biometrics: Theory, Applications and Systems

[15] Jebarra T (2004) Machine Learning: Discriminative and Generative. Kluwer Academic, Publishers, Boston

[16] Maio D, Maltoni D, Cappelli R (2002) Fingerprint Verification Competition. http://bias.csr.unibo.it/fvc2002/. Accessed 19 Jane 2010

[17] Mardia K V, Jupp P E (2000) Directional Statistics. Wiley, New York

[18] McDowell M A, Fryar C D, R H et al (2005) Anthropometric Reference Data for Children and Adults: United States Population, 1999 to 2002. Advance data from Vital and Health Statistics; Number 361

[19] Pankanti S, Prabhakar S, Jain A K (2002) On the Individuality of Fingerprints. IEEE Transactions on Pattern Analysis and Machine Intelligence, 24(8):

[20] Ralph M A S, D'Agostion. B (1986) Goodness-of-fit Techniques. CRC Press, Boca Raton

[21] Roxburgh T (1933) Galton's work on the Evidential Value of Fingerprints. Indian Journal of Statistics, 1: 62

[22] Saks F, Koehler J J (2008) The Individualization Fallacy in Forensic Science Evidence. Vanderbilt Law Review, 61: 199 − 219

[23] Srihari S N, Cha S, Arora H (2002) Individuality of Handwriting. Journal of Forensic Sciences, 47(4): 856 − 872

[24] Srihari S N, Cha S, Arora H (2008) Discriminability of Fingerprints of twins. Journal of Forensic Identification, 58(1): 109 − 127

[25] Srihari S N, Huang C, Srinivasan H (2008) On the Discriminability of the Handwriting of twins. Journal of Forensic Sciences, 53(2): 430 − 446

[26] Stoney D A (2001) Measurement of Fingerprint Individuality. In: Lee H, Gaensslen R (eds) Advances in Fingerprint Technology. CRC Press, Boca Raton

[27] Su C, Srihari S N (2008) Generative Models for Fingerprint Individuality Using Ridge Models. In Proceedings of International Conference on Pattern Recognition. IEEE Computer Society Press

[28] Tabassi E, Wilson C L, Watson C I. Fingerprint image quality.NISTIR 7151, National Institute of Standards and Technology, August 2004.

[29] Trauring M (1963) Automatic Comparison of Finger-ridge Patterns. Nature, p 197

22 Feature Mining and Pattern Recognition in Multimedia Forensics — Detection of JPEG Image Based Steganography, Double-Compression, Interpolation and WAV Audio Based Steganography

Qingzhong Liu[1], Andrew H. Sung[2,3,*], Mengyu Qiao[2],
Bernardete Ribeiro[4], Zhongxue Chen[5]

Abstract Steganography, the ancient art for secretive communications, has revived on the Internet by way of hiding secret data, in completely imperceptible manners, into a digital file. Thus, the steganography has created a serious threat to cyber security due to the covert channel it provides that can be readily exploited for various illegal purposes. Likewise, image tampering or forgery, which has been greatly facilitated and proliferated by photo processing tools, is increasingly causing problems concerning the authenticity of digital images. JPEG images constitute one of the most popular media on the Internet; yet they can be easily used for the steganography as well as easily tampered by, e.g., removing, adding, or splicing objects without leaving any clues. Therefore, there is a critical need to develop reliable methods for steganalysis (analysis of multimedia for the steganography) and for forgery detection in JPEG images to serve applications in national security, law enforcement, cybercrime fighting, digital forensics, and network security, etc. This article presents some recent results on detecting JPEG steganograms, doubly compressed JPEG images, and resized JPEG images based on a unified framework of feature mining and pattern recognition approaches. At first, the neighboring joint density features and marginal density features of the DCT coefficients of the JPEG image are extracted; then learning classifiers are applied to the features for the detection. Experimental results indicate that the method prominently improves the detection performances in JPEG images when compared to a previously well-studied method. Also, it

1 Department of Computer Science, Sam Houston State University, Huntsville, TX 77341, U.S.A. E-mail: qxl005@shsu.edu.

2 Department of Computer Science and [3]Institute for Complex Additive Systems Analysis, New Mexico Tech, Socorro, NM 87801, U.S.A. E-mail: sung@cs.nmt.edu.

3 Department of Computer Science, New Mexico Tech, Socorro, NM 87801, U.S.A. E-mail: myuqiao@cs.nmt.edu.

4 Department of Informatics Engineering, University of Coimbra, 3030-290 Coimbra, Portugal. E-mail: bribeiro@dei.uc.pt.

5 Center for Clinical and Translational Sciences, The University of Texas Health Science Center at Houston, Houston, TX 77030, U.S.A. E-mail: Zhongxue.Chen@uth.tmc.edu.

* Corresponding author.

is demonstrated that detection performance deteriorates with increasing image complexity; hence, a complete evaluation of the detection performance of different algorithms should include image complexity — in addition to other relevant factors such as hiding ratio or compression ratio — as a significant and independent parameter.

Compressed and uncompressed audios can also be easily used as carriers of the steganography; and this article also presents a new derivative-based method for steganalyzing uncompressed audios in WAV format. The method exploits Mel-cepstrum coefficients and Markov transition features from the second order derivative and employs pattern recognition techniques. Compared to a recently proposed signal stream based Mel-cepstrum method, the method exhibits considerable advantage under all categories of signal complexity-especially for audio streams with high signal complexity that are generally most challenging in steganalysis-and thereby significantly improves the state of the art in steganalysis of uncompressed audio streams.

22.1 Introduction

Steganography is the art and science of hiding information by embedding it into an innocuous digital media such as images, audios, videos, documents, network packets, etc. The innocuous digital media or files are called carriers or covers, the covers embedded with hidden data are called steganograms. Steganographic tools nowadays readily produce steganograms that are perceptibly indistinguishable from the cover. On the Internet digital steganography provides an easy way for covert communications; and compared to cryptography it has the advantage that the media do not attract attention to themselves, ensuring some additional degree of "security through obscurity". Consequently, steganography has created a serious threat for national and cyber security and law enforcement as terrorists, criminals, hackers and adversaries can all use it for a conceivable variety of nefarious purposes; as an example, recent news media reports [1–6] and court documents released by the U.S. Justice Department [7, 8] have revealed the first confirmed use of steganography for espionage. Therefore, there is a heightened need for implementing effective countermeasures for steganography, and developing steganalysis techniques with improved efficiency and reliability will be crucial to the effort.

Currently, most of steganography techniques adopt digital multimedia files, most often digital images, as carriers. Image steganography can roughly be divided into two types: space-hiding directly embeds data in pixel values, for instance, LSB matching [9], improved LSB matching steganography [10], and highly undetectable steganography [11]; transform-hiding hides data in the transform coefficients, such as compressed DCT compressed domain [12, 13], modified matrix encoding [14] and the improved version [15], Outguess [16]. Some other information hiding techniques include spread spectrum steganography [17], statistical steganography, and cover generation steganography [18].

Steganalysis generally employs techniques of signal processing, feature mining, and pattern recognition, and aims at detecting the existence of hidden material. To this date, a few popular steganographic systems such as LSB embedding, LSB matching, spread spectrum steganography, etc., have been successfully steganalyzed [19 – 36].

JPEG image is one of the most popular media and several steganographic systems for hiding data in JPEG images are available on the Internet, including commercial hiding software CryptoBola, well-known hiding algorithms/tools Outguess [16], F5 [13], Steghide [37], YASS [12], and the improved versions [38, 39], minimal distortion steganography by using modified matrix encoding (MME) [14] and the improved version based on BCH syndrome coding [15], etc.

In addition to steganography, today's technology allows digital media to be easily altered and manipulated. While we are routinely exposed to huge volumes of digital media, our traditional confidence in the integrity of these media has also been eroded since doctored pictures, video clips, and voices are appearing with growing frequency and increased sophistication in mainstream media outlets, scientific journals, political campaigns, and courtrooms. For example, a recent state-run newspaper in Egypt published a doctored picture, shown in Fig. 22.1, apparently attempting to create the impression that its country's president was leading the group in Middle East peace talks in Washington DC [40 – 43].

Fig. 22.1 An image forgery example with the doctored image (left) and original image (right) from CBS News [40,41]. The doctored image shows Egyptian President Hosni Mubarak leads Israeli Prime Minister Benjamin Netanyahu, President Obama, Palestinian Authority President Mahmoud Abbas and King Abdullah II of Jordan toward the East Room of the White House on the first day of the Middle East peace talks, Sept. 1, 2010 in Washington. The original image shows the same group with President Obama in the lead.

Generally, tampering manipulation on a JPEG image involves several different basic operations, such as image resize, rotation, splicing, double compression, etc. While we decode the bit stream of a JPEG image and implement the manipulation in spatial domain, and then compress the modified image back to JPEG format, if the quantization matrices are different between the original JPEG image and the modified image, we say the modified JPEG image has undergone a double JPEG compression. Although JPEG based double compression does not by itself prove malicious or unlawful tampering, it is an evidence of image manipulation.

Figure 22.2 shows two JPEG source images and a tampered image composited from the two source images, from worth1000.com. The quantization matrices affiliated with the luminance parts of these three JPEG images are given with different quantization values. In addition to double JPEG compression, the tampering also involves image resampling, with up-scaled size on doctored image (actual size is not shown in Fig. 22.2).

To detect audio steganograms, Ozer et al. constructed the detector based on the characteristics of the denoised residuals of the audio file [33]; Johnson et al. set up a statistical model by building a linear basis that captures certain statistical properties of audio signals [44]; Kraetzer and Dittmann

$$\begin{pmatrix} 11 & 7 & 7 & 11 & 16 & 26 & 34 & 40 \\ 8 & 8 & 9 & 13 & 17 & 38 & 40 & 36 \\ 9 & 9 & 11 & 16 & 26 & 38 & 46 & 37 \\ 9 & 11 & 15 & 19 & 34 & 57 & 53 & 41 \\ 12 & 15 & 24 & 37 & 45 & 72 & 68 & 51 \\ 16 & 23 & 36 & 42 & 53 & 69 & 75 & 61 \\ 32 & 42 & 51 & 57 & 68 & 80 & 79 & 67 \\ 48 & 61 & 63 & 65 & 74 & 66 & 68 & 65 \end{pmatrix}$$

$$\begin{pmatrix} 2 & 2 & 2 & 2 & 3 & 4 & 5 & 6 \\ 2 & 2 & 2 & 2 & 3 & 4 & 5 & 6 \\ 2 & 2 & 2 & 2 & 4 & 5 & 7 & 9 \\ 2 & 2 & 2 & 4 & 5 & 7 & 9 & 12 \\ 3 & 3 & 4 & 5 & 8 & 10 & 12 & 12 \\ 4 & 4 & 5 & 7 & 10 & 12 & 12 & 12 \\ 5 & 5 & 7 & 9 & 12 & 12 & 12 & 12 \\ 6 & 6 & 9 & 12 & 12 & 12 & 12 & 12 \end{pmatrix}$$

(a) Source1 and the quantization matrix   (b) Source2 and the quantization matrix

$$\begin{pmatrix} 6 & 4 & 4 & 6 & 10 & 16 & 20 & 24 \\ 5 & 5 & 6 & 8 & 10 & 23 & 24 & 22 \\ 6 & 5 & 6 & 10 & 16 & 23 & 28 & 22 \\ 6 & 7 & 9 & 12 & 20 & 35 & 32 & 25 \\ 7 & 9 & 15 & 22 & 27 & 44 & 41 & 31 \\ 10 & 14 & 22 & 26 & 32 & 42 & 45 & 37 \\ 20 & 26 & 31 & 35 & 41 & 48 & 48 & 40 \\ 29 & 37 & 38 & 39 & 45 & 40 & 41 & 40 \end{pmatrix}$$

(c) Tampered image and the quantization matrix

Fig. 22.2 Image tampering involves JPEG based double compression and resampling (actual image sizes are not shown here). The images are downloaded from worth1000.com.

recently proposed a Mel-cepstrum based analysis to detect hidden messages [45]; Zeng et al. presented new algorithms to detect phase coding steganography based on analysis of the phase discontinuities [46] and to detect echo steganography based on statistical moments of peak frequency [47]. Of all these methods, Kraetzer and Dittmann's Mel-cepstrum audio steganalysis is particularly noteworthy since it is the first to utilize Mel-frequency cepstral coefficients-which are widely used in speech recognition — for audio steganalysis.

Without doubt, multimedia forensics has emerged as a new discipline as it has important applications in protecting public safety and national security, as well as impacts to our daily life. In what follows, we discuss some well-known work in JPEG based steganalysis and double compression and Mel-cepstrum audio steganalysis in Section 22.2. Statistical models of DCT coefficients and the modification caused by the manipulations are described in Section 22.3. Section 22.4 presents the marginal density and neighboring joint density features for JPEG image forensics, followed by derivative based audio steganalysis in Section 22.5. Section 22.6 briefly introduces the pattern recognition techniques in our study, and Section 22.7 shows the experiments. We give conclusions in Section 22.8.

22.2 Related Works

The literature contains much work on image steganalysis and some recent work on audio steganalysis; this section briefly reviews the previous papers that are related to our own research.

22.2.1 JPEG-Based Image Forensics

To detect the information-hiding in JPEG images using the steganographic algorithm F5, the authors in [20] designed a method to estimate the cover-image histogram from the steganogram. To break a recently designed JPEG steganography YASS, several detection arts have been proposed: Li et al. proposed a method by extracting the statistical features from locations which are possible to hold embedding host blocks [25]; Yu and Babaguchi performed the detection via pixel and DCT coefficient analysis [36]; Kodovsky et al. conducted the study by using different feature sets [24]. To design a universal steganalysis method, Fridrich presented a feature-based steganalytic method for JPEG images [21]. Shi et al. applied a DCT intra-block-based Markov approach to effectively attacking JPEG steganography, which demonstrated a remarkable advantage over the prior methods [35]. Based on the work proposed in reference [35], Pevny and Fridrich merged their proposed DCT features and calibrated Markov transition probabilities to improve the detection performance [34]. Similarly, based on the Markov approach, the authors in [19] and [29, 31] individually expanded the original intra-block Markov approach to inter-block approach.

Although Markov approach and the improved detection methods are very promising for the detection of JPEG based steganograms, unfortunately, the authors did not elaborate on the reason why the approach is successful. To study this issue, we explored the statistical property of DCT coefficients and found that, most JPEG steganographic systems modify the DCT coefficients and change the correlation of neighboring DCT coefficients. Specifically, embedding of the secret information changes the neighboring joint distribution and, therefore, Markov transition probabilities are affected. Our study also shows that, that Markov approach does not completely represent the neighboring joint relation, that is, it falls short of fully exploring the modification caused by information-hiding [30].

In image forgery detection, researchers have proposed several different methods, which are included in a very nice survey by Farid [48]. Regarding detection of double JPEG compression, two methods were recently proposed by Pevny and Fridrich [49] and Chen et al. [50], respectively. Based on the fact that double JPEG compression changes the compressed DCT coefficients and hence modifies the histogram at a certain frequency in DCT 2D array, Pevny and Fridrich designed a feature set consisting of low-frequency DCT coefficients. An 8×8 DCT block has 64 frequency coefficients, the frequency coordinates are paired from $(1, 1)$ to $(8, 8)$, corresponding to upper-left corner to right-bottom corner. Pevny and Fridrich [49] extracted the histogram at each location from the following coordinate pair:

$$S = \{(2, 1), (3, 1), (4, 1), (1, 2), (2, 2), (3, 2), (1, 3), (2, 3), (1, 4)\}. \qquad (22.1)$$

The feature set consists of the following probability values

$$X = \left\{ \frac{1}{C_{ij}} (h_{ij}(0), h_{ij}(1), ..., h_{ij}(15)) | (i, j) \in S \right\}, \text{ s.b. } C_{ij} = \sum_{m=0}^{15} h_{ij}(m), \qquad (22.2)$$

where $h_{ij}(m)$ stands for the histogram of the DCT coefficient at frequency coordinate (i, j) with the value m.

To identify double JPEG compression, Chen et al. [50] designed a feature set consisting of Markov transition probability on the difference 2D array. The feature set is extracted as follows: Let matrix $F(u, v)$ represent the absolute values of DCT coefficients of the image. Four difference arrays are calculated along four directions: horizontal, vertical, diagonal, and minor diagonal, denoted $F_h(u, v)$, $F_v(u, v)$, $F_d(u, v)$, and $F_m(u, v)$, respectively.

$$F_h(u, v) = F(u, v) - F(u + 1, v), \qquad (22.3)$$
$$F_v(u, v) = F(u, v) - F(u, v + 1), \qquad (22.4)$$
$$F_d(u, v) = F(u, v) - F(u + 1, v + 1), \qquad (22.5)$$
$$F_m(u, v) = F(u + 1, v) - F(u, v + 1). \qquad (22.6)$$

The Markov transition matrix is constructed by

$$M_f(i, j) = \frac{\sum_{u=1}^{M-1} \sum_{v=1}^{N} \delta(F_f(u, v) = i, F_f(u + 1, v) = j)}{\sum_{u=1}^{M-1} \sum_{v=1}^{N} \delta(F_f(u, v) = i)}. \qquad (22.7)$$

Here $f \in \{h, v, d, m\}$ and F_f is the difference 2D array with the size of $M \times N$. The range of i and j is $[-4, 4]$, for a total of 324 features to be used in the detection.

In our study on detection of JPEG-based steganography, double compression and interpolation, we explore the statistical property of DCT coefficients and find that, the operations in steganography, double compression, and interpolation actually modify some DCT coefficients, and hence modify the marginal density at each specific frequency band or change the correlation of neighboring DCT coefficients; accordingly, we design a method to detect JPEG-based steganography, double compression and interpolation, based on feature mining on the DCT coefficients and pattern classification techniques.

22.2.2 Signal-Based Mel-cepstrum Audio Steganalysis

In speech processing, the Mel-frequency cepstrum is a representation of the short-term power spectrum of a sound. Mel-frequency cepstral coefficients (MFCCs) are coefficients that collectively make up a Mel-frequency cepstrum. Mel-cepstrum is commonly used for representing the human voice and musical signals [51]. Inspired by the success in speech recognition, Kraetzer and Dittmann proposed Mel-cepstrum based speech steganalysis, including the following two types of Mel-cepstrum coefficients [45]:

1. Mel-frequency cepstral coefficients (MFCCs), $sf_{mel1}, sf_{mel2}, \ldots, sf_{melM}$, where M is the number of MFCCs; the value of M is 29 for a signal with a sampling rate of 44.1 kHz. MFCCs can be calculated by the following equation, where MT indicates the Mel-scale transformation:

$$MelCepstrum = FT(MT(FT(f))) = \begin{bmatrix} sf_{mel1} \\ sf_{mel2} \\ \ldots \\ sf_{smlM} \end{bmatrix}. \tag{22.8}$$

2. Filtered Mel-frequency cepstral coefficients (FMFCCs), $sf_{mel1}, sf_{mel2}, \ldots, sf_{melM}$, which can be calculated by the following equation:

$$FilteredMelCepstrum = FT(\text{SpeechBandFiltering } (MT(FT(f))))$$

$$= \begin{bmatrix} sf_{mel1} \\ sf_{mel2} \\ \ldots \\ sf_{melM} \end{bmatrix}. \tag{22.9}$$

In Eq. (22.9), the role of speech band filtering is to remove the speech relevant bands (the spectrum components between 200 and 6819.59 Hz) [45].

22.3 Statistical Characteristics and Modification

Generalized Gaussian distribution (GGD) is widely used in modeling probability density function (PDF) of a multimedia signal. It is very often applied to transform coefficients such as discrete cosine transform (DCT) or wavelet ones. Experiments show that adaptively varying two parameters of the GGD [52, 53] can achieve a good PDF approximation, for the marginal density of transform coefficients. The GGD model is described as

$$\rho(x; \alpha, \beta) = \frac{\beta}{2\alpha\Gamma(1/\beta)} \exp\{-(|x|/\alpha)^{\beta}\}, \tag{22.10}$$

where $\Gamma(\cdot)$ is the Gamma function, scale parameter α models the width of the PDF peak, and shape parameter β models the shape of the distribution.

An 8×8 DCT block has 64 frequency coefficients, our study shows that the marginal density of DCT coefficients at each specific frequency approximately follows the GGD distribution and some manipulation, for instance, double JPEG compression, changes the density. Figure 22.3 demonstrates a singly compressed JPEG image with quality factor '75' (a), doubly compressed JPEG images with the first compression quality factor '55' (b) and '90' (c) respectively, followed by the second compression quality factor '75', and the marginal densities at frequency coordinates (2, 1), (2, 2), and (1, 3). Compared to the marginal density of the single compression, Fig. 22.3 (d), (g), and (j), the modification caused by the double compression from the low quality factor '55', shown in Fig. 22.3 (e), (h), and (k), is noticeable. However, the modification caused by the double compression from the high quality factor '90', Fig. 22.3 (f), (i), and (l), is less obvious.

Although there does not appear to exist a generally agreed-upon multivariate extension of the univariate generalized Gaussian distribution, researchers have defined a parametric multivariate generalized Gaussian distribution (MGGD) model that closely fits the actual distribution of wavelet coefficients in clean natural images, exploited the dependency between the estimated wavelet coefficients and their neighbors or other coefficients in different subbands based on the extended GGD model, and achieved good image denoising [54]. The MDDG model is as follows:

$$
p(x) = \gamma \exp \left\{ - \left(\frac{(x - \mu)^t \displaystyle\sum_x^{-1} (x - \mu)}{\alpha} \right)^\beta \right\},
\tag{22.11}
$$

where γ indicates a normalized constant defined by α and β, $\displaystyle\sum_X$ is the covariance matrix and μ is the expectation vector.

To exploit the dependency between the compressed DCT coefficients and their neighbors, we study the neighboring joint density of the DCT coefficients, and postulate that certain manipulations such as JPEG double compression, information hiding, etc., will modify the neighboring joint density. Let the left (or upper) adjacent DCT coefficient be denoted by random vector X_1 and the right (or lower) adjacent DCT coefficient be denoted by random vector X_2; let $X = (X_1, X_2)$. The DCT neighboring joint density will be modified by the manipulation, and the change hence leaves a track for the manipulation. An example of the modification of the joint density caused by JPEG double compression is illustrated by Fig. 22.4. Figure 22.4 (a), (b), and (c) show the neighboring joint density of the singly compressed JPEG image of Fig. 22.3 (a), the doubly compressed JPEG image of Fig. 22.3 (b), and the doubly compressed JPEG image of Fig. 22.3 (c), respectively. The differences of the neighboring joint density between the double compression and

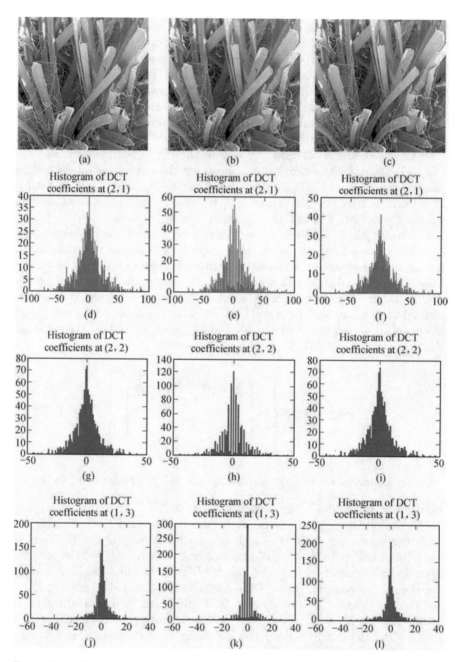

Fig. 22.3 Marginal densities of the singly compressed JPEG image with quality factor '75'(a) at the frequency coordinates (2, 1) (d), (2, 2) (g), and (1, 3) (j), and the doubly compressed JPEG images (b; c) and the marginal densities (e, h, and k; f, i, and l). X-axis shows the values of the DCT coefficients and Y-axis shows the occurrences.

the single compression are given by Fig. 22.4 (d) and (e). It verifies our postulation that the neighboring joint density has been modified by the double compression.

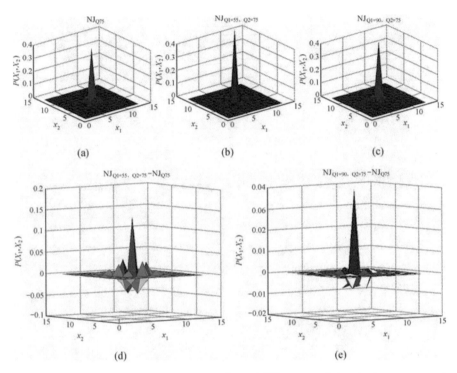

Fig. 22.4 Neighboring joint densities of the DCT arrays of the singly compressed JPEG image (Fig. 22.3 (a)) and the doubly compressed JPEG images (Fig. 22.3 (b) and Fig. 22.3 (c)) and the differences.

To illustrate the modification of neighboring joint density caused by information hiding, Fig. 22.5 (a) and (d) show a JPEG cover and the DCT neighboring joint density probability, respectively. Figure 22.5 (b) and (e) give the F5 steganogram carrying some hidden data and the neighboring joint density distribution. Figure 22.5 (c) and (f) are the Steghide steganogram and the joint density, respectively. Although the three images look identical, the neighboring joint densities are different. The differences are given in Fig. 22.5 (g) and (h), indicating that information hiding modifies the neighboring joint density. Figure 22.5 (c), (d), and (e) also demonstrate that the neighboring joint density is approximately symmetric about the origin. Considering this property, we anticipate that the neighboring joint density of the absolute values of the DCT coefficient array will also be changed by the information hiding, which is validated by Fig. 22.5 (i) and (j).

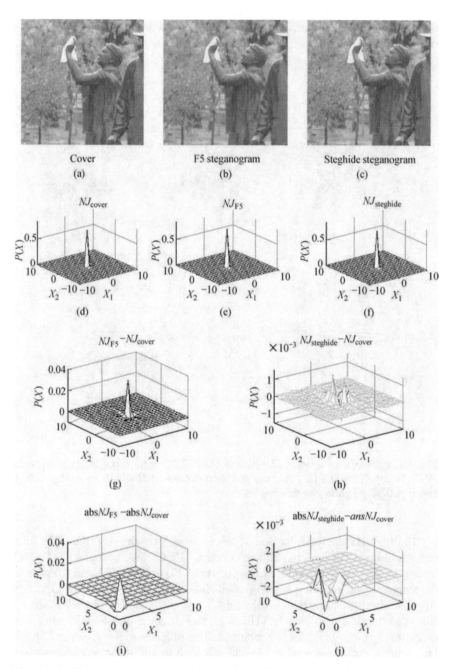

Fig. 22.5 Neighboring joint densities of the DCT arrays and the absolute arrays and the density differences.

22.4 Feature Mining for JPEG Image Forensics

In this section, we describe marginal density features and neighboring joint density features that can be explored to expose the difference between JPEG covers and steganograms or between original and doctored images.

22.4.1 Marginal Density Features

Since manipulations such as double JPEG compression modify the marginal density of DCT coefficients at each specific frequency coordinate [49], it will also modify the marginal density of the absolute DCT coefficients. To reduce the number of features and speed up the detection process, we design the following marginal density features at the low frequency of the absolute DCT coefficients.

An 8×8 DCT block has 64 frequency coefficients, the frequency coordinates are paired from $(1, 1)$ to $(8, 8)$, corresponding to upper-left low frequency to right-bottom high frequency. Let F denote the DCT coefficient array of a JPEG image, which consists of $M \times N$ blocks, F_{ij} ($i = 1, 2, \ldots, M$; $j = 1, 2, \ldots, N$). We extract the histogram at each location from the following coordinate pair:

$$S = \{(2, 1), (1, 2), (1, 3), (2, 2), (3, 1), (1, 4), (2, 3), (3, 2), (4, 1)\}. \qquad (22.12)$$

The feature set consists of the following probability values:

$$X = \left\{ \frac{1}{MN}(h_{kl}(0), h_{kl}(1), h_{kl}(2), h_{kl}(3), h_{kl}(4), h_{kl}(5),)|(k, l) \in S \right\}, \qquad (22.13)$$

where $h_{kl}(m)$ stands for the histogram of the absolute DCT coefficient at frequency coordinate (k, l) with the value m. In our experiments, there are 54 features in the marginal density set.

22.4.2 Neighboring Joint Density Features

In our detection algorithm, the neighboring joint features are extracted on intra-block and inter-block from the DCT coefficient array and the absolute array [55], described as follows.

DCT Coefficient Array Based Feature Extraction

1) Neighboring Joint Density on Intra-block

Let F denote the compressed DCT coefficient array of a JPEG image, consisting of $M \times N$ blocks $F_{ij}(i = 1, 2, \ldots, M; j = 1, 2, \ldots, N)$. Each block has a size of 8×8. The intra-block neighboring joint density matrix on horizontal direction NJ_{1h} and the matrix on vertical direction NJ_{1v} are constructed as follows:

$$NJ_{1h}(x, y) = \frac{\sum_{i=1}^{M} \sum_{j=1}^{N} \sum_{m=1}^{8} \sum_{n=1}^{7} \delta(c_{ijmn} = x, c_{ijm(m+1)} = y)}{56MN}, \quad (22.14)$$

$$NJ_{1v}(x, y) = \frac{\sum_{i=1}^{M} \sum_{j=1}^{N} \sum_{m=1}^{7} \sum_{n=1}^{8} \delta(c_{ijmn} = x, c_{ij(m+1)n} = y)}{56MN}. \quad (22.15)$$

where c_{ijmn} stands for the compressed DCT coefficient located at the mth row and the nth column in the block F_{ij}; $\delta = 1$ if its arguments are satisfied, otherwise $\delta = 0$; x and y are integers. For computational efficiency, we define NJ_1 as the neighboring joint density features on intra-block, as follows:

$$NJ_1(x, y) = \{NJ_{1h}(x, y) + NJ_{1v}(x, y)\}/2. \quad (22.16)$$

In our experiment, the values of x and y are in the range of $[-6, +6]$, so NJ_1 has 169 features.

It is worth noting that if the ranges of x and y increase, the number of features will increase greatly. As shown by Fig. 22.5, the modification caused by information hiding is negligible while the values of x and y are away from the origin; in such case, the feature extraction and classification take more time but the detection may not be better. On the contrary, if the ranges of x and y are too small, say, in the range of $[-1, 1]$, the features will be unable to completely reflect the modification caused by information hiding.

2) Neighboring Joint Density on Inter-block

The inter-block neighboring joint density matrix on horizontal direction NJ_{2h} and the matrix on vertical direction NJ_{2v} are constructed as follows:

$$NJ_{2h}(x, y) = \frac{\sum_{m=1}^{8} \sum_{n=1}^{8} \sum_{i=1}^{M} \sum_{j=1}^{N-1} \delta(c_{ijmn} = x, c_{i(j+1)mn} = y)}{64M(N-1)}, \quad (22.17)$$

$$NJ_{2v}(x, y) = \frac{\sum_{m=1}^{8} \sum_{n=1}^{8} \sum_{i=1}^{M-1} \sum_{j=1}^{N} \delta(c_{ijmn} = x, c_{(i+1)jmn} = y)}{64(M-1)N}. \quad (22.18)$$

We define NJ_2 as the neighboring joint density features on inter-block, calculated as follows:

$$NJ_2(x, y) = \{NJ_{2h}(x, y) + NJ_{2v}(x, y)\}/2. \quad (22.19)$$

Similarly, the values of x and y are in $[-6, +6]$ and NJ_2 has 169 features.

Absolute DCT Coefficient Array Based Feature Extraction

1) Neighboring Joint Density on Intra-block

Let F denote the compressed DCT coefficient array as before. The intra-block neighboring joint density matrix on horizontal direction $absNJ_{1h}$ and the matrix on vertical direction $absNJ_{1v}$ are given by

$$absNJ_{1h}(x,y) = \frac{\sum_{i=1}^{M}\sum_{j=1}^{N}\sum_{m=1}^{8}\sum_{n=1}^{7} \delta(|c_{ijmn}| = x, |c_{ijm(n+1)}| = y)}{56MN}, \quad (22.20)$$

$$absNJ_{1v}(x,y) = \frac{\sum_{i=1}^{M}\sum_{j=1}^{N}\sum_{m=1}^{7}\sum_{n=1}^{8} \delta(|c_{ijmn}| = x, |c_{ij(m+1)n}| = y)}{56MN}, \quad (22.21)$$

where c_{ijmn} is the DCT coefficient located at the mth row and the nth column in the block F_{ij}; $\delta = 1$ if its arguments are satisfied, otherwise $\delta = 0$; x and y are integers. For computational efficiency, we define $absNJ_1$ as the neighboring joint density features on intra-block, calculated as follows:

$$absNJ_1(x,y) = \{absNJ_{1h}(x,y) + absNJ_{1v}(x,y)\}/2. \quad (22.22)$$

In our algorithm, the values of x and y are in the range of $[0, 5]$, so $absNJ_1$ consists of 36 features.

2) Neighboring Joint Density on Inter-block

The inter-block neighboring joint density matrix on horizontal direction $absNJ_{2h}$ and the matrix on vertical direction $absNJ_{2v}$ are constructed as follows:

$$absNJ_{2h}(x,y) = \frac{\sum_{m=1}^{8}\sum_{n=1}^{8}\sum_{i=1}^{M}\sum_{j=1}^{N-1} \delta(|c_{ijmn}| = x, |c_{i(j+1)mn}| = y)}{64M(N-1)}, \quad (22.23)$$

$$absNJ_{2v}(x,y) = \frac{\sum_{m=1}^{8}\sum_{n=1}^{8}\sum_{i=1}^{M-1}\sum_{j=1}^{N} \delta(|c_{ijmn}| = x, |c_{(i+1)jmn}| = y)}{64(M-1)N}. \quad (22.24)$$

We define $absNJ_2$ as the neighboring joint density features on inter-block, calculated as follows:

$$absNJ_2(x,y) = \{absNJ_{2h}(x,y) + absNJ_{2v}(x,y)\}/2. \quad (22.25)$$

Similarly, the values of x and y are in $[0, 5]$ and $absNJ_2$ has 36 features.

22.5 Derivative Based Audio Steganalysis

As a widely used method of voice recognition, Mel-cepstrum analysis was adopted to successfully steganalyze WAV audio [45]. In JPEG image steganalysis, Markov approach was employed and achieved good porformance [35]. In this section, we propose derivative based spectrum and Mel-cepstrum analysis as well as derivative based Markov approach to minimize individual characterastics of WAV audio and further improve the accuracy of detection [74, 81].

22.5.1 Spectrum and Mel-cepstrum Analysis

In image processing, second order derivative is widely used to detect isolated points and edges [56]. Exploiting its great usefulness in detecting various objects, we design a scheme of second order derivative based audio steganalysis, the details of which are described as follows.

An audio signal is denoted $f(t)(t = 0, 1, 2, ..., N - 1)$. The second order derivative $D_f^2(\bullet)$ is defined as follows:

$$D_f^2(t) \equiv \frac{d^2 f}{dt^2} = f(t+1) - 2 * f(t) + f(t-1), \quad t = 1, 2, ..., N - 2. \quad (22.26)$$

Similar to the additive noise model proposed by Harmsen [22], a stego-signal is denoted $s(t)$, which can be modeled by adding a noise or error signal $e(t)$ to the original signal $f(t)$,

$$s(t) = f(t) + e(t). \quad (22.27)$$

Second order derivatives of $e(t)$ and $s(t)$ are denoted $D_e^2(\bullet)$ and $D_s^2(\bullet)$, respectively. Thus,

$$D_s^2(\bullet) = D_f^2(\bullet) + D_e^2(\bullet). \quad (22.28)$$

The Discrete Fourier Transforms (DFTs) of $D_s^2(\bullet), D_f^2(\bullet)$, and $D_e^2(\bullet)$, are denoted F_k^s, F_k^f, and F_k^e, respectively.

$$F_k^s = \sum_{t=0}^{M-1} D_s^2(t) e^{-\frac{j2\pi}{M}kt}, \quad (22.29)$$

$$F_k^f = \sum_{t=0}^{M-1} D_f^2(t) e^{-\frac{j2\pi}{M}kt}, \quad (22.30)$$

$$F_k^e = \sum_{t=0}^{M-1} D_e^2(t) e^{-\frac{j2\pi}{M}kt}, \quad (22.31)$$

where $k = 0, 1, 2, \ldots, M-1$ and M is the number of samples of the derivatives. So we have

$$F_k^s = F_k^f + F_k^e. \tag{22.32}$$

Assume θ is the angle between the vectors F_k^f and F_k^e, then

$$|F_k^s|^2 = |F_k^f|^2 + |F_k^e|^2 + 2|F_k^f| \cdot |F_k^e| \cdot \cos \theta. \tag{22.33}$$

For most steganographic systems, the hidden message or payload does not depend on the cover, that is, $e(t)$, the signal that approximates the payload signal, is irrelative to $f(t)$, the cover signal. Therefore, θ is an arbitrary value in the range $[0, \pi]$, the expected value of $|F_k^s|^2$ is calculated as follows:

$$E(|F_k^s|^2) = \frac{\int_0^\pi (|F_k^f|^2 + |F_k^e|^2 + 2|F_k^f| \cdot |F_k^e| \cdot \cos \theta) d\theta}{\int_0^\pi d\theta} = |F_k^f|^2 + |F_k^e|^2. \tag{22.34}$$

So, we have

$$E(|F_k^s|^2)/|F_k^f|^2 = 1 + |F_k^e|^2/|F_k^f|^2. \tag{22.35}$$

The expected value of the variance is obtained by the following equation:

$$E[(|F_k^s|^2 - E(|F_k^s|^2))^2] = \frac{\int_0^\pi 4|F_k^f|^2 \cdot |F_k^e|^2 \cdot \cos^2 \theta d\theta}{\int_0^\pi d\theta} = 2|F_k^f|^2 \cdot |F_k^e|^2. \tag{22.36}$$

Digital audio streams, especially speech audio clips, are normally band-limited. On the other side, regarding the low and middle frequency components, the power spectrum of audio signal (second order derivative) is much stronger than the power spectrum of the error signal or hidden data (second order derivative); that is, $|F_k^e|^2/|F_k^f|^2$ is almost zero. Based on Eq. (22.35), the difference between the spectrum of the cover and the stego-signal at low and middle frequency is negligible; however, the situation is very different at the high frequency components. As frequency increases, $|F^e|$ increases, and $|F^f|$ may decrease, the change of the spectrum resulted from embedding hidden data is no longer negligible, hence the statistics extracted from the high frequency components may be the clue to detecting the information-hiding behavior.

Figure 22.6 shows the spectra of the second order derivatives of a cover (left) and the correlated stego-signal (right) over the whole frequency range (first row) and over the high frequency region (second row). It clearly shows that the stego-signal has higher magnitude than the cover-signal in the derivative spectrum for high frequency components.

We may directly take the derivative based spectrum statistics in high frequency regions as features for audio steganalysis. In real-world detection,

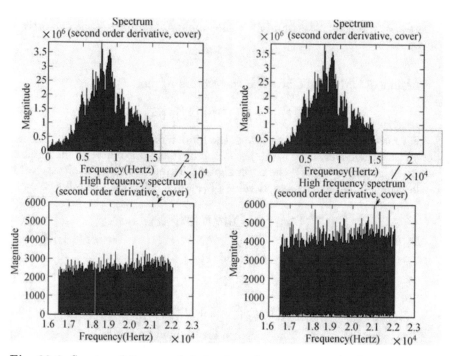

Fig. 22.6 Spectra of the second derivatives of a cover signal (left) and the stego-signal (right). Both figures in the first row show half magnitude values due to symmetric characteristics of Fourier transforms [80].

however, the cover reference shown in Fig. 22.6 is not available for steganalysis. Due to the fact that different audio streams have different spectrum characteristics, the detection derived from Eq. (22.35) may not be practical without a comparison with the original cover. In such case, Eq. (22.36) shows that the rate of power change in different spectrum bands of the stego-audio is quite different from the original. Based on Kraetzer and Dittmann proposed signal based Mel-cepstrum audio steganalysis [45], we formulate the second order derivative based MFCCs and FMFCCs, obtained by replacing the signal f in Eqs. (22.8) and (22.9) in Section 22.2 with the second order derivative $D_f^2(\bullet)$.

$$MelCepstrum = FT(MT(FT(D_f^2))) = \begin{bmatrix} sf_{mel1} \\ sf_{mel2} \\ \cdots \\ sf_{melM} \end{bmatrix}, \qquad (22.37)$$

and

$$FilteredMelCepstrum = FT(\text{SpeechBandFiltering } (MT(FT(D_f^2))))$$
$$= \begin{bmatrix} sf_{mel1} \\ sf_{mel2} \\ \cdots \\ sf_{melM} \end{bmatrix}. \tag{22.38}$$

Second order derivative based Mel-cepstrum coefficients, calculated by Eqs. (22.37) and (22.38), form the first type of features in our WAV audio steganalysis.

22.5.2 Markov Approach

The Markov approach has been widely used in different areas. In steganalysis, Shi et al. [35] presented a Markov process to detect the information-hiding behaviors in JPEG images based on the first order derivative of the quantized DCT coefficients. Since second order derivatives perform better than first order derivatives in detecting isolated points and edges [56], we design a Markov approach for audio steganalysis based on second order derivative of audio signals, described as follows:

An audio signal is denoted $f(t)(t = 0, 1, 2, ..., N-1)$, the minimal interval of the magnitude is 1. $D_f^2(t)(t = 1, 2, ..., N - 2)$ denotes the second order derivative, the Markov transition probability is calculated as follows:

$$M_{D_f^2}(i,j) = \frac{\displaystyle\sum_{t=1}^{N-3} \delta(D_f^2(t) = i, D_f^2(t+1) = j)}{\displaystyle\sum_{t=1}^{N-3} \delta(D_f^2(t) = i)}, \tag{22.39}$$

where $\delta = 1$ if its arguments are satisfied, otherwise $\delta = 0$. The range of i and j is $[-6, 6]$, so we have a 13×13 transition matrix, consisting of 169 features.

22.5.3 Signal Complexity

Our work in image steganalysis [27 − 29] has demonstrated that the information-hiding ratio cannot be used as the sole parameter in a complete and objective performance evaluation; this is because at the same hiding ratio, different image complexities are associated with different detection accuracy in that higher image complexity leads to lower detection accuracy, and vice versa. We measured the image complexity by using the shape parameter of the generalized Gaussian distribution (GGD) of the discrete wavelet/cosine transform coefficients.

We may employ the same metric of GGD shape parameter to calculate the audio signal complexity. For a more efficient computation, we instead utilize the following formula involving the second order derivative to measure the signal complexity:

$$C(f) = \frac{\dfrac{1}{N-2} \displaystyle\sum_{t=1}^{N-2} |D_f^2(t)|}{\dfrac{1}{N} \displaystyle\sum_{t=0}^{N-1} |f(t)|}. \tag{22.40}$$

$C(f)$ measures the ratio of the mean absolute value of the second order derivative to the mean absolute value of the signal. One may of course adopt several different metrics for signal complexity, $C(f)$ is introduced here as our measure as it can be computed much faster than, say, GGD, and still captures all essential elements of measures for signal complexity. Figure 22.7 shows six audio signal samples with different complexity values of $C(f)$. If we hide the same message into these different audio clips, the expectation of detection performance ought to be different: it should be easier to detect information hiding in the audios with lower signal complexity.

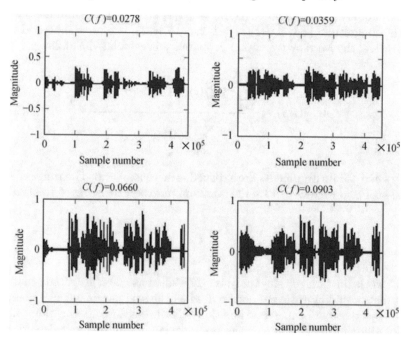

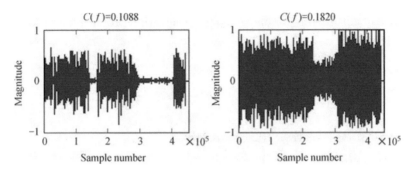

Fig. 22.7 Audio signal samples with different measurements of signal complexity, $C(f)$.

22.6 Pattern Recognition Techniques

The representative pattern recognition algorithms can be divided into: (1) classification, e.g., Naïve Bayes classifier, support vector machine; (2) clustering, e.g., K-means clustering, Hierarchical clustering; (3) regression, such as linear regression, neural networks, Gaussian process regression; (4) sequence labeling, e.g., hidden Markov models, maximum entropy Markov models, particle filters; (5) parsing or predicting tree structured labels, e.g., probabilistic context free grammars; (6) predicting arbitrarily-structured labels, e.g., Bayesian networks, Markov random fields, and (7) ensemble learning, e.g., bootstrap aggregating, boosting, etc. Pattern recognition algorithms have been extensively researched and used in various applications in diverse fields [57 – 70]. We briefly survey in this section two pattern recognition methods that were used in our study.

22.6.1 Introduction to Support Vector Machines

Support Vector Machine (SVM) is a classification and regression prediction tool that uses machine learning theory to maximize predictive accuracy while automatically avoiding over-fit to the data. The foundations of SVM have been developed by Vapnik [66] and gained popularity due to many promising features such as better empirical performance. The formulation uses the Structural Risk Minimization (SRM) principle. Linear SVM is a maximum margin classifier; to achieve the maximum margin $M = 2/||w||$, where w is weighting vector, the quadratic function with linear constraints should be optimized. The solution involves constructing a dual problem and where a Lagrange's multiplier α_i is associated. We need to find w and b such that $\Phi(w) = 1/2|w'||w|$ is minimized.

The quadratic programming (QP) optimization problem for SVM can be formulated as

1) SV classification:

$$\min_{f,\xi_i} \|f\|_K^2 + C\sum_{i=1}^{l} \xi_i \text{ and } y_i f(\boldsymbol{x}_i) \geqslant 1 - \xi_i, \text{ for all } i\xi_i \geqslant 0;$$

2) SVM classification, Dual formulation:

$$\min_{\alpha_i} \sum_{i=1}^{l} \alpha_i - \frac{1}{2}\sum_{i=1}^{l}\sum_{j=1}^{l} \alpha_i\alpha_j y_i y_j K(\boldsymbol{x}_i,\boldsymbol{x}_j), \text{ for all } i, 0 \leqslant \alpha_i \leqslant C,$$

$$\sum_{i=1}^{l} \alpha_i y_i = 0;$$

where the ξ_i are slack variables and they measure the error made at point (x_i, y_i).

As the data is far from linear and the datasets are inseparable, transforming the data into feature space makes it possible to define a similarity measure based on dot product, by using kernel functions. If the feature space is chosen suitably, pattern recognition becomes easy. Frequently used SVM kernel functions are:

1) Polynomial:

$$K(x_i, x_j) = <x_i, x_j>^d, \tag{22.41}$$
$$K(x_i, x_j) = (<x_i, x_j>+1)^d. \tag{22.42}$$

2) Gaussian Radial Basis Function

$$K(x_i, x_j) = \exp(-\|x_i - x_j\|^2/2\sigma^2). \tag{22.43}$$

3) Exponential Radial Basis Function

$$K(x_i, x_j) = \exp(-\|x_i - x_j\|/2\sigma^2). \tag{22.44}$$

4) Multi-Layer Perceptron

$$K(x_i, x_j) = tanh(\rho <x_i, x_j>+e). \tag{22.45}$$

22.6.2 Introduction to DENFIS

Neuron-fuzzy inference systems consist of a set of rules and an inference method that are embodied or combined with a connectionist structure for better adaptation. Evolving neuron-fuzzy inference systems are such systems, where both the knowledge and the mechanism evolve and change in time, with more examples presented to the system. The dynamic evolving neuron-fuzzy inference system, or DENFIS [71], uses the Takagi-Sugeno type of fuzzy inference method. The inference used in DENFIS is performed on m fuzzy rules indicated as follows:

$$If \ x_1 \ is \ R_{11} \ and \ x_2 \ is \ R_{12} \ and \ ... \ and \ x_q \ is \ R_{1q},$$
$$then \ y \ is \ f_1(x_1, \ x_2, \ ..., \ x_q)$$
$$If \ x_1 \ is \ R_{21} \ and \ x_2 \ is \ R_{22} \ and \ ... \ and \ x_q \ is \ R_{2q},$$
$$then \ y \ is \ f_2(x_1, \ x_2, \ ..., \ x_q)$$

$$\vdots$$

$$If \ x_1 \ is \ R_{m1} \ and \ x_2 \ is \ R_{m2} \ and \ ... \ and \ x_q \ is \ R_{mq},$$
$$then \ y \ is \ f_m(x_1, \ x_2, \ ..., \ x_q)$$

where "x_j is R_{ij}", $i = 1, 2, \ldots, m; j = 1, 2, \ldots, q$ are $m \times q$ fuzzy propositions that form m antecedents for m fuzzy rules respectively; $x_j, j = 1, 2, \ldots, q$, are antecedent variables defined over universes of discourse $X_j, j = 1, 2, \ldots, q$, and $R_{ij}, i = 1, 2, \ldots, m; j = 1, 2, \ldots, q$ are fuzzy sets defined by their fuzzy membership functions $\mu_{R_{ij}}: X_j \to [0, 1], i = 1, 2, \ldots, m; j = 1, 2, \ldots, q$. In the consequent parts of the fuzzy rules, y is the consequent variable, and crisp functions $f_i, i = 1, 2, \ldots, m$, are employed.

In the DENFIS model, all fuzzy membership functions are triangular type functions defined by the three parameters, $a, b,$ and c, as given below:

$$\mu(x) = mf(x, a, b, c) = \max(\min((x - a)/(b - a), (c - x)/(c - b)), 0), \quad (22.46)$$

where b is the value of the cluster centre on the x dimension, $a = b - d \times D_{thr}, d = 1.2$ to 2. The threshold value, D_{thr}, is a clustering parameter.

For an input vector $x^0 = [x_1^0 x_2^0 \cdots x_q^0]$, the result of the inference, y^0, or the output of the system, is the weighted average of each rule's output indicated as follows:

$$y^0 = \frac{\sum\limits_{i=1}^{m} w_i f_i(x_1^0, x_2^0, \cdots, x_q^0)}{\sum\limits_{i=1}^{m} w_i}, \quad (22.47)$$

where in the above, $w_i = \prod_{j=1}^{q} R_{ij}(x_j^0); \ i = 1, 2, \cdots, m; \ j = 1, 2, ..., q$.

In the DENFIS on-line model, the first-order Takagi-Sugeno type fuzzy rules are employed. In the DENFIS off-line models, the first-order and an extended high-order Takagi-Sugeno inference engines are used, corresponding to a linear model and an MLP-based model, respectively.

22.7 Experiments

This section describes in detail the experimental setups and the results of JPEG image steganalysis and forgery detection, and WAV audio steganalysis.

22.7.1 JPEG Image Steganalysis

A large dataset was constructed and numerous experiments were conducted to evaluate the performance of our method in comparison with previous work.

Experimental Setup

The 5000 original TIFF raw format digital images used in the experiments are 24-bit, 640×480 pixels, lossless true color and never compressed. We cropped these original images into 256×256 pixels in order to eliminate the low complexity parts and converted the cropped images into JPEG format with the default quality. The following six types of steganograms are generated by hiding different data into the 5000 JPEG images with different hiding ratios.

1) CryptoBola   CryptoBola is a commercial information-hiding software (http://www.cryptobola.com/) which determines the parts (bits) of the JPEG-encoded data that play the least significant role in the reproduction of the image, and it replaces those bits with the bits of the secret message.

2) JPHS (JPHIDE and JPSEEK)   The design objective of JPHS was not simply to hide a file but rather to do this in such a way that it is impossible to prove that the host file contains a hidden file. JPHS for Windows (JPWIN) is available at http://digitalforensics.champlain.edu/download/jphs_05.zip/.

3) Steghide   Hetzl and Mutzel [37] designed a graph-theoretic approach for information-hiding based on the idea of exchanging rather than overwriting pixels. Their approach preserves first-order statistics, and the detection on the first order doesn't work.

4) F5   Westfeld proposed the algorithm F5 that withstands visual and statistical attacks, yet it still offers a large steganographic capacity [13].

5) Model Based steganography without deblocking (MB1) and with deblocking (MB2)   Sallee presented an information-theoretic method for performing steganography [72]. Using the model-based methodology, an example steganography method is proposed for JPEG images which achieves a higher embedding efficiency and message capacity than previous methods while remaining secure against first order statistical attacks.

Some steganograms created by using these steganographic systems are shown in Fig. 22.8.

Comparison of Detection Performance

The authors of Ref. [19] designed an expanded Markov transition prob-

Fig. 22.8 Steganogram examples produced by CryptoBola (row 1), F5 (row 2), JPHS (row 3), Steghide (row 4), MB1 (left two, row5), MB2 (right two, row 5).

ability based feature set, and their study shows that the feature set outperforms other popular feature sets. Specifically, the method applies Markov approach to the differential neighboring coefficients on intra-block and inter-block. In our study, the Markov approach based feature set, DCT coefficient array based neighboring joint density feature sets, defined by Eqs. (22.16) and (22.19), and DCT coefficient absolute array based neighboring density feature sets, defined by Eqs. (22.22) and (22.25), are compared with respect to their performance in detecting JPEG steganography. Our approaches are abbreviated as NJ and absNJ, respectively. Table 22.1 lists the feature sets in our experiments.

Table 22.1   Feature sets tested in our detection

Detection method	The number of features		
	Intra-block feature set	Inter-block feature set	Intra-&inter-block feature set
Markov	324	162	486
NJ	169	169	338
absNJ	36	36	72

SVMlight is an implementation of Vapnik's Support Vector Machine [66] for the problem of pattern recognition, for the problem of regression, and for the problem of learning a ranking function. The optimization algorithms used in SVMlight are described in [62–64]. The algorithm has scalable memory requirements and can handle problems with many thousands of support vectors efficiently. In our experiments, we apply SVMlight with Radial Basis Function (RBF) kernel as the learning classifier to the features for classification, the kernel parameter is 0.01. In detection of each type of stego-images, three types of feature sets are compared: intra-block, inter-block and both. 100 experiments are performed on each type of feature set. In each experiment, 30% samples are randomly chosen for training and other 70% samples are used for testing. The testing results consist of true positive (TP), false positive (FP), false negative (FN), and true negative (TN). The detection performance is evaluated by the classification accuracy, $w \times TP/(TP+FN) + (1 - w) \times TN/(FP+TN)$, where w is a weighting factor in the range of [0, 1]. Without losing generality, w has been set to 0.5 in our experiments. The mean values over the 100 testing are shown in Table 22.2.

Table 22.2   Detection performance by using intra-block features, inter-block features, and total features (M: Markov transition feature set; NJ: Neighboring Joint density feature set) [55]

Steganographic systems	AES	Classification accuracy, 0.5*TP/(TP+FN)+0.5*TN/(TN+FP)								
		Intra-block features			Inter-block features			Intra-&inter-block features		
		M	NJ	absNJ	M	NJ	absNJ	M	NJ	absNJ
CryptoBola	0.11	**99.2%**	99.1	98.9	96.9	**99.4**	98.9	96.3	**99.4**	99.1
	0.18	99.5	**99.9**	99.8	97.9	**99.9**	99.8	95.7	**99.9**	**99.9**
JPHS	0.10	67.9	**76.8**	76.1	53.2	**75.6**	74.1	57.6	77.1	**78.2**
	0.12	76.5	82.8	**92.8**	62.2	81.2	**90.7**	66.3	83.3	**93.7**
Steghide	0.04	73.8	80.1	**82.8**	66.8	**75.1**	62.3	65.3	81.9	**89.5**
	0.06	83.9	92.7	**95.4**	75.3	**88.7**	76.7	76.0	93.4	**97.1**
F5	0.12	67.2	**82.3**	80.8	56.4	62.2	**74.8**	63.1	**87.6**	85.2
	0.22	83.9	**95.3**	93.2	67.3	78.0	**89.4**	82.4	**97.2**	95.3
MB1	0.09	82.6	89.2	**89.6**	58.2	**76.2**	75.3	78.0	88.6	**89.1**
	0.18	94.6	96.9	**97.1**	74.0	**93.2**	92.7	90.6	**97.4**	97.1
MB2	0.12	86.5	93.3	**94.0**	59.5	**78.7**	76.1	80.9	92.6	**93.5**
	0.24	95.5	98.1	**98.3**	74.5	**93.3**	93.1	92.5	98.1	**98.2**

In our study, the ratio of the number of modified DCT coefficients to the total number of non-zero DCT coefficients is used to measure the embedding strength; the ratio of the number of modified DCT coefficients to the total number of DCT coefficients is used as modification ratio. With the same amount of hidden data embedded into different JPEG images, the embedding ratio will likely be different since different JPEG images have different number of non-zero DCT coefficients. The average embedding strengths shown in Table 22.2 were used for experiments to demonstrate the detection performance under different embedding strength. In comparison of the three types of features, the best result is highlighted in bold. The results show that neighboring joint density based approaches outperform the Markov process based approach; some improvements are as high as over 20%.

Figure 22.9 demonstrates ROC curves in detecting the six types of steganograms by using absNJ, NJ, and Markov intra- and inter-block features. These results show that the advantage of the neighboring joint density based approaches over the Markov approach is significant and quite noticeable.

Detection Performance, Image Complexity, and Hiding Amount

As shown in our previous work on steganalysis of LSB matching steganography, even when the same amount of hidden data was embedded in images with the same spatial size, the detection performances under different image complexities will be different. To study the detection performance under different image complexities for image steganalysis, we previously adopted the shape parameter β of the GGD to measure the image complexity and found that image complexity is a significant factor in the evaluation of detection performance. To decrease the computational cost in measuring the image complexity, here we simply take the ratio of the number of non-zero DCT coefficients to the number of total DCT coefficients, including non-zero and zero DCT coefficients, to measure the image complexity. Additionally, we roughly represent the amount of covert message by the ratio of the number of modified DCT coefficients to the total number of DCT coefficients. Due to different hiding methods and different hiding capacity associated with different image complexity, in addition to investigating the relation between detection performance and image complexity, it would be more comprehensive and rigorous to explore the relation among detection performance, image complexity, and amount of hidden message.

In machine learning, Matthews Correlation Coefficient (MCC), which can be calculated by $(TP \times TN - FP \times FN)/sqrt((TP+FP) \times (TP+FN) \times (TN+FP) \times (TN+FN))$, is generally used as a balanced measure even if the classes are of very different sizes regarding the quality of binary classification. In Fig. 22.10, the figures on the left column demonstrate the MCC values in our detection by using Support Vector Machine; the figures in the middle column show the hiding statistics of the average modification ratio of the DCT coefficients in the steganograms and the covers under different image complexities; the figures on the right column exhibit the prediction error

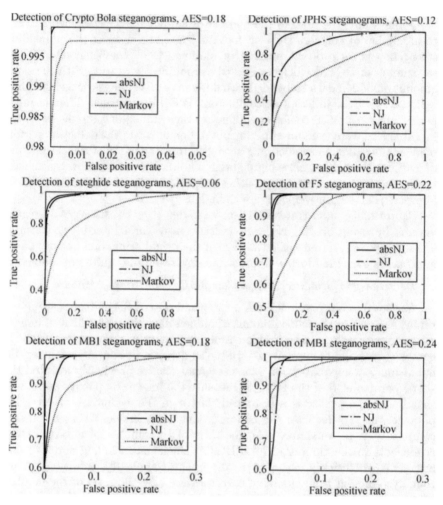

Fig. 22.9 ROC curves by using total features in absNJ, NJ, and Markov approaches [55].

of the modification ratio with the use of Dynamic Evolving Neural-Fuzzy Inference System (DENFIS) that was presented in Ref. [71].

It is clear that neighboring joint density based approaches are generally superior to the Markov approach. The results in Fig. 22.10 also demonstrate the relation among detection performance, image complexity, and the modification ratio. At low image complexity, as the modification ratio apparently increases with the increasing of image complexity, the detection performance measured by MCC values increases, which shows that, information hiding strength is the significant factor for the evaluation of detection performance. At middle image complexity, as the modification ratio increases with the increasing of image complexity, there is no noticeable increasing of the MCC

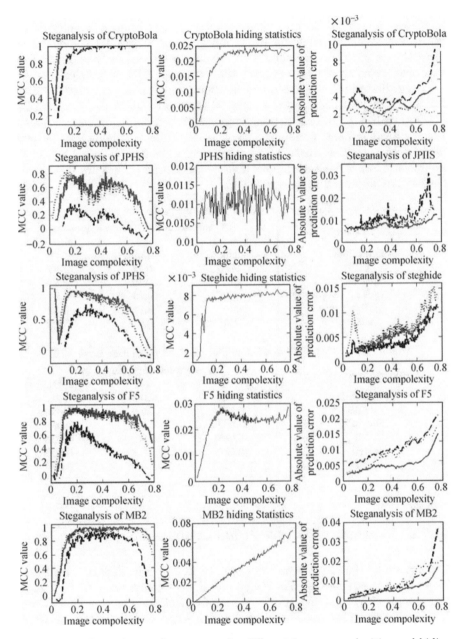

Fig. 22.10 Detection performance under different image complexities and hiding ratios by using total features in absNJ (solid line), NJ (dot line) and Markov (dash) approaches [55].

values, some detection performance even drops, which shows that image complexity is a significant factor for the detection evaluation. At high image complexity, when the modification ratio increases as the image complexity increases, the detection performance deteriorates, and image complexity is a crucial factor for the detection evaluation. Because the results of steganalysis of MB1 steganograms are similar to the results in steganalysis of MB2, to save space, Fig. 22.10 does not show the results on MB1 steganalysis.

Regardless of the specifics of information hiding ratio and image complexity, Table 22.3 lists the means of the MCC values and Table 22.4 gives the absolute values of the prediction errors on the modification ratio, wherein the mean values corresponding to the best detection are highlighted in bold. The results show that absNJ approach performs the best, followed by the NJ approach. Both neighboring joint density based approaches are superior to the Markov approach in most cases.

Table 22.3   Mean value of MCCs [55]

Steganographic system	MCC values under the three approaches		
	Markov	NJ	absNJ
CryptoBola	0.92	**0.99**	**0.99**
JPHS	0.14	0.54	**0.57**
Steghide	0.44	0.75	**0.80**
F5	0.42	0.85	**0.87**
MB1	0.67	**0.86**	0.86
MB2	0.75	0.91	**0.92**

Table 22.4   Mean values of absolute values of prediction errors [55]

Steganographic systems	Absolute value of prediction errors		
	Markov	NJ	absNJ
CryptoBola	0.016	**0.009**	0.011
JPHS	0.037	0.032	**0.028**
Steghide	**0.011**	0.020	0.015
F5	0.032	0.023	**0.015**
MB1	0.017	0.014	**0.012**
MB2	0.021	0.022	**0.014**

22.7.2   Detection of Double JPEG Compression

The original 5 150 raw images are obtained in 24-bit lossless true color and never compressed format. The single and double compressed JPEG images are generated by applying JPEG compression to these images with different quality factors. The first and second compression quality factors in JPEG double compression are denoted "Q1" and "Q2", respectively.

Table 22.5 shows the detection accuracy in the binary classification. The results at the first row are gained by using the 324 Markov transition probability features presented in reference [50], and the results at the second row are obtained by using the integration of 54 Marginal density features, defined in Eq. (22.13), and the 36 neighboring joint density features on the intra-blocks, defined in Eq. (22.22), total 90 features. The experimental results show that our approach obtains the higher detection accuracy. Due to the less number of features, our approach also gains the advantage of low computational cost.

Table 22.5 Average accuracy using Markov transition probability feature set (first row), marginal & neighboring joint density feature set (second row), in binary classification. Each value is the mean over 100 testing [82]

Q2 \ Q1	40	45	50	55	60	65	70	75	80	85	90
40		94.6	97.6	98.1	98.0	96.8	93.5	96.4	59.8	82.4	63.9
		96.4	**97.8**	**98.5**	**98.5**	**96.9**	**97.8**	**97.2**	**91.3**	**95.4**	**82.5**
45	96.1		86.6	96.6	97.3	97.9	96.8	94.2	90.6	88.9	72.9
	96.9		**92.8**	**97.3**	**98.3**	**98.5**	**97.2**	**98.2**	**96.0**	**94.5**	**89.9**
50	98.6	91.0		85.5	97.2	98.3	97.9	93.0	96.1	82.4	53.9
	98.6	**95.3**		**92.4**	**97.6**	**98.6**	**98.3**	**95.3**	**97.2**	**95.4**	**85.0**
55	99.1	98.3	90.2		91.2	97.6	98.4	97.6	95.2	66.3	83.8
	99.1	**98.4**	**94.7**		**95.8**	**98.4**	**98.7**	**98.1**	**97.2**	**94.5**	**94.7**
60	99.2	99.1	98.6	94.8		94.7	97.7	98.3	92.8	94.0	81.3
	99.4	**99.1**	**98.5**	**96.9**		**97.6**	**98.6**	**98.9**	**97.0**	**97.4**	**93.0**
65	99.3	99.4	99.2	98.9	97.1		94.7	97.9	98.2	95.5	88.6
	99.6	**99.6**	**99.3**	**99.1**	**98.1**		**97.4**	**98.6**	**98.5**	**98.5**	**94.4**
70	99.4	99.4	99.4	99.3	**99.3**	97.2		96.3	98.5	95.1	72.5
	99.7	**99.7**	**99.7**	**99.5**	99.2	**98.1**		**97.6**	**99.0**	**97.2**	**95.5**
75	99.5	99.4	99.4	99.4	99.5	**99.3**	98.2		97.1	98.6	94.8
	99.8	**99.8**	**99.8**	**99.8**	**99.7**	**99.3**	**98.3**		**98.9**	**99.1**	**96.8**
80	99.6	99.6	99.6	99.5	99.5	99.5	99.5	99.0		97.6	94.7
	99.8	**99.9**	**99.8**	**99.9**	**99.8**	**99.8**	**99.7**	**99.6**		**99.0**	**97.2**
85	99.6	99.6	99.6	99.6	99.7	99.6	99.6	99.6	99.4		98.5
	100.0	**100.0**	**100.0**	**99.9**	**99.9**	**100.0**	**99.9**	**99.9**	**99.5**		**99.4**
90	99.8	99.8	99.8	99.8	99.8	99.8	99.7	99.8	99.9	99.6	
	100.0	**100.0**	**100.0**	**100**	**100.0**	**100.0**	**100.0**	**100.0**	**100.0**	**99.9**	

Tables 22.6 and 22.7 show the detection accuracy in multi-class classification in the single compression and double compression with the JPEG

Table 22.6 Average accuracy of prediction of Q1 by using marginal & neighboring joint density feature set in multi-class classification (Q2=75). Q1 = 75 means the testing on single compression otherwise the testing on double compression

Prediction \ Q1	75	95	90	85	80	70	65	60	55	50	45	40
75	**44.9**	37.9	0.5	0.1	0.1	0.2	0	0	0	0	0	0
95	40.5	**47.4**	0.3	0.1	0.1	0.2	0	0.8	0	0	0	0
90	8.6	8.8	**97.4**	0	0.1	0.1	0.1	0.2	0	0	0	0
85	1.8	2.0	0.3	**99.5**	0.3	0.2	0	0	0	0	0	0
80	1.3	1.5	0.4	0.2	**99.2**	0.1	0	0	0	0	0	0
70	2.2	1.9	0.8	0.1	0	**98.9**	0	0	0	0	0	0
65	0.6	0.5	0.4	0	0.2	0.1	**99.6**	0	0	0	0	0
60	0	0	0	0	0	0.1	0.2	**99.4**	0.3	0.1	0	0
55	0	0	0	0	0	0	0	0	**99.3**	0.1	0.1	0.1
50	0	0	0	0	0	0	0	0	0.1	**98.6**	0.6	0
45	0	0	0	0	0	0	0	0	0.1	1.0	**99.0**	0
40	0	0	0	0	0	0	0	0.2	0.2	0.2	0.2	**99.8**

compression quality factors '75' and '90', respectively, by using the integration of marginal density and the neighboring joint density on the intra-block of the absolute DCT coefficient array, total 100 features. The results also show the good performance in recognition of the first compression history in the multiple types of double compression and single compression.

Table 22.7 Average accuracy of prediction of Q1 by using marginal & neighboring joint density feature set in multi-class classification (Q2=90). Q1 = 90 means the testing on single compression otherwise the testing on double compression

Q1 Prediction	90	95	85	80	75	70	65	60	55	50	45	40
90	**98.0**	4.8	0.2	0	0	0	0	0	0	0	0	0
95	2.0	**95.2**	0	0	0	0	0	0	0	0	0	0
85	0	0	**99.8**	0	0	0	0	0	0	0	0	0
80	0	0	0	**100.0**	0.1	0	0	0	0	0	0	0
75	0	0	0	0	**99.9**	0.3	0	0	0	0	0	0
70	0	0	0	0	0	**99.6**	0.1	0	0	0	0	0
65	0	0	0	0	0	0	**99.8**	0.1	0	0	0	0
60	0	0	0	0	0	0	0	**99.7**	0.2	0	0	0
55	0	0	0	0	0	0	0	0.2	**99.7**	0.2	0	0
50	0	0	0	0	0	0	0	0	0	**99.6**	0.3	0.1
45	0	0	0	0	0	0	0	0	0	0.1	**99.5**	0.1
40	0	0	0	0	0	0.1	0	0	0.1	0	0.1	**99.8**

22.7.3 Detection of JPEG Resampling

The original 5150 TIFF raw format digital images are obtained in 24-bit lossless true color and never compressed format. These images are converted into JPEG format with the default quality factor '75'. The resized images are produced by using the scale factors 0.3, 0.5, 0.7, 1.5, 2, and 2.2 with each following interpolation method: nearest-neighbor interpolation; bilinear interpolation; bicubic interpolation; interpolation with a box-shaped kernel; interpolation with a Lanczos-2 kernel; and interpolation with a Lanczos-3 kernel. These resized images are stored as JPEG images with the same quality factor '75'. Table 22.8 lists the detection accuracy in classification of original JPEG images and rescaled JPEG images by using marginal density features and the neighboring joint density features on the intra-block of the absolute DCT array. Each testing value in Table 22.8 is the mean over thirty experiments, which shows the promising performance by our approach. The results also indicate that the detection performance also depends on the scale factor. The detection on the rescaled JPEG images with scale factor 2 is much more reliable than the detection on the rescale at scale factor 0.7.

It should be noted that, in detecting double JPEG compression and resized JPEG images, similar to the detection of JPEG-based steganography, image complexity is also an important parameter to evaluate the detection

Table 22.8 Testing accuracy using marginal & neighboring joint density feature set to distinguish rescaled JPEG images w/ different scale factors and interpolation methods from original images

Interpolation Method \ Scale factor	0.3	0.5	0.7	1.5	2	2.2
Nearest	96.5	82.7	64.4	88.9	99.1	96.1
Bicubic	96.2	85.0	73.4	93.7	98.7	94.6
Bilinear	96.4	87.6	78.6	90.5	98.1	93.2
Box	96.2	83.9	64.6	89.0	99.1	96.0
Lanczos2	96.2	84.9	73.4	93.7	98.5	94.7
Lanczos3	96.2	84.0	70.2	95.4	98.7	95.5

accuracy. Generally speaking, as image complexity increases, the performance of steganalysis algorithms deteriorates.

22.7.4 WAV Audio Steganalysis

This subsection describes the experiments and results of the derivative-based audio steganalysis method presented in Section 5 for uncompressed WAV audio.

Experimental Setup

We have 19 380 mono 44.1 kHz 16 bit quantization in uncompressed, PCM coded WAV audio files, covering digital speeches and songs in several languages, e.g., English, Chinese, Japanese, Korean, and several types of music (jazz, rock, blue), etc. Each audio has the duration of 10 seconds. We produced audio steganograms by hiding different message into these audio files. The hiding tools/algorithms include Hide4PGP V4.0, available at http://www.heinzrepp.onlinehome.de/Hide4PGP.htm, Invisible Secrets, available at http://www.invisiblesecrets.com/, LSB matching [9], and Steghide [37]. The hidden data include voice, video, image, text, executable codes, random bits, etc., and the hidden data in any two audio files are different. The amounts of audio steganograms are:19380 produced by using Hide4PGP with 25% maximal hiding; 17158 and 17596 by Steghide with maximal and 50% maximal hiding; 18766 and 19371 by Invisible Secrets with maximal and 50% maximal hiding; 19000 and 19000 by using LSB matching with maximal and 50% maximal hiding; respectively.

Additionally, we have 6 357 mono 44.1 kHz 16 bit quantization in uncompressed, PCM coded WAV audio files, and most are on-line broadcast in English. Each audio has the duration of 19 seconds. We produced the same amount of the watermarking audio files by hiding randomly-produced 2 hexadecimal or 8 binary watermarking digits in each audio (maximal hiding) with the use of spread spectrum audio watermarking [73], which displays solid robustness against traditional signal processing, including arbitrary limited pitch bending and time-scaling.

Comparison of Signal and Derivative Based Audio Steganalysis

We compare signal based Mel-cepstrum audio steganalysis (S-Mel) with 58 Mel-cepstrum coefficients [45], with second order derivative based Mel-cepstrum steganalysis (2D-Mel) with the 58 features described in Eqs. (22.37) and (22.38), second order derivative based Markov approach (2D-Markov) with the 169 features calculated by Eq. (22.39), and combined derivative based detection containing all features described in Eqs. (22.37), (22.38), and (22.39), abbreviated as 2D-MM, in the four categories of signal complexity: low complexity (C < 0.04); middle complexity (0.04 ⩽ C < 0.08); middle-high complexity (0.08 ⩽ C < 0.12), and high complexity (C ⩾ 0.12). In Kraetzer and Dittmann's work, signal based Mel-cepstrum coefficients and several other statistical features form an AMSL Audio Steganalysis Tool Set (AAST) that is also tested in our experiments. To compare the detection performance, one hundred experiments are performed on each feature set under each category of signal complexity in each detection. In each experiment, 30% of the audio files are randomly assigned to the training group and 70% are used for testing for steganalysis of Hide4PGP, Invisible Secrets, LSB matching, and Steghide; 70% training and 30% testing are randomly grouped in steganalysis of spectrum spread audio watermarking. Support vector machines (SVM) with RBF kernels are used for classification. The results consist of true positive (TP), false positive (FP), false negative (FN), and true negative (TN). The classification accuracy is calculated as $w \times TP/(TP+FN) + (1-w) \times TN/(FP+TN)$, where $w \in [0, 1]$ is a weighting factor. Without losing generality, w was set to 0.5 in our experiments. Mean values of classification accuracy are listed in Table 22.9. For the comparison of the five feature sets, the highest mean testing values are highlighted in bold.

Regarding the relation of detection performance to signal complexity, as shown in Table 22.9, for signal and derivative based Mel-cepstrum and AAST feature sets, as signal complexity increases, the detection performances generally decrease. However, there is no obvious performance deterioration of the derivative based Markov approach in high signal complexity. In comparison of the five feature sets, second order derivative based Mel-cepstrum steganalysis improves the detection performance of signal based Mel-cepstrum set in each category of signal complexity. Especially noticeable regarding the detection of audio streams with high signal complexity, derivative based Mel-cepstrum improves the testing accuracy by about 15% to 20% for the steganalysis of Hide4PGP, Invisible Secrets, LSB matching, and Steghide. Compared to signal based Mel-cepstrum approaches, second order derivative based Markov approach also gains significant advantage, the improvements are about 23% to 34% in detecting audio steganograms produced by Hide4PGP, Invisible Secrets, LSB matching, and Steghide in high signal complexity.

Table 22.9 Testing accuracy of audio steganalysis over 100 experiments, with signal based Mel-cepstrum (S-Mel), AAST, and derivative based 2D-Mel, 2D-Markov, and 2D-MM [74]

Hiding method	Hiding ratio	Complexity C	Testing accuracy				
			S-Mel	AAST*	2D-Mel	2D-Markov	2D-MM
Invisible	100%	low	97.8	89.1	98.9	**99.6**	99.2
		middle	97.2	79.0	98.8	**99.9**	99.6
		middle-high	90.6	86.2	97.3	**99.9**	99.6
		high	76.4	65.9	91.5	**99.9**	99.6
	50%	low	93.8	78.2	96.7	**99.0**	98.0
		middle	89.0	71.2	96.5	**99.2**	98.9
		middle-high	78.7	74.9	88.9	**99.1**	98.7
		high	61.8	60.5	77.3	**99.3**	99.0
Hide4PGP	25%	low	97.8	86.1	98.9	**99.6**	99.2
		middle	97.2	80.1	98.9	**99.9**	99.7
		middle-high	90.6	86.2	97.4	**99.9**	99.7
		high	76.2	64.3	91.5	**99.9**	99.7
LSB matching	100%	low	97.8	86.8	98.9	**99.5**	99.2
		middle	97.2	80.1	98.9	**99.8**	99.6
		middle-high	90.8	87.1	97.3	**99.7**	99.6
		high	76.2	63.9	91.5	**99.9**	99.7
	50%	low	95.9	80.4	98.1	**99.2**	98.4
		middle	94.6	67.1	98.1	**99.5**	99.3
		middle-high	85.1	81.1	94.0	**99.3**	99.0
		high	66.1	60.1	84.8	**99.6**	99.4
Steghide	100%	low	97.0	89.6	**98.6**	97.6	97.7
		middle	96.4	81.8	**98.6**	98.6	**98.6**
		middle-high	87.4	83.6	96.2	**98.6**	98.3
		high	71.8	63.2	89.9	**99.1**	98.5
	50%	low	94.3	73.6	**97.2**	94.6	95.7
		middle	91.9	73.6	**97.3**	96.5	96.6
		middle-high	80.8	76.0	91.8	**96.6**	96.0
		high	64.0	59.8	84.4	**98.2**	97.1
Spread spectrum audio watermarking	100%	low	91.0	76.0	**92.6**	90.6	90.4
		middle	86.0	80.3	**92.9**	86.7	92.2
		middle-high	81.5	56.1	**87.2**	79.9	86.4
		high	67.5	51.0	70.7	**85.3**	81.8

* There are training failures with the use of AAST even when we adopt different kernels and kernel parameters. We calculate the mean testing accuracy based on the results obtained from the correct learning models.

Additionally, derivative based Markov approach is better than derivative based Mel-cepstrum steganalysis for detecting Hide4PGP, Invisible Secrets, LSB matching, and Steghide in high signal complexity. Although AAST includes all signal-based Mel-cepstrum features and several other statistical features, the detection performance is not as high as signal based Mel-cepstrum audio steganalysis. Our study also shows that the standard deviation value of the testing results by using AAST is high, that is, the testing performance is not stable. We surmise that some statistical feature design of AAST is not ideal, which is verified by the statistical analysis of each individual feature in AAST.

We note that in steganalysis of steganographic systems, derivative based Markov approach takes the lead in testing accuracy, followed by derivative based Mel-cepstrum method. However, in the steganalysis of audio watermarking, derivative based Mel-cepstrum performs the best except under high signal complexity. By combining derivative based Mel-cepstrum and Markov approaches, the testing results are very close to the best in each category of signal complexity; therefore, an effective detection system can be developed by incorporating both approaches.

In addition to the comparison shown in Table 22.9, the Receiver Operating Characteristic (ROC) curves by using S-Mel, 2D-Mel, and 2D-MM are also given in Fig. 22.11, for the steganalysis of Invisible (50% max-hiding), LSB matching (50% max-hiding), Steghide (50% max-hiding), and spread spectrum audio watermarking (abbreviated as SSAW in the figure, max-hiding) under the four categories of signal complexity (the ROC curves on Hide4PGP are similar to the curves on Invisible; to save space, the results are not included in Fig. 22.11).

Generally, derivative based Mel-cepstrum steganalysis outperforms signal based Mel-cepstrum approach, and the integration of derivative based Mel-cepstrum and Markov approaches delivers the best detection performance, the superiority is especially remarkable for steganalysis of audio streams with high signal complexity.

Second order derivative based methods have the advantage over the signal based Mel-cepstrum audio steganalysis. Our explanation is that audio signals are generally band-limited, while the embedded hidden data is likely broadband and most information-hiding inclines to randomly modify audio signals and tends to increase the high frequency information. Derivative based detections first preprocess signals by extracting the derivative information and it is relatively easy to expose the existence of hidden data; consequently, derivative-based methods are more accurate in comparison with signal based Mel-cepstrum audio steganalysis.

Derivative based Markov approach obtains remarkable detection performance even in high signal complexity, due to the fact that the transition features reflect the modification on the smooth parts of audio streams, not on the high complexity parts. Even an audio is associated with high signal complexity, there are many smooth parts or the sub-audio streams with low signal complexity, the difference between the magnitudes over the temporal neighborhood in these sub-audio streams is not so big, and Markov transition features are correlated to these sub-audio streams. On the other hand, in most audio steganographic systems, payload embedding is not correlated to audio signal, that is, these systems do not consider the signal complexity of the audio streams for adaptive hiding, the information hiding also modifies the magnitude values in the sub-audio streams with low signal complexity and high signal complexity; in such case, Markov transition features extracted from the low complexity sub-streams obtain impressive detection accuracy in the audio signals with high signal complexity.

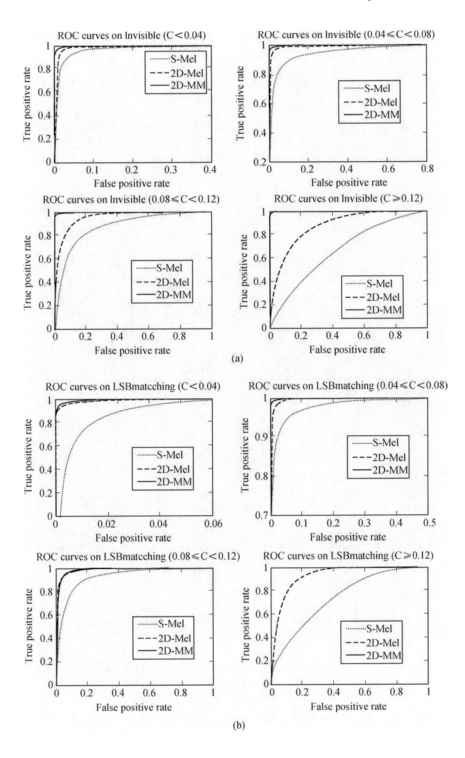

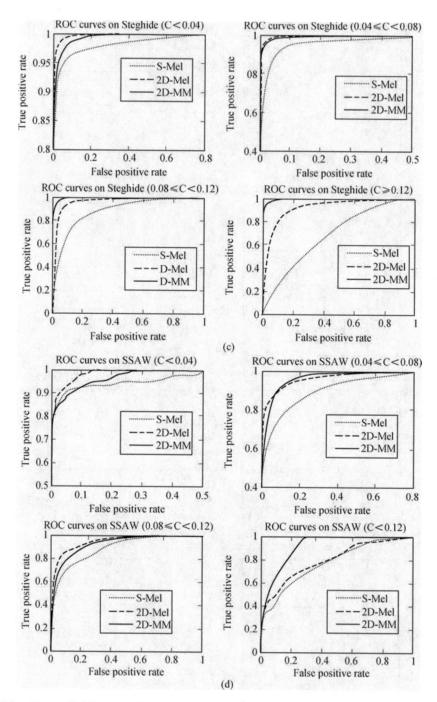

Fig. 22.11 ROC curves for the steganalysis of Invisible Secret (50% max-hiding, a), LSB matching (50% max-hiding, b), Steghide (50% max-hiding, c), and spread spectrum audio watermarking (abbreviated as SSAW, max-hiding, d) [74].

The advantage of derivative based Markov approach in steganalysis of spread spectrum watermarking is not so noticeable due to different emphasis of watermarking that focuses on robustness against traditional signal processing. The merged derivative based Mel-cepstrum and Markov approach still delivers good performance in different categories of signal complexity. Although AAST includes all signal-based Mel-cepstrum features and several additional statistical features, the detection performance is not as well as signal based Mel-cepstrum audio steganalysis. It indicates that feature selection is also an important issue in steganalysis, which have been conducted by our previous image steganalysis study [27, 31].

The proposed steganalysis method has just been tested on WAV uncompressed audio streams. To detect the information-hiding in compressed domain, for example, the steganalysis of MP3 audio streams, we utilize the statistics (mean, standard deviation, skewness, kurtosis) on the second order derivative of the modified discrete cosine transform (MDCT) coefficients, and/or combine the statistics with the inter-frame MDCT statistics, a MP3-based audio steganographic system has been successfully developed [75, 76]. A further study on predicting the embedded strength in MP3 audio has been conducted to quantitatively analysis MP3 steganography [77, 78].

We can use a high-frequency filter such as Daubechies wavelet analysis [79] instead of second order derivative and then obtain the Mel-cepstrum features, which is also better than signal-based Mel-cepstrum audio steganalysis, validated by our previous study [80]. In general, this alternative approach is not better than second derivative based Mel-cepstrum solution, which has been verified and discussed by our previous experiments [80, 81]. Our analysis indicates that the application of high-frequency filter such as Daubechies wavelet will produce the high-frequency signal that is similar to white noise and the spectrum is almost equally distributed over whole frequency band. However, the second derivative suppresses the energy in low frequency and amplifies the energy in high frequency, the spectrum does not equally distribute over whole frequency band. Figure 22.12 (a) shows the spectrum of the second derivative of the hidden data, called error signal in the figure, in an audio steganogram. Figure 22.12 (b) plots the spectrum of the detail wavelet subband of the same hidden data, filtered by using 'db8'. Based on Eq. (22.36) in Section 22.2, as the error spectrum increases, the expected value of the variance of the audio steganogram will prominently increases, that is, the rate of power change in different spectrum bands will dramatically change, since the Mel-cepstrum coefficients are used to capture the information for power change, in such case, the advantage of derivative-based Mel-cepstrum approach is noticeable.

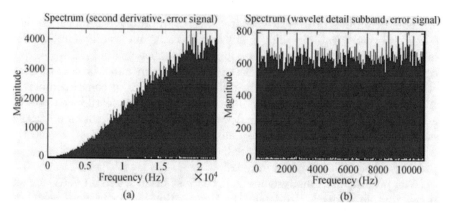

Fig. 22.12 Spectrum of the second order derivative of a hidden data in a 44.1 kHz audio steganogram, shown in (a), and spectrum of the detail wavelet sub-band of the same hidden data, filtered by using 'db8', shown in (b). The frequency shown in the x-axis (b) is reduced due to down-sampling of wavelet decomposition [80].

22.8 Conclusions

We presented in this article a method to detect JPEG-based steganography, double compression, and rescale interpolation, based on the framework of feature mining and pattern recognition techniques. The developed features include marginal density and the neighboring joint density features on the DCT coefficients. Compared to a recently proposed and well-known detection method, our method remarkably improves the detection performance, with respect to either detection accuracy or computational cost. Our study also shows that the detection performance is related not only to the information-hiding ratio, the compression quality factors, or the scale factor, but also to image complexity, which is an important parameter that seems to have been so far overlooked by the research community in conducting performance evaluation of steganalysis and forgery detection methods. In either steganalysis or forgery detection, the accuracy for high complexity mages is not as good as that for low complexity images.

To improve the prior WAV audio steganalysis methods, we also described a recently developed derivative based feature mining method that involves second derivative based Mel-cepstrum coefficients and Markov transition features, to discover the existence of covert message in the audio streams with the use of machine learning and pattern recognition techniques. Results indicate significant improvement over the previous methods; and the relation of detection performance versus signal complexity was experimentally explored to show that, as in the case of image steganalysis, a negative relation exists between the two.

Therefore, to formally study the performance evaluation issues pertaining to steganalysis and forgery detection, the image complexity or audio signal complexity-in addition to other relevant factors such as the compression quality, scale factor, data acquisition by different equipments, processing history, format, etc.-should be included in future research work.

Acknowledgements

The authors would like to acknowledge the support for this research provided by the Institute for Complex Additive Systems Analysis (ICASA), a research division of New Mexico Tech. Support from Sam Houston State University for the first author is also greatly appreciated.

We are grateful to Dr. Dan Ellis of Columbia University for his insightful discussions and invaluable suggestions, to Dr. Malcolm Slaney of Yahoo! research and Professor Haojun Ai for their valuable discussions, to Dr. Yun Q. Shi and Dr. Chunhua Chen for providing us their Markov feature extraction code, and to Dr. Professor Jana Dittmann and Christian Kraetzer for kindly providing us their AAST document.

References

[1]   http://www.wired.com/dangerroom/2010/06/alleged-spies-hid-secret-messages-on-public-websites/. Accessed 9 December 2010

[2]   http://www.msnbc.msn.com/id/38028696/ns/technology_and_science-science/. Accessed 9 Dec 2010

[3]   http://www.washingtonpost.com/wp-dyn/content/article/2010/06/30/AR2010063003108.html. Accessed 9 December 2010

[4]   http://www.newscientist.com/article/dn19126-russian-spy-ring-hid-secret-messages-on-the-web.html. Accessed 9 December 2010

[5]   http://www.nj.com/news/index.ssf/2010/06/accused_russian_spies_in_nj_us.html. Accessed 9 December 2010

[6]   http://digitalforensicsmagazine.com/blogs/?p=76. Accessed 9 December 2010

[7]   http://www.justice.gov/opa/documents/062810complaint1.pdf. Accessed 9 December 2010

[8]   http://www.justice.gov/opa/documents/062810complaint2.pdf. Accessed 9 December 2010

[9]   Sharp T (2001) An Implementation of Key-based Digital Signal steganography. LNCS, vol 2137: 13−26

[10]   Mielikainen J (2006) LSB Matching Revisited. IEEE Signal Processing Letters,13(5) : 285−287

[11]   Pevny T, Filler T, Bas P (2010) Using high-dimensional Image Models to Perform Highly Undetectable Steganography. Proc 12th Information Hiding, pp 161−177

[12]   Solanki K, Sarkar A, Manjunath B (2007) YASS: Yet Another Steganographic Scheme That Resists Blind Steganalysis. LNCS, vol 4567: 16−31

[13]   Westfeld A (2001) High Capacity Despite Better Steganalysis (F5−a Steganographic Algorithm). LNCS, vol 2137: 289−302

[14]   Kim Y, Duric Z, Richards D (2006) Modified Matrix Encoding Technique for Minimal Distortion Steganography. LNCS, vol 4437: 314−327

[15]   Sachnev V, Kim H J, Zhang R (2009) Less Detectable JPEG Steganography Method Based on Heuristic Optimization and BCH Syndrome Code. Proc of 11th ACM Multimedia & Security Workshop, pp 131−140

[16]   Provos N (2001) Defending Against Statistical Steganalysis. Proc 10th USENIX Security Symposium, vol 10: 323−335

[17]   Marvel L, Boncelet C, Retter C (1999) Spread Spectrum Image Steganography. IEEE Trans. Image Processing, 8(8): 1075 – 1083

[18]   Katzenbeisser S, Petitcolas F (2000) Information Hiding Techniques for Steganography and Digital Watermarking. Artech House Books

[19]   Chen C, Shi Y (2008) JPEG Image Steganalysis Utilizing Both Intrablock and Interblock Correlations. Proc of IEEE International Symposium on Circuits and Systems, pp 3029 – 3032

[20]   Fridrich J, Goljan M, Hogeam D (2002) Steganalysis of JPEG Images: Breaking the F5 Algorithm. Proc of 5th Information Hiding Workshop, pp 310 – 323

[21]   Fridrich J (2004) Feature-based Steganalysis for JPEG Images and its Implications for Future Design of Steganographic Schemes. LNCS, vol 3200: 67 – 81

[22]   Harmsen J, Pearlman W (2003) Steganalysis of Additive Noise Modelable Information Hiding. Proc of SPIE, vol 5020: 131 – 142

[23]   Ker A (2005) Improved Detection of LSB Steganography in Grayscale Images. LNCS, vol 3200: 97 – 115

[24]   Kodovsky J, Pevny T, Fridrich J (2010) Modern Steganalysis can Detect YASS. Proc of SPIE Electronic Imaging, Media Forensics and Security XII, pp 0201 – 0211

[25]   Li B, Shi Y, Huang J (2009) Steganalysis of YASS. IEEE Trans. Information Forensics and Security, 4(3): 369 – 382

[26]   Liu Q, Sung A H (2007) Feature Mining and Neuron-fuzzy Inference System for Steganalysis of LSB Matching Steganography in Grayscale Images. Proc of 20th International Joint Conference on Artificial Intelligence, pp 2808 – 2813

[27]   Liu Q, Sung AH, Chen H et al (2008) Feature Mining and Pattern Classification for Steganalysis of LSB Matching Steganography in Grayscale Images. Pattern Recognition, 41(1): p 56 – 66

[28]   Liu Q, Sung A H, Ribeiro B M et al (2008) Image Complexity and Feature Mining for Steganalysis of Least Significant bit Matching Steganography. Information Sciences, 178(1): 21 – 36

[29]   Liu Q, Sung A H, B. Ribeiro et al (2008) Steganalysis of Multi-Class JPEG Images Based on Expanded Markov Features and Polynomial Fitting. Proc of 21st International Joint Conference on Neural Networks, pp 3351 – 3356

[30]   Liu Q, Sung A H, Qiao M (2009) Improved Detection and Evaluation for JPEG Steganalysis. Proc of 17th ACM International Conference on Multimedia, pp 873 – 876

[31]   Liu Q, Sung A H, Qiao M et al (2010) An Improved Approach to Steganalysis of JPEG Images. Information Sciences, vol 180(9): 1643 – 1655

[32]   Lyu S, Farid H (2005) How Realistic is Photorealistic. IEEE Trans. Signal Processing, 53(2): pp 845 – 850

[33]   Ozer H, Sankur B, Memon N et al (2006) Detection of Audio Covert Channels Using Statistical Footprints of Hidden Messages. Digital Signal Processing, 16(4): 389 – 401

[34]   Pevny T, Fridrich J (2007) Merging Markov and DCT Features for Multi-class JPEG Steganalysis. Proc of SPIE, vol 650503, DOI:10.1117/12.696774

[35]   Shi Y, Chen C, Chen W (2007) A Markov Process Based Approach to Effective Attacking JPEG Steganography. LNCS, vol 4437: 249 – 264

[36]   Yu X, Babaguchi N (2008) Breaking the YASS Algorithm via Pixel and DCT Coefficients Analysis. Proc of 19th International Conference on Pattern Recognition, pp 1 – 4.

[37]   Hetzl S, Mutzel P (2005) A Graph-theoretic Approach to Steganography. LNCS, vol 3677: 119 – 128

[38]   Sarkar A, Nataraj L, Manjunath B S et al (2008) Estimation of Optimum Coding Redundancy and Frequency Domain Analysis of Attacks for YASS: A Randomized Block Based Hiding Scheme. Proc of 15th IEEE International Conference on Image Processing, pp 1292 – 1295

[39]   Sarkar A, Solanki K, Manjunath B S (2008) Further Study on YASS: Steganography Based on Randomized Embedding to Resist Blind Steganalysis. Proc of SPIE Security, Steganography, and Watermarking of Multimedia Contents (X), pp 681917.1 – 681917.11

[40]   http://www.cbsnews.com/8301-503543_162-20016679-503543.html. Accessed 9 December 2010

[41]   http://www.cbsnews.com/stories/2010/09/17/world/main6876519.shtml. Accessed 9 December 2010

[42]   http://news.yahoo.com/s/ap/20100917/ap_on_re_mi_ea/ml_egypt_doctored_photo. Accessed 9 December 2010

[43] http://www.npr.org/blogs/thetwo-way/2010/09/17/129938169/doctored-photograph-hosni-mubarak-al-ahram-white-house-obama-mideast-peace-talks. Accessed 9 December 2010

[44] Johnson M, Lyu S, Farid H (2005) Steganalysis of Recorded Speech. Proc of SPIE, vol 5681: 664 – 672

[45] Kraetzer C, Dittmann J (2007) Mel-cepstrum Based Steganalysis for VOIP-steganography. Proc of SPIE, vol 6505: 650505.1 – 650505.12

[46] Zeng W, Ai H, Hu R (2007) A Novel Steganalysis Algorithm of Phase Coding in Audio Signal. Proc of 6th International Conference on Advanced Language Processing and Web Information Technology, pp 261 – 264

[47] Zeng W, Ai H, Hu R (2008) An Algorithm of Echo Steganalysis Based on Power Cepstrum and Pattern Classification. Proc of International Conference on Information and Automation, pp 1667 – 1670

[48] Farid H (2009) Image Forgery Detection, a Survey. IEEE Signal Processing Magazine, 26(2): 16 – 25

[49] Pevny T, Fridrich J (2008) Detection of Double-compression in JPEG Images for Applications in Steganography. IEEE Trans. Information Forensics and Security, vol 3(2): 247 – 258

[50] Chen C, Shi Y, Su W (2008) A Machine Learning Based Scheme for Double JPEG Compression Detection. Proc of 19th International Conference on Pattern Recognition, pp 1 – 4

[51] McEachern R (1994) Hearing It Like It Is: Audio Signal Processing the Way the Ear does it. DSP Applications, pp 35 – 47

[52] Ohm JR (2004) Multimedia Communication Technology, Representation, Transmission and Identification of Multimedia Signals. Springer

[53] Sharifi K, Leon-Garcia A (1995) Estimation of Shape Parameter for Generalized Gaussian Distributions in Subband Decompositions of Video. IEEE Trans Circuits and Systems for Video Technology, vol 5: 52 – 56

[54] Cho D, Bui T (2005) Multivariate Statistical Modeling for Image Denoising Using Wavelet Transforms. Signal Processing: Image Communication, vol 20: 77 – 89

[55] Liu Q, Sung A H, Qiao M (2011) Neighboring Joint Density Based JPEG Steganalysis. ACM Trans. Intelligent Systems and Te chnology, in press (to appear in vol 2(2), 2011), vol 2(2),article 16

[56] Gonzalez R, Woods J K (2008) Digital Image Processing. 3rd edn, Prentice Hall

[57] Bayram S, Dirik AE, Sencar HT et al (2010) An Ensemble of Classifiers Approach to Steganalysis. Proc of 20th International Conference on Pattern Recognition, pp 4376 – 4379

[58] Bayram S, Sencar H T, Memon N D (2009) An Efficient and Robust Method for Detecting Copy-move Forgery. Proc of 34th IEEE International Conference on Acoustics, Speech and Signal Processing, pp 1053 – 1056

[59] Chang M S, Chou J H (2010) A Robust and Friendly Human-robot Interface System Based on Natural Human Gestures. International Journal of Pattern Recognition and Artificial Intelligence, 24(6): 847 – 866

[60] Chen Y, Ding X, Wang P S P (2009) Dynamic Structural Statistical Model Based Online Signature Verification. International Journal of Digital Crime and Forensics, 1(3): 21 – 41

[61] Dirik A E, Memon ND (2009) Image Tamper Detection Based on Demosaicing Artifacts. Proc of 16th Image Processing, pp 1497 – 1500

[62] Joachims T (1999) Making Large-scale SVM Learning Practical. Advances in Kernel Methods: Support Vector Learning, MIT Press

[63] Joachims T (2000) Estimating the Generalization Performance of a SVM Efficiently. Proc of 17th International Conference on Machine Learning, pp 431 – 433

[64] Joachims T (2002) Learning to Classify Text Using Support Vector Machines. Dissertation, Kluwer.

[65] Sutcu Y, Coskun B, Sencar H T et al (2007) Tamper Detection Based on Regularity of Wavelet Transform Coefficients. Proc of 14th Image Processing, pp 397 – 400

[66] Vapnik V (1998) Statistical Learning Theory, Wiley, New York

[67] Wang P S P, Yanushkevich S N (2007) Biometric Technologies and Applications. Proc of IASTED International Multi-Conference: Artificial intelligence and Applications, pp 249 – 254

[68] Yanushkevich S N, Gavrilova M L, Wang PSP et al (2007) Image Pattern Recognition: Synthesis and Analysis in Biometrics. World Scientific Publishing, Singapor

[69] Yanushkevich S N, Hurley D, Wang P S P (2008) Special Issue: Pattern Recognition

and Artificial Intelligence in Biometrics. International Journal of Pattern Recognition and Artificial Intelligence, 22(3): 367–369

[70] Zhang L, Chen J, Lu Y et al (2008) Face Recognition Using Scale Invariant Feature Transform and Support Vector Machine. Proc of 9th International Conference for Young Computer Scientists, pp 1766–1770

[71] Kasabov N K, Song Q (2002) DENFIS: Dynamic Evolving Neural-Fuzzy Inference System and its Application for Time-series. IEEE Trans Fuzzy Systems, 10(2): 144–154

[72] Sallee P (2004) Model Based Steganography. LNCS, vol 2939: 154–167

[73] Kirovski D, Malvar HS (2003) Spread Spectrum Watermarking of Audio Signals. IEEE Trans on Signal Processing, 51(4): 1020–1033

[74] Liu Q, Sung A H, Qiao M. Derivative Based Audio Steganalysis. ACM Trans Multimedia Computing, Communications and Applications, in press (to appear 2011)

[75] Qiao M, Sung A H, Liu Q (2009) Steganalysis of MP3stego. Proc of 22nd International Joint Conference on Neural Networks, pp 2566–2571

[76] Qiao M, Sung A H, Liu Q (2009) Feature Mining and Intelligent Computing for MP3 Steganalysis. Proc of International Joint Conference on Bioinformatics, Systems Biology and Intelligent Computing, pp 627–630

[77] Qiao M, Sung A H, Liu Q (2010) Predicting Embedding Strength in Audio Steganography, Proc of 9th IEEE International Conference on Cognitive Informatics, pp 925–930

[78] Qiao M, Sung A H, Liu Q (2010) Revealing Real Quality of Double Compressed MP3 audio. Proc of 18th ACM International Conference on Multimedia, pp 1011–1013

[79] Daubechies I (1992) Ten Lectures on Wavelets. Society for Industrial and Applied Mathematics

[80] Liu Q, Sung A H, Qiao M (2009) Temporal Derivative-based Spectrum and Mel-cepstrum Audio Steganalysis. IEEE Trans Information Forensics and Security, 4(3): 359–368

[81] Liu Q, Sung A H, Qiao M (2009) Novel Stream Mining for Audio Steganalysis. Proc of 17th ACM International Conference on Multimedia, pp 95–104

[82] Liu Q, Sung A H, Qiao M (2011) A Method to Detect JPEG-Based Double Compression. ISNN (2): 46–47

Part IV: Biometric Authentication

23 Biometric Authentication

Jiunn-Liang Lin[1,2], Ho-Ling Hsu[3], Tai-Lang Jong[4], and Wen-Hsing Hsu[5]

Abstract   This chapter will introduce several biometric recognition technologies, for example, fingerprint, person's face, eye pupil, sound, etc., and, in the meantime, also will introduce the standardization development of biometrics technology, recent projects in all countries for various applications, and some important issues while large-scale deploying biometric system. Especially for the gap to 100% correct of recognition for high security application requirements, we introduce the multi-press enrollment process to improve current fingerprint recognition performance. Also, we introduce the architecture of multiple biometric servers with minutiae-based classification to meet the quick response in large scale deployment of biometric system. By the way, wolf-attack problem is another one important issue needed to be faced and solved. At the end, due to the development of cloud networking technology during these years, what is changing in the biometrics service and system architecture will be discussed.

Among the daily life, we usually need to identify who you are in order to get a certain kind of service, for example getting into a building, accessing a computer or network system, obtaining banking services, reaching ATM machine, etc. Currently, most of the using certification ways is a simple password. The difference for everyone only is password length and the mixture of numeral and letter of alphabet. The biggest problem in password form is that everyone needs to commit to memory excessive passwords, and the passwords will be easily hacked. The recognition technology of biometric features provides a significantly higher security level and convenience than the password form. The biometrics recognition identifies individuals automatically according to the characteristic of physiology and/or behavior. Users would not need to commit to memory each password for the dissimilarity application. Users also would not need to worry that he/she can not get into a certain building or network system because of forgetting a password, or can not obtain a

1   Department of Electrical Engineering, Tsing Hua University, 30013, Hsinchu, Taiwan, China. E-mail: s9961819@m99.nthu.edu.tw
2   Industrial Technology Research Institute, 31040, Hsinchu, Taiwan, China. E-mail: Allan.Lin@itri.org.tw.
3   STARTEK Engineering Inc., 30075, Hsinchu, Taiwan, China. E-mail: holing@mail.startek-eng.com.
4   Department of Electrical Engineering, Tsing Hua University, 30013, Hsinchu, Taiwan, China. E-mail: tljong@ee.nthu.edu.tw.
5   Department of Electrical Engineering, Tsing Hua University, 30013, Hsinchu, Taiwan, China. E-mail: whhsu@ee.nthu.edu.tw.

certain financial service, for example, retrieving cash from the ATM machine.

23.1 Introduction

Generally speaking, the way to recognize something of a human being Mankind is to recognize sound, face, gait, odor, etc. Through a person's face or sound afar, we are able to identify who he/she is. The biometric recognition technology basically is imitating the human being's recognition to carry on same recognition, for example, the face recognition technique and speech recognition technique. More over, because of human beings having some features of physiology, which own the anatomy uniqueness, the biometric technology is developing and can quickly recognize an individual through these features, for example, fingerprint, eye pupil, finger vein, DNA, etc. These features come from human's body and are easily obtained by self and do not be forged. The biometrics recognition is very convenient on the application. Consequently, various applications needs to identify individuals adopt the biometrics recognition gradually. For example the passport inspection has already adopted the face recognition technique and fingerprint recognition technique to carry on the identification work at the border check. Currently, the development of biometrics technology in recognition has been rather mature. At the court, the biometrics technology also has been used to find or recognize the suspecting person through the collected fingerprint or DNA at the crime scene.

The biometric features have a few following characteristics [1]:

1) Universality: everyone has the features.

2) Distinctiveness: the features could be used to distinguish different identities.

3) Permanence: the features would not change with time.

4) Collectability: the features could be collected via a biometrics system, and formed uniqueness codes to process the identification and verification.

Applying biometric features to carry on identifying individuals are usually depended on the convenience of obtaining the biometric features. Take an example, most of the biometric features used to control at gate system is person's face, fingerprint and eye pupil, because these three kinds of biometric features are easily to provide at any one time for everyone without help of any instrument equipment. Most of the collected biometric features at crime scene are DNA and fingerprint because these two kinds of biometric features are easily left behind from human beings on the spots among daily life or behaviors. We will introduce some biometric features used in current most applications.

23.1.1 Fingerprint

Fingerprint has been one of the most popular biometric features. Many kinds of applications have been derived from the recognition of fingerprint feature, such as door security, laptop computer security, elevator control, sign-in of workers and students, identification of individual in court, etc. Actually, the related applications using recognition of fingerprint have appeared in our daily life. Currently, many countries, such as U.S.A., Hong Kong/Shenzhen, Europe, Japan, Australia, and Singapore, use fingerprint in border control systems.

Basically, the process of fingerprint recognition has three steps: fingerprint retrieve, feature extraction, feature matching. The quality of retrieved fingerprint will influence the result of fingerprint recognition seriously. Furthermore, the retrieved fingerprint is a 2D image. The surface of a finger is a 3D image. There exists a transformation problem between the 3D image and 2D image. After the image located, the fingerprint features (minutiae) will be extracted. Minutiae include ridge, ending and bifurcations. The matching mechanism will use these minutiae to find a matched template in database or to make a comparison with the other one. The recognition result could identify or verify someone to be a specific individual, as shown in Fig. 23.1.

Fig. 23.1 Fingerprint and minutiae.

The biometric system using fingerprint has been commercialized. Friendly human interface and accurate recognition have been accepted in popular. The feature extraction and matching algorithm have been near mature. The faced problems, currently, are how to improve the enrollment process to get a fingerprint with quality guarantee, and the challenges while large-scaled deployment. These problems will be discussed at the end of this chapter.

23.1.2 Face [2, 3]

Face recognition has been applied to some verification and identification systems. Usually, it is cooperated with the other biometrics technology, such as fingerprint, to form a multimodal biometrics system. The major algorithms of face recognition include principal components analysis (PCA), linear discriminant analysis (LDA) and elastic bunch graph matching (EBGM).

Face recognition is like fingerprint recognition. The first step of recognition process is to catch the image of human face via camera. The caught image will be sensitively influenced by the environment light. The strength of light will cause a large variance in the caught images of human face. The background of human face is an important and effective factor, too. Therefore, the operation place usually has to be restricted in order to lower the influence of light and background. Border control system will request a person stand still and face to a certain direction at a specific position to catch picture. All the light and background will be controlled and simplified under these conditions. It would have great improvement in the accurate rate of recognition. Otherwise, the face recognition still exist the problem of recognition accuracy.

The 3-Dimension image technology is gradually mature. The input image could be the 2-Dimension or 3-Dimension for face recognition. The recognition algorithm could be different for 2D and 3D images. the 2D image will be more sensitive on the light of environment than the 3D image. The light source and face angle will influence the recognition result. However, during 3D process, there is not existed the problems of light source and face angle. Currently, the key problem of the 3D image recognition is how to catch a proper 3D image, which will depend on the maturity of the 3D imaging technology. Currently, some vendors have developed the 3D face scanners. The performance gets to be 1.2 second in average for each scanning time (refer to http://www.space-vision.jp), as shown in Figs. 23.2 and 23.3.

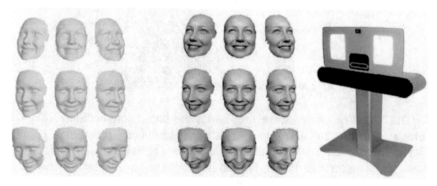

Fig. 23.2 The Cartesia Series 3D Face Scanner can produce high-resolution 3D images of a person's face in just 1.2 seconds (source: http://www.space-vision.jp/EP-Face_Scanner.html).

The Face Recognition Vendor Test (FRVT) 2006 evaluated face recognition technology by National Institute of Standards and Technology (NIST) include still face recognition and 3D face recognition. The FRVT 2006 is the

Fig. 23.3   3D   FastPass^{TM} Face Reader uses for access control (source: http:// www.l1id.com/).

latest evaluations for the face recognition that began in 1993. Years of evaluation by FRVT are shown in Fig. 23.4. It showed great improvement has been made under different light conditions. The best performers for 3D have a FRR interquartile range of 0.5% to 1.5% at a FAR of 0.1% for the Viisage normalization algorithm and a FRR interquartile range of 1.6% to 3.1% at a FAR of 0.1% for the Viisage 3D one-to-one algorithm (Fig. 23.5) [4].

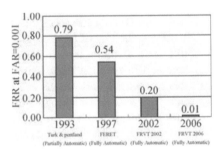

Fig. 23.4 Years of evaluation by FRVT [4].

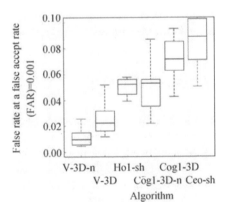

Fig. 23.5   Summary of performance for 3D face recognition algorithms [4].

23.1.3 Iris [5, 6]

Iris (Fig. 23.6) has high accuracy in recognition, but operation and cost are the problem while realizing iris recognition.

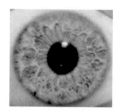

Fig. 23.6 Iris [7].

Iris is the muscles which surround the pupil and regulate the size of pupil to control the amount of light entering eye. The distribution of the muscles (iris texture) is so complicate and fine. The uniqueness of iris of an individual has been well accepted. Especially, iris is extremely hard to be forged. Even the twins have different irises.

An iris will be coded by 256 bytes using a polar coordinate system, which is called IrisCode. During iris recognition process, two IrisCodes will be compared. Therefore, the key of iris recognition is to create a qualified iris image which could present enough rich details. To catch a fine enough or qualified iris image, the resolution of a iris image is requested minimum of 70 pixels in iris radius. 80–130 pixels are more typical in common iris recognition systems. Therefore, iris recognition system needs a digital camera with high resolution. Besides, the commercial iris camera uses near-infrared light (700–900 nm) to illuminate for imaging, currently.

Although the near-infrared light is non-intrusive to a person, it still causes insecurity for a user because that user's eye is requested to be close to iris camera while operating. Obviously, it is an obstacle for the development of iris recognition system. However, the digital camera technology has a rapid progress in recent years. It could provide a way to move away the obstacle about near-infrared light.

23.1.4 Voiceprint [8, 9]

Voiceprint (Fig. 23.7) recognition is like the other biometric recognition, such as fingerprint, used to verify or identify individuals' identification. During the process of voiceprint recognition, the content of voice will not be analyzed.

Voiceprint is a kind of digital representation of vocal features. Voiceprint consists of some physical characteristics such as vocal cords, cadence of speech and duration of a vocal pattern. These characteristics are measured and translated into digital format called spectrogram. Voiceprint recognition is not like the other biometric recognition, such as fingerprint, which could match two

Fig. 23.7 Voiceprint.

samples directly and get high accuracy rate in recognition. Before the process of voiceprint recognition, the voiceprint template of a person needs to be built up through monitoring over time, which could not be abstracted from a short speech. With the template, the voiceprint recognition could identify whether the right person is speaking through the voice which could be from telephone system or computer system.

The voiceprint recognition system could be operated in real time, but in distance. The application scenario of voiceprint recognition is different with the other biometrics technologies. The modeling algorithms of voiceprint include vector quantization, neural networks, Gaussian mixture model, speaker-specific mapping, and kernel classifiers.

23.1.5 Vein [10, 11]

Vein recognition uses the image of vascular distribution of hand, palm or finger through the scan of near-infrared light. It is a new biometrics technology during these years, and has been deployed in financial system, such as ATM (Fig. 23.8), to identify an individual's ID. Finger vein could be acquired thru near infrared light. The vein pattern will be shown as a shadow. After the enrollment process, the vein pattern will be stored in ID card. Then, it will be used as a template for matching while identifying the user's identification.

Fig. 23.8 ATM using finger vein authentication [12].

The vein is beneath human's skin and hard to be forged. Therefore, some physical entry which needs high security will adopt this kind of biometrics technology. However, veins will be extending gradually as human's growing. Being as a biometric, the characteristics of biometric features, such as uniqueness and persistence, are still going on research.

Besides the mentioned biometrics technologies above, there still are some biometrics technologies under development for recognition, include hand geometry, DNA of human being for relations identification, steps appearance as gait recognition, face temperature as thermal facial recognition, etc.

Biometrics technology is a convenient and useful way for individual identification. However, different biometrics technology would result different accuracy of the recognition. Fingerprint and iris are well-known and higher accuracy of the recognition than voiceprint recognition. But the application of long-distance recognition, the technology of voiceprint recognition is the most suitable applied technology, because the telephone system can truly deliver a voice signal in a long distance and seamlessly integrate voice recognition technique to provide individual identification. According to various biometric features, there are Table 23.1.

Table 23.1 Comparison of various biometric technologies [1] High, Medium and Low are denoted by H, M, and L, respectively

Biometric identifier	Universality	Distinctiveness	Permanence	Collectability	Performance	Acceptability	Circumvention
fingerprint	M	H	H	M	H	M	M
face	H	L	M	H	L	H	H
iris	H	H	H	M	H	L	L
voiceprint	M	L	L	M	L	H	H
vein	M	M	M	M	M	M	L

Currently, there are many national projects going with biometrics technology. In 2008, President Bush signed Homeland Security Presidential Directive (HSPD-12). The NIST (National Institute of Standards and Technology) Computer Security Division initiated FIPS 201 program, entitled "Personal Identity Verification (PIV) of Federal Employees and Contractors", for improving the identification and authentication of Federal employees and contractors for access to Federal facilities and information systems (Fig. 23.9). Biometric datum, such as fingerprints and facial images, are required and stored in PIV cards. [13, 14] Currently, U.S. Homeland Security has adopted biometrics technology for border control to against terrorists. All the visitors entering America will be requested for fingerprints and facial images.

Hong Kong/Shenzhen adopted face, using near infrared face (NIR) and visible light (VL), and fingerprint biometric features in the biometrics system for the border control since 2005 (Fig. 23.10). There are 400 000 persons across the border every day. Currently, the system has enrolled over 1 800 000 people. The verification speed is about 6 seconds for each crossing [15].

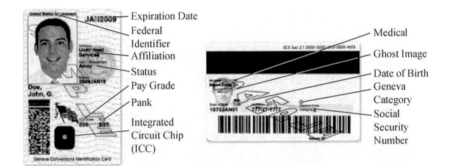

Fig. 23.9 PIV card and it's function. (source: http://www.cac.mil/CardInfoSecurity.html).

Fig. 23.10 Biometric Border Crossing: Shenzhen/Hong Kong [16].

Japan has implemented biometric identification system (J-BIS) for the border control since November, 2007. The system adopts fingerprint and face recognition technologies and processes approximately 25 000 visitors every day (Fig. 23.11) [17]. According to the Immigration Control and Refugee Recognition Law, all foreigners are required to provide their biometric datum which includes fingerprints and photographs for personal identification. The goal of the new border control policy is as same as the U.S. and includes the prevention of terrorism, illegal immigration and other crimes at the border. However, the requests shall not apply to the person who falls under any of the following items [18]:

Fig. 23.11 E-gate used for registered Japanese and registered re-entry visa holders [17].

1) a special permanent resident;

2) a person who is under 16 years of age;

3) a person who seeks to engage in Japan in an activity under "Diplomat" or "Official";

4) a person who is invited by the head of any national administrative organ;

5) a person provided for by a Ministry of Justice ordinance as equivalent to a person listed in any of the two items immediately preceding this item.

Border control is an important application of biometric technologies just like Hong Kong/Shenzhen and Japan cases described above. E-ID card is an another important application of biometric technologies like U.S. PIV case. E-ID card could be applied as civil ID, criminal ID, access control ID. Thailand, Singapore, Malaysia, Philippines and India have initiated the implementation of civil ID with fingerprint data. The United Kingdom National Identity Card is a personal identification document and European Union travel document (Fig. 23.12). Furthermore, the e-ID could be implemented for the personal

 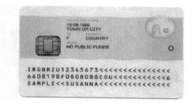

Fig. 23.12   NBIS (UK) e-ID project (source: http://www.telegraph.co.uk).

identification on consumer market. Users could have the e-ID for buying goods or services and remote transactions. For this kind of broad-applying service, users could access services after the e-ID authentication. WSA is an important architecture and will be discussed later.

23.2 Basic Operations of a Biometric System

A biometric system is basically a matching system of biometric features. Through the matching of caught biometric features and stored template in the database, identification of individuals will be accomplished. A biometric system consists of four main subsystems: Biometric Reader and Extracting Features, Matching Features, and Database of Features, as shown in Fig. 23.13.

1) Biometric Reader. It is usually the input interface of biometric system to input biometric image or voice signal, for example the fingerprint image, iris image, person face image, the image of the finger vein, talking signal, etc. Through this system input interface, the system provides the basic identification function according to the input biometric features. The different input

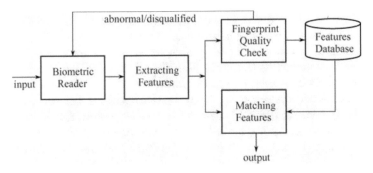

Fig. 23.13 Architecture of a biometric system.

interface will influence the possible applied scenario of this system. Take an example, the input of an iris image needs the biometric reader close and contact with person, the infrared rays scans on eyes to takes iris image. The images of fingerprint and the finger vein need to be apart from with biometric reader around $20-40$ cm in order to obtain image. The human face image can be caught by the camera with different focal length. Talking signal can be received through the close sound deliver or the long-distance telephone system, which almost is without the limitation of distance.

2) Extracting Features. The so-called features must represent a certain kind of the specialty. These features after particular assembling will have the uniqueness needed for identifying individuals. The features include type of ridges, minutiae type, minutiae location, direction, type of core, etc. After obtaining biometric image or voice signal, then Biometric Reader start carrying on the retrieval of biometric features. The retrieved biometric features will build a connection with this individual who owns the image or voice signal. The retrieved biometric features will be able to represent this individual who owns the image or speech signal. Therefore, the retrieved biometric features will consequently start the identification match with templates of biometric features which stored in database.

3) Matching Features. The retrieved biometric features will be able to represent this individual who owns the image or speech signal. Therefore, if the retrieved biometric features are matched with the template of biometric features which stored in database. Then it means the individual who owns the retrieved biometric features and the individual who owns the template of biometric features are the same one. The comparison could be classified into one-to-one and one-to-many according to the need of applications. The following will have a further discussion.

4) Features Database. While enrollment procedure, system will get the biometric image or voice signal through Reader of Biometric. After features extracting, the extracted biometric features will build a connection with this individual who owns the image or voice signal. The extracted biometric features and the information of this individual will be stored in a database at

the same time. This database is called Features Database. While starting recognition procedure, the subsystem of Matching Features will compare the input biometric features with templates stored in the database, if matched, then it will be judged as the input biometric features have the same identity with the matched template stored in database.

The biometric system is basically used to provide the services of identification and verification. Identification service mainly applies to security, financial system, passport, border control, etc. Applying the biometric recognition technology judges whether the applicant and the true identity match or not. This kind of service scenario is relatively simple and the integrity of the processed biometric datum is better than others. Therefore, the system process is quick with high accuracy, which is belonged to one to one process. The size of feature database is depended on the requirement of the application. Take an example. The number of restricted persons of a controlled area is limited normally. The size of feature database could be very small and stored in flash memory. As to the application of passport or border control, the size of feature database could be large considerable and need to store millions or billions biometric datum. The biometric system must be a super computer system with a complete database system in order to accomplish biometric features matching in an acceptable time period.

Biometric images or voiceprints will be different if the time or places of capturing them are different, even from identical person. Take fingerprint as an example. The factors influenced results include the fingerprint reader, moist degree of a finger, depth of finger press, location of a finger, angle of a finger, rotate of a finger, etc. The different conditions of formed images, changing environments and user's operations will cause the variance within these different images. Therefore, there will not be retrieved exact consistent biometric features even come from the identical one.

Biometric system will compute a matched score for a matching result. High matched score means the similar degree of these two sets of biometric features is high and the probability which these two sets of biometric features come from the identical is relative high. If the matched score is over a preset threshold, it will be judged as come from the same one. The distribution of the matched scores, which the matching features are retrieved from different individuals with the template features, is called impostor distribution. The distribution of the matched scored, which the matching features are retrieved from same person with the template features but different time, is called genuine distribution. The biometric system error rates are shown in Fig. 23.14.

In general, there is some system parameters used to measure the system performance [19]:

- FRR(false rejection rate)/FNMR(false non-match rate)—the rate of identifying the biometrics features of a person, which were sampled at different time, to be from different owners (Fig. 23.15);
- FAR(false acceptance rate)/FMR(false match rate)—the rate of identifying the biometrics features of different persons to be identical person

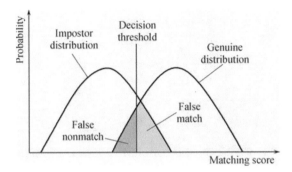

Fig. 23.14 Biometric system error rates [19].

(Fig. 23.16).

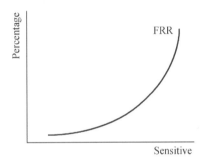

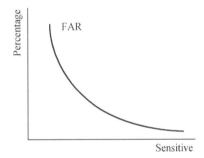

Fig. 23.15 The higher sensitivity with the higher FRR/FNMR.

Fig. 23.16 the higher sensitivity with the lower FAR/FMR.

From Fig. 23.16, we find FNMR and FMR are depend on decision threshold. Lower the decision threshold is, more the tolerance of influenced factors is (Fig. 23.17). The possibility of that system recognizes the matching features and template features which come from the same person to be different persons will be relative low. However, the possibility of that system recognizes the matching features and the template features which come from different persons to be the same one will be relative high.

Although the fingerprint recognition system has high recognition rate currently, in general, to get a proper fingerprint image still is an issue for the fingerprint recognition system. The forming of a fingerprint image will be effected by two factors, quality of a reader and situation of finger pressing. The quality of a reader includes geometric distortion of a fingerprint image, brightness contrast of a fingerprint image and uniformity of background of fingerprint image. In order to get the guaranteed quality of a reader, FBI and GSA have built up the certification process. The situation of finger pressing will cause the varieties of minutiae. Everyone's fingerprint could be deep, shallow, bold, thin, wet, dry, etc. At the enrollment process, many situations

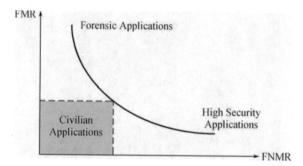

Fig. 23.17 System sensitivity is depend on the decision threshold and will cause different FMR and FNMR, respectively.

will be different each time, such as size of finger image, location of a finger in reader, rotate and distortion of a fingerprint while pressing, etc. Furthermore, the surface of a finger is a soft and bending image. How to translate the bending image into a flat image? All the factors will influence the result of the retrieved fingerprint. System will get the different fingerprint images for an identical individual at different times or different places. That is why the fingerprint recognition system may have a wrong judgment for a same person. The qualified fingerprint while enrollment is the key for this kind of problem.

How to get a qualified fingerprint under every different pressing situation? It could not be solved by system itself but also the enrollment process. 3 – 5 times pressing on a certified reader at multi-press enrollment process [20] could improve and get the qualified fingerprint image. It will also improve the variance of the situation factors, include the size of finger image, while finger pressing on a reader. Sometimes, there is abnormal press at enrollment process, too. System has to filter out the fingerprints while abnormal pressing.

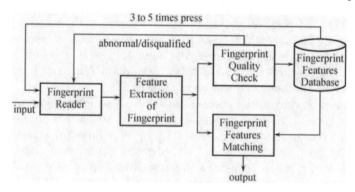

Fig. 23.18 Fingerprint recognition system with multi-press enrollment process [20].

In general, the fingerprint recognition system reads the fingerprint from the reader, than extracts out the features from the fingerprint and store in database. The features have been decided after caught the finger image. A complete fingerprint recognition system should include the enrollment process. Through the 3 – 5 presses while enrollment, system will get multiple images associated with different conditions. These images also present the custom of a register. After the combination of the extracted fingerprint features from these images, the set of fingerprint features will be stored in database, which will have wide range and more qualified features associated with the registered individual. The set of fingerprint features is not extracted form a single finger image only, but multiple finger images associated with different operation conditions. Therefore, system tolerance will be enlarged and able to process more complicate pressing situations. After the experimentation, the system performance will be improved up to around 3 – 4 times in identification if user pressed 3 – 5 times on a fingerprint reader while enrollment.

23.3 Biometrics Standardization

The international organization that work on the standardization of biometrics includes ITU (International Telecommunications Union), ISO(International Organization for Standardization) and IEC(International Electro-Technical Commission). ISO and IEC have a joint technical committee for information technology standards called JTC1. In 2002, ISO/ IEC JTC1 established a panel SC37 (http://isotc.iso.org/livelink/livelink/fetch/2000/2122/327993/2262372/customview.html?func=ll& objId=2262372& objActions=browse& sort=name) to build up biometrics standards. The SC37 establishes 6 technique working groups [21]:

1) WG1 – Harmonized Biometric Vocabulary
2) WG2 – Biometric Technical Interfaces
3) WG3 – Biometric Data Interchange Formats
4) WG4 – Biometric Profiles
5) WG5 – Biometric Performance Testing and Reporting
6) WG6 – Cross-Jurisdictional and Societal Aspects of Biometrics

The working group (WG3) of biometric data interchange formats has worked out a complete series standards since 2005 [22].

1) ISO/IEC 19794-1:2006 — Part 1: Framework
2) ISO/IEC 19794-2:2005 — Part 2: Finger minutiae data
3) ISO/IEC 19794-3:2006 — Part 3: Finger pattern spectral data
4) ISO/IEC 19794-4:2005 — Part 4: Finger image data
5) ISO/IEC 19794-5:2005 + A2:2009 — Part 5: Face image data
6) ISO/IEC 19794-6:2005 — Part 6: Iris image data
7) ISO/IEC 19794-7:2007 — Part 7: Signature/sign time series data
8) ISO/IEC 19794-8:2006 — Part 8: Finger pattern skeletal data
9) ISO/IEC 19794-9:2007 — Part 9: Vascular image data

10) ISO/IEC 19794-10:2007 — Part 10: Hand geometry silhouette data
Till now, the standardization of fingerprint has been complete, include features standards.

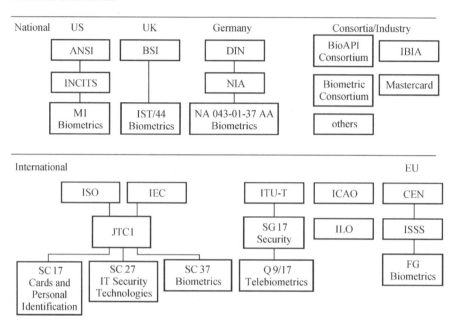

Fig. 23.19 Landscape of standardization in biometrics [7].

The standards of 3D face images are amended into ISO/IEC $19794-5$ since 2009. This amendment is intended to establish a standard data format for the interchange of 3D face images within systems or subsystems. Some new image types have been introduced, which combines 2D face images and associated 3D shape information.

Because the biometrics technology has been applied to many areas, for example, finance and economics (TC68), card and personal ID (JTC1 SC17), and IT security technology (SC27), SC37 is also responsible for the coordination of the needs for all the applications areas at biometrics standards. The International Committee on Information Technology Standards (INCITS) technical committee M1(http://www.incits.org/tc_home/m1.htm) is also responsible to standardize the biometric technology in the United States. Landscape of standardization in biometrisc is shown in Fig. 23.19.

23.4 Certification of Biometric System

A real biometrics system has to satisfy the need of purchase, deployment and usage, for example, the process speed, identification accuracy, database capacity, convenience in operation, open system.

The fingerprint identification system is very widespread use in many application areas, such as door security, computer security, passport ID and border check. According to the mentioned considerations above, the format of fingerprint features is one of the standardization items at the beginning. After the format of fingerprint features been standardized, the transaction and storage format of Biometric Reader, Extracting Features, Matching Features and Feature Database will have a consistent form. The message exchange within subsystems will also be solved because of the standardization of data format. Each subsystem does not need to be restricted by the same vendor. The subsystems produced by different manufacturer can combine each other and form an integral recognition system. Certainly, the recognition system will have different system parameters because of the constitution of different subsystems.

Because an integral recognition system may be consisted in some subsystems produced by different manufactures, each subsystem needs to identify the support toward the standard format of biometric feature and the recognition accuracy after a certain standard procedure, to ensure the integration degree within a recognition system.

MINEX (Minutiae Interoperability Exchange Test) is to approve the interoperability based on INCITS 378 and ISO/IEC 19794-2 fingerprint minutia standards within multi-vendors. MINEX is designed to evaluate the compatibility and interoperability within encoding schemes, probe templates, gallery templates, and fingerprint matchers from different vendors. There are parts of programs, include MINEX II, MINEX 04 and Ongoing MINEX. MINEX II evaluates the capabilities of fingerprint minutia matchers running on ISO/IEC 7816 smart cards. MINEX 04 evaluates the interoperability of minutiae data (rather than image data) between different fingerprint matching systems. Ongoing MINEX is an ongoing evaluation of the INCITS 378 fingerprint template which specifies finger minutiae format for data interchange, to evaluate template encoding and matching software. All the biometric software modules and systems of vendors will be testing through MINEX evaluation processes. MINEX will have reports about measurements of performance and interoperability of core template encoding and matching capabilities [23].

IREX (Iris Exchange) is a program of NIST to support interoperability of iris biometrics. IREX has addressed standards, formats and compression for data interchange, and intends to define and measure the quality of an iris image [24]. IREX conducted ISO/IEC 197946 standard which is initiated by the Working Group 3 of the ISO SC 37 committee in July 2009. Furthermore, IREX II-IQCE (Iris Quality Calibration and Evaluation) identifies iris image properties and evaluates iris image quality assessment algorithms. IREX II-IQCE quantifies and examines the effects of iris image properties to improve and speed up the maturity of iris recognition technology [24, 25].

23.5 Cloud Service — Web Service Authentication

Cloud service is a new kind of network service. As the network is gradually stable and connected easily, light client is the trend for network service. Users could not need to install heavy applications and proceed complicated computing on local computer. All the computing process could be moved to the network. Users are able to access the services from networks just through a simple portal connection.

Authentication through network (web service) is one of the cloud services. Usually, authentication will be proceeded at anytime and anywhere. Take an example, policemen identify or verify someone's identification at roadside. Fingerprint reader is the light client to get input fingerprint. However, the authentication needs the server to proceed matching for the input fingerprint with the templates stored in the database. If the authentication servers stay in the cloud network, authentication should be more convenient from the service angle. The system architecture of Startek's WSA for authentication cloud service is shown in Fig. 23.20.

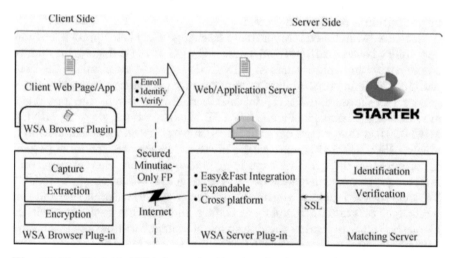

Fig. 23.20 Startek's WSA for authentication cloud service (source: startek.com. tw).

With the cloud architecture, the authentication system will have some advantages: [26]

1) Easy to install. Feature extraction of fingerprint and data encryption are built in the reader. Plugin program will be installed automatically while been used first time. After retrieved fingerprints, they will be extracted features, encrypted and transmitted to the network server for the further matching computing. The client is very simple and light without complicated computing.

2) Platform independent. The format of fingerprint feature will be fol-

lowed the definition of standards. Therefore, the network server could be platform independent. Moreover, if there are multiple servers existed in the network to provide the authentication service together, the servers could adopt different platforms. It will not affect the integrality of authentication service.

3) Easy to manage. All the network servers are stayed in the cloud network. Any change in the servers will not affect the user client. Therefore, the number, locations and architecture of servers could be adapted according to the current situation of authentication service. Moreover, the cloud could be public cloud, private cloud and hybrid cloud. The network design will have greater flexibility and security than before.

4) Easy to develop related applications. Because of the loose connection between clients and servers, the user interface on web could be easily developed and changed. It will be more user-friendly in use and meet the requirements.

Cloud service makes possible for fingerprint identification to substitute passwords as a new way of personalized authentication in daily applications, such as e-ATM and e-transaction. Window-based biometric framework has been built in all notebooks. These notebooks also comply to security framework, ISO19092-2008, for fingerprint authentication, It makes the electronic transactions over network more secure and reliable. Over 60 millions notebooks have already equipped with fingerprint sensors.

23.6 Challenges of Large Scale Deployment of Biometric Systems [26, 27, 28]

As the maturity of biometric technology, many national-wide projects have been implemented. In general, most of these projects are belonged to the application of border control, such as European Union, U.S. and Japan. Many important issues in large scale deployment of biometric systems have been addressed [29]:

1) Privacy and Security

(1) proportionality of the measures.

(2) restricted usage of the collected data.

2) Enrollment

quality control of the collected data during biometric capturing process (e.g., human factors, environment conditions, operation processes).

3) Database

management and architecture of storage of biometric data.

4) Accuracy

type of biometrics used in documents.

5) Standardization

(1) interoperability of databases.

(2) international data exchange.

(3) testing and certification.

Basically, large scale deployment is a complicate engineering process. Privacy and security issues are not related to biometric technology directly. Here will ignore these two issues temporary. The following concern is about standardization. Standardization happens in any open system network. A large system normally will be divided into many subsystems. All the interfaces between subsystems have to be standardized for the integration and purchase issues. Many biometric standards have been widely adopted in these deployments mentioned above. All the fundamental standards have been worked out by the standard organizations. Interoperability and certification are also prepared ready. The rest challenges return to the technology itself.

Currently, biometrics system could not provide 100% correct on recognition. The recognition rate FRR could be near 1% – 2%. However, this gap may lead to misjudgment. Certain high security applications could not accept any misjudgment. How to solve this problem? Well designed enrollment could be the key to approach it. A complete fingerprint recognition system should include the enrollment process. Actually, most of the misjudgments are caused by the different finger-pressing situations. System will get the different fingerprint images even from an identical individual at different time or different places. A well designed enrollment process, multi-press enrollment process [30], may collect all possible minutiae features and improve the recognition accuracy up to 3 – 4 times. The other way is to adopt multiple biometric technologies formed a multimodal biometric system.

In practice, there are two ways in enrollment process to improve FRR. One is multi-press enrollment process to relax the variance of the touch between finger and reader. The other way is using large field size, for example $0.8'' \times 0.8''$ with image size 400×400 (see Fig. 23.21) in enrollment process and medium field size, for example $0.5'' \times 0.64''$ with image size 256×325 (or $0.5'' \times 0.5''$ with image size 256×256, see Fig. 23.22), in matching process. There is a relative proportion between field size of a reader and minutiae features. The more minutiae features will get the more accurate recognition results (FRR/FAR).

Table 23.2   Average minutiae number in a fingerprint image

Image size	No of minutiae (average)
400×400 or above	51
256×325	35
256×256	28

Figures 23.23 and 23.24 show the fingerprints caught from FC320 (400×400) and FM210 (256×256). We can find the fingerprints caught by the reader FC320 with large field size are complete and keep the whole minutiae information. As to the fingerprints caught by the reader (FM210) with medium field size, they could be cut and lose a certain minutiae information.

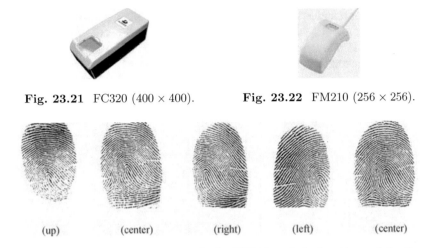

Fig. 23.21 FC320 (400 × 400).   **Fig. 23.22** FM210 (256 × 256).

(up)   (center)   (right)   (left)   (center)

Fig. 23.23 Fingerprint images caught by FC320 and the locations of the finger pressed on reader are up, center, right, left, center (from left to right).

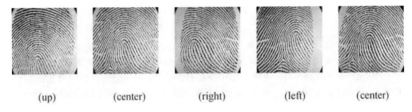

(up)   (center)   (right)   (left)   (center)

Fig. 23.24 Fingerprint images caught by FM210 and the locations of the finger pressed on reader are up, center, right, left, center (from left to right).

Through the multi-press enrollment process, system will get multiple images of a finger associated with different operation conditions. These images also present the operation custom of a register. After the combination of the extracted minutiae from these images, system will get wider range and more qualified minutiae in the enrollment process. Compare the readers FC320 and FM210. We found there are 51 minutiae in average in a single finger image caught by FC320. Moreover, after the combination of five finger images, we could get 63 minutiae for a single finger in average. For FM210, we found there are 28 minutiae in average in a finger image. After the combination of five images, we could get 42 minutiae for a single finger in average.

The set of minutiae is not extracted form a single finger image only, but multiple finger images associated with different operation conditions. Therefore, system tolerance will be enlarged and able to process more complicate pressing situations. With more minutiae, hereafter, system could have better matches naturally. After the experimentation, the system performance will be improved up to around 3−4 times in identification if user pressed 3−5 times on a fingerprint reader while enrollment.

To find matched minutiae from millions, even billions, template features is absolutely not an easy job. Multiple servers architecture are needed for the large scale deployment of biometric system. The system architecture is shown in Fig. 23.25.

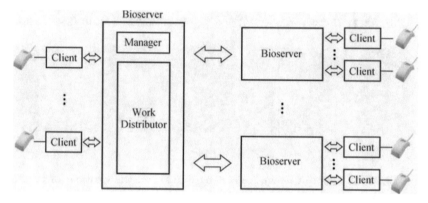

Fig. 23.25 Multiple servers architecture for large scale deployment of a biometric system.

Fingerprint as an example, based on the multiple servers architecture, the system could have high matching speed 60 000 – 80 000 templates/sec for each bio-server node with Pentium 3 GHz. At the same time, the minutiae stored in database must be classified. Classification will reduce most of invain matches and size of searching database to 1/100 to speed up the response up to 50 times (reference startek.com.tw)

The last big challenge is about the wolf attack in biometric system [30]. Due to the standardization of biometric technology and network connections within subsystems, there exist so many vulnerabilities and easy to be attacked. The attacker has complete knowledge of the biometric system to attack the system. He/She will present an artificial feature (wolf) to impersonate a victim and get false accept. The wolf attack will cause serious security problems. How to resist this kind of attack is a quite important issue for the large scale deployment of biometric systems. Sometimes, there are a few strong wolves in a biometric system (wolf accept probability ≫ false accept rate) . In system design, wolf attack problem must be and can be solved for the large scale deployment of biometric system.

23.7 Conclusion

Biometric technologies have been developed for near 30 years. Currently, they are progressing into development, even large scale development, stage. The border control and e-ID card are the important applications of biometric technologies. The related projects all have been large scale development. Such as there are accumulate 40 000 000 persons crossings the Hong Kong/Shenzhen border since June 2005. The system has enrolled over 1 800 000 people. Furthermore, the e-ID card with fingerprint data has been applied as civil ID, criminal ID, access control ID. For the future, the e-ID could be implemented for the personal identification on consumer market. Users could have the e-ID for buying goods or services and remote transactions.

Although there are some issues in system deployment, such as standardization and interoperability, and the identification and authentication capability of biometric system have satisfied the basic requirements of some applications, there are some fundamental problems still need to be improved, such as the system reliability.

A complete fingerprint recognition system should include the enrollment process. Through the multi-press enrollment process, system will get multiple images associated with different operation conditions. The four-finger reader could not be used in the multi-press enrollment process. After the experimentation, the system performance will be improved up to around $3-4$ times in identification if user pressed $3-5$ times on a fingerprint reader while enrollment. System tolerance has got the great improvement and capability to process more complicate situations of a finger. Following this enrollment process, the biometrics system will get higher accuracy in identification and authentication.

For a large scale deployment of biometrics system, there is a problem of response time except the issues addressed above. How to accomplish a fingerprint match with the database stored millions template fingerprints within seconds is a very challenge issue. Features classification is needed for the quick match. While enrollment process, the template fingerprints will be classified first, according to the extracted features, and then stored in features database. With the classification mechanism, the matching fingerprint could be matched with the limited template fingerprints in the features database. Therefore, the matching time could be reduced and meet the requirement of quick response under large scale deployment. Last, wolf attack is another big challenge for the large scale deployment of biometric system, and it must be and can be solved for the large scale deployment of biometric system.

Biometrics technologies are mature and ready for the daily security applications in life, even for large scale deployment. A biometric system could provide a more convenient security service than the others, such as keys and passwords, without doubt. Although the biometric system could not have 100% correct of recognition till now, multimodal biometrics could be an alternative way to meet the high accuracy requirement of real applications. Just like the case of fingerprint recognition technology, the accuracy of fingerprint recognition is quite high and used in daily applications. However, for a certain highest security applications, fingerprint and face recognitions will be selected by most of the multimodal biometric systems, at the same time, in order to compensate the little misjudgment percentage of each biometric technology. Otherwise, the enrollment process could be an important key to help fingerprint recognition technology to approach 100% correct.

References

[1]   Jain A K, Ross A, Prabhakar S (2004) An Introduction to Biometric Recognition. IEEE Transactions on Circuits and System for Video Technology, 14(1): 4–20
[2]   NTSC subcommittee on Biometric (2006) Face Recognition. http://www.biometrics. gov/Documents/FaceRec.pdf. Accessed 8 December 2010
[3]   Katakdound S (2004) Face Recognition: Study and Comparison of PCA and EBGM Algorithms. A thesis, the Department of Computer Science of Western Kentucky University
[4]   Phillips P J, Scruggs W T, O'Toole A J et al (2007) FRVT 2006 and ICE 2006 Large-Scale Results. NISTIR 7408. http://www.frvt.org/FRVT2006/docs/FRVT2006andI CE2006LargeScaleReport.pdf. Accessed 8 December 2010
[5]   NTSC subcommittee on Biometric (2006) Iris Recognition. http://www.biometrics. gov/Documents/IrisRec.pdf. Accessed 8 December 2010
[6]   Daugman J (2004) How Iris Recognition Works. IEEE Transactions on Circuits and Systems for Video Technology, 14(1): 21–30
[7]   ITU-T Technology Watch Report (2009) Biometrics and Standards. http://www.itu. int/dms_pub/itu-t/oth/23/01/T230100000D0002PDFE.pdf. Accessed 8 December 2010
[8]   NTSC subcommittee on Biometric (2006) Speaker Recognition. http://www. biometrics.gov/Documents/speakerrec.pdf. Accessed 8 December 2010
[9]   Haizhou L (2007) Overview of Voiceprint Technology. The Inaugural Asian Biometrics Conference & Exhibition
[10] NTSC subcommittee on Biometric (2006) Vascular Pattern Recognition. http:// www.biometricscatalog.org/NSTCSubcommittee/Documents/Vascular%20Pattern %20Recognition.pdf. Accessed 8 December 2010
[11] Yanagawa T, Aoki S, Ohyama (2007) Human Finger Vein Images are Diverse and its Patterns are Useful for Personal Identification. MHF Tech Rep MHF 2007-12, Kyushu University 21st Century COE Program, Development of Dynamic Mathematics with High Functionality
[12] Masahiro MIMURA (2007) Finger Vein Based Authentication. Inaugural Asian Biometrics Conference & Exhibition
[13] National Oceanic and Atmospheric Administration of U.S. Department of Commerce (2007) Homeland Security Presidential Directive HSPD-12 PIV-1 Implementation And Suitability Processing. http://www.wfm.noaa.gov/pdfs/WFM_HSPD12.pdf. Accessed 8 December 2010
[14] Computer Security Resource Center of Computer Security Division, NIST (2010) About Personal Identity Verification (PIV) of Federal Employees and Contractors. http://csrc.nist.gov/groups/SNS/piv. Accessed 8 December 2010
[15] Tan T (2007) Biometrics in China. Inaugural Asian Biometrics Conference & Exhibition
[16] Wen-Hsing Hsu (2008) Biometrics Market Trend in Asia (The Impact of US PIV Program). Asian Biometrics Conference 2008
[17] Wong T K (2008) The Border Control System That Utilized Biometrics in Japan. Asian Biometrics Conference 2008
[18] Japan Cabinet Secretariat. (2006) Japan Immigration Control and Refugee Recognition Act (Cabinet Order No. 319 of 1951). http://www.cas.go.jp/jp/seisaku/hourei/ data/icrra.pdf. Accessed 8 December 2010
[19] Delac K, Grgic M (2004) A Survey of Biometric Recognition Methods. 46th International Symposium Electronics
[20] Wen-Hsing Hsu (2010) A Multi-Presses Enrollment Method to Get Full Minutiae for Fingerprint Authentication. Asian Biometric Consortium Conference and Exhibition 2010
[21] ISO (2010) ISO JTC 1/SC 37 Biometrics. http://www.iso.org/iso/jtc1_sc37_home. Accessed 8 December 2010
[22] ISO (2010) ISO Standards—Browse by TC. http://www.iso.org/iso/iso_catalogue/ catalogue_tc.htm. Accessed 8 December 2010

[23] NIST (2010) MINEX Overview. http://www.nist.gov/itl/iad/ig/minex.cfm. Accessed
8 December 2010
[24] NIST (2010) IREX Overview. http://iris.nist.gov/irex/index.html. Accessed 8 December 2010
[25] NIST IREX II IQCE (2010) Iris Quality Calibration and Evaluation 2010. http://
biometrics.nist.gov/cs_links/iris/irexII/IQCE_conops_API_v44_26apr10.pdf. Accessed
8 December 2010
[26] Teng C H, Hsu H L, Hsu W H (2010) Biometric Authentication. Pattern Recognition
and Machine Vision: In: Honor and Memory of Professor King-Sun Fu, Wang P S P
(ed), River Publisher, pp: 335 – 348
[27] Teng C H, Hsu W H (2009) Large Scale System Design. Encyclopedia of Biometrics,
pp: 884 – 889
[28] Goldstein J, Angeletti R, Manfredbach et al (2008) Large-scale Biometrics Development in Europe: Identifying Challenges and Threats. JRC Scientific and Technical
Reports
[29] Goldstein J, Angeletti R, Holzbach M et al (2008) Large-scale Biometrics Deployment
in Europe: Identifying Challenges and Threats. JRC Scientific and Technical Report.
http://www.a-sit.at/pdfs/biometrics_report.pdf. Accessed 8 December 2010
[30] Une M, Otsuka A, Imai H (2008) Wolf Attack Probability: A Theoretical Security
Measure in Biometric Authentication System. IEICE - Transactions on Information
and Systems, ISSN:0916-8532 EISSN:1745-1361, vol E91-D Issue 5: 1380 – 1389

24 Radical-Based Hybrid Statistical-Structural Approach for Online Handwritten Chinese Character Recognition

Chenglin Liu[1], Longlong Ma[1]

Abstract The hierarchical nature of Chinese characters has inspired radical-based recognition, but radical segmentation from characters remains a challenge. This chapter describes a new radical-based online handwritten Chinese character recognition approach which combines the advantages of statistical methods and radical-based structural methods. We first establish a radical model database from the viewpoint of computer segmentation and recognition. The radicals are categorized into non-special (horizontally-vertically separable) ones and special ones. The parameters of statistical models for non-special radicals are estimated in automatic learning embedding radical segmentation on training character samples. Meanwhile, special radicals are modeled by binary classifiers, each of which detects a type of special radical from candidate radicals hypothesized from the input character. Characters either with or without special radicals are recognized in an integrated recognition framework, where the recognition performance is further improved using three strategies, namely, direct recognition in confident region, preferred recognition in confusing region and rejection of error detection. Our experimental results demonstrate the effectiveness of the proposed approach.

24.1 Introduction

With the emergence of digitizing tablets, tablet PCs, digital pens, pen-based PDAs, and mobile phones, on-line handwritten Chinese character recognition (OLHCCR) is gaining renewed interests. In the last decades, many approaches have been proposed and the recognition performance has been improved constantly [1]. To be implemented in hand-held devices with limited computation and storage capability, researchers are working towards high accuracy recognition methods with lower complexity.

The hierarchical nature of Chinese characters has inspired radical-based recognition methods, which model a much smaller number of radical classes instead of characters. Hierarchical character representation has also been used in Hangul character recognition [2–4], where components (graphemes) and

---

1 National Laboratory of Pattern Recognition (NLPR), Institute of Automation, Chinese Academy of Sciences 95 Zhongguancun East Road, Beijing 100190, China. E-mails: {liucl, longma}@nlpr.ia.ac.cn.

the relationships between them are statistically represented. Such hierarchical representation-based methods have three benefits. First, the model complexity is reduced by modeling radical shapes instead of holistic character shapes. Second, by focusing on radicals with simpler structures than characters, the recognition accuracy can be improved. Third, the classification of a small number of radicals needs a small set of training samples.

The methods of OLHCCR can be roughly grouped into two categories: statistical and structural. Statistical methods generally represent the holistic character shape as a feature vector and use a statistical classifier for classification. The feature vector representation, e.g., the directional feature density or the so-called direction histogram feature [5, 6], enables stroke-order and stroke-number free recognition. Statistical methods have yielded high accuracies but suffer from high complexity because of the large number of character classes. Structural methods are based on stroke analysis and radical analysis. To tolerate stroke-order variations, a character or radical is often modeled as a relational graph, with strokes or sub-strokes as primitives. Hidden Markov models (HMMs) are frequently used to model strokes and radicals [7, 8]. Discriminative training has been applied to decrease the HMM-based radical recognition error [9]. Since the HMM is stroke-order dependent, multiple models per radical/character are needed for stroke-order variations.

Radical-based structural methods usually decompose characters into radicals, which are extracted either from character strokes or character skeleton. Some rule-based or fuzzy rule-based methods are used to extract radicals. Rule-based radical detection using the prior knowledge of character structure and radical position [10] is likely to fail in cases of large shape variation. Some works have used fuzzy rules to model radicals and the relationships among them for dealing with divisible characters [11, 12]. A learning process was introduced to learn radical shape fuzzy prototypes and positioning fuzzy prototypes, and a fuzzy inference system aggregates these rules for decision making [12]. Fuzzy rule-based methods, however, have not been validated on large data sets.

Some methods use stable radicals to direct the holistic character recognition and speed up the character recognition process. Xiao and Dai defined 45 basic radicals (head or tail radicals) for 3 755 Chinese characters, and extracted reference templates of these radicals dynamically from stroke segments [13]. Chou et al. defined 64 top-level and 24 bottom-level radicals for coarse classification [14]. Via radical classification by neighboring segment matching, the number of candidate classes can be reduced substantially. Lay et al. introduced 453 front radicals and 273 rear radicals for filtering candidate radicals from characters using prior knowledge and geometric features [15].

To take advantage of the hierarchical structure of Chinese characters, Wang and Fan proposed a recursive hierarchical method to extract and match radicals in offline character recognition based on character structure information and prior knowledge about radicals [16]. For stroke-order-free on-line character recognition, a stroke-based cube graph model was proposed to decompose the character into intra-radical cube graphs and an inter-radical cube graph, and optimal paths are searched by two-level dynamic programming (DP) [17]. Based on a network representation of radical and ligature HMMs [8], radicals can be segmented by dynamically matching the radical models with sub-sequences of strokes. This approach is dependent on stroke order, however. Based on hierarchical representation and stroke segment matching, Kitadai and Nakagawa applied discriminative prototype learning to optimize radical models [18].

In all the above radical-based recognition methods, the problem of radical segmentation has not been solved. The previous methods are either dependent on stroke order or susceptible to stroke shape variance. A method avoids radical segmentation by radical location detection and location-dependent radical classification using neural networks on whole character images [19], but without radical segmentation, it suffers from the large number of location-dependent radical models and the low radical classification accuracy. To overcome the difficulty of radical segmentation, we proposed a new radical-based hybrid statistical-structural approach for online handwritten Chinese characters, and preliminarily applied to characters of left-right structures [20]. We deal with mixed left-right and up-down structures using a nested recursive radical segmentation method [21]. For special radicals that are not separable horizontally or vertically, we use binary classifiers for special radical detection from candidate radicals hypothesized from the input character [22].

Although some techniques of our system have been presented previously at conferences [20 – 22], this chapter aims to provide a comprehensive description of complete recognition system for recognizing characters of arbitrary structures. Particularly, we propose an integrated recognition framework to extract both non-special and special radicals, and improve the whole character recognition performance using three strategies, namely, direct recognition in confident region, preferred recognition in confusing region and rejection of error detection.

In the remainder of this chapter, Section 24.2 gives an overview of our radical-based recognition approach; Section 24.3 describes our method of radical modeling; Section 24.4 describes the integrated radical-based recognition framework; Section 24.5 presents our experimental results and Section 24.6 offers concluding remarks.

24.2 Overview of Radical-Based Approach

In previous works, radicals have been represented as sets of stroke segments, sequential models (such as HMMs) or relational graphs. These methods either depend on stroke order or rely on stroke segmentation. Our approach combines statistical and structural methods by representing radicals as statistical models (feature vectors, probability density models or statistical classifiers) while the hierarchical structure of Chinese character is exploited by hierarchical radical matching. We overcome the ambiguity of radical segmentation by over-segmentation of stroke segments and radical matching directed by character-radical dictionary. Compared to ordinary statistical methods which learn holistic models for each character class, our approach learns models of radicals of much fewer classes than characters. Compared to previous structural radical models, our radical model is based on feature vector representation, totally independent of stroke order and is easy to learn. Thus, our approach combines the merits of statistical and structural methods. Figure 24.1 summarizes the merits of our radical-based approach.

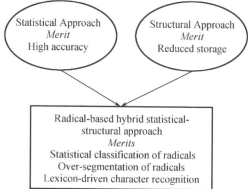

Fig. 24.1 Merits of radical-based hybrid statistical-structural approach.

Our hybrid approach takes advantage of appearance-based classification of radicals. The approach is similar to character string recognition [23] in the senses of candidate radical segmentation and tree representation of character compositions. It integrates appearance-based radical recognition and geometric context into a principled framework using a character-radical dictionary to guide radical segmentation and recognition during path search. To overcome the connection of strokes between radicals, corner points are detected to extract sub-strokes. For character recognition, we use two dictionary representation schemes and accordingly different search algorithms. The effectiveness of the proposed approach has been demonstrated on Chinese characters of non-special structures [21].

To recognize special-structure characters, we proposed a special radical detection method with radical hypotheses and verification using statistical binary classifiers [22]. When special radical is detected from the input character pattern, the remaining part is recognized using the approach for non-special structures. Otherwise, the input character is treated as non-special structure by default. We design 19 binary classifiers for detecting special radicals of 19 classes.

Finally, we introduce an integrated recognition framework based on the recognition approaches for non-special and special structures. To deal with the imperfection of special radical detection, we propose to improve the whole recognition performance by adopting three strategies: direct recognition in confident region, preferred recognition in confusing region and rejection of error detection. We applied the recognition framework to 6 763 character classes and achieved promising recognition performance in experiments.

24.3 Formation of Radical Models

Many Chinese characters share common sub-structures called radicals. Most radicals have semantic meanings and often a radical is also a single character. It is beneficial to use radicals as the units of classification because the number of radical classes is much smaller than the number of characters and the radicals have simpler structures. For character recognition, character models are formed by assembling radical models in a hierarchical structure, which guides the segmentation of radicals in the input character. Such radical model-guided recognition scheme is helpful to overcome the ambiguity of radical segmentation caused by character shape variation and stroke connection.

The formation of radical models plays an important role in overall recognition framework. The quality of the model database influences the recognition performance. In the following, we describe the basics for radical modeling of Chinese characters: Chinese character structure, character-radical dictionary representation and radical self-learning.

24.3.1 Chinese Character Structure

Chinese characters can be categorized into four rough structure types, single-element (SE), left-right (LR), up-down (UD) and special-structure. Special structures have been categorized in different ways: three types in [24], seven types in [25] and more detailed classes in [26]. We adopt the categorization of [25] but remove a structure type which involves very few character classes (we treat the characters of this type as single-element type). The nine types of structures considered in our system are shown in Fig. 24.2. For a character set of 6 763 Chinese characters (in standard GB2312–80), the number of characters belonging to nine structure types, the percentage and some examples of the characters are shown in Table 24.1.

Characters of non-special structure has recursively hierarchical characteristic and radicals in them are horizontally or vertically separable in recursion. While those of special structure have stable peripheral shape and the remaining part except the special radical is usually a non-special structure. Based on different characteristics and the relationship between non-special and special structures, we introduce an integrated radical-based recognition framework by combining the recognition approaches for non-special and special structures.

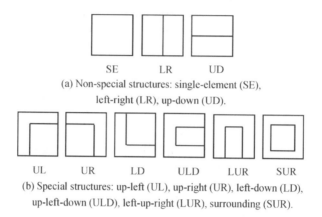

SE   LR   UD

(a) Non-special structures: single-element (SE),
left-right (LR), up-down (UD).

UL   UR   LD   ULD   LUR   SUR

(b) Special structures: up-left (UL), up-right (UR), left-down (LD),
up-left-down (ULD), left-up-right (LUR), surrounding (SUR).

Fig. 24.2 Nine structure types of Chinese characters.

Table 24.1   Statistics and examples of nine structure types

Type	#character	Percent(%)	Examples
SE	488	7.2	白本车册乘
LR	4 284	63.3	败帐挣保知
UD	1 489	22.0	曹罢恐息享
UL	240	3.6	疯雇层厄房
UR	28	0.4	氨甸氮旬氧
LD	141	2.1	迸趣毯魅飑
ULD	14	0.2	臣匿匹匣医
LUR	55	0.8	闭风冈闱闳
SUR	24	0.4	囤固国回困
Total	6 763	100	-

24.3.2   Character-Radical Dictionary Representation

The composition of characters with radicals as components can be represented in two schemes: sequential representation (Fig. 24.3 (a)) and hierarchical representation (Fig. 24.3 (b)).

By sequential representation, each character is represented as a string of radicals. The order of radicals is determined according to writing order (we assume that the writing order of radicals is relatively stable but allow stroke-order variation within radicals). Based on the representation, the character-radical dictionary is stored in a trie structure (Fig. 24.4) [20, 21]. The purpose of sequential representation is to take advantage of radical string matching and tree (trie) structure of the whole dictionary.

By hierarchical representation, each character class is represented as a tree structure indicating the radicals and the relationships between them.

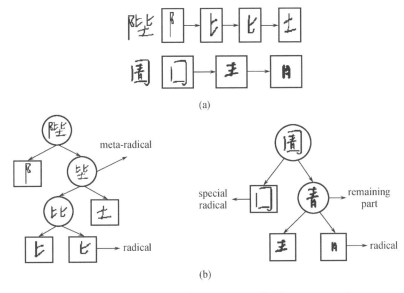

(a)

(b)

Fig. 24.3 Two schemes of character-radical representation.

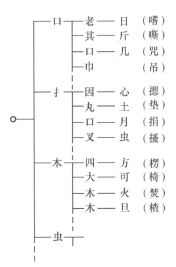

Fig. 24.4 Trie structure of character-radical dictionary.

Hierarchical representation is more informative and is totally free of radical order. However, it is not easy to unify the tree structures of different characters into a single data structure as the trie for sequential representation. In recognition, the input pattern is matched with the character models one by one, unlike the sequential representation that allows simultaneous matching of the input character with the whole character-radical dictionary.

24.3.3 Radical Modeling

From linguistic viewpoint, each radical has a semantic meaning. Some linguistic radicals, however, are hard to separate from characters by computer algorithms. We remove such radicals and define some new ones that appear frequently and easy to segment. The selection of radicals for modeling is influential to radical segmentation and character recognition.

Radical Selection

According to the characteristics of handwritten Chinese character, incorporating the radical selection criterion in [27] and our radical over-segmentation approach, we select radicals based on the following considerations:

1) According to writing custom and space information among the radicals within a character, cross strokes or strokes with large overlapping degree should belong to one radical.

2) A character or sub-character with large stroke-order variation is taken as a radical.

3) Select frequently used radicals when a character can be represented as different radical strings.

We take the character "陛" as an example. In Fig. 24.5, (a) denotes a character pattern, (b) shows the horizontally separable radicals as used in [20], and (c) shows the radicals used in nested segmentation [21]. We applied the criterion to 4,284 left-right (LR) structure Chinese characters and obtained 1,118 and 913 shared radical models according to the segmentation schemes of (b) and (c), respectively.

Fig. 24.5 A character pattern decomposed into different radicals.   (a)   (b)   (c)

Considering the hierarchical characteristic of Chinese characters, we use nested segmentation scheme to over-segment characters into radicals.

The selected radicals are further categorized into two types: special radicals and general (non-special) radicals. Special radicals are usually located at the beginning, end or both the beginning and end of stroke sequence in special-structure characters. General radicals appear in non-special structure characters and the part of a special-structure character excluding the special radical (remaining part). We applied the above radical selection criteria and the following radical self-learning method and finally, obtained 19 special radicals (Table 24.2) and 1 253 general radicals. Table 24.3 gives a few instances of general radicals. Note that some of the general radicals, such as 009, 044, 049, 064, and 069, contain special radicals. This is because they also appear in left-right (LR) and up-down (UD) structures and cannot be further segmented horizontally or vertically.

Table 24.2 Special radical models

Type	#Special radical	Radical models
UL	6	疒 广 厂 疒 户 尸
LD	6	辶 走 风 毛 九 鬼
LUR	3	冂 门 几
UR	2	勹 气
SUR	1	凵
ULD	1	乚

Table 24.3 Part of general radical models

Index	Radical	Index	Radical	Index	Radical	Index	Radical	Index	Radical
000	安	001	白	002	本	003	八	004	巴
005	贝	006	百	007	卑	008	半	009	包
010	恭	011	不	012	扁	013	步	014	丹
015	丙	016	并	017	卜	018	布	019	必
020	币	021	办	022	卞	023	表	024	虫
025	寸	026	才	027	车	028	仓	029	斥
030	产	031	成	032	采	033	厨	034	出
035	垂	036	春	037	串	038	差	039	臣
040	齿	041	丑	042	叉	043	册	044	辰
045	乘	046	川	047	长	048	丞	049	刍
050	处	051	从	052	大	053	电	054	歹
055	旦	056	单	057	刀	058	东	059	多
060	丁	061	当	062	岛	063	氏	064	店
065	弟	066	斗	067	度	068	豆	069	达

Among the 1 253 general radicals, 1 236 are from characters of non-special structures. Figure 24.6 gives the numbers of shared radicals among three types of non-special structures. Furthermore, for 6 261 characters of non-special structures, the number of characters containing different numbers of radicals and their percentages are shown in Table 24.4.

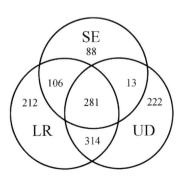

Fig. 24.6 The number of shared radicals for non-special structure.

Table 24.4 Statistics of different radical numbers for non-special structure

#Radical	#character	Percent (%)
1	726	11.6
2	3,361	53.7
3	2,010	32.1
4	154	2.5
5	10	0.1

Self-Learning of General Radicals

On determining the radical classes, the parameters of radical models are estimated in automatic learning on training character samples. Figure 24.7 shows the process of learning and corresponding examples.

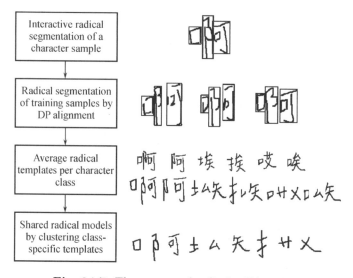

Fig. 24.7 The process of radical self-learning.

First, each character class has a sample correctly segmented into a sequence of radicals by human interaction. The remaining samples are matched with the correct radical sequence by DP to segment into radicals. The feature vectors of the segmented radicals of a radical class, from the training samples of a character class, are averaged to obtain the class-specific radical templates. In the second stage, the radical templates of all classes are clustered to obtain shared radical models. We used agglomerative clustering to obtain a hierarchy of radical partitions and then selected a partition by human judge. It is hard to determine an appropriate cluster number automatically, yet hierarchical clustering followed by human selection of partition works effectively.

During radical self-learning, we also obtain radical samples for modeling unary geometric features (outline shape and position of radicals) and binary geometric features (relationships between radicals in a character) [21].

Special radicals are modeled in a different way from general radicals. They are detected from candidate radicals hypothesized from input characters using binary classifiers. More details will be given in Section 24.4.1.

24.4 Radical-Based Recognition Framework

The recognition framework integrating special radical detection and nested segmentation-based recognition of non-special structures is diagramed in Fig. 24.8. The input character pattern is a sequence of strokes (each represented as a sequence of (x,y) coordinates). At the special radical detection stage, radical-specific heuristics (geometric rules, details in [22]) are used to group strokes for generating candidate radicals, which are classified using the special radical detectors (binary classifiers). If all the candidate radicals are rejected by each radical detector, the character is treated as a non-special structure

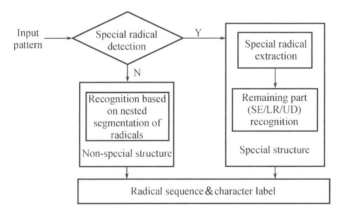

Fig. 24.8 Framework of radical-based recognition.

and is recognized using general radical models based on nested segmentation. If multiple candidate radicals are accepted by a radical detector, the one of maximum similarity is retained. After special radical extraction, the remaining part of the character is assumed to be SE/LR/UD structure and is recognized using the nested horizontal/vertical segmentation method.

If multiple special radicals are detected by different detectors, each radical is respectively extracted from the input character and the remaining part is recognized by the corresponding non-special structure (SE/LR/UD) module. The results of special radical detection and remaining part recognition are fused to give the final result of whole character recognition. We introduce three strategies to further improve the recognition accuracy. Finally, the optimal segmentation of radicals as well as their class labels (the radical classes in turn decide the character class) is given by the recognition framework.

In the following, we describe the methods for special radical detection, non-special structure recognition, special structure recognition, and integration strategies in sequel.

24.4.1 Special Radical Detection

The motivation of the special radical detection module is to detect whether the input character contains special radicals or not. The performance of detection decides the recognition accuracy of overall recognition.

Fig. 24.9 shows an example of special radical detection. We consider the character as a circular sequence of strokes (the last stroke is followed by the first stroke). Candidate radicals consist of sequences of consecutive strokes

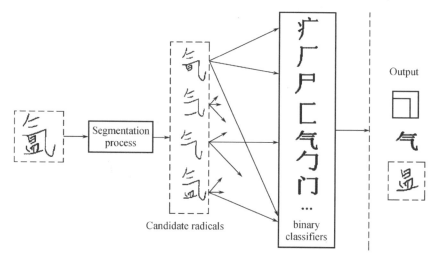

Fig. 24.9 An example of special radical detection.

in the circular sequence. All the possible candidate radicals can be generated using some prior knowledge on the segmentation points and important geometric properties. Specifically, candidate radicals are filtered using three rules: (1) the eligibility of cutting points between strokes, (2) the eligibility of special radical class (class-specific rules), and (3) the separation between the radical and the remaining part. Mostly, a special radical consists of start strokes of a character, end strokes, or both start and end strokes. Each class of special radical has relatively stable number of strokes and position in characters. A special radical and the remaining part of character are spatially apart to some extent.

After hypothesis and filtering, each retained candidate radical is classified by 19 binary classifiers (each for a special radical class) to decide whether it is a special radical or not. We use the support vector machine (SVM) [28] for binary classification. For a special radical class, a SVM is trained with positive radical samples and negative samples (of different classes or non-radicals) segmented from training character samples. Negative samples are collected using the bootstrap strategy. Whenever an SVM outputs a positive response, a special radical of specific class is detected. The detection module may output multiple detected special radicals because there are 19 detectors. More details of special radical detection can be found in [22].

24.4.2 Non-Special Structure Recognition

If no special radical is detected in the input character, we use the radical models of non-special (SE/LR/UD) structures to recognize the character based on nested horizontal/vertical segmentation and integrated segmentation-recognition directed by character-radical dictionary. When a special radical is detected from the input character, the remaining part is also treated as a non-special structure and recognized using this scheme. Figure 24.10 gives the process of non-special structure recognition. The method is outlined as follows, and more details can be found in [20, 21].

Character models (or remaining parts) are represented as nested horizontal/vertical strings of radicals (common sub-structures) in up to three hierarchies (horizontal-vertical-horizontal). Each radical is represented as a feature vector template or Gaussian density model (statistical radical model).

For segmentation, we first split the input strokes at corner points of high curvature to overcome the stroke connection between radicals. Candidate radicals are hypothesized from input character by grouping strokes according to the horizontal/vertical (depending on the hierarchy) overlapping degree. The candidate radicals are classified by statistical radical models to give matching scores. The combination of candidate radicals with highest score (minimum distance) gives the result of radical segmentation and character recognition. Radical hypothesis and classification are guided by a lexicon-driven tree search strategy, similar to lexicon-driven character string recognition [23]. Depending on the representation scheme of character-radical dictionary, we use two different search algorithms for radical segmentation-recognition: beam search for trie structure of sequential representation and nested DP matching for hierarchical representation.

In nested segmentation, the stroke sequence of input character is segmented into candidate radicals of up to three layers. For beam search with sequential dictionary representation, nested segmentation is fulfilled before search. For nested DP search with hierarchical dictionary representation, the input character is first segmented at the first layer and matched with the first hierarchy of radical/meta-radical sequence of a specific character class. The segmentation of the second layer (only for meta-radicals) depends on the matching result of the first layer, and the segmentation of the third layer depends on the second-layer matching. Figure 24.11 shows an example of nested segmentation.

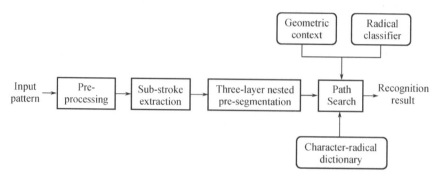

Fig. 24.10 Recognition process for non-special structure.

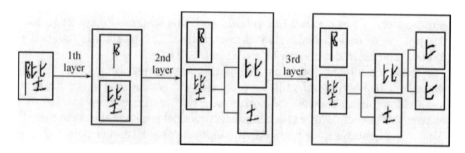

Fig. 24.11 Nested segmentation of a character.

For sequential dictionary representation, the segmented sequence of candidate radicals is matched with the radical sequence of all character classes stored in the trie structure of dictionary by beam search. This process is similar to lexicon-driven character string recognition [23]. An example of beam search is shown in Fig. 24.12, where each round box denotes a node (state) in the search space. In a round box, the left box denotes a candidate radical, the right tree denotes a node in the trie dictionary and the children of the node. On matching a candidate radical with a node of trie, the succeeding candidate radical is matched with the children of the trie node. A path in the search space gives a result of radical segmentation and character recognition, e.g., the path with edges of thick lines in Fig. 24.12 indicates that the input character is segmented into three radicals corresponding to character class "慚 (Can)".

More details of nested segmentation and lexicon-driven non-special structure recognition can be found in [21]. Note that for integrated recognition of special and non-special structures, there are 20 nested segmentation-recognition modules, one for SE/LR/UD whole character recognition and 19 for the remaining parts of 19 classes of special radicals.

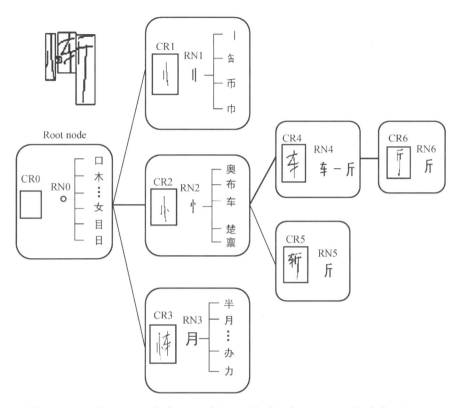

Fig. 24.12 Beam search for matching with the character-radical dictionary.

24.4.3 Special Structure Recognition

After special radical detection, characters with special radicals detected are recognized using special-structure models. Figure 24.13 gives an example of special-structure character recognition.

In special radical detection, one or more special radicals of different classes are detected by 19 detectors. The detected special radicals are extracted from the input character, and for each detected special radical, the remaining part is verified using the corresponding non-special structure recognition module specific to special radical class. When the output of remaining part recognition, the matching distance of the remaining part (minimum over all the remaining part classes for a special radical) exceeds an empirical threshold, the detected special radical is rejected. After verification for all the detected special radicals, if exactly one special radical is accepted, the whole character recognition result is immediately obtained. If all the detected radicals are rejected, the input pattern is treated as a non-special (SE/LR/UD) structure and recognized by the whole-character non-special structure recognition module. On the other hand, if multiple detected special radicals are

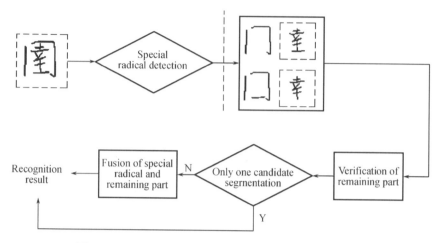

Fig. 24.13 Recognition process for special structure.

accepted after remaining part recognition, each detected radical is given a whole-character similarity by fusing the radical detection similarity and remaining part matching distance (transformed to sigmoidal confidence score), and the one of maximum similarity gives the character recognition result.

24.4.4 Strategies for Improving Recognition Accuracy

The accuracy of special radical detection influences the whole-character recognition performance. To overcome the imperfection of special radical detection by binary classifiers, we propose some strategies to improve the overall recognition performance.

For special radical detectors, it is desired that special radicals are detected in special-structure characters while on non-special structure characters, all the candidate radicals are rejected by the 19 special radical detectors. However, due to the imperfection of detectors (though the binary SVM classifiers perform over 99% accurately), some (false positive) candidate radicals are mis-detected as special radicals in non-special structure characters. We observed that the classifier outputs (confidence scores) of the false positives are relatively low compared to those of correct special radicals. To enhance the performance of special radical detection, we partition the confidence values of each special radical detector into high confident region and confusing region. When the confidence of special radical detection is sufficiently high (greater than a threshold $T2$) or low (smaller than a threshold $T1$), it is confident that the candidate radical is a special radical (positive) or rejected (negative). When the confidence is between $T1$ and $T2$, the detection is ambiguous. The confidence region $[T1, T2]$ is called confusing region, while $(-\infty, T1)$

and $(T2, \infty)$ are called confident regions. Different recognition strategies are adopted for confident and confusing regions.

Direct Recognition in Confident Region. The confident region $(-\infty, T1)$ indicates that the candidate radical is definitely not a special radical, i.e., the detector rejects the candidate. If all the 19 special radical detectors reject all the candidate radicals, the input character is treated as non-special structure and recognized using the whole-character non-special structure recognition module. The confident region $(T2, \infty)$ indicates that the candidate radical is definitely a special radical of a specific class. The remaining part is then recognized using the corresponding remaining-part non-special structure recognition module.

Preferred Recognition in Confusing Region. The confusing confidence region $[T1, T2]$ indicates that the detected special radical is not confident enough. We consider the special radical class of the highest confidence of the 19 detectors. If this highest confidence falls in the confusing region, we run both non-special structure recognition and special structure recognition on the input character, and obtain two whole-character similarity scores. The higher similarity gives the final result of character recognition. The fusing of two recognition modules thus can give more reliable recognition than either recognition module (special structure or non-special structure).

Rejection of Error Detection. For characters of non-special structures, some candidate radicals (non-special radicals) in the stroke sequences are similar to one of special radicals in shape and the special radical detector may falsely classify it as a special radical with confident greater than $T2$. This error can be corrected by remaining part verification using the non-special structure recognition module corresponding to the detected special radical class. If the remaining-part recognition module outputs low similarity, the input character is rejected as a special structure and is still recognized using the non-special structure recognition module. Figure 24.14 shows some characters with special radical detected and in fact these characters belong to non-special structures. Such mis-detected special radicals can be rejected in the subsequent recognition of remaining parts.

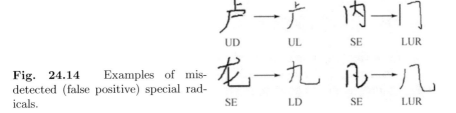

Fig. 24.14 Examples of mis-detected (false positive) special radicals.

24.5 Experiments

To evaluate the performance of the proposed radical-based recognition approach, we have experimented on a dataset of online handwritten Chinese characters of 6 763 classes (in the standard GB2312-80). The performance of the proposed approach is compared with a state-of-the-art holistic statistical recognition method.

24.5.1 Experiment Setting

In our dataset, each class has 60 samples produced by 60 writers. We used 50 samples per class for training classifiers, and the remaining 10 samples per class for evaluating the recognition performance.

For training radical models and recognizing candidate radicals, each radical sample undergoes the same procedures of trajectory normalization and direction feature extraction as done for holistic character recognition [29]. Specifically, a moment normalization method is used to normalize the coordinates of pen trajectory points, and 8-direction histogram features (512D) are extracted directly from pen trajectory. For radicals of non-special-structure characters, the 512D feature vector is reduced to 160D by Fisher linear discriminant analysis (LDA) for accelerating classification; while for special radical detection, the 512D feature vector is input to the SVM for classification.

For radical classification of non-special (SE/LR/UD) structures, we used a modified quadratic discriminant function (MQDF) classifier [30] with 20 principal eigenvectors per class. For special radical detection, we tested SVM classifiers with two types of kernels: polynomial kernel (SVM-poly, order 4) and Gaussian kernel (SVM-rbf). In training SVMs, the upper bounds of multipliers were set to 10.

Our recognition system was implemented in Microsoft Visual C++ 6.0 and all experiments were run on a PC with Intel Dual Core CPU and 2GB RAM.

24.5.2 Results and Analysis

In the following, we first give the results of single modules for radical detection, special and non-special structures recognition, and finally, give the results of integrated recognition of whole characters.

Performance of Special Radical Detection

We evaluates the performance of special radical detection in terms of the rates of Recall (R) and Precision (P), which are defined as

$$R = \frac{\text{\# correctly detected radicals}}{\text{\# true radicals}}, \tag{24.1}$$

$$P = \frac{\text{\# correctly detected radicals}}{\text{\# detected radicals}}. \tag{24.2}$$

Table 24.5 shows the performance of special radical detection, evaluated on samples of 502 Chinese characters containing special radicals. We can see that the detection rate (Recall) is high enough such that special-structure characters will be passed to the recognition modules of special-structures. For characters of non-special structures, some candidate radicals (non-special radicals) in the stroke sequences (Fig. 24.14) are misclassified as special radicals. Such mis-detected special radicals can be rejected in the subsequent recognition of remaining parts.

Table 24.5 Performance of special radical detection

Type	#character	SVM-poly(%)		SVM-rbf(%)	
		R	P	R	P
Special	502	99.83	97.34	99.65	97.29

Performance of Non-special Structure Recognition

We evaluated two methods of radical-based recognition for non-special-structure characters (6 261 classes): sequential representation with beam search and hierarchical representation with nested DP search. Radical patterns extracted from training character samples were also used for estimating the Gaussian PDFs of single-radical (unary) geometry and between-radical (binary) relationships, which were combined with the radical classification (MQDF) scores for evaluating candidate radical segmentation-recognition paths. The test accuracies of these two methods are listed in Table 24.6. We can see that radical-based recognition yields comparable character recognition accuracy to holistic recognition (MQDF classification based on feature vector representation of whole characters).

The main advantage of radical-based recognition is the smaller number of radical classes (1 236) than the number of character classes (6 261). Radical-based recognition, however, does not save computation time because it involves multiple candidates of radical segmentation. The test accuracies for different structure types are listed in Table 24.6. Nested DP with hierarchical representation is even more computationally intensive than beam search with sequential representation (about three times) because the character classes are matched one by one.

Binary geometric model measure the relationships between neighboring radicals within a character. Using binary geometric features, character pairs of different classes with the same sequential representation can be discriminated. Table 24.7 lists these character pairs from different structure types.

Table 24.6 Test accuracies (%) for non-special structures

Structure type	#character	Beam search (%)	DP search (%)
LR	4 284	97.24	97.19
UD	1 489	97.18	97.17
SE	488	98.15	98.04
Holistic	6 261	97.30	97.16

Table 24.7 Character pairs with the same sequential representation

吧—邑	标—奈	呗—员
晾—景	晚—冕	叻—另
晖—晕	旴—旱	屺—岜

For radical-based character recognition with two search methods, recognition errors are mainly due to pre-segmentation error of strokes or confusion
between similar characters of the same structure type. In pre-segmentation,
though we split connected strokes between radicals by corner point detection, errors still remain with skewed characters or heavily overlapping radicals
(Fig. 24.15). The confusion between similar characters (Fig. 24.16) often lies
in similar radicals. The confusion error can be reduced if the radical classifier is improved to better discriminate similar radicals. Table 24.8 gives some
similar radicals.

Fig. 24.15 Examples of radical pre-segmentation failure.

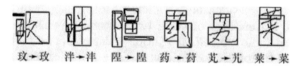

玟→玫   泮→泮   厘→厘   药→荮   芄→芄   莱→菜

Fig. 24.16 Examples of confusion between similar characters.

Table 24.8 Some groups of similar radicals

日曰	由申甲	目且	天夭
己己巳	人入八	大太犬	子孑孓
土士	七匕	无元	末未

Performance of Special Structure Recognition

Based on special radical detection and recognition of remaining parts, we
obtained recognition results on special-structure characters. Using SVM-poly
and SVM-rbf for special radical detection, MQDF for remaining parts recognition, the test accuracies on 502 special-structure characters are shown in
Table 24.9. In comparison, the accuracy of holistic recognition (by MQDF)

of 502 classes is 97.43%. Again, it is shown that radical-based recognition yields comparable character recognition accuracy with holistic recognition. The best recognition performance was obtained using sequential representation with beam search for remaining parts recognition and SVM-poly for special radical detection.

Table 24.9 Test accuracies on special-structure characters

Special radical detector	Search method	Accuracy (%)
SVM-poly	Beam	97.83
SVM-poly	DP	97.53
SVM-rbf	Beam	97.61
SVM-rbf	DP	97.49

Performance of Integrated Recognition

The above experimental results are produced by independent modules for special radical detection, non-special structure recognition and special structure recognition, respectively. When integrating these modules for whole-character recognition of arbitrary structures (6 763 classes), the test accuracies are shown in Table 24.10. In comparison, the test accuracy of holistic character recognition by MQDF is 97.19%. The accuracies of radical-based recognition are about 1% lower than that of holistic recognition though the accuracies of independent modules on specific structure types are fairly high. This is because the integration of special and non-special structures brings new confusion. We observed that the confusion between special-structure characters and non-special-structure characters is truly considerable. How to fuse radical similarity scores from special radical detectors and general radical classifier is also influential to the overall recognition accuracy.

Table 24.10 Test accuracies of integrated recognition framework for all structures

Special radical detector	Search method	Accuracy (%)
SVM-poly	Beam	96.03
SVM-poly	DP	96.07
SVM-rbf	Beam	96.19
SVM-rbf	DP	96.21

24.6 Concluding Remarks

We proposed a hybrid statistical-structural approach for radical-based online handwritten Chinese character recognition. Radical classes are selected for the ease of radical segmentation and radical model learning. Radical models for non-special radicals are learned from training character samples by automatic radical segmentation and clustering. Special radicals are detected by binary classifiers, one for each class of special radical. For non-special structure (LR/UD/SE) character recognition, two schemes, sequential and hierarchical, are adopted for character-radical dictionary representation, and two search methods, beam search and nested DP, are used accordingly. An integration framework was proposed to combine the special radical detectors, non-special and special structure recognition modules for whole-character recognition of arbitrary structures. Our experimental results show that the independent modules of non-special and special structure recognition perform comparably well as holistic character recognition by MQDF. The accuracy of integrated radical-based whole-character recognition is slightly lower than that of holistic recognition. Consider that most previous methods of radical-based Chinese character recognition were evaluated only on small number of character classes, our approach was evaluated on a large number of classes and resulted in competitive performance. The performance can be improved in future works by discriminative learning of radical models and elaborate fusion of radical similarity scores.

Acknowledgements

This chapter was supported by the National Natural Science Foundation of China (NSFC) under grants no. 60775004 and no. 60825301. Adrien Delaye of IRISA – INSA (France) participated in the work of special radical detection when he was an intern at the NLPR.

References

[1] Liu C L, Jaeger S, Nakagawa M (2004) Online Handwritten Chinese Character Recognition: The State of the Art. IEEE Trans Pattern Analysis and Machine Intelligence, 26(2): 198–213

[2] Kwon J, Sin B, Kim J H (1997) Recognition of On-line Cursive Korean Characters Combining Statistical and Structural Methods. Pattern Recognition, 30(8): 1255–1263

[3] Kang K W, Kim J H (2004) Utilization of Hierarchical Stochastic Relationship Modeling for Hangul Character Recognition. IEEE Trans Pattern Analysis and Machine Intelligence, 26(9): 1185–1196

[4] Lee S H, Kim J H (2008) Complementary Combination of Holistic and Component Analysis for Recognition of Low-resolution Video Character Images. Pattern Recognition Letters, 29: 383–391

[5] Kawamura A, Yura K, Hayama T et al (1992) On-line Recognition of Freely Handwritten Japanese Characters Using Directional Feature Densities. Proc 11th ICPR, Hague, 2: 183–186

[6] Hamanaka M, Yamada K, Tsukumo J (1993) On-line Japanese Character Recognition Experiments by an Off-line Method based on Normalization-cooperated Feature Extraction. Proc. 2nd ICDAR, Japan, pp 204–207

[7] Kim H J, Kim K H, Kim S K et al (1997) On-line Recognition of Handwritten Chinese Characters Based on Hidden Markov Models. Pattern Recognition, 30(9): 1489–1499

[8] Nakai M, Akira N, Shimodaira H et al (2001) Substroke Approach to HMM-based On-line Kanji Handwriting Recognition. Proc 6th ICDAR, Seattle, WA, pp 491–495

[9] Zhang Y D, Liu P, Soong F K (2007) Minimum Error Discriminative Training for Radical-based Online Chinese Handwriting Recognition. Proc 9th ICDAR, Brazil, pp 53–57

[10]   Liu Y J, Zhang L Q, Dai J W (1993) A new Approach to On-line Handwritten Chinese Character Recognition. Proc. 2nd ICDAR, Tsukuba, Japan, pp 192 – 195
[11]   Lee H M, Huang C W, C C Sheu (1998) A fuzzy Rule-based System for Handwritten Chinese Characters Recognition Based on Radical Extraction. Fuzzy Sets and Systems, 100 (1—3): 59 – 70
[12]   Delaye A, Mace S, Anquentil E (2008) Hybrid Statistical-structural On-line Chinese Character Recognition with Fuzzy Inference System. Proc. 19th ICPR, Tampa, FL
[13]   Xiao X H, Dai R (1998) On-line Handwritten Chinese Character Recognition Directed by Components with Dynamic Templates. Int J PRAI, 12 (1): 143 – 157
[14]   Chou K S, Fan K C, Fan T I (1996) Radical-based Neighboring Segment Matching Method for Online Chinese Character Recognition. Proc 13th ICPR, Vienna, pp 84-88
[15]   Lay S R et al (1996) On-line Chinese Character Recognition with Effective Candidate Radical and Candidate Character Selection. Pattern Recognition, 29(10): 1647 – 1659
[16]   Wang A B, Fan K C (2001) Optical Recognition of Handwritten Chinese Characters by Hierarchical Radical Matching Method. Pattern Recognition, 34(1): 15 – 35
[17]   Cai W J, Uchida S C, sakoe H (2006) An Efficient Radical-based Algorithm for Stroke-order-free Online Kanji Character Recognition. Proc. 18th ICPR, Hong Kong, pp 86 – 89
[18]   Kitadai A, Nakagawa M (2007) Prototype Learning for Structured Pattern Representation Applied to Online Recognition of Handwritten Japanese Characters. Int. J. Document Analysis and Recognition, 10(2): 101 – 112
[19]   Chellapilla K, Simard P (2006) A New Radical Based Approach to Offline Handwritten East-Asian Character Recognition. Proc. 10th IWFHR, La Baule, France, pp 261 – 266
[20]   Ma L L, Liu C L (2008) A New Radical-based Approach to Online Handwritten Chinese Character Recognition. Proc 19th ICPR, Tampa, FL
[21]   Ma L L, Liu C L (2009) On-line Handwritten Chinese Character Recognition Based on Nested Segmentation of Radicals. Proc of 2009 CCPR & First CJKPR, Nanjing, China, pp 929 – 933
[22]   Ma L L, Delaye A, Liu C L (2010) Special Radical Detection by Statistical Classification for Online Handwritten Chinese Character Recognition. Proc 12th ICFHR, Kolkata, India
[23]   Liu C L, Koga M, Fujisawa H (2002) Lexicon-driven Segmentation and Recognition of Handwritten Character Strings for Japanese Address Reading. IEEE Trans. Pattern Analysis and Machine Intelligence, 24(11): 1425 – 1437
[24]   Lin Z T, Fan K C (1994) Coarse Classification of On-line Chinese Characters Via Structure Feature-based Method. Pattern Recognition, 27(10): 1365 – 1377
[25]   Wang A B, Fan K C, Wu W H (1997) Recursive Hierarchical Radical Extraction for Handwritten Chinese Characters. Pattern Recognition, 30(7): 1213 – 1227
[26]   Cao H, Kot A C (2004) Online Structure Based Chinese Character Pre-classification. Proc 17th ICPR, Cambridge, UK, pp 395 – 398
[27]   Zhao W, Zhao C D (2006) A Continuous-recognition-oriented Handwritten Chinese Character Radical Set and its Statistical Rule. Journal of Chinese information processing, 20(5): 58 – 64 (in Chinese)
[28]   Burges C J C (1998) A Tutorial on Support Vector Machines for Pattern Recognition. Knowledge Discovery and Data Mining, 2(2): 1 – 43
[29]   Liu C L, Zhou X D (2006) Online Japanese Character Recognition Using Trajectory-based Normalization and Direction Feature Extraction, Proc. 10th IWFHR, La Baule, France, pp 217 – 222
[30]   Kimura F, Takashina K, Tsuruoka S et al (1987) Modified Quadratic Discriminant Functions and the Application to Chinese Character Recognition. IEEE Trans Pattern Analysis and Machine Intelligence, 9(1): 149 – 153

25 Current Trends in Multimodal Biometric System — Rank Level Fusion

Marina L. Gavrilova and Md. Maruf Monwar[1]

Abstract Biometric identification referes to idetifying an individual based on his or her physilogical or beavioral characteristics. The use of more than one biometric identfiers in a biometric system, called the multimodal biometric system, increases the overall system accuracy and hence increase security, as well as reduce the enrollment problems. An effective and appropriate fusion strategy is needed to integrate different biometric information in such multimodal systems. This chapter provides an in-depth overview of traditional multimodal biometric systems and current trends in multimodal biometric fusion. Various approaches of rank level fusion, which is an not heavily investigated by researchers yet, are also illustrated in details in this chapter. Pros and cons of these rank fusion approaches are discussed which can be helpful for large scale multimodal biometric system deployment.

25.1 Introduction

Controlling access to prohibited areas and protecting important national or public information are some of the main activities of security and intelligence services of many countries of the world. Often, to decide if a person is allowed to access a prohibited area, a biometric system is employed. A biometric system is an automated method of recognizing a person based on a physiological (face, iris, ear etc.) and/or behavioral (signature, voice, typing patterns etc.) characteristics [1]. Other applications of biometrics systems include e-commerce, access to computer networks, online banking, border control, parenthood determination, medical records management, and welfare disbursement.

The optimal biometric recognition method is one having the properties of distinctiveness, universality, permanence, acceptability, collect ability, and resistance to circumvention [2]. No existing biometric system simultaneously meets all of these requirements; however the use of more than one biometric can help to develop a system which approaches those goals [3]. The advantages of multimodal systems stem from the fact that there are multiple sources of information. The most prominent implications of this are increased accuracy, fewer enrolment problems and enhanced security. All multimodal biometric systems need a fusion module that takes individual data and combines it

1 University of Calgary, AB, Canada. E-mails: {mgavrilo, mmmonwar}@ucalgary.ca.

in order to obtain the authentication result: impostor or genuine user. This chapter provides a snapshot of the current trends of information fusion in multimodal biometric system and provides an in-depth information on rank level fusion.

25.1.1 Brief History of Biometric

Biometric originates from biometry — an area of biology dealing with human physiological characteristics. According to Latifi and Solayappan, the explorer Joao de Barros reported the first known example of biometrics in practice which was originated in China in the 14th century [4]. He wrote that Chinese merchants were stamping children's palm prints and footprints on paper with ink to distinguish the young children from one another. However, the first modern biometric device was introduced on a commercial basis only 35 years ago. This device was called *Identimat* [5] and it measured the shape of the hand and looked particularly at finger length. Although its production ceased in the late 1980s, the use of the *Identimat* set a path for biometric technologies as a whole. Later, in the mid 1980s, the first system to analyze the unique pattern of the retina was introduced [5]. This migration from research and development to commercialized use has been relatively slow, but nonetheless, still continues today. As mentioned earlier, the public sector-particularly military and law enforcement-were the early adopters of biometric technology. Today, it is becoming more common to see such devices in computer rooms, vaults, research labs, day care centers, blood banks, ATMs amusement parks, and military installations.

25.1.2 Overview of a Biometric System

A typical biometric system consists of four main components, namely, sensor module, feature extraction module, matching module, and decision module (see Fig. 25.1). The sensor module is responsible for acquiring the biometric data from an individual. The feature extraction module processes the acquired biometric data and extracts only the salient information to form a new representation of the data [2]. Ideally, this new representation should be unique for each person and also relatively invariant with respect to changes in the different samples of the same biometric collected from the same person. The matching module compares the extracted feature set with the templates stored in the system database and determines the degree of similarity (dissimilarity) between the two [2]. The decision module either verifies the identity claimed by the user or determines the user's identity based on the degree of similarity between the extracted features and the stored template(s).

There are some other modules not shown in this figure are data trans-

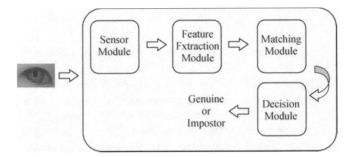

Fig. 25.1 Block diagram of a typical unimodal biometric system.

portation module, which is responsible for secure data transfer between modules, and the administration module which allows the management of the biometric system.

The functionalities provided by a biometric system can be categorized as verification and identification [2]. In verification, the user claims an identity and the system verifies whether the claim is genuine. In this scenario, the query is compared only to the template corresponding to the claimed identity. If the user's input and the template of the claimed identity have a high degree of similarity, then the claim is accepted as "genuine". Otherwise, the claim is rejected and the user is considered an "impostor".

Identification functionality can be classified into positive and negative identification. In positive identification, the user attempts to positively identify himself or herself to the system without explicitly claiming an identity. In negative identification, systems prove that you are not who you say you are not. The user in a negative identification application is considered to be concealing his true identity from the system. Negative identification is also known as *screening* [2] which is often used at airports to verify whether a passenger's identity matches with any person on a "watch-list". Screening can also be used to prevent the issue of multiple credential records (e.g., driver's license, passport) to the same person [2]. Negative identification is also critical in applications such as welfare disbursement to prevent a person from claiming multiple benefits under different names. In both positive and negative identification, the user's biometric input is compared with the templates of all the persons enrolled in the database and the system outputs either the identity of the person whose template has the highest degree of similarity with the user's input or a decision indicating that the user presenting the input is not an enrolled user.

25.1.3 Performance of Biometric System

The most common performance metrics of biometric systems are false accept rate, false reject rate and failure-to-enroll rate [3].

False accept rate (FAR) is the percentage of the likelihood that an impostor will be accepted by the biometric system due to large inter-user similarity. False reject rate (FRR) is the percentage of the likelihood that the genuine individual will be rejected by the system due to large intra-class variations. A FRR of 5% indicates that on average, 5 in 100 genuine attempts do not succeed. Similarly, A FAR of 0.1% indicates that on average, 1 in 1000 impostor attempts are likely to succeed.

Sometimes another term, Genuine Acceptance Rate (GAR) is used which is the percentage of the likelihood that a genuine individual is recognized as a match [2]. GAR can be obtained by

$$GAR = 1 - FRR. \tag{25.1}$$

Usually the above performance metrics are expressed using different graphs such as Score Histogram (SH), Receiver Operating Characteristic (ROC), and Cumulative Match Characteristic (CMC). SH plots the frequency of the scores for non-matches and matches over the match score range. ROC plots FRR against FAR at various thresholds on a linear or logarithmic or semi-logarithmic curve. When these error rates are plotted on a normal deviate scale, the resulting graph is called the Detection Error Tradeoff (DET). Sometimes, another measure, Equal Error Rate (EER), is used to summarize the performance of a biometric system which refers to that point in a DET curve where the FAR equals the FRR. A lower EER value, therefore, indicates better system's performance [2]. Cumulative Match Characteristic (CMC) curve is used in biometric identification to summarize the identification rate at different rank values [2].

The rest of the chapter is organized as follows. Section 25.2 puts an overview of multimodal biometric system and its advantages over a single biometric system. Section 25.3 describes different levels of biometric fusion. Section 25.4 illustrates rank level fusion strategies and its advantages over other fusion levels. Section 25.5 concludes the chapter.

25.2 Multimodal Biometric System

Multimodal biometric system becomes increasingly common in current and future real-world biometric system deployment, which is a system utilizing more than one biometric identifiers for biometric recognition [3]. For example, a multimodal biometric system may use face recognition, iris recognition and ear recognition to confirm the identity of a user. These systems address some of the problems faced by single biometric system through the utilization of multiple information sources.

25.2.1 Advantages

Recognition accuracy: The most immediate advantage of multimodal biometric system is recognition accuracy. As the combination of each of the biometric identifiers offers some additional evidence about the authenticity of an identity claim, one can have more confidence in the result. For example, two persons may have the similar signature patterns, in which case, the signature verification system will produce large FAR for that system. Addition of face recognition system with the signature verification system may solve the problem and reduce the FAR.

Biometric data enrollment: Multimodal biometric systems can address the problem of non-universality. In case of unavailability or poor quality of a particular biometric data, other biometric identifier of the multimodal biometric system can be used to capture data. For example, a face biometric identifier can be use in a multimodal system (involves fingerprint of general labors with lots of scars in the hand).

Privacy: Multimodal biometric systems increase resistance to certain type of vulnerabilities. For example, an attacker would have to spoof three different biometric identifiers which would be more challenging.

25.2.2 Architecture

Figure 25.2 shows the block diagram of a sample multimodal biometric system. A new module is added here (from the block diagram of unimodal biometric system), which is the fusion module. In this figure, the fusion module

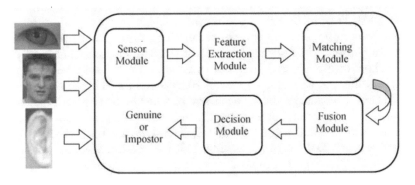

Fig. 25.2 Block diagram of a multimodal biometric system.

integrates information obtained from matching module.

25.2.3 Different Multimodal Biometric Systems

Several approaches have already been proposed and developed for multimodal biometric authentication system with different biometric traits and with different fusion mechanisms. In 1998, a bimodal approach was proposed by Hong and Jain for a PCA-based face and a minutiae-based fingerprint identification system with a fusion method at the match score level [6]. At a FAR of 0.01%, the monomodal systems obtained a FRR of 61.2% and 10.6% for face and fingerprint respectively. For the same FAR, the fusion approach obtained a FRR of 6.6%.

In 2000, Frischholz and Dieckmann developed a commercial multimodal approach, BioID, for a model-based face classifier, a VQ-based voice classifier and an optical-flow-based lip movement classifier for verifying persons [7]. Lip motion and face images were extracted from a video sequence and the voice from an audio signal. Weighted sum rule and majority voting approaches of decision level fusion method were used for fusion. Their experiments on 150 persons for three months demonstrated that the system can reduce the FAR significantly below 1%, depending the level of the security.

In 2003, Ross and Jain proposed a multimodal system for face, fingerprint, and hand geometry, with three fusion methods at the matching score level [8]. Sum-rule, decision trees and linear discriminant function are used and by their experiments, they showed that the sum rule performs better than the decision tree and linear discriminant classifiers. The FAR and FRR of the tree classifier are 0.036% and 9.63% respectively. The FAR and FRR of the linear discriminant classifier is 0.47% and 0.00% respectively. The FRR value in this case is a consequence of overfitting the genuine class as it has fewer samples in both the test and training sets. The sum rule that combines all three scores has a corresponding FAR of 0.03% and a FRR of 1.78%.

In 2004, Feng et al. developed a system using face and palmprint using feature level fusion strategy [9]. For feature level fusion, they use feature concatenation approach. They used two algorithms for classification — PCA and ICA. Their results showed that the standalone PCA-based face and palmprint classifier got an accuracy rate of 70.83% and 85.83% respectively while the feature level fusion got an accuracy of 95.83%. In their ICA-based implementation, the standalone face classifier and palmprint classifier got an accuracy rate of 85.00% and 92.50% respectively, while the combined system showed a nearly perfect result with 99.17% recognition accuracy rate.

In 2005, Jain et al. proposed a multimodal approach for face, fingerprint and hand geometry, with fusion at the score level [10]. The matching approaches for these modalities are minutiae-based matcher for fingerprint, which has similarity scores as output, PCA-based algorithm for face recognition, which has Euclidean distance as output, and a 14-dimensional features vector for hand-geometry, which also has Euclidean distance for output. Seven score normalization techniques (simple distance-t-similarity transformation with no change in scale, min-max normalization, z-score normalization, median-MAD normalization, double sigmoid normalization, tanh normalization and Parzen normalization) and three fusion techniques on the normalized scores (simple-sum-rule, max-rule and min-rule) were tested in this study. Except for one normalization technique (the median-MAD), all fusion approaches outperform the monomodal approaches. For example, the fingerprint system, which is the best monomodal system in this study, obtained a genuine acceptance rate of 83.6% at a FAR of 0.1%, while the multimodal approach obtained a genuine acceptance rate of 98.6% at a FAR of 0.1% when the z-score normalization and the sum-rule were used. At low FARs, the tanh and min-max normalization techniques outperforms the other techniques, while at higher FARs, the z-score normalization performs better than the other techniques.

In the next year, in 2006, Li et al. introduced a new feature metric — handmetric, which is a combination of palmprint, hand shape and knuckle print [11]. They integrated the features of these three biometrics using feature level fusion based on Kernel Principle Component Analysis (KPCA), which is a combination of kernel projection and PCA dimension reduction. They used four fusion rules for combining feature — sum, product, min and max. The system gets a FAR of 0.29% and a FRR of 0.20% for using the sum rule for 4 training samples of each user which is better than the authentication performance using pamprint or using finger.

In 2008, Nandakumar et al. presented a multimodal biometric system using likelihood ratio based match score level fusion [12]. They proposed a framework for optimal combination of match scores that is based on the likelihood ratio test which models the distributions of genuine and impostor match scores as finite Gaussian mixture model. At 0.01% FAR, the system gets 99.1% GAR for NIST multimodal database and 98.7% GAR for XM2VTS database.

In 2009, Monwar and Gavrilova developed a multimodal biometric system utilizing face, ear and signature biometric identifiers [13]. They proposed rank fusion approaches for biometric fusion using fisherimage method as matching algorithm and logistic regression and Borda count for fusing rank information.

In 2010, the same authors (Monwar and Gavrilova) introduced Markov chain rank fusion for multimodal biometric system [14]. They utilized face, ear and iris biometric identifiers for the multimodal system and examined various rank fusion methods including the novel Markov chain approach. They showed that the Markov chain approach for biometric rank fusion satisfies the Condorcet criterion which is essential for any fair rank aggregation process.

From the above discussion, it can be concluded that many multimodal biometric systems with various methods and strategies have been proposed over the last decade to achieve higher accuracy rate. Table 25.1 summarizes some of the multimodal biometric systems.

Table 25.1 Some Multimodal biometric systems

Year	Modalities Fused	Authors	Fusion Level	Fusion Approach
1998	Face and fingerprint	Hong and Jain	Match score	Product rule
2000	Face, voice and leap movement	Frischholz and Dieckmann	Decision	Weighted sum rule; Majority voting
2003	Face, fingerprint and hand geometry	Ross and Jain	Match Score	Sum-rule, decision tree and linear discriminant function
2004	Face and palmprint	Feng et al.	Feature	Feature concatenation
2005	Face, fingerprint and hand geometry	Jain et al.	Match score	Simple-sum-rule, max-rule and min-rule
2006	Palmprint, hand shape and knuckleprint	Li et al.	Feature	Sum, product, min and max
2008	Fingerprint, face and hand geometry	Nandakumar et al.	Match score	Likelihood ratio
2009	Face, ear and signature	Monwar and Gavrilova	Rank	Logistic regression, Borda count, highest rank
2010	Face, ear and iris	Monwar and Gavrilova	Rank	Markov chain, logistic regression, Borda count

25.2.4 Soft Biometrics

Multimodal biometric system that utilizes a combination of biometric identifiers like face, ear and iris can alleviate some of the problems associated
with unimodal biometric systems and can obtain better recognition accuracy.
However, using multiple biometric identifiers in a single system will increase
the identification or verification times and hence cause more inconvenience
to the users and increase the overall cost of the system. Thus, soft biometric is introduced in 2004 [15] to obtain the same recognition performance
without causing any additional inconveniences to the users [2] by incorporating it (soft biometric identifiers) to the primary multimodal systems. Soft
biometric identifiers include gender, ethnicity, height, weight, eye color, skin
color, hair color etc. Two key challenges need to be addressed in order to
incorporate soft biometrics into the traditional multimodal biometric framework. The first challenge is the automatic and reliable extraction of the oft
biometric information without causing inconveniences to the users and the
second challenge is to optimally combine this information with the primary
biometric identifier to achieve the best recognition performance.

25.3 Fusion in Multimodal Biometric System

Information fusion in multimodal biometric systems presents an elegant way
to enhance the matching accuracy of a biometric system without restoring to

non-biometric alternatives [3]. The fundamental issue of information fusion is to determine the type of information that should be fused and the selection of method for fusion. The goal of fusion is to devise an appropriate function that can optimally combine the information rendered by the biometric subsystems [2]. Evidences in a multimodal biometric system can be integrated in several different levels (Fig. 25.3), but the process can be subdivided into two main categories - prior-to-matching fusion and after matching fusion. The accuracy of a multimodal biometric system can actually be lower than that of the unimodal biometric system if an appropriate fusion method is not followed for combining the evidences provided by different sources.

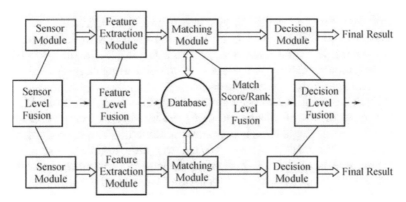

Fig. 25.3 Various fusion levels possibilities.

25.3.1 Prior to Matching Fusion

Fusion in this category integrates evidences before matching. This category fuses the information of a multimodal biometric system in the following levels:

1) Sensor level: The raw data acquired from multiple sensors can be processed and integrated to generate new data from which features can be extracted. For example, Ratha et al. described a fingerprint mosaicing scheme to integrate multiple snapshots of a fingerprint as the user rolls the finger on the surface of the sensor [16].

2) Feature level: The feature sets extracted from multiple data sources can be fused to create a new feature set to represent the individual. For example, Feng et al. developed a feature level fusion based multimodal biometric system using face and palmprint [9]. They used PCA (Principal Component Analysis) and ICA (Independent Component Analysis) as classification algorithms and feature concatenation as fusion approach.

25.3.2 After Matching Fusion

Fusion in this category integrates evidences after matching. The category includes:

1) Match score level: In this case, multiple classifiers output a set of match scores which are fused to generate a single scalar score. As an example, the match scores generated by the face, fingerprint and hand modalities of a user may be combined via the simple sum rule in order to obtain a new match score which is then used to make the final decision [8].

2) Rank level: This type of fusion is relevant in identification systems where each classifier associates a rank with every enrolled identity (a higher rank indicating a good match). Thus, fusion entails consolidating the multiple ranks associated with an identity and determining a new rank that would aid in establishing the final decision. For example, in [13], a rank level fusion based multimodal biometric system is developed using highest rank, Borda count and logistic regression approaches.

3) Decision level: When each matcher outputs its own class label (i.e., accept or reject in a verification system, or the identity of a user in an identification system), a single class label can be obtained by employing techniques, such as "AND"/"OR", majority voting, weighted majority voting etc [2]. For example, Frischholz and Dieckmann developed a commercial multimodal biometric system using weighted sum rule and majority voting approaches of decision level fusion method for a model-based face classifier, a vector quantification-based voice classifier and an optical flow-based lip movement classifier for person identification [7].

25.3.3 Data Available for Fusion

Typically, the amount of information available to the system for fusion decreases as one proceeds through the biometric system (Fig. 25.4).

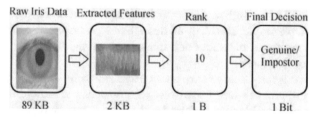

Fig. 25.4 Example of how information available for fusion decreases in every levels of a biometric verification system (adopted from [2]).

Biometric data processing at different stages are expected to decrease the intra-user variability and the amount of noise that is contained in the available information. In many practical multibiometric systems, higher levels of information, such as the raw images or feature sets are either not available (e.g., proprietary feature sets used in commercial-off-the-shelf systems) or the information available from the different sources is not compatible (e.g., fingerprint minutiae and eigenface coefficients). For this unavailability and incompatibility of desired information, sensor level and feature level fusion are not possible in all cases. Decision level fusion approaches are well investigated for biometric systems but are too rigid and only consider single information for fusion, which has a high probability of produce wrong recognition result [8]. Match score level fusion is the most investigated fusion method so far which considers the match or similarity/distance score for fusion. But, the similarity/distance scores need to be normalized before fusion as they can be in different ranges. Rank level fusion, which remains significantly understudied, also uses the similarity/distance scores, but not directly. Instead, this method uses the corresponding ordering of identities based on those similarity/distance but does not need any normalization procedure.

25.4 Rank Level Fusion

Rank-level fusion is used in identification systems and is applicable when the individual matcher's output a ranking of the "candidates" in the template database. The system is expected to assign a higher rank to a template that is more similar to the query. Very few methods for consolidation of biometric rank information can be found in the literature, as it is still an understudied problem. Three methods described by Ho, Hull, and Srihari [2] for making the final decision in a general multiple classifier system, can be used for rank level fusion in multimodal biometric systems. These three methods are highest rank, Borda count and logistic regression methods. Recently, Nandakumar and others introduced Bayesian approach for rank level fusion [18]. All of these methods for rank level fusion are briefly discussed in the next subsections.

25.4.1 Highest Rank Method

The highest rank method is good for combining a small number of specialized matchers and hence can be effectively used for a multimodal biometric system where the individual matchers are the best. In this method, the consensus ranking is obtained by sorting the identities according to their highest rank.

Suppose that we have m classifiers which assign ranks to all classes. Then the consensus rank of a particular class is obtained by

$$\text{Consensus rank, } R_c = \min_{i=1}^{m} R_i. \tag{25.2}$$

The final identity authentication ranking is then obtained by sorting the consensus ranking of each class in the ascending order.

The advantage of this method is the ability to utilize the strength of each matcher. Even if only one matcher assigns the highest rank to the correct user, it is still very likely that the correct user will receive the highest rank after reordering. The disadvantage of this method is that the final ranking may have many ties (which can be broken randomly). The number of classes sharing the same ranks depends on the number of classifiers used. Due to these properties, this method cannot be a good choice for a security critical multimodal biometric system.

25.4.2   Borda Count Method

The Borda count method [19] is the most widely used rank aggregation method and uses the sum of the ranks assigned by individual matchers to calculate the final rank. Suppose we have m classifiers. Then the consensus rank of a particular class is obtained by

$$Consensus\ rank,\ R_c = \sum_{i=1}^{m} R_i. \tag{25.3}$$

The final identity authentication ranking is then obtained by sorting the consensus ranking of each class in the ascending order.

This method assumes that the ranks assigned to the users by the individual matchers are statistically independent and the performances of all three matchers are equal. The magnitude for the Borda count for each class measures the strength of agreement by the three matchers that the input pattern belongs to that class. The advantage of this method is that it is easy to implement and requires no training stage. These properties made the Borda count method feasible to incorporate in multimodal biometric systems. The disadvantage of this method is that it does not take into account the differences in the individual matcher's capabilities and assumes that all the matchers perform equally, which is usually not the case in most real biometric systems.

25.4.3   Logistic Regression Method

The logistic regression method, which is a variation of the Borda count method, calculates the weighted sum of the individual ranks. In this method, the final consensus rank is obtained by sorting the identities according to the summation of their rankings obtained from individual matchers multiplied by the assigned weight.

Suppose we have m classifiers which assign ranks to all classes. Then the consensus rank of a particular class is obtained by

$$Consensus\ rank,\ R_c = \sum_{i=1}^{m} W_i R_i, \qquad (25.4)$$

where W_i is the weight assigned to the i-th classifier. The final identity authentication ranking is then obtained by sorting the consensus ranking of each class in the ascending order.

The weight to be assigned to the different matchers is determined by a "logit" function using logistic regression [20]. This method is very useful when the different matchers have significant differences in their accuracies but requires a training phase to determine the weights which can be computationally expensive. Also one of the key factors that have direct effect on the performance of a biometric system is the quality of the biometric samples. Hence the single matchers' performances can vary with different sample sets which make the weights allocating process more challenging. Inappropriate weight allocation can eventually reduce the recognition performance of this multimodal biometric system (using logistic regression) compared to unimodal matchers. So, in some cases, logistic regression method cannot be employed for rank aggregation.

Figure 25.5 illustrates the highest rank, the Borda count and the logistic regression rank fusion approaches in a face, ear and signature based multimodal biometric system.

In this Figure, it is considered that the less the value of the rank, the more accurate the result is. The rank for "Person 1" is 1, 2 and 2 respectively from the face, ear, and signature matchers. For the Highest rank method, the fused score (highest rank) is 1 for "Person 1". Similarly, for "Person 2", "Person 3", "Person 4" and "Person 5", the highest ranks are 1, 3, 2, and 3 respectively. There is a tie between "Person 1" and "Person 2" and between "Person 3" and "Person 5". These ties are broken arbitrarily. So, in the final reordered ranking, "Person 1" gets the top position in the reordered rank list whereas, "Person 2" gets the second position. For the Borda count method, the initial ranks are first added. Thus, 5, 7, 13, 9, and 11 can be found as the Borda scores for "Person 1" to "Person 5" respectively. So, "Person 1" gets the top position in the reordered list due to his/her lowest Borda score. For the logistic regression method, the matchers need to be assigned weights which are determined by the recognition performance of the matchers. For this system, face matcher is assigned a weight of 0.1, ear matchers assigned a weight of 0.5 and signature matcher is assigned a weight of 0.4. For this system, it is assumed that the matcher with the minimum weight performs better than other matchers. So face matcher works better than the ear matcher or signature matcher. The fused scores (weighted Borda scores) are 2.2, 1.4, 4.8, 3.4, and 3.5 for "Person 1" to "Person 5" respectively. So, "Person 2" is on the top position in the reordered ranking list.

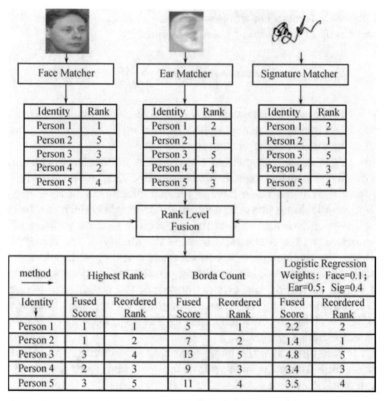

Fig. 25.5 Example of rank level fusion with different methods (adopted from [13]).

25.4.4 Bayesian Approach

Bayesian approach for biometric rank fusion is based on Bayes decision theory. This approach uses the rank distribution (probability that a identity is assigned a rank by an individual matcher is a true identity), which can be estimated when the marginal genuine and impostor match score densities are known [18]. This estimation requires two assumptions: (1) match scores of the individual users are independent and (2) match score distributions of different users are identical. The consensus rank is obtained as the product of the posterior probabilities of the individual matchers.

If we have m classifiers, then the consensus rank of a particular class is obtained by the following equation:

$$Consensus\ rank,\ R_c = \prod_{i=1}^{m} P_i(R_i), \tag{25.5}$$

where $P_i(R_i)$ is the posteriori probability that the identity (class) which is assigned rank R_i by the i-th classifier is the true identity. The final identity authentication ranking is then obtained by sorting the consensus ranking of each class in the descending order.

25.4.5 Markov Chain Approach

Markov chain is a novel biometric rank fusion approach for multimodal biometric system proposed by the authors in 2010 [14] Markov chain is a random process or set of states in which the probability that a certain future state will occur depends only on the present or immediately preceding state of the system, and not on the events leading up to the present state. In the Markov chain biometric rank aggregation method, it is assumed that there exists a Markov chain on the enrolled identities and the order relations between those identities in the ranking lists (obtained from different biometric matchers) represent the transitions in the Markov chain [21]. The stationary distribution of the Markov chain is then utilized to rank the entities. The construction of the consensus ranking list from the Markov chain can be summarized as below:

Step 1   Map the set of ranked lists to a single Markov chain, where one node of the chain represents single identity in the initial ranking lists.

Step 2   Compute the stationary distribution on the Markov chain.

Step 3   Rank the identities based on the stationary distribution. That is, the node with the highest score in the stationary distribution is given the top rank, and so on until the node with the lowest score in the stationary distribution which is given the last rank.

Fig. 25.6 shows a Markov chain with its transition matrix. There are five

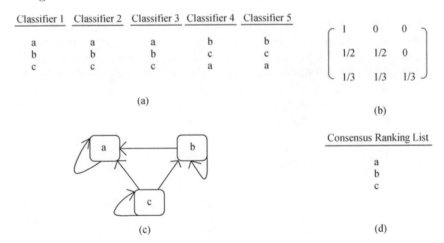

Classifier 1   Classifier 2   Classifier 3   Classifier 4   Classifier 5

a	a	a	b	b
b	b	b	c	c
c	c	c	a	a

(a)

$$\begin{bmatrix} 1 & 0 & 0 \\ 1/2 & 1/2 & 0 \\ 1/3 & 1/3 & 1/3 \end{bmatrix}$$

(b)

(c)

Consensus Ranking List

a
b
c

(d)

Fig. 25.6   (a) Initial ranking lists. (b) Transition matrix. (c) Markov chain. (d) Consensus list.

matchers which outputs five different ranking lists. Based on the ranking list, a Markov chain is constructed and the consensus ranking list is obtained from the Markov chain.

As only the top ranked identities (partial ranking lists) will be used for consensus ranking, there can be situations where some identities are in some initial ranking lists but are absence in other lists. These situations can be handled by finding the position of the absence identities in a list using their relative positions in the lists where they are present.

25.5 Conclusion

The design of a multimodal biometric system is a challenging task due to heterogeneity of the biometric sources in terms of the type of information, the magnitude of information content, correlation among the different sources and conflicting performance requirements of the practical applications. In addition, there are some disadvantages of multimodal biometric systems, as they may be more expensive and complicated due to the requirement of additional hardware and matching algorithms, and there is a greater demand for computational poser and storage. Also, from a user's point of view, the systems may be more difficult to use, leading to longer enrollment and verification times. Furthermore, there are inter-operability challenges related to the integration of products from different vendors. Despite these challenges and shortcomings, the use of multimodal biometric systems for enhancing system performance at both an algorithmic and system levels is becoming increasingly important due to its numerous benefits over unimodal systems.

One of the key differences in the design of unimodal and multimodal biometric systems is information fusion. Which information needs to be fused and in what level to obtain the maximum recognition performance is still the main focus of many biometric researchers. In this chapter we presented an overview of the current trends in recent multimodal biometric fusion research and illustrated in details an under-investigated fusion strategy — rank level fusion. From the discussion, it is clear that rank level fusion can be a better choice in multimodal identification systems and hence can be effectively used in any security critical biometric applications.

References

[1] Jain A K, Ross A, Prabhakar S (2004) An Introduction to Biometric Recognition, IEEE Transacations on Circuits and Systems for Video Technology, Special Issue on Image- and Video-Based Biometrics, 14 (1): 4 – 20
[2] Ross A, Nandakumar K, Jain A K (2006) Handbook of Multibiometrics. Springer, New York
[3] Dunstone T, Yager N (2009) Biometric System and Data Analysis: Design, Evaluation, and Data Mining. Springer, New York

[4]   Latifi S, Solayappan N (2006) A Survey of Unimodal Biometric Methods. In: Proc of Int Conf on Security & Management, 57 – 63, Las Vegas, USA

[5]   Bowman E (2000) Everything you Need to Know About Biometrics. Identix Corporation. http://www.ibia.org/EverythingAbout Biometrics, 14 October 2000

[6]   Hong L, Jain A K (1998) Integrating Faces and Fingerprints for Personal Identification. IEEE Transacations on Pattern Analysis and Machine Intelligence, 20(12): 1295 – 1307

[7]   Frischholz R, Dieckmann U (2000) BioID: A Multimodal Biometric Identification System, IEEE Computer, 33(2): 64 – 68

[8]   Ross A, Nandakumar K, Jain A K (2003) Information Fusion in Biometrics. Pattern Recognition Letters, 24: 2115 – 2125

[9]   Feng G, Dong K, Hu D et al (2004) when Faces Re-combined With Palmprints: A Novel Biometric Fusion Strategy. In proc of 1st Int Conf on Biometric Authentication, 701 – 707, Hong Kong, China

[10] Jain A K, Nandakumar K, Ross A (2005) Score Normalization in Multimodal Biometric Systems. Pattern Recognition, 38: 2270 – 2285

[11] Li Q, Qiu Z, Sun D (2006) Feature-level Fusion of Hand Biometrics for Personal Verification Based on Kernel PCA, In: Zhang D, Jain A K (eds) LNCS 3832, ICB 2006, 744 – 750

[12] Nandakumar K, Chen Y, Dass S C et al (2008) Likelihood Ratio-based Biometric Score Fusion. IEEE Trans on Pattern Analysis and Machine Intelligence, 30 (2): 342 – 347

[13] Monwar M M, Gavrilova M L (2009) A Multimodal Biometric System Using Rank Level Fusion Approach. IEEE Trans SMC-B: Cybernetics, 39(4): 867 – 878

[14] Monwar M M, Gavrilova M L (2010) Secured Access Control Through Markov Chain Based Rank Level Fusion Method. In: Proc of 5th Int Conf on Computer Vision Theory and Applications (VISAPP), pp 458 – 463, Angers, France

[15] Jain A K, Nandakumar K, Lu X et al (2004a) Integrating Fingerprint, Faces and Soft Biometric Traits for User Recognition. In proc ECCV Int Workshop on Biometric Authentication, LNCS 3087: 259 – 269, Springer

[16] Ratha N K, Connell J H, Bolle R M (1998) Image Mosaicing For Enrolled Fingerprint Construction. In proc of 14th Int Conf on Pattern Recognition, 2: 1651 – 1653, Brisbane, Australia

[17] Ho T K, Hull J J, Srihari S N (1994) Decision Combination in Multiple Classifier Systems. IEEE Trans on Pattern Analysis and Machine Intelligence, 16 (1): 66 – 75

[18] Nandakumar K, Jain A K, Ross A (2009) Fusion in Multibiometric Identification Systems: What about the Missing Data. LNCS 5558: 743 – 752, Springer

[19] Borda J C (1781) M'Emoire sur Les 'Elections au Scrutin. Histoire de l'Acad'emie Royale des Sciences, France

[20] Agresti A (2007) An Introduction to Categorical Data Analysis, 2nd edn. Wiley-Interscience

[21] Dwork C, Kumar R, Naor M et al (2001) Rank aggregation methods for the web. In: Proc of 10th Intl WWW Conf, pp 613 – 622, Hong Kong, China

26 Off-line Signature Verification by Matching with a 3D Reference Knowledge Image — From Research to Actual Application

Maan Ammar[1]

Abstract This chapter introduces a method for off-line verification of signatures. The method can be used to verify signatures and detect skilled forgeries with outstanding performance. It is based on matching the questioned signature with a 3D reference knowledge image (RKI) using ammar matching technique (AMT). The AMT which was developed and modified through years (1989 – 2010) is introduced in detail, and the 3D RKI derived from the methodology of the forensic document examiner (FDE) of working is elaborated. The features extracted using the new knowledge representation method and the AMT has been found to be highly effective in signature verification, and distinctly outperform the classical features like slants, baseline, and contour based ones on the data used. Experimental results of using RKIs built from binary, high pressure, and thinned images for feature extraction and verification are also presented and discussed.

26.1 Introduction

Image understanding techniques like matching, similarity, features extraction, and learning have their applications not only to detect and recognize objects using cameras, but also to be applied to forensic science issues like signature verification and recognition [1].

Due to the importance of off-line signature in our daily life, and the fact that it is the target of forgers, it has received, and yet is receiving a considerable attention from researchers to computerize the process of its verification. One should bear in mind that, similar to medical image diagnosis cases, the final decision concerning the document authenticity, especially in valuable documents (high value bank checks, criminal cases documents, etc.), will be made by the authorized person (FDE, for example). The benefit of higher automatic signature verification (ASV) performance (lesser error rates) will be very important for: (1) reducing the large volume of scanned documents for verification to a handful of documents containing the suspect ones, and (2) giving a higher confidence in the decision to be made in special cases like criminal ones.

---

1 Biomedical Engineering Department, Damascus University, P. O. Box 86, Damascus, Syria. E-mail: maan_ammar@yahoo.com.

Different feature extraction techniques were used for off-line signature verification: Nagel R. and Rosenfeld A. used handwriting model parameters as a basis for feature extraction [2]. Ammar M. et al. used high pressure features as well as parametric shape ones [3 – 7]. Among those used transforms, Deng P. S. P. et al. used wavelet transform for verification [8]. While Almudena G. et al. extracted features from the contours of the signature [9], Alessandro Z. and Lee L. L. used on-line signature information for segmentation of off-line signature in a hybrid ASV system [10]. Recently, Jesus F. et al. used the histogram of gray level image [11], and Larkin L. used statistical information [12]. The above mentioned methods for feature extraction used between 1973 and 2009 (36 years) are naturally not exhaustive but, to a good extent, representative. The common factor in most off-line research works is that all features extracted can be classified as parametric features (or shape descriptors) like slopes and slants extracted from contours and thinned images, aspect ratio, baseline, etc., measured globally or locally on the signature image. On the other hand, one-dimensional reference patterns (from projections) and matching were used for feature extraction [13], and Katsuhiko U. used a selected genuine signature image as a pattern and used matching for off-line Japanese signature verification [14]. Ammar M. analyzed the performance of matching-based features using horizontal and vertical projections, and finished to the conclusion that the 2D reference pattern based features (RPBFs) should be more effective in signature verification [15], and reported outstanding results in off-line ASV by using Ammar Matching Technique (AMT), and a multi-valued reference image [16]. In connection with this methodology (2D pattern representation), and 5 years later (1997), Sabourin R. et al. used a 2D pattern built from binary skeletons of genuine signatures to extract granulometric features for verification, and tested it on random forgeries [17]. Ten years later (2002), Parker J. used the same idea of representation to construct a 2D image from skeletons of genuine signatures obtained from a tablet, and used the representation to compute a simple distance for signature recognition [18, 19].

In this Chapter, RKI and its relation to FDE methodology of working are explained in detail, AMT with its relation to the previous work are elaborated, the performance of the proposed entire method is evaluated using binary, thinned, and high pressure images of the same signature data, and finally some conclusions are reached.

26.2 Used Signature Data

The signature data used in this Chapter consists of 200 genuine signatures obtained from 20 writers, 10 from each one, and 200 forgeries produced by 10 imitators after good practice so that the forgeries came with a good degree of skill. Each imitator imitated the signatures of the 20 persons. Each image was digitized into a gray image of 256×1024 pixels and 256 gray levels. This signature data was prepared by Fujitsu Labs, Japan, using a TV camera. The signature images contained shadows, and the background was not homogeneous so that considerable preprocessing was necessary to be developed and applied to the signature data to insure that clean images are used for feature extraction.

26.2.1 Preprocessing

Figure 26.1 shows an example of the signatures in this data and the preprocessing stages applied in sequence, and their results. Figure 26.1 (a) shows a signature image of 256×1024 pixels and 256 gray levels with non-homogeneous background. Figure 26.1 (b) shows the result of thresholding this image using the well-known Otsu's method. Naturally, this result is not useful at all. Figure 26.1 (c) is the result of equalizing the original image of Figure 26.1(a) by subtracting from each pixel value the average of its column. Figure 26.1(d) shows the result of background reduction by eliminating the negative values in Figure 26.1(c). The result of averaging pixels within 3×3 window is shown in Figure 26.1(e). Figure 26.1(f) is the result of semi-thresholding of Figure 26.1(e) using Otsu's method to find the threshold. Figure 26.1(g) shows the result of restoring the original values of pixels into the extracted signature body in Fig. 26.1(f). Finally, Fig. 26.1(h) is the result of recovering the last

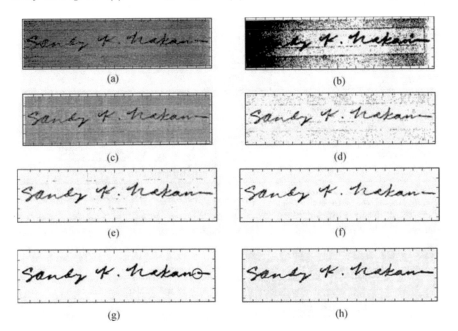

Fig. 26.1 The preprocessing phases.

part of the signature (surrounded by a circle in Fig. 26.1(g)) which was lost
during preprocessing phases due to its very low gray levels. The reconstruction was achieved by adding the average value of the row of each pixel to its
value before thresholding. More details about these preprocessing phases can
be found in Refs. [20, 24], and to a lesser extent in [6].

As can be seen, this sequence of preprocessing phases was very effective
in extraction of clean images to be used for feature extraction.

26.3 Image Types Used for Feature Extraction and Evaluation

Three types of images are used for feature extraction: binary, thinned, and
high pressure regions (HPRs) images.

26.3.1 Binary and Thinned Signature Images

Binary images are those extracted at the end of the preprocessing phases
(Fig. 26.1 (f)) with "1" value assigned to signature pixels and "0" value assigned to background ones.

Thinned images are obtained from the binary ones using Hilditch's
algorithm. Figure 26.2 shows four examples of the thinned signatures with
different global slants.

Fig. 26.2 Four thinned signatures with different global slants. (a) positively
slanted, (b) less positively slanted, (c) vertically slanted, and (d) negatively slanted
[5].

26.3.2 HPRs Images

These images are obtained from the extracted signature images with their original densities restored by thresholding their gray levels so that those higher than 0.7. g_{max} are considered as HP pixels (g_{max} is the maximum gray level), and the others are considered as low pressure ones. Naturally, the HPRs image has three values ("zero" for background, "High" for high pressure pixels, and "Low" for low pressure ones). More details about HPRs can be found in [6, 20, 24]. In Fig. 26.3 (a) shows an original signature image with non-homogenous background and shades, and Fig. 26.3 (b) shows the signature extracted from background with HPRs (black ones) extracted. The gray areas are the low pressure areas of the signature. This case gives an idea of the effectiveness of the preprocessing phases in extraction of signature images from background. Figure 26.3 (c) shows a genuine signature extracted from background and displaycd as contour image with HPRs extracted, Fig. 26.3 (d) shows a forgery signature similar to Fig. 26.3 (c) but the HPRs are very different. This difference which cannot be compensated by shape features provides a new ability to detect skilled forgeries in off-line cases. Figure 26.4 shows two examples: (a) and (b) are two HPRs images of the signature of the same person; (c) and (d) are two forgeries of the same signature from the data used. This figure shows clearly that: (1) HPRs are very much similar in genuine samples. (2) They are quite different in size, location, and shape between forgery samples and genuine ones, although the shape similarity is very high. This finding highlight the merit of HPRs over shape features in case of skilled forgery cases.

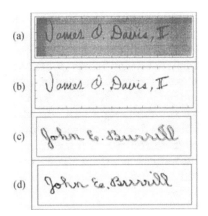

Fig. 26.3 (a) An original signature image. (b) The signature extracted from background with HPRs extracted. (c) A genuine signature extracted from background and displayed as contour image with HPRs extracted. (d) A forgery signature similar to that of (c) but HPRs are very different allowing efficient detection of forgeries.

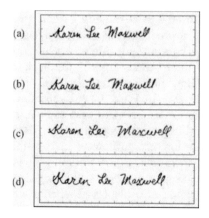

(a) *Karen Lee Maxwell*

(b) *Karen Lee Maxwell*

(c) *Karen Lee Maxwell*

(d) *Karen Lee Maxwell*

Fig. 26.4 HPRs of two genuine signatures (a) and (b), and two forgery ones, (c) and (d).

26.4 Skills of Forgery Creation of Used Forgeries

Forged signatures were classified in several ways according to the method used for creation and the date the term appeared; however, the most common and recent terms used are: skilled forgery and unskilled forgery. The unskilled forgery refer to: (1) random forgery, a completely different signature is used instead of the genuine one, which is commonly used to evaluate different signature verification algorithms and methods since it is already available as the signatures of other persons in the data. (2) simple forgery which refers to forgery samples created with little care and practice so that it resembles the genuine signature to some extent, but can be easily detected by a human or by computerized algorithms [20]. The skilled forgery refers to forgery samples created with sufficient practice so that the forger is convinced that he/she made a convincing forgery, which can be accepted by inspectors or even by the person to whom the signature belongs, himself[2]. Those skilled forgeries are created either by imitation, tracing, or by free hand movement after memorizing the signing process with sufficient practice and producing it from memory with natural handwriting. This is the most dangerous kind of forgeries. In fact, and as a final comment about forgery creation, the author faced also some cases in which the forgery signature was created with some scaling of a genuine sample and partial tracing so that the detection becomes very difficult. In general we can say, **skilled** and **unskilled** forgery. The degree of forgery creation skill will, no doubt, affect the performance of the algorithm of the system tested. Since there is no standard measure of the skill of forgery creation, several examples of the genuine and forgery

---

2   In fact, during the author's work as a questioned document examiner, he faced this
    kind of forgeries in which the victim says: yes, this is my signature but I did not sign
    it!

samples of some persons from the data used are displayed so that the reader can appreciate how skillful the forgeries have been made. Figure 26.5 shows two handwritten genuine signatures of two different persons and their two corresponding forgeries from the data used.

Genuine
Forgery
Genuine
Forgery

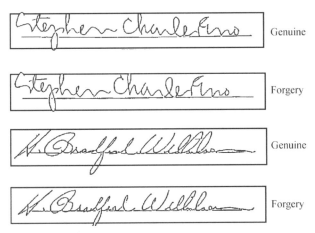

Fig. 26.5 Two different genuine signatures and their imitations.

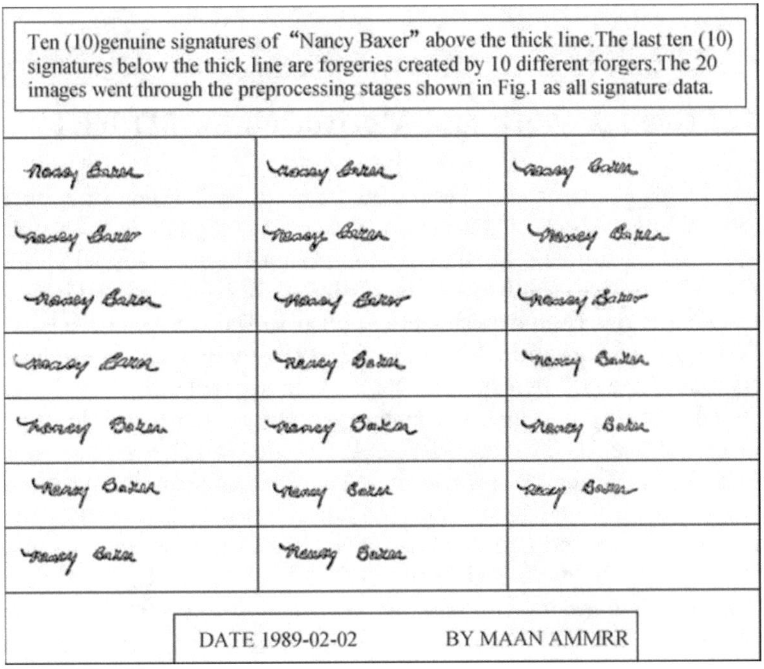

Ten (10)genuine signatures of "Nancy Baxer" above the thick line.The last ten (10) signatures below the thick line are forgeries created by 10 different forgers.The 20 images went through the preprocessing stages shown in Fig.1 as all signature data.
DATE 1989-02-02
BY MAAN AMMRR

Fig. 26.6 The set of 10 genuine samples and 10 forgeries of Nancy's signature.

Figure 26.6 and Fig. 26.7 show 2 sets of signatures. Each one consists of
10 genuine samples and 10 forged others. It is believed that these forgeries
are skilled to a good extent. These sets of signatures are displayed as contour-
images for the convenience of display and printing.

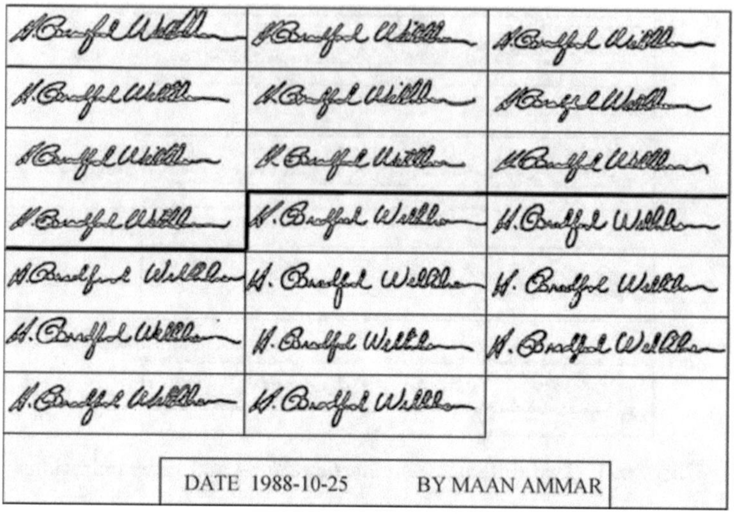

DATE 1988-10-25   BY MAAN AMMAR

Fig. 26.7 Ten genuine samples and 10 forgeries of the same signature.

26.5 Previous Work and Motivation for 3D RKI

Ammar M. et al. reported a successful work on verification of skilled forgeries in off-line systems in mid-eighties using the signature data described
in Sections 26.2 and 26.4 [3, 4]. They adopted a threshold-based verification
decision with a pre-established value of the verification threshold VTH estimated using verification experiments applied to the whole data available [5,
6, 20]. They used this approach because in the practical situation there is no
forgery class for every person. The verification approach they used measures
the distance of the test sample from the genuine cluster using the Euclidean
distance measure. Then if the distance exceeded VTH, the test sample is
considered as a forgery, otherwise, it is accepted as genuine. Details of this
approach can be found in [5, 20]. They developed also a feature selection
algorithm that can find and display the curves of the best feature set and an
(X, O) graph (to be explained in this section), as well as the result matrices
of n^2 feature sets among the possible $n!$ feature sets of "n" given features.
This facility, which has been developed recently to be used as an interactive feature selection technique (IFST) [22], was behind the motivation for
3D RKI-based signature verification. In order to give an idea about the fea-

ture selection technique, and understand how it was behind the technique presented in this paper, it is described very briefly here.

26.5.1 Feature Selection Technique (FST)

FST is based on the circulant matrix idea to generate n^2 feature sets among the possible $n!$ feature set of "n" given features [5, 20, 21]. For every feature set, it computes the percentage of correct acceptance (PCA), the percentage of correct rejection (PCR), and the system reliability (SR), where SR=(PCA+PCR)/2. At the end of each run, it displays 3 result matrices, PCA, PCR and SR, matrices. Each one contains n^2 values. A practical explanatory three results matrices for $n = 4$ case, is shown in Fig. 26.8.

SR

	1	2	3	4
1	78.75	85.25	93.75	94.25
2	80.50	93.00	95.75	94.25
3	85.75	93.50	93.25	94.25
4	86.00	89.75	93.00	94.25

PCA

	1	2	3	4
1	92.00	96.00	93.50	96.50
2	88.00	94.00	96.00	96.50
3	93.00	94.50	94.50	96.50
4	95.00	90.50	94.00	96.50

PCR

	1	2	3	4
1	65.50	74.50	94.00	92.00
2	73.00	92.00	95.50	92.00
3	78.50	92.50	92.00	92.00
4	77.00	89.00	92.00	92.00

Fig. 26.8 A practical explanatory example of three result matrices for $n = 4$.

26.5.2 Performance of Parametric Shape Features

Figure 26.9 shows SR result matrix obtained from running FST using 7 parametric shape features ($n = 7$). The 7 features and their initial order were as follows: global baseline, percentage of vertically, positively and horizontally slanted pixels, length, global area, percentage of negatively slanted pixels. The best feature set is that of the entry (1, 3) with SR=86.00 [5].

Figure 26.10 shows the PCA, PCR, and SR curves belonging to the best

i j	1	2	3	4	5	6	7
1	79.75	84.50	86.00	85.75	85.25	84.50	83.75
2	69.25	70.75	77.50	82.00	79.75	78.25	83.75
3	68.50	75.75	78.75	76.50	76.50	82.75	83.75
4	68.50	74.25	73.00	73.75	81.00	83.50	83.75
5	66.00	65.00	67.50	79.00	82.25	83.00	83.75
6	61.50	65.00	78.75	83.50	83.50	83.75	83.75
7	63.25	79.00	83.00	83.25	83.25	84.75	83.75

Fig. 26.9 The result matrix of seven chosen shape features. The best feature set
is that of entry (1, 3), from [5].

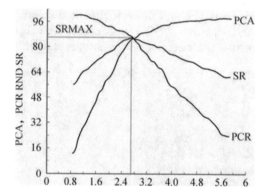

Fig. 26.10 SR, PCA, and PCR curves of the best shape feature set, entry (1, 3)
of SR result matrix in Fig. 26.9. The horizontal axis is VTH [5].

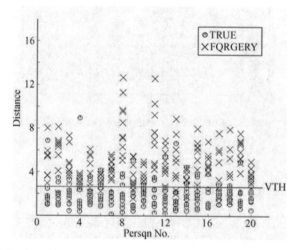

Fig. 26.11 The (X, O) graph with VTH for the best shape feature set in Fig. 26.9
[5].

feature set, and Fig. 26.11 shows the (X, O) graph of this feature set at SR= 86.00 (SR_{max}). The (X, O) graph shows the result of each signature sample of 400 samples database. Every genuine sample is shown by "O" and every forgery one is shown by "X" in the graph with the vertical axis representing the distance measure, and the horizontal axis representing persons denoted by their number $(1-20)$. By a glance on the graph, one can evaluate the performance of the used feature set where each "O" below VTH is a correct decision, and each "X" above VTH is also a correct decision. Examining Fig. 26.11 reveals the fact that although SR=86.00, and all the forgeries of several persons in the data $(1, 7, 8, 11, 16, 19)$ are correctly detected, (X is above VTH), PCR for some persons, like persons No. 17 and 18, is distinctly low (PCR=50.00) which means that 50% of the forgeries are accepted. In [5], they tried to overcome this problem by using a combination of HPR features as well as the shape features and got SR=91.00 by using the set of 3 features (global percentage of Vertically and Positively slanted pixels, and Global Baseline) of which the results are shown in Figs. 26.10 and 26.11, augmented by the global high pressure factor. The overall performance improved with this feature set but the problem of unpleasant forgery detection for specific persons remained. Ammar M. tried to alleviate this problem by introducing and using the one-dimensional reference pattern based features by matching the vertical and horizontal projections of the test signature with the reference projections built from the training samples of the concerned person [13, 15]. This attempt and its results are explained in the next section.

26.5.3 One-dimensional RPBFs

Figure 26.12 shows the horizontal projection of a signature and its corresponding binary reference pattern. Figure 26.13 (a) shows a genuine signature with its HPRs and its horizontal projection HP, Fig. 26.13 (b) the HPRs reference pattern, and Fig. 26.13 (c) a forgery signature with its HPRs and their horizontal projection. By extracting features using matching between horizontal and vertical projections and their corresponding reference patterns, computing the distance measure and making the verification decision, Ammar M. found that the one-dimensional reference pattern-based shape and HPRs features, in general, improves the performance. And through an analytical discussion, he finished to a conclusion that 2D RPBFs must be more effective in signature verification, and indicated that this conclusion motivated conducting real experiments on using 2D RPBFs with surprising results [15].

At this stage, and besides the motivation emanated from the analytical conclusion said above, there was a second motivation towards how to build the 2D Reference Pattern and use it for feature extraction. This second motivation which is as important as the previous one, or even more, is explained

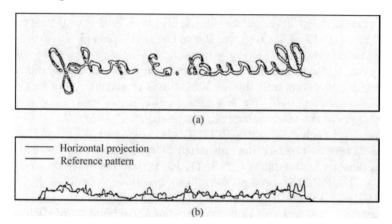

Fig. 26.12 The HP of a signature and the binary reference pattern [5].

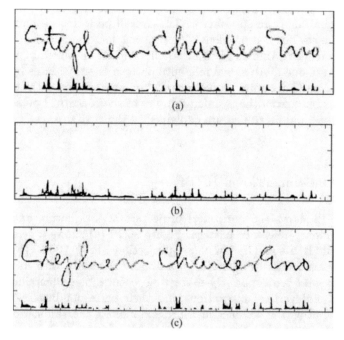

Fig. 26.13 (a) HPRs genuine signature with its HP, (b) HPRS reference pattern, and (c) HPRs forgery signature with its HP [5].

as follows.

If we are to use the verification method based on parametric features explained in Section 26.5.2, or on one-dimensional RPBFs explained in Section 26.5.3 in a practical system, we will face real problems with persons like No. 17 and No. 18 in Fig. 26.11. It turns out that improving the ability of forgery detection becomes a necessity if we are aiming at an actual ASV system. Therefore, the author reinvestigated carefully the samples of those persons, as well as others and found that an experienced human eye can do better than a computer with the results reported in the previous sections. This finding oriented the research towards studying carefully the way of FDE works and modeling it for implementation via computer.

In 1992, Ammar M. reported outstanding results in signature verification on the signature data explained in Section 26.2 using a matching technique that matches the test signature with a 2D multi-valued Reference Image (RI) and called the technique "ammar matching technique" (AMT) [16]. That extended abstract-paper concentrated on the experimental results and their discussion rather than elaborating on details of the reference image and the matching technique explained briefly in the paper. Figure 26.14 shows an example of RI built for a specific person from the same signature data [16], where this multi-valued RI was displayed as contour-image for the convenience of printing facilities available that time. The innermost contour represents the highest level in the multi-valued image, which will be explained in detail in the following sections.

Fig. 26.14   2D multi-valued reference image, from [16], 1992.

26.5.4 FDE Working Methodology and Knowledge Acquisition Model

Knowing that FDE may retrieve up to 3 dozens of genuine signatures before giving a decision [2], and some times retrieves all genuine samples available, as the experience of the author according to FDE for Syrian Justice Ministry since 1995, the way by which FDE acquires knowledge of a specific signature will be modeled. Then, the model will be used in building some off-line signature knowledge to be used for ASV. The diagram in Fig. 26.15 below summarizes this model. In this diagram, FDE compares the input signature (IS) which is the questioned one, with the first genuine signature available. If an authorized person can give a decision about the authenticity of IS, he gives his decision with explanations. If he cannot give a decision, he retrieves another genuine sample and compare with these two samples according to his experience. If he cannot give a decision, he continues to retrieve more genuine samples and compare until he can decide, or no more genuine samples

remain. In all cases, he must give a decision with explanations. In actuality:
(1) he rarely deals with one reference genuine signature, but this case may
exist, and (2) he rarely reach a decision by comparing with only one reference
genuine signature.

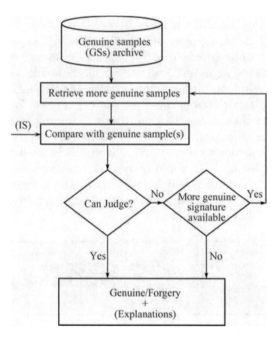

Fig. 26.15 A model for acquiring the knowledge of a signature for verification by
FDE.

In fact, when FDE retrieves one genuine sample after another and every
time he compares its general shape, way of writing, pen movement, letter
design, spontaneity of the stroke, specific marks, etc., with the questioned
(test) signature, he accumulates his knowledge of this person's signature by
investigating one genuine sample after another. Finally, with the knowledge
he could acquire, and comparison with the questioned signature, he gives his
decision, almost with explanations. During knowledge building (acquisition),
when a feature or an aspect of the genuine signature is repeated in all or most
samples, it is given higher weight in the decision, and this weight is based on
2D shape characteristics. Moreover, if a shape feature is developing from one
sample to another through time within some or all samples, it is considered.
In brief, FDE learns through the successive genuine sample investigation
not only for decision making, but also for feature extraction before decision
making.This is the key point of the above discussion. Therefore, in order
to approach FDE methodology, we should learn two-dimensionally during
feature extraction.

26.6 3D Reference Knowledge of Signature

The 3D reference knowledge of a signature derived from FDE methodology of signature-knowledge acquisition is represented by a reference knowledge image (RKI). RKI represents the knowledge of 2D shape aspects of the genuine signature, the spatial relations between its parts in different samples, and the way they change from one sample to another. It is simply built by superimposing the binary genuine samples centered around their gravity center (GC, this is a primitive process for those working with image processing). The number of occurrences of pixels from different genuine samples at the same place in RKI is represented on the third dimension. Figure 26.16 shows RKI of "Karen Maxwell" signature shown in Fig. 26.4. Here in Fig. 26.16: (a) is a 3D display; Fig. 26.16 (b) is a level-contour display; Fig. 26.16 (c) is a gray level display. RKI in this figure is built from 10 genuine samples.

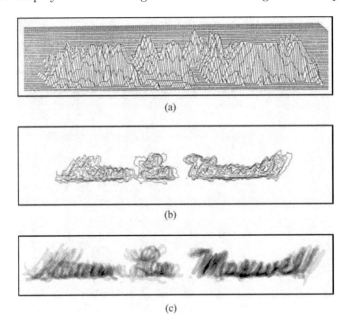

(a)

(b)

(c)

Fig. 26.16 RKI of "Karen Maxwell" displayed as (a) 3D image, (b) level-contours, and (c) gray levels.

Before commenting on these 3 types of display of RKI, the accumulation process (build-up) of RKI from "n" genuine samples will be shown below. Figure 26.17 shows 8 binary genuine samples of a specific person, and Fig. 26.18 shows the process of building RKI from these samples, step by step, displayed as gray level image. The resultant RKI in this case is an image of 8 gray levels, since the gray level cannot exceed the number of used samples.

Returning to Fig. 26.16, we may feel that (a) gives the best visual impression, however, there are hidden surfaces; (b) gives the full details of RKI

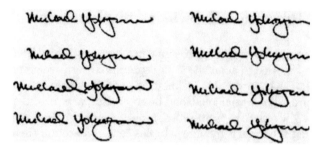

Fig. 26.17 Eight binary genuine samples of the same person.

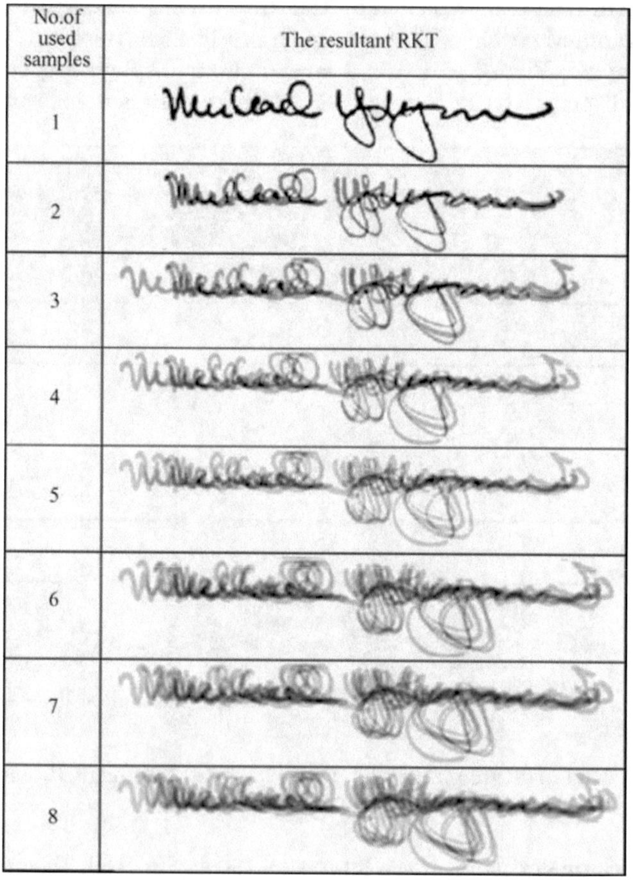

No.of used samples	The resultant RKT
1	
2	
3	
4	
5	
6	
7	
8	

Fig. 26.18 Building up RKI by accumulating 8 genuine samples centered on their GC, step by step.

since every level is represented by its contour within RKI. We can feel the clarity of this representation if we use a larger image as shown in Fig. 26.19. The innermost contour represents the highest level and it should given the largest weight when features are extracted. RKIs, shown in [16], and shown in Fig. 26.14 from [16], are displayed in this way as level-contours. Figure 26.16 (c) represents the number of occurrences of genuine samples at the same point in the RKI as gray level. Eventually, the preference of the display method among 3 ones shown above will depend on the viewer (subjective impression).

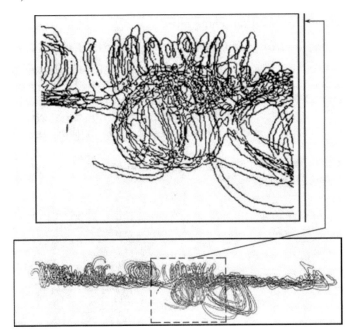

Fig. 26.19 RKIs of the signature in Fig. 26.18 displayed as contours with an enlarged portion.

If the samples of RKI is produced by a machine, RKI will have one gray level that equals to the number of genuine samples used to construct it, or it will be the outline of the signature as one contour. However, since two genuine samples can not be exactly the same (a forensic rule), RKI will always have values form 0 to n, where n is the number of genuine samples used. Examining carefully Fig. 26.16 or Fig. 26.19 can show how much 2D characteristics are lost when only parametric features are used. In other words, parametric features do not learn about 2D signing process characteristics, and the discrimination power of this learning is lost.

26.7 Ammar Matching Technique

Ammar Matching Technique (AMT) matches the test signature with RKI and produces two main features: matching and mismatching [16], which are measured using a method that takes in consideration of the signing process itself. Since, up to the author's information, there was no known matching technique done in the way that this matching method works, and the referees of the 11ICPR agreed, it was called AMT [16].

In fact, mismatching features, to be explained, has been proved to be little bit more effective than matching ones with the used signature data for forgery detection, and increase considerably the discrimination power of the features when used with matching ones.

Computing Matching and Mismatching from RKI Using AMT

Computing matching and mismatching using RKI will depend on the philosophy of building it. As an introduction to this concept, let us consider the two patterns in Fig. 26.20 as a general representation to explain the principle of AMT. The two patterns differ to some extent from each other:

Pattern 1: the reference pattern drawn by bold thick line (stands for RKI).

Pattern 2: the test pattern drawn by thinner ordinary line (stands for test signature, "questioned").

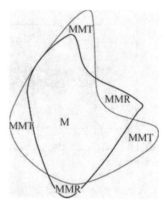

Fig. 26.20 Matching and mismatching areas in a general reference pattern and a test pattern: M represents matching areas, and MMT and MMR represent mismatching areas.

In the general case, we will have areas of matching and mismatching in both of the patterns if superimposed on top of each other as shown.

– Matching area "M". This area includes all parts of the test pattern superimposed over nonzero reference pattern parts. In signature terms: test signature pixels superimposed over nonzero RKI pixels.

– Mismatching areas "MM". Those are two types:

Type 1 Areas of test pattern that do not superimpose over reference pattern parts "MMT". In signature terms: test signature pixels that do not superimpose over RKI pixels.

Type 2 Areas of reference pattern that do not underlay any test pattern parts "MMR". In signature terms: RKI pixels that do not underlay any pixel of the test signature image.

Now, let "n" be the number of binary genuine signature samples used to build RKI represented as gray levels. In this case, RKI will have $g_{max} = n$, and $g_{min} = 0$, where g_{max} and g_{min} are the maximum and minimum gray levels, respectively. The test signature (TS) as a binary image will have the values 1, and 0. Computing matching and mismatching, will be done as follows:

Matching (MA) since M represents the areas of TS that matches some areas of RKI, each pixel of matching areas of TS will be given a weight proportional to the number of pixel occurrences from the training samples in RKI's matching position, which is simply, the gray level in this representation. In other words, the matching feature will be the summation of RKI pixel values that match (underlay) TS pixels in the area M, according to

$$MA = \sum_{i=0}^{M} \sum_{j=0}^{N} \text{RKI}(i,j) * \text{TS}(i,j). \tag{26.1}$$

Where: MA is the matching result; M, N is the dimensions of RKI image. It is always kept larger than the largest TS to accommodate lt; RKI is reference knowledge image and TS is test signature.

Mismatching (MM) since we have two kinds of mismatching areas (MMR and MMT), the mismatching feature will have two components:

MMR: forensically, among n genuine signing trials, no pixel of TS matched any pixel in these areas. Therefore the degree of mismatch between TS and training samples will be proportional to the number of pixel occurrences of training samples in these areas. Simply, MMR will be the summation of RKI pixel values in these areas:

$$\text{MMR} = \sum_{i=0}^{M} \sum_{j=0}^{N} \text{RKI}(i,j) * \text{NOT}(\text{TS}(i,j)). \tag{26.2}$$

MMT: forensically also, among n genuine signing trials, there are no occurrences in RKI similar to TS pixels in these areas. Therefore, pixels of TS in these areas will be given a weight proportional to the number of the genuine trials they missed. Simply, the weight is n for all pixels in these areas, and MMT will be n times the summation of TS pixels in these areas:

$$\text{MMT} = \sum_{i=0}^{M} \sum_{j=0}^{N} \text{NOT}(\text{RKI}b(i,j)) * \text{TS}(I,J) * n, \tag{26.3}$$

where $\text{RKI}b(i,j)$ is the binary image of the RKI, constructed by replacing all nonzero RKI pixels by 1, and leaving the others "0", and n is the number of genuine samples used to build up RKI.

Finally, MM is computed by

$$\text{MM} = \text{MMR} + \text{MMT}. \tag{26.4}$$

26.8 Feature Extraction

In this Section, Four global features and 22 local ones will be extracted.

26.8.1 Global Features

The global features (GFs) are those computed on the whole signature image.
Four global features are extracted. Two of them are the global MA and MM:

$$\text{GF}(1) = \text{MA}, \tag{26.5}$$
$$\text{GF}(2) = \text{MM}. \tag{26.6}$$

Two others are X and Y coordinates of GC:

$$\text{GF}(3) = \text{X},$$
$$\text{GF}(4) = \text{Y}.$$

26.8.2 Local Features

Local MA and MM Measured on RKI and TS Sixths

All these features are of MA and MM types. They are computed on TS
and RKI images divided into six parts, three equal to width parts the right of
GC of RKI, and three equal width others to the left of GC of RKI, according
to the drawing shown in Fig. 26.21. This way of segmentation was found to be
stable and reliable. We must keep in mind that some parts of TS go beyond
RKI, and some RKI parts extend beyond TS ones, therefore, RKI image
space must take this fact into account (kept larger enough).

MA and MM are computed according to Eqs. 26.1 and 26.4, respectively on each one of the six parts. Therefore, we have six local MA features
$\text{LMF}_6(1), \ldots, \text{LMF}_6(6)$, measured on the six parts shown in Fig. 26.21, and
six local MM ones $\text{LMMF}_6(1), \ldots, \text{LMMF}_6(6)$, measured on the same parts.

Local MA and MM Measured on RKI and TS Thirds

In order to study the effect of decreasing the locality, MA and MM are
measured on three parts of RKI and TS as follows:

$$\text{LMF}_6(1) + \text{LMF}_6(2) = \text{LMF}_3(1), \tag{26.7}$$
$$\text{LMF}_6(3) + \text{LMF}_6(4) = \text{LMF}_3(2), \tag{26.8}$$
$$\text{LMF}_6(5) + \text{LMF}_6(6) = \text{LMF}_3(3). \tag{26.9}$$

Similarly,

$$\text{LMMF}_6(1) + \text{LMMF}_6(2) = \text{LMMF}_3(1), \tag{26.10}$$

$$\mathrm{LMMF}_6(3) + \mathrm{LMMF}_6(4) = \mathrm{LMMF}_3(2), \qquad (26.11)$$

$$\mathrm{LMMF}_6(5) + \mathrm{LMMF}_6(6) = \mathrm{LMMF}_3(3). \qquad (26.12)$$

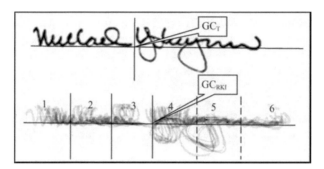

Fig. 26.21 An explanatory example of dividing the left side of RKI (left to the $\mathrm{GC_{RKI}}$) into 3 equal width parts. The same thing is done to the right side part (dotted).

Local MA and MM Measured on RKI and TS halves

$$\mathrm{LMF}_6(1) + \mathrm{LMF}_6(2) + \mathrm{LMF}_6(3) = \mathrm{LMF}_2(1), \qquad (26.13)$$

$$\mathrm{LMF}_6(4) + \mathrm{LMF}_6(5) + \mathrm{LMF}_6(6) = \mathrm{LMF}_2(2). \qquad (26.14)$$

Similarly,

$$\mathrm{LMMF}_6(1) + \mathrm{LMMF}_6(2) + \mathrm{LMMF}_6(3) = \mathrm{LMMF}_2(1), \qquad (26.15)$$

$$\mathrm{LMMF}_6(4) + \mathrm{LMMF}_6(5) + \mathrm{LMMF}_6(6) = \mathrm{LMMF}_2(2). \qquad (26.16)$$

In this way, we have $(6 \times 2) + (3 \times 2) + (2 \times 2) = 22$ local AMT-based features. Adding the 4 global features explained above, the number of used features for performance evaluation becomes 26 features, 24 of them are AMT-based.

26.9 Distance Measure and Verification

In a way similar to the previous works [4–9], the distance measure of the test sample is computed using the weighted Euclidean distance according to

$$\mathrm{dist} = \left(1/n \sum_{i=1}^{n} ((F_i - \mu_i)/\sigma_i)^2 \right)^{1/2}, \qquad (26.17)$$

where F_i is the value of the ith feature measured on the TS,

μ_i is the mean of ith feature computed on the set of genuine(training) samples of the same person,

σ_i is the standard deviation of the ith feature computed on the set of training samples of the same person,

n is the number of training samples.

A verification threshold VTH on this distance gives the decision: If the distance exceeded VTH, the signature is judged to be a forgery, otherwise, it is accepted as genuine. The leave-one-out method is used to verify the genuine samples. Naturally, when computing MA and MM features for genuine samples, the genuine sample for which the features are being computed, is excluded from RKI. For the purpose of results evaluation, terms PCA, PCR, and SR are used, where, PCA is percentage of correct acceptance, PCR is percentage of correct rejection, and SR is system reliability = (PCA+PCR)/2.

26.10 Experimental Results and Discussion

Three types of RKI are used for performance evaluation: (1) Built from binary images, (2) built from HPRs images, and (3) built from thinned images.

26.10.1 RKI from Binary Images

Performance of MA and MM globally and on signature halves

Extracting all the features mentioned in Section 26.8 from the binary images of the signature data after the necessary preprocessing explained in Section 26.2, and using the circulant matrix idea based FST explained in Section 26.2 with the following features: X and Y are coordinates of GC, and MA and MM measured globally and on signature halves in this initial order:

$$n = 8 = \{X, Y, \mathrm{MA}, \mathrm{LMF}_2(1),\ \mathrm{LMF}_2(2), \mathrm{MM}, \mathrm{LMMF}_2(1), \mathrm{LMMF}_2(2)\}$$

gave the result matrices shown in Fig. 26.22.

SR result matrix shows that:

1) MA and MM measured globally and on signature halves give SR=94.25 (entry $(3, 6)$) which exceeds the best result ever obtained on the same signature data using any possible selection of parametric or one-dimensional RPBFs. This result practically confirms the expectations based on the analysis done in [15] that the 2D RPBFs must give better performance.

2) MA and MM measured globally and on signature halves, with Y coordinate of GC, give SR=95.75 with very high PCR=97.5 (entry $(2, 7)$). This result can be thought of as: the forgers can imitate the horizontal distribu-

SR

	1	2	3	4	5	6	7	8
1	78.75	85.25	93.00	94.25	94.25	96.00	96.00	95.50
2	80.50	91.50	92.25	93.75	96.50	96.25	(95.75)	95.50
3	86.00	87.50	89.50	94.75	94.50	(94.25)	94.50	95.50
4	84.00	89.75	95.50	94.50	93.75	94.00	95.25	95.50
5	85.25	92.25	91.00	91.25	92.50	94.25	95.00	95.50
6	85.75	84.50	86.25	89.75	92.75	94.75	95.00	95.50
7	82.50	85.75	89.75	93.25	94.50	95.00	95.75	95.50
8	84.75	88.75	93.25	94.50	95.50	95.75	96.25	95.50

PCA

	1	2	3	4	5	6	7	8
1	92.00	96.00	94.00	96.50	95.00	96.50	94.50	93.00
2	88.00	96.00	97.50	94.00	95.50	95.50	94.00	93.00
3	95.00	94.00	93.00	96.00	94.50	93.50	94.00	93.00
4	94.50	91.50	95.50	94.50	92.50	93.50	93.50	93.00
5	89.50	94.50	93.00	92.50	93.00	95.00	94.00	93.00
6	93.00	92.50	93.00	90.50	91.00	93.50	94.00	93.00
7	91.50	92.00	90.50	93.00	93.00	93.50	94.00	93.00
8	85.50	90.00	93.00	95.50	94.00	96.00	95.50	93.00

PCR

	1	2	3	4	5	6	7	8
1	65.50	74.50	92.00	92.00	93.50	95.50	97.50	98.00
2	73.00	87.00	87.00	93.50	97.50	97.00	97.50	98.00
3	77.00	81.00	86.00	93.50	94.50	95.00	95.00	98.00
4	73.50	88.00	95.50	94.50	95.00	94.50	97.00	98.00
5	81.00	90.00	89.00	90.00	92.00	93.50	96.00	98.00
6	78.50	76.50	79.50	89.00	94.50	96.00	96.00	98.00
7	73.50	79.50	89.00	93.50	96.00	96.50	97.50	98.00
8	84.00	87.50	93.50	93.50	96.00	95.50	97.00	98.00

Fig. 26.22 The result matrices of MA and MM measured globally and on signature halves, and X and Y coordinates of the GC.

tion of the signature components (which is easier to recognize) more than the vertical distribution, since X is less effective as the result shows.

Performance of MA and MM globally and on signature sixths

Measuring MA, MM globally and on signature sixths and generating the result matrices for this initial order of 16 features: $n=16=\{X, Y, \text{MA}, \text{LMF}_6(1), \ldots, \text{LMF}_6(6), \text{MM}, \text{LMMF}_6(1), \ldots, \text{LMMF}_6(6)\}$ gave SR matrix shown in Fig. 26.23.

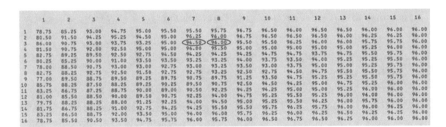

Fig. 26.23 SR result matrix of X and Y coordinates of GC and (MA and MM) measured globally and on signature sixths.

This SR matrix shows that:

1) global MA and local MA measured on signature sixths (7 features of entry (3,7)) give SR=94.5 which is much better than the performance

of global MA and MA on halves (SR=89.5, in Fig. 26.22): It means that
increasing the localization of matching gives better performance,

2) The performance of MA and MM measured globally and MA measured
on signature sixths (8 features, entry (3,8), gives the best SR = 96.00.

3) Several other meaningful results can be inferred from Fig. 26.23, but
only these two indicative ones are mentioned.

As a conclusion:

1) AMT with 3D RKI built from binary images, gives much higher performance than any combination of parametric and one-dimensional RPBFs,
with an improvement exceeding 5%.

2) MA and MM when measured locally on signature sixths (as segmented)
appreciably improve the performance.

26.10.2   RKI from Thinned Images

Principally, when a subject signs one's signature by different writing instruments, the line width will change. When a signer or an imitator uses the pen
with different pressure habits, the line width will also differ. In this section,
the performance of AMT with RKI using thinned images will be tested in detail because thinning will eliminate the role of line width. Figure 26.24 shows
the complete result matrices (SR, PCA, and PCR) obtained by using RKI
and AMT-based global and local features extracted from thinned images.
Here, all matrices are shown to discuss clearly details of the performance of
AMT-based features, that cannot be reached by using only SR matrix or
(X,O) graph. The result matrices were generated from this initial order of
features:

$$N = 16 = \{X, Y, \mathrm{MM}, \mathrm{MA}, \mathrm{LMF}_6(1), \ldots, \mathrm{LMF}_6(6),$$
$$\mathrm{LMMF}_6(1), \ldots, \mathrm{LMMF}_6(6)\}.$$

Anatomy of Effectiveness of AMT-Based Features of Thinned images

1) Global MM. This global MM alone gives the performance:
SR = 89.50, PCA = 96.50, PCR = 82.50 (entry 3, 1). This result is better
than the best parametric shape feature set.

2) Global MA and MM. The global MA and MM give this performance:
SR=92.50, PCA = 91.00, PCR = 94.00 (entry 3, 2). This result is better than
any feature set that could be obtained using parametric and one-dimensional
RPBFs. It is rather in favor of correct rejection, which is preferred in the
practical application. It is obvious that PCR has improved a lot when MM
was used with MA.

3) Local MM. The performance of local MM 6 features is SR = 96.00,
PCA = 96.00, PCR = 96.00 (entry 11, 6). This result is far better than that
of parametric features, and better than that of global AMT-based features,

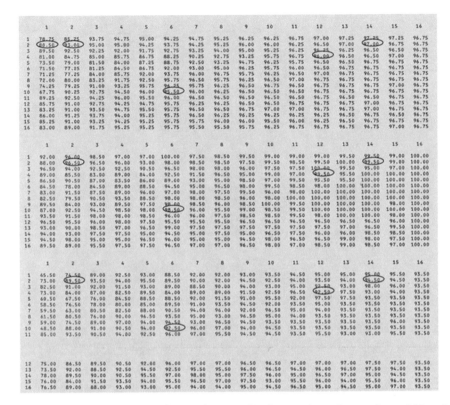

Fig. 26.24 SR (top), PCA (middle) and PCR (bottom) result matrices of X and Y coordinates of GC and (MA and MM) measured globally and locally on thinned signature sixths.

especially if we remember that the result now is above SR = 95.00. The advantage of this result is that it is obtained at an equal type I/type II error, and with the elimination of the writing instrument role.

4) Local MM and MA. The performance of local MA and MM (12 features, entry 11, 12) is SR = 96.50, PCA = 95.50, PCR = 97.50. This results is great in comparison with parametric and one-Dimensional RPBFs.

At this stage, it is meaningful to see how (X, O) graph and SR, PCA and PCR curves have been shown. Figure 26.25 shows the curves and Fig. 26.26 shows (X, O) graph. Taking a look at (X, O) graph reveals quickly the following facts:

- The problem of distinctly low PCR for some persons have disappeared.
- The genuine samples have clustered nicely below VTH and the vast majority of the forgeries have moved above VTH with excellent separation. This fact is reflected on SR curve where the peak around the equal two-type error point is almost flat.
- The preceding two points reflects the high discrimination power of 3D RKI and AMT-based features.

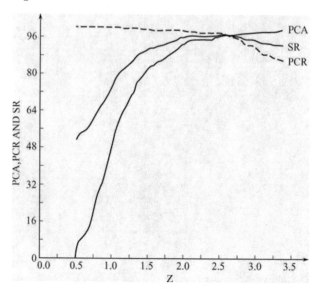

Fig. 26.25 SR, PCA and PCR curves of local MA and MM.

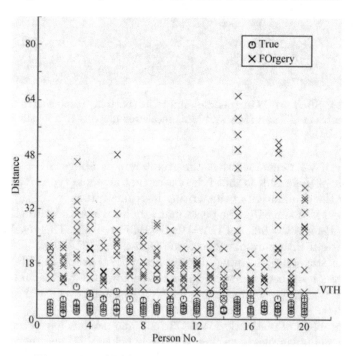

Fig. 26.26 (X,O) graph corresponding to Fig. 26.25.

5) Global and local AMT-based features. The performance of 14 AMT-based features (2 global and 12 local) is SR=96.50, PCA=95.00, PCR=98.00

(entry 3, 14). This is also good that PCR has improved 0.5% at the cost of PCA. This means that using both global and local AMT-based features gives the best forgery detection.

6) AMT-based features and the coordinates of the GC. These are all the features used to generate the result matrices in Fig. 26.24 (16 features, entry (1,16)). Here, we find also another interesting result: SR has improved 0.25 to become 96.75, PCA became perfect, (PCA=100), however, at some cost of PCR which fall down to 93.50. As a conclusion, augmenting AMT-based features by coordinates of GC made the performance PCA-biased. However, such result could be preferred in the practical reality when the convenience of the customer has the highest priority. This could be possible for low-value checks.

7) Ignored high performance entries. The three entries (1, 13), (1, 14), and (1, 15) in Fig. 26.24 gave SR=97.25, however, these entries consider mismatching on some sixths, and exclude others, therefore this result is ignored.

26.10.3 RKI from HPRs Signature Images

In order to study the efficiency of using AMT with RKI built from HPRs images, the same procedure used with binary and thinned images was repeated but using HPRs images, and SR result matrix of the same features used in 26.10.1 with HPRs images was computed and shown in Fig. 26.27. RKI of HPRs of the same person of Fig. 26.14 is shown in Fig. 26.28 displayed as level-contours, and (X, O) graph of the feature set of the entry (4, 6) is shown in Fig. 26.29.

	1	2	3	4	5	6	7	8	9	10	11	12	13	14	15	16
1	78.75	85.25	93.00	94.75	95.00	95.50	95.50	95.75	96.75	96.00	95.75	94.75	93.75	93.50	93.00	91.75
2	80.50	91.50	94.25	95.25	94.50	95.00	96.25	96.00	96.00	95.00	94.25	93.75	93.50	92.75	91.75	91.75
3	86.00	90.75	93.00	93.75	93.25	95.00	94.50	95.00	93.75	92.75	92.25	92.00	91.00	90.25	91.50	91.75
4	81.50	90.75	92.00	92.50	95.00	95.00	95.50	93.50	92.25	91.75	92.00	91.00	90.00	90.50	91.50	91.75
5	82.75	89.25	89.50	92.50	92.75	92.75	90.75	89.25	88.25	88.50	87.75	87.00	87.75	89.75	90.75	91.75
6	80.25	85.25	90.00	91.00	91.50	90.00	87.50	86.50	87.00	86.50	86.00	87.25	88.25	90.50	91.00	91.75
7	78.00	88.50	90.75	90.75	89.00	86.50	86.00	84.25	85.00	84.75	86.00	89.00	88.75	90.75	91.50	91.75
8	82.75	88.25	89.00	86.50	85.00	84.75	85.25	84.25	83.75	84.75	86.75	88.00	89.75	90.50	91.25	91.75
9	77.00	82.00	79.75	78.50	78.00	78.50	77.25	76.25	80.75	84.25	87.25	88.75	89.50	90.75	91.00	91.75
10	71.25	69.50	69.25	70.50	71.75	70.75	69.75	77.00	81.25	86.75	87.25	89.00	89.50	90.50	91.25	91.75
11	61.50	64.75	69.75	70.75	70.75	69.75	76.00	81.00	86.75	87.50	89.00	89.50	90.50	91.50	92.00	91.75
12	64.75	70.50	70.75	70.25	70.00	77.00	82.50	87.50	88.50	90.00	90.75	91.00	92.00	92.50	92.25	91.75
13	66.00	70.00	70.25	69.00	77.75	83.75	88.00	90.00	91.25	92.00	92.25	93.00	93.50	93.00	93.00	91.75
14	58.50	61.25	67.75	78.00	84.00	89.75	91.50	93.00	93.25	93.50	94.00	94.75	94.25	93.75	92.75	91.75
15	62.00	67.50	78.25	84.25	90.50	91.50	93.00	93.25	93.50	94.00	95.00	94.50	93.75	93.00	92.00	91.75
16	67.50	80.75	86.00	92.00	92.75	94.25	94.50	95.00	95.00	96.25	95.25	94.75	93.50	92.50	92.25	91.75

Fig. 26.27 The result matrix of X and Y coordinates of GC and(MA and MM) measured globally and locally on HPRs signature sixths.

SR result matrix of Fig. 26.27 reveals the following facts:

1) The best performance that can be obtained from AMT with HPRS RKI is by using MA measured on the signature sixths with SR=95.00 (6 features of entry (4, 6), marked by ellipse).

2) MM features are not effective at all with this kind of images, since the performance deteriorates when used.

Fig. 26.28 RKI of HPRs of the same person of Fig. 26.14, displayed as contours.

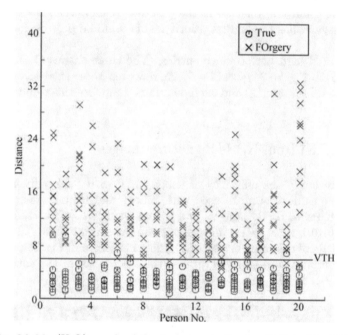

Fig. 26.29 (X,O) graph of the feature set of entry (4,6) in Fig. 26.27.

26.11 Limited Results are Shown and Discussed

As can be seen from the previous presentation through using binary, thinned
and HPRs images with the circulant matrix-based FST used, the possible
experimental cases and combinations that can be shown and discussed are
huge. Therefore, some limited but representative cases were selected and
discussed. They are believed to be indicative and give a very good insight
into the work done using the introduced method in off-line ASV field. The
improvement that can be provided by AMT and 3D knowledge representation
is seen to be outstanding in comparison with other methods.

26.12 AMT Features and Signature Recognition

Experimental results on using AMT-based features in signature recognition using minimum distance classifier gave excellent results.

26.13 AMT and Closely Related Works

Since the idea of 2D RPBFs for signature verification appeared in 1992 [16], there have been a few works used this idea. Table 26.1 summarizes these works.

Table 26.1   AMT and closely related works

Author	Reference	Signature representation used
Ammar M., 1991	IJPRAI [15]	One-dimensional representation with the conclusion that 2D representation must be more effective in ASV.
Ammar M., 1992	11ICPR [16]	2D RPBFs using AMT with multi-valued 2D reference pattern for representation: 2D *multi − valued* Reference Image displayed as contours.
Sabourin R. et al., 1997	PAMI[17].	2D representation from binary skeletons of signatures. 2D *binary* pattern.
Parker J. R., 2002	Vision Interface 2002 [18,19]	2D image representation from binary skeletons. 50 signatures aligned and centered from data tablet.
Ammar M., 2010 [this paper]	this book	3D representation from binary, HPRs, and thinned signature images. *3D reference pattern* (same as multi-valued but displayed as 3D instead of contours)

As Table 26.1 and the details of papers shown, while Ammar M. in 1992 (11ICPR) computed the distance measure of the test samples (skilled forgeries) using the Euclidean distance from matching and mismatching features extracted using AMT and a multi-valued reference image, Sabourin R. et al., 1997 [17] used binary 2D representation to extract local features and tested the performance using random forgeries, and Parker R. J. used the same representation but on-line signatures collected via a tablet, and used a simple distance measure for signature recognition.

Concerning segmentation to capture the local behavior of the signing process, Sabourin R. et al. used the grid, while Ammar M. used segmentation of the signature in the horizontal direction based on GC as a basis [16]. This way is believed to be more stable, and it captures efficiently the local behavior, and exploit its effectiveness in verification as the experimental results have shown.

Concerning the weights given to mismatching pixels, Parker gave an arbitrary weight (penalty weight = 5), while they are given in this work a weight equals to the number of genuine signing processes they missed.

This is a brief comparison between 3 main works dealt with 2D representation of the signature as a priori stage to feature extraction. This representation was first proposed by the author in 1992 [16]. The work for each of those 3 ones has its specialty. The reader can benefit from each one according to his/her interest.

26.14 Transition from Research to Prototyping then Pilot Project and Actual Use

Transition between these stages (research, prototyping, actual environment pilot project, and actual use system) that the author passed until a commercial ASV system serving many banks in the USA and others in different countries have been a reality [21], was not easy. At each transition between these stages, one might face a surprise, according to the difference in data nature and work environment. For example, from research to prototyping, the data changed from fixed writing format to free format, from one language to several languages of signature writing [22], from a main frame computer to a Personal Computer, and so on. Figure 26.30 shows examples of the free format RKIs from the prototyping signature data. The biggest surprise in this transition was the decrease in the performance of AMT-based features by about 10% in comparison with the research stage due to the nature of the data. For the next stage (Pilot Project with actual banks data), the author had to introduce the Spiral AMT.

In the transition from prototyping to actual environment pilot project, we had to change from clean signature background to actual checks background including all possible symbols, background print and lines, and unstable writing format. From gray-level images to only binary data, and finally, sometimes to badly scanned checks. Concerning reference data, we had to deal with sin-

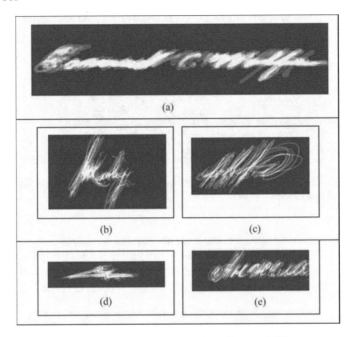

Fig. 26.30 Five different free format RKIs.

gle reference signature instead of the one-genuine-class used in research and
prototyping stages with all what this means of drastic changes in the ASV
process.

During the transition from pilot project to actual use, we faced different situations in which one have to respond appropriately to the customer
feedback and deal wisely with his complains. Some times, there was a necessity to change completely a processing method. As an example, although
the preprocessing stages presented in this chapter are 25 years old, they were
the basis for the highly sophisticated procedures used for signature extraction from check background. They do not lose importance, especially for
new researchers. Figure 26.31 shows examples of the check images and the
automatically extracted signatures. If good extraction fails, no system may
work at all, whatever the feature extraction or classification approaches are
efficient.

Another example of actual use problems is shown in Fig. 26.32 which
contains two single reference signature verification cases where "s" stands for
"genuine", and "t" stands for "test". These two cases complain one problem
(False Rejection). This is not a dangerous decision but may be inconvenient.
However, many customers liked it (pleased to be asked whether they signed
that specific check or not, which means that the bank takes care of their
checks, especially low value ones). Such decision should be avoided by further
development and investigations. As an idea of the amount of false rejections

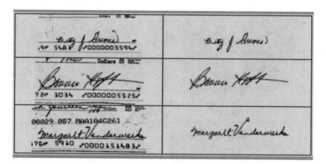

Fig. 26.31 An example of *Binary* Signature Areas from Checks and Automatically Extracted Signatures.

in our commercial system, through actual use of about 10 years, we kept false acceptance=0, while false rejections are within 10% $(5-10; PCR = 90-95)$ depending on the daily batch. Number of signatures investigated in batches are in the range $(3\,000-50\,000)$ depending on the bank of providing the batch.

116s	116t	116as	116at

Fig. 26.32 Two examples of actual single reference signature cases. Both test samples are false rejections.

In fact, the above cases are two examples from many cases that appear among the questioned cases and come out every day from about $2\,000\,000$ checks pass the system in all its working copies.

One final comment about the transition to actual use: Until now, there is no one method for ASV that gives the really desired performance. Several methods are used together in order to reach the final decision about a signature [22, 23].

26.15 Conclusions

This chapter has introduced the motivation and details of a new method for representing the knowledge of off-line signatures as a 3D reference knowledge image based on the forensic document examiner knowledge acquisition model of a signature. The details of AMT and its using of 3D knowledge were explained in detail. Their efficiency was explored through extensive experiments using binary, thinned and HPRs images, at the global and local levels. The results have shown that the improvement in performance is huge in comparison with parametric and with one-dimensional RPBFs. A comparison of AMT and 3D knowledge in signature verification was made with the closely related works in this field. Finally, an overview of the transition process between research, prototyping, pilot project, and actual use stages in building an actual ASV system, was summarized.

Acknowledgements

A part of this research was prepared during the sabbatical period of Professor Ammar M. at Watanabe-Lab, Nagoya University according to Damascus University delegation decision NO. 5301/DC for a scientific research period (27/10/2009 – 1/3/2010).

References

[1]   Ammar M, (2003) Application of Artificial Intelligence and Computer Vision Techniques to Signatory Recognition. Pakistan Journal of Information and Technology, 2(1): 44–51
[2]   Nagel R N, Rosenfeld A (1997) Computer Detection of Freehand Forgeries. IEEE Transacations on Computers, 26(9): 895–905
[3]   Ammar M, Yoshida Y, Fukumura T (1986) A New Effective Approach for Automatic Off-line Verification of Signatures by Using Pressure Features, Proceedings of the 8ICPR, Paris, pp 566–569
[4]   Ammar M, Yoshida Y, Fukumura T (1986) Automatic Off-line Verification of Signatures Based on Pressure Features. IEEE Trans on Systems Man and Cybernetics, 16(3): 39–47
[5]   Ammar M, Yoshida Y, Fukumura T (1989) Feature Extraction and Selection for simulated Signature Verification, Plamondon R, Suen C Y, Simner M L(edS), Computer Recognition and Human Production of Handwriting(1989), World scientific Publishing, pp 61–76
[6]   Ammar M, Yoshida Y, Fukumura T (1989) Off-line Preprocessing and Verification of Signatures. International Journal of Pattern Recognition and Artificial Intelligence, 2(4): 589–602
[7]   Ammar M (2002) Method and apparatus for verification of signatures. United States Patent: No 6424728, 07/23/2002.
[8]   Deng S P, Liao H Y, Ho C et al (1999) Wavelet-based Off-line Handwritten Signature Verification. Computer Vision Image Understanding, 76: 173–190
[9]   Almudena G, et al (2008) Off-line Signature Verification Using Contour Features. In Proceedings, ICFHR, pp 19–21
[10] Alessandro Z, Lee L L. (2003) A Hybrid On/Off Line Handwritten Signature Verification System, Proc 7th 7ICDAR, vol 1: 424–429
[11] Jesus F, Bonilla V, Miguel A et al (2009) Off-line signature Verification Based on Pseudo-Capstral Coefficients. 10th ICDAR, pp 126–130
[12] Larkin L (2009) Off-line Signature Verification, Doctor Thesis, University of Waikato.
[13] Ammar M (1990) Performance of Parametric and Reference Pattern Based Features in Static Signature Verification: A Comparative Study. In IEEE Proceedings of the 10th International Conference on Pattern Recognition. (Atlantic City, New Jersey, USA), IEEEE Computer Society Press, 646–649
[14] Katsuhiko U (2003) Investigation of Off-line Japanese Signatures Verification Using Pattern Matching, 7th ICDAR, vol 2, p 951
[15] Ammar M (1991) Progress in Verification of Skillfully Simulated Handwritten Signatures. International Journal of Pattern Recognition and Artificial Intelligence, 5(1&2): 337–351
[16] Ammar M (1992) Elimination of Skilled Forgeries in Off-line Systems: A Breakthrough, Proceedings, 11ICPR, the Netherlands, pp 415–418, Sept 1992

[17]   Sabourin R et al (1997) Off-line Signature Verification by Local Granulometric Size
Distributions, PAMI, 19(9): 976 – 988
[18]   Parker J R (2002) Simple Distances Between Handwritten Signatures. Vision Interface, pp 218 – 223, Calgary, Alberta, 27 – 29 May 2002
[19]   Parker J R (2007) Composite Systems for Handwritten Signature Recognition. In:
Yanushkevich S N, Gavrilova M L, Wang P S P (eds) Image Pattern Recognition:
Synthesis and Analysis in Biometrics, pp 159 – 182, WSP
[20]   Ammar M (1989) Signature Verification and Description. Doctoral dissertation,
Nagoya University, Nagoya, Japan, 1989.
[21]   http://www.asvtechnologies.com, Accessed 1 October 2010
[22]   Ammar M (2011) Raising the Performance of Automatic Signature Verification above
that Obtainable by the best Feature Set. IJPRAI. IJPRAI, 25(2): 183 – 206
[23]   Ammar M (2010) Using Multisets of Features and Interactive Feature Selection to
Get the Best Qualitative Performance for Automatic Signature Verification (this
book)
[24]   Ammar M (2011) Intelligent Signature Verification and Analysis, Lambert Academic
Publishing

27 Unified Entropy Theory and Maximum Discrimination on Pattern Recognition

Xiaoqing Ding[1]

Abstract Pattern Recognition is a very urgent research area in intelligent information processing and computer intelligent perception, such as computer vision, content-based retrieval, etc. In general, the research on pattern recognition is carried out partial separately as feature extraction, classification, etc. in which the global optimum could not been achieved. In this chapter the unified entropy theory on Pattern Recognition is presented firstly, in which the information procedures in learning and recognition and the determine role of Mutual Information have been discovered. Secondly the Maximum MI Discrimination-based subspace pattern recognition is presented to get optimum recognition performance, which is crucial for difficult pattern recognition problems. Experiments on handwritten character recognition prove their effective and efficient.

27.1 Introduction

Pattern recognition is important and growing fast in intelligent information processing, and efficiently and effectively affect today's progresses on computer vision, biometrics, video surveillance, etc. because Pattern Recognition is the fundamental of intelligent activities.

There are two important problems in pattern recognition: one is classifier design, and the other is feature extraction/selection. More papers have been published for classifier design and for feature extraction independently, but few for the relation between them, i.e., few from whole learning and recognition. More papers published for the research on this area partial separately, such as machine learning, classifier design, and feature selection, post processing, etc., but few papers to studied feature extraction/selection globally together with classifier design, or with learning and recognition. In fact, pattern recognition is close relative with both feature extraction/selection and classifier design, and with both learning procedure and recognition procedure. The research on pattern recognition on the global and relative way rather is a very urgent problem.

Designing a good performance recognition system is still an urgent issue in pattern recognition research. The researchers know that the classifier design

---
1   Dept. of Electronic Engineering, Tsinghua University Beijing, 100084, China. E-mail: dingxq@tsinghua.edu.cn.

and the feature extraction are the most important, but they need to know
how the recognition performance is affected by both feature and classifier.

In this chapter, first, we extend the Information Theory into Pattern
Recognition area. An unified entropy theory of Pattern Recognition is represented, in which pattern recognition can be described by an information
entropy procedure. Also the mutual information generated from learning procedure and depended on the feature extraction is crucial affecting and determining the recognition performance. Second, based on the maximum mutual
information principle, a MI Discrimination-based Sub-space recognition approach is proposed, which is crucial for get optimum recognition performance
and is an urgent step for solving difficult pattern recognition problems. Experiments on handwritten character recognition prove the effective and efficient
to achieve the excellent pattern recognition.

Even now more statistical learning theory and algorithms have been studied and made more progresses, which are more direct based on the samples
learning for pattern classification, such as SVM, Adaboosting, etc. The statistical learning theory studies the learning methods for classifier design, avoiding the most difficult probability distribution estimation problems in general
pattern recognition model. Even though, there still could not regardless the
existence of various probability distributions of random variables. Therefore,
the unified information entropy theory is useful to describe the whole pattern
recognition procedure from information point of view, including learning and
recognition procedure, the feature extraction relative with classifier performance, etc. Despite of various learning methods, the entropy analysis and
unified Entropy frame is useful for deeper analysis and comprehensive understanding of pattern recognition.

Because mutual information is the discriminate entropy for recognition,
a maximum MI discrimination Subspace Pattern Recognition approach is
also proposed, which will reduce the indiscriminate components and noise
to achieve the good recognition performance for difficult pattern recognition
problems.

This chapter is organized as follows: Section 27.2 introduces proposed
Unified Entropy Theory in Pattern Recognition; Section 27.3 discusses the
Discriminate Entropy in Pattern Recognition; Section 27.4 discusses MI Discrimination Analysis; Section 27.5. introduces Maximum MI Principle; Section 27.6 describes Maximum MI Discrimination Sub-space Recognition in
handwritten Chinese character recognition; and Section 27.7 is the conclusion

27.2 Unified Entropy Theory in Pattern Recognition

Recognition is a procedure to determine the category of an unknown testing
sample based on some known category samples, which can be described by

an information entropy procedure. The information entropy system of pattern recognition is composed by feature entropy $H\left(F\right)$, system ntropy $H\left(E\right)$, conditional entropy $H\left(F|E\right)$, a posterior entropy $H\left(E|F\right)$, and mutual information $I\left(F,E\right)$, which described in the Appendix.

The unified entropy procedure including:

1) Learning information procedure, $I\left(F,E\right) = H\left(F\right) - H\left(F|E\right)$,

2) Recognition information procedure, $H\left(E|F\right) = H\left(E\right) - I\left(E,F\right)$,

in which a whole information procedure happens in pattern recognition

27.2.1 Information Procedures in Recognition

Pattern recognition is to identify the category or index of an unknown sample from a category probability space $\Omega = \left(\omega_1,\ldots,\omega_n; P\left(\omega_i\right),\ldots,P\left(\omega_n\right)\right)$, which has preliminary category uncertainty described by the system entropy

$$H\left(E\right) = -\sum_{i=1}^{n} P\left(\omega_i\right) \log P\left(\omega_i\right), \quad \sum_{i=1}^{n} P\left(\omega_i\right) = 1.$$

For example, for Chinese character recognition, H(E)=12–16 bits.

When the feature X is extracted from sample in the feature probability space,

$$F = \left(X; P\left(X\right)\right).$$

Therefore, for the total sample set, the sample feature matrix could be represented as

$$\boldsymbol{X} = \left[X_1, X_2, \cdots, X_L\right]. \tag{27.1}$$

The probability distribution of sample feature X could be estimated from more samples, but it is a very difficult problem forever.

The mean feature vector and the feature covariance matrix can be estimated by the training samples.

The mean feature vector is estimated as

$$\hat{M} = \frac{1}{L}\sum_{i=1}^{L} X_i.$$

Then the feature covariance matrix can be estimated as

$$\hat{\Sigma}_X = \frac{1}{L}\boldsymbol{X}\boldsymbol{X}^{\mathrm{T}} - \hat{M}\hat{M}^{\mathrm{T}}. \tag{27.2}$$

Provided that the feature probability density $p\left(X\right)$ is a Gaussian distribution, then $p\left(X\right)$ could be estimated by the mean feature vector and the feature covariance matrix only:

$$p\left(X\right) = \frac{1}{\left(2\pi\right)^{\frac{N}{2}}\left|\hat{\Sigma}_X\right|^{\frac{1}{2}}} \exp\left\{-\frac{1}{2}\left(X - \hat{M}\right)^{\mathrm{T}} \hat{\Sigma}_X^{-1}\left(X - \hat{M}\right)\right\}.$$

Based on the feature probability density $p(X)$, the feature differential entropy $h(F)$ could be calculated as

$$\max_{\Sigma_X = E\{XX^T\}} h(F) = \frac{1}{2} \log (2\pi e)^N |\Sigma_X|, \quad \text{with equality iff } X \in N(M_X, \Sigma_X).$$

And the feature entropy will be

$$H(F) = \frac{1}{2\ln 2} \left[\ln (2e\pi)^N + \ln \left| \hat{\Sigma}_X \right| \right].$$

Learning information procedure

In the training procedure, the features probability distribution and feature conditional probability distribution could be estimated from training samples, then both feature entropy $H(F)$ and category conditional entropy $H(F|E)$ are obtained. The leaning entropy reduction $H(F) - H(F|E)$, and the same as the mutual information $I(F, E) = H(F) - F(F|E)$ will be obtained too.

Based on the training samples, the ith category training sample set is represented as a sample matrix

$$\boldsymbol{X}_i = [X_{i1}, X_{i2}, \cdots, X_{iL}], \quad i = 1, 2, \cdots, n. \tag{27.3}$$

Then the ith category mean vector and its covariance matrix can be estimated as

$$\hat{M}_i = \frac{1}{L_i} \sum_{j=1}^{L_i} X_{ij}, \quad i = 1, 2, \cdots, n.$$

$$\hat{\Sigma}_i = \frac{1}{L} \boldsymbol{X}_i \boldsymbol{X}_i^T - M_i M_i^T, \quad i = 1, 2, \cdots, n. \tag{27.4}$$

Provided that ith class-conditional probability density $p(X|\omega_i)$ $(i = 1, 2, \cdots, n)$ is a Gaussian distribution, it will be represented as

$$p(X|\omega_i) = \frac{1}{(2\pi)^{\frac{N}{2}} \left| \hat{\Sigma}_i \right|^{\frac{1}{2}}} \exp \left\{ -\frac{1}{2} \left(X - \hat{M}_i \right)^T \hat{\Sigma}_i^{-1} \left(X - \hat{M}_i \right) \right\},$$

$$i = 1, \cdots, N. \tag{27.5}$$

The class-conditional feature entropy $H(F|E)$ will be calculated as

$$H(F|E) = \sum_{i=1}^{n} P(\omega_i) H(F|\omega_i) \leqslant \frac{1}{2\ln 2} \left[\ln (2e\pi)^N + \sum_{i=1}^{n} P(\omega_i) \ln \left| \hat{\Sigma}_i \right| \right]$$

with equality iff

$$X \in N(M_X, \Sigma_X) \tag{27.6}$$

Therefore, the learning entropy reduction or the mutual information in feature space of Gaussian feature will be

$$
I\left(F, E\right) = H\left(F\right) - H\left(F \mid E\right) = \frac{1}{2\ln 2}\left[\ln\left|\Sigma_X\right| - \sum_{i=1}^{n} P\left(\omega_i\right)\ln\left|\Sigma_i\right|\right]. \quad (27.7)
$$

The learning entropy reduction in feature space obtained from machine learning represents the acquired information from training samples, and it presents the relation between feature and category regardless the learning methods, even it is a statistical parameter estimation of probability distributions, or any statistical learning methods.

The smaller the Class Feature Entropy $H\left(F \mid E\right)$ is, the bigger the learning entropy reduction $I(F, E)$ will be. The class feature entropy $H\left(F \mid E\right)$ represents the feature variation about the category, or the feature instability of the pattern, which is injuring information for recognition, and will weaken the feature recognition ability.

In general, the information gotten from any machine learning algorithm is the learning entropy reduction $I(F, E)$, and is equal to the mutual information in feature space, which could be estimated from learning procedure.

Recognition information procedure

In the recognition procedure, the posteriori entropy $H\left(E \mid F\right)$ and the cognition entropy reduction $H\left(E\right) - H\left(E \mid F\right)$ will happen from the features of an unknown category samples, which equals to the mutual information $I(F, E)$.

In pattern recognition the mutual information is acquired from learning procedure and transferred to the recognition procedure to reduce the category uncertainty.

A different classifier $G_i\left(X\right)$ is used to determine the category of unknown testing sample based on its feature. By the category recognition, the category uncertainty of recognition system could be reduced from the system information entropy $H(E)$ to the smaller known feature survivals information entropy, that should be the posterior entropy $H\left(E \mid F\right)$.

If the classifier is the optimum Bayes classifier, that is

$$
\omega_i = \arg\max_i P\left(\frac{\omega_i}{X}\right). \quad (27.8)
$$

In this case, it will be proved in Section 27.2, the recognition error will satisfy

$$
Pe \leqslant \frac{1}{2}H\left(E \mid F\right). \quad (27.9)
$$

That is to show that $H\left(E \mid F\right)$ represents the smallest survivals information entropy and close related with the recognition error, as the upper limit of error by the best recognition Bayes classifier. Therefore, the uncertainty of system has been reduced from the system entropy $H(E)$ to the reserved posteriori system entropy $H(E|F)$, so that the recognition entropy reduction $I(E, F)$ in the category space is

$$I(E, F) = H(E) - H(E|F). \qquad (27.10)$$

And it also equals to the mutual information $I(F, E)$.

Obviously, the bigger $I(E, F)$ is, the smaller $H(E|F)$ and P_e will be, and the better recognition will be achieved.

The pattern recognition entropy procedure can be described by Fig. 27.1

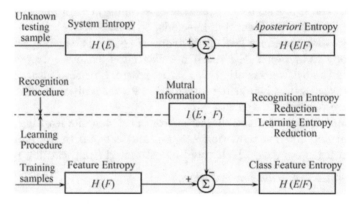

Fig. 27.1 Unified entropy procedure diagram on pattern recognition.

It should be noted that the mutual Information is transferred from mutual information in feature space into mutual information in class space when the cognition procedure happens.

27.2.2 A Posterior Entropy Upper limited Bayes Classifier Error

It's well known that the error P_e of a posterior Bayes classifier is

$$P_e = E_X \left\{ 1 - \max_i P\left(\frac{\omega_i}{X}\right) \right\} = 1 - E_X \left\{ \max_i P\left(\frac{\omega_i}{X}\right) \right\}, \qquad (27.11)$$

there is an inequality

$$1 - \max_i P\left(\frac{\omega_i}{X}\right) \leqslant 1 - \sum_{i=1}^{n} P^2\left(\frac{\omega_i}{X}\right), \qquad (27.12)$$

because $\sum_{i=1}^{n} P^2\left(\frac{\omega_i}{X}\right) \leqslant \left[\max_j P\left(\frac{\omega_i}{X}\right)\right] \left[\sum_{i=1}^{n} P\left(\frac{\omega_i}{X}\right)\right] = \max_j P\left(\frac{\omega_i}{X}\right)$

also $1 - \sum_{i=1}^{n} P^2\left(\frac{\omega_i}{X}\right) = \sum_{i=1}^{n} P\left(\frac{\omega_i}{X}\right) \left[1 - P\left(\frac{\omega_i}{X}\right)\right]$

$$\leqslant -\frac{1}{2}\sum_{i=1}^{n} P\left(\frac{\omega_i}{X}\right) \log P\left(\frac{\omega_i}{X}\right),$$

$$1 - \sum_{i=1}^{n} P^2\left(\frac{\omega_i}{X}\right) \leqslant -\frac{1}{2}\sum_{i=1}^{n} P\left(\frac{\omega_i}{X}\right) \log P\left(\frac{\omega_i}{X}\right).$$

Take the expectation in both sides of the above inequality,

$$Pe \leqslant \frac{1}{2} H\left(E|F\right) \text{ or } P_e \leqslant \frac{1}{2}[H(E) - I(E, F)]. \tag{27.13}$$

It claims that the a *posteriori* entropy $H\left(E|F\right)$ represents the upper limit of error rate P_e of the Bayes classifier.

If required, the error P_e will be smaller, then the a *posteriori* entropy $H\left(E|F\right)$ should be near to zero.

For a practical classifier, the classifier error rate ε is composed of two parts, that is

$$\varepsilon = P_e + P_{\bar{e}}, \tag{27.14}$$

here $P_{\bar{e}}$ represents classifier error caused by apart from the Bayes classifier, which often happen in practice.

27.3 Mutual-Information — Discriminate Entropy in Pattern Recognition

From the analysis above, we can see that the learning entrpy reduction and the recognition entropy reduction are both eiqual to the relative entrop mutual information (or Kullback leibler distance) definated on the intersection space $E \cap F$, which is called the discriminate entropy because it definitely determines the recognition performance.

$$I\left(F, E\right) = H\left(F\right) - H\left(F|E\right)$$
$$= \sum_{i=1}^{n} \int_{R^N} p\left(X, \omega_i\right) \log_2 \frac{p\left(X|\omega_i\right)}{p\left(X\right)} dX, \tag{27.15}$$

$$I\left(E, F\right) = H\left(E\right) - H\left(E|F\right)$$
$$= \sum_{i=1}^{n} \int_{R^N} p\left(X, \omega_i\right) \log_2 \frac{p\left(\omega_i|X\right)}{P\left(\omega_i\right)} dX, \tag{27.16}$$

$$I\left(F, E\right) = I\left(E, F\right) = \sum_{i=1}^{n} \int_{R^N} p\left(X, \omega_i\right) \log_2 \frac{p\left(X, \omega_i\right)}{P\left(\omega_i\right) p\left(X\right)} dX, \tag{27.17}$$

1) The mutual information determines the error upperlimit of optimal Bayes classifier, $P_e \leqslant \frac{1}{2}\left[H(E) - H\left(E|F\right)\right]$, i.e., determines the optimal

classification performance. The bigger the mutual information is, the better recognition performance will be.

2) The key and decision factor for pattern recognition is to extract better features, which should be as closer as possible related with classified categories and bring higher mutual information. It is absolutely necessary for successful pattern recognition.

3) Mutual information represents the discriminate ability of features, therefore, which can be used as the mearsure for feature selection, i.e.,

$$I\left(F, E\right) = H\left(F\right) - H\left(F \mid E\right).$$

Theorem 27.1   The mutual information is unchanged after reviseable feature linear transformation.

If $Y = f\left(X\right)$ is an one to one corresponded linear transformation, the relation between two probability density function is

$$p_Y\left(Y\right) \mathrm{d}Y = p_X\left(X\right) \mathrm{d}X \quad \text{or} \quad p_Y\left(Y\right) = p_X\left(X\right)/|J|,$$

here $|J|$ is the Jacobin determine.

Then the mutual imformation for transformed random features Y and the mutual imformation for original random features X are related equially each other by

$$I_Y\left(F, E\right) = I_X\left(F, E\right). \tag{27.18}$$

It is good for the analysis of mutual information regardless any unsingular linear transforms processed on the features.

Theorem 27.2   The mutual information of a high dimensional features is equal to the sum of the mutual Informations of its separated complementary subspaces.

If a feature vector X in N-dimensional feature space could be separated as the sum which are in d-dimensional feature subspace F_d and in its complementary $N - d$ dimensional feature subspace F_{N-d},

$$F_d \cup F_{N-d} = F, \quad F_d \cap F_{N-d} = \phi.$$

Then it can be drived:

$$I\left(F, E\right) = I\left(F_d, E\right) + I\left(F_{N-d}, E\right). \tag{27.19}$$

4) The bigger mutual information can be gotten in different way, such as in integrated features as a whole vector, or in discrete multiple steps like Adaboost. By multiple weaker features and its classifiers with smaller mutual information, the whole bigger feature mutual information and better performance of classifier could be obtained at the same time.

5) Summary for mutual information (MI in short for mutual information)

(1) MI represents the correlation between the class space E and the feature space F,

$$I\left(F, E\right) = H\left(F\right) + H\left(E\right) - H\left(F, E\right).$$

(2) MI also represents the learning entropy reduction and transferred as the recognition entropy reduction,

(3) MI determines the Bayes recognition error upper limit.

(4) MI could be as the optimum measure of feature extracted/selected, and as the guide for feature extraction and selection.

The maximum mutual information principle will be proposed as the best principle for feature selection or feature dimension compression in recognition, because the bigger mutual information is absolute nessasary to get the better recognition performance.

When the feature probabilities are Gaussian distributions, maximum mutual information principle can be presented by MI discrimination analysis in pattern recognition

27.4 Mutual Information Discrimination Analysis in Pattern Recognition

When an optimum feature with the high mutual information has been extracted/selected, there has a strong possibility to get higher recognition performance, but how to realize/implement an approached optimul Bayes classifier is still a seriouse and challenge problem.

In some practical pattern recognition problems, the extracted high dimensional features are involved with broad noises. When an Euclidean distance is used to measure the similarity between the patterns,

$$\bar{\varepsilon}^2 = E\left\{\sum_{i=1}^{N}\left\{(y_i - b_i)^2\right\}\right\} = \sum_{i=1}^{m} E\left\{(y_i - b_i)^2\right\} + \sum_{i=m+1}^{N} E\left\{(y_i - b_i)^2\right\}.$$

If the white noise is often appeared:

$$y_i \Rightarrow (y_i + n_i),$$

so

$$\bar{\varepsilon}^2 = E\left\{\sum_{i=1}^{N}\left\{((y_i + n_i) - b_i)^2\right\}\right\}$$

$$= \left[\sum_{i=1}^{N} E\left\{(y_i - b_i)^2\right\} + \sum_{i=1}^{N} E\left\{(n_i)^2\right\}\right]. \tag{27.20}$$

When feature demension has been increased too much, the distance will be more distroed by noisy components $\sum_{i=1}^{N} E\left\{(n_i)^2\right\}$, which even make the trure pattern matching hardly be obtained.

Therefore, except "the curse of dimension" problem caused by the lack of training samples, the better the feature dimension should also be reduced.

The best way of feature dimension reduction is to delete some feature components, but still reserve the biggest mutual information at the same reduced dimensions. Due to so, a maximum mutual information discrimination

subspace recognition algorithm is proposed and has shown to realize the best
performance of recognition.

Mutual Information Discrimination Analysis

In order to realize the feature dimension reduction based on maximum
mutual information principle, the mutual information discrimination analysis
is proposed as following:

Definition 27.10   Because the mutual information of a homogeneous
Gaussian distribution is equal to

$$I\left(F, E\right) = \log_2 \left(\frac{|S_w + S_b|}{|S_w|}\right)^{1/2},$$

the mutual information discriminate matrix is definited as

$$\frac{S_t}{S_w} = \frac{S_w + S_b}{S_w}, \tag{27.21}$$

where $\Sigma_x = S_t$ is the whole feature scatter matrix, S_w is the within-class
scatter matrix, and S_b is the between-classes scatter matrix.

Definition 27.11   The mutual information discrimination analysis is
defined by a transformation Φ, which makes the diagonalization of the mutual
information discriminate matrix S_t/S_w,

$$\left[\frac{S_t}{S_w}\right]\Phi = \Phi\Lambda. \tag{27.22}$$

That is $S_t\Phi = \Phi\Lambda$ and $S_w\Phi = \Phi I$

where $\Phi = [\varphi_1 \cdots \varphi_N]$ is a $N \times N$ eigenvector matrix of the mutual
information discriminate matrix S_t/S_w, and is consist of its N eigenvectors,
also,

$$\Lambda = \begin{bmatrix} \lambda_1 & 0 & 0 \\ 0 & \ddots & 0 \\ 0 & 0 & \lambda_N \end{bmatrix}$$

is the diagonal eigenvalue matrix of S_t/S_w.

Therefore, $$\left[\frac{S_t}{S_w}\right]\varphi_i = \lambda_i\varphi_i, \qquad i = 1, \cdots, N. \tag{27.23}$$

Let the N dimensional random feature vector $X = \begin{bmatrix} x_1 & \cdots & x_N \end{bmatrix}^{\mathrm{T}}$ is
transformed by the eigenvector matrix Φ^{T} of the MI discriminate matrix
S_t/S_w. That is

$$Y = \Phi^{\mathrm{T}}X, \quad Y = \begin{bmatrix} y_1 & \cdots & y_n \end{bmatrix}^{\mathrm{T}},,$$

here $$y_i = \varphi_i^{\mathrm{T}}X, \quad i = 1, \cdots, n. \tag{27.24}$$

We can find following important performances of the MI discrimination
analysis:

1) MI discrimination analysis is to diagonalize the MI discriminate matrix S_t/S_w as the diagonalized MI discriminate matrix $\Lambda = \mathrm{diag}(\lambda_1, \cdots, \lambda_N)$, λ_i $i = 1, \cdots, N$, which represents the discriminate components ordered from the bigger to the smaller.

2) After MI discrimination analysis by linear transformation Φ^{T}, the mutual information of feature is the same as before, and reserves constant.

3) Based on the MI discriminetion analysis, the mutal information MI can be calculated as

$$I\left(F, E\right) = \frac{1}{2} \log_2 \left| \frac{S_t}{S_w} \right| = \frac{1}{2} \sum_{i=1}^{N} \log_2 \lambda_i. \qquad (27.25)$$

4) Based on the MI discrimination analysis, if the smallest $N-d$ dimension components is deleted, which are corresponding to the smallest eigenvalues of MI discrimination matrix, then the reserved MI will be as bigger as possible as follows:

$$I_d\left(F, E\right) = \left\{ \frac{1}{2} \log_2 \left| \frac{S_t}{S_w} \right| \right\}_d = \frac{1}{2} \sum_{i=1}^{d} \log_2 \lambda_i.$$

It is proved that the maximum mutual information dimension reduction based on MI discrimination analysis is the best than any other dimension reduction method in pattern recognition.

27.5 Maximum MI Principle

As discussed in MI discrimination analysis, the maximum **MI** principle should satisfy the following conditions:

$$\max \sum_{i=1}^{d} e_i^{\mathrm{T}} \frac{S_t}{S_w} e_i \qquad (27.26)$$
$$constrain : e_i^{\mathrm{T}} e_i = c_i.$$

Using Lagrange multiplier,

$$u = \sum_{i=1}^{d} e_i^{\mathrm{T}} \frac{S_t}{S_w} e_i - \lambda \left(\sum_{i=1}^{d} (e_i^{\mathrm{T}} e_i - c_i) \right),$$

maximize MI, from

$$\frac{\partial u}{\partial e_i} = 2 \frac{S_t}{S_w} e_i - 2\lambda e_i = 0, \qquad i = 1, \cdots, d.$$

We can get
$$\frac{S_t}{S_w} e_i = \lambda e_i, \quad 1 \leqslant i \leqslant d, \qquad (27.27)$$

where e_i $1 \leqslant i \leqslant d$ are the vectors which correspond the biggest eigenvalues of the MI discriminate matrix S_t/S_w.

The subspace $\boldsymbol{W}_d = [\boldsymbol{e}_1 \quad \boldsymbol{e}_2 \quad \cdots \quad \boldsymbol{e}_d]$ is spanned by the eigenvectors that with the biggest d eigenvalues of the MI discriminate matrix S_t/S_w, and is called the maximum MI discriminate subspace.

Definition 27.4 The feature dimension reduction based on maximum MI principle is the optimization, which will be mapping the feature into $d-$dimensional sub-space spanned by d eigenvectors $\boldsymbol{W}_d = \{\boldsymbol{w}_1 \quad \boldsymbol{w}_2 \quad \cdots \quad \boldsymbol{w}_d\}$, which correspond with the biggest d eigenvalues of the MI discriminate matrix S_t/S_w

Therefore, based on the d-dimensional maximum MI discriminate subspace $\boldsymbol{W}_d$, the feature dimensional reduction is the maximum mutual Information reserved and it will be able to get the best recognition performance.

27.5.1 Maximum MI Discriminate Sub-space Pattern Recognition

Definition 27.4 Maximum MI principle is the feature mapping from N dimensional feature space into d dimensional maximum MI discrimination sub-space $\boldsymbol{W}_d$, which is spanned by d dimensional the biggest discriminate eigenvectors,

$$Y_d = \boldsymbol{W}_d^{\mathrm{T}} X,$$

where, $\boldsymbol{W}_d$ is $N \times d$ matrix,

$$\boldsymbol{W}_d = \{\boldsymbol{w}_1 \quad \boldsymbol{w}_2 \quad \cdots \quad \boldsymbol{w}_d, \} \tag{27.28}$$

and

$$\frac{S_t}{S_w} \boldsymbol{W}_d = \boldsymbol{W}_d \Lambda_d, \quad \Lambda_d = \begin{bmatrix} \lambda_1 & 0 & 0 \\ 0 & \ddots & 0 \\ 0 & 0 & \lambda_d \end{bmatrix}. \tag{27.29}$$

At this time, the mutual information is reduced by$I\left(F_{N-d}, E\right)$, because

$$I\left(F, E\right) = \frac{1}{2}\log_2\left|\frac{S_t}{S_w}\right| = \frac{1}{2}\sum_{i=1}^{N}\log_2\lambda_i = \frac{1}{2}\sum_{i=1}^{d}\log_2\lambda_i + \frac{1}{2}\sum_{i=d+1}^{N}\log_2\lambda_i$$

$$= I\left(F_d, E\right) + I\left(F_{N-d}, E\right). \tag{27.30}$$

Preserved mutual information $I\left(F_d, E\right)$ vs. total mutual information rate is

$$\eta = \sum_{i=1}^{d}\log_2\lambda_i \Big/ \sum_{i=1}^{N}\log_2\lambda_i. \tag{27.31}$$

Because reduced mutual information eigenvalues λ_i, $i = d+1, \cdots, N$ are the smallest, $I\left(F_{N-d}, E\right) = \frac{1}{2}\sum_{i=d+1}^{N}\log_2\lambda_i$ will be the smallest than any other $d-$dimension reduction algorithm. Therefore the dimension reduction in MI discriminate sub-space preserves the more of the mutual information than any other d-dimension reduction algorithm. When d is bigger enough, we can get $I(F,E) \approx I(F_d, E)$. That is to say, even the dimension has been reduced, it does not affect the discriminate ability seriously.

27.5.2 The Best Maximum MI Discrimination Sub-space dimension reduction

The dimension reduction is a crucial problem in pattern recognition, especially because of the "curse of dimension" problem. Compare with different dimensional reduction approaches, we get:

1) Dimensional reduction in original feature space. In this case, the mutual information for every original signal would be distributed equally. So dimension reduction may cause mutual information proportional reduced, and then the cognition error would be increased.

2) Dimensional reduction in PCA subspace. In this case the dimension reduction could be implemented by deleting the feature components with the smallest feature eigenvalues. So, the deleted feature components have the smallest variances. It is best for data compression, but may not be better for recognition, because compressed components may have the bigger mutual information.

3) Dimensional reduction in Maximum MI Discrimination Subspace. In this case, the dimension reduction is implemented by deleting the feature components with the smallest discriminate feature components with the smallest mutual information. Therefore, MI discrimination reduction has preserved the biggest mutual information to achieve the best cognition results than any other algorithm.

The conclusion can be gotten that maximum MI discriminate sub-space recognition brings us the biggest possibility to design better classifiers.

27.6 Maximum MI Discriminate SubSpace Recognition in Handwritten Chinese Character Recognition

Because there are large scale of pixel data in an image, more complex and bigger variance of image objects will be in the image that make more difficulties in object detection and recognition from image, such as character in nature scenery or face in complex background, etc. People will hardly to find the effective and efficient features to discriminate the different object patterns, especially for the patterns, which are embedded in the broader noisy or in pattern-useless background, for example, the character independent font recognition on a single Chinese character and so on.

For the Chinese character recognition with larger scale numbers (more than thousands) , at least the minimum mutual information 15 bits are necessery required. It has been proved in practice that Chinese character recognition was failed by the algorithms based on the features extracted from limited character strokes, because the low quantity of mutual information could be obtained only. But in contract, the recognitions have been succeeded by much higher dimentional features (more than hundreds) extracted from character textures with enough higher mutual information for Chinese character recognition. The same situations also have met in the face recognition by the features extracted from few face geometric parameters vs. extracted higher dimension of face texture features.

The problems, such as selection/extraction of optimum features with higher mutual information, reduction of the feature dimention and design of the optimum classifier, are still remained to get higher performance recognition.

Handwritten Chinese Character Recognition

Because of the large scale numbers (more than 4 000) of Chinese characters and bigger variances of handwritten Chinese characters, it makes handwritten Chinese character recognition to be one of the most difficult recognition problems. The essential ways for recognition is to choice and extract the features possessing enough higher quantity of multual information. It has been proved in TH-OCR practice that for some successful handwritten Chinese character recognition algorithms two kinds of available features, such as directional element feature DEF and gradient feature, have been extracted, they both have large enough of mutual information for recognition.

The mutual information of directional element feature DEF or gradient feature GF for handwritten Chinese character recognition are calculated from

$$I(E, F) = \frac{1}{2} \log_2 \frac{|S_t|}{|S_w|} = \frac{1}{2} \left(\sum_{i=1}^{N} \log_2 \lambda_{ti} - \sum_{i=1}^{N} \log_2 \lambda_{wi} \right), \qquad (27.32)$$

there the parameters of (S_t, S_w) are evaluated on a training sample group with 1800 handwritten Chinese set (4000 different Chinese characters per set) and the feature dimension $N=392$.

After the maximum MI feature discriminate analysis $Y_d = W_d^T X$, the features are mapped into the d-dimension discriminate subspace. The mutual information will be changed with the reduced dimension d as shown in Fig. 27.2.

We can see for the handwritten Chinese character recognition that there are big enough MI in both DEF and gradient features and MI of the DEF feature is little smaller than the MI of the gradient feature. It has been proved by experiments in Fig. 27.4, that the recognition preferment on the GF gradient features is better than the recognition preferment on the DEF features.

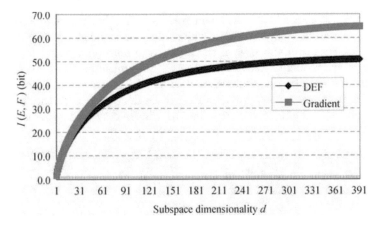

Fig. 27.2 Mutual information vs. subspace dimensionality in handwritten Chinese character recognition (feature dimensionality N=392, subspace dimensionality d varies from 1 to 392).

Fig. 27.4 shows that the MI of every discriminate component in the feature vector is dramatically attenuated with the feature sequence order increased. It is obvious that MI decreased with the increasing feature sequence order, and the truncation of the highest order of the discriminate components will delete the smallest in the whole feature MI.

$$I(E, F) = \sum_{i=1}^{N} l_i (E, F), \tag{27.33}$$

$$l_i (E, F) = \frac{1}{2} \log_2 \lambda_{Ti} - \log_2 \lambda_{Wi}, \quad i = 1, \cdots, N. \tag{27.34}$$

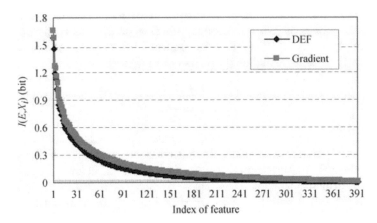

Fig. 27.3 Mutual information provided by single-dimensional feature.

Figure 27.4 shows the experiment results of handwritten Chinese character recognition on the test set, in which the handwritten Chinese character
recognition rate is changed with the reduced dimension d of the feature discriminated components. We can see that the recognition rate will be increased
with the increase of the dimension d, and will reach the maximum when
$d=182$. After that the recognition rate will be decreased with the increase of
the dimension d, at the same time the MI will be continuing increased.

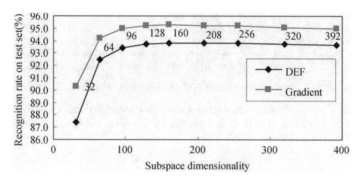

Fig. 27.4 Comparison of recognition performance on test set.

These phenomena have been pointed out, because when the dimension d
is continually increased, MI is increased slowly, but the effects of the "curse
of dimension" and of discriminate-useless noisy components are relatively
enhanced. The results make the recognition performance is not increased,
even the reduction happens at all.

From now on, we can see that maximum MI discrimination sub-space
discriminate recognition is very important to achieve the best recognition
performance of classifiers, especially, for the difficult recognition problems, in
which the discriminate features are smaller and hard to be extracted directly
from a huge embedded noise.

It should be pointed out that there will be more approximate because
the feature probability distribution of practical samples is really apart from
Gaussian distribution. Therefore the mutual information of non-Gaussian
features will be much smaller than the calculated:

$$I(E, F) \leqslant \frac{1}{2} \log_2 \frac{|S_T|}{|S_W|} = \frac{1}{2} \left(\sum_{i=1}^{N} \log_2 \lambda_{Ti} - \sum_{i=1}^{N} \log_2 \lambda_{Wi} \right).$$

Although, it is impossible to exactly estimate the non-Gaussian feature probability distribution and calculate its mutual Information, it is still very useful
for the relatively feature comparison for feature selection and recognition performance estimation.

From Figs. 27.2 and 27.3, two kinds of features for handwritten Chinese
character recognition, directional element feature DEF and gradient feature
GF, have been compared from their mutual information: MI of gradient

feature GF is bigger than that MI of directional element feature DEF. Unimpeachable, it has been proved by experiments shown in Fig. 27.4, that two kinds of features compared with their recognition performance: the recognition rate of gradient feature classifier is higher than that of directional element feature DEF classifier, Both classifiers are in the same Euclidean distance classifiers.

From above discussing the mutual information is a good measure for feature merit and recognition algorithm and the maximum MI discrimination sub-space recognition is the best for feature dimension reduction.

27.7 Conclusion

The unified entropy theory in pattern recognition and maximum MI discrimination based sub-space pattern recognition are presented in this chapter The information procedures in learning procedure and recognition procedure in pattern recognition are discovered. The mutual information is the finally determination for the recognition performance, which has been shown is the best merit for feature selection. Also the maximum MI discrimination subspace dimensional reduction is proposed and proved, that is the best for pattern recognition.

Acknowledgements

The author is indebted to the National Basic Research Program of China (973 program) under Grant No. 2007CB311004 for supporting this work, to Dr. Hailing Liu for his works on handwritten Chinese character experiments in this Chapter.

Appendix

A.1   Two Probability Spaces and Their Entropies in Recognition System

In pattern recognition, either learning procedure or cognition procedure all concern the features extracted from samples and their corresponding categories attribution. There are two kinds of uncertainties in feature space and class space. Therefore two different probability spaces and their entropies are defined first.

Definition A.1   Pattern sample ξ has two attributes $\{X, \omega\}$, here X is its observable feature vector, $X = (x_1, x_2, \cdots, x_N)^{\mathrm{T}}$, $\omega \in \Omega$ is its class attribution, and Ω is the class set.

Definition A.2   The probability space of pattern classes E is composed of pattern class set Ω and its probability $P(\omega_i)$, that is

$$E = \left\{ \begin{array}{cccc} \omega_1 & \omega_2 & \cdots & \omega_n \\ P(\omega_1) & P(\omega_2) & \cdots & P(\omega_n) \end{array} \right\}, \tag{A.1}$$

where $\Omega = \{\omega_1, \omega_2, \cdots, \omega_n\}$, n is the number of pattern classes,

$$P(\omega_i), \quad i = 1, 2, \cdots, n. \quad \sum_{i=1}^{n} P(\omega_i) = 1.$$

Definition A.3   The system entropy H(E) is defined as an entropy on the category space E of pattern classes. That is

$$H(E) = -\sum_{i=1}^{n} P(\omega_i) \log P(\omega_i). \tag{A.2}$$

System entropy H(E) represents the uncertainty of the space E of pattern classes, or the capability of a recognition system. The bigger H(E) is, the bigger uncertainty is in the recognition system.

When $P(\omega_i) = \frac{1}{n}, i = 1, 2, \cdots, n$ the biggest system entropy H(E) is

$$H_{\max}(E) = \log_2 n, \quad H(E) \leqslant \log_2 n.$$

For two classes classification problem, $H(E) \leqslant 1$ bit

For the Chinese character recognition problem, when $n = 6763$, then $H(E) \approx 16$ bits.

Definition A.4   The feature probability space F is composed of N-dimensional random vector X and its probability distribution density $p(X)$. That is

$$F = \left\{ \begin{array}{c} X \\ p(X) \end{array} \right\} = \left\{ \begin{array}{c} (x_1, x_2, \cdots, x_N)^{\mathrm{T}} \\ p(x_1, x_2, \cdots, x_N), \end{array} \right\}, \tag{A.3}$$

where X is the N-dimensional feature vector of sample ξ, $X = (x_1, x_2, \cdots, x_N)^{\mathrm{T}}$ and its probability density function P(X) is

$$p(X) = p(x_1, x_2, \cdots, x_N),$$

$$\int_{R^N} p(X) \, \mathrm{d}X = 1.$$

Definition A.5   Feature entropy $H(F)$ is defined as an entropy on the feature probability space F,

$$H(F) = -\int_{R^N} p(X) \log p(X) \, \mathrm{d}X. \tag{A.4}$$

The feature entropy H(F)* represents the information contained in the feature space F.

When N-dimensional random vector X is an independent random vector, that is

$$p(X) = p(x_1, x_2, \cdots, x_N) = \prod_{i=1}^{N} p(x_i).$$

Then the feature entropy is

$$H(F) = \sum_{i=1}^{N} H(F_i) = -\sum_{i=1}^{N} \int_{R^N} p(x_i) \log p(x_i)\, \mathrm{d}x_i.$$

When $p(X)$ is a Gaussian distribution function of the random vector X,

$$p(X) = \frac{1}{(2\pi)^{\frac{N}{2}} |\Sigma_X|^{\frac{1}{2}}} \exp\left\{ -\frac{1}{2}(X - M_X)^{\mathrm{T}} \Sigma_X^{-1}(X - M_X) \right\}, \qquad \text{(A.5)}$$

$$M_X = E\{X\}, \quad \Sigma_X = E\left\{ (X - M_X)(X - M_X)^{\mathrm{T}} \right\}.$$

Then Gaussian feature differential entropy $h(F)$ is

$$h(F) = \frac{1}{2\ln 2} \left[\ln(2\pi e)^N + \ln|\Sigma_X| \right]$$

or
$$h(F) = \frac{1}{2}[N + N \log_2 2\pi e + \log_2 |\Sigma_X|]\text{bits}. \qquad \text{(A.6)}$$

The discrete Gaussian feature entropy is

$$H(F) = \frac{1}{2\ln 2} \left[\ln(2\pi e)^N + \ln|\Sigma_X| \right] + \ln\Delta, \quad \Delta \to 0. \qquad \text{(A.7)}$$

Therefore, the bigger N and Σ_X have, the bigger $H(F)$ is.

There need to be noted that the entropy is satisfied with

$$\max_{\Sigma_X = E\{XX^{\mathrm{T}}\}} h(F) \leqslant \frac{1}{2} \log(2\pi e)^N |\Sigma_X|,$$

with equality iff $X \in N(M_X, \Sigma_X)$. \qquad \text{(A.8)}

That is to say, for the non-Gaussian distribution, the feature entropy is smaller than Gaussian feature entropy.

The feature space and its feature entropy represent the observable features and its uncertainty respectively. The class space and the system entropy represent the inherence category attribute and its uncertainty, respectively. The cognition is obtained from observed feature of testing sample to find its unknown category attribute.

---

*   This $H(F)$ is really a differential entropy. The corresponding discrete entropy should be $H^\Delta(F) = H(F) - \log\Delta$, Δ *is the length of quantization bin, here we use* $H^\Delta(F)$ $\cong H(F)$

A.2   Relative Entropies Between the Feature Space and Class Space

Both learning procedure and recognition procedure concern the relations between the feature space and the class space. Relative entropies between them clearly describe these relations.

Definition A.6   Class-conditional feature entropy $H\left(F\,|E\right)$ is defined as

$$H\left(F\,|E\right) = -\sum_{i=1}^{n} \int_{R^N} p\left(X, \omega_i\right) \log_2 p\left(X\,|\omega_i\right) \mathrm{d}X. \tag{A.9}$$

Class-conditional feature entropy $H\left(F\,|E\right)$ represents the mean feature entropy under certain class. If features are closely related with the classes, the class-conditional feature entropy $H\left(F\,|E\right)$ becomes smaller. These features are beneficial to recognition. Inversely, when features are weakly related with the classes, the class-conditional feature entropy $H\left(F\,|E\right)$ becomes larger, these features will injure the recognition.

Lemma A.1   It could be proved that

$$H\left(F\,|E\right) \leqslant H\left(F\right).$$

When both features and classes are unrelated, then

$$p\left(X\,|\omega_i\right) = p\left(X\right), \quad i = 1, 2, \cdots, n,$$

$$H\left(F\,|E\right) = H\left(F\right).$$

For the Gaussian feature distribution,

$$p\left(X\,|\omega_i\right) = \frac{1}{\left(2\pi\right)^{\frac{N}{2}} \left|\Sigma_i\right|^{\frac{1}{2}}} \exp\left\{-\frac{1}{2}\left(X - M_i\right)^{\mathrm{T}} \Sigma_i^{-1}\left(X - M_i\right)\right\}, \tag{A.10}$$

where $M_i = E\left\{X_i\right\}$, $\Sigma_i = E\left\{\left(X_i - M_i\right)\left(X_i - M_i\right)^{\mathrm{T}}\right\}, i = 1, 2, \ldots, n$, Gaussian discrete class-conditional feature entropy $H\left(F\,|E\right)$ is

$$H\left(F\,|E\right) = \sum_{i=1}^{n} P\left(\omega_i\right) h\left(F\,|\omega_i\right) \tag{A.11}$$

$$= \frac{1}{2\ln 2}\left[N + N\ln 2\pi + \sum_{i=1}^{n} P\left(\omega_i\right)\ln\left|\Sigma_i\right|\right] + \ln \Delta \quad \text{when } \Delta \to 0$$

Definition A.7   A posterior entropy $H\left(E\,|F\right)$ is defined as

$$H\left(E\,|F\right) = -\sum_{i=1}^{n} \int_{R^N} p\left(X, \omega_i\right) \log_2 p\left(\omega_i\,|X\right) \mathrm{d}X. \tag{A.12}$$

The mutual information from discrete feature entropies is

$$I\left(F, E\right) = H\left(F\right) - H\left(F \,|E\right) = \frac{1}{2 \ln 2} \left[\ln |\Sigma_X| - \sum_{i=1}^{n} P\left(\omega_i\right) \ln |\Sigma_i| \right].$$

A posteriori entropy $H\left(E \,|F\right)$ is feature-conditional system entropy, which represents the average system entropy after the features have been extracted. $H\left(E \,|F\right)$ represents the leftover uncertainty after obtaining the observed features of sample, which is closely related with the error probability of recognition system.

A *posteriori* entropy of the Gaussian feature is

$$H\left(E \,|F\right) = H\left(E\right) - H\left(F\right) + H\left(F \,|E\right)$$

$$= \frac{1}{2 \ln 2} \left[\ln |\Sigma_X| - \sum_{i=1}^{n} P\left(\omega_i\right) \ln |\Sigma_i| \right] - \sum_{i=1}^{n} P\left(\omega_i\right) \log P\left(\omega_i\right). \quad \text{(A.13)}$$

Lemma A.2   It could be proved that *a posteriori* entropy is no bigger than the system entropy,

$$H\left(E \,|F\right) \leqslant H\left(E\right).$$

When both the features and classes are statistical independent each other, then

$$P\left(\omega_i \,|X\right) = P\left(\omega_i\right), \quad i = 1, 2, \cdots, n, \quad H\left(E \,|F\right) = H\left(E\right).$$

Definition A.8   Combined entropy $H\left(E, F\right)$ on the product space $E \otimes F$ of feature space F and class space E is defined as

$$H\left(F, E\right) = -\sum_{i=1}^{n} \int_{R^N} p\left(X, \omega_i\right) \log_2 p\left(X, \omega_i\right) \mathrm{d}X$$

$$H\left(F, E\right) = -\sum_{i=1}^{n} \int_{R^N} p\left(X, \omega_i\right) \log_2 p\left(X, \omega_i\right) \mathrm{d}X = H\left(E\right) + H\left(F \,|E\right).$$

$$\text{(A.14)}$$

Lemma A.3   When both the class space and feature space are independent each other, that is $P\left(\omega_i, X\right) = P\left(\omega_i\right) p\left(X\right), \quad i = 1, 2, \cdots, n.$
Then $\qquad\qquad\qquad H\left(E, F\right) = H\left(E\right) + H\left(F\right).$
In general, $\qquad\qquad\quad H\left(E, F\right) \leqslant H\left(E\right) + H\left(F\right). \qquad\qquad \text{(A.15)}$

Lemma A.4   Because $\qquad p\left(X, \omega_i\right) = p\left(X \,|\omega_i\right) P\left(\omega_i\right)$
or $\qquad\qquad\qquad\qquad\quad p\left(X, \omega_i\right) = p\left(\omega_i \,|X\right) P\left(X\right),$
it will easily be proved,

$$H\left(E, F\right) = H\left(E\right) + H\left(F \,|E\right), \qquad\qquad \text{(A.16)}$$

$$H\left(F, E\right) = H\left(F\right) + H\left(E \,|F\right). \qquad\qquad \text{(A.17)}$$

Definition A.9   The mutual information between the class space E and
the feature space F can bedefined as

$$I\left(F,E\right) = \sum_{i=1}^{n} \int_{R^N} p\left(X,\omega_i\right) \log_2 \frac{p\left(X \mid \omega_i\right)}{p\left(X\right)} \mathrm{d}X, \qquad (A.18)$$

$$I\left(E,F\right) = \sum_{i=1}^{n} \int_{R^N} p\left(X,\omega_i\right) \log_2 \frac{p\left(\omega_i \mid X\right)}{P\left(\omega_i\right)} \mathrm{d}X, \qquad (A.19)$$

or $$I\left(E,F\right) = I\left(F,E\right) = \sum_{i=1}^{n} \int_{R^N} p\left(X,\omega_i\right) \log_2 \frac{p\left(X,\omega_i\right)}{P\left(\omega_i\right) p\left(X\right)} \mathrm{d}X, \qquad (A.20)$$

$$I\left(F,E\right) = H\left(F\right) - H\left(F \mid E\right), \qquad (A.21)$$

$$I\left(E,F\right) = H\left(E\right) - H\left(E \mid F\right), \qquad (A.22)$$

$$I\left(F,E\right) = H\left(F\right) + H\left(E\right) - H\left(F,E\right),$$

$$I\left(F,E\right) \leqslant \min\left[H\left(F\right), H\left(E\right)\right].$$

The mutual information is the correlation information between the feature
space and the class space, which is the most important and the decision factor
in pattern recognition,

Lemma A.5   The combined entropy is related with the mutual information:

$$H\left(E,F\right) = H\left(E\right) + H\left(F\right) - I\left(F,E\right).$$

Lemma A.6   The mutual information MI for the Gaussian feature distribution equals to

$$I\left(F,E\right) = \frac{1}{2\ln 2}\left[\ln\left|\Sigma_X\right| - \sum_{i=1}^{n} P\left(\omega_i\right)\ln\left|\Sigma_i\right|\right]$$

with equality iff $X \in N\left(M_X, \Sigma_X\right)$. $\qquad (A.23)$

Because $H\left(F\right) = \dfrac{1}{2\ln 2}\left[N + N\ln 2\pi + \ln\left|\Sigma_X\right|\right] + \ln\Delta, \quad$ when $\Delta \to 0$
and

$$H\left(F \mid E\right) = \frac{1}{2\ln 2}\left[N + N\ln 2\pi + \sum_{i=1}^{n} P\left(\omega_i\right)\ln\left|\Sigma_i\right|\right] - \ln\Delta, \quad \text{when } \Delta \to 0.$$

Lemma A.7   When the *a piori* probabilities for every class are equal,
the heteroscedastic Gaussian mutual information is

$$I\left(F,E\right) = \frac{1}{2\ln 2}\ln\frac{\left|\Sigma_X\right|}{\displaystyle\prod_{i=1}^{n}\left|\Sigma_i\right|^{\frac{1}{n}}}.$$

Lemma A.8 When all within-class feature scatter matrix for every class are equal, $\sum_i = S_w$, and $\sum_X = S_t = S_b + S_w$, then homoscedastic Gaussian mutual information is

$$I\left(F,E\right) = \frac{1}{2\ln 2}\ln\frac{|\Sigma_X|}{|S_w|},$$

$$I\left(F,E\right) = \frac{1}{2}\log_2\left(\frac{|S_w + S_b|}{|S_w|}\right)^{1/2}. \tag{A.24}$$

References

[1] Cover T M, Thomas J A (1991) Elements of Information Theory. Wiley, New York

[2] Ding X, Wu Y (1993) Unified Information Theory in Pattern Recognition. ACTA Electronica Sinica, 21(8): 1–8

[3] Biem A, Katagiri S, Juang B H (1997) Pattern Recognition Using Discriminative Feature Extraction, IEEE Tranc on Signal Processing, 45(2): 500–504

[4] Watanabe H, Yamaguchi T, Katagiri S (1997) Discriminative Metric Design for Robust Pattern Recognition, Transaction on signal Processing, 45(11): 2655–2662

[5] Maes F, vandermeulen D, Suetens P (2003) Medical Image Registration Using Mutual Information. Proceeding of IEEE, 91(10): 1699–1722

[6] Ding X Q, Chen L, Wu T (2007) Character Independent Font Recognition on a Single Chinese Character, IEEE Transaction on Pattern Recognition and Machine Intelligence, 29(2): 195–204

[7] Ding X (1993) Information Theory Application in Pattern Recognition, '93' National Conference on Communication Theory and Information Theory, Yichang, Hubei

[8] Ding X (1991), Pattern Recognition Integrated Entropy Theory Based on Information Theory, NJC_ACTAI'91, Proceeding of NJC_ACTAI'91, Beijing, China. pp 339–344

[9] Ding X (1996), Pattern Recognition Feature & Information Entropy Principal of Feature Selection, Software Transaction, Special Issue for Intelligent Computer Term of 863 High-Tech Plan, pp 394–400

[10] Escolano F, Suau P, Bonev B (2009), Information Theory in Computer Vision and Pattern Recognition, Springer, New York

[11] Ding S, Shi Z (2005), Studies on Incidence Pattern Recognition Based on Information Entropy, Journal of Information Science, 31(6): 497–502

[12] Ding S, Zhang Y, et al (2009) Research on a Principal Components Decision Algorithm Based on Information Entropy, Journal of Information Science, 35(1): 120–127

[13] Dasa K, Nenadic Z (2008), Approximate Information Discriminant Analysis: A Computationally Simple Heteroscedastic Feature Extraction Technique, Pattern Recognition, 41(5): 1548–1557

[14] Keysers D, Och F J, Ney H (2002), Efficient Maximum Entropy Training for Statistical Object Recognition, Informatiktage 2002 der Gesellschaft fur Informatik, pp 342–345

[15] Keysers D, Och F J, Ney H (2002), Maximum Entropy and Gaussian Models for Image Object Recognition, Proceedings of the 24th DAGM Symposium on Pattern Recognition, Zurich, Switzerland, pp 498–506

[16] Normandin Y (1992), Hidden Markov Models, Maximum Mutual Information Estimation, and the Speech Recognition Problem, Doctoral Thesis, McGill University, Montreal, Que, Canada

[17] Matton M, Wachter M D, Compernolle D V, Cools R (2005), Maximum Mutual Information Training of Distance Measures for Template Based Speech Recognition. Proc. International Conference on Speech and Computer, Patras, Greece, pp 511–5145

[18] Kim H Y, Kim J H (2000), Minimum Entropy Estimation of Hierarchical Random Graph Parameters for Character Recognition. ICPR 2000: 6050–6053

[19] Minh T A, Victor D (2004), Minimum Entropy Estimation as a Near Maximum-Likelihood Method and its Application in System Identification with Non-Gaussian

Noise. 2004 IEEE International Conference on Acoustics, Speech, and Signal Processing, pp 545 – 548

[20] Wang Q R, Suen C Y (1984), Analysis and design of a decision tree based on entropy
      reduction and its application to large character set recognition, IEEE Transacations
      on Pattern Analysis and Machine Intelligence, 6(4): 406 – 417

[21] Tong C S, Shing Y M (2000), Two-Stage Entropy-Enhanced Chinese Character
      Recognition, Proceedings of ICCLC2000, 163 – 167

[22] Zhu S C, Wu Y N, Nian Z, Mumford D (1997), Minimax Entropy Principle and Its
      Application to Texture Modeling, Neural Computation, 9(8): 1627 – 1660

[23] Srikantan G, Lam S W, Srihari S N (1996), Gradient-Based Contour Encoder for
      Character Recognition, Pattern Recognition, 29(7): 1147 – 1160

28 Fundamentals of Biometrics — Hand Written Signature and Iris

Radhika K R[1] and Sheela S V[2]

Abstract   Biometric is a metric of apparent nontransferable uniqueness provided by user's presence. Biometrics are extremely convenient form of providing identity and cannot be lent to another individual. Due to digital impersonation security techniques are predominantly using biometrics. To eliminate identity theft, a measurable physical characteristic or behavioral trait is more reliable. In factual scenario, blend of biometric identification and a keypad code provides virtually unbreakable security. The consequential statistical error rates of biometric security systems are calculated for a large population. The choice of biometric is application specific. The number of users, technical implementation and operating environment will influence the selection of distinguishing biometric trait. The hand written signature characteristics and iris texture variations form an occurrence vector to provide biohashing.

28.1 Prologue

Biometric characteristics are broadly grouped into two categories, namely physiological and behavioral. Physiological biometrics are based on the measure of anatomical biocharacteristics such as fingerprints, facial thermogram, vein structure, iris pattern, retina pattern, hand geometry, and facial recognition. Behavioral biometrics are based on the measure of trait during a period of time such as speech, signature, handwriting, keystrokes, and mouse dynamics. A biometric security system user should first endure process called as enrollment. In enrollment a user provides a biometric trait to the system. After certain processing stages of system compatibility, the system stores biometric trait as a template in database. To authenticate a user later, the template is used to compare the given biometric trait. The extracted pertinent features are compared with stored templates. Information labeling process is known as pattern recognition. This forms an important component of machine intelligence domain. Associating an identity with an individual is called personal identification. There are two modes to use biometrics in

---

1   Department of Information Science and Engineering, B M S College of Engineering, Bull Temple Road, Bangalore-560019, India. E-mail: radhikakr@ieee.org.
2   Department of Information Science and Engineering, B M S College of Engineering, Bull Temple Road, Bangalore-560019, India. E-mail: sheelasv@ieee.org.

personal identification. They are authentication and identification. The authentication and identification are shown in Fig. 28.1.

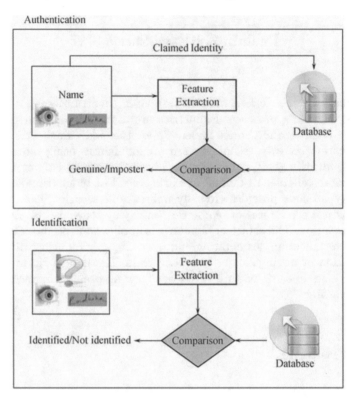

Fig. 28.1 Authentication vs identification.

Authentication or verification refers to the problem of confirming or denying a person's claimed identity. The two types of authentications are personal identity and web service authentication. Recognition or identification refers to the problem of establishing a subject's identity from a set of already known identities [1]. The primary application focus of biometric technology is verification and identification of humans using their possessed biological properties [2]. Last two decades have witnessed an explosive growth in biometric authentication systems. The premise is that, biometric is a measurable trait which forms a reliable indicator of identity when combined with legacy systems such as passwords. A blend of acquisition, feature extraction and classification has evidently become technology oriented in e-commerce and m-commerce applications.

28.2 Fundamentals of Handwritten Signature

The first informatization with signature, occurred in Sumerian era, which led to transition from a purely verbal communication to writing. It is the culture of writing that made it possible to communicate binding declarations of intent over a distance. It is the tool used to establish relations between parties absent. The signature is of importance not only in the legal sense but also as an expression of a social state of trust between the parties. By providing the signature, an identity-conclusion function is observed. Signer allows herself/himself to be identified. The conclusion function signals the finality of providing authentication [3]. Signature is user-friendly, ubiquitous and non-invasive. For many years it is acquired in a number of written information systems. It is socially and legally well accepted, providing data integrity and non-repudiation. Authentication performance is stable across all evaluated age groups proving the ability of signature system to be deployed for use, within a general population. It is non-intrusive and suitable for hospitality transactions. A handwritten signature is biologically linked to a specific individual. The digital signature provided by cryptographic authentication systems binds signatures to individuals through technical and procedural mechanisms. Handwritten signatures are under the direct control of the signer and digital signatures are applied by a computer command. A variant method which shares some of the benefits of signatures, is to generate cryptographic keys from handwritten passphrases. Passwords and ID cards are shared by others. People forget to bring the ID cards or forget the passwords. Passwords are revoked or reissued.

Handwritten signature authentication is the process used to recognize an individual's signature which is natural and intuitive. In the case of signature biometrics, it is well known that no two genuine signatures of a person are

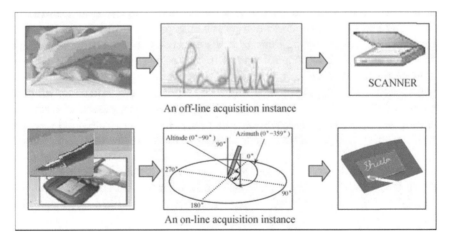

An off-line acquisition instance

An on-line acquisition instance

Fig. 28.2 Acquisition system.

precisely the same. Signature experts note that, if two signatures written on paper are same, then they might be considered forgery by tracing [4]. Two systems of signature authentication are Off-line Signature Authentication System (OFAS) and On-line Signature Authentication System (ONAS). The acquisition instances are shown in Fig. 28.2. In OFAS, the signatures are treated as gray level images. ONAS refers to matching the time functions of the signing process as shown in Fig. 28.3.

Fig. 28.3 Time Warping Process.

28.2.1   Challenges Faced by OFAS and ONAS

The signature pattern is strongly affected by user-dependencies. Discriminative power of the features that are extracted is affected by the physical and emotional state of the user. This leads to difficulty in detecting true signature. The authentication system considers statistical and elastic matching methods with template and decision adaptation methods. These methods are required as different features have different thresholds for different users. The signature varies from one signing instance to another in a known way — causing a duplicate to be rejected.

In OFAS, the one dimensional parameterization of the signature curve is unavailable. The appropriate formatting and pre-processing techniques are required. The angle at which people sign is different due to seating position or support taken by hand on the writing surface. The dynamic information regarding the signing process such as velocity, pressure and stroke order are unavailable. The ability of human beings to recognize patterns is superior to that of a machine but, when it comes to processing speed and management of large data sets with consistency, machines are far superior. Dependency on automation is due to the difficulty faced in visual assessment for different types and sizes of signatures such as simple, cursive, graphical and segmental signatures. The different types of forgeries with respect to OFAS are freehand, simulated and traced forgery. Freehand forgeries are written in forger's own handwriting without the appearance knowledge of the genuine signature. They are identified by document examiners on the basis of differences between the handwriting characteristics of the forger and the genuine writer. Simulated forgery is achieved by practice. Tracing a signature is the easiest way of forgery. The amount of off-line input data is two orders of magnitude higher compared to on-line. This has a negative effect on necessary storage required and the time needed for authentication. ONAS faces the challenges to achieve discriminability power due to low permanence, as the handwritten signature tends to vary along time and it is vulnerable to direct attacks by forgers. The types of forgeries with respect to ONAS are random, simple, skilled and unskilled as shown in Fig. 28.4.

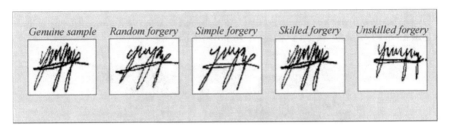

Fig. 28.4 Types of forgeries, courtesy: GVGD database.

In random forgery, the forger has no knowledge about the original signature and does not try to imitate the shape of the signature. In simple forgery, a forged signature is produced with the knowledge of genuine signer's name. In simulated forgery, a forged signature imitating reasonably a genuine signature is captured. In unskilled forgeries, inexperienced forger imitates signature after observing genuine specimens closely for some time. The spelling of the name is unknown. Skilled forgeries are forgeries where a forger sees the genuine signature and has time to practice the imitations. In timed forgery, forger is provided with information about average genuine writing time. The signers vary in coordination and consistency of the signatures. The selection process of pertinent parameters is challenging. Necessity still exists to build an automated tool that is trustworthy, based on implicit knowledge acquired from the latest acquisition devices.

28.2.2 Goal

The behavioral pattern of the user is mined for authentication. The objective of the biometric security system is to extract useful, implicit and novel knowledge from the handwritten signature patterns. These patterns are classified using advanced machine learning techniques in the field of pattern recognition to authenticate the user. Using handwritten signature biometric, the goal is:
- to track and analyze the signature of the user;
- to develop off-line authentication algorithm;
- to develop on-line authentication algorithm.

28.2.3 Methodology

Features are useful measurements obtained by observing the objects. These provide discriminating and characterizing information to identify an object. Selected features are invariant for transformations. We can infer that, there are certain subparts of signature, which vary even in genuine samples as shown in Fig. 28.6 for the sample shown in Fig. 28.5.

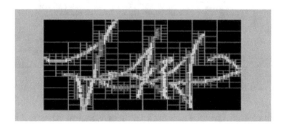

Fig. 28.5 Genuine sample of a subject.

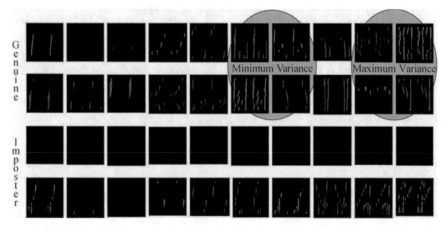

Fig. 28.6 Minimum variance and maximum variance parts in samples.

Off-line features are horizontal difference $x_{max} - x_{min}$, vertical difference $y_{max} - y_{min}$, curvature measurements, ratio of long to short stroke, segment length, vertical midpoint $(y_{min}/(y_{max} - y_{min}))$, number of vertical midpoint crossings, total pen writing distance per signature area, modified direction features, maximum pixel change, maximum rotation angle, centre of signature gravity and histograms. $x_{max}, x_{min}, y_{max}$ and y_{min} are maximum and minimum co-ordinates of X, Y axis respectively. The intersection points provide significant details as shown in Fig. 28.7.

The zero pressure segments of both genuine and imposter are depicted in Fig. 28.8.

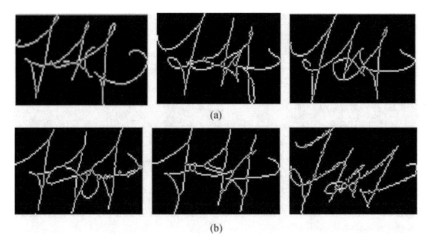

Fig. 28.7 (a) Intersection points of genuine samples. (b) Intersection points of imposter samples.

The prominent on-line features for signature authentication are xy coordinates, pressure, azimuth and inclination as shown in the Fig. 28.9. Azimuth (θ) is the angle between the z-axis and the radius vector connecting

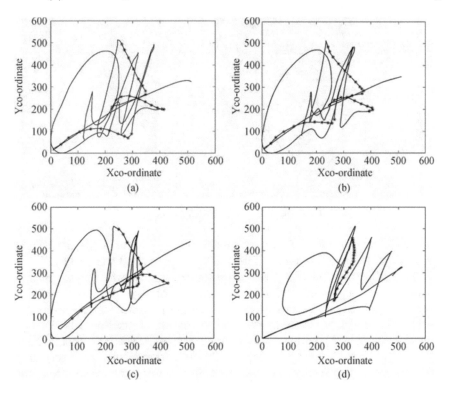

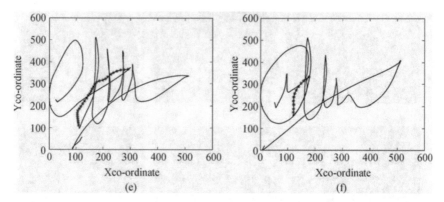

Fig. 28.8 (a)–(c) Zero pressure plot of genuine samples. (d)–(f) Zero pressure plot of imposter samples.

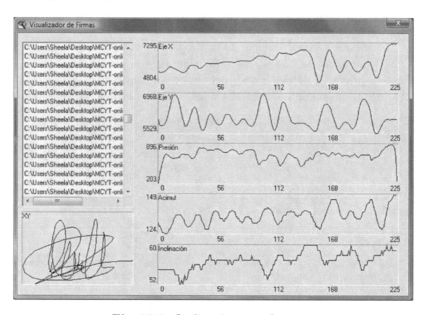

Fig. 28.9 On-line signature features.

the origin to any point of interest. Inclination (φ) is the angle between the projection of the radius vector onto the $x - y$ plane and the x-axis.

The other features are jitter, aspect ratio, normalized length, mean of pixels in a sliding computation window, center of mass, torque, moments of inertia, weighted cross-correlation, extrema points, total time, pen-up time, total path length, and two dimensional histogram of pixel distributions as a function of angle. The other novel on-line features are extrinsic energy, intrinsic energy, neuro-motor trajectory attribute, instantaneous trajectory angle, instantaneous displacement, curvature radius, centripetal acceleration, number of pendowns, and number of pen-ups. The signature is in various lengths and sizes. The positional concept specifies what percentage of signature has what velocity value. This provides a basis for segmentation. From Fig. 28.10 (a) − (b), it is analysed that the velocity scatter plot looks similar for two genuine samples. From Fig. 28.11 it is analysed that, the velocity scatter plot of imposter sample differ from velocity scatter plot of a genuine sample. Acceleration values provide the same inference. The shape of acceleration plot of two genuine samples are similar. The shape of acceleration plot of the imposter sample differs from that of a genuine sample.

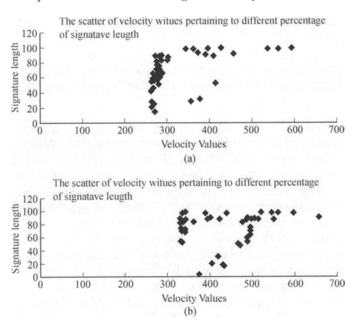

Fig. 28.10 Velocity plot of two genuine samples.

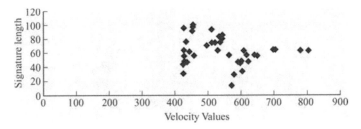

Fig. 28.11 Velocity plot of an imposter sample.

Fig. 28.12 (a) shows the shape similarity of acceleration plot between the two genuine samples of same person and Fig. 28.12 (b) shows enormous variation in shape of acceleration plot when compared between a genuine and an imposter sample of the same person.

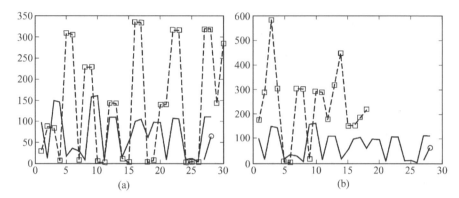

Fig. 28.12 Acceleration Plot. (a) Two Genuine signature samples. (b) A genuine sample and an imposter sample (Number of on-line pixels vs acceleration values).

Some of the on-line derived features are number of sign changes in x and y accelerations, path tangent angle, duration of velocity in x-plane>0, duration of velocity in x-plane<0, duration of velocity in y-plane>0 and duration of velocity in y-plane<0. The other derived features are ratio of total pen-down duration to total signing duration, ratio of maximum pen velocity occurrence time to total pen-down duration, ratio of minimum pen velocity occurrence time to total pen-down duration, ratio of maximum velocity occurrence time in the x-plane to total pen-down duration, ratio of minimum velocity occurrence time in the x-plane to total pen-down duration, ratio of mean velocity to maximum velocity, ratio of minimum velocity in x plane to average velocity in x-plane, ratio of minimum velocity in y-plane to average velocity in y-plane, ratio of average velocity to maximum velocity in the x-plane, ratio of average velocity to maximum velocity in the y-plane and total number of samples with zero velocity value. The rotation invariant features can be extracted. For instance applying Zernike moment for sample, leads to rotation invariant values as shown in the Fig. 28.13.

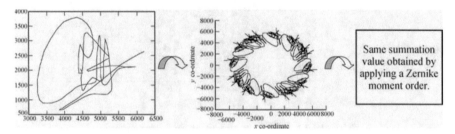

Fig. 28.13 Rotation Invariant property.

28.2.4 Preprocessing

Normalization of co-ordinates and base-line rotation of the sample are achieved. The goal is to condition the acquired data so that noise from various sources is removed. In OFAS, the images are preprocessed using median filter to achieve uniform background. The outliers are removed using rough set processing after the preprocessing steps such as binarization, filtering and smoothing, normalisation, thinning, base-line detection, slant correction and segmentation. The preprocessing is applied for training samples and also for testing samples. The standard sample is obtained for feature extraction.

Binarization

Binarization process converts a gray scale image into bi-level image as shown in the Fig. 28.14.

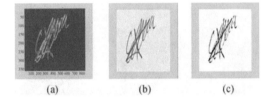

Fig. 28.14 Binarization.   (a)   (b)   (c)

The histogram of the gray values of the image is generated. The signature images in rgb format are converted to grayscale. The resulting intensity image is converted to binary. The gray value at peak level in histogram, is used as cut-off point. All the gray values less than are equal to cut-off point are changed as black. All the gray values greater than cut-off point are changed to white.

Filtering and Smoothing

Filtering and smoothing remove noise variations in the input image. Filtering and smoothing can be achieved either with spatial techniques or morphological techniques. A 3×3 mask is applied pixel-by-pixel on the 2D image. Morphology is an alternative technique for smoothing. The closing operation eliminates minute openings and fill gaps on the curve endings. The opening operation break narrow curve endings and sharp peaks. Gray-scale manipulation maps each pixel to a gray value. This improves the contrast of the images. Median filter is applied to reduce noise and preserve edges as shown in the Fig. 28.15.

Normalization

Normalization is used to have images of fixed size. If the sizes of the images are different, comparison becomes tedious process as shown in Fig. 28.16.

$$\boldsymbol{y}_n(t_i) = \frac{(\boldsymbol{y}(t_i) - min_y)Q}{(max_y - min_y)}, \ i \in \{1, 2, ..., length(\boldsymbol{y}(t))\} \ and \ Q = 512 \quad (28.1)$$

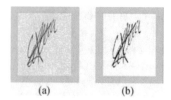

Fig. 28.15 Filtering and Smoothing.   (a)   (b)

$$\boldsymbol{x}_n(t_i) = \frac{(\boldsymbol{x}(t_i) - min_x)Q}{(max_x - min_x)}, \ i \in \{1, 2, ..., length(\boldsymbol{x}(t))\} \ and \ Q = 512 \quad (28.2)$$

where min_x, max_x, min_y and max_y are minimum and maximum values of $\boldsymbol{x}(t)$ and $\boldsymbol{y}(t)$ respectively. The normalized image of size 512×512 is obtained.

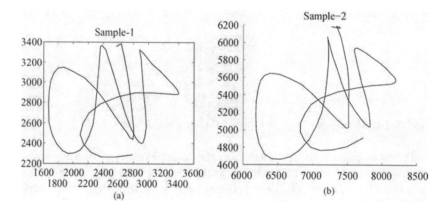

Fig. 28.16 Sample-1 and sample-2.

Thinning

Thinning eliminates the thickness differences of pen by creating the image one pixel thick as shown in the Fig. 28.17. Thinning provides good connectivity. Thinning algorithms are sequential and parallel. Sequential algorithm deals with only one pixel at a time. Parallel algorithm activate on all the pixels. The thinning algorithms are based on an edge attrition. A window or mask is applied to the image as shown in the Fig. 28.18. Morphological closing provides thinning. Edge detection is adopted in order better connectivity.

(a)   (b)   (c)

Fig. 28.17 Thinning.

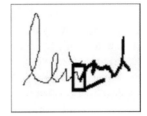

Fig. 28.18 Thinning with mask.

Base-line Detection

Base-line is defined as the line of signing orientation. Horizontal projection histogram is considered as one of the methods for fixing the base-line. Operations are performed, which lead to a smooth connected signature. The minimum bounding box is found and signature is extracted.

Slant Correction

The angle between longest stroke in a signature and the vertical direction is referred as slant. Slanted strokes are identified. To normalize signature patterns, slant correction is done.

Segmentation

The signature sample can be subdivided, before feature extraction. For instance, quad tree components can be extracted as shown in Fig. 28.19. A region quadtree with the subregion size of f is used to represent a subimage consisting of $f \times f$ pixels, where each pixel value is 0 or 1. The root node represents the entire binary image. Let R represent the entire normalized binary signature image. Quadtree partitions R into 4 subregions, R_1, R_2, R_3, R_4, such that (a) $\bigcup R_i = R$. (b) R_i is a connected region, $i = 1, 2, 3, 4$. (c) $R_i \cap R_j = \phi$ for all i and $j, i \neq j$. For further subdivisions, same clauses are applicable.

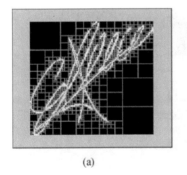

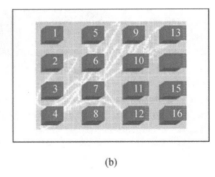

(a)   (b)

Fig. 28.19 Segmentation.

The quadtree structure is selected because in a real time scenario, storing external files are simpler since every node is either a leaf or it contains four children. The binary tree involves a number of traversals for different encode levels. Quadtrees have proven to be very helpful for visualization without being time consuming. Subregions represent the presence of critical information of the signature.

In on-line scenario, an instance of preprocessing is to minimize the effect of spatial resolution obtained by the input device. The features are normalized to achieve zero mean and unit variance. The assumptions for measuring the dynamic features with respect to pen-tip should be stated. Different signatures of the same individual have variability in their form and characteristics. Time alignment algorithms such as equispacing by linear interpolation, time normalization, location normalization and size normalization are used to minimize the time differences among the user's samples.

28.2.5 Normalisation of feature values

The different types of score normalization procedures are min-max, z-score, tanh, median, sigmoid, parzen and simple sum of scores fusion normalization methods [5]. z-score normalization techniques are sensitive to outliers in the data. If many tail-points influence the tanh estimate, then estimate is not robust. The min-max procedure provides robust and efficient normalization.

The score normalization technique is carefully chosen depending on the amount of robustness required which in turn depends on the estimate of the amount of noise in the available training data. Min-max estimate is used to transform the scores into a common domain. If the location and scale parameters of the matching scores are known in advance, then simple normalization techniques like min-max would suffice. For instance, the normalized acceleration profiles to a length of 100 subsections, are shown in Fig. 28.20.

$$a_{s1}(t) = \frac{\dfrac{a_1(t) - \mathrm{mean}(a_1(t))}{\sigma(a_1(t))}}{\mathrm{max}(a_1(t))}, \quad a_{s2}(t) = \frac{\dfrac{a_2(t) - \mathrm{mean}(a_2(t))}{\sigma(a_2(t))}}{\mathrm{max}(a_2(t))}. \tag{28.3}$$

$$a_k(t_{i+1}) = 1|(a_k(t_{i+1}) > a_k(t_i)) \& (a_k(t_{i+1})$$
$$\geqslant a_k(t_{i+2})) \& (a_k(t_{i+1}) > 0), \quad k \in \{s1, s2\}. \tag{28.4}$$

The scaling transformation is applied for further enhancing the shape characteristics using analytical Eqs. 28.3 and 28.4 as shown Fig. 28.21 (a) – (b). These normalisations aid in warping process as shown in Fig. 28.21 (c).

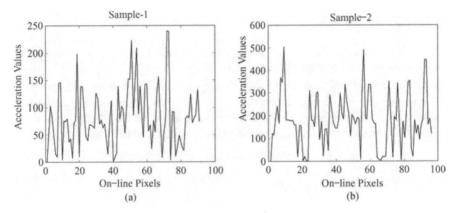

Fig. 28.20 Acceleration profiles.

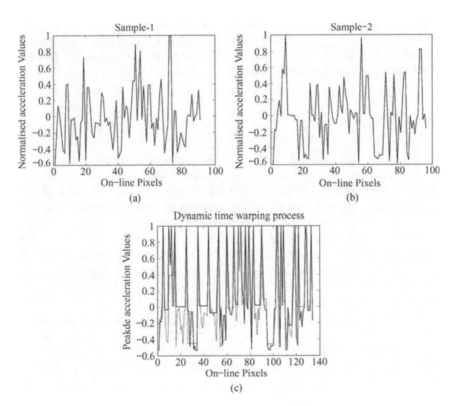

Fig. 28.21 (a) – (b) Shape characteristics, (c) warping process.

28.2.6 Prediction

To predict the value for a specific attribute temporal and spatial assumptions play a major role. Temporal assumption states that, an instance of an event occurs at roughly the same fraction of time for a signature regardless of the overall signing speed. Spatial assumption works on linear scaling of the horizontal and vertical displacements by considering different scaling constants. Histogram aids in predicting the value for a specific attribute.

28.2.7 Model Visualization

To make the discovered knowledge representable, a model is created. Three pattern recognition models are statistical pattern recognition, syntactic pattern recognition and neural network pattern recognition. Statistical model is a formal and precise decision theoretic approach. The premise of syntactic model is that, the structure of an entity is paramount and it is used for description. The neural network model involves large interconnected networks of relatively simple non-linear units.

28.2.8 Classification

To determine the class label of a sample, training procedure maps a relationship between features and corresponding class labels. The samples of distinguishing biometric trait of a person is termed as data set. The data set is divided into two different subsets as training set and testing set. The set of training samples are the samples with known class labels. Class label specify the class category of a sample. Classifier is a model to estimate the correct class label for testing samples as shown in Fig. 28.22. The testing samples are the samples which are used to test the classifier. In supervised training, the class labels of testing samples are known. In unsupervised training, the class labels of testing samples are not known.

The quantitative measures that represent the classification error are considered. Incorrect labeling of the testing samples by classifier is termed as classification error. They are False Acceptance Rate (FAR) and False Rejection Rate (FRR). The other measures which represent error in OFAS are slant measure and variability measure. Biometric systems utilize a user specific threshold value to tune the system for the best possible performance. The Fig. 28.23 illustrates the relationship between the FAR and FRR. FRR value decrease and FAR value increase with increase in training samples. Both the measures are trade off against one another. Equal Error Rate (EER) is the value of error where FAR is equal to FRR.

The general structure of a biometric authentication system is shown in

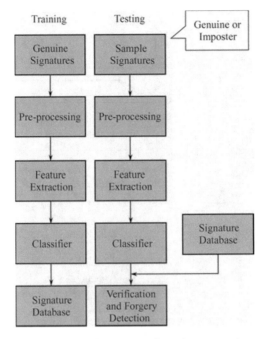

Fig. 28.22 Training and testing scenario.

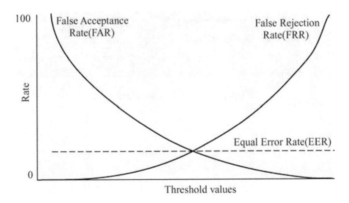

Fig. 28.23 Biometric characteristics.

Fig. 28.24.

Factors which affect the performance are user-dependent decision thresholds, posterior threshold alignment measures and number of training samples in multi-sessions. The factors of on-line acquisition hardware are quality of pen tablet compared to paper, tablet resolution and pen tablet without visual feedback. A common agreement in the research community on benchmark databases and protocols for performance evaluation is vital.

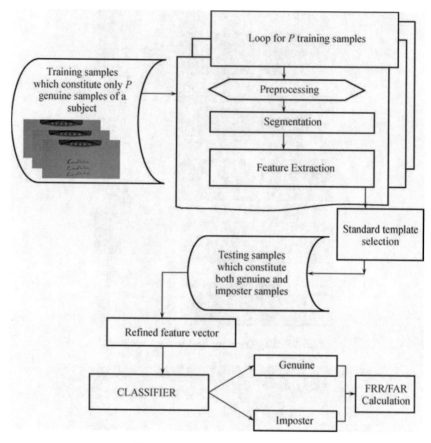

Fig. 28.24 General structure.

28.3 Acquisition

One of the most important requirements for designing a successful pattern recognition system is to acquire sufficient number of samples and learn the decision boundary. Decision boundary is a functional mapping between the features and correct class labels. The main goal of acquisition is to provide meaningful variations of data instances to form training and test data.

In off-line scenario, the card reader, electronic pen with miniature cameras, electronic signature pad and scanner are the input devices used to extract signature samples. The resolution of the resulting binary image justifies the storage capacity required. The resolution is dependent on the monitor or digital camera used. In a well controlled environment, thin signature samples are extracted from a uniform background. In an uncontrolled environment, samples are occluded with other textual information and spread out value of ink is inconsistent. In on-line scenario, the devices are used to extract on-line features of a signature sample. Input devices used are digitizing tablets, smart pens and hand gloves. Electro-magnetic sensing area of tablet acts as a paper and stylus acts as a pencil to capture handwritten signature.

Machine readable medium is required at the stage of enrollment as shown in Fig. 28.25 (a). The system includes a digitizing tablet with stylus, input/output devices, processor, memory and network interface as shown in Fig. 28.25 (b). A sensing device is used to keep track of pen-up/pen-down positions.

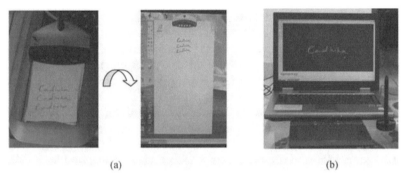

(a)   (b)

Fig. 28.25 (a) Writing pad to display, (b) bamboo tablet.

28.4 Databases

– Biometric Recognition Group, ATVS is devoted to research in the areas of biometrics, pattern recognition, image analysis, and speech. The group maintains European public projects and holds contracts with companies [6]. MCYT database is from Ministerio de Ciencia Y Tecnologia, Spanish ministry of science and technology project. In on-line scenario, samples are acquired in multi-session procedure by incorporating intrinsic short-term signature variability, signature size variability and time constrained signature forgeries. The 25 client signatures and 25 highly skilled forgeries (with natural dynamics) are obtained for each individual. The sub-corpus consists of $100 \times (25 + 25)$ samples. In MCYT on-line corpus the acquisition device used is a Wacom pen tablet, model INTUOS A6 USB. The tablet resolution is 2 540 lines per inch (100 lines/mm) and the precision is $+/- 0.25$ mm. The maximum detection height is 10 mm for pen-up movements. The capture area is 127 mm $\times$ 97 mm. This tablet provides the following discrete-time dynamic sequences: (i) position in x-axis, x_t: $[0 - 12\,700]$, corresponding to $0 - 127$ mm; (ii) position in y-axis, y_t: $[0 - 9700]$, corresponding to $0 - 97$ mm; (iii) pressure, p_t applied by the pen:

$[0-1024]$; (iv) azimuth angle θ_t of the pen with respect to the tablet, corresponding to $0° - 360°$ and (v) altitude angle φ_t of the pen with respect to the tablet, corresponding to $30° - 90°$ [6]. In off-line scenario, image of the written signature is considered. The bitmap images are of size 850 × 360 in rgb format.

- Graphics Visualization and Games Development [GVGD] database is from malaysia. GVGD provides samples of 40 subjects. It provides 5 genuine training samples, 5 genuine testing samples, 5 skilled imposter samples, and 5 unskilled imposter samples per subject.
- Dolfing database: The data set contains 4 800 signatures from 51 writers. For each writer, there are 15 training signatures, 15 genuine test signatures, and 60 skilled forgeries. Each of these signatures contain static and dynamic information captured at 160 samples per second (London).
- Unipen project : The international Unipen Foundation is installed to safeguard the distribution of a large database of on-line handwritten samples, collected by a consortium of 40 companies and institutes, from more than 2 200 writers (The Netherlands).
- GPDS database (Grupo de Procesado Digital de Seales): It contains 160 signature sets of 24 genuine and 30 targeted signatures for each set (Spain).
- BioSecure: The database consists of samples from 600 individuals (France).
- Caltech Signature Database: The on-line database consists of samples from 56 subjects (California Institute of Technology).
- US-SIGBASE: Database consists static and dynamic signatures of 51 people. All signatures were recorded on paper placed on a Wacom UD-0608-R digitizing tablet with 50 mm × 20 mm signing area. The static binary signatures are sampled at 600 dpi.

Providers of Signature Verification Solutions:
1) Communication Intelligence Corporation,
2) Cyber-SIGN Japan Inc,
3) DATAVISION corporation,
4) SOFTPRO,
5) Security Biometrics Inc,
6) WonderNet.

28.5 Signature Analysers

The theoretical foundation of the existing signature authentication techniques are linked to milestones. A milestone is an event signifying the completion of a major research work or a set of related activities. The listing is done for the past four decades. The related activities like, signature verification competitions and reviews are listed. Usually a milestone is used as a research checkpoint to validate how the preprocessing methods, feature extractors and classifiers are progressing. Milestones are used as high-level conclusions according to the researcher's point of interest. The sacrosanct knowledge obtained by the milestones are listed in Table 28.1.

Table 28.1 Milestones

Milestone	Sacrosanct Knowledge obtained.
1975	Pressure is reduced to few measures via mathematical transformations (Strenberg, Automated signature verification, WESCON).
1977	To assume that for a given writer, the parameter values for a specific stroke in particular context are less variable than those strokes of the same class (Nagel and Rosenfeld, Off-line system, IEEE T. Comp).
1977	The fine structure of the muscle forces exerted during the writing of a signature is constant and well defined for most of the people (Liu and Herbst, On-line system, IBM J.Res. Dev).
1989	Gradient of the edges are significantly different in an original signature from that in a signature forged by tracing. Perceptually important points are determined during the segmentation process (Plamondon and Lorette, State-of-the art, PR).
1994	Most fraudulent duplications of human signatures occurred in commercial transactions are simple forgeries (Leclerc and Plamondon, State-of-the-art, IJPRAI).
1995	To develop a feature vector which is individually optimized (Fairhurst, Structural modularisation, IEEE Security and detection).
1995	For each signature a Markov model is constructed, using a set of sample signatures described by the normalized directional angle function along the signature trajectory (Prasad et al., Regional-HMM, PR)
1996	Orthogonalizing features in accordance with the availability of training data and the level of system complexity is stated (Lee et al., Holistic, PAMI).
1997	Simplified version of skeletonisation technique (Papamarkos and Baltzakis, Off-line system, IEEE DSP).
1997	To declare a system to work, it is necessary to handle self fraud. Self fraud is fraud by the genuine signer. At feature extraction level this condition is taken care, else it opens up the possibility of a genuine signer authorizing a transaction with the apriori intent of later denying authorization (Nalwa, Local-DTW, Proc. IEEE).
2000	Fundamental characteristics of handwritten pattern are three-fold. The purpose is achieved by virtue of the Mark's conventional relation to language (Plamondon and Srihari, State-of-the-art, PAMI).
2002	More enrollment data is required to reliably estimate the writer-dependent threshold. An alignment between two signatures is found. The difference between the number of strokes in the two signatures are incorporated as the dissimilarity measure (Jain et al., Local-DTW, PR).
2003	User-specific decision thresholds produce results that in average usually outperform the global user-independent decision approach (Ortega-Garcia et al., Regional-HMM, AVBPA).
2004	First International Signature Verification Competition. To achieve objective evaluation and comparison of algorithms (Yeung et al., SVC, ICBA).
2005	Complete list of global features is provided. Results in four common conditions are compared (few/many training signatures and skilled/random forgeries). A good working point of the combined system in four conditions are depicted. Signature trajectories are first preprocessed by subtracting the center of mass followed by a rotation alignment based on the average path tangent angle (Fierrez et al., Fusion of local and global information, IEEE).
2006	A signature is schematized in the human motor memory as an action map made up of a sequence of virtual targets connected by discrete strokes. When a person is about to write, this plan is voluntarily activated to produce a sequence of vectorial commands (Plamondon et al., Interactive trajectory synthesizer, IEEE).

Continued

Milestone	Sacrosanct Knowledge obtained.
2007	The application of a target-centric score normalization approach provided a performance improvement for skilled forgeries while maintaining the performance for random forgeries (Fierrez et al., HMM based, PRL).
2008	The Discrete Cosine Transform is used to reduce the approximation coefficients vector obtained by wavelet transform to a feature vector (Nanni L et al., Novel local method, PRL).
2009	Features are ranked according to the inter-user class separability. Applicability of symbolic data analysis for signature verification is explored. System considers the test sample described by a set of feature values and compares with the interval type features (Guru et al., Symbolic Representation, PAMI).
2009	The influence of multi-sessions, environmental conditions and signature complexity on the performance is stated (Sonia Garcia-Salicetti et al., Competition, ICB).

The factors useful for identifying the milestones are:
- Related publications ascertained, depending on required system characteristics.
- Publications pertaining to usage of selected databases.
- Details of activities which provide the state-of-art.

28.6 Off-line Methods

The broad categories are template matching methods, hidden Markov models, structural techniques, and feature based techniques.

28.6.1 Template Matching Methods

Warping is the method to map one of the curves onto another curve while attempting to preserve its original shape.

Order of co-ordinates are used to match the co-ordinates of two exterior curves. Exterior curves are formed by top and bottom parts of the template signature and test signature [7]. To achieve similarity metric, the curve is represented in mass-spring system. The nodes in the graph represent unit mass particles and springs represent edges. Disconnected curve parts form a separate graph. The first ring neighbors are the neighboring nodes connected to node by a single edge. These are sorted by specifying a criterion on the angle to provide structural springs. The structural springs provide a basis for comparison between template and test signature samples. Classifier measures intrinsic and extrinsic energy. Derivation of intrinsic energy is by structural curve constraints and extrinsic energy is by forces of attraction between the nodes of the template signature and test signature.

Deformable templates based on multi-resolution shape features use chain code contours of extremas along the contour for convexity and rotated versions [8]. Deformation is measured by point to point matching using isomorphic projection between local extrema sets of template and test signature. The thin-plate spline mapping function is considered as an objective function for gradient structural concavity algorithm to achieve region matching. Stroke characterization is achieved by matching contour directional chain codes. Adaptive feature thresholding is achieved by converting signature feature vector to a binary feature vector. The creation of a binary feature vector is based solely on the gradient direction of each pixel across a signature.

28.6.2 Hidden Markov Model

The common characteristics of hidden Markov model methods are listed:
- Markov statistical learning theory has the ability to absorb both variability and similarity between the patterns. The learning theory is strictly causal.
- It is based on empirical risk minimization principle where decision rule is based on finite number of training samples. The number of states depends on the signature length. Left-to-right topology depends on learning probability. The training patterns form visible states.
- For specific number of states, the best validation probability is defined by maximum likelihood or Bayesian approach. The medium threshold is defined by learning probability normalized by signature length.

28.6.3 Structural Techniques

Structural features are extracted from the signature contour as modified direction feature. It utilizes location of transitions from background to foreground pixels in vertical and horizontal directions with respect to signature. For each transition, location and direction of transition are stored. The division of the signature image into three equal parts lead to tri-surface structural features. Dividing each component based on horizontal line calculated from centre of gravity, leads to sixfold surface. This leads to 6 features of interest. The centroid feature relating to dominant angle of the signature is defined as the angle between the horizontal axis and the line obtained by linking the centre of gravity points.

28.6.4 Feature Based Techniques with Global Features

Global features are computed in subregions of the signature image. The classical method and morphological method for outline extraction exist. Fuzzy vault biometric cryptosystems consider maxima and minima from upper and lower envelopes of the signature. The upper envelope is extracted as first non-zero pixel for each column in a normalized binary image. The lower envelope is extracted as last non-zero pixel for each column in a normalized binary image. Using moving average method of span 35, smoothing envelopes are obtained. Fuzzy vault input key is formed by a set of quantized envelope values [9]. The location where a template and a key are combined into a unique token allows key reconstruction. The extracted peripheral features describe internal and external structural changes of signatures periodically.

28.6.5 Feature Based Techniques with Local Features

Local features provide better localized characterization with a difficulty of computing them more reliably. The technique is based on fuzzy Takagi-Sugeno model, which uses distance and angle features extracted by box method. The different formulations are stated depending on the number of rules, such as single rule (for all input features) or multiple rules (as many rules as number of features) [10]. In order to reduce the number of fuzzy sets, feature values are distributed into intervals. The fuzzification is achieved by defining membership function for each interval. Equal horizontal density approximation method is considered for small data sets. Preprocessed image is partitioned into 8 portions. The rule for partition constitutes equal number of dark pixels obtained after scanning the image horizontally left to right and right to left. The summation of the angles of all points in each portion with respect to bottom left corner is calculated. The normalized summation is obtained by dividing summation with number of pixels in the partition. These parameters track the variations.

28.7 On-line Methods

The existing methods are broadly categorised into four classes as global parametric feature based approach, function-based approach, hybrid method and trajectory construction method.

28.7.1 Feature Based Approach

In these methods, a holistic vector representation of global features is derived from the signature trajectories. The parameter model provides a maximin distance for a subject, by ordering features. The best features for which distance defined by a criterion is the largest from the rest of the entire population are selected. The criterion is based on sample mean and variance of the feature. In symbolic representation, signature features are stored in compact form. Enrollment data size is considerably small and constant. It is a statistical parametric approach [11]. Symbolic representation is stable against the variations in the local regions. It captures the variation of features in the form of interval type data. The representation records only parameters. Using parameters, original signature pattern cannot be reconstructed. All n samples of a person with m features form a class. Each feature value is represented in the form of an interval with the aid of their respective mean and standard deviation values. The m intervals forming symbolic vector represent the entire class. The verification is done by a threshold called acceptance count. It is the count of features mapped into intervals by the symbolic vector. The entropy associated to a given portion of a signature is represented by group of outcomes. To achieve right parameterization, time variability of signatures are considered.

28.7.2 Function Based Local Method

Function based approaches are methods in which time sequences describing local properties of the signature are used. In local method, the time functions of different signatures are directly matched using elastic distance measures. Dynamic time warping is a template matching method. The reference and test pattern are sampled with the same sampling rate. The goal is to find an optimal time alignment between reference and test pattern by extracting form and motion features. In large datasets, apriori knowledge about the relative importance of different parts of the patterns is obtained to reduce over-sampling. In differential geometric shape analysis method, the extrema points such as peaks and valleys are selected for warping [4]. The extrema of both x and y profiles together and the pen-up events from left to right are represented as a string. The four basic symbols forming a string are local maxima and minima of x profile, local maxima and minima of y profile with respect to time.

28.7.3 Function Based Regional Method

In regional methods, time functions are converted to a sequence of vectors describing regional properties [12]. The basic challenge for signature authentication is, the allowance of certain variabilities within the original class and simultaneously detecting significant differences between the original and the forged class [13]. The continuous models lead to better authentication rates on smaller databases because of the interpolating effect. The larger the database, the better it fits a discrete model. Five time sequences such as horizontal position trajectory, vertical position trajectory, azimuth, altitude of pen with respect to tablet and pressure signal are used to find extended features like path tangent angle, path velocity magnitude, log curvature radius and total acceleration magnitude. These form function based instantaneous feature set. The set is normalized to form discrete time sequences. Sampling rate affects the performance.

28.7.4 Hybrid Approach

Local and global predicted values are applied to a method which is both feature based and function based. Based on the normalized arc-length of the signature curve with weighted cross correlation, shape and dynamics related features are extracted. The general distribution of the feature is observed in whole database. The general distribution of the feature and a person's distribution of the feature, constitute the parameter for DPF. The comparison of distance measure between the test sample feature and reference provides the Degree of Authenticity (DoA) of the feature. The DoA of shape aspect and the DoA of dynamics aspect are combined by fuzzy inference system to yield the DoA of the overall test signature. Representing only one possible temporal occurrence of feature sequence leads to the use of one distinct sample as a reference or a personalized template. The method is shown in Fig. 28.26.

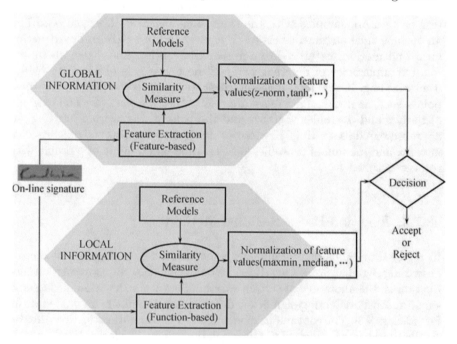

Fig. 28.26 Hybrid Approach.

28.7.5 Trajectory Generation Methods

Trajectory construction methods produce and control complex two-dimensional synergistic movements. Contemporary techniques are developed in two directions such as analysis and synthesis of biometric information. Two general methodologies exist for generating trajectory using synthesis of biometric information. The first methodology aims to replicate features of handwritten pattern. Some of the models of this type are oscillatory model, second order linear model, minimum-jerk model, minimum-torque model and delta-lognormal synergic model. The second methodology focuses on psychologically descriptive features. Some of the models of this type are, feedback error learning neural network model, dynamic optimization model, forward inverse relaxation model and Beta-elliptical Model (BM). A neuromuscular synergy is composed of two parallel and global activities called agonist and antagonist activities. Complex human movements are segmented into basic and simple strokes. Each stroke is described by a set of parameters that characterize the hand movement both in the kinematic and the static domains [14]. The kinematic theory describes the basic properties of a single stroke and how strokes are added vectorially to generate a stroke from a given series of input commands. A handwritten pattern is generated, in terms of velocity vs time signal or position vs time signal.

The Lognormal Model (LM) deals with the study of strokes to characterize a stroke velocity profile. The study of outlier patterns for interactive comparative analysis uses rapid human movements in its vectorial version [15].

$$v_i(t; t_{0i}, \mu_i, \sigma_i^2) = \frac{\|D_i\|}{\sigma_i \sqrt{2\pi}(t - t_{0i})} exp\left(\frac{-[ln(t - t_{0i}) - \mu_i]^2}{2\sigma_i^2}\right). \qquad (28.5)$$

The value $i = 1$ depicts agonist and $i = 2$ depicts antagonist neuromuscular systems in reaction to two simultaneous input commands D_1 and D_2, respectively. The vector D_i describes the amplitude and the direction of the input command. t_{0i} is the time of occurrence of the input command. μ_i is logtime delay which depicts time delay of the neuromuscular system on a logarithmic time scale. σ_i, is logresponse time which depicts the response time of neuromuscular system on a logarithmic time scale in Eq. 28.5. The direction of the velocity vector is determined by starting θ_{is}, and ending θ_{ie} directions of the curve's input command. These represent trajectory in cartesian co-ordinates. The trajectory is recovered by integration. The system combines kinematic and geometry of the trajectory, using stroke numbers based on velocity signal extremum. The application of BM deals with the inherent mechanisms which govern cursive strokes. Basic strokes are formed from algebraic summation of beta velocity profiles. The method describes handwritten pattern in context of kinematic points and static points that obey the beta law. A trajectory is planned according to arm position in an intrinsic kinematic space with an assumption that any handwritten object is partially programmed in advance.

$$\beta(t, t_0, t_1, p, q) = \frac{(t_1 - t)^p}{(t_1 - t_c)^p} \cdot \frac{(t - t_c)^q}{(t_c - t_0)^q}. \tag{28.6}$$

The parameters of dynamic domain are given by the Beta function for approximating the original curvilinear velocity signal as shown in Eq. 28.6. t_0 is the starting time. $t_c = (pt_0 + qt_1)/(p + q)$, t_c is the instant where the curvilinear velocity reaches its maximum and t_1 is the ending time. p and q are intermediate parameters. The critical points to split the pen path into smaller entities are local extrema and points of inflexion in velocity profiles.

28.7.5.1 Comparison

A comparison is made between LM and BM depending on characteristics such as representation schemes, neurophysiological point of view used to estimate the angular parameters, visuality, constraint on shape, extractor, optimization, extraction consistency evaluation methods and parameter variability restriction. These methods provide sacrosanct knowledge for signature reconstruction. The question for which any reconstruction model answers is *"Does the extracted solution form a plausible action plan for the movement realized?"*.

28.8 Fundamentals of Iris

The iris is highly protected, non-invasive and ideal for handling applications requiring management of large user groups, like voter ID management. The iris recognition techniques potentially prevent unauthorized access to ATMs, cellular phones, desktop PCs, workstations, buildings and computer networks. The accuracy of iris recognition systems is proven to be much higher compared to other types of biometric systems like fingerprint, handprint and voiceprint. The system provides audit trail, socially acceptable and cost effective. In the year 1885, a French ophthalmologist, Alphonse Bertillon first proposed iris pattern as a basis for personal identification. In 1987, an unimplemented concept of automated iris biometrics system was obtained. A report on iris patterns was published by Johnston in 1992 without any experimental results. Iris is the annular ring between the pupil and the sclera of the eye. The potency of iris recognition lies in its textual information. The structural formation in the human iris is fixed from about one year in age and remains constant over time. Iris exhibits long-term stability and infrequent re-enrollment. The variations in the gray level intensity values effectively distinguish two individuals using left or right eye images. The difference persists between identical twins.

The sclera is the white region of the connective tissues surrounding the iris. The outer membrane that covers the iris and the pupil is the cornea.

The pupil is the darkest region in the eye. The anterior surface of the iris is composed of two regions, the central pupillary zone and the outer ciliary zone. The collarette is the circular boundary between these two regions. The pit like oval structures surrounding the collarette are called crypts. The radial band in the pupilary zone appear straightened when pupil is constricted and appear wavy during pupil dilation. This band is called as radial furrows. The average diameter of the iris is approximately 12 mm with an average thickness of about 0.5 mm. The dilation and constriction of the pupil regulate the amount of light. The anatomy of iris is shown in Fig. 28.27.

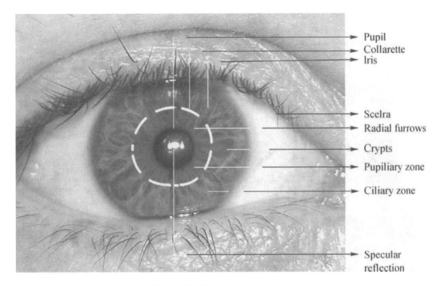

Fig. 28.27 Iris anatomy.

Color and Texture

The iris vary in color from light blue to dark brown. The variations are found in different ethnic races, left and right eyes and different regions in the iris. The color is due to the cellular density and the pigment contained in the iris stroma, which is a connective tissue in the iris [16]. The color gradings are blue, grey, light brown and dark brown as shown in Fig. 28.28. The texture of the iris and pupil of an individual remains constant irrespective of the image quality. Texture provides randomness in the iris region, which form unique patterns for each individual. The textures can be graded as fine, fine/medium, medium/coarse and coarse as shown in Fig. 28.29.

Considering the ethnicity aspect it can be stated that the iris patterns from different races show distinctive characteristics. The diameter of the pupil and iris, color of the iris and shape of the eyelid are significant features for ethnic classification. The Fig. 28.30 illustrates iris patterns.

A typical iris recognition system involves four main modules as shown in

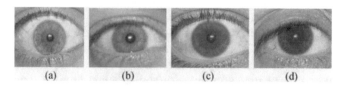

Fig. 28.28 Color gradings. (a) Blue. (b) Grey. (c) Light brown. (d) Dark brown.

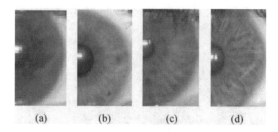

Fig. 28.29 Texture gradings. (a) fine. (b) fine/medium. (c) medium/coarse. (d) coarse.

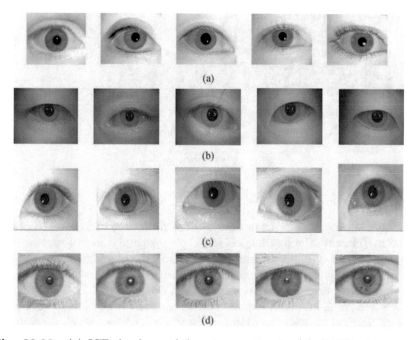

Fig. 28.30 (a) ICE database of American nationals. (b) CASIA database of Chinese nationals. (c) MMU database of Malaysian nationals. (d) UBIRIS database of Portugal nationals.

Fig. 28.31.

The first module, image acquisition deals with capture of a sequence of

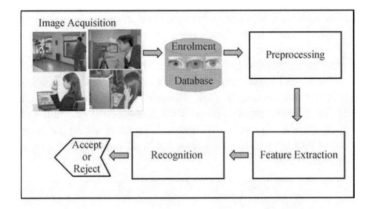

Fig. 28.31 Iris recognition system.

iris images from the subject using a specifically designed sensor. Iris based security applications thrive on infrared cameras and video cameras for logins. Many iris recognition systems require stern cooperation of the user for image acquisition. The benchmark databases use cameras like Nikon E5700, Canon EOS 5D, OKI and Iridian LG EOU2200 [17 – 19]. The design of an image acquisition apparatus deals with illumination, position and physical capture system. The entire sequence of images can be acquired during enrollment. The best workable images are selected to increase flexibility. Strong identity management solutions begin at enrollment.

The second module is preprocessing which includes detection of iris liveness, pupil, iris boundary, eyelid and eyelashes. Iris liveness detection ensures that the image is of a live subject instead of an iris photograph, a video playback, a glass eye or other artifacts. It is possible that biometric features are forged. Several methods like Hough transformation, integrodifferential operator, gradient based edge detection are used to localize the portion of iris and the pupil from the eye image. The contours of upper and lower eyelids are fit using the parabolic arcs resulting the eyelid detection and removal. The variation in the distance between the camera lead to iris images of unequal size. Illumination causes iris to dilate and contract. It is essential to map the extracted iris region to a normalized form. The iris localization methods are based on spring force, morphological operators, gradient, probability and moments.

The third module is feature extraction which identifies the most prominent features for classification. The features are radius of pupil-iris, shape-size of the pupil, intensity values, resolution and rotational measurements of the pupil ellipse. The features are encoded to a suitable format for recognition.

The fourth module is recognition which match features with known patterns in the database. In biometric security systems, the major issue lies in inter-class and intra-class variability. In most of the approaches, beginning with the acquisition, a sequence of preprocessing, edge detection, iris extraction, normalization and feature extraction are performed to generate a unique iris code to identify an individual. The sequence is shown in Fig. 28.32. The preprocessing includes filtering operations to eliminate noise. The operators like Canny, Sobel and Prewitt can be used for edge detection based on threshold. The circular boundaries are identified based on Hough transform, connected components and integrodifferential operators. The iris is extracted by converting the pixels to polar coordinates. The unwrapped iris is normalized. A binary code can be calculated based on local entropy for the rectangular iris forms the iris code.

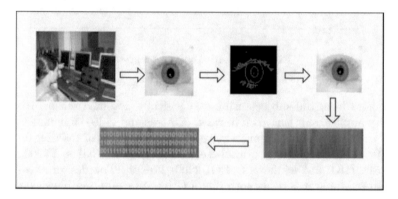

Fig. 28.32 General Approach.

The major challenges are based on iris localization, normalization, occlusion, segmentation, liveness detection and large scale identification. The iris recognition algorithms need to be developed and tested in diverse environment and configurations. It is required to achieve lowest false rejection rate and fastest composite time for template creation.

28.9 Feature Extraction

Feature extraction identifies the most significant features for identification. Some of the features are wavelet decomposition coefficients, texture difference, log-Gabor wavelet components, moment summation values, 2D Gabor filter response, velocity of the gray level values and shape of acceleration plot. A 2D wavelet transform applied on the image produces the approximation coefficients and the details. The detail coefficients are obtained along the three orientations, horizontal, vertical and diagonal. The summation, average and the magnitude of the detail coefficients form useful features. The wavelet types that are commonly used are Haar, Daubechies, biorthogonal, symlets and coiflets. The coefficients for three levels of wavelet decomposition is shown in Fig. 28.33.

The gradient identifies the edges or the change overs on the image. The intensity gradient from pupil region to iris marks the pupil boundary. The gradient from iris to sclera detects the iris boundary. The texture difference are the features. Iris localization is based on intensity gradient and texture difference [20]. The gradient along the circular boundary of the pupil and the iris is calculated using the integrodifferential operator. In the boundary region, the texture difference between the outer zone and the inner zone are computed. The Kullback-Leibler divergence method is used to measure the distance between two probability distributions derived from the inner and the outer zones. The iris portion can be restricted to the left and the right quadrants at an angle between −45 deg to +45 deg and 225 deg to 135 deg as shown in Fig. 28.34. The local binary pattern is used to extract the features by analyzing the texture.

Iris segmentation based on collarette area localization is less sensitive to pupil dilation and eliminates occlusion by eyelid and eyelashes [21]. The data points along the radial lines of the pupiliary zone form significant features.

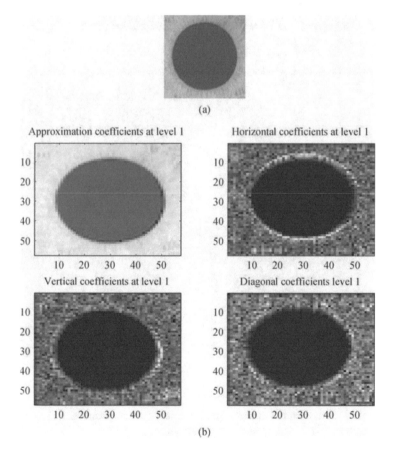

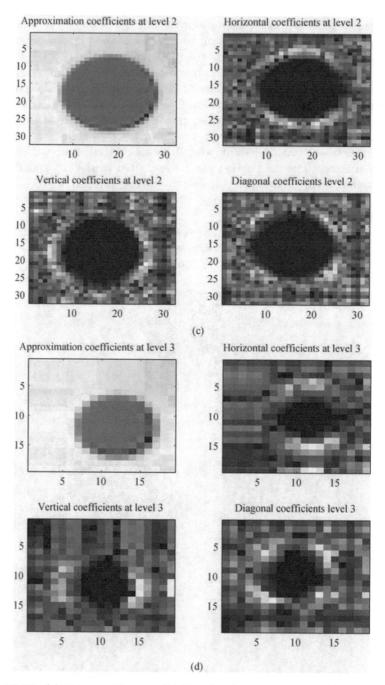

Fig. 28.33 (a) Extracted Image. (b) First level decomposition. (c) Second level decomposition. (d) Third level decomposition.

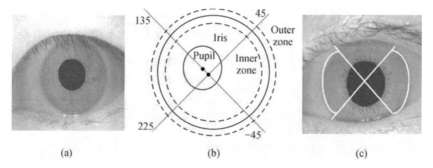

Fig. 28.34 Texture difference.

The feature vector is obtained by convolution of the normalized collerette region with the 1D log-Gabor wavelets. The collerette segmentation is shown in Fig. 28.35.

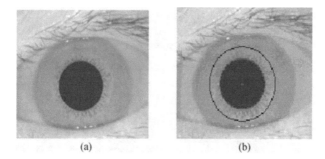

Fig. 28.35 Iris segmentation based on collarette.

The iris and the pupil region can be decomposed into subparts as shown in Fig. 28.36. Considering Fig. 28.36 (b), the moment values computed on minimum variance subregions form the feature vector. The minimum variance subparts correspond to the pupil region. The shape of the iris in each of the subparts forms distinguishing characteristics for each subject. The variations are prominent in the subparts consisting of the iris part. The subparts and minimum variance parts are shown in Fig. 28.37.

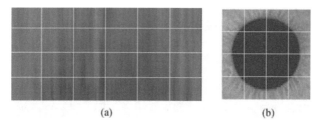

Fig. 28.36 (a) Iris subparts. (b) Iris and pupil subparts.

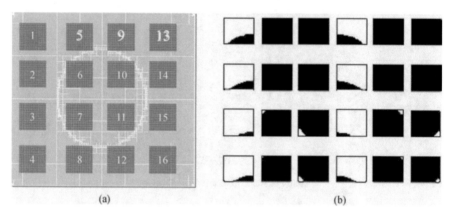

Fig. 28.37 (a) Subparts. (b) Minimum variance subparts.

The peak in the histogram of the eye image gives the highest number of pixels. The intensity values less than the peak value corresponds to the pupil and the eyelashes. The intensity values greater than the peak value corresponds to the sclera and the eyelids. For the pixels less than or equal to the peak value, the connected components using 8-connectivity objects are labeled.

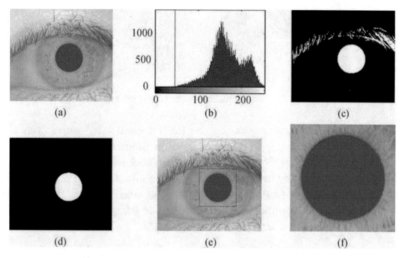

Fig. 28.38 (a) Eye image. (b) Histogram. (c) Connected components. (d) Maximum area connected component. (e) Pupil-iris detection. (f) Pupil-iris frame.

The area is determined for the identified components. It is observed that there are a number of connected components with the smaller area corresponding to eyelashes and one component with large area corresponding to pupil. The area of the pupil is always the maximum. The maximum and minimum values in the x and y direction are used to determine the center of the pupil and its radius. From the center of the pupil, a window size is determined as a function of the pupil radius. A normalized bounding box extracts the pupil and the iris region for further processing. The sequence of steps is shown in Fig. 28.38. The iris region obtained consists of the part that forms unique characteristics for an individual.

The normalization technique removes the concentricity of the iris and pupil. The iris is converted into a rectangular region. With the detection of iris boundaries and pupil center as the origin, the iris image $I(x, y)$ is mapped into polar coordinates $I(r, \theta)$ [22]. The process is performed in a circular manner from the pupil boundary until the iris boundary is reached. The r value is computed for each value of θ in the range $[0, 2\pi]$. $I(x(r, \theta), y(r, \theta)) \leftarrow I(r, \theta)$. The mapped image is generally normalized to the size $90 \times 360, 64 \times 256, 80 \times 256$ and 128×256. The steps are shown in Fig. 28.39.

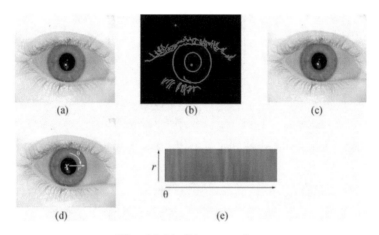

Fig. 28.39 Iris extraction.

Considering region wise division, the iris part is divided into 6 regions. Starting from an angle of -45 deg, Regions 1 to 4 are marked in successive quadrants. Regions 5 and 6 are marked as outer and inner concentric parts. This approach of extracting iris eliminates the reflections or illumination spots and the occlusion due to eyelids. Feature extraction using 2D Gabor filters is performed to each region and independent biometric signatures are obtained [23]. The regions are shown in Fig. 28.40.

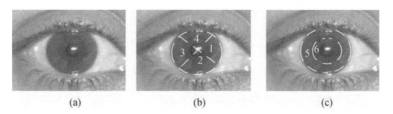

Fig. 28.40 Region wise iris extraction.

The velocity and acceleration are the kinematic characteristics applied to the samples of the iris patterns. The texture of the pupil is identified by extracting the 2D static features to calculate the rate of change of gray values. The velocity and acceleration in two-dimension is calculated for each pixel of the iris images. The velocity scatter plots are similar for genuine samples and different for imposter samples. The acceleration plots for a person show a similar variation pattern. It varies when compared to acceleration plots of a different person. The shape of the acceleration plot is similar for genuine samples. The velocity plot is shown in Fig. 28.41.

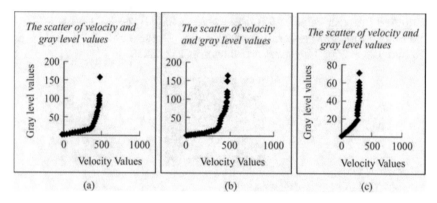

Fig. 28.41 Velocity scatter plot. (a) – (b) Two genuine samples. (c) Imposter sample.

The acceleration plot is shown in Fig. 28.42.

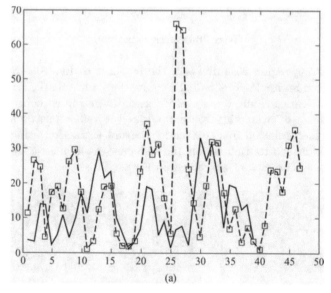

(a)

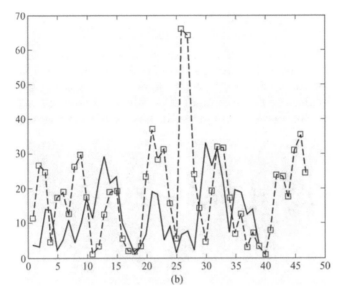

Fig. 28.42 Acceleration plot. (a) Genuine samples. (b) Genuine and imposter.

28.10 Preprocessing

Preprocessing involves operations to eliminate specular reflections, iris localization and removal of noise, eyelid and eyelashes.

28.10.1 Elimination of Specular Reflections

The specular reflections are caused by the illumination during image acquisition. It appears as a bright spot in the pupil region. The elimination method can use the intensity values of the neighboring pixels. The reflection is removed based on a suitable threshold value as shown in Fig. 28.43.

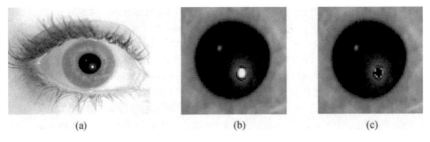

Fig. 28.43 (a) Eye image. (b) Segmented image. (c) Image after eliminating reflection.

28.10.2 Noise Removal

The types of noise that could be present in an image are Gaussian white noise, Poisson noise, salt and pepper noise and multiplicative noise. Noise elimination process use median and Wiener filters as shown in Fig. 28.44. Some of the image enhancement techniques like contrast stretching and histogram equalization can be used to improve the quality of the images.

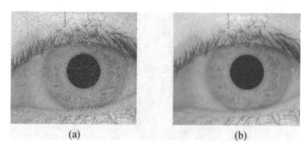

Fig. 28.44 (a) Noise in the image. (b) Image after noise elimination.

28.10.3 Elimination of Eyelid and Eyelash Occlusion

Occlusion is the measure of the iris texture covered by the eyelids and eyelashes. Occlusion reduce the accuracy of iris recognition. Eyelids can be identified using edge detection operators. The horizontal multiscale edges in the binary image represent the portion of eyelid and eyelashes. The eye image and iris segmented image is shown in Fig. 28.45 (a) and (b). Detection of eyelid in normalised rectangular image is performed by scanning for the white pixels. The length of white pixels determine the noise due to eyelids [24]. This is shown in Fig. 28.45 (c) and (d).

Eyelids are detected using arcuate curves with spline fitting, parabolic curve model and a linear model. The eyelids form a dome shaped representation that is detected using approximation of the circular model. The eyelids are eliminated using rectangular blocks in the region. The intensity of the eyelash near the curve boundaries are either lightened or smoothed. This is shown in Fig. 28.45 (e) – (g). A statistical model is developed with three Gaussians based on occlusion free images. The probability of the noisy images is lower compared to the well focused, non occluded images [25].

Filters help to retain useful information in the image. The types of filters that are used during preprocessing are the averaging filter, circular averaging filter, Gaussian low pass filter, filter to estimate linear motion and contrast enhancement filter. The result of applying these filters is shown in Fig. 28.46.

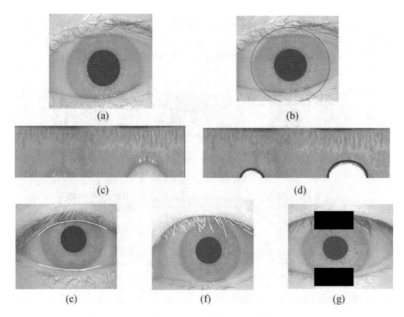

Fig. 28.45 Eyelid and Eyelash Elimination.

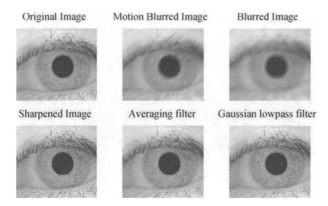

Fig. 28.46 Filtering operations.

28.10.4 Boundary Detection Operators

Edge detection operator applied on the grayscale image produces a binary image with the edges identified. The edges correspond to the pupil, iris, and eyelid boundaries.

The edge represents the change over form one intensity range to the other. The operators return the edges at the points where the gradient in the image is maximum. The types of edge detection operators are Sobel, Prewitt, Roberts, Laplacian of Gaussian, Canny and zero crossings. The result of applying the operators to the eye image is shown in Fig. 28.47.

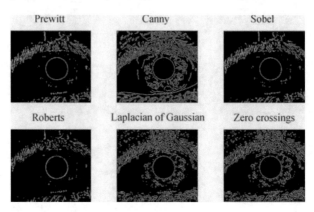

Fig. 28.47 Edge detection operators.

Morphology operations extract image components useful in the representation and description of region shape such as boundaries. The two operations commonly used in edge detection are dilation and erosion. Some of the structural elements used for operations are square, disk, line and octagon. The effect of dilation and erosion is prominent in the image where the intensity changes rapidly. The edges are clearly identified when the eroded image is subtracted from the dilated image. The operations are shown in Fig. 28.48.

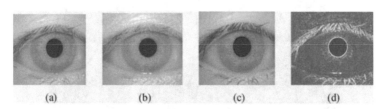

Fig. 28.48 (a) A original eye image. (b) Image after dilation. (c) Image after erosion. (d) subtraction of eroded image from dilated image.

Gradient operators use the first order derivative of the gray level. The Gaussian filter for the gradient operator is given by $y = 1/\sigma\sqrt{2\pi}exp(-x^2/2\sigma^2)$. The first order derivative of Gaussian is used to obtain the gradient images in the horizontal and vertical directions. The absolute summation value of the gradients is used to detect the edges. Figure 28.49 shows the gradient images for $\sigma = 0.5$.

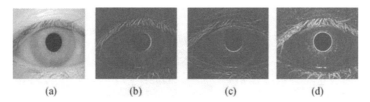

Fig. 28.49 (a) An eye image. (b) Gradient along x direction. (c) Gradient along y direction. (d) Absolute value of gradient.

28.10.5 Pupil Dilation

Pupil dilation is measured by ratio of the pupil radius to the iris radius. The dilation ratio lies in the range [0,1]. For ICE database, the dilation ratios are between 0.213 7 and 0.700 9 [26]. For the best performance the degree of dilation during enrollment must be similar to that during verification. Large differences in pupil dilation increases FRR. The samples with dilated pupil are shown in Fig. 28.50. Due to large size of the pupil, the iris portion is reduced which could be insufficient for recognition.

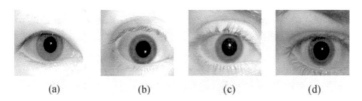

Fig. 28.50 Dilated Pupil samples.

28.10.6 Liveness Detection

Liveness detection can be performed during acquisition or during processing (27). It can be implemented by additional hardware, by using information present in the device or liveness information. The variation in the average pixel value is used to determine the liveness data. The fake iris does not show variations and has a constant pattern. This helps to reject the iris with cosmetic lenses. An eye image with lens is shown in Fig. 28.51 [28].

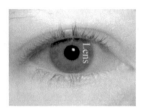

Fig. 28.51 Image with contact lens.

28.11 Iris Image Databases

The specifications of benchmark iris databases are listed in Table 28.2.

Table 28.2 Database specifications

Database	Version	Camera	Images	Subjects	Resolution
UBIRIS, SOCIA	V1	Nikon E5700	1877	241	400 × 300
Lab, Portugal	V2	Canon EOS 5D	11,102	261	800 × 600
	V1	Self-developed	756	108	320 × 280
	V2	Self-developed	1200	60	640 × 480
CASIA, NLPR, China	V3-Interval	Self-developed	2655	249	320 × 280
	V3-Lamp	OKI	16213	411	640 × 480
	V3-Twins	OKI	3183	200	640 × 480
ND 2004-2005, USA	–	Iridian LG EOU 2200	64,980	356	640 × 480
University of Bath, UK	Iris DB 400	AD-100 Iris Guard	8,000	200	1280 × 960
	Iris DB 800	AD-100 Iris Guard	16,000	400	1280 × 960
	Iris DB 1600	AD-100 Iris Guard	32,000	800	1280 × 960
UPOL, Olomouc	–	SONY DXC 950P	384	64	576 × 768
MMU, Malaysia	MMU1	LG Iris Access 2200	450	100	320 × 280
	MMU2	Panasonic BM ET100US	995	100	320 × 280

28.12 Iris Analyzers

Iris recognition methods are based on phase information, texture, zero-crossings, local intensity variations and kinematic characteristics.

28.12.1 Phase-based Method

The phase based method deals with recognising iris patterns based on phase information. Phase information is independent of imaging contrast and illumination [22, 29]. The eye images are captured with image focus assessment performed in real time. The pupil and iris boundaries are determined using integrodifferential operator. The representation of iris texture is binary coded by quantizing the phase response of a texture filter using quadrature 2D Gabor wavelets. The recognition is based on the test of statistical independence involving degrees of freedom. The dissimilarity measure between two iris patterns is computed using Hamming distance.

28.12.2   Texture-analysis Based Method

The limbus and pupil are modeled with circular contours which is extended to upper and lower eyelids with parabolic arcs. The particular contour parameter values x, y and radius r are obtained by the voting of the edge points using Hough transformation. The largest number of edge points represents the contour of the iris. The Laplacian of Gaussian is applied to the image at multiple scales and Laplacian pyramid is constructed. The iris code was produced using wavelet packets. Classification is performed using Fisher's linear discriminant function [30].

28.12.3   Zero-Crosssing Representation Method

The method represents features of the iris at different resolution levels based on the wavelet transform zero-crossing. The algorithm is translation, rotation and scale invariant [31]. The input images are processed to obtain a set of 1D signals and its zero crossing representation based on its dyadic wavelet transform. The wavelet function is the first derivative of the cubic spline. The centre and diameter of the iris is calculated from the edge-detected image. The virtual circles are constructed from the center and stored in circular buffers.

28.12.4   Approach Based on Intensity Variations

In the method characterized by local intensity variations, the sharp variation points of iris patterns are recorded as features [32]. In the iris localization phase, the centre coordinates of the pupil are estimated by image projections in horizontal and vertical directions. The exact parameters of the pupil and iris circles are calculated using Canny edge detection operator and Hough transform. The iris in cartesian coordinate system is projected into a doubly dimensionless pseudopolar coordinate system.

28.12.5   Approach using Independent Component Analysis

The iris recognition system adopts independent component analysis to extract iris texture features. Image acquisition is performed at different illumination and noise levels [33]. The iris localization is performed using integrodifferential operator and parabolic curve fitting. From the inner to outer boundary of iris, fixed number of concentric circles n with m samples on each circle is obtained. This is represented as a matrix $n \times m$ for a specific iris image which is invariant to rotation and size. The independent components are uncorrelated and determined from the feature coefficients. The independent components are estimated and encoded. The centre of each class is determined by competitive learning mechanism which is stored as the iris coding of a given person. The average Euclidean distance classifier is used to recognize iris patterns and match bits are used to measure reliability of match result.

28.12.6   Iris authentication Based on Continuous Dynamic Programming

The iris authentication method is based on kinematic characteristics, acceleration [34]. Continuous dynamic programming is used with the concept of comparing shape characteristics part wise. The acceleration plot is segmented and only certain parts of acceleration curve are used to verify with input's acceleration curve. For iris samples, rate of change of gray level intensities within bounding box forms acceleration feature plot. The implementation is based the concept of accumulated minimum local distances between a reference template and input sample. The reference template is obtained using Leave one out method. The distance measure is the count of directional changes in acceleration plot. The local distances are directional changes in respective segmented slot of acceleration plot. The methods are summarized in Table 28.3.

Table 28.3   Iris recognition methodologies

Group	Methodology	Size of Database
Daugman, 1994	Phase-based	4258 images
Wildes et al., 1994	Texture-analysis based	60 images
Boles and Boashash, 1998	Zero-crossings representation	Real images
Ya-Ping Huang, 2002	Independent Component Analysis	Real images
Li Ma, 2004	Local intensity variations	2245 images (CASIA)
Masek	Daugman-like algorithm with 1D log Gabor filters	624 images
Xiaomei Liu, 2005	Masek's method	12000 images (ICE)
H. Proenca and L.A.Alexandre, 2005	Moment-based	1877 images (UBIRIS)

Group	Methodology	Continued Size of Database
Tisse and Martin, 2002	Gradiet Hough Transform, 2D Hilbert transform	300 images
Jinyo Zuo, 2008	Pupil and Iris Segmentation	450 images (MMU1)
Jong Gook Ko et al., 2007	Cumulative-sum-based change point analysis	(i) 820 images from 82 individuals(ii) 756 images (CASIA)
N.Tajbakhsh, 2009	2D-Discrete Wavelet Transform	1877 images(UBIRIS)
Karen Hollingsworth et al., 2009	Improving recognition rates using (a) fragile bit masking, (b) signal-level fusion, (c) detect local distortions in iris texture and (d) analyze effects of pupil dilation	(a) 1226 images from 24 subjects(ICE). (b) 1061 videos from 296 eyes. (c) ICE database (d) 1263 images from 18 subjects(ICE)
Radhika et al., 2009	Continuous Dynamic Programming	(a)1205 images (UBIRIS)(b)1200 images (CASIAv2)

Iris Solution Providers:
1) LG Electronics Inc.
2) Oki Electric Industry Co. Ltd.
3) Iritech
4) Iridian
5) IBM Technologies

28.12.7 Products and Solutions

Iris authentication products are used in significant applications like civilian identification management programs. The iris recognition system by L-1 Identity Solutions is based on Daugman method. The algorithm has been used in the National Institute of Standards and Technology Iris Exchange testing program. The Offender Identification System supports identification of prisoners in jail environment. PIER 2.4 provides mobile identification with iris technology in a real time environment. The Handheld Interagency Identity Detection Equipment is a multi-biometric handheld device. It is used in defense agencies and in remote or centralized enrollments. The LG IrisAccess, Panasonic BM-ET200, Oki, IBM, IrisGuard IG-AD100, Sagem, Securimetrics and Argus systems work by analyzing the iris patterns and converting them into digital templates.

Biometric standards are system integrators to allow the interoperability of many technologies. Different manufacturers reduce the technical risk the consumer face, by implementing the standards in a scenario of complicated recognition algorithms. With standards, biometric service providers need only focusing on the products that are complaint with standards and purchasers will not be locked-in by a specific vendor due to technical limitations [35]. International organization for standardization, International Electrotechnical commission and InterNational Committee for information technology standards are some of the biometric standard providers. A common programming interface is provided by BioAPI consortium. To evaluate the performance of a biometric system, the following standard procedures are required: Modality-specific testing, Testing methodologies, Interoperability performance testing, and A cross-validation test.

28.13 Conclusion

The biometric security system rely on FRR and FAR values of system, hashing mechanism to retrieve template from database, liveliness detection mechanisms, anti-spoofing mechanisms and efficiency of communication media. Anti-spoofing identify fake properties. The fake aspects in iris detection are plastic fingers. Liveness detection process identifies alive properties. The utilization of external physiological biometrics with behavioral biometrics increases the reliability of authentication system. An application with different biometric information on a single algorithm of authentication, creates adaptiveness of the interactions. The factors for security with convenience lead to proper biometric selection. With varying performance of devices to extract features, multimodal methods create trustworthy system for large population sectors. Anatomical biometric is a constant signal. Behavioral biometric provides self-certification. With common algorithm, a method for userdependent threshold decisions are achieved for both biometrics in uniform fashion. The integration of static iris information and signature dynamic information can be achieved at decision level, to improve response time of the security system as compared to feature level and score level. In feature level, all features of both biometrics are integrated which increases the dimensionality. Score-level integration is at the classifier level. At decision level, inferences are drawn using voting techniques. Cancelable biometrics incorporates protection and replacement of features. In general, using a trajectory generation method comprising data compression and mining technique, a suitable authentication method for web and m-commerce applications can be provided. A server is a computer in network that is used to provide services to other computers. A client is a computer in network that uses the services provided by server. In on-line scenario, the authentication from server side can be performed by passing only the parameters of biometric sample from the client side. The biometric sample can be reconstructed at server-side. A good parameter mining method using unique biometric vault generation can be developed for authentication purpose. These methods search large databases for augmentation of stroke parameters. Instead of providing an image of constant size to recognizer, a numerically augmented parameter value with topological ordering is provided to make the authentication process faster in on-line database search scenario. The abundance applications on PDAs and smartphones represent a new scenario for automatic signature verification.

Acknowledgements

The authors would like to thank J.Ortega-Garcia for the provision of MCYT Signature database from Biometric Recognition Group, B-203, Universidad Autonoma de madrid SPAIN [6], Proenca H and Alexandre L.A, Portugal for UBIRIS database [17], Patrick J.Flynn, Department of Computer Science and Engineering, University of Notre Dame for ICE database [19] and CASIA-V1 by the Chinese Academy of Sciences Institute of Automation (CASIA) [7].

References

[1]   Jain A K, Bolle R, Pankanthi S (1998) Biometrics: Personal Identification in Networked Society. Kluwer Academic Publishers, Dordrecht
[2]   Yanushkevich S, Shmerko V, Stoica A et al (2007) Introduction to Synthesis in Biometrics. In: Yanushkevich S, Wang P, Gavrilova M et al (eds), Image Pattern Recognition: Synthesis and Analysis in Biometrics, Machine Perception and Artificial Intelligence, 67: 5 – 29. World Scientific Publishing, Singapore
[3]   Coats W S, Bagdasarian A, Helou T et al (2007) The Practitioners Guide to Biometrics. ABA Publications, Illinois
[4]   Gupta G K, Joyce R C (2007) Using Position Extrema Points to Capture Shape in Online Handwritten Signature Verification. Pattern Recognition, 40: 2811 – 2817
[5]   Jain A K, Nandakumar K, Ross A (2005) Score Normalization in Multimodal Biometric Systems, Pattern Recognition, 38: 2270 – 2285
[6]   Ortega-Garcia J, Fierrez-Aguilar J, Simon D et al (2003) MCYT Baseline Corpus: a Bimodal Biometric Database. In: IEEE Proceedings of Vision, Image and Signal Processing, 150(6): 395 – 401
[7]   Agam G, Suresh S (2007) Warping-Based Offline Signature Recognition, In: IEEE Transactions on Information Forensics and Security, Chicago, September 2007, p 430
[8]   Chen S, Srihari S (2006) A New Off-line Signature Verification Method Based on Graph Matching. In: IEEE International Conference on Pattern Recognition, 24 – 26 August 2006, Hong Kong, China
[9]   Freire M R, Fierrez J, Martinez-Diaz M et al (2007) On the Applicability of Off-line Signatures to the Fuzzy Vault Construction. In: IEEE International Conference on Document Analysis and Recognition, 23 – 26 September 2007, Parana, Brazil
[10] Hanmandlu M, Murthy, O R (2007) Fuzzy Model Based Recognition of Handwritten Numerals. Pattern Recognition, 40: 1840 – 1854
[11] Guru D S, Prakash H N (2007) Symbolic Representation of On-line Signatures. In: IEEE International Conference on Computational Intelligence and Multimedia Applications, 13 – 15 December 2007, Tamilnadu, India
[12] Fierrez J, Ortega-Garcia J, Ramos D et al (2007) HMM-based on-line Signature Verification: Feature Extraction and Signature Modeling. Pattern Recognition Letters, 28: 2325 – 2334.
[13] Kashi R, Hu J, Nelson N L et al (1997) On-line Handwritten Signature Verification Using Hidden Markov Model Features. In: IEEE International Conference on Document Analysis and Recognition, 18 – 20 August 1997, Ulm, Germany
[14] Bezine H, Alimi A M, Sherkat N (2007) Generation and Analysis of Handwriting Script with the Beta-Elliptic Model. International Journal of Simulation, 8: 45 – 65
[15] Djioua M, OReilly C, Plamondon R (2006) An Interactive Trajectory Synthesizer to Study Outlier Patterns in Handwriting Recognition and Signature Verification. In: IEEE International Conference on Pattern Recognition, 20 – 24 August 2006, Hong Kong, China
[16] Boyce C, Ross A, Monaco M, et al (2006) Multispectral Iris Analysis: A Preliminary

Study. In: Conference on Computer Vision and Pattern Recognition Workshop, 17–22 June 2006, New York, USA

[17]   Proenca H, Alexandre L A (2005) UBIRIS: A Noisy Iris Image Database. In: International Conference on Image Analysis and Processing, 3617: 970–977, 6–8 September 2005, Cagliari, Italy. Lecture Notes in Computer Science, vol 3617, p 970, Springer

[18]   Tieniu Tan (2009) CASIA-IrisV3, http://www.cbsr.ia.ac.cn/IrisDatabase.htm. Accessed 19 Jane 2010

[19]   Liu X, Bowyer K W, Flynn P J (2005) Experimental Evaluation of Iris Recognition. In: IEEE Computer Society Conference on Computer Vision and Pattern Recognition Workshops, pp 158–165, June 2005, San Diego, USA

[20]   Guodong G, Micheal J Jones (2008) Iris Extraction Based on Intensity Gradient and Texture Difference. In: IEEE Workshop on Applications of Computer Vision, Washington DC, January 2008, pp 1–6

[21]   Kaushik R (2008) Prabir Bhattacharya, EURASIP Journal on Image and Video Processing, pp 1–20

[22]   Daugman J (1993) High Confidence Visual Recognition of Persons by a Test of Statistical Independence. IEEE Transactions on Pattern Analysis and Machine Intelligence, 15(11): 1148–1161

[23]   Proenca H, Alexandre L A (2007) Toward Noncooperative Iris Recognition: A Classification Approach Using Multiple Signatures. In: IEEE Transactions on Pattern Analysis and Machine Intelligence, 29(4): 607–612

[24]   Passi A, Kumar A (2007) Improving iris Identification Using User Quality and Cohort Information. In: IEEE Conference on Computer Vision and Pattern Recognition, Minneapolis, 18–23 June 2007, Minneapolis, Minnesota, USA

[25]   Krichen E, Garcia-Salicetti S, Dorizzi B (2007) A new Probabilistic Iris Quality Measure for Comprehensive Noise Detection. In: IEEE Conference on Biometrics. Theory and Applications, 27–29 September 2007, Washington DC, USA

[26]   Hollingsworth K, Bowyer K W, Flynn P J (2009) Pupil Dilation Degrades Iris Biometric Performance. Computer Vision and Image Understanding, 113(1): 150–157

[27]   Deloitte T B (2005) Biometrics Liveness Detection, Information Security Bulletin, 10: 291–297

[28]   Kanematsu M, Takano H, Nakamura K (2007) Highly Reliable Liveness Detection Method for Iris Recognition. SICE Annual Conference, SICE Annual Conference, 17–20 September 2007, Takamatsu, Japan

[29]   Daugman J (2004) How Iris Recognition Works. In: IEEE Transactions on Circuits and Systems for Video Technology, 14(1): 21–30

[30]   Wildes R, Asmuth J, Green G et al (1996) A Machine-vision System for Iris Recognition. Machine Visual Application, 9: 1–8

[31]   Boles W W, Boashash B (1998) A Human Identification Technique Using Images of the Iris and Wavelet Transform. IEEE Transactions on Signal Processing, 46(4): 1185–1188

[32]   Ma L, Tan T, Wang Y et al (2003) Personal Identification Based on Iris Texture Analysis. IEEE Transactions on Pattern Analysis and Machine Intelligence, 25(12): 1519–1533

[33]   Huang Y P, Luo S W, Chen E Y (2002) An Efficient iris Recognition System. In: International Conference on Machine Learning and Cybernetics, Beijing, November 2002, pp 450–456

[34]   Radhika K R, Sheela S V, Venkatesha M K et al (2009) Multimodal Authentication Using Continuous Dynamic Programming. In: International Conference on Biometric ID Management and Multimodal Communication, 16–18 September 2009, Madrid, Spain, Springer LNCS vol 5707, pp 228–235

[35]   Wang P S P (2010) Pattern Recognition and Machine Vision. River Publishers

[36]   Kherallah M, Bouri F, Alimi A M (2009) On-line Arabic Handwriting Recognition System Based on Visual Encoding and Genetic Algorithm. Engineering Applications of Artificial Intelligence, 22: 153–170

[37]   Martinez-Diaz M, Fierrez J, Ortega-Garcia J (2007) Universal Background Models for Dynamic Signature Verification. In: IEEE International Conference on Biometrics: Theory, Applications and Systems, pp 1–6, 27–29 September 2007, Washington DC, USA

[38]   Mingming M, Wijesoma W, Sung E (2000) An Automatic On-line Signature Verification System Based on Three Models. In: IEEE Canadian Conference on Electrical and Computer Engineering, pp 890–894, March 2000, Halifax, Canada

[39]   Ramesh V E, Murty M N (1999) Off-line Signature Verification Using Genetically

Optimized Weighted Features. Pattern Recognition, 32: 217 – 233

[40] Sen A, Ananthakrishnan G, Sundaram S et al (2009) Dynamic Space Warping of Strokes for Recognition of Online Handwritten Characters. International Journal of Pattern Recognition and Artificial Intelligence, 23: 925 – 943

[41] Wen J, Fang B, Tang Y Y et al (2009) Model-based Signature Verification With Rotation Invariant Features. Pattern Recognition, 42: 1458 – 1466

29 Recent Trends in Iris Recognition

Lenina Birgale[1] and Manesh Kokare[2]

Abstract   Security has been an issue these days. Every nation all over the world is very much concerned about its data security says cables obtained by WikiLeaks. Generally, one can identify oneself with a system by three basic methods based on Knowledge (what you know), Possession (what you have), and Reality (who you are). Both knowledge (passwords, PINs etc) and possession (e-tokens, ID cards etc) based methods are theft prone. Hence only the third type reality based (biometrics) are the options to rely on. Biometrics classified as physiological and behavioural traits are in use. Physiological traits like facial features, voice patterns, hand geometry, retinal patters, vein patterns, facial thermography, DNA matching, nailbed identification, ear shape recognition, finger prints and behavioural traits like signature dynamics, voice verification, gait analysis, keystroke dynamics etc all explored as biometric identifiers with varying levels of success. However, they have their own limitations. Nevertheless, iris has unique patterns, as it is a fact that no two-iris patterns are alike. That is one cannot be enrolled with the right eye and authenticated with the left. Uniqueness of iris motivates oneself to sustain it as a biometric authentication technique. After rigorous review of hundreds of papers on iris recognition systems, the authors are presenting this chapter. It contributes for the recent trends in iris recognition methodologies.

29.1 Introduction

An ocean of information with unlimited expanse and unfathomable depth got created and made accessible to all human beings through huge databases. Even if the ocean of information is available, only a small drop of this ocean, which is appropriate for the specific event, is the requirement of the day. Thus in the vast ocean of available databases picking the drop that quenches our thirst is the current field of research as the level of ocean of database is increasing day by day. The curse of e-wars in all the e-systems is calling for more and more secure systems and protection of data. At the same time, the matter of border security has become a matter of concern. Hence, the researchers have focused their attention on the secure biometric systems. Biometrics like the one fingerprints [1 – 2], facial features [3], voice patterns, hand geometry, retinal patters, vein patterns, signature dynamics, iris etc

---

1, 2   Department of Electronics Communication Engineering, Shri Guru Gobind Singhji Institute of Engineering and Technology, Vishnupuri, Nanded, Maharashtra State, 431606 (India). E-mails: lenina2003_2003@yahoo.com; mbkokare@sggs.ac.in

are used. Of the entire biometrics iris, found to be as a unique one. Iris as an identifier has been developing since 1994. The uniqueness of iris patterns had identified since then. This property of iris can be quoted in the words of Daugman [4,5] as, "Advantage of the iris shares with fingerprints is the chaotic morphogenesis of its minutiae. The iris texture has a chaotic dimensions because its details depend on initial conditions in embryonic genetic expression, yet the limitation of partial genetic penetrance (beyond expression of form, function, color and general textural quality), ensures that even identical twins have uncorrelated iris minutiae. Thus the uniqueness of every iris, including the pair possessed by individual, parallels the uniqueness of every fingerprint."

The main motivation of this chapter is the guarantee assured by the iris as a lifelong and unique password as analysed in [6 – 10]. Biometric [11] based personal identification methods have recently gained more interests with an increasing emphasis on security. A lot of research is going on in this area since last two decades [12 – 28]. This work tries to give a consolidated view of recent trends in iris recognition to identify the recent trends in iris recognition systems. Iris in an eye is identified as shown in Fig. 29.1.

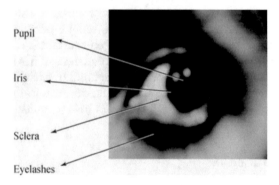

Fig. 29.1 A front-on view of the human eye.

Rest of the chapter is organised as follows. Section 29.2 gives basic modules of iris recognition. Section 29.3 highlights the performance measures used in iris recognition systems. Section 29.4 discusses limitations of available techniques of iris recognition systems. Lastly Section 29.5 elaborates the future scope of technique.

29.2 Basic Modules of Iris Recognition

Almost all the work on iris recognition revolves around five basic modules for decision that is acquisition, segmentation, normalisation, encoding and matching as shown in Fig. 29.2.

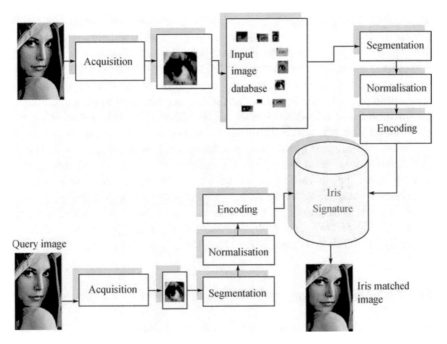

Fig. 29.2 Basic modules in iris recognition.

29.2.1 Acquisition

As long as the user is cooperative in image acquisition, iris recognition is a cakewalk. As the user becomes non-cooperative and situations are, non-ideal the work becomes quite difficult. Many researchers started working on this. Zhou et al. [29], Zuo et al. [30], Matey et al. [31], Proenca et al. [32], Medioni et al. [33], He et al. [34,35], Jinyu et al. [30,36], Krichen et al. [37], Hollingsworth et al. [38,39], Perez et al. [40], Thornton et al. [41], Schuckers et al. [42], and others also have worked to address this issue [42 – 47]. They gave new methods for the compensation of non-ideal iris images in order to improve iris recognition performance. However, some researchers have used Extended-Depth-of-Field to avoid restoration process [48]. Fast search algorithms [49], cascaded classifiers [50] and composite classifiers [51] are used for Large Fuzzy Databases after image acquisition.

29.2.2 Segmentation

The segmentation module identifies the pupillary and limbus boundaries. It detects and identifies the regions where the eyelids and eyelashes interrupt the limbus boundary's contour. The traditional detection mechanism is integro-differential operator. More recent works has promoted the use of active contours to account for nonconic boundary attributes. Shah et al. [52] proposed geodesic active contours (GACs) to extract the iris from the surrounding structures. Proenca [43, 45], presented a segmentation method that can handle degraded images acquired in less constrained conditions. Smereka [53] proposed a methodology with a capability of reliably segmenting nonideal imagery, which is simultaneously affected with such factors as specular reflection, blur, lighting variation, occlusion, and off-angle images. The adaptive fuzzy leader clustering (AFLC) architecture is a hybrid neural-fuzzy system, which learns on-line in a stable and efficient manner. It is proposed by Iskander et al. [54]. Vatsa et al. [55], improved recognition performance using segmentation, quality enhancement, match score fusion and indexing. He et al. [35] are the first to build and extract a rough position of the iris centre by an Adaboost-cascade iris detector.

29.2.3 Normalisation

The segmentation module has estimated the iris boundary. It is followed by the normalization module that uses a rubber-sheet model to transform the iris texture from cartesian to polar coordinates. The process is, often called iris unwrapping. It is also called as Daugmans rubber sheet model. Generally all methodologies [56–77] are normalization based except [7, 8]. Normalisation is generally, followed by encoding.

29.2.4 Encoding

Most systems first use a feature extraction routine to encode the iris textural content. A recognition system can use the unwrapped iris directly to compare two irides by using filters like the correlation filters. But, this is not suggestible as it requires more time and memory. Encoding algorithms generally perform a multiresolution analysis of the iris by applying wavelet filters and examining the ensuing response. In a commonly used encoding mechanism, 2D Gabor wavelets are first used to extract the local phasor information of the iris texture. The mechanism then encodes each phasor response using two bits of information, resulting in an Iris Code [78]. Schmid and Nicolo [79] worked on Principle Component Analysis (PCA) and Independent Component Analysis (ICA) Encoding for iris recognition.

29.2.5 Matching

The matching module generates a match score by comparing the feature sets of two iris images. Most widely used technique for comparing two iris codes is the Hamming distance, a number corresponding to the bits that differ between the two iris codes. The binary mask computed in the normalization module ensures that the technique compares only bits corresponding to valid iris pixels. The two iris codes need aligning before computing the Hamming distance through a registration procedure. While a simple translation operation could suffice in most cases, schemes that are more sophisticated can account for the elastic changes in iris texture. Feng et al. [80] proposed a novel method called classifier combination and its application in iris recognition, for matching. According to them in the context of verification, the methods for information fusion at the matching score level can be classified into two categories.

1) The classification approach.
2) The score combination approach.

In the first approach, the matching scores obtained from different modalities are treated as a feature vector, and a classifier, such as a support vector machine, is adopted to assign the input pattern to a certain category. The other category is the score combination approach. The matching scores are combined to generate a single score, and the final decision is made according to this score. Their study focuses on the latter, and they investigated two different combination rules (i.e., the max rule and the min rule). They concluded that, the max-combined classifier and min-combined classifier performances are conditional.

Researchers have also designed other types of encoding and matching schemes, based on discrete cosine transforms [81], ordinal features [82], phase based [83, 84] and scale-invariant feature transforms.

29.3 Performance Measures

Generally used performance measures in iris recognition are:
- FAR   False Acceptance Rate.
- FRR   False rejection Rate.
- ROC   Receivers Operating Curves.
- DET   Detection of Error Trade-off.

29.3.1 False Acceptance Rate

False acceptance rate (FAR), the fraction of access attempts by an un-enrolled individual that are nevertheless deemed a match. The FAR must be very, very low to provide any confidence in the technology.

$$FAR = \frac{NFA}{NIVA}.$$ (29.1)

- NFA   Number of False Acceptances.
- NIVA   Number of Imposter Verification Attempts.
- FAR   The measure of likelihood that the biometric security system would incorrectly accept that an attempt by an unauthorized user.

29.3.2 False rejection Rate

False rejection rate (FRR), the fraction of access attempts by a legitimately enrolled individual that are nevertheless rejected. The FRR must be sufficiently low that the users will not abandon the technique due to frustration.

$$FRR = \frac{NFR}{NEVA}.$$ (29.2)

- NFR   Number of False Rejections.
- NEVA   Number of Imposter Verification Attempts.
- FRR   The measure of likelihood that the biometric security system would incorrectly reject that an attempt by an authorized user.

29.3.3 Receivers Operating Curves

ROC is a plot of genuine acceptance rate against false acceptance rate for all possible system operating points (that is matching threshold) and measure the overall performance of the system. Each point on the curve represents a particular decision threshold. In ideal case, both the error rates that is FAR and FRR should be zero and imposter distribution should be disjoint. Ideal ROC in such case is a step function at zero FAR. On the other hand if genuine and imposter distributions are exactly the same, then ROC is a line segment with a slope of $45°$ (forty five degrees) with an end point at zero FAR. In practice, these curves behave in between these two extremes. An equal error rate (ERR) is defined as the operating point where the two types of errors (FAR and FRR) are equal.

29.3.4 Detection of Error Trade-off

Modified ROC curve is known as DET curve. It plots error rates on both axes.

It gives uniform treatment to both types of errors.

29.4 Limitations of Current Techniques

Most research had been focused on the development of new iris processing and recognition algorithms for frontal view iris images. However, a few challenging directions in iris research have been identified, including processing of a non-ideal iris and iris at a distance. Schukers et al. [42], describe two non-ideal iris recognition systems and analyze their performance. The word "non-ideal" is used in the sense of compensating for off-angle occluded iris images. Their system is designed to process non-ideal iris images in two steps:

1) Compensation for off-angle gaze direction;

2) Processing and encoding of the rotated iris image.

Two approaches presented by them account for angular variations in the iris images. In the first approach, they use Daugmans integro differential operator as an objective function to estimate the gaze direction. After the angle is estimated, the off-angle iris image undergoes geometric transformations involving the estimated angle and is further processed as if it were a frontal view image. The encoding technique developed for a frontal image is based on the application of the global independent component analysis. Their second approach uses an angular deformation calibration model.

The major limitation of the approach resides in linearization of the angle-estimation problem. Because of a convex shape of the eye and the difference of indexes of refraction characterizing the interior and exterior portions of the eye, an iris pattern imaged at an off-angle undergoes highly nonlinear transformations. In their approach to this problem, they assumed that the nonlinear distortions can be linearly approximated. In addition, in the algorithm, they have made an initial guess in estimating the angle. They use the "assumed" angle as the initial guess in this implementation. In a real-world application, an automatic method in determining an initial guess is needed. The method is also computationally intensive and, thus, may look impractical. However, knowledge of approximate gaze direction, for example, evaluated by some other means, will limit the search of optimum angle to a narrow range of angles and will make the computations feasible. Thus, these two limitations identify directions for the future work. Their present implementation is not capable of handling varying distances between the eye and the capturing device. Another drawback of their model is that the training is semi automated.

Kalka et al. [12], proposed a method for estimation and fusion of quality factors for iris image. The main limitation of their approach is the requirement of segmentation. Failed localization/segmentation will result in inaccurate quality scores. However, this would have negative consequences only if the matching algorithm applied to the same iris image performs segmentation successfully. Therefore, as long as the segmentation algorithm used for quality evaluation is as sophisticated as the one used in quality evaluation, it is unlikely that quality scores will be misleading.

Eyelids, eyelashes and shadows are three major challenges for effective iris segmentation.

Also not much work is found on spectral analysis of iris.

29.5 Future Scope

All the algorithms proposed are on the ideal databases obtained in controlled conditions. These databases are created in ideal condition with cooperative users at predefined distances. Now the challenge lies if the range of iris acquisition is extended from centimeters to meters. That is when the conditions turn out to be non-ideal. Human intervention is necessary in the present systems. Fully automated training systems without human intervention are the future scope of the iris recognition systems. At the same time, computational cost needs to be maintained as low as possible.

All the spectral components of the image may be used for recognition process if image acquisition cameras support for colored images. The need to deploy segmentation within the quality assessment algorithm makes Kalka et al.'s [12] approach unsuitable for real-time applications in which a quality factor would be used for the selection of the "best" frame from a sequence (e.g., streaming video). Future work includes perfecting the estimation techniques for the described quality factors, along with experimenting with the new quality scores that incorporate correlation. Furthermore, their proposed framework is open for the inclusion of new iris quality factors that will undoubtedly emerge through further research or through further relaxation of acquisition constraints (e.g., distance, motion, and non-uniform lighting). He et al. [34, 35], proposed contactless auto feedback Iris Capture Design. It can be further improved by focusing on the following four issues:

1) Designing a more presentable interface;

2) Considering occlusion of the eyelids and eyelashes, which may cause FRR, in future work, which they did not do in their experiments, and adding the iris image quality evaluating algorithm to the device to automatically capture better iris images;

3) Embedding the live iris detection algorithm to the system;

4) Conducting experiments on a large number of iris databases in various environments for the system to be more stable and reliable.

Thus, an iris recognition system which supports all time, all environments is the need of the day.

Acknowledgements

We grateful to all the anonymous reviewers of this chapter for the efforts they have put towards rigorous review of this chapter. We would also thank the editors who have brought this chapter to this form. Last but not the least we would like to acknowledge our regards to all the researchers who are actively working in this area and formed the source of this chapter.

References

[1] Sanchez-Reillo R (2001) Smart Card Information and Operations Using Biometrics. IEEE Aerospace and Electronic Systems Magazine, 16(4): 3–6

[2] Conti V, Militello C, Sorbello F et al (2010) Frequency-based Approach for Features Fusion in Fingerprint and Iris Multimodal Biometric Identification Systems. IEEE Transactions on Systems, Man, and Cybernetics, 40(4): 384–395.

[3] Moriyama T, Kanade T, Jing Xiao et al (2006) Meticulously Detailed Eye Region Model and its Application to Analysis of Facial Images. IEEE Transactions on Pattern Analysis and Machine Intelligence, 28(5): 738–752

[4] Daugman J (1993) High Confidence Visual Recognition of Persons by a Test of Analysis of Statistical Independence. IEEE Transactions on Pattern Analysis and Machine Intelligence, 15(11): 1148–1161

[5] Daugman J (2006) Probing the Uniqueness and Randomness of Iris Codes: Results From 200 Billion Iris Pair Comparisons. Proc of the IEEE, 94(11): 1927–1995

[6] Lenina Birgale and Manesh Kokare (2009) Iris Recognition Using Discrete Wavelet Transform. Proceedings of IEEE International Conference ICDIP09, 7–9 March 2009, Bangkok, Thailand

[7] Lenina Birgale and Manesh Kokare (2010) Iris Recognition Without Iris Normalisation. Journal of Computer Science. Science Publications, 6(9): 1042–47

[8] Lenina Birgale and Manesh Kokare (2009) Iris Recognition Without Iris Normalisation. In: Proceedings of Indian Conference on Computer Vision, Graphics, Image and Video. Nagpur a blank India, 13–14 Mar 2009

[9] Birgale L, Manesh K (2009) A Survey on Iris Recognition. The Icfai University Journal of Electrical & Electronics Engineering (IUJEEE), 2(4): 7–25

[10] Williams G O (1997) Iris Recognition Technology. In IEEE Aerospace and Electronics. Systems Magazine, 12(4): 23–29

[11] Leung M K H, Fong A C M, Siu C H (2007) Palmprint Verification for Controlling Access to Shared Computing Resources. IEEE Pervasive Computing, 6(4): 40–47

[12] Kalka N D, Jinyu Zuo, Schmid N A et al (2010) Estimating and Fusing Quality Factors for Iris Biometric Images, IEEE Transactions on Systems, Man and Cybernetics, Part A: Systems and Humans, 40(3): 509–524

[13] Wildes R P (1997) Iris Recognition: An Emerging Biometric Technology. In: Proceedings of the IEEE, 85(9): 1348–1363

[14] Weaver A C (2006) Biometric Authentication. Computer, 39(2): 96–97

[15] Negin M, Chmielewski T A, J, Salganicoff M et al (2000) An Iris Biometric System for Public and Personal Use. Computer, 33(2): 70–75

[16] Sanchez-Avila C, Sanchez-Reillo R, De Martin-Roche D (2002) Iris-based Biometric Recognition Using Dyadic Wavelet Transform. IEEE Aerospace and Electronic Systems Magazine, 17(10): 3–6

[17] Daugman J (2004) How Iris Recognition Works. IEEE Transactions on Circuits and Systems for Video Technology, 14(1): 21–30

[18] Sasse M A (2007) Red-Eye Blink, Bendy Shuffle, and the Yuck Factor: A User Experience of Biometric Airport Systems. IEEE Security & Privacy, 5(3): 78–81

[19] Miller B (1994) Vital Signs of Identity. IEEE Spectrum, 31(2): 22–30

[20] Fu L M, Hsu H H, Principe, J C (1996) Incremental Back Propagation Learning Networks, IEEE Transactions on Neural Networks, 7(3): 757–761

[21] Phillips P J, Newton E M (2009) Biometric Systems: The Rubber Meets the Road. I Proceedings of IEEE, 97(5): 782–783

[22] Wayman J L (2008) Biometrics in Identity Management Systems. IEEE Security & Privacy, 6(2): 30–37
[23] Wang Z F, Han Q, Li Q et al (2009) Complex Common Vector for Multimodal Biometric Recognition. Electronics Letters, 45(10): 495–496
[24] Dass S C, Yongfang Zhu, Jain A K (2006) Validating a Biometric Authentication System: Sample Size Requirements. IEEE Transactions on Pattern Analysis and Machine Intelligence, 28(12): 1902–1319
[25] Kroeker K L (2002) Graphics and Security: Exploring Visual Biometrics. IEEE Computer Graphics and Applications, 22(4): 16–21
[26] Yager N, Dunstone T (2010) The Biometric Menagerie. IEEE Transactions on Pattern Analysis and Machine Intelligence, 32(2): 220–230
[27] Shen W C, Khanna R (1997) Prolog To Iris Recognition: An Emerging Biometric Technology. Proceedings of the IEEE, 85(9): 1347–1347
[28] Scotti F, Piuri V (2010) Adaptive Reflection Detection and Location in Iris Biometric Images by Using Computational Intelligence. IEEE Transactions on Techniques Instrumentation and Measurement, 59(7): 1825–1833
[29] Zhi Zhou, Yingzi Du and Belcher C (2009) Transforming, Traditional Iris Recognition Systems to Work in Nonideal Situations. IEEE Transactions on Industrial Electronics, 56(8): 3203–3213
[30] Jinyu Zuo, Natalia A Schmid, Xiaohan Chen (2007) On Generation and Analysis of Synthetic Iris Images. IEEE Transactions on Information Forensics and Security, 2(1): 77–90
[31] Matey J R, Naroditsky O, Hanna K et al (2006) Iris on the Move: Acquisition of Images for Iris cognition in Less Constrained Environments. Proceedings of the IEEE, 94(11): 1936–1947
[32] Proenca H, Filipe S, Santos R et al (2010) The UBIRIS. v2: A database of Visible Wavelength Iris Images Captured On-the-Move and At-a-Distance. IEEE Transactions on Pattern Analysis and Machine Intelligence, 32(10): 1529–1535
[33] Medioni G, Choi J, Cheng-Hao Kuo et al (2009) Identifying Noncooperative Subjects at a Distance Using Face Images and Inferred Three-Dimensional Face Models. IEEE Transactions on Systems, Man and Cybernetics, Part A: Systems and Humans, 39(1): 12–24
[34] Xiaofu He, Jingqi Yan, Guangyu Chen et al (2008) Contactless Autofeedback Iris Capture Design. IEEE Transactions on Instrumentation and Measurement, 57(7): 1369–1375
[35] Zhaofeng He, Tieniu Tan, Zhenan Sun et al (2009) Toward Accurate and Fast Iris Segmentation for Iris Biometrics. IEEE Transactions on Pattern Analysis and Machine Intelligence, 31(9): 1670–1684
[36] Jinyu Zuo, Schmid N A (2010) On a Methodology for Robust Segmentation of Nonideal Iris Images. IEEE Transactions on Systems, Man, and Cybernetics, Part B: Cybernetics, 40(3): 703–718
[37] Krichen E, Garcia-Salicetti S, Dorizzi B (2009) A New Phase-Correlation-Based Iris Matching for Degraded Images. IEEE Transactions on Systems, Man, and Cybernetics, Part B: Cybernetics, 39(4): 924–934
[38] Hollingsworth K, Peters T, Bowyer K W et al (2009) Iris Recognition Using Signal-Level Fusion of Frames From Video. IEEE Transactions on Information Forensics and Security, 4(4): 837–848
[39] Hollingsworth K P, Bowyer K W, Flynn P J (2009) The Best Bits in an Iris Code. IEEE Transactions on Pattern Analysis and Machine Intelligence, 31(6): 964–973
[40] Perez C A, Lazcano V A, Estevez P A (2007) Real-Time Iris Detection on Coronal-Axis-Rotated Faces. IEEE Transactions on Systems, Man, and Cybernetics, Part C: Applications and Reviews, 37(5): 971–978
[41] Thornton J, Savvides M, Kumar V (2007) A Bayesian Approach to Deformed Pattern Matching of Iris Images. IEEE Transactions on Pattern Analysis and Machine Intelligence, 29(4): 596–606
[42] Schuckers S A C, Schmid N A, Abhyankar A et al (2007) On Techniques for Angle Compensation in Nonideal Iris. Recognition. IEEE Transactions on Systems, Man, and Cybernetics, Part B: Cybernetics, 37(5): 1176–1190
[43] Proenca H, Alexandre L A (2007) Toward Noncooperative Iris Recognition: A Classification Approach Using Multiple Signatures. IEEE Transactions on Pattern Analysis and Machine Intelligence, 29(4): 607–612
[44] Rakvic R N, Ulis B J, Broussard R P et al (2009) Parallelizing Iris Recognition. IEEE Transactions on Information Forensics and Security, 4(4): 812–823

[45]   Proença H (2010) Iris Recognition: On the Segmentation of Degraded Images Acquired in the Visible Wavelength. IEEE Transactions on Pattern Analysis and Machine Intelligence, 32(8): 1502–1516

[46]   Kang Ryoung Park, Jaihie Kim (2005) A Real-time Focusing Algorithm for Iris Recognition Camera. IEEE Transactions on Systems, Man, and Cybernetics, Part C: Applications and Reviews, 35(3): 441–444

[47]   Schonberg D, Kirovski D (2006) EyeCerts. IEEE Transactions on Information Forensics and Security, 1(2): 144–153

[48]   Boddeti V N, Kumar B V K V (2010) Extended-Depth-of-Field Iris Recognition Using Unrestored Wavefront-Coded Imagery. IEEE Transactions on Systems, Man and Cybernetics, Part A: Systems and Humans, 40(3): 495–508

[49]   Feng Hao, Daugman J, Zielinski P (2008) A Fast Search Algorithm for a Large Fuzzy Database. IEEE Transactions on Information Forensics and Security, 3(2): 203–212

[50]   Zhenan Sun, Yunhong Wang, Tieniu Tan et al (2005) Improving Iris Recognition Accuracy via Cascaded Classifiers. IEEE Transactions on Systems, Man, and Cybernetics, Part C: Applications and Reviews, 35(3): 435–441

[51]   Dasarathy B V, Sheela B V (1979) A Composite Classifier System Design: Concepts and Methodology. Proceedings of the IEEE, 67(5): 708–713

[52]   Shah S, Ross A (2009) Iris Segmentation Using Geodesic Active Contours. IEEE Transactions on Information Forensics and Security, 4(4): 824–836

[53]   Smereka J M (2010) A New Method of Pupil Identification. IEEE Potential, 29(2): 15–20

[54]   Iskander D R, Collins M J, Mioschek S et al (2004) Automatic Pupillometry From Digital Images. IEEE Transactions on Biomedical Engineering, 51(9): 1619–1627

[55]   Vatsa M, Singh R, Noore A (2008) Improving Iris Recognition Performance Using Segmentation, Quality Enhancement, Match Score Fusion, and Indexing. IEEE Transactions on Systems, Man, and Cybernetics, Part B: Cybernetics, 38(4): 1021–1035

[56]   Kang B J, Park K Ry (2007) Real-Time Image Restoration for Iris Recognition. IEEE Transactions on Systems, Systems, Man, and Cybernetics, Part B: Cybernetics, 37(6): 1555–1566

[57]   Chou C T, Shih S W, Chen W S et al (2010) Non-Orthogonal View Iris Recognition. IEEE Transactions on System, Circuits and Systems for Video Technology, 20(3): 417–430

[58]   Belcher C, Yingzi Du (2008) A Selective Feature Information Approach for Iris Image-Quality Measure. IEEE Transactions on Information Forensics and Security, 3(3): 572–577

[59]   Kong A W K, Zhang D, Kamel M S (2010) An Analysis of IrisCode. IEEE Transactions on Image Processing, 19(2): 522–532

[60]   Ma L, Tan T, Yunhong Wang et al (2003) Personal Identification Based on Iris Texture Analysis. IEEE Transactions on Pattern Analysis and Machine Intelligence, 25(12): 1519–1533

[61]   Daugman J (2007) New Methods in Iris Recognition. IEEE Transactions on Systems, Man, and Cybernetics, Part B: Cybernetics, 37(5): 1167–1175

[62]   Poh N, Bourlai T, Kittler J et al (2009) Benchmarking Quality-Dependent and Cost-Sensitive Score-Level Multimodal Biometric Fusion Algorithms. IEEE Transactions on Information Forensics and Security, 4(4): 849–866

[63]   Rakshit S, Monro D M (2007) An Evaluation of Image Sampling and Compression for Human Iris Recognition. IEEE Transactions on Information Forensics and Security, 2(3): 605–612

[64]   Luengo-Oroz M A, Angulo J (2009) Cyclic Mathematical Morphology in Polarogarithmic. IEEE Transactions on Image Processing, 18(5): 1090–1096

[65]   Chernyak D A (2005) Iris-based Cyclotorsional Image Alignment Method for Wavefront Registration. IEEE Transactions on Biomedical Engineering, 52(12): 2032–2040

[66]   Bhanu B, Ratha N K, Kumar V, et al (2007) Guest Editorial: Special Issue on Human Detection and Recognition. IEEE Transactions on Information Forensics and Security, 2(3): 489–490

[67]   Daugman J G (1993) High Confidence Visual Recognition of Persons by a Test of Statistical Independence. IEEE Transactions on Pattern Analysis and Machine Intelligence, 15(11): 1148–1161

[68]   Phillips P J, Bowyer K W, Flynn P J (2007) Comments on the CASIA Version 1.0 Iris Data Set. IEEE Transactions on Pattern Analysis and Machine Intelligence,

29(10): 1869–1870

[69] Lee Y J, Park K R, Lee S J et al (2008) A New Method for Generating an Invariant Iris Private Key Based on the Fuzzy Vault System. IEEE Transactions on Systems, Man, and Cybernetics, Part B: Cybernetics, 38(5): 1302–1313

[70] Daugman J (1997) Face and Gesture Recognition: Overview. IEEE Transactions on Pattern Analysis and Machine Intelligence, 19(7): 675–676

[71] Boles W W, Boashash B (1998) A Human Identification Technique Using Images of the Iris and Wavelet Transform. IEEE Transactions on Signal Processing, 46(4): 1185–1188

[72] Newton E M, Phillips P J (2009) Meta-Analysis of Third-Party Evaluations of Iris Recognition. IEEE Transactions on Systems, Man and Cybernetics, Part A: Systems and Humans, 39(1): 4–11

[73] Velisavljevic V (2009) Low-Complexity Iris Coding and Recognition Based on Directionlets. IEEE Transactions on Information Forensics and Security, 4(3): 410–417

[74] Sudha N, Puhan N B, Xia H et al (2009) Iris Recognition on Edge Maps. IET Computer Vision, 3(1): 1–7

[75] Jaiyen S, Lursinsap C, Phimoltares S (2010) A Very Fast Neural Learning for Classification Using Only New Incoming Datum. IEEE Transactions on Neural Networks, 21(3): 381–392

[76] Joni-Kristian Kamarainen and Ville Kyrki (2006) Invariance Properties of Gabor Filter-based Features-overview and Applications. IEEE Transactions on Image Processing, 15(5): 1088–1099

[77] Ma L, Tan T, Wang Y et al (2004) Efficient Iris Recognition by Characterizing key Local Variations. IEEE Transactions on Image Processing, 13(6): 739–750

[78] Ross A (2010) Iris Recognition: The Path Forward. Computer, 43(2): 30–35

[79] Schmid N A, Nicolo F (2008) On Empirical Recognition Capacity of Biometric Systems Under Global PCA and ICA Encoding. IEEE Transactions on Information Forensics and Security, 3(3): 512–528

[80] Feng X, Ding X, Wu Y et al (2008) Classifier Combination and Its Application. In Iris Recognition, IJRAI, 22(3): 617–638

[81] Monro D M, Rakshit S, Zhang D (2007) DCT-Based Iris Recognition. IEEE Transactions on Pattern Analysis and Machine Intelligence, 29(4): 586–595

[82] Sun Z, Tan T (2009) Ordinal Measures for Iris Recognition. IEEE Transactions on Pattern Analysis and Machine Intelligence, 31(12): 2211–2226

[83] Schmid N A, Ketkar M V, Singh H et al (2006) Performance Analysis of Iris-based Identification System at the Matching Score Level. IEEE Transactions on Information Forensics and Security, 1(2): 154–168

[84] Miyazawa K, Ito K, Aoki T et al (2008) An Effective Approach for Iris Recognition Using Phase-Based Image. IEEE Transactions on Matching Pattern Analysis and Machine Intelligence, 30(10): 1741–1756

30 Using Multisets of Features and Interactive Feature Selection to Get Best Qualitative Performance for Automatic Signature Verification

Maan Ammar[1]

Abstract Traditionally, Automatic Signature Verification (ASV) that uses threshold-based decision, depends on using one feature set for verification. Recently, the Multi-Sets of Features (MSF) approach has been proposed in the field of ASV to get higher performance than that obtained by the best feature set. This chapter reports the results obtained by using the MSF technique in case of using large number of feature sets. It introduces also the concept of qualitative performance in ASV and using Interactive Feature Selection to evaluate it.

30.1 Introduction

Writing instrument and the data capture used, determine the type of signature. Essentially, signatures are two types. The first type is on-line signatures like those obtained from a three axis writing pen (x, y, and pressure) [1], digitizing tablets [2], or hand gloves [3], and the second is off-line ones like those we usually find on letters, contracts, and bank checks. The research introduced in this paper is related to the second type.

Ammar et al. reported in 1986 the first successful work on verification of skilled forgeries [4, 5]. Their principle of extracting High Pressure Regions (HPRs) in signatures was adopted later by other researchers for further study [6, 7]. It has also motivated others to explore other ways of determining the threshold used to extract the HPRs [8, 9]. In 1989 Ammar et al. investigated using shape features, HPR features, and both of them for ASV [10]. At the same time, they investigated the effectiveness of individual shape features, different shape feature sets, and mixed ones with the implications of automatic determination of the verification threshold VTH, using a feature selection algorithm they developed [11]. In 1990, Ammar used signature projections and matching for extracting new features [12]. He investigated the performance of the new features and all previous ones using the same feature selection algorithm, and reached new results [13]. Later, other researchers

1 Biomedical Engineering Department, FMEE, Damascus University, P. O. Box 86 Syria.
  E-mail: maan_ammar@yahoo.com.

used projections and shape features introduced previously by Ammar et al.
[11] like baseline, area, and the ratio of height and width for off-line ASV using Support Vector Machine (SVM) and ANN techniques [14, 15]. Recently,
some research works attempt to practically evaluate published approaches
[16], and others are reattempting to explore the potential effectiveness in the
gray level image [17]. More ASV-related works can be found in related review
papers [18 – 20] and recent publications.

In 1995, Ammar et al. reported the realization of a portable software for
off-line ASV usable with PCs under DOS (SIGVA 1.0) [Technical Report TR
2/95 (1995), SigSoft Company, Damascus, Syria]. Later in 2002, an off-line
ASV system that can be used in the Interbank Check Imaging (ICI) environment in USA banks was realized and put in actual use [21, 22]. Developing
this system required solving several problems like extracting signature image
from noisy documents, recognizing the signatory in multi-signatory accounts
and verification in case of Single Reference Signature (SRS) even with signature resolution as low as 80 dpi [21]. Figure 30.1 shows two examples from the
ICI environment. In this environment, raising the performance of ASV using
several techniques was necessary [23 – 26]. The MSF-based decision making
technique used to conduct the MSF related study reported in this paper was
one of these techniques.

SVM classifier introduced in 1992 [27] has been popular in Pattern Recognition applications research. It has been used widely in ASV research works
[3, 14, 15, 28 – 31]. Some reported results showed that it outperforms other
classification methods (Naive Bayes, and Distance Statistics), when training on the two classes (genuine and forgery). However, when training on the
genuine class (16 samples) only, which is the practical case of ASV, *SVM
was the worst among four methods* (SVM, Naive Bayes, Distance Statistics,
and Distance Threshold) with 46% error rate on research data [31]. E. J. R.
Justino et al. reported False Rejection (FR)=32% (type I error) at the best
False Acceptance (FA) [30]. Considering these results and the others in the
above mentioned SVM works, and the intrinsic nature of the SVM technique,
the SVM was considered as inappropriate for actual ASV system, where the
majority of the accounts are SRS ones, only the genuine class is available for
training in the rest of the accounts, *and the FA should be "zero" , and FR
should be within 10%.*

In this chapter, the MSF-based decision making technique for off-line
ASV is explained in brief. Its efficiency in case of using medium and large
number of features, and in case of using small and large number of feature
sets is evaluated and discussed. The concept of Qualitative Performance (QP)
in ASV and using Interactive Feature Selection (IFS) to evaluate it, is also
introduced.

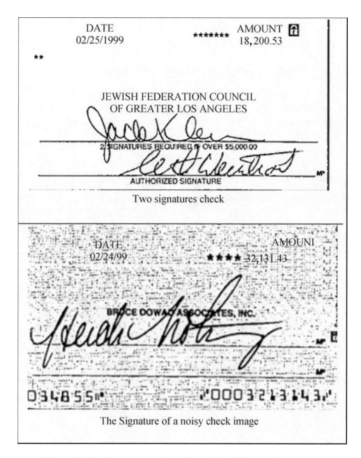

Two signatures check

The Signature of a noisy check image

Fig. 30.1 Two examples from the ICI environment.

30.2 Signature Data

The genuine samples of this signature data were collected from documents from the daily life activities so that the evaluation experiments give a good indication of the behavior of the system when put in actual applications. The signature data consists of 560 genuine and forgery signatures belong to 26 writers. The signatures are written in different languages by people of different nationalities including Arabic, Japanese, Koreans, German, and Americans. The number of genuine signatures and forgeries differ from one person to another. Moreover, the documents from which the signatures were extracted vary from white paper, business documents, to bank checks so that the signature data is naturally written under widely different conditions. Forgeries were created with a good attention in order to have convincing

forgeries, and some forgeries are real ones obtained from actual caseworks.
Figure 30.2 shows the complete set of genuine and forgery samples of 4 per-

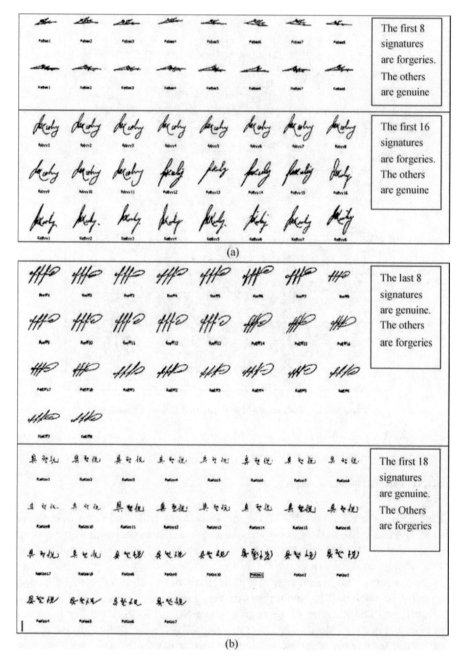

(a)

(b)

Fig. 30.2 Four complete sets of the signatures of 4 persons. Each set consists of
unequal number of genuine and forgery samples (specified in the figure).

sons, and Fig. 30.3 shows examples of the signatures available in the signature database.

Fig. 30.3 Examples of the signatures in the database.

30.3 ASV Systems Using Threshold-Based Decision

ASV systems using threshold-based decision usually follow the general approach shown in Fig. 30.4 in order to give a decision about an input signature, whether it is genuine or an attempted forgery. Since the MSF and the IFS technique introduced in this paper are using threshold-based ASV approach, the processes in Fig. 30.4 as used in the system will be explained first.

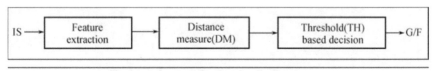

IS: input signature; G: Genuine; F: Forgery.

Fig. 30.4 The general approach of threshold-based ASV system.

30.3.1 Feature Extraction

This stage may follow a preprocessing one [10]. While dynamic time wrapping
and matching are used for feature extraction in recent on-line works [29, 32],
a new matching technique has been in use for feature extraction in off-line
systems [33]. Features in off-line systems are essentially two types: (1) shape
features like handwriting slants (positive, vertical, negative, and horizontal),
relative measures of signature height and width, middle zone width and signature width, and (2) pseudo-dynamic features like High Pressure Factor.
Those features can be extracted globally on the signature as a whole, and
locally on the signature divided into specific parts [10, 11, 33].

Features used in this research: The features used in this paper
are a modified version of the previous ones reported in [11]. They are the
four slants: percentage of (positively, negatively, vertically and horizontally
slanted pixels) [10, 11] measured locally on the contour-detected-signature
divided horizontally into six parts as shown in Fig. 30.5, and globally on the
image as a whole. The six parts are determined as 3 equal width parts (Xleft)
to the left of the Gravity Center (GC) of the signature, and 3 equal width others to its right (Xright); the "x" and "y" coordinates of the Gravity Center;
effective length: the length containing 80% of the area of the signature after
omitting 10% to the left and 10% to the right; effective width computed in a
way similar to the length; the baseline; and the area of the signature in each
one of the six parts computed as a percentage of the total area. Measuring
slants and areas in a relative manner makes them scale invariant.

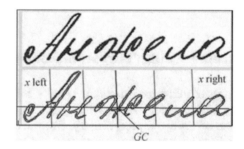

Fig. 30.5 Signature segmentation for feature extraction.

30.3.2 Distance Measure (DM)

DM measures the similarity between the input signature and the reference
one(s). The Euclidean distance is used for this purpose in this research. It is
computed from the features using

$$DM = \left(1/n \sum_{i=1}^{n}(f_i - \mu_i/\sigma_i)^2\right)^{1/2},$$ (30.1)

where f_i is the ith feature ($1 \leqslant i \leqslant n$), n is the number of used features, μ_i is the ith feature computed on the set of genuine (training) samples of the related person, σ_i is the standard deviation of the ith feature computed on the same set.

30.3.3 Verification Decision

The verification decision is made as follows:

If DM > VTH, the input signature is judged to be "genuine", otherwise, it is judged to be "an attempted forgery". VTH is the Verification Threshold.

The value of the VTH is usually determined based on some evaluation experiments using a reference signature data so that it minimizes the error rate (maximizes the correct decisions).

Determining the used features is usually done either based on the developer experience (not very accurate, but works), or based on a feature selection technique that selects the best feature set (BFS). BFS is the feature set that gives the highest performance. Ammar et al. [11] developed a feature selection technique based on the principle of the "Circulant Matrix" (Circulant Matrix-Based Feature Selection Technique CMBFST) to generate n^2 feature sets among the possible $n!$ feature sets of n given features, and found that evaluating the signature data available using these n^2 feature sets will lead to the BFS after, at most, one or two shuffling processes of the initial order of the features $(f_1, f_2, \ldots, f_n)$. This CMBFST is a very fast one and gives a clear idea of the effectiveness of the individual features, and their contribution to the effectiveness of the different feature sets if augmented by to form new ones. It is also used to develop the MSF technique for ASV, and allows using the IFS. Therefore, it will be briefed below with some experimental results.

The CMBFST

For a given primary feature set $\{f_1, f_2, \ldots, f_n\}$, the n^2 feature sets are formed as follows:

Step 1   A matrix of $n \times n$ entries is formed as follows: the first row is made to be the primary feature set. Each following row is formed from the preceding one by shifting its contents one entry to the right (could be to the left). $n - 1$ shifts are made, as shown below for $n = 4$ (in actual systems n may reach tens of features but taken here 4 for simplicity of explanation):

$$
\begin{array}{cccc}
f_1 & f_2 & f_3 & f_4 \\
f_4 & f_1 & f_2 & f_3 \\
\\
f_3 & f_4 & f_1 & f_2 \\
f_2 & f_3 & f_4 & f_1
\end{array}
$$

Step 2 From each row, n different feature sets are formed starting from the first row as follows:

$$S_1 = \{f_1\};\ S_2 = \{f_1, f_2\};\ S_3 = \{f_1, f_2, f_3\};\ S_4 = \{f_1, f_2, f_3, f_4\}.$$

In this way, we generate $n \times n$ feature sets, but since the last feature set generated from each row is essentially the same, $n \times n - (n - 1)$ different feature sets are actually generated.

Now, in order to find the BFS among the $n \times n - (n-1)$ sets, the signature data is verified using all generated feature sets forming 3 result matrices: SR, PCA, and PCR, where SR is the system reliability $=$ (PCA+PCR)/2, PCA is the percentage of correct acceptance (percentage of genuine signatures accepted as genuine samples), PCR is the percentage of correct Rejection (percentage of forgeries rejected and classified as attempted forgeries).

Figures 30.6–30.8, show an *illustrative practical example* of the three result matrices of 144 (12 × 12) entries respectively, formed by the CMBFST using the shape features appearing in the screen shot containing the result matrix, and explained as follows:

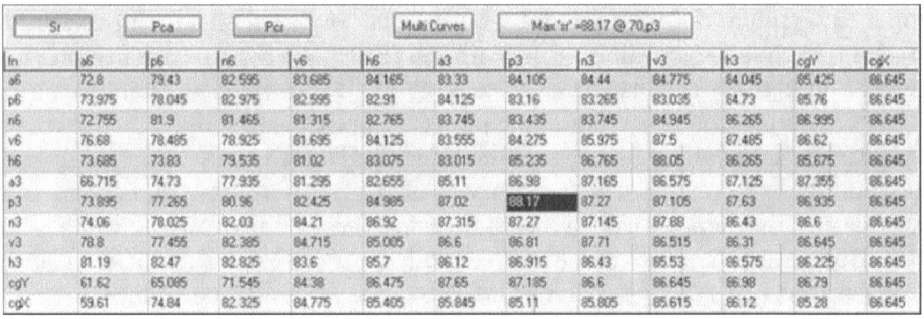

| Sr | Pca | Pcr | Multi Curves | Max 'sr' =88.17 @ 70.p3 |

In	a6	p6	n6	v6	h6	a3	p3	n3	v3	h3	cgY	cgX
a6	72.8	79.43	82.595	83.685	84.165	83.33	84.105	84.44	84.775	84.045	85.425	86.645
p6	73.975	78.045	82.975	82.595	82.91	84.125	83.16	83.265	83.035	84.73	85.76	86.645
n6	72.755	81.9	81.465	81.315	82.765	83.745	83.435	83.745	84.945	86.265	86.995	86.645
v6	76.68	78.485	78.925	81.695	84.125	83.555	84.275	85.975	87.5	87.485	86.62	86.645
h6	73.685	73.83	79.535	81.02	83.075	83.015	85.235	86.765	88.05	86.265	85.675	86.645
a3	66.715	74.73	77.935	81.295	82.655	85.11	86.98	87.165	86.575	87.125	87.355	86.645
p3	73.895	77.265	80.96	82.425	84.985	87.02	88.17	87.27	87.105	87.63	86.935	86.645
n3	74.06	78.025	82.03	84.21	86.92	87.315	87.27	87.145	87.88	86.43	86.6	86.645
v3	78.8	77.455	82.385	84.715	85.005	86.6	86.81	87.71	86.515	86.31	85.645	86.645
h3	81.19	82.47	82.825	83.6	85.7	86.12	86.915	86.43	85.53	86.575	86.225	86.645
cgY	61.62	65.095	71.545	84.38	86.475	87.65	87.185	86.6	86.645	86.98	86.79	86.645
cgX	59.61	74.84	82.325	84.775	85.405	85.845	85.11	85.805	85.615	86.12	85.28	86.645

Fig. 30.6 SR Matrix of 144 entries.

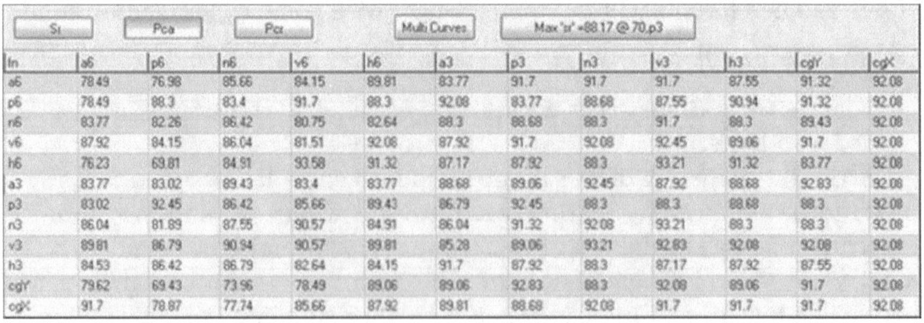

| Sr | Pca | Pcr | Multi Curves | Max 'sr' =88.17 @ 70.p3 |

In	a6	p6	n6	v6	h6	a3	p3	n3	v3	h3	cgY	cgX
a6	78.49	76.98	85.66	84.15	89.81	83.77	91.7	91.7	91.7	87.55	91.32	92.08
p6	78.49	88.3	83.4	91.7	88.3	92.08	83.77	88.68	87.55	90.94	91.32	92.08
n6	83.77	82.26	86.42	80.75	82.64	88.3	88.68	88.3	91.7	88.3	89.43	92.08
v6	87.92	84.15	86.04	81.51	92.08	87.92	91.7	92.08	92.45	89.06	91.7	92.08
h6	76.23	69.81	84.51	93.58	31.32	87.17	87.92	88.3	93.21	91.32	83.77	92.08
a3	83.77	83.02	89.43	83.4	83.77	88.68	89.06	92.45	87.92	88.68	92.83	92.08
p3	83.02	92.45	86.42	85.66	89.43	86.79	92.45	88.3	88.3	88.68	88.3	92.08
n3	86.04	81.89	87.55	90.57	84.91	86.04	91.32	92.08	93.21	88.3	88.3	92.08
v3	89.81	86.79	90.94	90.57	89.81	85.28	89.06	93.21	92.83	92.08	92.08	92.08
h3	84.53	86.42	86.79	82.64	84.15	91.7	87.92	88.3	87.17	87.92	87.55	92.08
cgY	79.62	69.43	73.96	78.49	89.06	89.06	92.83	88.3	92.08	89.06	91.7	92.08
cgX	91.7	78.87	77.74	85.66	87.92	89.81	88.68	92.08	91.7	91.7	91.7	92.08

Fig. 30.7 PCA Matrix of 144 entries.

a6, p6, n6, v6, and h6 are area and percentage of positively, negatively, vertically, and horizontally slanted pixels in the 6 parts of the signature explained in Section 30.3.1.

| St | Pca | Pcr | Multi Curves | Max 'sr' -88.17 @ 70,p3 |

fn	a6	p6	n6	v6	h6	a3	p3	n3	v3	h3	cgY	cgX
a6	67.11	81.88	79.53	83.22	78.52	82.89	76.51	77.18	77.85	80.54	79.53	81.21
p6	69.46	67.79	82.55	73.49	77.52	76.17	82.55	77.85	78.52	78.52	80.2	81.21
n6	61.74	81.54	76.51	81.88	82.89	79.19	78.19	79.19	78.19	84.23	84.56	81.21
v6	65.44	72.82	71.81	81.88	76.17	79.19	76.85	79.87	82.55	85.91	81.54	81.21
h6	71.14	77.85	74.16	68.46	74.83	78.86	82.55	85.23	82.89	81.21	87.58	81.21
a3	49.66	66.44	66.44	79.19	81.54	81.54	84.9	81.88	85.23	85.57	81.38	81.21
p3	64.77	62.08	75.5	79.19	80.54	87.25	83.89	86.24	85.91	86.58	85.57	81.21
n3	62.08	74.16	76.51	77.85	88.93	98.59	83.22	82.21	82.55	84.56	84.3	81.21
v3	67.79	68.12	73.83	78.86	80.2	87.52	84.56	82.21	80.2	80.54	81.21	81.21
h3	77.85	78.52	78.86	84.56	87.25	80.54	85.91	84.56	83.89	85.23	84.3	81.21
cgY	43.62	60.74	69.13	90.27	83.89	86.24	81.54	84.9	81.21	84.9	81.38	81.21
cgX	27.52	70.81	86.91	83.89	82.89	81.88	81.54	79.53	79.53	80.54	78.86	81.21

Fig. 30.8 PCR Matrix of 144 entries.

a3, p3, n3, v3, and h3 are the same features above but computed on the signature thirds resulting from combining each two sixths starting from left to the right.

cgX and cgY are x and y coordinates of the center of gravity.

It is worth noting here that "a6" means the 6 local areas measured on the six parts of the signature, and "p6" means the percentage of the positively slanted pixels in the six parts of the signature (6 features). Since no part of the signature can be omitted, the result matrix considered the local features as groups (6 local areas, 6 percentage of positively slanted pixels, 3 local areas, and so on.) Therefore, the total number of features actually used in the result matrix at the last column is: $(5 \times 6) + (5 \times 3) + (2) = 47$ features. It is obvious that the last column of each matrix has the same value because the feature set is essentially the same.

Since the BFS is chosen to maximize SR, it will be that of the entry (7, 7) which gives SR=88.17 with corresponding PCA=92.45 and PCR=83.89.

Published works on signature verification usually mention total evaluation results like the curves shown in Fig. 30.9 or as total performance expressed as percentage of correct decisions (or type I and type II error rates). Ammar et al. [11] displayed the result as (X, O) graph which gives a clear insight into the evaluation process enabling to see what is happening to each sample, and the performance of the verification approach for individual persons.

Figure 30.9 shows the SR, PCA and PCR curves corresponding to (7, 7) entry in the SR matrix (the BFS that can be formed from the mentioned features), and Fig. 30.10 shows its corresponding (X, O) graph. In Fig. 30.10, persons included in the signature database are displayed as "symbolic names" on the horizontal axis, and the distance measure of each sample is displayed on the vertical axis. Each forgery sample is represented by "X", and each genuine sample is represented by "O". Each X above the verification threshold VTH is a correct decision, and each "O" below VTH is a correct decision. The clear view provided by the (X, O) graph shown in Fig. 30.10, which was used by Ammar [23] to explain the motivation for the MSF technique, will be used here to explain the QP and the IFS.

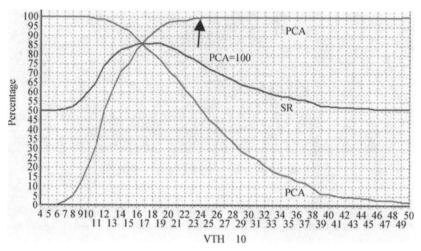

Fig. 30.9 SR, PCA and PCR curves of the entry (7, 7) of the SR result matrix in Fig. 30.6.

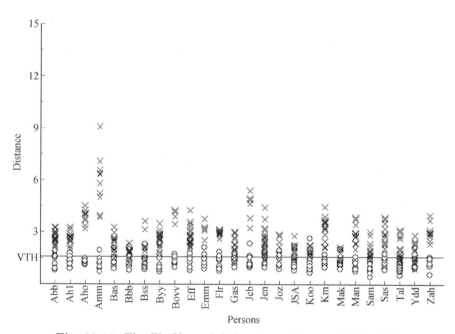

Fig. 30.10 The (X, O) graph belonging to SR_{max} of Table 30.1.

30.4 MSF and Its Performance

Ammar [23] explained the motivation for the MSF technique shown in Fig. 30.8 where f is the feature; $S_1, \ldots, S_m$ are the Effective Feature Sets(EFS); ThBD is Threshold-Based Decision; F is Forgery; G is Genuine. In this figure, the CMBFST is used to find the EFS (close in performance to the BFS), then for every input signature, it is verified using all EFS. If rejected by any one of them, it is rejected as a forgery, otherwise, it is accepted as genuine. In [23], Two distinct cases were discussed: (1) using MSF alone, and (2) using the MFS with the BFS. These two cases are explained below before reporting the results of using the MSF with large number of feature sets, and using the IFS.

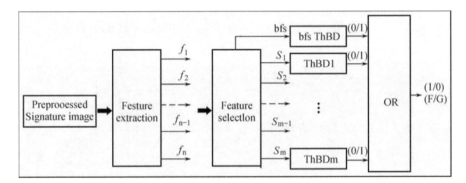

Fig. 30.11 The MSF-based ASV.

30.4.1 Performance of MSF with Medium Number of Features (144 Entries SR Matrix and 47 Features)

Figure 30.12 shows the SR, PCA and PCR curves of the MSF-based verification using 18 EFS formed from the primary feature set, used to produce Figs. 30.6 – 30.8, with SR over 87.0, as appears in the screen shot in Fig. 30.12.

Figure 30.13 shows the curves of both BFS and MSF. In this figure, the merit of the MSF over the BFS alone appears clearly, where the thick curves are those of the MSF, and the thin ones are those of the BFS. Examining these curves shows that:

1) There is a considerable gain in PCR between VTH=1.9 and VTH=4.0 ranging from 10% to 15%. This improvement is the effectiveness that can be provided by the EFS, but can not be captured by the BFS. This gain has been provided by the new MSF technique. It is also a real reflection of the fact that every feature can detect some aspects of the variability of the signature that can not be completely compensated by other features.

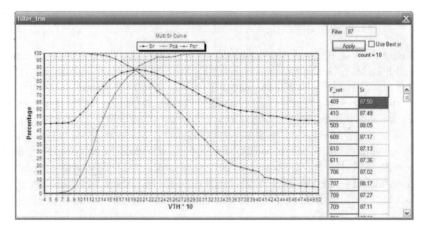

Fig. 30.12 SR, PCA, and PCR curves of the MSF using 18 EFS of SR>87.0.

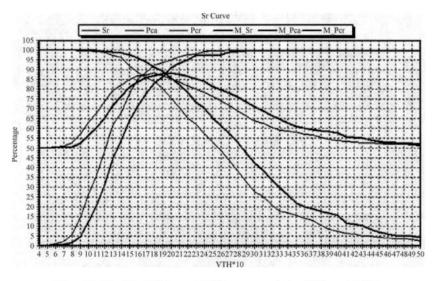

Fig. 30.13 SR, PCA, and PCR curves of the BFS and MSF.

2) The gain explained above is pure for VTH>2.9, since there is no loss in PCA.

3) From VTH = 2.9 down, the loss in PCA starts to appear gradually, but remains less than the improvement in PCR so that the total change remains positive until VTH=1.85. Below this value, the loss in PCA becomes larger than the gain in PCR, and consequently, a loss in SR, as the curves show. In fact, this result is natural because this region below VTH 1.85 in the (X, O) graph is the genuine samples zone. VTH may not be used in this zone at all.

4) The highest SR (SR$_{max}$) obtained by the MSF is little bit higher than

that of the BFS (88.26 in comparison with 88.17), as Fig. 30.14 shows.

5) In general, by using the MSF we can get at the same SR a PCR higher than that obtained by the BFS (Fig. 30.14).

6) The importance of the improvement in performance gained by the MSF technique is that it can not be obtained by the one feature set based approach. It is usually lost.

Sr Curve	Sr Values	Xo Map	VTCH							
Distance	1.4	1.5	1.6	1.7	1.8	1.9	2	2.1	2.2	2.3
707 SR	82.12	84.85	86.94	87.3	88.17	86.89	85.3	83.71	81.59	80.1
707 Pca	67.92	77.74	85.28	88.68	92.45	93.58	95.09	96.6	97.74	98.1
707 PCR	96.31	91.95	88.59	85.91	83.89	80.2	75.5	70.81	65.44	62.0
Multi SR	76.31	80.9	83.65	85.85	87.15	87.84	88.26	87.71	86.64	85.4
Multi Pca	53.96	64.15	71.32	77.74	83.02	86.42	90.94	93.21	95.09	97.3
Multi Pcr	98.66	97.65	95.97	93.96	91.28	89.26	85.57	82.21	78.19	73.4

Fig. 30.14 SR, PCA, and PCR values for BFS and MSF around the peak.

30.4.2 Performance of MSF with Large Number of Feature Sets

The curves and results shown and discussed in Section 30.4.1 above, were produced by 18 feature sets having over SR=87.0 performance. It is interesting to investigate the effect of increasing the number of EFS used for MSF.

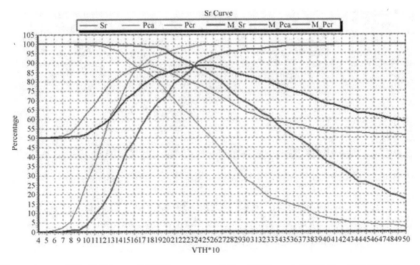

Fig. 30.15 SR, PCA, and PCR curves of the BFS and MSF for 108 EFS having SR>80.0.

Figure 30.12 shows the result obtained by 108 feature sets having performance over SR=80.0. This result is very interesting because the MSF gave SR approaching 89.0 in comparison with 88.26 for 18 feature sets. The gain in forgery detection is very distinctive here. It is in the range of 30%–200% compared with that of the BFS, over a wide range of VTH = 2.0 − 4.0.

30.4.3 Performance of the MSF with Large Number of Features (625 Entries SR Matrix and 63 Features)

In order to investigate the behavior of the MSF technique with larger number of features, it was tested with 625(25 × 25) entries result matrices and 63 features. A part of the SR matrix is shown in the screen shot with the curves of the obtained BFS in Fig. 30.16. The BFS obtained is marked at (1, 20) entry with SR=90.9.

The MSF curves corresponding to this FS and obtained by using 15 EFS with SR>90.5 as well as the curves of the BFS are shown in Fig. 30.17. In these curves, we find that:

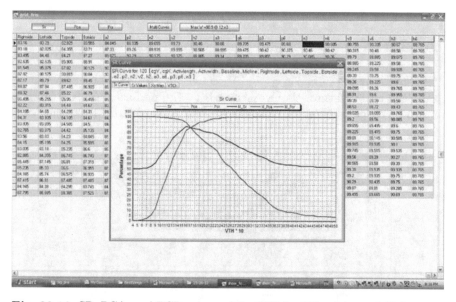

Fig. 30.16 SR, PCA, and PCR curves of the BFS for 625 entries and 63 features.

1) The performance of the BFS obtained from the 625 entries result matrix and 63 features is considerably better than that obtained from 144 entries and 47 features in Fig. 30.6 (90.9 in comparison with 88.17).

2) The MSF performance is also considerably better than that obtained from the 144 entries and 47 features (90.31, as appears in Fig. 30.17, in comparison with 88.26).

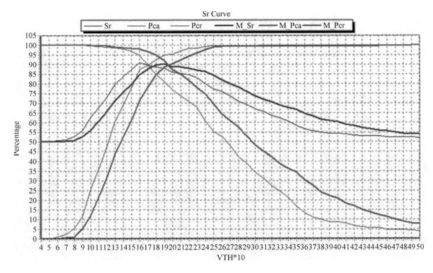

Fig. 30.17 SR, PCA, and PCR curves of the BFS and MSF obtained from a primary feature set containing 63 features.

3) In Fig. 30.17, although the peak of MSF curve is little bit lower than that of the BFS (90.31 in comparison with 90.9), the peak of the MFS is more flat. Consequently, it is more convenient for selection of the VTH for practical use.

4) The PCA curve of the MSF is more smooth and approached better the PCA curve of the BFS. Consequently, it is more convenient for making a "Zero false alarm" decision. In Fig. 30.17, we can get a "Zero false alarm" decision at PCR=70, in comparison with PCR= 50 in Fig. 30.13. This result leads us to the fact that: increasing the number of used features in the primary feature set, and with the proper selection, we can get better performance.

5) The relative pure gain in forgery detection remained the same: about 15% on the PCR scale.

6) In general, and as the curves in Fig. 30.17 shows in comparison with Fig. 30.13, the gained performance in forgery detection with the MSF is better. This finding reflects the fact that the MSF technique from a feature set containing larger number of properly selected features, provide better forgery detection.

30.4.4 Performance of the MSF with Large Number of Features and Large Number of Feature Sets

In Section 30.4.2, the performance of the MSF with large number of feature sets(108) and medium number of features (47) was tested. It is meaningful to conduct similar test with large number of features(63) used in Section 30.4.3, and compare between the two cases. Figure 30.18 shows the performance of the MSF when the initial feature set contained 63 features and the result matrix of 625 entries was created (Section 30.4.3 test), and 445 EFS, having a performance over SR=85.0, were used. It is worth reminding here that when we specify SR=85.0, the program automatically selects all feature sets with a performance exceeding this limit, like in Fig. 30.11. Here, the number came 445. Examining Fig. 30.18 reveals that:

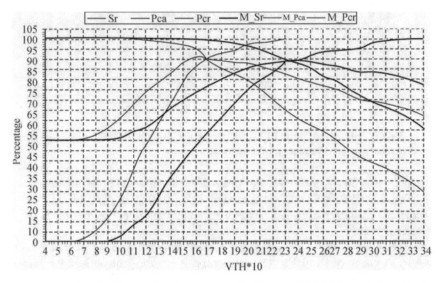

Fig. 30.18 SR, PCA, and PCR curves of the BFS and MSF for 445 EFS having SR>85.0.

1) PCA=100 is realized at VTH=3.1 with PCR=65.0 in comparison with VTH=4.1 and PCR=35.0 in Fig. 30.12. This important improvement should be a result of the fact that the performance of the individual feature sets here is higher than that of those used for Fig. 30.15.

2) SR curve is more flat and stable around the peak VTH=2.3 and convenient for comfortable selection of a global VTH.

3) Several other things can be concluded from the comparison of the curves in Fig. 30.15 and Fig. 30.15, according to the interest of the reader, but, in general, there is an improvement over the BFS in this studied case, and over the MSF of the case of medium number of features and 108 feature sets. The only negative point could be the computation time needed here. However, with nowadays computers, it should not be a problem.

30.4.5   Performance of the MSF with the BFS

In order to check the efficiency of using the MSF with the BFS, the verification procedure of Fig. 30.8 (including the bold block of BFS) was tested for 15 EFS with SR>90.5 used to produce Fig. 30.17. The result came interesting as shown in Fig. 30.16. Where:

  1) We got higher $SR_{max}= 91.28$ in comparison with 90.9 for the BFS, and 90.31 for the MSF.

  2) This higher SR is obtained at distinctly high PCR approaching 95%.

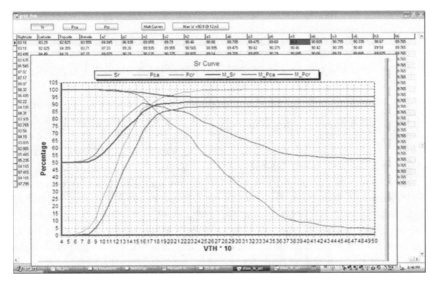

Fig. 30.19   The performance of BFS and the MSF together.

In fact, the MSF technique and the feature selection procedure used, the number of EFS used, the number of features of the initial feature set, using MSF with the BFS, and their varieties give a huge space for experimentation and analysis in the field of ASV. What was reported in this paper is only some selected aspects of interest here.

30.5   IFS and QP

A few researchers used feature selection in ASV and reported it explicitly [23]. The feature selection is used to find the BFS which gives the highest performance on the used signature data. However, this BFS might give unsatisfactory results for some persons, the situation that should be avoided in a practical system. Moreover, the individual performance of features might be desired to be known, and the feature sets that give the best performance for PCR, or PCA, could be a matter of interest. In order to explore and investigate the previous issues we need the IFS that gives the performance of feature sets as PCA, PCR and SR curves or values, or in detail like the (X, O) graph in Fig. 30.7. Ammar et al. developed a feature selection technique (the CMBFST) reported in [4] and explained and used in this paper. The CMBFST provides great advantages like: (1) finding the BFS, (2) giving the ability of using the IFS, (3) selecting a feature set with a specific performance, (4) studying the effectiveness of individual features and their impact on the overall performance, or (5) finding a feature set with the best qualitative performance. In the following, some of these topics will be explained by practical examples.

30.5.1   Selecting a Feature Set with a Specific SR, PCA or PCR

Figure 30.6 – 30.8 will be used for explanation of this topic. If we use the CMBFST to give the BFS, the feature set is given automatically as the one belonging to the entry (7, 7) with SR=88.17, PCA=92.45, and PCR=83.89 (marked blue in the SR matrix). However, this result may not be the one preferred because it is considerably PCA-biased at its maximum performance. A feature set with almost PCA = PCR at its maximum performance could be preferred. In this case, the feature set of the entry (8, 6) with SR=87.315, PCA=86.04, and PCR=88.59, for example, would be better. This feature set can be found visually from the three result matrices in Tables 30.1 – 30.3. Needless to say, such a process can be automated, but visual inspection leads some times to interesting findings.

30.5.2   Finding a Feature Set with the Best Qualitative Performance

When we say: a feature set with SR=88.17, PCA=92.45, and PCR=83.89, or with SR=87.315, PCA=86.04, and PCR=88.59, this is a *quantitative* performance. It does not give any information about how the values of SR, PCA, and PCR change from one person to another. Sometimes, although the global SR is high, it is low (bad) for some persons in the data, or PCA or PCR is distinctly low for a specific person, and consequently not good in practical use if the approach is used in a real application. *Such cases should be avoided in actual systems*, but how? Using the IFS with the (X, O) graph will enable us to do that. For example, consider the SR result matrix for an *arbitrary* initial feature arrangement (only a part of the SR matrix is displayed) shown in Fig. 30.17. The best feature set in this case is that of the entry (5, 10) or (bv6, cgY) with SR=88.57, PCA=93.58, PCR=83.56. Directly below the SR matrix, the (X, O) graph is displayed. Now, let us focus on the following points:

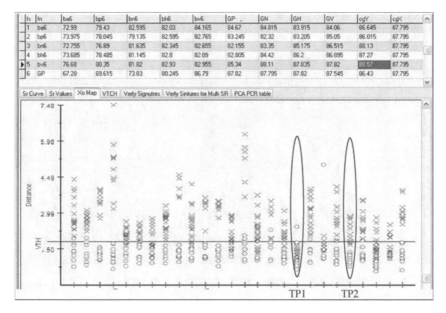

Fig. 30.20 The SR result matrix with BFS of SR=88.57 and its correspond (X, O) graph.

1) If we displayed only SR, then we will not know that this feature set is PCA biased.

2) If we do not display the (X, O) graph, we will not know that for the person named "TP1" and surrounded by an ellipse, PCR=0.0, *which is absolutely bad*, but this fact is hidden in the global performance in the SR result matrix, and in the SR, PCA, PCR curves shown in Fig. 30.22.

3) In fact, "TP1", and "TP2" (the person surrounded by the second ellipse), are two signatures of the same person, but the first signature is signed in Japanese, and the second one is signed in English. Here, the (X, O) graph highlighted the fact that the BFS is *language-sensitive*.

4) Now, we will use the IFS which allows us to display the (X, O) graph for any feature set in the SR matrix by clicking on it to search for better feature set. Doing that concentrating on EFS, and in a minute or so, we found the feature set of the entry (6, 6) or (GP, GP) in Fig. 30.21, with SR=87.82, PCA=86.04, PCR=89.6, which avoids the unsatisfactory situation where:

– Although SR of this feature set is little bit lower than that of the BFS, it is more balanced between PCA and PCR, and little bit in favor of PCR which is preferred in ASV systems.

– It is almost insensitive to signature language where its performance for "TP1" became PCA=100 and PCR= 90.0, and for "TP2" became PCA=96.0 and PCR=100.

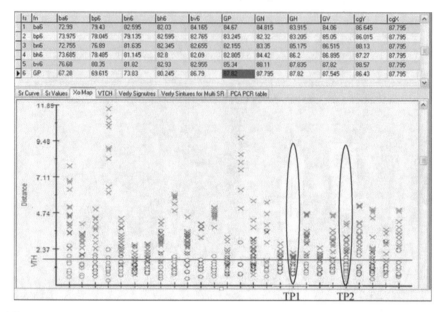

ts	fn	ba6	bp6	bm6	bh6	bv6	GP	GN	GH	GV	cgY	cgX
1	ba6	72.99	79.43	82.595	82.03	84.165	84.67	84.815	83.915	84.06	86.645	87.795
2	bp6	73.975	78.045	79.135	82.595	82.765	83.245	82.32	83.205	85.05	86.015	87.795
3	bm6	72.755	76.89	81.635	82.345	82.655	82.155	83.35	85.175	86.515	88.13	87.795
4	bh6	73.685	78.485	81.145	82.8	82.09	82.805	84.42	86.2	86.895	87.27	87.795
5	bv6	76.68	80.35	81.82	82.93	82.955	85.34	88.11	87.835	87.82	88.57	87.795
6	GP	67.28	69.615	73.83	80.245	86.79	87.82	87.795	87.82	87.545	86.43	87.795

Fig. 30.21 The SR result matrix with BFS of SR=88.57 and the (X,O) graph of
the feature set of SR=87.82.

– If we compare the SR curves of the BFS and those of the second one in
Fig. 30.19, we find that the curve of the second one is more flat and wider
around the peak. Naturally, this is better than that of the BFS.

This example demonstrates the fact that the global evaluation is **"blind"**
and may lead to very bad situations, however, the IFS in the way used, is
very fast and effective to avoid such situation, and consequently, may lead to
successful practical system.

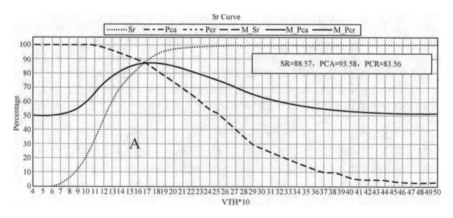

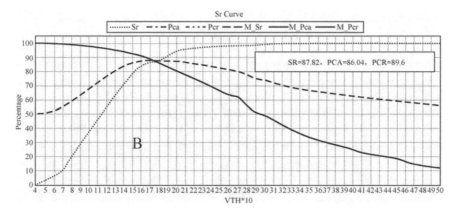

Fig. 30.22 "A" belongs to the BFS in Fig. 30.21, and "B" belongs the feature set of the (X,O) graph in Fig. 30.21.

30.5.3 Deeper Qualitative Performance Evaluation

In the (X,O) graph, we can see the distance of every sample without knowing its identity. In order to evaluate the performance accurately, it is necessary to know the decision for every signature for every feature set. A "0"/"1" table like that in Fig. 30.23 where every sample is displayed as "0" if accepted, and as "1" if rejected. Genuine signatures are shown in green, and forgeries are shown in red. Eventually, in a glance, we can evaluate the overall situation of

SR	PCA	PCR	Multi Curves	Max 'SR' =88.57 @ 5 , 10 [bv6 , GP , GN , GH , GV , cgY , cgX , ba6 , bp6 , bn6

fs	fn	ba6	bp6	bn6	bh6	bv6	GP	GN	GH
5	bv6	76.68	80.35	81.82	82.93	82.955	85.34	88.11	87.835
▶6	GP	67.28	69.615	73.83	80.245	86.79	97.82	87.795	87.82
7	GN	70.2	78.235	83.33	81.86	83.435	87.605	87.44	87.375
8	GH	60.745	67.83	74.02	74.94	84.06	86.415	87.125	86.6
9	GV	62.23	67.31	69.49	83.54	85.51	87.145	86.79	87.33
10	cgY	61.92	66.135	75.8	84.38	86.475	85.82	87.02	88.05

Sr Curve	Sr Values	Xo Map	VTCH	Verfy Signutres	Verfy Sintures for Multi SR	PCA PCR table

Abb	0 0 1 0 1 0 0 0 1 1 1 1 1 1 1 1 1 1 1 1 1 1 1 1 1
Ahl	1 0 0 1 0 0 0 0 1 1 1 1 1 1 1 1 1 1 1 1 1
Aho	0 0 0 0 0 0 1 0 1 0 1 1 1 1 1 1 1 1 1 1
Amm	1 0 0 0 0 0 1 0 0 0 0 1 1 1 1 1 1 1 1 1 1
Bas	1 0 0 0 0 1 0 0 1 1 1 1 1 1 1 1 1 1 1 1 1 0

Fig. 30.23 The "0"/"1" table that shows the decision for every sample of every person in the signature database for the selected feature set (matrix entry).

the data. For example, the left bottom and right bottom samples (of "Bas"
person) in Fig. 30.20 are misclassified. The result of 5 persons is shown in the
figure.

At a lower level, the IFS software built, enables us to display for every feature set (among the n^2 sets) every signature image of every person
and its status (accepted or rejected). If accepted, it is displayed in green,
and if rejected, it is displayed in red (the signature image version of the
"0"/"1" table). Figure 30.24 shows the result for the Target Person (TP1)
and Fig. 30.25 shows the result for (TP2), for the feature set of Fig. 30.21.
TP1 is the Japanese signature and TP2 is the English signature of the Target
Person. Accepted signatures are displayed in "green" and rejected signatures
are displayed in "red". Red genuine signatures and green forgery signatures

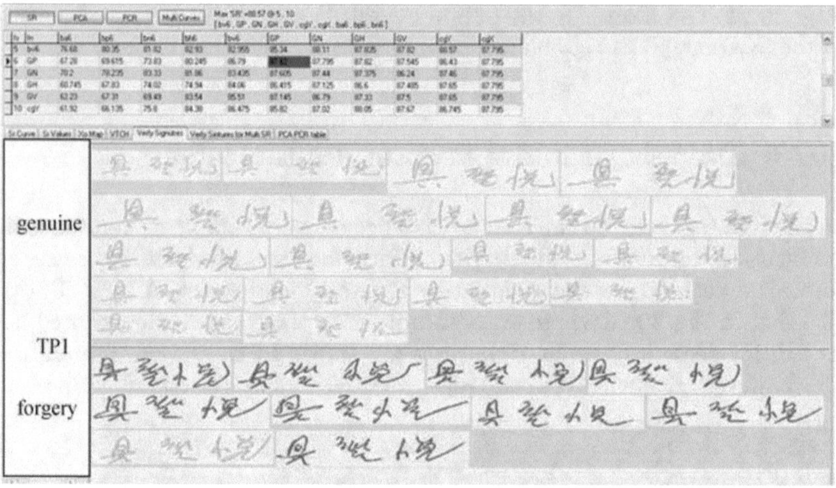

Fig. 30.24 The verification results of TP1.

are misclassified. For example, the bottom left forgery Japanese signature
sample is misclassified, and all the other genuine and forgery Japanese samples are correctly classified in Fig. 30.24. For the English signature, only one
genuine signature is misclassified, and displayed in "red". It is good to remind
here that the two signatures belong to the same person.

The advantage of displaying every sample with its decision is as follows:
when we feel that the result is not convincing, we display an enlarged signature image, go back to the values of extracted features and their computed
reference statistics, and figure out what could be wrong. This facility provided
by the IFS is very useful when aiming a practical system.

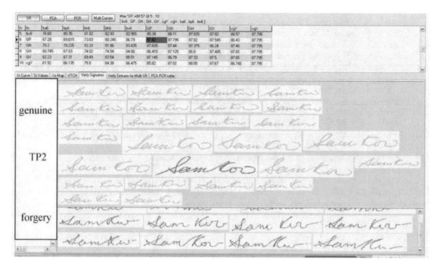

Fig. 30.25 The verification results of TP2.

30.6 Conclusion

This chapter has presented the new MSF technique for ASV and reported
the new results obtained by using large number of feature sets of hundred-
order. The new results have shown important improvement in comparison
with using small number of feature sets of ten-order. It proposed also two
new concepts in ASV field: the Qualitative Performance and the Interactive
Feature Selection, and showed by a practical example that when the new
concepts are used, some bad performance cases that may occur in the real
application can be avoided, and with the same features, better qualitative
performance can be obtained.

References

[1] Crane H D, Ostrem J S (1983) Automatic Signature Verification Using a Three-axis-
     force-sensitive Pen. IEEE Trans SMC 12: 329 – 337
[2] Alonso-Fernandez F, Fierrez-aguilar F, del-valle F et al (2005) On-line Signature
     Verification Using Tablet. In: Abstracts of the Fourth Int. Symposium on Image and
     Signal Processing and Analysis, ISPA, 2005
[3] Kamel N S, Sayeed S (2008) SVD-based Signature Verification Technique Using Data
     Glove. IJPRAI, 22(3): 431 – 443
[4] Ammar M , Yoshida Y, Fukumura T (1986) A New Effective Approach for Automatic
     Off-line Verification of Signatures by Using Pressure Features. In: Abstracts of the
     IEEE 8th International Conference on Pattern Recognition, Paris, France
[5] Ammar M, Yoshida Y, Fukumura T (1986) Automatic Off-line Verification of Signa-
     tures Based on Pressure Features. IEEE Trans on SMC, 16(3): 39 – 47

[6]   Huang K, Hong Y (1997) Off-line Signature Verification Based on Geometric Feature
       Extraction and Neural Network Classification. Patten Recognition 30(1): 9 – 17
[7]   Sansone C, Vento M (2008) Signature Verification: Increasing Performance by a
       Multi-stage System. Pattern Analysis and Applications, 3: 169 – 181
[8]   Francisco J V, Carlos M T, Jesus B A (2008) Off-line Signature Verification Based on
       High Pressure Polar Distribution. In: Abstracts of the 11th International Conference
       on Frontiers in Handwriting Recognition
[9]   Mitra A, Banerjee P, Ardil C (2005) Automatic Authentication of Handwritten Docu-
       ments via Low Density Pixel Measurements. International Journal of Computational
       Intelligence, 2(4): 219 – 223
[10]   Ammar M, Yoshida Y, Fukumura T (1989) Off-line Preprocessing and Verification
       of Signatures. IJPRAI, 2(4): 589 – 602
[11]   Ammar M, Yoshida Y, Fukumura T (1989) Feature Extraction and Selection for
       Simulated Signature Verification. In: Plamondon R, Suen C Y, Simner ML(eds)
       Computer Recognition and Human Production of Handwriting. World scientific, 1989
[12]   Ammar M (1990) Performance of Parametric and Reference Pattern Based Features
       in Static Signature Verification: A Comparative Study. In: Abstracts of the IEEE
       10th International Conference on Pattern Recognition, Atlantic City, New Jersey,
       USA, 1990. IEEE Computer Society Press
[13]   Ammar M (1991) Progress in Verification of Skillfully Simulated Handwritten Sig-
       natures. IJPRAI, 5(1 – 2): 337 – 351
[14]   Frias-Martinez E, Sanchez A, Velez J F (2006) Support Vector Machines Versus
       Multi-layer Perceptrons for Efficient Off-line Signature Verification. Engineering Ap-
       plications of Artificial Intelligence, 19(6): 693 – 704
[15]   Nguyen V, Blumenstein M, Muthukkumarasamy V et al (2007) Off-line Signature
       Verification Using Enhanced Modified Direction Features in Conjunction with Neural
       Classifiers and Support Vector Machines. In: Abstracts of the ICDAR
[16]   Larkin L (2009) Off-line Signature Verification. Doctor thesis, University of Waikato
[17]   Jesus F, Bonilla V, Miguel A et al (2009) Off-line Signature Verification Based on
       Pseudo-capstral Coefficients. In: Abstracts of the 10th ICDAR
[18]   Leclerc F, Plamondon R (1994) Automatic Signature Verification: the State of the
       Art–1989 – 1993. IJPRAI, 8(3): 643 – 660
[19]   Plamondon R, Lorette G (1989) Automatic Signature Verification and Writer Iden-
       tification: the State of the Art. Pattern Recognition, 22(2): 107 – 131
[20]   Plamondon R, Srihari S (2000) On-line and Off-line Handwriting Recognition: a
       Comprehensive Survey. IEEE Transactions on PAMI, 22(1): 63 – 84
[21]   Ammar M (2002) Method and Apparatus for Verification of Signatures. US Patent
       6424728, 23 July 2002
[22]   http://www.asvtechnologies.com. Accessed 30 August 2010
[23]   Ammar M (2010) Raising the Performance of Automatic Signature Verification over
       That Obtainable by Using the Best Feature Set, Accepted, IJPRAI
[24]   Ammar M, Aquel M (2005) Verification of Signatures of Bank Checks at Very Low
       Resolutions and Noisy Images. Jordan Journal of Applied Science University, 7(1):
       1 – 23
[25]   Ammar M (2003) Application of Artificial Intelligence and Computer Vision Tech-
       niques to Signatory Recognition. Pakistan Journal of Information and Technology,
       2(1): 44 – 51
[26]   Aquel M, Ammar M (2005) Functions, Structures and Operation of Modern Systems
       for Authentication of Signatures of Bank Checks. Information Technology Journal,
       4(1): 96 – 105
[27]   Boser B E, Guyon I M, Vapnik V N (1992) A Training Algorithm for Optimal Margin
       Classifiers. In D Haussler (ed). Proceedings of the 5th Annual ACM Workshop on
       COLT, Pittsburgh, PA. ACM Press, Boston
[28]   Fauziyah S, Mardiana B, Zahariah M et al (2009) Signature Verification System
       Using Support Vector Machine. MASAUN Journal of Basic and Applied Sciences,
       1(2): 291 – 294
[29]   Hong Q D, Chen C, Xian J (2005) Signature Verification by Support Vector Machine
       with ALCSL &LS kernel. Journal of Fudan University, 43(5): 805 – 814
[30]   Justino E J R, Bortolozzi F, Sabourin R (2005) A Comparison of SVM and HMM
       Classifiers in the Off-line Signature Verification. Pattern recognition letters, 26(9):
       1377 – 1385
[31]   Srihari S N, Xu A, Kalera M K (2004) Learning Strategies and Classification Methods
       for Off-line Signature Verification. In: Abstracts of the IWFHR, IEEE Computer

Society, 2004

[32] Chen Y, Ding X, Wang P S P (2009) Dynamic Structural Statistical Model Based Online Signature Verification. IJDCF, 1(3): 21 – 41

[33] Ammar M (2011) Off-line Signature Verification by Matching with a 3D Reference Knowledge Image: From Research to Actual Application

31 Fourier Transform in Numeral Recognition and Signature Verification

Giovanni Dimauro[1]

Abstract Digital transforms have been intensively used in the field of pattern recognition with the advance of computers. In the field of handwriting recognition, applications of such mathematical transformations ranges from character and numeral recognition, to word recognition and signature verification. In this chapter we present the fundamentals of digital transforms and their use in handwriting recognition. This chapter is divided into two parts. The first part deals with the fundamental concepts and an overview of digital transforms. The second part deals with the implementation of digital transforms. Specifically the Discrete Fourier Transform (DFT) is explained in detail. Some computational aspects of the DFT are evaluated and some algorithms for Fast Fourier Transform are presented. In particular the "Radix-2 FFT" algorithms are discussed in detail. In the second part of the chapter some applications are presented: the DFT represents one of the most powerful tools for plane curve analysis and recognition. This is useful in handwriting recognition where often most of the information for the description and the classification of the pattern can be found in its boundary.

31.1 Concepts of Digital Transforms

Frequently the understanding of a signal or image informative contents, or the recover of a part of theirs, is better done if, instead of a temporal or spatial description, a different one is used. It must be underlined that temporal or spatial description is imposed by the phenomena physics. Infact since a signal is generated in time, it is suitable, or even inevitable, to acquire and describe it in the temporal reference; the same can be said for an image that, occuping a physical space, can be naturally described using an Euclidean reference system. The tools used to describe a signal or an image in a different way are offered by mathematics. Even if these tools seem complex, mostly it is a matter of reference system changes and therefore of linear tranformations. The understanding difficulties usually rise from the fact that we deal with many dimensions spaces, even of infinite dimensions. Really, so that the new signal or image description is efficient, it must be complete at least as the original one, in the sense that from an informative point of view nothing

---

1 Università degli Studi di Bari 'Aldo Moro', Dipartimento di Informatica, Bari, Italy.
   E-mail: dimauro@di.uniba.it.

is lost passing from the first description to the second one. In maths we talk about completeness of a representation system to say that all elements of a set can be described using that system. Such as we also talk about equipower of two systems to say that the set of all elements which can be described using the first system is equal to the set of all elements which can be described using the second. Two fundamental problems connected to the element description change exist: the first concerns the simple way of characterize the elements for which the different representation is permitted, the second instead concerns the way of identifing the transforms. Generally, the sets of the elements that show the property of being well represented by different systems are characterized through general properties such as continuity, periodicity, absolute integrability, etc. Regarding transforms it must be said that they exist of different kinds and they depend on either the nature of the elements that we want to transform or the specific properties of the same elements that we want to point out.

31.2 Orthonormal System of Trigonometric Functions

The problem of the completeness, as presented in the previous section, was faced by Weierstrass and Stone. They looked at this as a density problem and they defined the conditions under which a lower-algebra is dense in the set of real (and complex), continuous and in a compact space defined functions. One of the most important corollary of their famous "Weierstrass-Stone Theorem" is the one which shows that the set of trigonometric polynomial with coefficients in a vectorial seminormated space G, i.e., of the kind

$$P(\vartheta) = \sum_{n \in Q} c_n e^{j\vartheta n} \tag{31.1}$$

is dense in the space of continuous 2π-periodic functions and with values in G. The method proposed by Weierstrass and Stone, gives the conditions to verify the density, but does not give the method to calculate the coefficients of polynomials that approximate an element of the algebra. Really, the problem of coefficient calculation is much more wide and was solved by Fejer.

We can prove that if we consider the space L_2 of square integrable and defined in $[-\pi, +\pi]$ functions with the ordinary Lebesgue measure, the system of the functions:

$$\frac{1}{2\pi}, \quad \frac{\cos nx}{\sqrt{\pi}}, \quad \frac{\sin nx}{\sqrt{\pi}}, \quad n = (1, 2, ...) \tag{31.2}$$

is a complete orthogonal system of L_2. In fact we can verify through direct calculation that the two following relations exist:

$$\int_{-\pi}^{\pi} \cos nx \cos mx \, \mathrm{d}x, \quad \forall n \neq m \tag{31.3}$$

$$\int_{-\pi}^{\pi} \frac{\cos^2 nx}{\pi}\,dx = 1, \quad \int_{-\pi}^{\pi} \frac{\sin^2 nx}{\pi}\,dx = 1, \quad \forall n \in N^*, \qquad (31.4)$$

and then the system in (31.2) is orthonormal.

Now considering the generic element f of L_2, and calculating the coefficients:

$$a_0 = \frac{1}{\pi} \int_{-\pi}^{\pi} f(x)\,dx, \qquad (31.5)$$

$$a_n = \frac{1}{\pi} \int_{-\pi}^{\pi} f(x)\cos nx\,dx, \quad \forall n \in N^*, \qquad (31.6)$$

$$b_n = \frac{1}{\pi} \int_{-\pi}^{\pi} f(x)\sin nx\,dx, \quad \forall n \in N^*. \qquad (31.7)$$

We can prove that the series

$$\frac{a_0}{2} + \sum a_n \cos nx + b_n \sin nx, \quad \forall n \in N^* \qquad (31.8)$$

converges in quadratic mean to f. This result can be easily extended to the case that the functions f are defined into the interval $[-1,1]$ instead of being defined in $[-\pi, +\pi]$. For this purpose it is enough to set

$$t = \frac{l}{\pi} x \qquad (31.9)$$

and to consider the coefficents

$$a_0 = \frac{1}{l} \int_{-l}^{l} f(t)\,dt, \qquad (31.10)$$

$$a_n = \frac{1}{l} \int_{-l}^{l} f(t)\cos\frac{n\pi t}{l}\,dt, \quad \forall n \in N^*, \qquad (31.11)$$

$$b_n = \frac{1}{l} \int_{-l}^{l} \sin\frac{n\pi t}{l}\,dt, \quad \forall n \in N^*, \qquad (31.12)$$

It must be also observed that existing the following equalities

$$\frac{a_0}{2} + \sum_{n=1}^{+\infty} a_n \cos nx + b_n \sin nx$$

$$= \frac{a_0}{2} + \sum_{n=1}^{+\infty} \left(a_n \frac{e^{inx} + e^{-inx}}{2} - ib_n \frac{e^{inx} - e^{-inx}}{2} \right)$$

$$= \frac{a_0}{2} + \sum_{n=1}^{+\infty} \frac{a_n - ib_n}{2} e^{inx} + \sum_{n=1}^{+\infty} \frac{a_n + ib_n}{2} e^{-inx}$$

$$= \sum_{n=-\infty}^{+\infty} c_n e^{inx}, \qquad (31.13)$$

where
$$c_0 = \frac{a_0}{2} \tag{31.14}$$

$$\forall n \in \mathrm{N}^* : \begin{cases} c_n = \dfrac{a_n - ib_n}{2}, \\ c_{-n} = \dfrac{a_n + ib_n}{2}. \end{cases} \tag{31.15}$$

It results also that the complex trigonometric Fourier series

$$\sum_{n=-\infty}^{+\infty} c_n e^{inx} \tag{31.16}$$

converges in quadratic mean to f. Then, through (31.5), (31.6), (31.7), and (31.16) the formula to calculate directly the coefficients C_n can be extracted. In fact we can observe that multipling $f(x)$ and (31.16) by e^{-imx} and integrating each of the two expressions we obtain:

$$\int_{-\pi}^{\pi} f(x) e^{-inx} \mathrm{d}x = \sum_{n=-\infty}^{+\infty} c_n \int_{-\pi}^{\pi} e^{inx} e^{-imx} \mathrm{d}x. \tag{31.17}$$

Then it easy to verify that

$$\int_{-x}^{x} e^{inx} e^{-imx} \mathrm{d}x = \begin{cases} o, & for \ n \neq m, \\ 2\pi, & for \ n = m \end{cases} \tag{31.18}$$

and therefore it results

$$c_m = \frac{1}{2\pi} \int_{-x}^{x} f(x) e^{-imx} \, \mathrm{d}x, \quad m = 0, \pm 1, \pm 2, \ldots. \tag{31.19}$$

The series development in Eq. (31.16) is still valid for complex and square module integrable defined into the $[-\pi,+\pi]$ interval functions. In other terms the family $(e^{inx})_{n \in Z}$ constitutes a base for the L_2 space of complex, square module integrable defined into the $[-\pi,+\pi]$ interval functions. So the expression in Eq. (31.19) represents the scalar product of f by e^{imx}. If these functions are defined into the interval $[-1, 1]$, then the base of L_2 becomes the family $[e^{(in\pi t)/l}]_{n \in Z}$. We can prove that the system

$$(\cos nx)_{n \in N} \tag{31.20}$$

constitutes a complete orthogonal system for the square integrable functions defined into the $[0, \pi]$ interval. The othogonality between the elements of the sequence in Eq. (31.20) can be proved by a direct calculation. In order to prove the completeness, called f any square integrable on $[0, \pi]$ function, we can consider the function defined as

$$f^*(x) = \begin{cases} f(x), & if \ 0 \leqslant x \leqslant \pi, \\ f(-x), & if \ -\pi \leqslant x \leqslant 0. \end{cases} \tag{31.21}$$

It is obvious that f^* is square integrable in $[-\pi,+\pi]$ and so can be developed in Fourier series in accordance with Eq. (31.8). But we must observe that all the b_n coefficients to be calculated using Eq. (31.7) are identically null due to the effect of the parity of f^*. Therefore it is proved that f^* and, even more f, can be developed in series of only cosines. It can also be proved that the system

$$(\sin nx)_{n \in N} \qquad (31.22)$$

forms a complete orthogonal system for square integrable defined in $[0, \pi]$ functions. The demonstration can be done as in the previous case but defining f^* in the following way:

$$f^*(x) = \begin{cases} f(x), & if\ 0 \leqslant x \leqslant \pi, \\ -f(-x), & if\ -\pi \leqslant x \leqslant 0. \end{cases} \qquad (31.23)$$

The most typical example of a set of functions of those here described is the sine and cosine set as shown in Fig. 31.1.

It is well known that sinusoidal functions play an important role in synthesizing smoothly varying plane curves. This is reflected in the effectiveness of the Fourier Transform applied in the synthesis of handwriting. This topic, with particular attention on the application of the Discrete Fourier Transform, will mainly treated in the following sections.

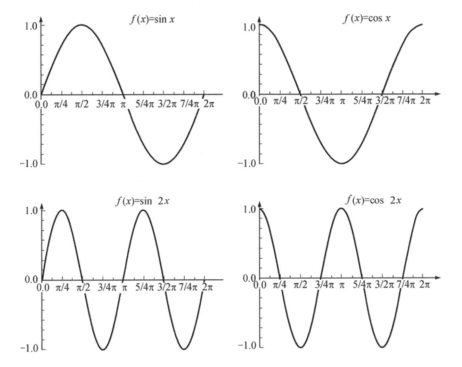

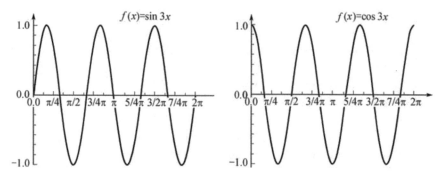

Fig. 31.1 Functions in the sine and cosine set.

31.3 Introduction to Discrete Fourier Transform

In Section 31.3 it has been shown that every 2π-periodic function, can be developed in trigonometric polynomial series and that if the function x(t) is continue, periodic of period T_0, that is

$$x\left(t + mT_0\right) = x\left(t\right), \quad \forall m \in \text{Z}. \tag{31.24}$$

Using the trigonometric system of complex exponentials we obtain

$$x\left(t\right) = \sum_{n=-\infty}^{+\infty} c_n e^{jn\omega_0 t}, \tag{31.25}$$

where $\omega_0 = 2\pi/T_0$ is the fundamental harmonic, ω_0 is the harmonic of order n, and further:

$$c_n = \frac{1}{T_0} \int_{-T_0/2}^{T_0/2} x\left(t\right) e^{-jn\omega_0 t} dt, \quad \forall n \in \text{Z}. \tag{31.26}$$

Let now $x_p(t)$ be a periodic function of period T_0, and t_0 a real not null number, let us consider the $x_p(nt_0)$ sequence that represents a sampled version of the $x_p(t)$ function, that is

$$x_p\left(nt_0\right) = x_p\left(t\right)|_{t=nt_0}, \quad \forall n \in \text{Z} \tag{31.27}$$

To keep the periodicity we suppose here, more specifically that T_0 is multiple of t_0 and, for notational simplicity, also that t_0 is unitary. So let $x_p(n)$ be the sampled version of the $x_p(t)$ function, that is following Eq. (31.27) it results:

$$x_p(n) = x_p(t)|_{t=n}, \quad \forall n \in \text{Z} \tag{31.28}$$

being $x_p(t)$ periodic, for Eq. (31.25) it results:

$$x_p(t) = \sum X_p^1(m) e^{j\frac{2\pi}{T_0}mt} \tag{31.29}$$

(where now the $X_p^1(m)$ substitute the c_n) and then:

$$
\begin{aligned}
x_p(n) &= x_p(t)|_{t=n} \\
&= \left(\sum_{m=-\infty}^{+\infty} X_p^1(m) e^{j\frac{2\pi}{T_0}mt} \right) \Big|_{t=n} \\
&= \sum_{m=-\infty}^{+\infty} X_p^1(m) e^{j\frac{2\pi}{N}mn} \tag{31.30} \\
&\overset{(1)}{=} \sum_{K=-\infty}^{+\infty} \left(\sum_{m=0}^{N-1} X_p^1(m+KN) e^{j\frac{2\pi}{N}(m+KN)n} \right) \\
&= \sum_{m=0}^{N-1} \left(\sum_{k=-\infty}^{+\infty} X_p^1(m+KN) e^{j\frac{2\pi}{N}(m+KN)n} \right),
\end{aligned}
$$

where in the (1) equality it has been considered that

$$e^{j\frac{2\pi}{N}m} = e^{j\frac{2\pi}{N}(m+kN)}, \quad \forall k \in \mathbb{Z}. \tag{31.31}$$

That is, under the enunciated conditions, only N complex exponentials can be discriminated. Still by the observation in Eq. (31.31), from Eq. (31.30) we obtain

$$x_p(n) = \sum_{m=0}^{N-1} e^{j\frac{2\pi}{N}mn} \sum_{k=-\infty}^{+\infty} X_p^1(m+kN), \tag{31.32}$$

thus, placed

$$\sum_{k=-\infty}^{+\infty} X_p^1(m+kN) = -\frac{X_p(m)}{N}.$$

we obtain

$$x_p(p) = \frac{1}{N} \sum_{m=0}^{N-1} X_p(m) e^{j\frac{2\pi}{N}mn}. \tag{31.33}$$

We call Eq. (31.33) expression "Inverse Discrete Fourier Transform" (IDFT). If now we consider the expression

$$\sum_{n=0}^{N-1} X_p(n) e^{j\frac{2\pi}{N}nm}, \tag{31.34}$$

we can see that its values are exactly the X_p (m) in Eq. (31.34), that is

$$X_p(m) = \sum_{n=0}^{N-1} X_p(n) e^{j\frac{2\pi}{N}nm}. \tag{31.35}$$

In fact:

$$\sum_{n=0}^{N-1} x_p(n) e^{-j\frac{2\pi}{N}nm} = \sum_{n=0}^{N-1} \left(\frac{1}{N} \sum_{k=0}^{N-1} x_p(k) e^{-j\frac{2\pi}{N}kn} \right) e^{-j\frac{2\pi}{N}nm}$$

$$= \sum_{k=0}^{N-1} \frac{X_p(k)}{N} \sum_{n=0}^{N-1} e^{-j\frac{2\pi}{N}n(k-m)} \qquad (31.36)$$

$$= \sum_{k=0}^{N-1} \frac{X_p(k)}{N} N\mu_0(k-m) = X_p(m),$$

where $u_o(n)$, known in literature as digital impulse, represents a sequence that is anywhere null except for $n = 0$ where it is equal to 1. The expression

$$X_p(m) = \sum_{n=0}^{N-1} x_p(n) e^{-j\frac{2\pi}{N}nm} \qquad (31.37)$$

is called "Discrete Fourier Tranform" (DFT).

The following couple is called Discrete Fourier Transform relations exists

$$X(k) = \sum_{n=0}^{N-1} x(n) e^{-j\frac{2\pi}{N}nk} \quad \text{(DFT)}, \qquad (31.38)$$

$$X_p(n) = \frac{1}{N} \sum_{n=0}^{N-1} X(k) e^{-j\frac{2\pi}{N}nk} \quad \text{(IDFT)}. \qquad (31.39)$$

We can observe that Eqs. (31.38) and (31.39) relations create a correspondence between N points of the variable with index n and N points of the variable with index k. This can be done either with periodic $x(n)$, $X(k)$ sequences with N points in each period, or with finite sequences with N elements. That is why in Eqs. (31.38) and (31.39) the sub-index "p" has been omitted.

31.4 Properties of DFT

Let N be a natural integer and let's consider the vectorial space C^N on the field C; it must be underlined here that the DFT introduced by Eq. (31.38) can be considered as an application of C^N on C^N. There are several properties of the DFT that play an important role in practical techniques for signal processing. In the following subsections we will summarize some of these properties.

31.4.1 Relations between Z-Transform, Fourier Transform and DFT

If $x(n)$ is a sequence, the expression

$$X(z) = \sum_{n=-\infty}^{+\infty} x(n)z^{-n}, \quad \text{where } z \in C, \tag{31.40}$$

if exists, is called "Z- Transform" of $x(n)$. In [1] the properties of the Z-transform of a sequence are shown. Using the Z-Transform, making suitable assumptions on $x(n)$, it can be easily shown that under certain conditions the DFT is nothing but a sampled version of the Fourier Transform of $x(n)$. This gives a method to pass from the Z-transform of a sequence, or from its Fourier Transform, to its Discrete Transform. Now there is the problem to find directly the Z-transform and/or the Fourier Transform of a sequence when the values of its DFT are known. For this purpose the following interpolating functions can be used

$$\frac{1 - z^{-N}}{1 - e^{j\frac{2\pi}{N}k}z^{-1}}. \tag{31.41}$$

In fact it is easy to demonstrate that it results

$$X(z) = \sum_{k=0}^{N-1} \frac{X(k)}{N} \frac{1 - z^{-N}}{1 - e^{j\frac{2\pi}{N}k}z^{-1}}, \tag{31.42}$$

$$X(e^{j\omega}) = \sum_{k=0}^{N-1} \frac{X(k)}{N} \frac{1 - e^{-j\omega N}}{1 - e^{-j\omega}e^{j\frac{2\pi}{N}k}}. \tag{31.43}$$

31.4.2 How to Increase Frequency Resolution

The observations made in the previous section allow to develop some very useful considerations. One of these is the way of increase the frequency resolution for a given sequence. In other words, it can be shown that to obtain the values in L harmonics, of an N points sequence, it is sufficient to add L-N zeros to the original sequence and then to calculate the DFT of the new sequence obtained. As we will see in the following sections, the most common algorithms for the computation of the DFT work with sequences constituted by a number of points that is a power of 2; consequently the method shown here is also an useful artifice to calculate the DFT of every sequence using these algorithms.

31.4.3 Linearity Property

If $x(n)$ and $y(n)$ are two sequences of C^N and α and β are two elements of the field C, then the DFT of the sequence $\alpha x(n) + \beta y(n)$ is equal to the linear combination, with the same coefficients α and β, of the DFT of $x(n)$ and respectively $y(n)$. Being the DFT a linear application, it can have a matrix representation. In fact, let $W_N = e^{-j(2\pi/N)}$,

$$X(k) = \sum_{n=0}^{N-1} x(n)W_N^{kn} \tag{31.44}$$

just represents the following product rows by columns

$$
\begin{bmatrix} X(0) \\ \vdots \\ X(k) \\ \vdots \\ X(N-1) \end{bmatrix}^{transp.}
=
\begin{bmatrix}
W_N^{0,0} & \cdots & W_N^{n,0} & \cdots & W_N^{N-1,0} \\
\vdots & & \vdots & & \vdots \\
W_N^{0,k} & \cdots & W_N^{n,k} & \cdots & W_N^{N-1,k} \\
\vdots & & \vdots & & \vdots \\
W_N^{0,N-1} & \cdots & W_N^{n,N-1} & \cdots & W_N^{N-1,N-1}
\end{bmatrix}
\begin{bmatrix} x(0) \\ \vdots \\ x(n) \\ \vdots \\ x(N-1) \end{bmatrix}. \tag{31.45}
$$

31.4.4 Translation Property

Let $x(n)$ be an N-elements sequence and n_o a natural integer. If $X(k)$ is the DFT of $x(n)$, it can be shown that the DFT of $x(n - n_o)$ is

$$e^{-j\frac{2\pi}{N}kn_o} X(K). \tag{31.46}$$

31.4.5 Symmetry Property

Let $x(n)$ be an N real points sequence and $X(k)$ its DFT; if $\text{Re}[X(k)]$ is the real part of $X(k)$ and $\text{Im}[X(k)]$ its imaginary part, it can be shown that:
a) $\text{Re}[X(k)] = \text{Re}[X(N-k)]$,   b) $\text{Im}[X(k)] = -\text{Im}[X(N-k)]$.

31.4.6 Direct Calculation of DFT of two Real Sequences

Since one generally deals with real sequences, the symmetry property may be conveniently used to obtain the DFT's of two sequences using a single DFT. Let $x(n)$ and $y(n)$ two N real sequences, and $X(k)$ and $Y(k)$ their respective discrete Fourier transforms. Considered the complex sequence

$$z(n) = x(n) + \mathrm{j}y(n) \tag{31.47}$$

and denoted by $Z(k)$ its DFT, from the linearity property we know that results

$$Z(k) = X(k) + \mathrm{j}Y(k). \tag{31.48}$$

It can be shown that

$$\mathrm{Re}[X(k)] = \frac{\mathrm{Re}[Z(k)] + \mathrm{Re}[Z(N-k)]}{2}, \tag{31.49}$$

$$\mathrm{Im}[X(k)] = \frac{\mathrm{Im}[Z(k)] - \mathrm{Im}[Z(N-k)]}{2}, \tag{31.50}$$

$$\mathrm{Im}[Y(k)] = \frac{\mathrm{Re}[Z(N-k)] - \mathrm{Re}[Z(k)]}{2}, \tag{31.51}$$

$$\mathrm{Re}[Y(k)] = \frac{\mathrm{Im}[Z(N-k)] + \mathrm{Im}[Z(k)]}{2}. \tag{31.52}$$

Thus, a single N-point DFT can effectively transform two N-point real sequences at the same time.

31.5 DFT Calculation Problem

N^2 complex multiplications and $N(N-1)$ additions are required to calculate Eq. (31.44). Since we can ignore the time necessary to calculate additions in respect to multiplications, we can say that the computation cost depends almost fully on the N^2 multiplications, then

$$C(N) = N^2. \tag{31.53}$$

To reduce the number of operations and consequently the computation cost of Eq. (31.44) many algorithms, identified as FFT (Fast Fourier Transform), were proposed. The first among these was that proposed by Cooley, Lewis and Welch in 1967 and 1968 [2,3]. The most important variations of this algorithm are the Decimation In Time algorithm (DIT) and the Decimation In Frequency algorithm (DIF). Subsequently, using the polynomial transformations, a systematic approach to the algorithms formulation for the fast DFT calculation was proposed. Another algorithm that we particulary point out is the Winograd-Fourier's one [4, 5].

31.5.1 Decimation in Time Algorithm

In this section we examine a particular FFT algorithm, known in literature as Decimation in Time Algorithm (DIT). Let us suppose that $x(n)$ has $N = 2^\beta$ elements, with β natural integer. Let

$$x_1(n) = x(2n), \quad n = 0, 1, \dots, N/2 - 1 \tag{31.54}$$

and

$$x_2(n) = x(2n + 1), \quad n = 0, 1, \dots, N/2 - 1 \tag{31.55}$$

be two sub-sequences of $x(n)$, each one having $N/2$ elements, the first constituted by all the even index elements of $x(n)$, and the second by the odd index ones. Let us denote with $X(k)$, $X_1(k)$, $X_2(k)$ respectively the DFT of $x(n), x_1(n), x_2(n)$. The following equations allow to find the relationship between $X(k)$ and the couple $(X_1(k), X_2(k))$

$$
\begin{aligned}
X(k) &= \sum_{n=0}^{N-1} x(n) W_N^{nk} \\
&= \sum_{n=0}^{\frac{1}{2}N-1} x(2n) W_N^{2nk} + \sum_{n=0}^{\frac{1}{2}N-1} x(2n+1) W_N^{(2n+1)k} \\
&= \sum_{n=0}^{\frac{1}{2}N-1} x_1(n) W_N^{2nk} + W_N^k \sum_{n=0}^{\frac{1}{2}N-1} x_2(n) W_N^{2nk} \\
&= \sum_{n=0}^{\frac{1}{2}N-1} x_1(n) W_{\frac{1}{2}N}^{2nk} + W_N^k \sum_{n=0}^{\frac{1}{2}N-1} x_2(n) W_{\frac{1}{2}N}^{nk} \\
&= X_1(k) + W_N^K X_2(k).
\end{aligned}
\tag{31.56}
$$

So Eq. (31.56) allows to transform the DFT calculation of an N points sequence into the calculation of two $N/2$ points DFT, reducing so the calculation cost from N^2 to $N^2/2$. It can be calculated dividing itself in two parts: in the first part we calculate the first N/2 values corresponding to $k = 0, 1, \dots, N/2 - 1$, in the second part the remaining $N/2$ corresponding to $k = N/2, \dots, N - l$; in this way we have

$$
X(k) = \begin{cases}
X_1(k) + W_N^k X_2(k), & for\ k = 0, \dots, \dfrac{1}{2}N - 1, \\[2mm]
X_1\left(k - \dfrac{1}{2}N\right) - W_N^{k - \frac{1}{2}N} X_2\left(k - \dfrac{1}{2}N\right), \\[2mm]
\qquad\qquad for\ k = \dfrac{1}{2}N, \dots, N - 1.
\end{cases}
\tag{31.57}
$$

The procedure used to formulate Eq. (31.56) and then Eq (31.57) can be repeated first on $X_1(k)$ and $X_2(k)$, and then on the sequences as obtained from these ones, till to obtain two elements sub-sequences. Let

$$X_{i_1,\ldots,i_{\beta-1}}(k), \quad \text{where } k = 0, 1 \tag{31.58}$$

be a generic sequence among these ones; following Eq. (31.57) we will be able to individuate the sequences that have been reduced to one element only

$$X_{i_1,\ldots,i_{\beta-1};1}(k), \quad \text{where } k = 0, \tag{31.59}$$

$$X_{i_1,\ldots,i_{\beta-1};2}(k), \quad \text{where } k = 0, \tag{31.60}$$

so that

$$X_{i_1,\ldots,i_{\beta-1}}(k) = \begin{cases} X_{i_1,\ldots,i_{\beta-1};1}(k) + W_2^0 X_{i_1,\ldots,i_{\beta-1};2}(k), & \text{if } k = 0, \\ X_{i_1,\ldots,i_{\beta-1};1}(k) - W_2^0 X_{i_1,\ldots,i_{\beta-1};2}(k), & \text{if } k = 1. \end{cases} \tag{31.61}$$

We can now observe that Eqs. (31.59) and (31.60) represents DFT of sequences of one point only. As the DFT of a point is the point itself, the X (upper) in the (31.61) can be replaced with the x (lower). More generally, if we suppose to denote these points with a and b, then Eq. (31.61) becomes:

$$X_{i_1,\ldots,i_{\beta-1}}(k) = \begin{cases} a + W_2^0 b, & for\ k = 0, \\ a - W_2^0 b, & for\ k = 1. \end{cases} \tag{31.62}$$

Looking at Eqs. (31.57) and (31.61) we can observe they have the same structure, and it can be shown that the structure in Eq. (31.62) is constant through the whole decomposition process; this structure can generally be represented by the butterfly in Fig. 31.2. The black arrow in the figure represents a multiplier and the circle an adder (or a subtracter) depending on, as we move along the butterfly branches from left to right, we get up or we get down. So the $X(k)$ calculation algorithm can be represented by a butterflies fall in which each level matches a decompositon step in the iteration shown before. In Fig. 31.3 it is reported the graphic scheme of the DIT algorithm applied to an 8 points sequence. The following two observations will help us in designing an algorithm for the computation of the butterflies in Fig. 31.2.

1) To obtain the output sequence from the pattern of Fig. 31.2 in natural order, it is necessary to change the order of the input sequence. An optimal

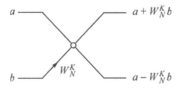

Fig. 31.2 Butterfly representation of the DIT algorithm.

algorithm for the solution of this problem was the "Bit Reversing algorithm" proposed by Rader.

2) Another important problem that raises in the transform calculation according to the pattern of Fig. 31.3 regards the phase factors. Observing that $W_2^0 = W_4^0 = W_8^0$ and also that $W_1^4 = W_8^2$ we can see that the phase factors in Fig. 6.3, can be grouped as $W_2^0 = W_4^0 = W_8^0$, $W_8^2 = W_4^1$, $W_8^1 = W_8^3$. So a recurrence exists in the value of the phase factors. Further considerations based on the fact that the phase factors are obtainable iteratively with the relation:

$$W_N^k = W_N^{k-1} W_N^1 \qquad (31.63)$$

led to the formulation of fast calculation algorithms.

In Fig. 31.3 it is reported the algorithm that provides both for the phase factors generation and the butterflies-fall calculation according to the algorithm derived by the iteration in Eq. (31.57) and exemplified in Fig. 31.2. Particularly we point out that the strategy used in this algorithm is that of calculating time after time all the butterflies that belong to the same level, and this procedure is repeated a number of times that equals the number of the levels that, as we know, is $M = \log_2 N$.

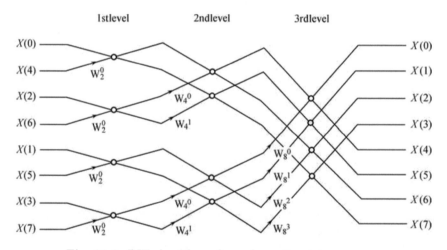

Fig. 31.3 DIT algorithm scheme for an 8 points sequence.

Fixed now a level, we can see that one phase factor can be associated to each butterfly, but, on the contrary, several butterflies are associated to each phase factor: for example, in the first level of Fig. 31.3, N/2 butterflies are associated to the phase factor W_2^0. For this reason, all the butterflies of the same level can be grouped so that the butterflies in which the same phase factor appears belong to the same group.

Firstly all the same group butterflies, and this group by group in each level, are calculated. For example, in the first level we calculate only the butterflies group in relation with the phase factor W_2^0, in the second level the two butterflies groups in relation with the phase factors W_4^0 and W_4^1 and so on. In Fig. 31.3 three nesting loops are evident: the inner performs the calculation of the same group butterflies, the intermediate instead scans the various butterflies groups inside each level. In fact in this loop the instruction $U = U * W$ appears; this instruction, fixed the initial phase factor U and the spacing factor W, allows to generate in iterative manner the phase factor of each group. It is the most external loop that scans the various levels and it provides, from time to time, to determine the initial phase factor and the spacing factor calculated with the sine and cosine standard routines.

Fig. 31.3 shows that, for sequences of $N = 2^\beta$ elements, the DIT algorithm foresees $\beta = \log_2 N$ levels, and as it is necessary to carry out $N/2$ multiplications for each level, we can deduce the total number of multiplications equals $1/2N\log_2 N$. Similar observations can be done also for the DIF algorithm. Naturally to determine the effective cost of the DFT calculation by one of these two algorithms, we must consider also other necessary operations as, for example, the bit-reversing one. However we can consider that the cost $c(N)$ is of $1/2N\log_2 N$ order. The reader is invited to confront the two sequences N^2 and $1/2N\log_2 N$, that is respectively the cost of the direct calculation of the DFT and its calculation through the DIT and DIF algorithms. The knowledges and the properties of the DFT reported in the preceding sections can be also extended to multidimensional sequences.

In the next sections will be presented some example of applications of Discrete Fourier Tranform to handwriting recognition.

31.6 Description of a Numeral Through Fourier Coefficents

Let us consider a plane, continue, of finite lenght curve, for example the one that we obtain writing a one-stroke character. Here we assume that a stroke is the writing from pen-down to pen-up.

As the points of the plane (in this case of the paper sheet) can be described by complex numbers considering the abscissa of a point as the real part and its ordinate as the imaginary part of the same complex number, the description of the traced curve can be described by the complex function

$$Z = z(t) = x(t) + jy(t), \tag{31.64}$$

where t could be a time parameter during the writing. Almost in the totality of the cases it is difficult to find a simple analytical expression of $z(t)$. In order to describe $z(t)$ it can be sufficient and useful to discretize it, taking N points distributed on all its lenght, i.e., the sequence of complex numbers

$$z(n) \text{ with } n = 0, 1, \ldots, N - 1 \tag{31.65}$$

that represents the description by points of the curve $z(t)$. By applying an FFT algorithm to $z(t)$ we obtain N new complex numbers: each of this complex number represents the module and the phase of a particular harmonic component of the curve $z(t)$. As an experiment we can write a numeral on a graphic tablet and extract the $z(n)$ sequence. Then, if necessary we should add as many zeros until the number of the points that describe the curve becomes a power of two. In this way we will be able to apply the DIT algorithm and calculate the Fourier coefficients. Repeating this procedure for many characters it can be verified that for most of them the value of the modules obtained is low: in fact most of the informative contents of an handwritten character stroke is carried by the lowest few order harmonics; then we can use only them to rebuild the curve using the inverse transformation process. In Figs. 31.4 and 31.5 are shown some numerals as acquired from a graphic tablet and their reconstruction (Fig. 31.6) using only four Fourier Coefficents.

```
PROCEDURE fft (VAR a:coordarray);
  VAR
    1,le,le1,ip,j,i          : integer;
    u1,u2,w1,w2,t1,t2,v1,v2: double;
  BEGIN
    FOR 1:=1 TO M DO
      BEGIN
        le:=trunc(exp(1*ln(2)));
        le1:=le div 2;
        u1:=1;
        u2:=0;
        w1:=cos(pi/le1);
        w2:=-sin(pi/le1);
        FOR j:=1 TO le1 DO
          BEGIN
            i:=j;
            while i<=n DO
              BEGIN
                ip:=i+le1;
                t1:=a[ip,1]*u1-a[ip,2]*u2;
                t2:=a[ip,1]*u2+a[ip,2]*u1;
                a[ip,1]:=a[i,1]-t1;
                a[ip,2]:=a[i,2]-t2;
                a[i,1]:=a[i,1]+t1;
                a[i,2]:=a[i,2]+t2;
                i:=i+le;
              END;
            v1:=u1*w1-u2*w2;
            v2:=u1*w2+u2*w1;
            u1:=v1;
            u2:=v2;
          END;
      END;
  END;
```

Fig. 31.4 DIT algorithm.

We can now conclude that Fourier Coefficents are good features and can be considered representative of the numeral/character that they describe. As we know pattern features are useful to classify pattern itself and then we wonder, as an example, is it possible to design a numeral recognition system basing on numerals description through Fourier Coefficents? In order to do this, we will firstly define a concept of similarity among numerals, then will show as it

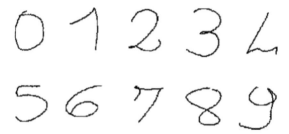

Fig. 31.5 A samples of mono-stroke handwritten numerals.

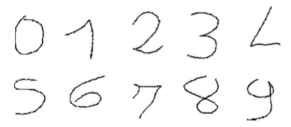

Fig. 31.6 Reconstruction of the samples in Fig. 31.4 using 4 Fourier Coefficents.

is possible to obtain a large number of numerals derived from a handwritten one, then will define a space useful to place all original and derived numerals. After having described all possible numeral shape variations we will think about numeral class definition: then we will show as it can be implemented a numeral recognition system based on Fourier Coefficents description of numerals. An example of this system will be shown in the next sections.

31.6.1 Similarity Between two Handwritten Numerals

Handwritten characters consist either of one or of a finite number of strokes. Here we assume that each stroke corresponds to a geometric plane curve. Hence, since the Euclidean plane can be considered isomorphic to set C of the complex numbers, each plane curve can be represented by means of a parametric equation of the kind:

$$z = z(t) = x(t) + jy(t), \qquad (31.66)$$

where t belongs to either a bounded and closed real interval, if the character consists of one stroke, or a finite set of intervals of the same kind, if it includes many strokes. Since a multi-stroke character can be investigated by analyzing one stroke at a time, here only the set $\{z(t)\}$ of the plane curves, defined in $[0, T]$ real interval, will be taken into consideration.

Distance

It can be proved that set $\{z(t)\}$ can become a metric space if the distance:

$$d(z_1, z_2) = ||z_1(t) - z_2(t)||, \qquad (31.67)$$

where $\forall z \in \{z(t)\} : ||z|| = (1/T \int_0^T z(t)\overline{z(t)}dt)^{1/2}$ is taken into consideration [6]. The distance in Eq. (31.67) represents the global difference between the plane curves $z_1(t)$ and $z_2(t)$.

Furthermore it can be proved that $\{z(t)\}$ is also a Hausdorff space thus, both the limit of a sequence of plane curves and the limit and the derivative of a function of the plane curves $M = M(z)$ may exist. Therefore the boundary of a subset of $\{z(t)\}$ can be considered. It can be also shown that $\{z(t)\}$ is an algebra.

The Weierstrass-Stone theorem proves that all the subalgebras of $\{z(t)\}$ which separate the points of $\{z(t)\}$ are dense in $\{z(t)\}$ (i.e., $\overline{A} = \{z(t)\}$, where $\overline{A}$ is the closure of the subalgebra A). Therefore the elements of $\{z(t)\}$ can be uniformly approximated by elements of A. The subalgebra of complex esponential polynomials is here taken into consideration.

Weierstrass has shown that this subalgebra separates the points of $\{z(t)\}$ and thus that it is dense in $\{z(t)\}$. In other words all the plane curves belonging to $\{z(t)\}$ set can be uniformly approximated by complex exponential polynomials. In addition, Fejer has shown that the coefficients of these polynomials can be obtained by using the Fourier coefficients computed by means of the scalar product between the plane curve and the elements of the complex exponential orthonormal system.

The measurement of the global difference between the two plane curves $z'(t)$ and $z''(t)$ given in Eq. (31.67) can thus be approximated by

$$d(z', z'') = \sum_{-M}^{+M} ((\text{Re}[Z_n''] - \text{Re}[Z_n'])^2 + (\text{Im}[Z_n''] - \text{Im}[Z_n''])^2)^{1/2}, \qquad (31.68)$$

where Z' and Z'' are the Fourier coefficients of $z'(t)$ and $z''(t)$ respectively.

Instead of the $z(t)$ sequence its even part $z_e(t)$ can be taken into consideration to eliminate the discontinuity effect of the end points. Another approach to eliminate the same effect could be the choice of the cosine orthonormal system.

To avoid the effect of size and orientation of the distance in Eq. (31.68), instead of the Fourier coefficients, the Fourier descriptors:

$$d_{1j} = \frac{Z_j}{Z_1}, \quad \text{where } j \in Z - \{0\} \qquad (31.69)$$

will be used from this point on [7–10]. Other types of descriptors can be found in [11–16].

A new distance between the two plane curves $z'(t)$ and $z''(t)$ that measures the difference in shape between them can thus be given by:

$$D(z', z'') = \sum_{n=-M}^{+M} ((\text{Re}[d''_{1n}] - \text{Re}[d'_{1n}])^2 + (\text{Im}[d''_{1n}] - \text{Im}[d'_{1n}])^2)^{1/2}, n \neq 0.$$
(31.70)

Obviously the distance in Eq. (31.70) is independent of position, size and orientation of the two plane curves. It allows to detect whether the two plane curves are equal or different and in the latter case it measures the difference between the two shapes.

Similarity

After having discussed about the measure of the difference between two plane curves it is now time to discuss as we can define the "similarity" between two plane curves, for example representing numerals or a part of them.

Preliminary it must be noted that the writing process can be considered as a stochastic process. Hence the organization degree of the handwritten numerals can be computed. It has been shown that this organization degree is high enough so that the shape of the plane curve:

$$z_m(t) = \sum_{n=-M}^{+M} (d_{1n} + \Delta d_{1n})e^{-jn(2\pi/T)t}$$
(31.71)

obtained from the handwritten 'numeral' $z(t)$, with a slight increment Δd_{1n} of the Fourier descriptors d_{1n} can be recognized by the writer as the same numeral $z(t)$. That is

$$M \vdots z_m(t) \vdots = M \vdots z(t) \vdots,$$
(31.72)

where M represents a recognition operator.

Consequently it can be assumed that $z(t)$ and $z_m(t)$ are similar [17]. Thus it can be stated that if $z_1(t)$ and $z_2(t)$ are two samples of the same numeral "j", then:

> $z_1(t)$ is similar to $z_2(t)$ if and only if there exists a line L joining $z_1(t)$ to $z_2(t)$ so that M recognizes each $z(t)$ point of the line L as the same numeral j represented by $z_1(t)$ and $z_2(t)$

(31.73)

It can be easily shown that this similarity definition is an equivalence relation and that it will divide the set of all the handwritten numerals into similarity classes. The hypotesis that each class is a simply connected set will be assumed, so that each class can be selected directly by detecting its boundary. Consequently if $z(t)$ is a handwritten sample of the numeral j, the problem of selecting all the samples of the numeral j, belonging to the class of $z(t)$, can be solved by detecting only the boundary of this class.

It has been shown that the contribution carried by the higher order harmonics is very small and thus it can be ignored in representing these numerals [7]. Here only the first four harmonics have been taken into consideration and experiments to detect class boundaries have been carried out by investigating the tridimensional complex space of the Fourier descriptors d_{12}, d_{13} and d_{14}.

31.7 Numeral Recognition Through Fourier Transform

In the character recognition field many efforts has been done in order to design reliable systems. In the field of printed characters excellent results, in terms of high speed and low misrecognition rate, has been achieved [18,19]. Unfortunately in the field of handwritten character recognition results are not so excellent, since the systems proposed up to now are affected by a misrecognition rate which is not negligible. This depends on the fact that different people write characters in a different way and even one man writes the same character in different ways depending on factors like age, physical and psychological states, the semantic contents of writing as well as environmental conditions [20]. It results that many plane curves represent the same character, i.e., more than one image description of the same pattern can be stored within the same family or class, and different families may be related to the same pattern. Notwithstanding man is not perfect as recognizer, in fact his 'recognition rate' changes with time and experience, and also he shows some uncertainty [21 – 23], he remains unchallenged in handwritten character recognition. Much more reliable is man's performance when also context is taken into consideration.

Many researches has been conducted about this point and the approach to create an extensible database of all possible numeral shapes is opposed to the approach that performs any possible type of analysis on patterns. Multiexpert systems follow the approach to integrate multi-type recognition systems results.

All approaches are welcome but up to now none is effectively able to solve the handwriting recognition problems definitively. Here we will show as it is possible to create a recognition system basing on Fourier Description of the curves representing numerals [24]. Maybe it could be not sufficient to solve the problem but it allows to extend its knowledge base of totally unconstrained handwritten numerals without any limit and then is an interesting approach to study.

31.7.1 Core System

Let be $z(t) = x(t) + \mathrm{j}y(t)$ (where t belongs to either a bounded and closed real interval) the description of a stroke that is here considered as belonging to a complex plane; then let denote by $\{z(t)\}$ the set of the plane curves, defined in the $[0, T]$ real interval. As we have shown these plane curves can be represented using Fourier coefficents and, to avoid the effect of position, size and orientation, some suitable Fourier Descriptors have been used.

In order to develop the learning phase of the system it must be noted that the shape of a plane curve $z_2(t)$ obtained with a slight increment of the Fourier descriptors of the handwritten "numeral" $z_1(t)$, is generally recognized by the writer as the same numeral represented by $z_1(t)$. Consequently it can be assumed that $z_1(t)$ and $z_2(t)$ are similar. This procedure can be iterated for $z_2(t)$ and so on, but not indefinetely.

We have seen also that to describe the strokes representing handwritten numerals only the first four Fourier coefficents can be used and experiments to detect class boundaries can be carried out by investigating the tridimensional complex space of the Fourier descriptors d_{12}, d_{13}, and d_{14}, where $d_{1j} = c_j/c_1$ for each j=2,3,4, being c_j for j=1,2,3, and 4 the Fourier coefficents.

This space can be investigated through the six dimensional real space R^6 and if the polar representation of complex numbers is used, since the phase is periodic, it can be investigated trough a tridimensional array of 'cubes' each one of size $(2\pi)^3$. In this space the boundary of each class is a hypersurface including "the subset of curves recognized similar to the handwritten sample". To detect one of these surfaces the tridimensional finite space of ternaries $(\Phi d_{12}, \Phi d_{13}, \Phi d_{14})$ must be investigated for each module ternary $(|d_{12}|, |d_{13}|, |d_{14}|)$.

Basing on the similarity definition, we can discretize this space. For this purpose, a quantum step "q" for Φd_{1j} has been selected. In this manner each block consists of $Q \times Q \times Q$ plane curves, where $Q = 2\pi/q$.

Similarly if "p_j" is the quantum step of $|d_{1j}|$ and if P_j is the number of the quanta levels for $|d_{1j}|$ then the tridimensional space of the modules will consist of an array of $P_2 \times P_3 \times P_4$ cubes, each of which includes $Q \times Q \times Q$ plane curves.

It has been shown [32] that for handwritten numerals the best quanta step values are respectively $q = (2\pi/16)$, P_2=1, P_3=0.1 and P_4=0.05, while the maximum value for descriptors modules are respectively $|d_{12}|$=50, $|d_{13}|$=30 and $|d_{14}|$=10. It follows that a tridimensional array of cubes including 50 × 300×200×4096 = 1. 2288×10^{10} plane curves must be taken into consideration to investigate our Fourier descriptors space. Figure 31.7 shows the discrete representation of this descriptor space.

We have now the possibility to describe the shape of all possible mono-stroke numerals and place them in our knowledge base [25 – 29].

How we can now fill this knowledge base and then use it to decide if any shape is a numeral and which class it belongs to ?

We could easily acquire a large quantity of numerals and begin to fill the base. As it allows to introduce new samples without any limit, the system learning phase can continue indefinitely.

In the following section we will show a different approach that is more

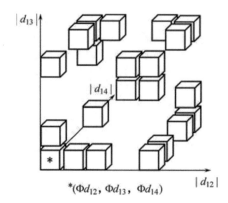

Fig. 31.7 Discrete representation of the descriptor space.

fascinating also if not easy to implement.

31.7.2 Numeral Class Detection

Let $z(t)$ be a handwritten element of $\{z(t)\}$ and $z_e(t)$ its even part. Let's consider:

- The application on $z(t)$ of a sampling function;
- The application of the Discrete Fourier Transform on the sampled sequence;
- The selection of the first four Fourier Coefficents and d_{12}, d_{13}, and d_{14} Fourier Descriptors calculation;
- The Inverse Discrete Fourier Transform;
- The new plan curve $z_1(t)$.

The new generated plane curve is described by only 4 Fourier coefficents; Fig. 31.5 that shows the results of the application of the above procedure to the even part of the handwritten samples in Fig. 31.5.

During the pre-processing phase the acquired sequence is normalized to N: N is chosen to be equal to 2α (α integer) to allow the use of the DIT algorithm to extract the features from the curves. In order to apply more conveniently the DFT transform we can duplicate the set $\{x(n)n = 0, \dots , N-1\}$ of the points describing the curve obtaining the new set $\{x'(n)\} = \{x(0), x(1), \dots ,$ $x(N-2), x(N-1), x(N-2), \dots , x(1)\}$: so we close the curves representing the numerals. Then it must be specified that N is chosen to be $2(N-1) = 2\alpha(\alpha$ integer). Alternatively the DCT (Discrete Cosine Transform) can be applied [30,31]. To create the structure of the knowledge base a space discretization has been realized as previously shown. In order to do this, a quantum step is taken into consideration for the modules and the phases of the Fourier descriptors. The steps $M_{12} = 0.125$ $M_{13} = 0.08$ $M_{14} = 0.05$ and $F_{12} = F_{13} = F_{14} = 2\pi/16$ (where M_{1n} is the modules quanta and F_{1n} the phases quanta)

were experimentally determined. A complete description of the structure of
the discrete Fourier Descriptors space can be found in [32].

Now by changing the value of the phases of the Fourier descriptors d_{12}
and d_{13} quantum step by quantum step, a map of the $Q \times Q$ plane curves
can be obtained. Since this can be done for each quantized value of the
descriptor d_{14} phase, it follows that Q maps of this kind can be obtained.
Figure 31.6 presents an example of map containing 16×16 numerals derived
from the handwritten one that is included (coordinates 8,8). The problem
of class selection basing on human knowledge is now met by submitting to
one or more person these maps and by asking them to discriminate the plane
curves representing numerals similar to handwritten $z(t)$ from the others.
The boundaries selected with this procedure are evident in the same map in
Fig. 31.8.

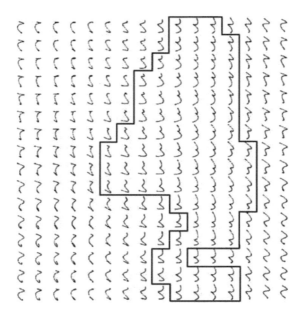

Fig. 31.8 Map containing 16×16 numerals derived from an handwritten one.

The composition of these boundaries in a tridimensional arrangement
inside the "cube", gives the surface which represents the boundary of the
subclass which includes the original handwritten numeral.

The value of the first Fourier coefficent of the original numeral has to be
recorded in order to save the orientation of the strokes and used for example
to distinguish the curves representing numerals like 6 and 9.

Obviously all the plane curves in this subclass exhibit the same descriptor
modules. An investigation similar to the previous one must be now repeated
for all the cubes neighbouring those which have been selected in the module space until none plane curves under examination will be recognizable as

similar to the handwritten sample.

To accomplish system training many handwritten samples written by different people can be employed as starting $z(t)$ "numeral".

The last problem concerns database recording. No requirements in this case: just as an example, looking at the 3D structure of each space (Φd_{12}, Φd_{13}, Φd_{14}), it could be easily represented through a cube of $16 \times 16 \times 16 = 4096$ bit linearized and compressed. The values of (Φd_{12}, Φd_{13}, Φd_{14}) must be recorded for each module ($|d_{12}|$, $|d_{13}|$, $|d_{14}|$) ternary value.

In order to classify an unknown sample, its descriptors are computed at first and its "module quantum level" (i.e., the point corresponding to the computed ternary in the module space) of belonging is selected.

Subsequently the cluster corresponding to the computed ternary of modules and describing the phases boundaries is considered, if it exists, and the inclusion of the unknown point into the class is controlled. If it is included, the sample is recognized otherwise a new class could be generated by the interactive process. A rejection occurs both if the quantum level in the module space is not classified yet or the sample is out of the boundaries selected in the phases space.

The method proposed allows the designing of a recognition system that can extend its knowledge base appreciably. As a matter of fact no knowledge base limit is exhibited by this system. If during the learning phase some classes are left out and thus an unknown handwritten sample can not be classified, its class can be immediately selected using the interactive system described. Another advantage of this method lies that very simple and fast systems can be designed to classify unknown samples. This is due to the fact that all the ternary of Fourier descriptors corresponding to handwritten numerals can be quickly found. The high speed recognition and the simplicity of the systems proposed depends on the fact that they have to control only the contents of a memory location addressed by the Fourier descriptors value of the unknown samples.

The method proposed may also be employed in other fields of pattern recognition. But in applications where more than four Fourier descriptors are required, the learning phase can become very expensive. As we said in advance, the limit of this type of system is the amount of work and time required to select all the boundaries in order to develop the knowledge base.

It must be finally pointed out that the discrimination carried out by the human observer during the boundary selection process depends on the observer and then the investigation about the handwritten numeral knowledge base of different observers is certainly interesting. Details about Pattern Recognition and applications of Fourier Transform can be found in [33–39].

31.8 Signature Verification Systems Trough Fourier Analysis

High-security access control is important in many applications and for this purpose, several systems for automatic personal verification has been proposed up to now [40]; among these systems many researches have been conducted on the use of signature with the aim of person identification.

In order to evaluate the effectiveness of an identification systems, many factors should be considered, such as the accuracy of the system, the verification speed, the simplicity of the enrolment process, the possibility to work on remote systems.

In this sense, an important advantage of signature is that it is a common way to do legal attestation or to identify an individual in daily operations such as banking transactions and fund transfers. Signature verification is a pattern recognition problem: it is not easy to solve and a wide community of scientists is involved in the improvement of algorithms for signature verification [41 – 47].

As other pattern recognition systems, a Signature Verification System (SVS) consists of several processing phases. After data acquisition, noise is removed during the preprocessing phase in order to obtain significant information from the raw data. In the feature extraction phase, the discriminant features are taken into consideration: they will be used to discriminate genuine signatures from forgeries. During the training process, the features extracted from the set of reference signatures are enrolled into the personal database. In the comparison phase, they are matched against those belonging to the input (test) signature in order to verify the authenticity of the input signature. During comparison process the authenticity of the test specimen is evaluated by matching its features against those of the reference database. This process produces a single response given in the form of a boolean value true if the test signature is considered genuine or false if it is considered a forgery one. For this purpose, several matching strategies can be adopted to compare the test signature with the reference signatures already included in the reference database [48, 49]:

- wholistic matching,
- regional matching,
- multiple regional matching.

The simplest strategy is based on the wholistic approach and it consists of matching the test signature against each one of the reference signatures considered as a whole. Of course this approach does not allow any regional evaluation of the signature.

A more reliable strategy is based on a regional matching approach. In this case the test signature S^t is split into n segments

$$(S_1^t, S_2^t, \dots S_k^t, \dots, S_n^t)$$

as well as S^r which is split into n segments

$$(S_1^r, S_2^r, \dots S_k^r, \dots , S_n^r)$$

The matching between S^t and S^r is performed by evaluating the local responses R_k^r obtained by matching S_k^t against S_k^r for $k = 1, 2, \dots , n$:

$R_k^r = $ *false* if and only if S_k^t results a forgery when compared with S_k^r,

$R_k^r = $ *true* if and only if S_k^t results genuine when compared with S_k^r.

This approach allows a regional analysis of the signature, but it is carried out in a one-by-one comparison process: i.e., the test signature is considered genuine if and only if a reference signature exists for which, in the comparison process, all or a suitable number of segments of the test signature are found to be genuine.

An excellent matching approach is the multiple regional matching. In this case each segment S_k^t of the test signature is matched against the entire set of the corresponding segments of the reference signatures. Again the test signature is considered genuine if a suitable number of segments are found to be genuine. This approach allows a regional evaluation of the signature without requiring a large set of reference signatures.

The procedure just described requires that each comparison technique is based on a suitable similarity (or dissimilarity) measure: two kinds of comparison techniques can be distinguished according to the type of features, parameters or functions, that are used.

When parameters are considered to be features, the signatures are described by vectors of parameters in a multidimensional feature space while when functions are considered to be features, the signatures are described as time functions which represent complete dynamic signals acquired directly from the acquisition device or derived from the input signals. In this case the matching techniques must take into account the variations of signal duration from one signature to another (even if the specimens are produced by the same writer). Furthermore, random variations, due to the writer's pauses or hesitations, can create portions of signals, such as deletions, additions and gaps, which complicate the problem of matching. Dynamic non-linear time warping is one of the most common matching methods used for this purpose [50].

In the next sections we will show the basic processing phases concerning signature verification and focus the contribution that Fourier analysis can carry to these type of systems.

31.8.1 Signature Acquisition and Preprocessing

Two different kinds of acquisition devices can be used for signature acquisition, on-line [51 – 56] and off-line [57 – 59]: here we will focus only on-line acquisition since will show as Fourier analysis can be used in the on-line signature verification systems basing on the coordinate points acquired during the writing process.

On-line acquisition devices such as graphic tablets produce signals representative of the signature trace during the writing process. Special tablets can also give different type of signals: coordinates, velocity, acceleration, pressure and force signals, pen-down and pen-up signals (a pen-down movement is the operation of pulling down the tip of the pen toward the writing plane, while a pen-up movement is the operation of lifting the tip of the pen away from the writing plane) and so on. Figure 31.9 shows an example of an on-line acquired signature described by means of its coordinates $(x(i), y(i)), i = 1, 2, \ldots, N$, where N is the number of samples acquired during the signing process. In Fig. 31.10 the pen-down and pen-up signals are also reported. They are marked with "*" and "0", respectively.

Fig. 31.9 An example of on-line signature.

Fig. 31.10 Pen-down and pen-up points of the signature in Fig. 31.9.

After noise removal and signature normalization, a critical preprocessing task is the segmentation of signature into strokes. This task is necessary when signature verification is carried out by the analysis of parts of the signature.

One of the approaches to signature segmentation is to consider a segment as a piece of the written trace between a pen-down and a pen-up movement. This approach is based on the consideration that the signature can be regarded as a sequence of writing units delimited by abrupt interruptions. Writing units are the regular parts of the signature, while interruptions are the singularities of the signature [60, 61]. Experimental evidence has shown that singularities occur in definite positions in the signature of an individual. Therefore only a finite set of writing units, called components or fundamental strokes, can be generated by each writer. As an example, Fig. 31.10 shows the components of the signature in Fig. 31.9. Furthermore, it has been shown that although signatures of the same writer may be composed of a different number of components, and also signatures with the same number of components may be very different from each other because of the different pen-lift distribution, the components that each writer uses to produce his signature are well defined and belong to a finite set of fundamental components.

An interesting segmentation technique is based on a dynamic splitting procedure [62]. The basic idea is to perform the splitting by using the information from both the reference signatures and the input signature, in fact, the reference signatures are segmented into basic strokes taking into consideration the characteristics of the input signature. This technique first detects the Candidate Splitting Points of each reference signature and of the input signature as the local maxima and minima in the vertical direction of each specimen. Then, an elastic matching procedure is applied and the best coupling between the Candidate Splitting Points of the input signature and of the reference signatures leads to the identification of the set of coupled strokes.

31.8.2 Discriminant Features

Whatever the means for separate the writing units (indeed signature could be considered a single writing unit), two types of features can be used for signature verification: parameters or functions [48]. In on-line signature verification, typical parameters used as personal features are the total duration of the signing process, the pen-down time ratio which is the ratio of pen-down time to total time, the average and the root-mean-square of the velocity and acceleration, the correlation between the magnitude of V_x and V_y, the number of pen lifts and specific coefficients derived from mathematical transforms [52, 63]. Specifically, based on the consideration that the signature generation process of each writer is constrained by the particular characteristics of the handwriting system (nerves and muscles of the harm, hand, fingers and so on), suitable Fourier descriptors can be used to describe the signature shape and used as discriminant features for on-line signature verification.

Let $(x(i), y(i)), i = 0, 1, \dots , N - 1$, be the sequence of coordinates representing a normalized signature, i.e., a signature with a number of samples $N = 2^\alpha$ (α integer). If we assume that the signature has been generated on a two-dimensional complex space, its description becomes $z(i) = x(i) + jy(i), i = 0, 1, ..., N - 1$. This complex sequence can be transformed by a Fast Fourier Transform algorithm, obtaining the Fourier coefficients $Z_i i = 0, 1, ..., N - 1$. As we have already seen, the Fourier descriptors are the complex values:

$$d_{1i} = Z_{i+1}/Z_1, \quad i = 0, 1, ..., N - 2.$$

It is easy to verify that these features are related to signature dynamics.

Many other features can also be considered which are related to the shape of the signature, like the length-to-width ratio, the direction histogram or the histogram of the direction changes.

31.9 On-line Signature Verification System Based on Fourier Analysis of Strokes

As we have just seen as in on-line signature verification systems the data acquisition phase is carried out during the writing process, hence, using a control signal (pen-up/pen-down signal), a signature can be acquired as a finite sequence of plane curves called strokes; each stroke consists in a finite sequence of consecutive points between two control signals [64].

Since we have seen that a signature can be considered a curve described through suitable Fourier descriptors [63, 65, 66], now we will see an on-line signature verification system (SVS) based on one-dimensional spectral analysis of the strokes considered separate 'writing units'. For this purpose we will show a structured knowledge-base of the SVS where the training signatures are grouped in suitable classes. In this way the verification process of an unknown signature will take into consideration only those signatures of the knowledge-base belonging to specific classes.

31.9.1 Stroke Based Knowledge Base: Segment and Group

Different samples of signatures of the same person can present not only strokes with different length but also a different number of strokes. Therefore, in order to develop a verification system more consistent with the human variability in signing, a structured knowledge-base has been designed.

The number of strokes which compose a signature can be obtained by counting the pen-up/pen-down signals during the acquisition process. In Fig. 31.11 a sampled signature and the strokes detected are shown while the characteristics about the j-th stroke of the i-th signature may be evaluated by detecting the number of minima m_{ij} and maxima M_{ij} in the stroke as the example in Fig. 31.12. In this case the organization of the knowledge-base could be the one reported in Fig. 31.13.

The First Level Classes (FLC) are obtained by grouping the signatures with the same number of strokes: this number represents the entry of the class. Within a FLC, a further classification is done. More specifically a signature belonging to the nth FLC (and then composed by n strokes) is grouped in a Second Level Class (SLC) if and only if for each signature of the SLC:

$$|m_{sk} - m'_k| < 2, \qquad k = 1, \ldots, n,$$
$$|M_{sk} - M'_k| < 2, \qquad k = 1, \ldots, n,$$
$$|(m_{sk} + M_{sk}) - (m'_k + M'_k)| < 2, \quad k = 1, \ldots, n,$$

Fig. 31.11 Strokes detected in the signature of Fig. 31.8.

Fig. 31.12 minima and maxima points.

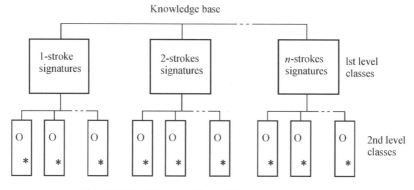

Fig. 31.13 Organization of the knowledge-base.

with m'_k number of minima in the k-th stroke of the signature to be classified; M'_k number of maxima in the k-th stroke of the signature to be classified; m_{sk} number of minima in the k-th stroke of the s-th signature of the SLC; M_{sk} number of maxima in the k-th stroke of the s-th signature of the SLC.

The entry of a SLC is then represented by the averages, computed over all the signatures of the class, of the number of the minima and of the maxima for each stroke.

31.9.2 Fourier Descriptors as Discriminant Features of Single Strokes

The aforesaid structure of the knowledge-base permits to partition the learning signatures in suitable classes but, in order to obtain global informations about the characteristics of the signatures in a given class, further features are needed. For this purpose, in this approach, we take into consideration the Fourier description of the strokes. Specifically if

$$z_i^s = x_i^s + y_i^s \quad i = 0, \ldots, N_s - 1$$

is the sequence of the N_s points of the s-th stroke of a n-stroke input signature, after a suitable standardization of the sequence length to N′ points ($N' = 2^\alpha$, α integer), a FFT algorithm is applied on z_i^s and the Fourier coefficients

$$c_i^s, \quad i = 0, \ldots, N', \quad s = 1, \ldots, n$$

and then the Fourier descriptors

$$d_{1-i}^s = c_i^s / c_1^s \quad i = 2, \ldots, N'$$

are computed for each stroke. Many experiments have shown that the greatest part of the information of each stroke is conveyed by the low-frequency harmonics, therefore only the first N low-order descriptors can be taken into consideration. N depends on the typical shape of the signature of each writer and can be defined during the training process, for example trough a statistical base [67]. Since this, let g be the number of signatures of a selected SLC of the n-th FLC and let

$$V^{sk} = (d_{1-2}^{sk}, \ldots, d_{1-N}^{sk}),$$

be the feature vectors of the Fourier descriptors moduli of the k-th stroke of the s-th signature in that class; a global information about the harmonical characteristics of the signatures of the class is conveyed by the vectors

$$V^k = (d_{1-2}^k, \ldots, d_{1-N}^k), \quad k = 1, \ldots, n$$

and

$$\sigma^k = (\sigma_{1-2}^k, \ldots, \sigma_{1-N}^k) \quad k = 1, \ldots, n,$$

where, for $i = 2, \ldots, N$:

$$d_{1-i}^k = 1/g \sum_{s=1}^{g} d_{1-i}^{sk}, \quad \sigma_{1-i}^k = \left[1/g \sum_{s=1}^{g} (d_{1-i}^k - d_{1-i}^{sk})^2 \right]^{1/2}.$$

31.9.3 Signature Verification Based on a Simple Decision

The hierarchical organization of the knowledge-base makes possible to compare an unknown input sample with the signatures of the same type only. More specifically let n be the number of strokes of an input signature and let m_k' and M_k' be the number of the minima and of the maxima of the k-th stroke; then the SLCes selected for the recognition procedure are all those belonging to the n-th FLC for which the following relations are true:

$$|m_k - m'_k| < 2,$$
$$|M_k - M'_k| < 2,$$
$$|(m_k + m'_k) - (M_k + M'_k)| < 2,$$

where m_k and M_k are respectively the averages of the number of the minima and of the maxima in the k-th stroke of the signatures of the SLC. If the above condition is satisfied for none of the SLCes of the n-th FLC, then the unknown signature will be labeled as inconsistent in that it cannot be verified.

When a SLC has been selected, a recognition algorithm based on the characteristics of the signatures in that class is used. Let

$$V'^k = (d'^k_{1-2}, \ldots, d'^k_{1-N}), \quad k = 1, \ldots, n$$

be the feature vectors of the unknown signature and V^k and σ^k, $k = 1, \ldots, n$, the statistical measures of the characteristics of the selected SLC, if we denote with p^k the percentage of features related to the k-th stroke for which

$$|d'^k_{1-i} - d^k_{1-i}| < 2\sigma^k_{1-i}, \quad \text{for } i = 2, \ldots, N;$$

then the decision rule can be the following:
- if $p^k > 60\%$ for $k = 1, \ldots, n$ then the signature is accepted;
- if $p^k < 40\%$ for $k = 1, \ldots, n$ then the signature is rejected;
- otherwise the signature is unclassified.

We have shown as it is possible to design and implement an on-line Signature Verification System based on Fourier descriptors and using a stroke approach. Moreover we have shown as the problem of variability in signatures may be successful faced by means of a local analysis of the signature trace. The real application of this type of system depends on a good training phase. It is important to underline that it has been experimentally proved that the personal variability in signing is not a random phenomenon but, more probably, each person signs in a finite number of different ways and therefore it is always possible to associate a signature to a specific class of signatures of that subject [68–70].

References

[1] Rabiner L R, Gold B (1975) Theory and Applications of Digital Signal Processing. Prentice Hall, Englewood Cliffs
[2] Cooley J W et al (1969) The Fast Fourier Transform and its Application. IEEE Trans Educ, E–12(1): 27–34
[3] Cooley J W, Tukey J W (1967) An Algorithm for the Machine Calculation of Complex Fourier Series. In: Rabiner LR, Rader CM (eds) Digital Signal Processing. IEEE Press

[4]   Dimauro G et al (1989) Integration of the Cooley, Rader and Winograd-Fourier Algorithms for a faster computation of the DFT. In: Cantoni V et al (eds) Lecture Notes in Computer Science, Springer, Berlin

[5]   Winograd S (1978) On Computing the Discrete Fourier Transform, Mathematics of Computation vol 32: 175 – 199

[6]   Taylor A E, Lay C (1980) Introduction to function analysis. Wiley, New York

[7]   Impedovo S et al (1978) A Fourier Descriptor Set for Recognizing Nonstylized Numerals. IEEE Trans Syst Man and Cybern, SMC 8-(8): 640 – 645

[8]   Krzyzak A et al (1988) Reconstruction of Two Dimensional Patterns by Fourier Descriptors. In: Proc of 9th ICPR, Rome, Italy, pp 555 – 558

[9]   Persoon E, Fu K S (1977) Shape Discrimination Using Fourier Descriptors. IEEE Trans Syst Man and Cybern, SMC, 7(3): 170 – 179

[10] Zahn C T et al (1972) Fourier Descriptors for Plane Closed Curves. IEEE Trans Comp, C 21 – 3: 269 – 281

[11] El Oirrak A et al (2002) Affine Invariant Descriptors Using Fourier Series. Pattern Recognition Letters, 23: 1109 – 1118

[12] Granlud G (1976) Fourier Preprocessing for Hand-printed Character Recognition. IEEE Trans on System, Man and Cybern, SMC 6 – 2

[13] Granlund G (1972) Fourier Preprocessing for Hand Printed Character Recognition. IEEE Trans Comput, C-21 (3): 195 – 201

[14] Iivari Kunttu I et al (2006) Multiscale Fourier Descriptors for Defect Image Retrieval. Pattern Recognition Letters, 27: 123 – 132

[15] Shridhar M, Badreldin A (1984) High Accuracy Character Recognition Algorithm Using Fourier and Topological Descriptors. Pattern Recognit, 17: 515 – 524

[16] Weiss I (1993) Geometric Invariants and Object Recognition. Internat J Comput Vision, 10: 207 – 231

[17] Impedovo S, Abbattista N (1982) Hand-written Numeral Recognition; the Organization Degree Measurement. In: Proc of 6th ICPR, Munich, Germany, pp 40 – 43

[18] Davis R H, Lyell J (1986) Recognition of Handwritten Characters: A Review. Image Vision Comput

[19] Suen C Y, Ahmed P (1987) Computer Recognition of Totally Uncostrained Handwritten Zip Codes. International Journal of Pattern Recognition and Artificial Intelligence, Word Scientific Publishing, 1(1): 1 – 15

[20] Duvernoy J, Charraut D (1979) Stability and Stationarity of Cursive Handwriting. Pattern Recognition, 11: 145 – 154

[21] Dimauro G et al (1989) Ambiguous Patterns: Investigation on their Properties. In: Cantoni V et al (eds) Progress in Image Analysis and Processing. World Scientific Publishing, Singapore

[22] Dimauro G et al (1992) Uncertainty in the Recognition Process: Some Considerations on Human Variable Behaviour. In: Impedovo S, Simon J C (eds) From Pixels to Features III: Frontiers in Handwriting Recognition, Elsevier North-Holland publishers, Amsterdam

[23] Suen C Y et al (1990) Classification of Confusing Handwritten Numerals by Human Subjects. Proc of First IWFHR, Montreal, Canada

[24] Persoon E, Fu K (1986) Shape Discrimination Using Fourier Descriptors. IEEE Trans Pattern Anal Machine Intell, PAMI – 8(8): 388 – 397

[25] Hand D J (1981) Discrimination and Classification. John Wiley and Sons

[26] Spath H (1980) Cluster Analysis Algorithms for Data Reduction and Classification of Objects Ellis Harwood

[27] Sobel I (1978) Neighborhood Coding of Binary Images for Fast Contour Following and General Array Processing C G and I P, pp 127 – 135

[28] Van Ryzien J (1977) Classification and clustering. Academic Press

[29] Yana Y, Zhanga Y J (2008) 1D Correlation Filter Based Class-dependence Feature Analysis for Face Recognition. Pattern Recognition, 41: 3834 – 3841

[30] Gelman L et al (2003) Signal Recognition: Fourier Transform vs Cosine Transform. Pattern Recognition Letters, 24: 2823 – 2827

[31] Rao K et al (2001) Discrete Cosine Transform: Algorithms, Advantages, Applications. Academic Press

[32] Impedovo S, Dimauro G (1990) An Interactive System for the Selection of Handwritten Numeral Classes. Proc of 10th International Conference on Pattern Recognition, IEEE Computer Society Press

[33] Brigham E O (1988) The Fast Fourier Transform and Applications. Prentice-Hall, New York

[34]   Fukunaga K (1990) Introduction to Statistical Pattern Recognition, 2nd edn. Academic Press, New York

[35]   Gelman L, Braun S (2001) The Optimal Usage of the Fourier Transform for Pattern Recognition. Mech Syst Signal Process, 15(3): 641–645

[36]   Gonzàlez E et al (2008) Active Object Recognition Based on Fourier Descriptors Clustering. Pattern Recognition Letters, 29: 1060–1071

[37]   Jing X Y et al (2006) Face Recognition Based on Discriminant Fractional Fourier Feature Extraction. Pattern Recognition Letters, 27: 1465–1471

[38]   Pavlidis T (1982) Algorithms for Graphics and Image Processing. Springer, Berlin

[39]   Tou J T, Gonzales R C (1981) Pattern Recognition Principles Reading. Addison-Wesley

[40]   Wirtz B (1998) Technical Evaluation of Biometric Systems. In: Proc of ACCU '98, Hong Kong

[41]   Fang B, Wen J, Tang Y Y et al (2010) Automatic Off-line Signature Verification by Computer. In: Fang B, Wen J, Tang Y Y, Wang P S P (eds) Pattern Recognition and Machine Vision, King-Sun Fu Memorial Book, River Pub Co, Denmark

[42]   Jain A K et al (2002) On-line Signature Verification. Pattern Recognition, 35: 2963–2972

[43]   Kamel N S, Sayeed S (2008) SVD-Based Signature Verification Technique Using Data Glove. IJPRAI (Int J of Pattern Recognition and Artificial Intelligence), 22(3): 431–443

[44]   Leclerc F, Plamondon R (1994) Automatic Signature Verification: The State of the Art 1989–1993. IJPRAI 8(3): 643–660

[45]   Nemcek W F, Lin W C (1974) Experimental Investigation of Automatic Signature Verification. IEEE T-SMC 4: 121–126

[46]   Plamondon R (1994)(ed) Progress in Automatic Signature Verification, World Scientific Publ, Singapore

[47]   Plamondon R, Lorette G (1989) Automatic Signature Verification and Writer Identification. The state of the art Pattern Recog, 22(2): 107–131

[48]   Dimauro G et al (2004) Recent Advancements in Automatic Signature Verification. Proc of 9th International Workshop on Frontiers in Handwriting Recognition (IWFHR-9 2004). IEEE Computer Society Press, Kokubunji, Tokyo

[49]   Fairhurst M C, Brittan P (1994) An Evaluation of Parallel Strategies for Feature Vector Construction in Automatic Signature Verification Systems. Int J Pattern Recog Artif Intell, 8(3): 661–678

[50]   Wirtz B (1995) Stroke-based Time Warping for Signature Verification. In: Proc IC-DAR. IEEE Press, pp 179–182

[51]   Chen Y, Ding X Q, Wang P S P (2009) Dynamic Structural Statistical Model Based Online Signature Verification. IJDCF (International J of Digital Crime and Forensics, 1(3)

[52]   Dimauro et al (1990) On-line Signature Verification System Through Stroke Analysis. In: New Concepts in Computer Science, Association Française pour la Cybernétique Économique et Technique (AFCET), Paris

[53]   Lorette G (1984) On-line Handwritten Signature Recognition Based on Data Analysis and Clustering. In: Proc 7th Int Conf Pattern Recognition, 2: 1284–1287

[54]   Qi Y, Hunt B R (1995) A multiresolution approach to computer verification of handwritten signatures. IEEE Trans Image Processing, 4(6): 870–874

[55]   Sato Y, Kogure K (1982) On-line Signature Verification Based on Shape, Motion and Handwriting Pressure. In proc of 6th ICPR, Munich

[56]   Tseng L Y, and Huang T H (1992) An On-line Chinese Signature Verification Scheme based on the ART1 neural network. In: proc of Int J Conf on Neural Networks, Maryland

[57]   Bajaj R, Chaudhury S (1997) Signature Verification using multiple neural classifiers. Pattern Recognition, 30(1): 1–7

[58]   Plamondon R, Srihari S N (2000) On-line and Off-line Handwriting Recognition: A Comprehensive Survey. IEEE T-PAMI, 22(1): 63–84

[59]   Sabourin R, Drouhard J P (1992) Off-line signature verification using directional PDF and neural networks. In: Proc of 11th ICPR

[60]   Dimauro G et al (1992) A Stroke-Oriented Approach to Signature Verification. In: Impedovo S, Simon JC (eds) From Pixels to Features III: Frontiers in Handwriting Recognition, Elsevier North-Holland publishers, Amsterdam

[61]   Dimauro G et al (1994) Component-Oriented Algorithms for Signature Verification.

Int J Pattern Rec and Artificial Intelligence, 8: 771–793

[62]   Dimauro G et al (1993) A Signature Verification System Based on a Dynamical Segmentation Technique. Proc of Third International Workshop on Frontiers in Handwriting Recognition, Partners' Press Inc

[63]   Impedovo S et al (1988) A Fourier Analysis Based Signature Verification System. Proc of 9th ICPR, IEEE Computer Society Press

[64]   Wirtz B (1997) Average Prototypes for Stroke-Based Signature Verification. In: Proc of ICDAR, IEEE Press

[65]   Lam C F, Kamins D(1989) Signature Recognition Through Spectral Analysis. Pattern Recog, 22(1): 39–44

[66]   Castellano M et al (1988) A Spectral Analysis Based Signature Verification System. In Goos G, Hartmanis J (eds) Lecture Notes in Computer Science: Recent Issues in Pattern Analysis and Recognition. Springer, Berlin

[67]   Dimauro G et al (1989) Decision Making Process in a Signature Verification System. In: Cantoni V et al (eds) Progress in Image Analysis and Processing. World Scientific Publishing, Singapore

[68]   Plamondon R (1995) A Kinematic Theory of Rapid Human Movements: Part I: Movement Representation and generation. Biological Cybernetics, 72 (4): 295–307

[69]   Plamondon R (1995) A Kinematic Theory of Rapid Human Movements: Part II: Movement Time and Control. Biological Cybernetics, 72 (4): 309–320

[70]   Plamondon R (1997) A Kinematic Theory of Rapid Human Movements: Part III: Kinetic Outcomes. Biological Cybernetics, 72 (4)

Index

Printed in the United States
By Bookmasters